Beat Ernst, Gerald W. Hart, Pierre Sinaÿ

Carbohydrates in Chemistry and Biology

Related Titles from WILEY-VCH

A. E. Stütz (ed.)
Iminosugars as Glycosidase Inhibitors
XIV, 397 pages with 152 figures, 24 in color, and 48 tables
plus a comprehensive 123-table of glycosidase inhibitors
1999. ISBN 3-527-29544-5. Cloth

Y. Chapleur (ed.)
Carbohydrate Mimics – Concepts and Methods
XXVIII, 604 pages with 375 figures and 52 tables
1998. ISBN 3-527-29526-7. Cloth

H. van Bekkum, H. Röper, A. G. Vorhaben (eds.)
Carbohydrates and Organic Raw Materials III
X, 315 pages with 156 figures and 68 tables
1996. ISBN 3-527-30079-1. Cloth

T. K. Lindhorst
Essentials of Carbohydrate Chemistry and Biochemistry
XIV, 323 pages with 244 figures and 11 tables
2000. ISBN 3-527-29543-7. Paperback

Beat Ernst, Gerald W. Hart, Pierre Sinaÿ

Carbohydrates in Chemistry and Biology

Part I Chemistry of Saccharides

Vol. 2 Enzymatic Synthesis of Glycosides and Carbohydrate-Receptor Interaction

Weinheim · New York · Chichester
Brisbane · Singapore · Toronto

Prof. Dr. B. Ernst
Institut für Molekulare
Pharmazie
Universität Basel
Klingenbergstrasse 50
4051 Basel
Switzerland

Prof. Dr. G. W. Hart
Dept. of Biological Chemistry
Johns Hopkins University
School of Medicine
725 N. Wolf St. Rm.
401 Hunterian
Baltimore, MD 21205-2185
USA

Prof. Dr. P. Sinaÿ
Dept. de Chimie, URA 1686
Ecole Normale Supérieure
24 rue Lhomond
75231 Paris Cedex 05
France

> This book was carefully produced. Nevertheless, editors, authors and publisher do not warrant the information contained therein to be free of errors. Readers are advised to keep in mind that statements, data, illustrations, procedural details or other items may inadvertently be inaccurate.

Cover picture: 3,6-dideoxyhexose binding to antibody se155.4
Courtesy of David Bundle, University of Alberta

Library of Congress Card No.: applied for

A catalogue record for this book is available from the British Library.

Die Deutsche Bibliothek - CIP Cataloguing-in-Publication-Data
A catalogue record for this publication is available from Die Deutsche Bibliothek

ISBN 3-527-29511-9

© WILEY-VCH Verlag GmbH, D-69469 Weinheim (Federal Republic of Germany). 2000
Printed on acid-free and chlorine-free paper.
All rights reserved (including those of translation in other languages). No part of this book may be reproduced in any form - by photoprinting, microfilm, or any other means - nor transmitted or translated into machine language without written permission from the publishers. Registered names, trademarks, etc. used in this book, even when not specifically marked as such, are not to be considered unprotected by law.
Composition: Asco Typesetters, Hongkong. Printing: betz-druck gmbH, D-64291Darmstadt.
Bookbinding: Wilhelm Osswald & Co., D-67433 Neustadt.
Printed in the Federal Republic of Germany.

Contents

Part I   **Chemistry of Saccharides**

Vol. 1   **Chemical Synthesis of Glycosides and Glycomimetics**

	List of Contributors	LV
	Abbreviations Used in Volumes 1 and 2	LXIII
I	**Chemical Synthesis of Glycosides**	1
1	**Introduction to Volumes 1 and 2**	3
2	**Trichloroacetimidates**	5
	Richard R. Schmidt and Karl-Heinz Jung	
2.1	Introduction	5
2.2	Methods	6
2.3	*O*-Glycosides	7
2.3.1	Synthesis of Oligosaccharides	7
	β-Glucosides, β-Galactosides, α-Rhamnosides, *etc.*	7
	Aminosugar Trichloroacetimidates	8
	β-Mannosides	13
	2-Deoxyglycosides	13
	Miscellaneous Compounds	14
	Complex Oligosaccharides	14
2.3.2	Inositol Glycosides	36
2.3.3	Glycosylation of Sphingosine Derivatives and Mimics	38
2.3.4	Glycosylation of Amino Acids	40
2.3.5	Polycyclic and Macrocyclic Glycosides	42

2.3.6	Glycosides of Phosphoric and Carboxylic Acids	44
2.3.7	Solid-Phase Synthesis	45
2.4	S-Glycosides	49
2.5	N- and P-Glycosides	51
2.6	C-Glycosides	51
2.7	Conclusion and Outlook	53
	References	53
3	**Iterative Assembly of Glycals and Glycal Derivatives: The Synthesis of Glycosylated Natural Products and Complex Oligosaccharides**	**61**
	Lawrence J. Williams, Robert M. Garbaccio, and Samuel J. Danishefsky	
3.1	Introduction	61
3.2	Ciclamycin 0	64
3.3	Allosamidin	66
3.4	KS-502 and Rebeccamycin	69
3.5	Extension to Thioethyl Donors	74
3.6	Lewis^y	76
3.7	Globo H	82
3.8	KH-1	86
3.9	Concluding Remarks	90
	Acknowledgments	90
	References	90
4	**Thioglycosides**	**93**
	Stefan Oscarson	
4.1	Introduction	93
4.2	Synthesis of Thioglycosides	94
4.2.1	From Anomeric Acetates	94
4.2.2	From Glycosyl Halides	95
4.2.3	Protecting Group Manipulations in Thioglycosides	96
4.3	Glycosylations with Thioglycoside Donors	97
4.3.1	A Two-Step Procedure: Transformation of Thioglycosides into Other Types of Glycosyl Donors	97
4.3.2	Direct Activation of Thioglycoside Donors	99
	Heavy Metal Salt Promoters	99
	Halonium, sulfonium and carbonium type promoters	100
	Single-Electron Activation	106
	Other Types of Donors With an Anomeric Sulfur	108
4.4	Applications of Thioglycosides	110
4.4.1	Block Syntheses, Orthogonal Glycosylations	110
	Thioglycosides as Acceptors	110
	Thioglycosides as Both Donors and Acceptors	111
4.4.2	Intramolecular Glycosidations	112
4.4.3	Solid Phase Synthesis	113
	References	113

5	**Glycosylation Methods: Use of Phosphites** 117	
	Zhiyuan Zhang and Chi-Huey Wong	
5.1	Introduction 117	
5.2	Preparation of Glycosyl Phosphites 118	
5.3	Glycosylation using Glycosyl Phosphites 119	
5.3.1	Mechanism 119	
5.3.2	Low Temperature-Dependent Stereoselectivity 121	
5.3.3	Glycosylation of Sialyl Phosphites 122	
5.3.4	Glycosylation of C-2-Acylated Glycosyl Phosphites............... 123	
5.3.5	Glycosylation with C-2-*O*-Benzylated Glycosyl Phosphites 124	
	Glycosylation using Glucosyl Phosphites with a Benzyl Group at C-2 124	
	Glycosylation using Galactosyl and Fucosyl Phosphites with a Benzyl Group at C-2... 125	
	Glycosylation using other Glycosyl Phosphites with a Benzyl Group at C-2... 126	
5.3.6	Glycosylation with 2-Deoxy Glycosyl Phosphites 127	
5.4	Other Applications of Glycosyl Phosphites 128	
5.4.1	Synthesis of CMP–NeuAc 129	
5.4.2	Synthesis of GDP–Fucose 129	
5.4.3	Formation of Glycosyl Phosphonate................................... 131	
5.4.4	Transformation to other Types of Glycosyl Donor 131	
	Phosphate 131	
	Phosphorimidate 131	
	References 132	
6	**Glycosylation Methods: Use of *n*-Pentenyl Glycosides** 135	
	Bert Fraser-Reid, G. Anilkumar, Mark R. Gilbert, Subodh Joshi, and Ralf Kraehmer	
6.1	Introduction 135	
6.2	Fundamental Reactions 135	
6.3	Determination of Relative Reactivities 138	
6.4	*n*-Pentenyl Orthoesters as Glycosyl Donors 141	
6.5	*n*-Pentenyl Orthoesters as Latent C2 Esters 144	
6.6	Protecting Groups 146	
6.7	Solid-Phase Iterative Couple–Deprotect–Couple Strategy 146	
	References 153	
7	**Glycosylidene Diazirines** 155	
	Andrea Vasella, Bruno Bernet, Martin Weber, and Wolfgang Wenger	
7.1	Introduction 155	
7.2	Synthesis of Glycosylidene Diazirines 155	
7.3	Stability of the Glycosylidene Diazirines 158	
7.4	Glycosidation by Glycosylidene Diazirines 158	
7.4.1	General Aspects 158

7.4.2	Glycosidation of Strongly Acidic Hydroxy Compounds	162
	Glycosidation of Phenols	162
	Glycosidation of Fluorinated Alcohols	163
7.4.3	Glycosylation of Weakly Acidic Hydroxy Compounds	163
	Glycosidation of Monovalent Alcohols	163
	Glycosidation of Diols and Triols	164
7.5	Synthesis of Spirocyclopropanes	168
7.6	Addition to Aldehydes and Ketones	170
7.7	Exploratory Use of Diazirines: Formation of Glycosyl Phosphines, Stannanes, N-Sulfonylamines, Esters, Boranes, and Alanes, and of 1,1-Difluorides	171
	Acknowledgments	174
	References	174
8	**Glycosylation Methods: Alkylations of Reducing Sugars**	**177**
	Jun-ichi Tamura	
8.1	Introduction	177
8.2	Anomeric O-Alkylation	177
8.2.1	Anomeric O-Alkylation of Ribofuranose with Primary Triflates: Effect of the Protecting Group at O-5 of Ribofuranose	178
8.2.2	Anomeric O-Alkylation of Mannofuranose with Primary Triflates: The Crown Ether Effect	179
8.2.3	Anomeric O-Alkylation of Gluco- and Galactopyranoses with Primary Triflates: High β-Selectivity as a Result of the Reactive Anomeric β-Anion	180
8.2.4	Anomeric O-Alkylation of Acyl-Protected Nucleophiles with Primary Triflates	181
8.2.5	Anomeric O-Alkylation of Mannopyranose with Primary Triflates: Possibility of Intramolecular Complexation of the Nucleophile	184
8.2.6	Anomeric O-Alkylation of KDO with Primary Triflates	185
8.2.7	Anomeric O-Alkylation of Some Protected Aldoses with Primary Triflate	186
8.2.8	Anomeric O-Alkylation of Unprotected Aldoses with Primary Triflate, Bromides, and Cyclic Sulfates	187
8.2.9	Anomeric O-Alkylation with Secondary Triflates and Nonaflate	188
8.3	Glycosylation *via* the Locked Anomeric Configuration	189
8.3.1	Synthesis of Methyl, Allyl, and Benzyl Glycosides *via* Stannylene Acetals	189
8.3.2	Epimerization at C-2 by the Locked Anomeric Configuration Method	189
8.3.3	The Locked Anomeric Configuration Method for Rhamnosyl Stannylene Acetal	190
8.3.4	The Locked Anomeric Configuration Method for Mannosyl Stannylene Acetal: Isomerization of Acetal [25, 26]	190
8.3.5	The Locked Anomeric Configuration Method for Stannylene Acetal with the Glucose Configuration [25, 26]	191

8.4	Conclusion	192
	References	193
9	**Other Methods of Glycosylation**	**195**
	Luigi Panza and Luigi Lay	
	Introduction and Summary	195
	Highlights	195
9.1	Enol Ethers	197
9.1.1	Endo-Enol Ethers	198
9.1.2	Exo–Enol Ethers	201
9.1.3	Endo–Glycals	202
9.1.4	Exo–Glycals	204
9.1.5	Vinyl Glycosides	206
9.2	1-Hydroxy Sugars	209
9.2.1	Acidic Activation	210
	Acidic Activation With Additional Reagents	211
9.2.2	Dehydrative Glycosylation	212
	In the Presence of the Acceptor From the Beginning	213
9.2.3	Mitsunobu Glycosylation	214
9.2.4	1-O-Silyl Glycosides	215
9.3	Esters and Related Derivatives	216
9.3.1	Esters	216
9.3.2	Sugar Carbonates and Derivatives	221
9.3.3	Orthoesters and Oxazolines	223
9.3.4	Phosphorus and Sulfur Derivatives	229
	References	233
10	**Polymer-Supported Synthesis of Oligosaccharides**	**239**
	Jiri J. Krepinsky and Stephen P. Douglas	
10.1	Introduction	239
10.2	General Reflections	240
10.3	Polymer Supports	246
10.4	One-Phase Systems (Syntheses in Solution)	247
	Polyethyleneglycol$_\omega$-monomethylether (MPEG)	248
	Linear Polystyrene	249
10.4.1	Linkers	250
	Succinoyl Diester	250
	Dioxyxylyl Diether (DOX)	252
10.4.2	Chemistry Investigations	254
10.5	Two-Phase Systems (Syntheses on Solid Supports)	255
	Controlled Pore Glass	256
	Cross-Linked Polystyrene	256
	Polyethylene Grafts on Cross-Linked Polystyrene	256
10.5.1	Linkers	259
	Dialkyl- or Diaryl-Silyl	259
	Thioglycoside Linkers	259

	Linkers Cleavable by Photolysis	260
10.6	Examples of Syntheses	260
10.7	Combinatorial Libraries	261
10.8	Capping	262
10.9	Concluding Remarks	262
	References	262

11 Glycopeptide Synthesis in Solution and on the Solid Phase 267
Horst Kunz and Michael Schultz

11.1	Introduction 267
11.1.1	Which Protecting Groups are Suitable for Carbohydrates (Table 2)? 269
11.1.2	Which Glycosylation Methods are Useful for the Formation of Glycopeptides? 271
	Formation of Asparagine *N*-Glycosides 271
	α-Fucosylation 271
	Formation of the β-Lactosamine Linkage 272
	α-Sialylation 272
11.1.3	Glycopeptides Containing Particularly Sensitive Linkages 272
	Acid Sensitivity 273
	Base Sensitivity 273
11.2	Synthesis of Glycopeptides in Solution 274
11.2.1	*O*-Glycopeptides 274
	Glycopeptides Carrying *N*-Acetylgalactosamine (Tn-Antigen) 274
	Glycopeptides Carrying the T-Antigen (Gal–GalNAc) 276
	Glycopeptides Carrying the Sialyl T Antigen (NeuAcα2,6[Galβ1,3]GalNAc) 278
	Glycopeptides Carrying *O*-GlcNAc 279
11.2.2	*N*-Glycopeptides 280
	N-Glycopeptides Carrying Natural Saccharide Side-Chains 280
	N-Glycopeptides with Lewis-Type Saccharide Side-Chains 285
11.3	Glycopeptide Synthesis on the Solid Phase 286
11.3.1	*O*-Glycopeptides 287
	Glycopeptides Carrying *N*-Acetylgalactosamine (Tn-Antigen) 287
	O-Glycopeptides Carrying the T Antigen (Gal–GalNAc) 290
	O-Glycopeptides Carrying the Sialyl Tn Antigen (NeuNAc–α2,6-GalNAc) 291
	O-Glycopeptides Carrying the 2,3-Sialyl T Antigen 293
	O-Glycopeptides Carrying *O*-GlcNAc Side-Chains 294
	O-Glycopeptides Carrying O-Linked Fucose 295
	O-Glycopeptides Carrying a Sialyl Lewis Antigen Structure 296
11.3.2	*N*-Glycopeptides 297
	The Construction of *N*-Glycopeptide Libraries on the Solid Phase 298
	Sequential *N*-Glycopeptide Synthesis on the Solid Phase with Oligosaccharides from Natural Sources 299

11.4	Conclusion	300
	References	300
12	**Glycolipid Synthesis**	305
	Hideharu Ishida	
12.1	Introduction	305
12.2	Synthesis of Ganglio-Series Gangliosides	305
12.2.1	Retrosynthetic Analysis of Ganglioside GD1a	305
12.2.2	Preparation of Sialylgalactose Donor as Building Block	306
12.2.3	Construction of Oligosaccharide	308
12.2.4	Transformation of Oligosaccharide into Glycolipid	310
12.3	Synthesis of Polysialo Ganglio-Series Gangliosides	311
12.3.1	Retrosynthetic Analysis of GQ1b	313
12.3.2	Preparation of Building Block	313
12.3.3	Construction of Oligosaccharide	314
12.4	Conclusion	315
	References	316
13	**Stereoselective Synthesis of β-Mannosides**	319
	Vince Pozsgay	
13.1	Introduction	319
13.2	Chemical Methods	320
13.2.1	Glycosylation with Mannosyl Donors	320
	Mannosylation using Insoluble Promoters	320
	The Sulfonate Approach	322
	Intramolecular Mannosylation	324
	Other Mannosyl Donor-Based Methods	327
13.2.2	Epimerization of β-Glucopyranosides at C-2	329
	The Oxidation–Reduction Approach	329
	Direct Inversion	329
13.2.3	The 2-Ulosyl Donor Method	331
13.2.4	Anomeric *O*-alkylation	332
	Alkylation of 1-*O*-Metal Complexes	332
	The Stannylene Acetal Method	332
13.2.5	Miscellaneous Methods	333
	Radical Inversion of the Anomeric Chirality of α-D-Mannopyranosides	333
	Reductive Cleavage of Cyclic Orthoesters	334
	De novo Syntheses	334
13.2.6	2-Acetamido-2-deoxy-β-D-mannopyranosides	335
13.2.7	Aryl β-D-mannopyranosides	336
13.2.8	1-Thio-β-D-mannopyranosides	336
13.2.9	β-D-Mannopyranosylamines	337
13.3	Enzymatic Synthesis	337
13.4	Conclusions	338
	References	338

14	**Special Problems in Glycosylation Reactions: Sialidations**	345
	Makoto Kiso, Hideharu Ishida and Hiromi Ito	
14.1	Introduction	345
14.2	Sialidation by the Koenigs–Knorr Method	345
14.3	Sialidation Using an Auxiliary Group at C-3	347
14.4	Sialidation Using 2-Thioglycosides, Xanthates, or Phosphites of Sialic Acids in Acetonitrile	349
14.4.1	Thioglycosides	349
14.4.2	Xanthates and Phosphites	356
14.4.3	Reaction Mechanism	359
14.5	Further Solutions to the Problem	359
14.5.1	Combination of C-3 Auxiliary and Sterically less Hindered Sugar Acceptors	359
14.5.2	Combination of C-3 Auxiliary and Specific Activation of the Anomeric Center C-2	360
14.5.3	Thioglycoside of N,N-Diacetylneuraminic Acid and Combination with C-3 Auxiliary	363
	References	364
15	**Special Problems in Glycosylation Reactions: 2-Deoxy Sugars**	367
	Alain Veyrières	
15.1	Introduction	367
15.2	Electrophilic Additions to Glycals: Mechanistic Aspects and Applications to the Synthesis of 2-Deoxyglycosides	368
15.2.1	Protonation of Glycals	369
15.2.2	Enzyme-Catalyzed Additions to Glycals	370
15.2.3	Halogenation of Glycals	370
15.2.4	Bromo- and Iodoalkoxylation of Glycals	372
15.2.5	Epoxidation of Glycals	377
15.2.6	Addition of Sulfur Based Electrophiles to Glycals	379
15.2.7	Addition of Selenium Based Electrophiles to Glycals	382
15.3	The Cycloaddition Way to Glycosyl Transfer	384
15.4	Fluoroglycosylation of Glycals	385
15.5	Glycosyl Donors with a C-2 Heteroatom	386
15.5.1	2-Bromo-2-deoxyglycosyl bromides	386
15.5.2	2-Deoxy-2-(thiophenyl)-glycosyl fluorides	387
15.5.3	2,6-Anhydro-2-Thio-Glycosyl Donors	388
15.5.4	1,2-Di-*O*-Acetyl-β-Hexopyranoses and *N*-Formylglucosamine Derivatives	392
15.6	2-Deoxyglycosyl Donors	393
15.6.1	2-Deoxy-Hexopyranoses	394
15.6.2	Tert-Butyldimethylsilyl 2-Deoxyglycosides	394
15.6.3	1-*O*-Acyl- and Acetimidyl-2-Deoxy-Hexopyranoses	394
15.6.4	2-Deoxyglycosyl Bromides and Fluorides	395
15.6.5	*S*-(2-Deoxyglycosyl)phosphorodithioates	396

15.6.6	2-Deoxyglycosyl Phosphates, Phosphoramidites and Phosphites....	397
15.6.7	2-Deoxy Thioglycosides	398
15.6.8	2-Deoxyglycosyl Sulfoxides...	399
15.7	Other Approaches to 2-Deoxyglycosides...........................	400
15.7.1	Cyclization of Acyclic Sugars	401
15.7.2	Use of Alkoxy-Substituted Anomeric Radicals	402
	References...	403
16	**Orthogonal Strategy in Oligosaccharide Synthesis**	407
	Osamu Kanie	
16.1	Introduction...	407
16.2	Analysis of the Strategic Aspects of Oligosaccharide Synthesis	408
16.2.1	General Aspects...	408
16.2.2	The Pursuit of Efficiency in Oligosaccharide Synthesis.........	408
16.3	The Introduction of the Orthogonal Glycosylation Strategy	410
16.3.1	Limitation of Current Concepts.....................................	410
16.3.2	The Orthogonal Coupling Concept	412
16.3.3	What is Orthogonality Anyway?	413
16.3.4	Orthogonal Glycosylation and Solid-Phase Oligosaccharide Synthesis	414
16.4	The Orthogonal Glycosylation Strategy	414
16.4.1	Orthogonal Chain Elongation of Homo-Oligosaccharides: Synthesis of Chito-Oligosaccharides [19].........................	414
16.4.2	Orthogonal Coupling for Hetero-Oligomer Synthesis [22].....	418
16.4.3	Application to Polymer-Supported Synthesis [26]	420
16.5	Conclusions and Prospects	421
	Acknowledgments	424
	References...	424
17	**Protecting Groups: Effects on Reactivity, Glycosylation Stereoselectivity, and Coupling Efficiency**	427
	Luke G. Green and Steven V. Ley	
17.1	Introduction...	427
17.2	Glycosidic Mechanism...	428
17.3	Electronic and Torsional Effects	430
17.4	Influence of Protecting Group on Donor Reactivity.............	431
17.5	Stereoselectivity..	436
17.5.1	Neighboring-Group Participation	436
17.5.2	Reactivity Control	437
17.6	Influence of the Protecting Group on the Acceptor	441
17.7	Steric Effects on Glycosylation	443
17.8	Conclusions	444
	Acknowledgments	445
	References...	446

18	**Intramolecular Glycosidation Reactions**	449
	Jacob Madsen and Mikael Bols	
18.1	Introduction	449
18.2	Reactions in which the Tether Participates in the Reaction	450
18.2.1	Tethering to the Glycosyl Donor	450
	Carbon Tethers	450
	Silicon Tethers	454
18.2.2	Tethering to the Leaving Group	459
18.3	Reactions in which the Tether does not Participate in the Reaction	459
18.4	Conclusion	464
	References	465
19	**Classics In Total Synthesis of Oligosaccharides and Glycoconjugates**	467
	Jean-Maurice Mallet and Pierre Sinaÿ	
19.1	Introduction	467
19.2	Syntheses of Nod factors	467
19.2.1	Introduction	467
19.2.2	The K. C. Nicolaou Synthesis (1992) [3]	468
19.2.3	The J.-M. Beau Synthesis (1994) [12]	471
19.2.4	The T. Ogawa Synthesis (1994) [16]	475
19.2.5	The Y. Z. Hui Synthesis (1992) [18]	477
19.2.6	Conclusion	480
19.3	Synthesis of the Antithrombin-Binding Pentasaccharide Sequence in Heparin (1984) [19, 20]	480
19.3.1	Introduction	480
19.3.2	An Overview of the Synthesis of the Protected Pentasaccharide 73	481
19.3.3	Synthesis of the Disaccharidic Bromide Donor 68	483
19.3.4	Synthesis of the Disaccharidic Acceptor 69	484
19.3.5	Synthesis of the Protected Pentasaccharide 73	484
19.3.6	Synthesis of the Active Site of Heparin	485
19.4	Total Synthesis of VIM-2 Ganglioside [31]	485
19.4.1	Introduction	485
19.4.2	The Total Synthesis of VIM-2—a General Strategy	486
19.4.3	Preparation of the Key Protected Octasaccharide 87	487
19.5	Epilogue	490
	References	491
II	**Synthesis of Oligosaccharide Mimics**	493
20	**Synthesis of C-Oligosaccharides**	495
	Troels Skrydstrup, Boris Vauzeilles, and Jean-Marie Beau	
20.1	Introduction	495
20.2	The Anionic Approach	496
20.2.1	C5-Alkynyl Anions	496

20.2.2	C1-Glycal Carbanions	500
20.2.3	Anomeric Samarium Species	502
20.2.4	*C*-Branched Carbanions	506
20.2.5	C6-Phosphoranes	508
20.3	The Radical Approach	511
20.3.1	Intermolecular Anomeric Radical Addition	511
20.3.3	Intramolecular Anomeric Radical Addition	513
20.4	The Partial *de Novo* Approach	518
20.5	The Cycloaddition and Rearrangement Approach	527
	References	528
21	**Synthesis of Oligosaccharide Mimics: S-Analogs**	**531**
	Jon K. Fairweather and Hugues Driguez	
21.1	Introduction	531
21.2	General Synthesis	532
21.2.1	Preparation of Thioglycoses	532
	1-Thioglycoses	532
	2-, 3-, 4-, 5-, or 6-Thioglycoses	532
	Selective S-Deprotection of Thioglycoses	533
	Glycosylation Methods	534
21.3	Establishment of 1,6-Thio Linkages	534
21.3.1	6-Thiodisaccharides	534
21.3.2	6-Thiooligosaccharides	538
21.3.3	Branched Thiocyclodextrins	538
21.4	Establishment of 1,4-Thio Linkages	541
21.4.1	1,4-Thiodisaccharides	541
	General Approaches	541
	S_N2-Displacement on Triflates	541
21.4.2	1,4-Thiooligosaccharides	546
	Conventional Approaches	546
	Chemoenzymatic Approaches	548
	Michael Addition to Unsaturated Acceptors	549
	Solid-Support Synthesis	550
21.5	Establishment of 1,3-Thio Linkages	551
21.5.1	1,3-Thiodisaccharides	551
	Conventional Methods	551
	Cyclic Sulfamidate and Aziridine	551
21.5.2	1,3-Thiooligosaccharides	552
21.6	Establishment of 1,2-Thio Linkages	553
21.6.1	1,2-Thiodisaccharides	553
	Conventional Methods	554
	Other Approaches	555
21.7	Establishment of 1,1-Thio Linkages	557
21.8	Establishment of Mixed Thio linkages	558
21.9	Thiooligosaccharides and Proteins	558
21.9.1	The Conformation of Thiooligosaccharides in Solution	558

21.9.2	Enzyme–Substrate Interactions	560
	α-Glucan-Active Enzymes	560
	β-Glucan-Active Enzymes	561
21.9.3	Lectin–Ligand Interactions	562
21.10	Conclusion	562
	Acknowledgments	562
	References	562
22	**Saccharide–Peptide Hybrids**	**565**
	Hans Peter Wessel	
22.1	Introduction	565
22.2	Carbohydrate Amino Acids	566
22.2.1	Natural Carbohydrate Amino Acids	566
22.2.2	Synthetic Carbohydrate Amino Acids	567
22.3	Amide-Linked Carbohydrate Polymers	572
22.4	Amide-Linked Carbohydrate Oligomers	574
22.4.1	Solution Synthesis	574
22.4.2	Solid-Phase Synthesis	578
22.4.3	Biological Activity	579
22.4.4	Conformational Properties	582
	References	583
	Index	**I 1**

Part I Chemistry of Saccharides

Vol. 2 Enzymatic Synthesis of Glycosides and Carbohydrate-Receptor Interaction

III	**Enzymatic Synthesis of Glycosides**	**587**
23	**On the Origin of Oligosaccharide Species—Glycosyltransferases in Action**	**589**
	Dirk H. van den Eijnden	
23.1	Introduction	589
23.2	Protein N-Glycosylation: Pre-assembly of Oligosaccharide–PP–Dolichol and en bloc Transfer	591
23.3	Trimming of the Polypeptide-Bound Oligosaccharide	592
23.4	Folding and Quality Control	593
23.5	Committed Steps in the Formation of Complex-Type Oligosaccharide Chains and Branching	594
23.6	Topology of the Reaction Catalyzed by a Typical GlcNAcT	596
23.7	Elongation and Termination Reactions in the *trans*-Golgi	596
23.8	Activity with Branched Substrates	598

23.9	Branch Specificity	600
23.10	Essential Requirements for Activity with LacNAc	601
23.11	Further Terminal Reactions in Complex-Type Oligosaccharide Synthesis	602
23.12	Specific Modifications of Polylactosaminoglycans	603
23.13	The Invariable Core of N-linked Oligosaccharide Chains, and Site- and Protein-Specific Processing	606
23.14	Comparison of the Synthesis of Type 1 (Gal(β1-3)GlcNAcβ–R) and Type 2 (Gal(β1-4)GlcNAcβ–R) Chains	607
23.15	The LacdiNAc Pathway of Complex-Type Oligosaccharide Synthesis	607
23.16	Protein O-Glycosylation	608
23.17	Glycosyltransferase Families	608
23.18	Sialyltransferase Family	610
23.19	α2-Fucosyltransferase Family	611
23.20	α3/4-Fucosyltransferase Family	612
23.21	α3-Galactosyl/N-Acetylgalactosaminyltransferase (Histo-Blood Group ABO) Family	613
23.22	β6-N-Acetylglucosaminyltransferase Family	613
23.23	Polypeptide N-Acetylgalactosaminyltransferase Family	614
23.24	β4-N-Acetylgalactosaminyltransferase Family	615
23.25	β4-Galactosyltransferase Family	615
23.26	β3-Galactosyltransferase Family	617
23.27	β3-Glucuronyltransferase Family	617
23.28	Glycosyltransferases Standing Alone	617
23.29	Concluding Remarks	618
	References	618
24	**Synthesis of Sugar Nucleotides**	**625**
	Reinhold Öhrlein	
24.1	Introduction	625
24.2	Synthesis of Sugar Nucleotides	626
24.2.1	Chemical Synthesis	626
	UDP-Activated Donors	626
	CMP–Activated Sugars	629
	GDP-Activated Donors	632
	Comments	634
24.2.2	Chemo–Enzymatic Synthesis	635
	Uridine Diphosphate-Activated Donor Sugars	635
	CMP–Activated Sugars	637
	GDP-activated sugars	639
	Comments	640
24.3	In situ Generation of Sugar Nucleotides	641
	Comments	641
24.4	Outlook	644
	References	644

25	**Enzymatic Glycosylations with Glycosyltransferases**	647
	Ossi Renkonen	
25.1	Introduction	647
25.2	In vitro Synthesis of the Core Region of O-Glycans	648
25.2.1	Initialization of O-Glycan Biosynthesis	648
25.2.2	Synthesis of Core 1	648
25.2.3	Synthesis of Core 2	649
25.2.4	Synthesis of Core 3 and Core 4	650
25.2.5	In vitro Extension of Core 1 Glycans	650
25.2.6	In vitro Extension of Core 2 Glycans	651
25.2.7	Extension of Core 3 and Core 4 Glycans	651
25.3	Enzymatic in vitro Synthesis of Polylactosamine Backbones	651
25.3.1	Enzymatic Synthesis of the Primary Chains of Blood Group i-Type	652
25.3.2	Distal Branching of i-Type Polylactosamine Backbones	653
25.3.3	Central Branching of i-Type Polylactosamine Backbones	654
25.3.4	β4-Galactosylation in Polylactosamine Backbones	656
25.4	α3-Sialylation of N-Acetyllactosaminoglycans at the Terminal Gal	656
25.5	α3-Fucosylation of Lactosamine Saccharides	657
	References	659
26	**Recycling of Sugar Nucleotides in Enzymatic Glycosylation**	663
	Kathryn M. Koeller and Chi-Huey Wong	
26.1	Introduction	663
26.2	Glycosyltransferases of the Leloir Pathway and their Sugar Nucleotide Substrates	663
26.3	Design of Regeneration Systems	665
26.4	Practical Regeneration Systems	666
26.4.1	UDP–Galactose	666
26.4.2	Other UDP–Sugars	669
26.4.3	CMP–NeuAc	671
26.4.4	GDP–Sugars	676
26.4.5	Other Carbohydrate-Based Regeneration Systems	680
26.5	Conclusion	682
	References	683
27	**Enzymatic Glycosylations with Non-Natural Donors and Acceptors**	685
	Xiangping Qian, Keiko Sujino, and Monica M. Palcic	
27.1	Introduction	685
27.2	Enzymatic Glycosylations	686
27.2.1	Galactosylations	686
	β1,4-Galactosyltransferase	686
	α1,3-Galactosyltransferase	688

27.2.2	Fucosylations	690
	Human Milk α1,3/4-Fucosyltransferase	690
	FucT III and VI	692
	FucT V	692
27.2.3	Sialylations	692
	α2,3-Sialyltransferase and α2,6-Sialyltransferase	692
27.2.4	N-Acetylglucosaminylation	696
	N-Acetylglucosaminyltransferase I, II, and III	696
	N-Acetylglucosaminyltransferase V	698
27.3	Summary	700
	Acknowledgments	700
	References	700

28 Solid-Phase Synthesis with Glycosyltransferases 705
Claudine Augé, Christine Le Narvor, and André Lubineau

28.1	Introduction	705
28.2	General Aspects	705
28.3	Enzymatic Synthesis on Insoluble Supports	707
28.3.1	Enzymatic Synthesis of Oligosaccharides	707
	Use of an Amino-Functionalized Water-Compatible Polyacrylamide Gel	707
	Use of a Sepharose Matrix	708
	Use of Controlled-Pore Glass	711
28.3.2	Enzymatic Synthesis of Glycopeptides	712
	Use of Controlled-Pore Glass	712
	Use of Polyethylene Glycol Polyacrylamide (PEGA)	715
28.4	Enzymatic Synthesis of Oligosaccharides and Glycoconjugates on Soluble Supports	715
28.4.1	Enzymatic Synthesis of Oligosaccharides	715
	Use of Water-Soluble Amino-Substituted Poly(vinyl alcohol)	715
	Use of Water-Soluble Glycopolymer Synthesized by Polymerization	717
28.4.2	Enzymatic Synthesis of Glycolipids on Water-Soluble Polyacrylamide–Poly(N-acryloxysuccinimide) (PAN)	718
	References	722

29 Glycosidase-Catalysed Oligosaccharide Synthesis 723
David J. Vocadlo and Stephen G. Withers

29.1	Introduction	723
29.2	Background on Glycosidases	723
29.3	Basic Mechanisms	724
29.4	Synthesis by the 'Thermodynamic' Approach	724
29.5	The Kinetic Approach	728
29.6	Recent Developments and New Directions	732
	References	838

30	**Production of Heterologous Oligosaccharides by Recombinant Bacteria (Recombinant Oligosaccharides)**	845
	Roberto A. Geremia and Eric Samain	
30.1	Introduction	845
30.2	Concept and Methodology of Heterologous ('Recombinant') Oligosaccharide Production in *E. coli*.	847
30.2.1	Biosynthesis of Nod Factors	847
30.2.2	Expression Systems and Cloning Strategy	849
30.2.3	High Cell-Density Cultivation	851
30.2.4	Purification of Recombinant Oligosaccharides	852
30.3	Examples of Recombinant Oligosaccharides	852
30.3.1	Production of Chitin Oligosaccharides in *E. coli* Expressing NodC	852
30.3.2	Production of Nod Factor Precursors	853
30.3.3	Production of Derivatives of *N*-Acetyllactosamine................	855
30.4	Conclusions and Future Perspectives	856
30.4.1	Production of Labeled Chitin Oligosaccharides to Study Their Interactions with Proteins	856
30.4.2	Improvement of Oligosaccharide Production, and Metabolic Engineering	858
30.4.3	Production of More Complex Oligosaccharides...................	858
	Acknowledgments	859
	References	859
IV	**Carbohydrate–Protein Interactions**...............................	861
31	**Protein–Carbohydrate Interaction: Fundamental Considerations** ...	863
	Nikki F. Burkhalter, Sarah M. Dimick, and Eric J. Toone	
31.1	Introduction	863
31.2	Association in Aqueous Solution	864
31.2.1	Gas Phase Non Covalent Interactions.............................	864
	Dipole–Dipole Interactions	864
	Dipole–Induced Dipole	866
	Dispersive Interactions	867
	Specific Forces: Hydrogen Bonding and *n*-σ Bonding..............	868
31.2.2	The Effect of Water on Intermolecular Interactions................	869
	Coulombic Stabilization...	870
	Hydrogen Bonding	871
	Dispersive Interactions	872
31.2.3	'Hydrophobic' Interactions..	872
31.3	The Evaluation of Protein–Carbohydrate Binding.................	876
31.3.1	Precipitin Assays...	877
31.3.2	Enzyme-Linked Lectin Assay (ELLA).............................	878
31.3.3	Isothermal Titration Microcalorimetry	878
31.4	The Interpretation of Calorimetric Data...........................	882

31.4.1	Solvation/Desolvation	882
	Solvation Entropy	883
	Translational/Rotational Entropy	884
31.4.2	Other Contributions to Thermodynamics of Association	885
	Proton Transfer	885
	Salt Effects/Binding Site Reorganization	885
31.4.3	van't Hoff versus Calorimetric Enthalpies	886
31.5	The Thermodynamics of Protein–Carbohydrate Interaction	887
31.6	The Role of Multivalency in Protein–Carbohydrate Interaction	901
31.6.1	Phenomenology	901
31.6.2	The Energetic Consequence of Ligand Linkage	905
	Enthalpic Contributions to ΔG_i	906
	Entropic Contributions to ΔG_i	907
31.6.3	A Molecular Basis for the Cluster Glycoside Effect	910
	Acknowledgments	911
	References	911
32	**Structural Analysis of Oligosaccharides: FAB-MS, ES-MS and MALDI-MS**	**915**
	Anne Dell, Howard R. Morris, Richard Easton, Stuart Haslam, Maria Panico, Mark Sutton-Smith, Andrew J. Reason, and Kay-Hooi Khoo	
32.1	Introduction	915
32.2	Fast Atom Bombardment-Mass Spectrometry (FAB-MS)	915
32.3	Matrix Assisted Laser Desorption Ionization-Time of Flight-Mass Spectrometry (MALDI-TOF-MS)	917
32.4	Electrospray-Mass Spectrometry (ES-MS)	918
32.5	Appearance of Mass Spectra Obtained in FAB-MS, MALDI-MS and ES-MS Experiments	919
32.6	Assignment of Mass Values	921
32.7	Derivatisation	921
32.8	Fragmentation Pathways	922
32.9	Protocols for MS Analysis	924
32.9.1	Protocol 1—Sample Loading for FAB-MS Analysis	924
32.9.2	Protocol 2—Sample Loading for NanoES-MS and MS-MS Analysis on the Q-TOF	925
32.9.3	Protocol 3—Sample Loading for LC-ES-MS and LC-ES-MS-MS on the Q-TOF	925
32.9.4	Protocol 4—Sample Loading for MALDI-MS Analysis	925
32.10	Applications of FAB-MS, MALDI-MS and ES-MS in Glycobiology	926
32.10.1	Case Study 1—Molecular Weight Profiling of Polysaccharides by MALDI-MS	926
32.10.2	Case Study 2—Analysis of Glycoproteins by LC-ES-MS and FAB-MS	927

32.10.3	Case Study 3—Characterization of a Novel *N*-Glycan by FAB-MS and FAB-MS-MS	930
32.10.4	Case Study 4—High Sensitivity Sequencing of a Novel Glycopeptide by Q-TOF ES-MS-MS and MALDI-MS	933
32.10.5	Case Study 5—FAB-MS Screening of Biological Samples for Glycan Content	935
32.10.6	Case Study 6—MS Analysis of Mycobacterial Glycoconjugates	942
32.11	Concluding Remarks	944
	References	945
33	**Conformational Analysis in Solution by NMR**	**947**
	S. W. Homans	
33.1	Introduction	947
33.2	Solution Conformations of Oligosaccharides	947
33.2.1	The NMR Technique	947
33.2.2	Conformational Parameters in Oligosaccharides	948
33.2.3	Conformational Restraints	949
33.2.4	^{13}C Isotopic Enrichment	949
33.2.5	Additional Conformational Restraints	950
	Exchangeable Protons	950
	Heteronuclear Overhauser Effects	952
	^{13}C–^{13}C Coupling-Constants	953
	Dipolar Couplings	954
33.3	Experimental Restraints in Conformational Analysis	955
33.3.1	Restraining Protocol	955
	Biharmonic Restraints	955
	Time-Dependent Restraints	957
33.3.2	Dynamical Simulated Annealing	957
33.4	Analysis of Oligosaccharide Dynamics	958
33.4.1	Monte-Carlo Simulations	959
33.4.2	Molecular Dynamics Simulations	959
33.5	A Case Study on Neu5Acα2-3Galβ1-4Glc	959
33.5.1	Resonance Assignments in Neu5Acα2-3Galβ1-4Glc	960
33.5.2	ROE Connectivities	960
33.5.3	'Global Minimum' Conformation of Neu5Acα2-3Galβ1-4Glc	961
33.5.4	Conformational Dynamics of Neu5Acα2-3Galβ1-4Glc	962
33.5.5	Short-range vs Long-range Restraints	963
33.6	Conclusions	966
	References	966
34	**Oligosaccharide Conformations by Diffraction Methods**	**969**
	Serge Pérez, Catherine Gautier, and Anne Imberty	
34.1	Introduction	969
34.2	General Analysis	970
34.3	Crystalline Conformations of Disaccharide Moieties	973
34.3.1	The Disaccharides	973

34.3.2	The Analogs (S, C, N, ...)	985
34.4	Hydrogen Bonding in Crystalline Oligosaccharides	987
34.5	Packing Features	988
34.6	Selected Examples	990
34.7	Crystalline Conformations of Oligosaccharides Complexed with Lectins	992
34.8	Concluding Remarks	996
	References	998

35	**Transfer NOE Experiments for the Study of Carbohydrate–Protein Interactions**	**1003**
	Thomas Peters	
35.1	Introduction	1003
35.2	The Transfer NOE Experiment	1004
35.3	Measurement of trNOEs	1006
35.4	Bioactive Conformations of Carbohydrate Ligands From trNOE Experiments	1008
35.5	Spin Diffusion may Generate Misleading Distance Constraints	1009
35.6	The Conformation of Sialyl Lewis^x Bound to E-selectin	1011
35.7	Interaction of Bacterial Lipopolysaccharide Fragments with Monoclonal Antibodies	1016
35.8	Conclusions and Future Directions	1019
	References	1021

36	**Carbohydrate–Protein Interactions: Use of the Laser Photo Chemically Induced Dynamic Nuclear Polarization(CIDNP)-NMR Technique**	**1025**
	Hans-Christian Siebert and Johannes F. G. Vliegenthart	
36.1	Introduction	1025
36.2	The CIDNP Method	1026
36.3	CIDNP-related Molecular Modelling	1027
36.4	Applications	1027
36.5	Hevein-like Lectins	1029
36.6	Galactoside-binding Lectins from Plant and Animal Origin	1032
36.7	Sialidase from *Clostridium Perfringens* (Wild Type and Mutants)	1037
36.8	CIDNP Analysis of Glycoproteins	1039
36.9	Conclusions	1040
	Acknowledgments	1041
	References	1042

37	**Biacore**	**1045**
	Wolfgang Jäger	
37.1	Introduction	1045
37.1.1	Real-time Analysis by Surface Plasmon Resonance	1045
37.1.2	Information in a Sensorgram	1047
37.2	Experimental Procedures	1048

37.2.1	Immobilization of Biomolecules at the Sensor Surface	1048
37.2.2	Surface Regeneration	1050
37.2.3	Interaction Analysis and Controls	1051
37.2.4	Determination of Kinetic Rate Constants	1052
37.2.5	Affinity Determination	1053
37.3	Application Areas	1054
37.3.1	Selectin Binding to a Glycoprotein Ligand	1054
37.3.2	Oligosaccharide Characterization	1055
37.3.3	*In situ* Modification of Immobilized Carbohydrates	1056
	References	1056
V	**Carbohydrate–Carbohydrate Interactions**	1059
38	**Carbohydrate–Carbohydrate Interactions**	1061
	Dorothe Spillmann and Max M. Burger	
38.1	Introduction	1061
38.2	From Structural Components to Cell Recognition	1063
38.2.1	Carbohydrate–Carbohydrate Interactions as Part of Structural Components	1063
	Extracellular Matrix of Seaweeds—Agarose, Carrageenan and Alginate	1063
	Cell Walls	1064
	Mammalian Extracellular Matrix Components	1066
38.2.2	Carbohydrate–Carbohydrate Interactions as Part of Recognition Keys?	1068
	Carbohydrate Interactions in Invertebrates—The Marine Sponge *Microciona prolifera* as a Model System	1069
	Carbohydrate Interactions in Vertebrates—Embryonal and Tumor Cells	1071
	Repulsive Carbohydrate–Carbohydrate Interactions	1072
38.3	Molecular Aspects of Carbohydrate Interactions	1074
38.3.1	Polyvalence to Inforce Weak Interactions	1074
38.3.2	Arrangement of Motifs and the Possibility to Control Specificity	1075
38.3.3	Molecular Basis of Carbohydrate–Carbohydrate Interactions	1076
38.4	Experimental Approaches	1078
38.4.1	General Considerations	1078
38.4.2	Affinity Interactions	1079
	Cell Binding Studies	1079
	Aggregation of *de novo* Complexes	1081
	Affinity Chromatography	1082
	Distribution between Compartments	1082
38.4.3	Microscopy	1083
	Electron Microscopy	1083
	Atomic Force Microscopy	1083
38.4.4	Crystallography	1084

38.4.5	Mass Spectrometry	1085
38.4.6	Nuclear Magnetic Resonance	1085
38.4.7	Molecular Modelling	1086
38.4.8	Tools	1086
	Synthetic Oligosaccharides	1086
	Antibodies against Carbohydrate Motifs	1087
	Cells	1088
	References	1088
VI	**Carbohydrate–Nucleic Acid Interactions**	1093
39	**Carbohydrate–Nucleic Acid Interactions**	1095
	Heinz E. Moser	
39.1	Introduction	1095
39.2	Carbohydrates Binding to DNA	1096
39.2.1	Ene-Diyne Antibiotics and Antitumor Agents	1096
	Esperamicins	1096
	Calicheamicins	1100
39.2.2	Anthracyclins	1106
39.2.3	Pluramycins and Aureolic Acids	1111
39.3	Carbohydrates Binding to RNA	1112
39.3.1	Aminoglycosides	1113
	References	1120
	Index	I 1

Part II Biology of Saccharides

Vol. 3 Biosynthesis and Degradation of Glycoconjugates

	Introduction to Volumes 3 and 4	V
	Abbreviations Used in Volumes 3 and 4	LV
I	**Biosynthesis of Glycoconjugates**	1
1	**Metabolism of Sugars and Sugar Nucleotides**	3
	Hudson H. Freeze	
1.1	Introduction	3
1.2	Basic Principles	3
1.3	Transporters Deliver Monosaccharides to Cells	4
1.4	Intracellular Sources of Sugars	5
1.4.1	Salvage	5

1.4.2	Activation and Interconversion of Monosaccharides	6
	Glycogen	6
	Glucose	7
	Glucuronic acid	8
	Iduronic acid	8
	Xylose	8
	Mannose	8
	Fucose	9
	Galactose	10
	N-Acetylglucosamine	10
	N-Acetylgalactosamine	10
	Sialic acids	11
1.5	Sugar Nucleotide Transporters	11
1.6	Control of Sugar Nucleotide Levels	13
1.7	Possible Future Directions	13
	References	14
2	**Nucleotide Sugar Transporters**	19
	Rita Gerardy-Schahn and Matthias Eckhardt	
2.1	Introduction	19
2.2	General Considerations	20
2.3	The Requirement for Nucleotide Sugar Transporters and Their Mechanism of Function: A Comprehensive Overview of the Last 20 Years	20
2.4	Molecular Cloning of Nucleotide Sugar Transporters	22
2.5	The Structure of Nucleotide Sugar Transporters	25
2.6	The Subcellular Distribution of Nucleotide Sugar Transporters	27
2.7	Molecular Defects that Cause Inactive UDP-Galactose and CMP-Sialic Acid Transporters	28
2.8	Association Between Defects in Nucleotide Sugar Transporters and Diseases	29
2.9	Involvement of Nucleotide Sugar Transporters in the Regulation of Glycosylation	29
2.10	Future Perspectives	30
	Acknowledgements	31
	References	32
3	**Biosynthesis of Oligosaccharyl Dolichol**	37
	Sharon S. Krag	
3.1	General Overview	37
3.2	Oligosaccharyl Dolichol	38
3.3	Key Enzymatic Steps in the Assembly Process	39
3.4	Topology of the Assembly Process	42
3.5	Utilization of Oligosaccharyl Dolichol	42
	Acknowledgment	43
	References	43

4	**Biochemistry and Molecular Biology of the *N*-Oligosaccharyltransferase Complex**..	45
	Roland Knauer and Ludwig Lehle	
4.1	Introduction..	45
4.2	Biochemistry of OST	46
4.2.1	Lipid–Saccharide Donor	47
4.2.2	Acceptor Specificity of OST..	48
4.2.3	Catalytic Mechanism of OST	49
4.2.4	Regulation of OST Activity	51
4.3	Isolation of OST Complexes from Different Sources	51
4.4	Molecular Biology of OST	52
4.4.1	WBP1/OST48...	54
4.4.2	SWP1/Ribophorin II	54
4.4.3	OST1/Ribophorin I...	55
4.4.4	OST3/OST6...	55
4.4.5	OST5	56
4.4.6	OST4	56
4.4.7	OST2/DAD1..	57
4.4.8	STT3	58
4.5	Structural Organization of the OST Complex	59
	Acknowledgments	60
	References..	60
5	**Processing Enzymes Involved in the Deglucosylation of *N*-Linked Oligosaccharides of Glycoproteins: Glucosidases I and II and Endomannosidase**	65
	Robert G. Spiro	
5.1	Introduction..	65
5.2	Glucosidase I..	66
5.3	Glucosidase II...	68
5.4	Endo-α-mannosidase..	70
5.5	Concerted Action of Deglucosylation Enzymes.......................	72
5.6	Mutants	74
5.7	Role of Monoglucosylated *N*-Linked Oligosaccharides and Glucose Trimming Enzymes in Regulating Quality Control of Glycoproteins	75
5.8	Effect of Glucosidase Inhibitors on Viral Proliferation..............	77
	Acknowledgments	78
	References..	78
6	**α-Mannosidases in Asparagine-linked Oligosaccharide Processing and Catabolism**...	81
	Kelley W. Moremen	
6.1	Overview	81
6.2	Introduction..	82

6.2.1	Roles of *N*- and *O*-Linked Glycans and Compartmentalization of Biosynthetic and Catabolic Reactions	82
6.2.2	Processing of Asn-Linked Oligosaccharides	82
6.2.3	Early Trimming Events: importance for quality control glycoprotein degradation and anteriograde transport	85
6.2.4	Glycoprotein Catabolism: multiple routes for glycoprotein breakdown	87
6.2.5	Consequences of Genetic Defects in Oligosaccharide Biosynthesis and Catabolism	88
6.3	Mannosidases in Glycoprotein Processing and Catabolism	89
6.3.1	Classification of Mannosidases	89
6.3.2	Class 1 Mannosidases: enzymes of the ER and Golgi	93
	ER mannosidase I subfamily	93
	Golgi mannosidase I sub-family	95
	Fungal secreted mannosidases	97
	New genes with unknown functions	98
6.3.3	Class 2 Mannosidases: enzymes of the cytosol, ER, Golgi, and Lysosomes	98
	Golgi mannosidase II	99
	Lysosomal mannosidase	101
	Epididymal/sperm mannosidase	103
	Heterogeneous cluster of mannosidase homologs among eukarya, eubacteria, and archaea	104
6.4	Conclusions and Future Prospects	106
	Acknowledgments	107
	References	107
7	**The Role of UDP-Glcyglycoprotein Glucosyltransferase as a Sensor of Glycoprotein Conformations**	119
	Armando J. Parodi	
7.1	Introduction	119
7.2	General Properties	120
7.3	GT Recognizes Glycoprotein Conformations	121
7.4	The Primary Structure of the UDP-Glcyglycoprotein Glucosyltransferase	122
7.5	The Role of Monoglucosylated Oligosaccharides in Glycoprotein Folding	123
	Acknowledgments	126
	References	127
8	**Mannosyltransferases**	129
	Peter Orlean	
8.1	Introduction	129
8.2	Occurrence of Covalently-linked Mannose	130

8.2.1	Eukaryotic Secretory Glycoproteins	130
8.2.2	Glycophospholipids	130
8.2.3	Eubacterial and Archaeal Mannose-containing Molecules	130
8.2.4	C-linked Mannose	130
8.3	Biochemistry of Mannosyl Transfer	131
8.3.1	Many Linkages, Two Donors	131
8.3.2	Donor Specificity	131
8.3.3	Acceptor Specificity	132
8.3.4	Structural Features of Man-T	132
8.4	Man-T Families and the Pathways They Participate in	133
8.4.1	Man-Ts of the ER [1–5]	134
	Alg1p	134
	Alg2p/Alg11p	134
	Dpm1p	135
	Alg3p	135
	Alg9p/PIG-Bp family	135
	Pmt1p family	136
8.4.2	Golgi Man-Ts and Fungal Mannan Synthesis	136
	Och1p family	137
	Mnn9p family	137
	Mnn10p/Mnn11p family	138
	Mnn1p family	138
	Ktr1p family	138
8.4.3	"Missing" Eukaryotic Man-T	138
8.4.4	Eubacterial and Archaeal Man-T	139
8.5	Coordinating Man Transfer with the Cell Cycle and Morphogenesis	139
8.6	Concluding Remarks	140
	Acknowledgments	140
	References	140
9	**Branching of *N*-Glycans: *N*-Acetylglucosaminyltransferases**	145
	Harry Schachter	
9.1	Introduction	145
9.2	Processing of *N*-Glycans within the Endomembrane Assembly Line	146
9.3	General Properties of the *N*-Acetylglucosaminyltransferases	148
9.3.1	Domain Structure	148
9.3.2	Targeting to the Golgi Apparatus	150
9.4	UDP-GlcNAc:Manα1-3R [GlcNAc to Manα1-3] β-1,2-*N*-Acetylglucosaminyltransferase I (GnT I, EC 2.4.1.101)	150
9.5	UDP-GlcNAc:Manα1-6R [GlcNAc to Manα1-6] β-1,2-*N*-Acetylglucosaminyltransferase II (GnT II, E.C. 2.4.1.143)	152
9.6	The Role of GnT I and II in Mammalian Development	153

9.7	UDP-GlcNAc:R_1-Manα1-6[GlcNAcβ1-2Manα1-3]Manβ1-4R_2 [GlcNAc to Manβ1-4] β-1,4-N-Acetylglucosaminyltransferase III (GnT III, E.C. 2.4.1.144)	155
9.7.1	Overexpression of GnT III Activity	156
9.7.2	GnT III Activity and Cancer	157
9.8	UDP-GlcNAc:R_1Manα1-3R_2 [GlcNAc to Manα1-3] β-1,4-N-Acetylglucosaminyltransferase IV (GnT IV, E.C. 2.4.1.145)	157
9.9	UDP-GlcNAc:R_1Manα1-6R_2 [GlcNAc to Manα1-6] β-1,6-N-Acetylglucosaminyltransferase V (GnT V, E.C.2.4.1.155)	158
9.9.1	GnT V Activity and Cancer	159
9.10	UDP-GlcNAc:R_1(R_2)Manα1-6R_3 [GlcNAc to Manα1-6] β-1,4-$N1$-Acetylglucosaminyltransferase VI (GnT VI)	161
9.11	GnT VII and GnT VIII	161
	References	162
10	**The Galactosyltransferases**	175
	Nancy L. Shaper, Martin Charron, Neng-Wen Lo, Jane R. Scocca, and Joel H. Shaper	
10.1	Introduction	175
10.2	Using the Databanks to Obtain Information on the Galactosyltransferases	177
10.2.1	Nomenclature	177
10.3	The Dual Role of β4-Galactosyltransferase-I (β4GalT-I) in Oligosaccharide and Lactose Biosynthesis: The Early Days	178
10.3.1	β4GalT-I: Isolation and Characterization of cDNA Clones	181
10.3.2	The Murine β4GalT-I Gene: Genomic Organization and Structure of the 5′-End	181
10.3.3	β4GalT-I and Lactose Biosynthesis	182
10.3.4	β4GalT-I and the Vertebrate β4GalT Gene Family	182
10.3.5	Evolution of the β4-Galactosyltransferase Gene Family	184
10.4	The Vertebrate β3Galactosyltransferase (β3GalT) Gene Family	185
10.4.1	General Characteristics of the β3-Galactosyltransferase Gene Family Members	186
10.4.2	β3GalT-IV: UDP-galactose:GM1 β3-galactosyltransferase (GM1 Synthase; GalT-3)	187
10.4.3	Other Vertebrate β-Galactosyltransferase Activities	187
10.4.4	UDP-Galactose:Ceramide β-Galactosyltransferase (CGalT; EC 2.4.1.45)	187
10.5	The Vertebrate α3-Galactosyltransferase Gene Family	188
10.5.1	α3-Galactosyltransferase (α3GalT: UDP-Gal:Galβ4GlcNAcα3-Galactosyltransferase; EC 2.4.1.87)	188
10.5.2	The Blood Group B α3-Galactosyltransferase (EC 2.4.1.37)	190
10.5.3	The Forssman Glycolipid Synthetase (EC 2.4.1.88)	191
10.5.4	Evolution of the α3GalT Gene Family	191

10.6	A UDP-Gal:Galβ3GalNAc α4Galactosyltransferase Activity	192
	Acknowledgments	192
	References	192

11 Fucosyltransferases ... 197
Ernesto T. A. Marques, Jr.

11.1	Introduction	197
11.2	General Characteristics	198
11.2.1	Nomenclature	198
11.2.2	Gene Structure	199
11.2.3	Sequence Peptide Motifs	199
11.2.4	Specificity	199
11.2.5	Protein Structure and Topology	200
11.2.6	Enzymatic Reaction Mechanism	201
11.2.7	Inhibitors	203
11.3	Specific Fucosyltransferases	203
11.3.1	GDP-Fucose: Fucα1(Fucα1,2Fuc)α2-fucosyltransferase	204
11.3.2	GDP-Fucose: Galβ1(Fucα1,2Gal)α2-fucosyltransferase	204
11.3.3	GDP-Fucose: Galβ1,4/3GlcNAc(Fucα1,3/4GlcNAc)α3/4-fucosyltransferases	204
	Blood group Lewis: FucT III, V and VI	204
	Myeloid enzyme: FucT IV	205
	Leukocyte enzyme: FucT VII	206
	Neuronal enzyme: FucT IX	206
11.3.4	GDP-Fucose: Galβ1,3GlcNAc(Fucα1,3GlcNAc) bacterial (*Helicobacter pylori*) α3-fucosyltransferase	207
11.3.5	GDP-Fucose: GlcNAc-N(Fucα1,6GlcNAc)α6fucosyltransferases	207
11.3.6	GDP-Fucose: *O*-Ser(Fucα1-*O*-Ser)GlcNAc polypeptide fucosyltransferases	207
11.3.7	Unconventional Types of Fucosylation: Fucβ1-P-Ser and cytoplasmic Fucα1,2-Galβ1,3-GlcNAc-Pro (*Dictyostelium discoideum*)	207
	Fucβ1-P-Ser	207
	Fucα1,2-Galβ1,3-GlcNAc-Pro	208
	Acknowledgments	208
	References	208

12 Sialyltransferases ... 213
Joseph T.Y. Lau and Sherry A. Wuensch

12.1	Introduction	213
12.2	General Features of Sialyltransferases	213
12.3	Cloning and Identification Strategies for Sialyltransferases	215
12.4	Sialyltransferase Classification and Nomenclature	216
12.5	The α2,3-ST Family	216
12.6	The α2,6-ST Family	217
12.7	The α2,8-ST Family	218

12.8	Regulation and Functionality of Sialyltransferases................	219
	References	221
13	**Biochemistry of Sialic Acid Diversity**...........................	227
	Roland Schauer	
13.1	Introduction	227
13.2	Occurrence and Biosynthesis.....................................	227
13.3	General Biological Functions	229
13.4	N-Glycolylneuraminic Acid	231
13.5	O-Acetylated Sialic Acids	234
13.6	O-Methylated and O-Sulfated Sialic Acids	238
	Acknowledgments	239
	References	239
14	**Carbohydrate Sulfotransferases**..................................	245
	Steven D. Rosen, Annette Bistrup, and Stefan Hemmerich	
14.1	Introduction	245
14.2	Basic Features of Sulfotransferase Reactions.....................	245
14.3	Tyrosine Sulfation	246
14.4	Diversity of Carbohydrate Sulfation..............................	246
14.5	Biochemical Demonstration of Carbohydrate Sulfotransferases ...	249
14.6	Molecular Cloning of Carbohydrate Sulfotransferases.............	250
14.7	Primary Structures of Carbohydrate Sulfotransferases.............	252
	Acknowledgments	256
	References	256
15	**Novel Variant Pathways in Complex-type Oligosaccharide Synthesis**	261
	Dirk H. van den Eijnden	
15.1	Introduction	261
15.2	The lacNAc Pathway of Complex-type Oligosaccharide Synthesis	261
15.3	Occurrence and Biology of lacdiNAc-based Complex-type Oligosaccharides...	262
15.4	Biosynthesis of lacdiNAc Backbone Units........................	263
15.5	The lacdiNAc Pathway of Complex-type Oligosaccharide Synthesis	264
15.6	Other Shared Properties of β4-GalT and β4-GalNAcT	266
15.7	Cloning of a snail UDP-GlcNAc:GlcNAcβ-R β4-N-acetylglucosaminyltransferase	266
15.8	The Chitobio Pathway of Complex-type Oligosaccharide Synthesis	267
15.9	Competition Between Pathways...................................	267
15.10	Concluding Remarks	269
	References	269

16	**Control of Mucin-Type *O*-Glycosylation: *O*-Glycan Occupancy is Directed by Substrate Specificities of Polypeptide GalNAc-Transferases**	273
	Helle Hassan, Eric P. Bennett, Ulla Mandel, Michael A. Hollingsworth, and Henrik Clausen	
16.1	Introduction..	273
16.2	The Mammalian UDP-GalNAc: Polypeptide GalNAc-Transferase Gene Family	274
16.3	The GalNAc-Transferase Gene Family is Evolutionarily Old	276
16.4	The Kinetic Properties of GalNAc-Transferase Isoforms are Different..	278
16.4.1	Lessons from in vivo Analysis of GalNAc-transferase Substrate Specificities...	279
16.4.2	Lessons from in vitro Analysis of the Acceptor Substrate Specificities of GalNAc-transferase Isoforms	280
	Isoforms may have distinct acceptor substrate specificities.........	281
	Isoforms may have overlapping substrate specificities..............	283
	Isoforms may act in different order on substrates with multiple acceptor sites...	283
	Isoforms may require prior (GalNAc) glycosylation	283
16.5	Expression of the GalNAc-Transferase Genes are Differentially Regulated	285
16.6	Predictive Value of in vitro *O*-glycosylation?	288
16.7	Conclusions and Future Perspectives..................................	288
	References..	289
17	**Glycosyltransferase Inhibitors**..	293
	Xiangping Qian and Monica M. Palcic	
17.1	Introduction..	293
17.2	Inhibitors of Glycosyltransferases	296
17.2.1	Inhibitors of Galactosyltransferases...................................	296
	Inhibitors of β1,4-galactosyltransferase	296
	Inhibitors of α1,3-galactosyltransferase	297
17.2.2	Inhibitors of Fucosyltransferases	298
	Inhibitors of α1,2-fucosyltransferases	300
	Inhibitors of α1,3-fucosyltransferases	300
17.2.3	Inhibitors of Sialyltransferases	301
	Inhibitors of α2,6-sialyltransferase	302
	Inhibitors of α2,3-sialyltransferase	304
17.2.4	Inhibitors of *N*-Acetylglucosaminyltransferases	305
17.2.5	Inhibitors of Human Blood Group A and B Glycosyltransferases.	306
17.3	Summary..	309
	Acknowledgments	309
	References..	309

18	**Biosynthesis of the *O*-Glycan Chains of Mucins and Mucin Type Glycoproteins**	313
	Inka Brockhausen	
18.1	Summary	313
18.2	Introduction	313
18.3	Structures of *O*-Glycans	314
18.4	Functions of Mucin Type *O*-Glycans	314
18.5	Primary *O*-Glycosylation	315
18.6	Synthesis of *O*-Glycan Core 1	315
18.7	Synthesis of *O*-Glycan Core 2	317
18.8	Synthesis of *O*-Glycan Core 3	319
18.9	Synthesis of *O*-Glycan Core 4	319
18.10	Synthesis of *O*-Glycan Cores 5–8	319
18.11	Elongation and Branching Reactions	320
18.12	Synthesis of Terminal Structures	321
	Acknowledgments	324
	References	324
19	**Glycosyltransferases in Glycosphingolipid Biosynthesis**	329
	Subhash Basu, Kamal Das, and Manju Basu	
19.1	Introduction	329
19.2	Fucosyltransferases in Glycolipid Biosynthesis	329
19.3	Galactosyltransferases in Glycolipid Biosynthesis	332
19.4	*N*-Acetylgalactosaminyltransferases in Glycolipid Biosynthesis	334
19.5	*N*-Acetylglucosaminyltransferases in Glycolipid Biosynthesis	336
19.6	Sialyltransferases in Glycolipid Biosynthesis	337
19.7	Glucuronyltransferases in Glycolipid Biosynthesis	340
	Acknowledgments	342
	References	342
20	**Biosynthesis of Glycogen**	349
	Peter J. Roach	
20.1	Summary	349
20.2	Introduction	350
20.3	Glycogenin and the Initiation of Glycogen Synthesis	351
20.3.1	History	351
20.3.2	Properties	351
20.3.3	Reaction Mechanism	352
20.3.4	Domain Structure	352
20.3.5	Function	354
20.4	Glycogen Synthase and the Bulk Synthesis of Glycogen	354
20.4.1	Properties	354
20.4.2	Structure of Glycogen Synthase	355
20.4.3	Branching Enzyme	356
20.5	Intermediates in the Biosynthesis of Glycogen	357

20.6	Conclusion	358
	Acknowledgments	359
	References	359

21 Biosynthesis of Hyaluronan ... 363
Paraskevi Heldin and Torvard C. Laurent

21.1	Introduction	363
21.2	Site of Biosynthesis	364
21.3	Biosynthetic Precursors	364
21.4	Hyaluronan Synthases	365
21.4.1	Microbial Enzymes	365
21.4.2	Vertebrate Synthases	366
21.5	Mechanism of Synthesis	367
21.5.1	Chain Elongation	368
21.5.2	Translocation	369
21.5.3	Shedding	369
21.6	Regulation of HA Synthesis	370
21.7	Concluding Remarks	371
	Acknowledgments	372
	References	372

22 Biosynthesis of Chondroitin Sulfate and Dermatan Sulfate Proteoglycans ... 375
Geetha Sugumaran and Barbara M. Vertel

22.1	Introduction	375
22.2	Proteoglycan Structure	379
22.2.1	Proteoglycans and Their Core Proteins	379
22.2.2	What Initiates GAG Chain Addition?	381
22.2.3	The Linkage Region	381
22.2.4	CS and DS Chains	382
22.3	Biosynthesis of CS and DS Proteoglycans	383
22.3.1	Biosynthesis of the Core Protein	383
22.3.2	Origin of Sugar and Sulfate Precursors	384
22.3.3	Addition of the Linkage Oligosaccharides	385
	Xylosylation	385
	Galactosylation	386
	Addition of GlcA and completion of the common tetrasaccharide linkage region	387
	Initiation of CS/DS chains by addition of the first GalNAc	388
22.3.4	Formation of the CS/DS Chains	388
	Addition of the repeating disaccharides	388
	Epimerization of GlcA to IdoA to form DS	389
	Sulfation of GalNAc	390
	Sulfation of uronic acid	391

22.4	Concluding Remarks/Perspectives	391
	Acknowledgments	392
	References	392

23	**Biosynthesis of Heparin and Heparan Sulfate Proteoglycans**	395
	Lena Kjellén and Ulf Lindahl	
23.1	Introduction	395
23.2	The Proteoglycans: Structure, Location and Functions	396
23.3	Biosynthesis of the Polysaccharide Backbone	396
23.4	Outline of Polymer-Modification Reactions	397
23.4.1	The *N*-Deacetylase/*N*-Sulfotransferases	399
23.4.2	The C5-Epimerase	399
23.4.3	The 2-*O*-Sulfotransferase	399
23.4.4	The 6-*O*-Sulfotransferases	400
23.4.5	The 3-*O* Sulfotransferases	400
23.5	The Products, Heparin and Heparan Sulfate	400
23.6	Interactions with Proteins	401
23.7	Regulation of HS Biosynthesis	402
	References	403

24	**Biosynthesis of Proteoglycans with Keratan Sulfates**	407
	James L. Funderburgh	
24.1	Introduction: Keratan Sulfate Renaissance	407
24.2	Keratan Sulfate Structure and Distribution	407
24.2.1	Corneal KS	408
24.2.2	Non-corneal KSI	409
24.2.3	KSII	409
24.2.4	KSIII	410
24.3	Keratan Sulfate Proteoglycans	410
24.3.1	SLRPs	410
24.3.2	Aggrecan	411
24.3.3	Cell-Associated KS	411
24.3.4	Brain	412
24.4	Enzymatic Reactions of KS Biosynthesis	412
24.5	Metabolic Control of KS Synthesis	413
	Acknowledgments	414
	References	414

25	**The Biosynthesis of GPI Anchors**	417
	Yasu S. Morita, Alvaro Acosta-Serrano, and Paul T. Englund	
25.1	Introduction	417
25.2	Structure of GPI Anchors	417
25.2.1	Glycan Core Modifications	417

25.2.2	Variations in Anchor Lipid Structure	419
25.3	GPI Precursor Synthesis	419
25.3.1	GlcNAc-PI Synthesis	420
25.3.2	GlcNAc-PI Deacetylation	421
25.3.3	Inositol Acylation	421
25.3.4	GPI Mannosylation	422
25.3.5	Transfer of EtN-P	423
25.3.6	Lipid Remodeling	423
25.3.7	Addition of Carbohydrate Side Chains	424
25.3.8	Topology of Biosynthetic Pathways	424
25.4	Attachment of the GPI Precursor to a Protein	425
25.4.1	Basic Features	425
25.4.2	Protein Machinery for GPI Addition	426
25.4.3	Signal Sequence for GPI Addition	426
25.5	Evolution of GPI Biosynthesis	426
25.6	Future Studies	427
	Acknowledgments	427
	References	427
26	***Escherichia coli* Lipid A: A Potent Activator of Innate Immunity**	**435**
	Teresa A. Garrett and Christian R. H. Raetz	
26.1	Introduction	435
26.2	Structure of Lipopolysaccharide	435
26.3	Lipid A Biosynthesis in ***E. coli***	437
26.3.1	Acylation of UDP-GlcNAc	439
26.3.2	Disaccharide Formation	440
26.3.3	Phosphorylation by the Lipid A $4'$ Kinase	440
26.3.4	Kdo Addition and the Late Acyltransferases	441
26.3.5	Other Acyltransferases	442
26.3.6	Transport of Lipid A and the Role of MsbA	442
26.4	Lipid A Activation of Signal Transduction in Animal Cells	444
26.5	Summary	447
	Acknowledgments	447
	References	447
II	**Glycosidases**	**453**
27	**Lysosomal Degradation of Glycolipids**	**455**
	Thomas Kolter and Konrad Sandhoff	
27.1	Summary	455
27.2	Introduction	455
27.3	Mechanisms of Lysosomal Glycolipid Degradation	456
27.3.1	Glycosidases	456

27.3.2	Topology of Endocytosis and Lysosomal Glycolipid Degradation	457
27.3.3	Sphingolipid Activator Proteins	458
	The GM2-activator and its role in lysosomal digestion	459
	SAP-A to SAP-D	460
27.3.4	Lateral Pressure	460
27.3.5	Lipid Composition	461
27.3.6	Membrane Curvature	462
27.4	Degradation of Selected Lipids	462
27.4.1	Ganglioside GM2	462
27.4.2	Lactosylceramide	464
27.4.3	Glucosylceramide	464
27.4.4	Ceramide	465
27.4.5	Sphingomyelin	465
27.4.6	Sulfatide	465
27.4.7	Galactosylceramide	466
27.5	Pathobiochemistry	466
27.5.1	Animal Models for Sphingolipidoses	467
27.5.2	Therapy	469
27.6	Future Directions	470
	References	470
28	**Lysosomal Degradation of Glycoproteins**	473
	Nathan N. Aronson, Jr.	
28.1	Summary	473
28.2	Introduction	473
28.3	Roles of Lysosomes	474
28.4	Lysosomal Degradation of *N*-Linked Glycoproteins	475
28.4.1	General Features	475
28.4.2	Carbohydrate Digestion	476
28.4.3	Protein and Linkage Hydrolysis	476
28.5	Formation of Thyroid Hormone via Lysosomal Degradation of Thyroglobulin	477
28.5.1	Synthesis of Thyroid Hormone	477
28.5.2	Carbohydrate Degradation	478
28.5.3	Proteolysis	479
28.6	Degradation of Free Polymannose-Type Oligosaccharides Derived from *N*-Linked Glycoproteins During Biosynthesis	479
	References	481
29	**Sialidases**	485
	Garry Taylor, Susan Crennell, Carl Thompson, and Marina Chuenkova	
29.1	Abstract	485
29.2	Introduction	485

29.3	Influenza Virus Neuraminidase	486
29.4	Paramyxovirus Hemagglutinin-Neuraminidase (HN)	487
29.5	Non-Viral Sialidases	487
29.6	Small Sialidases	490
29.7	Large Sialidases	491
29.8	T. cruzi Trans-Sialidase (TS)	491
29.9	Conclusion	493
	Acknowledgments	494
	References	494
30	**Microbial Glycosidases**	497
	Kenji Yamamoto, Su-Chen Li, and Yu-Teh Li	
30.1	Exo-Glycosidases	497
30.1.1	α-Glucosidase	497
30.1.2	β-Glucosidase	498
30.1.3	α-Galactosidase	498
30.1.4	β-Galactosidase	499
30.1.5	α-Mannosidase	500
30.1.6	β-Mannosidase	500
30.1.7	β-*N*-Acetylhexosaminidase	501
30.1.8	α-*N*-Acetylgalactosaminidase	501
30.1.9	α-L-Fucosidase	502
30.1.10	β-D-Fucosidase	503
30.1.11	Sialidase	503
30.1.12	KDNase	504
30.1.13	α-L-Rhamnosidase	504
30.1.14	β-Xylosidase	505
30.2	Endo-Glycosidases	505
30.2.1	Endo-β-*N*-acetylglucosaminidase	505
30.2.2	Peptide-N-glycanase F	506
30.2.3	Endo-α-*N*-acetylgalactosaminidase	506
30.2.4	Endo-β-galactosidase	507
30.2.5	Endoglycoceramidase	507
	References	508
31	**Glycoprotein Processing Inhibitors**	513
	Magid Osser and Alan D. Elbein	
31.1	Introduction	513
31.2	Structural Classification	515
31.3	Distribution of Glycosidase Inhibitors in the Plant Kingdom	515
31.4	Isolation and Structural Determination	516
31.5	Glycosidase Inhibitory Activity	517
31.6	Structure-Activity Relationships	518
31.7	*N*-Linked Oligosaccharide Processing	519
31.8	Inhibitors of *N*-Linked Oligosaccharide Processing	522

31.8.1	Glucosidase Inhibitors	522
31.8.2	Mannosidase Inhibitors	525
31.9	Summary and Future Prospects	528
	References	529
	Index	I 1

Part II Biology of Saccharides

Vol. 4 Lectins and Saccharides Biology

III	**Lectins**	533
32	**Plant Lectins**	535
	Marilynn E. Etzler	
32.1	Summary	535
32.2	Introduction	535
32.3	Carbohydrate Specificity	536
32.4	Other Activities	539
32.5	Structure	540
32.6	Biological Roles	543
	Acknowledgments	546
	References	547
33	**Interactions of Oligosaccharides and Glycopeptides with Hepatic Carbohydrate Receptors**	549
	Yuan C. Lee and Reiko T. Lee	
33.1	Summary	549
33.2	Introduction	550
33.3	Molecular Characteristics of Hepatic Lectins	551
33.4	Cellular Aspects of HL	552
33.5	Binding Specificity	553
33.6	Photoaffinity Labeling	557
33.7	Subunit Organization on Rat Hepatocyte Surface	558
33.8	Applications	559
	References	560
34	**P-Type Lectins and Lysosomal Enzyme Trafficking**	563
	Patricia G. Marron-Terada and Nancy M. Dahms	
34.1	Introduction	563
34.2	Intracellular Trafficking of the MPRs	564
34.3	Primary Structure and Biosynthesis of the MPRs	566

34.3.1	CI-MPR	566
34.3.2	CD-MPR	567
34.4	Lysosomal Enzyme Recognition by the MPRs	569
34.5	Structural Determinants of Man-6-P Recognition	571
34.5.1	Expression of Mutant Forms of the MPRs	571
34.5.2	Crystal Structure of the CD-MPR	572
34.6	Conclusions	574
	Acknowledgments	574
	References	575
35	**The Siglec Family of I-Type Lectins**	**579**
	Paul R. Crocker and Soerge Kelm	
35.1	Introduction	579
35.2	The Immunoglobulin Superfamily and Carbohydrate Recognition	579
35.3	Siglecs as a Family of Sialic Acid Binding Proteins	580
35.4	Biology of Siglecs	581
35.5	Sialic Acids in Cellular Recognition	583
35.6	Mode of Carbohydrate Recognition by Siglecs	584
35.7	Importance of Multivalent Binding	588
35.8	Sialic Acid Recognition by the Immunoglobulin Fold—Evolutionary Considerations	588
35.9	Role of *cis* Interactions in Modulating Adhesion to Other Cells in *trans*	589
35.10	Sialoadhesin as a Macrophage Adhesion Molecule	590
35.11	Signalling Versus Adhesion Mediated by Siglecs	591
35.12	Conclusions	592
	Acknowledgments	592
	References	592
36	**C-Type Lectins and Collectins**	**597**
	Russell Wallis	
36.1	Summary	597
36.2	Structure and Function of C-Type Animal Lectins	598
36.2.1	The Carbohydrate-Recognition Domain	599
36.2.2	Ligand Binding	600
36.3	Mannose-Binding Protein and Collectins	601
36.3.1	Domain Organization	601
36.3.2	MBPs as Prototype Collectins	603
36.3.3	Ligand Binding by Serum MBP	603
36.3.4	MBP and Innate Immunity	604
36.3.5	Liver MBP	607
36.3.6	Pulmonary Surfactant Proteins	608
36.3.7	Conglutinin and CL-43	608
36.4	Conclusions	609

	Acknowledgments	609
	References	609
37	**Selectins**	613
	Rodger P. McEver	
37.1	Introduction	613
37.2	Structure of Selectins	613
37.3	Selectin Ligands	614
37.4	Requirements for Selectins to Mediate Tethering and Rolling of Leukocytes under Hydrodynamic Flow	619
37.5	Functions of Selectins and their Ligands in vivo	621
37.6	Conclusions	621
	References	622
38	**Galectins**	625
	Douglas N.W. Cooper and Samuel H. Barondes	
38.1	Introduction	625
38.2	Galectin Structure	626
38.3	Novel Candidate Galectins	631
38.4	Unorthodox Subcellular Targeting	635
38.5	Regulation of Galectin Expression	637
38.6	Galectin Binding Specificity and Identified Ligands	639
38.7	Physiological Functions	640
38.8	Summary	642
	References	642
IV	**Saccharide Biology**	649
39	**Structures and Functions of Nuclear and Cytoplasmic Glycoproteins**	651
	Robert S. Haltiwanger	
39.1	Introduction	651
39.2	O-Linked N-Acetylglucosamine (O-GlcNAc)	652
39.2.1	O-GlcNAc Appears to be a Regulatory Modification much like Phosphorylation	653
39.2.2	Modulation of Protein Stability and Function by O-GlcNAc	655
39.3	Other Forms of Nuclear and Cytoplasmic Glycosylation	658
39.3.1	Unique Cytoplasmic Forms of Glycosylation	658
39.3.2	Conventional Forms of Glycosylation in the Nucleus and Cytoplasm	660
39.3.3	Nuclear and Cytoplasmic Lectins	661
39.4	Conclusions	662
	Acknowledgments	662
	References	662

40	**Structure and Functions of Mucins**	669
	Joyce Taylor-Papadimitriou and Joy M. Burchell	
40.1	Classification of Mucins	669
40.2	The Epithelial Mucins	670
40.3	Mucin Type *O*-Glycosylation Pathways	670
40.3.1	Initiation of *O*-Glycosylation	671
40.3.2	Chain Extension	671
40.3.3	Chain Termination	671
40.4	Expression of Epithelial Mucins	672
40.5	The Complex Gel-Forming Mucins: Processing and Function	672
40.6	Epithelial Membrane Mucins	674
40.7	Studies Related to the MUC1 Mucin	675
40.7.1	Changes in the Patterns of *O*-Glycosylation of MUC1 in Breast Cancer	675
	Differences in sites of glycosylation	675
	Changes in the composition of *O*-glycans added to MUC1 in Breast Cancer	676
	Correlation in changes of Glycosyltransferase activities with changes in *O*-glycan structure in Breast Cancer	676
40.7.2	Changes in Glycosylation of MUC1 in other Cancers	677
40.7.3	Effects of MUC1 Expression on the Behavioral Properties of Cancer Cells	677
	Effects on cell interactions and tumourogenicity	677
40.7.4	MUC1 Expression and Immune Responses	678
40.7.5	Active Specific Immunotherapy Based on MUC1	679
	Animal models	679
	Clinical studies	680
40.8	Comments	681
	References	681
41	**Biological Roles of Hyaluronan**	685
	Bryan P. Toole	
41.1	Introduction	685
41.2	Hyaluronan is a Biopolymer with Unusual Physical Properties	685
41.3	Hyaluronan Binds to Several Types of Proteins (Hyaladherins)	687
41.3.1	General Properties of Hyaladherins	687
41.3.2	Structural Hyaluronan-Binding Proteins	688
41.3.3	Hyaluronan Receptors	688
41.3.4	Intracellular Hyaluronan-Binding Proteins	689
41.3.5	Inter-α-Trypsin Inhibitor	689
41.4	Hyaluronan-Dependent Pericellular Matrices Assemble Around Several Cell Types	690
41.4.1	Hyaluronan-Dependent Cellular "Coats"	690
41.4.2	Assembly of Chondrocyte Pericellular Matrix	691

41.4.3	Tethering of Cell Surface Hyaluronan to Hyaluronan Synthase	691
41.5	Hyaluronan Influences Cell Behavior During Morphogenesis and Tissue Remodeling	693
41.5.1	Migratory and Proliferating Cells are Surrounded by Hyaluronan-enriched Matrices	693
41.5.2	Hydrated Pericellular Milieux Provide Cellular Pathways	693
41.5.3	Receptors Mediate Effects of Hyaluronan	693
41.5.4	Hyaluronan-Cell Interactions in Limb Development	694
41.5.5	Hyaluronan-Cell Interactions in Other Physiological and Developmental Systems	695
41.6	Hyaluronan Plays a Crucial Role in Cancer	696
	References	696

42	**Biological Roles of Heparan Sulfate Proteoglycans**	**701**
	Ofer Reizes, Pyong Woo Park, and Merton Bernfield	
42.1	Introduction	701
42.2	Heparan Sulfate Biosynthesis	701
42.3	Functions of Heparan Sulfate	702
42.4	Proteoglycans	703
42.5	Intracellular Proteoglycans	703
42.5.1	Serglycin and Heparin	704
42.6	Cell Surface Heparan Sulfate Proteoglycans	705
42.6.1	Syndecans	705
42.6.2	Glypicans	706
42.7	Part-time Cell Surface Heparan Sulfate Proteoglycans	707
42.7.1	Betaglycan	707
42.7.2	CD44	708
42.8	Functions of Cell Surface Heparan Sulfate Proteoglycans	708
42.8.1	Ligand Receptors	708
42.8.2	Ligand Coreceptors	709
42.8.3	Shed Effectors	709
42.9	Extracellular Matrix Heparan Sulfate Proteoglycans and Their Functions	710
42.9.1	Perlecan	710
42.9.2	Agrin	712
42.9.3	Other Extracellular HSPGs	713
42.10	Conclusions	713
	References	713

43	**Biological Roles of Keratan Sulfate Proteoglycans**	**717**
	Gary W. Conrad	
43.1	Introduction	717
43.2	Corneal Transparency	718

43.3	Nerve Growth Cone Guidance	719
43.4	Cell Adhesion	721
43.5	Other Possible Roles of KSPGs	722
	Acknowledgment	723
	References	723

44	**Developmental and Aging Changes of Chondroitin/Dermatan Sulfate Proteoglycans**	729
	J. Michael Sorrell, David A. Carrino, and Arnold I. Caplan	
44.1	Proteoglycans	729
44.2	Glycosaminoglycans	729
44.3	Core Proteins	731
44.3.1	Hyalectans	731
44.3.2	Small Leucine-rich Proteoglycans	733
44.4	Chondrotin/Dermatan Sulfate Proteoglycans in Development and Aging	735
44.4.1	Core Proteins in Development, Aging, and Pathologies	735
44.4.2	Chondroitin/Dermatan Sulfate Glycosaminoglycan Chains in Development, Aging, and Pathologies	736
44.5	Summary	740
	References	740

45	**Proteoglycans and Hyaluronan in Vascular Disease**	743
	Thomas N. Wight	
45.1	Introduction	743
45.2	Proteoglycans and Hyaluronan	744
45.3	Versican (CSPGs)	745
45.4	Hyaluronan	747
45.5	Decorin/Biglycan (DSPGs)	748
45.6	Perlecan/Syndecans (HSPGs)	749
45.7	Summary	750
	Acknowledgments	750
	References	750

46	**Functions of Glycosyl Phosphatidylinositols**	757
	Nikola A. Baumann, Anant K. Menon, and David M. Rancour	
46.1	Introduction	757
46.2	Parasite Coats: Extreme GPI-Anchoring	758
46.3	Yeast GPIs and the Cell Wall	758
46.4	Paroxysmal Nocturnal Hemoglobinuria (PNH): Disease and Defects in GPI-Anchoring of Proteins	759
46.5	GPIs in the Secretory and Endocytic Pathways	760
46.6	Organization of GPI Proteins in the Plasma Membrane	762

46.7	Association of GPI-Anchored Proteins with Caveolae	764
46.8	Detergent Insolubility and Signaling via GPI-Proteins	764
46.9	Membrane Release of GPI-Anchored Proteins	765
46.10	GPIs as Second Messenger Signaling Molecules	766
46.11	Summary	767
	Acknowledgments	767
	References	768
47	**Glycosphingolipid Microdomains in Signal Transduction, Cancer, and Development**	**771**
	Sen-itiroh Hakomori and Kazuko Handa	
47.1	Clustered GSLs as Functional Units	771
47.2	GSL Clusters, Associated with Signal Transducers, are Functional Units Separable from Caveolae	772
47.3	Cell Adhesion Coupled with Signal Transduction Initiated by GSL Microdomain: Concept of Glycosignaling Domain (GSD)	773
47.4	Role of GSLs in Control of Growth Factor and Hormone Receptors: Possible Relationship with GSL Microdomain	774
47.5	Functional Role of Developmentally-Regulated and Tumor-Associated GSLs	776
	References	778
48	**The Primary Cell Walls of Higher Plants**	**783**
	Jocelyn K. C. Rose, Malcolm A. O'Neill, Peter Albersheim, and Alan Darvill	
48.1	Introduction (What is a Cell Wall?)	783
48.2	Purification of Cell Walls and Isolation of Wall Components	784
48.3	The Structural Components of the Primary Cell Wall	786
48.3.1	Cell Walls and the Diversity of Flowering Plants	786
48.3.2	The Structural Components of the Primary Wall	786
48.4	Biosynthesis of Wall Components	791
48.5	Organization of the Plant Primary Cell Wall	793
48.6	Cellulose-Xyloglucan Interactions	793
48.7	Interactions Between Pectins and Other Cell Wall Components	794
48.8	Glycoproteins in the Cell Wall	796
48.9	Heterogeneity in the Primary Cell Wall	797
48.10	Function and Metabolism of Plant Primary Cell Walls	798
48.10.1	Mechanical Support	798
48.10.2	Regulation of Cell Expansion	798
48.10.3	Morphogenesis and Differentiation	800
48.10.4	Plant Cell Wall Oligosaccharides in Defense and Cell Signalling	801
48.11	Intercellular Transport and Storage	803
48.12	Biotechnology and Future Directions in the Commercial Applications of Plant Primary Cell Walls	803

	Acknowledgment	804
	References	804

49	**Glycolipids and Bacterial Pathogenesis**	809
	Clifford A. Lingwood	
49.1	Introduction	809
49.2	Modulation of Glycolipid Receptor Function	810
49.3	Stress Response and Glycolipid Receptors	812
49.4	Subcellular Gb_3 Trafficking	813
49.5	Model for Lipid Sorting Based on Chain Length	815
49.6	Glycosphingolipids and Signal Transduction	815
	Acknowledgments	817
	References	817

50	**Glycobiology of Viruses**	821
	Hildegard Geyer and Rudolf Geyer	
50.1	Summary	821
50.2	General Aspects	821
50.2.1	Functions of Viral Surface Glycoproteins	825
50.2.2	Biosynthesis	826
50.2.3	Function of Carbohydrate Substituents	826
50.2.4	Oligosaccharide Diversity	827
50.3	Examples	830
50.3.1	Friend Murine Leukemia Virus Complex	830
50.3.2	Marburg Virus (MBGV)	832
50.3.3	Hepatitis B Virus (HBV)	833
	Acknowledgments	836
	References	836

51	**The Glycobiology of Influenza Viruses**	839
	Stephen J. Stray and Gillian M. Air	
51.1	Introduction	839
51.2	Receptor Binding Proteins: Influenza A HAg and Influenza C HEF	840
51.2.1	Structure of Receptor Binding Domain and Mechanism of Sialic Acid Recognition	843
51.2.2	HEF Esterase Domain and Mechanism of Cleavage	844
51.3	Influenza NAm (types A and B)	844
51.3.1	Mechanism of Sialic Acid Cleavage	845
51.4	Function of Viral Receptor Destroying Enzymes	847
	Acknowledgments	847
	References	848

52	**Glycobiology of Aids**	851
	Ten Feizi	
52.1	Abstract	851
52.2	Introduction	851
52.3	The Repertoire of N-Glycans on the Envelope Glycoprotein of HIV of Human Immunodeficiency Virus Produced in Different Cell Types	853
52.4	Evidence for the Occurrence of O-Glycans on the Envelope Glycoproteins of HIV-1 Produced in Certain Cell Lines	854
52.5	Oligosaccharides of gp 120 and gp 41 at N-Glycosylation Sites and Their Possible Influence on Viral Infectivity	855
52.6	gp 120 Glycosylation Can Influence Antigenicity and Immunogenicity	856
52.7	Saccharides Recognized by Carbohydrate-binding Proteins and Antibodies as Potential Neutralization Epitopes on the Envelope Glycoprotein of HIV-1	857
52.7.1	Lectins and Antibodies with Mannose-related Specificities	857
52.7.2	Antibodies to O-Glycan Sequences	858
52.7.3	Antibodies to Blood Group A	858
52.7.4	Xeno-antibodies to Galα1-3Gal Sequence	859
52.7.5	Potential Medical Relevance	859
52.8	Does Viral Oligosaccharide Display Influence Tissue Tropism?	860
52.9	Concluding Remarks	862
	Acknowledgment	862
	References	863
53	**Glycobiology of Protozoan and Helminthic Parasites**	867
	Richard D. Cummings, and A. Kwame Nyame	
53.1	Introduction	867
53.2	General Classification of Parasites	867
53.3	The Major Protozoan Parasites	868
53.3.1	Malaria	868
53.3.2	Trypanosomiasis	873
53.3.3	Leishmaniasis	874
53.4	Other Protozoan Parasites	878
53.4.1	*Entamoeba histolytica*	878
53.4.2	*Acanthamoeba*	878
53.4.3	*Giardia lamblia*	878
53.4.4	*Cryotosporidium parvum*	878
53.4.5	*Sarcocystis* spp.	879
53.4.6	*Toxoplasma gondii*	879
53.4.7	*Pneumocystis carinii*	879
53.5	Helminthic Parasites	879
53.6	Carbohydrate-Binding Proteins in Parasitic Helminths	883

53.7	Unusual Glycans in Other Helminthic Parasites	883
53.8	Future Directions	885
	Acknowledgments	885
	References	886

54	**The Involvement of the Oligosaccharide Chains of Glycoproteins in Gamete Interactions at Fertilization**	895
	Noritaka Hirohashi and William J. Lennarz	
54.1	Introduction	895
54.2	Advantages of Marine Invertebrates as an Experimental System	895
54.3	Induction of the Acrosome Reaction	896
54.3.1	Studies in Sea Urchins	896
54.3.2	Studies in Starfish	899
54.4	Sperm–Egg Coat Binding	899
54.4.1	Studies in Mammals	900
54.4.2	Studies in Frog	900
54.4.3	Studies in Ascidians	902
54.4.4	Studies in Sea Urchins	904
54.5	Carbohydrate as a Species-Specific Determinant	906
	References	907

55	**Glycosylation and Development**	909
	Michèle Aubery and Christian Derappe	
55.1	Summary	909
55.2	Introduction	911
55.3	Lectins as Tools to Analyze Changes in Cell-surface Glycoconjugates During Development	911
55.4	Cell-adhesion Molecules	912
55.4.1	Neural Cell-adhesion Molecule	912
55.4.2	The Adhesion Molecule L1	914
55.5	Glycosyltransferases	914
55.6	Altered Expression of Endogenous Lectins During Development	915
55.6.1	Galectins	915
55.6.2	Selectins	917
55.6.3	Other Endogenous Lectins	917
55.7	Conclusion	918
	References	918

56	**Protein Glycosylation and Cancer**	923
	James W. Dennis and Maria Granovsky	
56.1	Introduction	923
56.2	Protein Glycosylation Generates Molecular Diversity	923
56.3	Cancer Initiation and Progression	926

56.4	Tumor Cell Proliferation	927
56.5	Cell Migration	930
56.6	Sialylation and Metastasis	933
56.7	Endogenous Lectins and Tumor Cell Adhesion	934
56.8	Carbohydrate Processing Inhibitors as Anti-Cancer Agents	935
56.9	Other Considerations	936
	Acknowledgments	937
	References	937
57	**Lysosomal Storage Diseases**	945
	Nathan N. Aronson, Jr.	
57.1	Summary	945
57.2	Introduction	946
57.3	Animal Models	947
57.4	Mucopolysaccharidoses	947
57.5	Cathepsin K Deficiency and Pycnodysostosis	949
57.6	Mouse Models for Tay-Sachs and Sandhoff Diseases	951
57.7	Impact of Lysosomal Diseases and Their Study	953
	References	954
58	**Genetic Diseases of Glycosylation**	959
	Tomoya Akama and Michiko N. Fukuda	
58.1	Introduction	959
58.2	CDGS	959
58.2.1	CDGS Type I	959
58.2.2	CDGS Type II	961
58.3	HEMPAS	963
	References	964
59	**Glycobiology of *Helicobacter pylori* and Gastric Disease**	967
	Karl-Anders Karlsson	
59.1	Introduction	967
59.2	The Bacterial Surface and Molecular Mimicry	968
59.3	Host Surfaces and *H. pylori* Recognition of Glycoconjugates: Unique Complexity	968
59.3.1	Sialic Acid	969
59.3.2	Sulfatide	970
59.3.3	Heparan Sulfate	970
59.3.4	Fucose-Dependent Binding (H-1 and Lewis b)	971
59.3.5	Gangliotetraosylceramide	971
59.3.6	Lactosylceramide	972
59.4	The Meaning of Multiple Binding Specificities	972
59.5	Aspects for the Future	973
	References	973

60	**Immunoglobulin G Glycosylation and Galactosyltransferase Changes in Rheumatoid Arthritis**	977
	John S. Axford	
60.1	Introduction	977
60.2	Oligosaccharide Synthesis	977
60.3	Galactosyltransferase	978
60.4	Immunoglobulin G	978
60.5	Rheumatoid Arthritis	979
60.6	Quantification of IgG sugars in RAr	979
60.7	RAr and Pregnancy	980
60.7.1	Galactosylation of IgG	980
60.7.2	α3-Fucosylation of α1-Acid Glycoprotein	981
60.8	Agalactosyl-IgG and Rheumatoid Factor Binding	982
60.9	Tissue-specific Galactosyltransferase Abnormalities in an Experimental Model of Rheumatoid Arthritis	983
60.10	Glycosylation Homeostasis within RAr Lymphocytes is Abnormal	985
60.11	Are the Rheumatoid Arthritis Associated Glycosylation Abnormalities Unique?	986
60.12	Sugar Printing Rheumatic Disease is Possible	990
60.13	Rapid Profiling of IgG N-Glycans by Fluorophore-coupled Oligosaccharide Electrophoresis has the Potential of Differentiating Rheumatic Diseases	992
60.14	In What Way could GTase Enzymatic Control be Abnormal?	992
60.15	Conclusion	993
	References	994
61	**Calnexin, Calreticulin and Glycoprotein Folding Within the Endoplasmic Reticulum**	997
	Michael R. Leach and David B. Williams	
61.1	Structure and Properties of Calnexin and Calreticulin	997
61.2	Biological Functions	1000
61.3	Mechanism of Action	1002
61.4	Functional Relationship Between Calnexin and Calreticulin	1005
61.5	Relationship with other ER Chaperones and Folding Catalysts	1007
	References	1008
62	**Glycobiology of The Nervous System**	1013
	Ronald L. Schnaar	
62.1	Introduction	1013
62.2	Nervous System Glycoconjugates—Overview	1013
62.3	Nervous System Glycolipids	1014
62.3.1	Galactosylceramides	1014

62.3.2	Gangliosides and Related Anionic Glycosphingolipids	1016
62.4	Nervous System Glycoproteins	1019
62.4.1	Polysialic Acid	1020
62.4.2	The HNK-1 Determinant	1020
62.5	Nervous System Glycosaminoglycans	1020
62.6	Lectins in the Brain	1021
62.6.1	Myelin-Associated Glycoprotein	1021
62.6.2	Other Nervous System Lectins	1022
62.7	Concluding Remarks	1023
	References	1023

63	**Glycobiology of the Immune System**	1029
	Elizabeth F. Hounsell	
63.1	Infection and Pathogenesis	1029
63.2	Control of the Immune Response	1033
63.3	Bacterial and Tumor Antigens, Mucins and Mucin-like Molecules	1034
63.4	Immunoglobulins and Pathology	1036
	References	1038

64	**Metabolic Engineering Glycosylation: Biotechnology's Challenge to the Glycobiologist in the Next Millennium**	1043
	Thomas G. Warner	
64.1	Introduction	1043
64.2	Recent Developments in Carbohydrate Biosynthesis	1044
64.2.1	Optimizing Sialylation of Recombinant Proteins by Metabolic Engineering Sialic Acid Biosynthesis	1044
64.2.2	Optimizing Galactosylation of Recombinant Proteins by Metabolically Engineering Galactose Biosynthesis	1049
64.2.3	Mannose Biosynthesis and Mannosylation of Recombinant Proteins	1052
64.3	Glycosylation Engineering Alternate Expression Hosts For Recombinant Protein Therapeutic Production	1053
64.3.1	Engineering Glycosylation of Recombinant Proteins Expressed in Baculovirus-Insect Cells	1053
	Genes needed to supplement glycosylation of recombinant proteins in insect cells	1054
	Deleterious genes may need to be deleted or inhibited to enhance recombinant glycoprotein biosynthesis in insect cells	1054
64.3.2	Engineering Glycosylation of Recombinant Proteins Expressed in Plants	1056
	Genetic addition and supplementation needed to improve plant recombinant protein glycosylation	1058
	Inhibition or deletion of plant glycosylation genes	1059

64.4	Summary		1059
	Acknowledgments		1060
	References		1060
	Index		I 1

Part I

Volume 2

III

Enzymatic Synthesis of Glycosides

23 On the Origin of Oligosaccharide Species–Glycosyltransferases in Action

Dirk H. van den Eijnden

23.1 Introduction

Glycoproteins and glycolipids, jointly called glycoconjugates, are macromolecules essential to eukaryotic life. The oligosaccharide chains carried by glycoconjugates often contain biological intelligence which is encoded by the oligosaccharide structure. These chains thus confer specific biological functions on the proteins and lipids carrying them. When embedded in the cellular plasma membrane they decorate the cell surface with specific oligosaccharide structures which can be crucial to cell function. The synthesis of protein- and lipid-linked oligosaccharides proceeds by the sequential addition of monosaccharide building blocks to a growing chain that is covalently linked to the protein or lipid aglycone. In a similar way oligosaccharides present in some secretions (*e.g.* milk) are built up from a single sugar (glucose).

Unlike the processes of protein and nucleic acid synthesis, in which the order of attachment of amino acids and nucleotides is read from a nucleic acid matrix, glycosylation is a non-template-directed process. This enables the formation of branches which are common to oligosaccharides but are not found in nucleic acids and proteins. The enzymes that are responsible for the assembly of the oligosaccharides on glycoconjugates are known as **glycosyltransferases** (in the early stages of N-glycosylation glycosidases also play a role—see later in this chapter). Being proteins, glycosyltransferases are primary products of their respective genes and, therefore, the oligosaccharide chains formed by their action can be considered as secondary gene products. Glycosyltransferases typically catalyze the following general reaction:

$$\text{nucleoside–P–P–sugar} + \text{acceptor} \rightarrow \text{sugar–acceptor} + \text{nucleoside–P–P}$$

In this reaction nucleotide–sugars are the donor substrates (for some glycosyltransferases involved in the early stages of protein N-glycosylation a sugar–phosphate–dolichol is the donor molecule; see later). In sequential reactions monosaccharides are attached, one at a time, through glycosidic linkages at non-reducing ends and at branching points in a specific order; the sugar–acceptor product of one reaction is the acceptor substrate in the next. In vitro glycosyltransferases generally have optimum activity at a pH of *ca* 7.0. In most instances Mn^{2+} ions are required for enzymatic activity (except for sialyltransferases and some β6-*N*-acetylglucosaminyltransferases which are cation-independent). Because glycosyltransferases are generally membrane-associated proteins, determination of their activity in vitro requires the solubilization of the enzymes with detergents. Some glycosyltransferases, however, occur in a soluble form in body fluids, probably as a result of proteolytic release of the catalytically active part of the enzyme.

Glycosyltransferases of eukaryotic organisms have high specificity for the sugar they transfer and the linkage they establish. This has enabled their classification in groups, such as the sialyltransferases which all exclusively utilize CMP–sialic acid as a donor substrate, and the galactosyltransferases which highly prefer UDP–Gal as a donor, with subgroups based on the linkage that is established. In many (but not all) instances the molecular cloning of the cDNAs of these enzymes has provided molecular support for this classification (see Section 23.17, *Glycosyltransferase families*). Glycosyltransferases are, furthermore, generally specific for the accepting sugar and its anomeric configuration, the linkage by which this sugar is linked to or is substituted by another sugar residue, and the nature of these other residues. Sometimes the entire underlying oligosaccharide structure and possibly its spatial conformation seems to be recognized. It also has been suggested that some glycosyltransferases recognize specific peptide motifs of the polypeptide aglycone, which could be one explanation for site specific glycosylation patterns on glycoproteins with more than one glycosylation site.

Very many different glycosidic linkages are known to occur in glycoconjugate-linked and free oligosaccharides. The formation of each linkage is catalyzed by at least one specific glycosyltransferase. As mentioned above, such an enzyme is generally not capable of catalyzing the formation of another linkage-type. A notable exception to the 'one enzyme–one linkage' concept [1], however, is the Lewis blood group α3/4-fucosyltransferase, which can establish Fuc(α1–4)[Gal(β1–3)]GlcNAc and Fuc(α1–3)[Gal(β1–4)]GlcNAc linkages in vivo and in vitro [2–4]. Also other glycosyltransferases, particularly when the substrate concentrations used are unphysiologically high, can establish more than one linkage by transferring to another acceptor sugar [5] or by transferring a monosaccharide from a different sugar–donor substrate (donor promiscuity) [6–11].

Even though glycosylation is a non-template directed process, oligosaccharide structures are generally produced with high fidelity. Firstly, the glycosylation potential of a cell (expression levels of the different glycosyltransferases; subcellular localization of the enzymes; supply of donor substrates) determines which pathways potentially can be followed. It should be noted that this potential is not constant but may vary with cell status (normal *versus* malignant), with the developmental stage of the cell, and with the environment wherein the cell grows (*e.g.* energy supply). Furthermore the specificities of the glycosyltransferases involved play a major controlling role in the formation of the correct final product structure(s). In such a process, however, it is inevitable that at intersections in the metabolic pathway, where enzymes compete with one another for common intermediate substrate structures, various directions can be taken. In practice this leads to the synthesis of an array of related oligosaccharide structures rather than of one single structure. Such variations in oligosaccharide structure at single glycosylation sites is a common feature of glycoproteins and is known as **microheterogeneity**. It causes glycoproteins to occur in different **glycoforms**, which are oligosaccharide variants of the same polypeptide.

Many excellent reviews have described the topography, mechanisms, and enzymology of protein and lipid glycosylation [12–25]. This chapter gives a concise introduction to the enzymatic properties and molecular relationships of glycosyltransferases with emphasis on protein N-glycosylation and the corresponding pathways of oligosaccharide assembly, in particular the late stages. Where appropriate, however, the glycosyltransferases involved in protein O-glycosylation, biosynthesis of the core of proteoglycans, and the glycosylation of sphingoglycolipids are also discussed.

23.2 Protein *N*-Glycosylation: Pre-assembly of Oligosaccharide–PP–Dolichol and en bloc Transfer

Protein N-glycosylation is a process that has been highly conserved during evolution. In this process several stages can be distinguished; the early ones are shared by all organisms from yeasts to mammals. The initial steps, in which an oligosaccharide precursor structure is synthesized on a dolichol–diphosphate (Dol–PP) lipid carrier, occur in the endoplasmic reticulum (ER). The most simple picture that complies best with current knowledge [12, 15, 18] is that depicted in Figure 1.

Several glycosyltransferases, embedded in the membranes of the ER with their catalytic domain facing the cytosol of the cell, transfer GlcNAc–P, GlcNAc, and Man to Dol–P from UDP–GlcNAc and GDP–Man, respectively, to yield a Dol–PP linked $Man_5GlcNAc_2$ oligosaccharide. The molecule then flip-flops over the membrane resulting in a luminal exposed oligosaccharide structure. Subsequently four additional Man and three Glc residues yielding $Glc_3Man_9GlcNAc_2$–PP–Dol are attached by glycosyltransferases that have their catalytic domain facing the lumen. The latter enzymes are unusual in that they utilize Man–P–Dol and Glc–P–Dol rather than GDP–Man and UDP–Glc as a sugar donor. The Dol–P linked sugars are synthesized from the nucleotide–sugars at the cytosolic face of the ER, whereafter they flip-flop over the membrane to become utilizable for the luminal glycosyltransferases.

In the next stage of N-glycosylation this pre-assembled oligosaccharide is transferred en bloc to an Asn–X–Ser/Thr sequon (where X can be any amino acid except Pro or Asp) in the nascent polypeptide chain of a glycoprotein being synthesized at the ribosome (Figures 1 and 2). This co-translational step is catalyzed by the oligosaccharyltransferase (OST) complex which consists of at least three proteins [26, 27] that seem to be intimately associated with membrane-bound ribosomes [19].

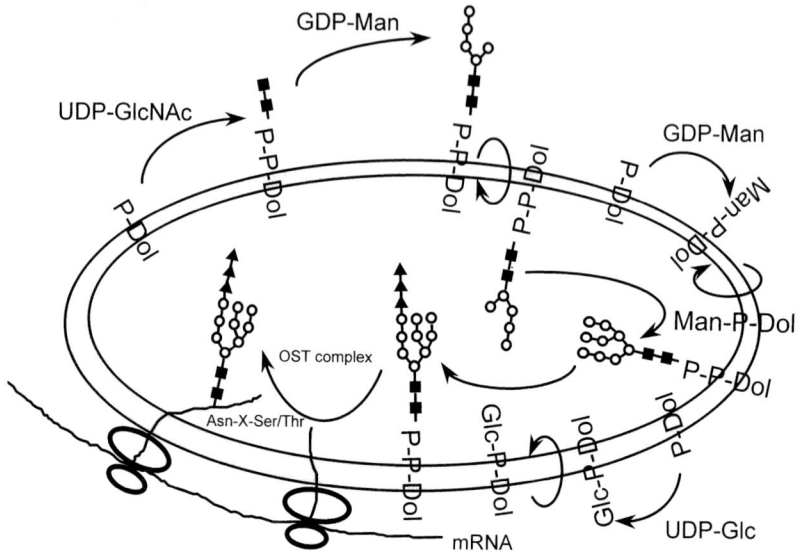

Figure 1. Initial steps in protein N-glycosylation. Pre-assembly of oligosaccharide–P–P–Dol and en bloc transfer to the protein in the endoplasmic reticulum. Dol, dolichol; OST, oligosaccharyltransferase; P, phosphate; ■, GlcNAc; ○, Man; ▲, Glc.

23.3 Trimming of the Polypeptide-Bound Oligosaccharide

Once the preformed oligosaccharide is attached to the polypeptide it is extensively trimmed before it is extended with more sugars. Three Glc and four Man residues are removed from the polypeptide-bound oligosaccharide to yield the common intermediate $Man_5GlcNAc_2$ structure (Figure 2). The first Glc (α2-linked) is removed by ER glucosidase I that might already act while the glycosylated polypeptide chain is still being synthesized on the ribosome. The other two Glc residues (that are α3-linked) are removed by ER glucosidase II (reviewed in [19, 23]). The catalytic sites of both enzymes face the lumen of the ER in which the glycosylated polypeptide resides after its vectorial translocation over the ER membrane. The first (α2-linked) Man may be removed in the ER by ER mannosidase to yield a unique $Man_8GlcNAc_2$ structure after which the glycoprotein is translocated to the *cis*-Golgi compartment (Figure 2) [19, 28]. Alternatively, the first Man residue is removed after this translocation by Golgi-localized (α2-)mannosidase IA and IB. The specificity of these enzymes differs from that of the ER mannosidase, and they yield the two other possible $Man_8GlcNAc_2$ isomeric structures. Subsequent repeated actions of mannosidase IA and/or IB on either $Man_8GlcNAc_2$ structure results in the aforementioned single intermediate $Man_5GlcNAc_2$ structure [28]. In yet another

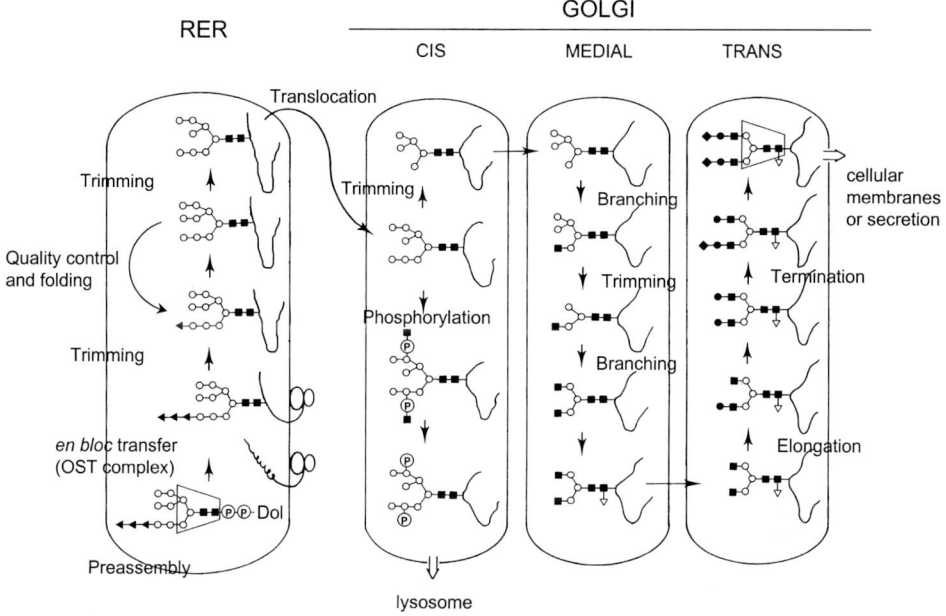

Figure 2. Topography of reactions involved in N-glycosylation in the rough endoplasmic reticulum and the Golgi membranes. Note that the boxed structure remains untouched during the entire process. The pathway leading to a di-antennary disialylated oligosaccharide is shown. Dol, dolichol; OST, oligosaccharyltransferase; P, phosphate; ■, GlcNAc; ○, Man; ▲, Glc; ●, Gal; ▽, Fuc; ◆, NeuAc.

pathway the last Glc and the Man residue to which it is attached are removed as Glc(α1–3)Man in one step by an ER localized *endo*-mannosidase yielding a Man$_8$GlcNAc$_2$ structure that is processed further in the *cis*-Golgi [29].

23.4 Folding and Quality Control

Although the synthesis of the Glc$_3$Man$_9$GlcNAc$_2$ oligosaccharide followed by its degradation to a Man$_5$GlcNAc$_2$ structure might seem a waste of energy, one important role of the oligosaccharide is that it aids the maturation of the polypeptide chain to a functional protein [30–32]. After the removal of the first two Glc residues by ER glucosidase I and II the glycosylated polypeptide is released from the ribosome. In the lumen of the ER it then associates with membrane bound calnexin or soluble calreticulin, chaperones which recognize the monoglucosylated oligosaccharide in a lectin like fashion. While transiently bound the polypeptide is assisted by these chaperones to fold properly and, for oligomeric glycoproteins, to assemble in oligomers. At this stage also disulfide bridges are formed.

After subsequent removal of the third Glc from the oligosaccharide by glucosidase II the glycoprotein is no longer bound and is released, and the oligosaccharide is further processed on the glycoprotein's journey *via* the Golgi to its cellular destination. In case the process of correct folding or assembly fails, however, the deglucosylated polypeptide is retained in the ER where it can be re-glucosylated by a glycoprotein glucosyltransferase specific for improperly folded glycoproteins [33, 34]. Re-glucosylated proteins can be bound again by calnexin or calreticulin and so have another chance of folding correctly. The glycoprotein glucosyltransferase (together with the chaperones) thus functions in the control mechanism by which only properly folded and assembled glycoproteins can leave the ER.

23.5 Committed Steps in the Formation of Complex-Type Oligosaccharide Chains and Branching

After the trimming reactions in the ER and the *cis*-Golgi the glycoprotein containing the resulting $Man_5GlcNAc_2$ oligosaccharide is translocated to the *medial*-Golgi compartment (Figures 2 and 3). Here the first committed step in the formation of

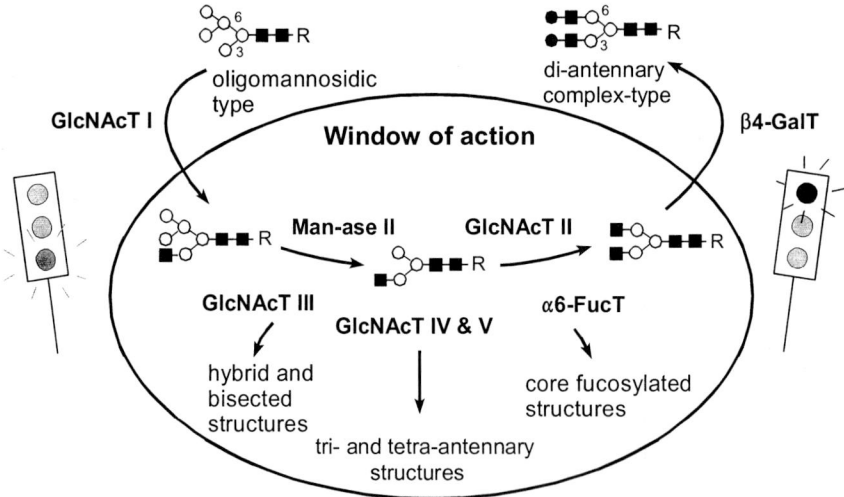

Figure 3. Committed steps in the synthesis of complex-type N-linked oligosaccharides. Action of GlcNAcT I on the oligomannoside structure opens the window for several possible reactions as catalyzed by α-mannosidase II (Man-ase II), GlcNAcT II–V, and core-specific α6-FucT. The window is closed by the action of β4-GalT as shown for a di-antennary structure. For key to the symbols see legend to Figure 2.

23.5 Committed Steps in the Formation of Complex-Type Oligosaccharide Chains

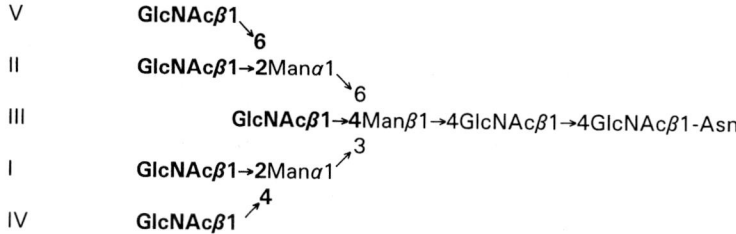

Figure 4. Structural representation of the linkages formed by GlcNAcT I–V. The Roman number refers to the GlcNAcT involved.

complex-type oligosaccharides takes place by the addition of a peripheral GlcNAc residue in (β1–2) linkage to the Man(α1–3)Manβ site as catalyzed by GlcNAcT I to form the first branch [13, 17, 20].

This GlcNAc functions as a green traffic light for several possible reactions catalyzed by α-mannosidase II, GlcNAcT II, III, IV and V (Figure 4), and N-linked core-specific α6-FucT that catalyzes reactions leading to di-, tri- and tetra-antennary oligosaccharides, hybrid and bisected structures, and core-fucosylated structures [13, 17] (Figure 3).

It is at this stage of oligosaccharide processing that there is an extensive regulation at the substrate level. Divergence in the pathway occurs here in particular and structures with varying number of branches (a branch or antenna is build up from a GlcNAc residue as introduced by GlcNAcT I, II, IV or V) might be formed [13, 21]. Evidently GlcNAcT II, yielding a di-antennary oligosaccharide, can only act after α-mannosidase II has trimmed two additional Man residues from the Man(α1–6)Manβ branch.

When GlcNAcT III acts before α-mannosidase II the oligosaccharide is frozen into a hybrid structure (combining the features of a complex-type structure at the Man(α1–3)Manβ branch and an oligomannosidic-type structure at the Man(α1–6)Manβ branch) because α-mannosidase II is blocked by the bisecting GlcNAc residue attached by GlcNAcT III action. When GlcNAcT III acts after GlcNAcT II a so-called bisected di-antennary oligosaccharide is formed. Action of GlcNAcT IV and V leads to the formation of tri- and tetra-antennary oligosaccharides and these might also become bisected by the subsequent action of GlcNAcT III.

The other route, in which the order of action of these GlcNAcTs is reversed, is impossible because GlcNAcT II, IV and V do not act on structures containing a bisected GlcNAc residue. A bisecting GlcNAc also inhibits core α6-fucosylation. It should be mentioned here that in plants and insects α3-fucosylation of the core GlcNAc might occur as catalyzed by a core-specific α3-FucT [35, 36, 188]. In addition, in plants an α2-xylosyltransferase has been described that can introduce an α-Xyl residue to OH-2 of the β-Man of the core [37]. All these reactions can no longer occur after in a second committed step of complex-type oligosaccharide formation, by β4-galactosyltransferase action, a Gal residue is attached in (β1–4) linkage to the GlcNAc(β1–2)Man(α1–3)Manβ branch (to which the first Gal preferentially is introduced [38, 39]) to yield a Gal(β1–4)GlcNAc (lacNAc) terminus [13, 17, 21]. This Gal residue thus functions as a red light and closes the stage of oligosaccharide branching and modification of the core (Figure 3). Occasionally a GalNAc rather than a Gal is introduced by a β4-N-acetylgalactosaminyltransferase to yield a terminal GalNAc(β1–4)GlcNAc (N, N'-diacetyllactosediamine, lacdiNAc) unit [40] (see also Section 23.15—*The lacdiNAc Pathway of Complex-Type Oligosaccharide Synthesis*).

23.6 Topology of the Reaction Catalyzed by a Typical GlcNAcT

All eukaryotic Golgi glycosyltransferases are typically membrane-bound enzymes with a type II membrane topology (cytosolic NH_2-terminus) as has become apparent from analysis of the deduced amino acid sequences of cloned enzymes [41, 42]. GlcNAcT I, II, III, IV and V form no exceptions in this respect [43–49]. Most glycosyltransferase polypeptides consist of *ca* 325–425 amino acids and have a very similar domain structure—a short cytoplasmic tail ($\approx$5–30 amino acids) attached to a transmembrane domain ($\approx$15–20 amino acids), which functions as a signal–anchor sequence by which the enzyme is targeted to and anchored in the Golgi membrane. This is followed by a so-called stem region ($\approx$35–45 amino acids) and a large catalytic domain (approximately 300 amino acids, containing the COOH terminus) extending into the lumen of the Golgi (Figure 5).

The immature protein-linked oligosaccharide chains are also exposed into the Golgi lumen during the glycoprotein's traverse through the cellular endo membranes. UDP–GlcNAc is thus utilized inside the Golgi lumen. The synthesis of this substrate by the enzyme UDP–GlcNAc pyrophosphorylase, however, occurs in the cytosol and, as nucleotide–sugars cannot freely diffuse through the Golgi membrane, it has to be transported across this barrier. This transport is facilitated by a specific UDP–GlcNAc transporter by a mechanism in which UDP–GlcNAc is exchanged for UMP that is formed from UDP, the other reaction product of the GlcNAc-transfer reaction [25] (Figure 5). Similar specific transporters for CMP–NeuAc, UDP–Gal, GDP–Man, and UDP–GalNAc have also been demonstrated and cloned [25, 50], and aberrant glycosylation patterns are, indeed, observed for cells with defects in transporter activity [25]. As yet, however, it is unclear if and to what extent the import of nucleotide–sugar donor–substrates might be limiting and can thus contribute to the regulation of protein glycosylation.

23.7 Elongation and Termination Reactions in the *trans*-Golgi

By the time the first Gal residue is introduced to the oligosaccharide chain, the glycoprotein has reached the *trans*-Golgi in which the elongation and termination

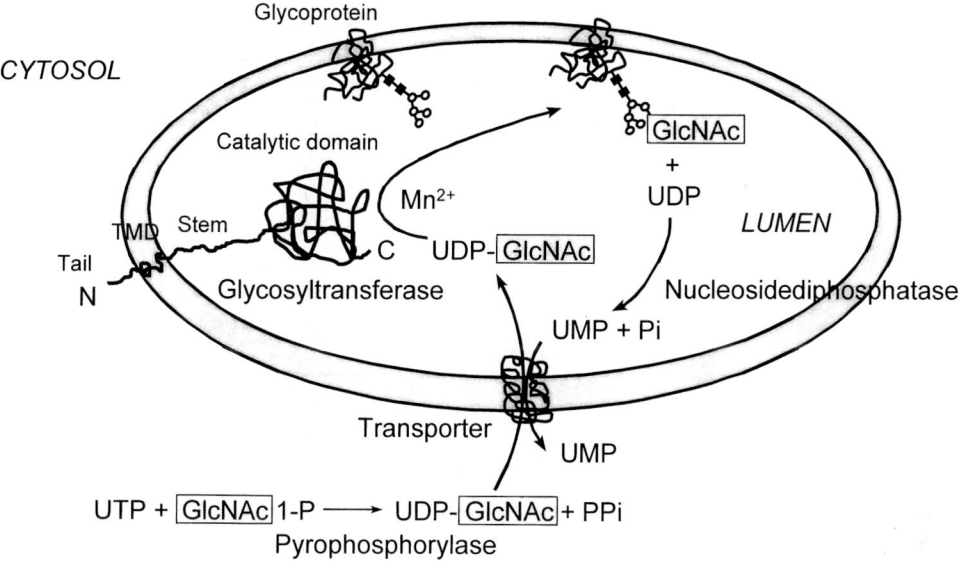

Figure 5. Organization of a typical glycosyltransferase in the Golgi membrane and topology of the reaction catalyzed as exemplified for a GlcNAcT. Note that the nucleotide sugar donor substrate is synthesized in the cytosol. Before it can be utilized by the glycosyltransferase it has to be translocated in exchange for a nucleotide monophosphate into the lumen of the Golgi by a specific transporter. Sugar transfer takes place to the oligosaccharide chain of a glycoprotein during its traverse through the secretory system. TMD, trans-membrane domain.

reactions occur (Figures 2 and 6). As mentioned above, β4-galactosylation disables the action of α-mannosidase II, GlcNAcT I–V, and N-linked core-specific α6-FucT. In contrast, it is the green traffic light for many glycosyltransferases which all act on an accepting lacNAc unit. The action of these enzymes yields the mature oligosaccharide structures. Here additional diversity (leading to microheterogeneity and different glycoforms of a glycoprotein; see Section 23.1, *Introduction*) is introduced because these enzymes potentially compete for common Gal(β1–4)GlcNAc acceptor sites on the protein-linked oligosaccharides. Six of the most common reactions, as catalyzed by α6-NeuAcT, α3-NeuAcT, α3-GalT, α3-FucT, α2-FucT, and β3-GlcNAcT, respectively, are shown in Figure 6 for a di-antennary glycan as acceptor. α6-NeuAcT, α3-NeuAcT, α3-GalT, and α2-FucT directly cap the Gal of the lacNAc units by introduction of a terminal sugar, whereas α3-FucT adds a fucose to the GlcNAc of these units. Introduction of a β3-GlcNAc residue to a lacNAc unit by β3-GlcNAcT enables a second β4-galactosylation reaction (Figure 6).

In this concerted action of β3-GlcNAcT and β4-GalT polylactosaminoglycan extensions are formed which also may be sialylated, fucosylated, or α-galactosylated. These reactions can in principle take place with any terminally exposed lacNAc unit irrespective of whether it is presented on di-, tri- or tetra-antennary, hybrid, or bisected N-linked oligosaccharides or on O-linked oligosaccharides (see later).

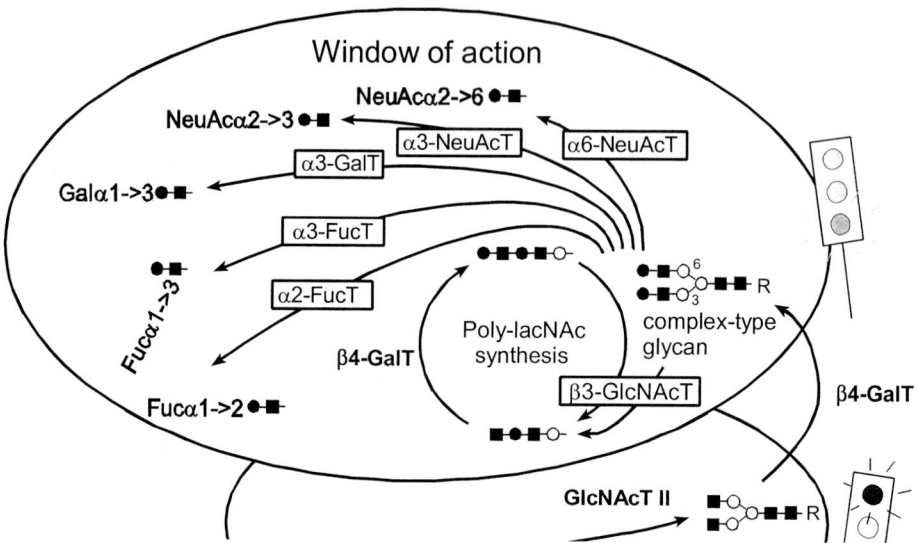

Figure 6. Elongation and termination reactions as enabled by the action of β4-GalT. β4-Galactosylation blocks the action of the branching GlcNAcTs, Man-ase II and core α6-FucT (*c.f.* Figures 2 and 3). By way of contrast, it opens the window of action for the elongating (polylactosaminoglycan-forming) and terminating enzymes shown. The reactions cannot only occur with the di-antennary substrate shown, but also with tri- and tetra-antennary, bisected, and hybrid structures.

Because of the substrate specificities of each of these enzymes, however, not all of the theoretically possible product structures are formed in nature in detectable amounts. By way of contrast, such structures may be formed in vitro where very active (recombinant) enzyme preparations can be used. In the next paragraphs the specificities of the enzymes are further described.

23.8 Activity with Branched Substrates

Not all N-linked oligosaccharides with terminal lacNAc units are acted upon by α6-NeuAcT, α3-NeuAcT, α3-GalT, α3-FucT, and β3-GlcNAcT at the same rate. Substrate specificity studies have shown that α6-NeuAcT (bovine colostrum; corresponds to ST6Gal I described below in Section 23.18) much prefers di-antennary substrates and is less active with tri- (having an additional Gal(β1–4)GlcNAc β4-linked at the Man(α1–3)Manβ branch as initiated by GlcNAcT IV action), tri'- (having an additional Gal(β1–4)GlcNAc β6-linked at the Man(α1–6)Manβ branch as initiated by GlcNAcT V action), tetra- (having both additional Gal(β1–

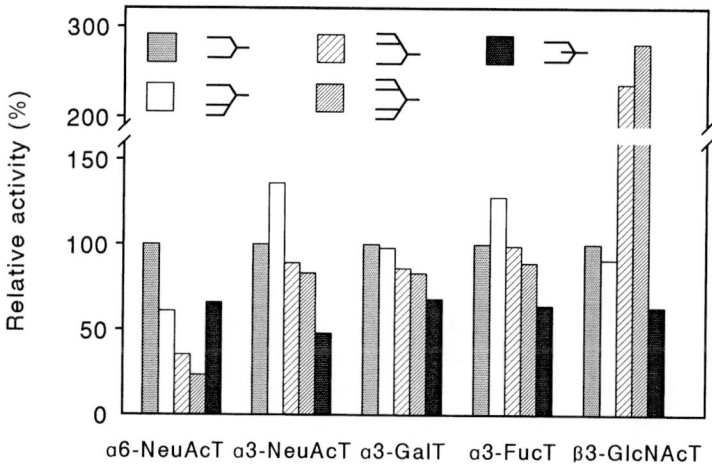

Figure 7. Relative activities of α6-NeuAcT, α3-NeuAcT, α3-GalT, α3-FucT, and β3-GlcNAcT with structures that form part of branched, N-linked oligosaccharides. The acceptor oligosaccharides are represented by symbolic structures the branch linkages of which correspond to those in the structures shown in Figures 4 and 8.

4)GlcNAc units), and bisected di-(GlcNAcT III action) antennary oligosaccharides [51] (Figure 7).

By way of contrast, α3-NeuAcT (human placenta; high preference for type-2 chain acceptors; corresponds to ST3Gal IV) prefers the tri-antennary substrate whereas a bisected acceptor is a poorer substrate [52]. Very similar specificity is shown by α3-FucT VI from human milk (D. H. Van den Eijnden and W. E. C. M. Schiphorst, unpublished results). α3-GalT (calf thymus) is relatively indifferent to the branching of the oligosaccharide substrates, but also acts at a lower rate on a bisected di-antenna [53]. Finally β3-GlcNAcT (Novikoff tumor cell ascites fluid) has the most unusual behavior of these enzymes, because it strongly prefers structures with a Gal(β1–4)GlcNAc(β1–6)Man sequence as occurs in tri'- and tetra-antennary oligosaccharides [54]. Consequently, changes in the activities of GlcNAcT III, IV and V, yielding a differently branched substrate population for α6-NeuAcT, α3-NeuAcT, α3-GalT, α3-FucT, and β3-GlcNAcT, by themselves will result in quantitative changes in the terminal structures on N-linked oligosaccharides. Increased GlcNAcT III activity will result in suppression of the elongation and termination reactions. Elevated activity of GlcNAcT IV will result in a shift from α6-sialylation to α3-sialylation [55] and α3-fucosylation (and consequently in increased expression of Lewis^x and sialyl-Lewis^x, see later), whereas increased GlcNAcT V activity will promote polylactosaminoglycan formation and will further suppress α6-sialylation. Elevated activities of the branching GlcNAcTs (III, IV, and V) have been reported [56–60] or have been concluded [61] to occur in cases of malignant transformation. Such changes thus form one cause of the altered terminal glycosylation of N-linked glycans which is often observed in malignant cells. These effects might be further amplified by changes in the activities of the glycosyltransferases acting on the lacNAc termini [58].

23.9 Branch Specificity

The different lacNAc termini on branched N-linked oligosaccharides are not acted upon at the same rates by all enzymes. The preference of an enzyme for a lacNAc unit on one particular branch has been termed 'branch specificity'. A summary of the branch specificities of four common glycosyltransferases acting on Gal(β1–4)GlcNAcβ termini is shown in Figure 8.

In particular α6-NeuAcT shows a strong preference for the Gal(β1–4)GlcNAc(β1–2)Man(α1–3)Manβ branch in di-, tri-, and tetra-antennary oligosaccharides [51, 62, 63]. The Gal(β1–4)GlcNAc(β1–2)Man(α1–6)Manβ branch is α6-sialylated at least 10 times more slowly, whereas the Gal(β1–4)GlcNAc(β1–6)Man(α1–6)Manβ branch as occurs in tri'- and tetra-antennary oligosaccharides is very resistant to α6-sialylation. Branch specificity, however, is only displayed by this sialyltransferase when the N-linked substrate contains at least one core GlcNAc residue and it has been proposed that the α6-NeuAcT recognizes this residue in a lectin like fashion aiding the correct positioning of the substrate on the enzyme [62]. By way of contrast α3-GalT prefers the Gal(β1–4)GlcNAc(β1–2)Man(α1–6)Manβ branch 4–5

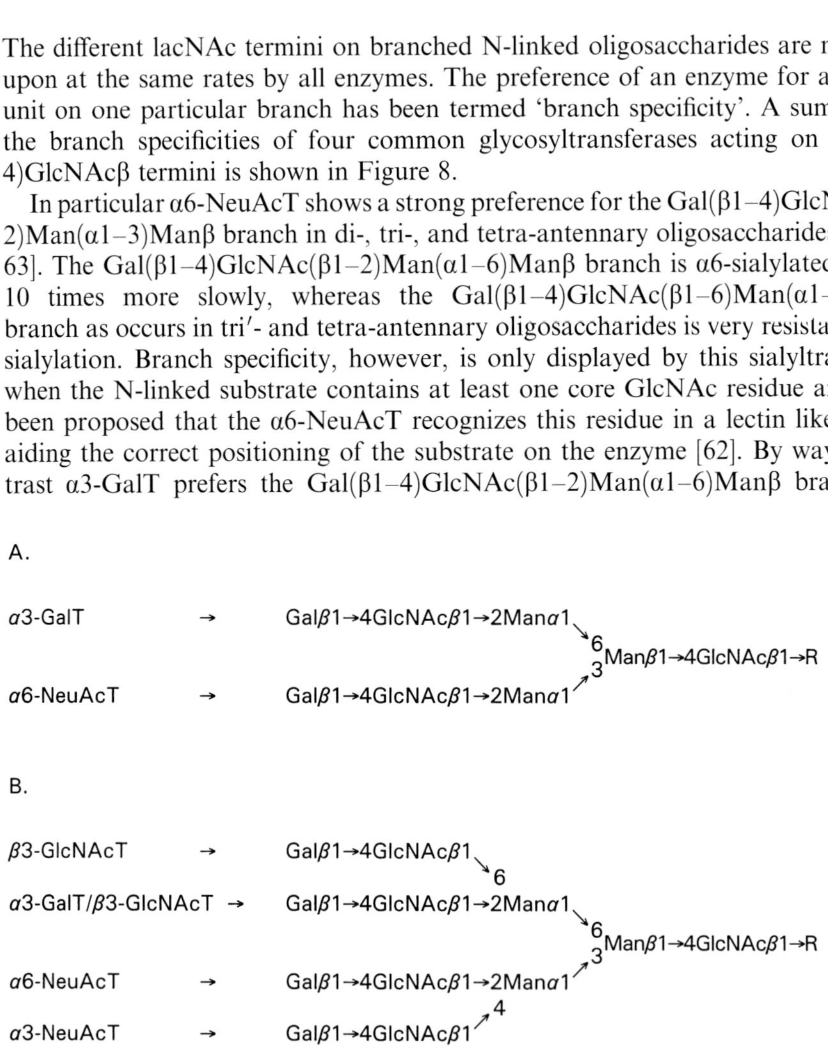

A.

α3-GalT → Galβ1→4GlcNAcβ1→2Manα1
                                                    ↘6
                                                       Manβ1→4GlcNAcβ1→R
                                                    ↗3
α6-NeuAcT → Galβ1→4GlcNAcβ1→2Manα1

B.

β3-GlcNAcT → Galβ1→4GlcNAcβ1
                                           ↘6
α3-GalT/β3-GlcNAcT → Galβ1→4GlcNAcβ1→2Manα1
                                                             ↘6
                                                                Manβ1→4GlcNAcβ1→R
                                                             ↗3
α6-NeuAcT → Galβ1→4GlcNAcβ1→2Manα1
                                                       ↗4
α3-NeuAcT → Galβ1→4GlcNAcβ1

Figure 8. Branch specificities of Gal(β1–4)GlcNAc-specific enzymes with (A) an N-linked di-antennary substrate and (B) a tetra-antennary substrate. The branches to which the enzymes preferentially introduce a sugar residue are indicated by →. Note that α3-NeuAcT and β3-GlcNAcT do not show branch specificity with the di-antennary structure.

times more than the Gal(β1–4)GlcNAc(β1–2)Man(α1–3)Manβ branch in a di-antennary substrate [64]. The Gal(β1–4)GlcNAc(β1–2)Man(α1–6)Manβ branch seems to be even more preferred by α3-GalT when it is present in tri- and tetra-antennary oligosaccharides [65]. Branch specificity is different yet again for α3-NeuAcT (human placenta). Although it shows no preference for either branch in a di-antennary substrate [52] it prefers the Gal(β1–4)GlcNAc(β1–4)Man(α1–3)Manβ branch in tri- and tetra-antennary substrates although this sialyltransferase can also act on all other branches (M. Nemansky, C. H. Hokke and D. H. Van den Eijnden, unpublished results). The typical α3-/α6-sialylation patterns as occur on N-linked glycans [51] therefore seem to be the result of the differential branch specificities of both sialyltransferases. Like α3-NeuAcT, the β3-GlcNAcT does not show branch preference with di-antennary oligosaccharides [64]. This enzyme, however, highly prefers the Gal(β1–4)GlcNAc(β1–2)Man(α1–6)Manβ and Gal(β1–4)GlcNAc(β1–6)Man(α1–6)Manβ branches equally well to the other branches of tri'- and tetra-antennary oligosaccharide structures [54]. Because of the strongly increased activity of β3-GlcNAcT on such substrates (Figure 7) polylactosaminoglycan chain formation is highly favored on such oligosaccharides at the branches on the Man(α1–6)Manβ unit. This further emphasizes the important role of GlcNAcT V (which initiates the formation of the Gal(β1–4)GlcNAc(β1–6)Man(α1–6)Manβ branch) in this process.

23.10 Essential Requirements for Activity with LacNAc

It will be clear from the previous paragraphs that the activity of the glycosyltransferases with each of the lacNAc units on branched N-linked oligosaccharide substrates can be strongly influenced by the structural environment (number and nature of neighboring branches, branch location) of the oligosaccharide substrate. In addition, studies with deoxygenated, substituted and otherwise modified lacNAc derivatives as acceptors for α6-NeuAcT, α3-NeuAcT (rat liver; preference for type 1 acceptors but also acting on type 2 structures; corresponds to ST3Gal III described below in Section 23.18, *Sialyltransferase Family*), α3-FucT and α3-GalT have revealed which modifications of the lacNAc acceptor site are allowed and which abolish the acceptor property for each of these enzymes [5, 39, 66–72]. Obviously the OH group to which a sugar is to be attached by the glycosyltransferase is essential and should be unsubstituted. The further requirements for activity explain why some enzymes can still act after another glycosyltransferase has introduced a sugar to the lacNAc unit (*e.g.* α3-fucosylation is still possible after α3-sialylation), whereas other enzymes are mutually exclusive (*e.g.* α6-sialylation precludes α3-fucosylation *vice versa*). On the other hand not all product structures of reactions that are possible in vitro are found in nature (*e.g.* no α6-sialylated form of the blood group H type 2 oligosaccharide has been reported even though α6-NeuAcT can act on Fuc(α1–2)Gal(β1–4)GlcNAc). A summary of the requirements for activity of these enzymes with lacNAc is given in Figure 9.

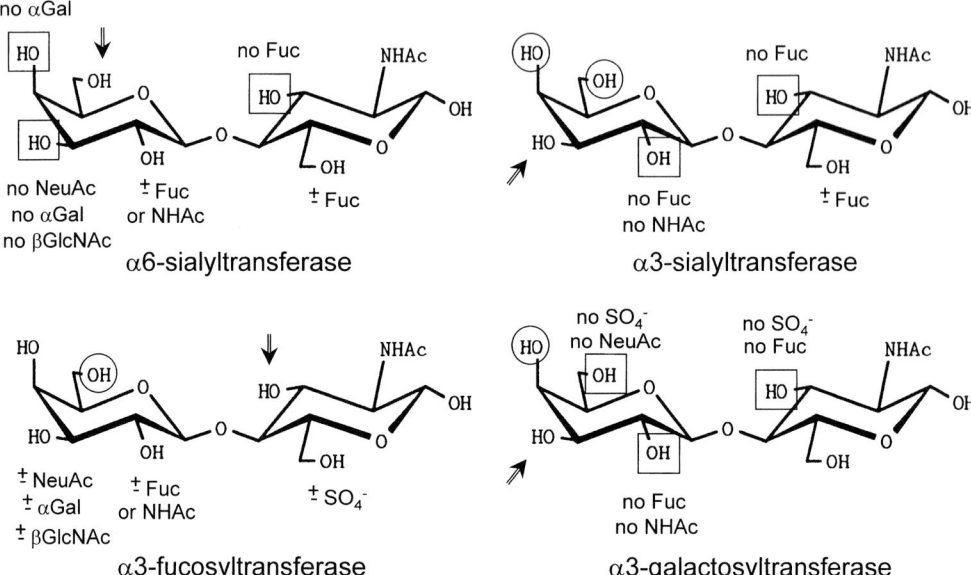

Figure 9. Summary of the structural requirements of α6-NeuAcT, α3-NeuAcT, α3-FucT and α3-GalT for activity with lacNAc and its derivatives. Essentially required OH groups are indicated by an arrow (site of sugar attachment) and by a circle. The latter OH groups are required for enzyme binding and catalysis (key polar groups). OH groups that can be deoxygenated without complete loss of activity but cannot be substituted by another sugar are indicated by a square. Other OH groups may be modified (deoxygenated, substituted, and/or replaced by an N-acetyl group) with only limited effect on the activity of the enzyme. All enzymes accept the replacement of the NHAc group of the GlcNAc residue by an OH moiety but are (much) less active with the resulting compound (lactose).

In particular α3-FucT seems to be the enzyme that is most easily satisfied, which explains why in metabolic routes α3-fucosylation is generally a late step. It clearly seems that each of the enzymes recognizes a lacNAc acceptor site in a unique way, suggesting that these enzymes do not share a common lacNAc recognition sequence motif. Indeed it has not yet been possible to identify such a motif in the deduced amino acid sequences of cloned lacNAc-specific enzymes.

23.11 Further Terminal Reactions in Complex-Type Oligosaccharide Synthesis

In addition to the glycosyltransferases mentioned in the previous paragraphs several other enzymes are known that can act on lacNAc units but catalyze less common reactions. Among these are a β3-GlcAT [73], a 3′-sulfotransferase (to Gal) [74], and a 6-sulfotransferase (to GlcNAc) [75], a β6-GlcNAcT [76], and a β3-GalNAcT [77] (Table 1). Furthermore, an α4-GalT acting on the lacNAc terminus of lacto-*N-neo*-tetraosylceramide [78] has been reported. Very recently a novel α4-GlcNAcT was cloned that preferentially acts on lacNAc present on O-linked core 2 [79]. After α6-sialylation or α3-fucosylation of lacNAc generally no further modifications are possible. The only exception known so far is the action of schistosomal α2-fucosyltransferase on the Lewis^x structure to form Gal(β1–4)[Fuc(α1–2)Fuc(α1–3)]-GlcNAcβ–R (difucosyllacNAc) [80]. Action of one of the other enzymes, however, often enables additional reactions to occur. This is particularly true of α2-fucosylation or α3-sialylation (Table 1). The first reaction yields the H-type 2 antigen that not only can be converted into the Lewis^y determinant by the action of α3-FucT, but also into the blood group A or B antigenic structures by the action of the A-enzyme (α3-GalNAcT) and B-enzyme (α3-GalT), respectively [24, 81]. α3-Sialylation in turn yields the structure that is intermediate in the synthesis of the sialyl–Lewis^x determinant by α3-FucT action, or polysialic acid through the action of one of the two known polysialyltransferases [82], or the Sd^a antigen by the action of a β4-GalNAcT [83] (Table 1).

For reasons of completeness the reactions catalyzed by three different sulfotransferases are also included in the figure. On the basis of the sequences of the two sulfotransferases of this group that have been cloned so far it seems that these enzymes have a domain structure that strongly resembles that of a Golgi glycosyltransferase [75, 84] (*c.f.* Figure 5).

23.12 Specific Modifications of Polylactosaminoglycans

The terminal lacNAc unit on polylactosaminoglycan chains can be capped like any other terminal lacNAc unit by the action of the glycosyltransferases described in the previous paragraphs, albeit that some of these enzymes act only at low rates on these glycans (*e.g.* α6-NeuAcT [55]). In addition to these reactions, however, several specific reactions can occur in which sugars are introduced to internal Gal or GlcNAc residues. Linear polylactosaminoglycan chains as synthesized by the concerted action of β3-GlcNAcT and β4-GalT (Figure 6) show blood group i-activity [24, 81]. They may be converted to branched, blood group I-active structures by the action of a β6-GlcNAcT and subsequent β4-galactosylation yielding a Gal(β1–4)GlcNAc(α1–3)[Gal(β1–4)GlcNAc(β1–6)]Gal branching point [85, 86].

Two different type of branching β6-GlcNAcTs have been described, one of which acts only on penultimate Gal residues capped by a single β3-linked sugar (GlcNAc or Gal) ('predistally acting') [87, 88], the other acting on any Gal residue along the linear chain except the terminal or penultimate ('centrally acting') [89, 90]. In a recent study, however, a recombinant β6-GlcNAcT was described which has both these activities [91]. Earlier a β6-GlcNAcT acting on terminal Gal residues ('terminally acting') has been reported [76]. A β3-GlcNAcT acting on resulting

Table 1. Modifications of lacNAc as shown by primary product structures and further terminal reactions in complex-type oligosaccharide synthesis. After the introduction of a sugar to lacNAc further additions are often, but not always, possible.

Primary product structure		Final structure	Name of (antigenic) structure
NeuAcα2→6Galβ1→4GlcNAcβ-R		no further reaction	
NeuAcα2→3Galβ1→4GlcNAcβ-R	↑	NeuAcα2→3Galβ1→4[Fucα1→3]GlcNAcβ-R	sialyl-Lewis^x
	↑	(NeuAcα2→8)$_n$NeuAcα2→3Galβ1→4GlcNAcβ-R	polysialic acid
	↗	NeuAcα2→3[GalNAcβ1→4]Galβ1→4GlcNAcβ-R	Sd^a
Galβ1→4[Fucα1→3]GlcNAcβ-R	↑	Galβ1→4[Fucα1→2Fucα1→3]GlcNAcβ-R	Lewis^x → difucosyllacNAc
Galα1→3Galβ1→4GlcNAcβ-R	↑	Galα1→3Galβ1→4[Fucα1→3]GlcNAcβ-R	α-galactosyl-Lewis^x
Fucα1→2Galβ1→4GlcNAcβ-R	↑	Fucα1→2Galβ1→4[Fucα1→3]GlcNAcβ-R	Lewis^y
	↖	Fucα1→2[GalNAcα1→3]Galβ1→4GlcNAcβ-R	H-type 2 ↗ blood group A
	↗	Fucα1→2[Galα1→3]Galβ1→4GlcNAcβ-R	↘ blood group B
GlcAβ1→3Galβ1→4GlcNAcβ-R	↑	SO$_4^-$-3GlcAβ1→3Galβ1→4GlcNAcβ-R	HNK-1
SO$_4^-$-3Galβ1→4GlcNAcβ-R	↑	SO$_4^-$-3Galβ1→4[Fucα1→3]GlcNAcβ-R	3'-sulfo-Lewis^x
Galβ1→4[SO$_4^-$-6]GlcNAcβ-R	↑	(NeuAcα2→3)$_\pm$Galβ1→4[Fucα1→3][SO$_4^-$-6]GlcNAcβ-R	6-sulfo-(sialyl-)Lewis^x
GlcNAcβ1→6Galβ1→4GlcNAcβ-R	↑	?	
GalNAcβ1→3Galβ1→4GlcNAcβ-R	↑	?	
Galα1→4Galβ1→4GlcNAcβ-R	↑	?	P1
GlcNAcα1→4Galβ1→4GlcNAcβ-R	↑	?	

23.12 Specific Modifications of Polylactosaminoglycans

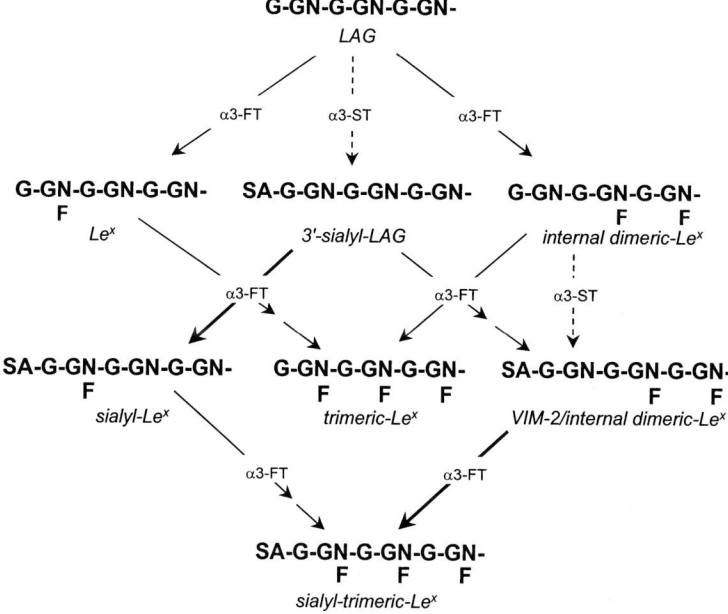

Figure 10. Cooperative action of α3-NeuAcT (α3-ST, dashed arrow), α3-FucT IV (myeloid enzyme; α3-FT, thin solid arrow), and α3-FucT VII (α3-FT, leukocyte enzyme; bold solid arrow) in the synthesis of sialyl-oligomeric Lewis^x and related structures on polylactosaminoglycan chains. F, Fuc; G, Gal; GN, GlcNAc; LAG, polylactosaminoglycan; SA, NeuAc.

GlcNAc(β1–6)Galβ1-R to yield a GlcNAc(α1–3)[GlcNAc(β1–6)]Galβ1-R branching point, however, has not yet been described. (See also Section 23.23, β6-N-*acetylglucosaminyltransferase family*).

Another specific modification of polylactosaminoglycan chains is the introduction of α3-Fuc residues to non-terminal GlcNAcs along the chain to yield a (sialyl-) oligomeric-Lewis^x structure. In leukocytes the formation of this structure, which is a ligand for E- and P-selectin, is catalyzed by two different α3-FucTs, the myeloid FucT (FucT IV) and the leukocyte FucT (FucT VII) [92, 93] (Figure 10) (see also Section 23.20, *α3/4-Fucosyltransferase Family* below).

β-GlcNAc cannot be introduced to position C-6 of Gal residues that form part of the Lewis^x structure [90]. In turn, as the OH at C-6 of Gal in lacNAc is a key polar group for α3-FucTs [69], α3-fucosylation of GlcNAc residues underlying blood group I branching points is impossible. It thus seems that the formation of oligomeric Lewis^x and of blood group I branching points are competing processes. Branched polylactosaminoglycans can, therefore, only carry Lewis^x groups at lacNAc units that are not engaged in a branch [94]. Similar to the β6-GlcNAcTs, different α3-FucTs show differential preferences for the penultimate and internal acceptor sites along the polylactosaminoglycan acceptor chain ("site specificity"). This specificity is further detailed under *α3/4-Fucosyltransferase family* described below.

23.13 The Invariable Core of N-linked Oligosaccharide Chains, and Site- and Protein-Specific Processing

Mature N-linked oligosaccharides all have the same basic core consisting of three Man and two GlcNAc residues. This core structure is already present in the preformed dolichol–PP–linked oligosaccharide precursor and remains untouched throughout the entire processing after its en bloc transfer to the polypeptide chain (Figure 2). Nevertheless, the oligosaccharide chains on different glycoproteins synthesized in one particular tissue or cell type are often structurally different. Apparently, the protein-linked chains can be processed to different extents.

When, for some reason, GlcNAcT I fails to act or when not all α2-linked Man residues are removed by α-mannosidase IA or IB, the oligosaccharide will remain as an oligomannosidic-type structure. Such a chain can occur at one Asn of a polypeptide even though the oligosaccharides at other Asn residues of the same glycoprotein have been processed to complex-type structures. Many examples of glycoproteins with this characteristic are known (*e.g.* tissue plasminogen activator [95]; glycodelin [96]). A differential accessibility of the various oligosaccharide chains to the processing enzymes as dictated by the polypeptide environment of the individual chains has been proposed to account for this feature [97]. However, the variations in structure at a single glycosylation site (microheterogeneity) rather seem to be determined stochastically. In other instances the N-linked oligosaccharides at different Asn residues differ in their amount of branching (*e.g.* human $α_1$-acid glycoprotein [98]) or in the presence or absence of a bisecting GlcNAc residue (human myeloma IgG_1 [99]). A hypothetical explanation of the latter differences might be that the polypeptide can interact with the oligosaccharide in a way that it is locked in a conformational state that is not acted upon by certain GlcNAcTs (*c.f.* [99]).

Finally, two classes of glycoprotein have been described the N-linked chains of which have a specific character because of the recognition of a peptide motif by a sugar-transferring enzyme. The first class is formed by the lysosomal enzymes. When α2-linked Man residues are still present in the oligosaccharide the structure is phosphorylated by the sequential action of an *N*-acetylglucosamine phosphotransferase and a phosphodiesterase resulting in the formation of Man 6-P determinants [100] (Figure 2). This structural element mediates the targeting of the lysosomal enzyme to the lysosome through its interaction with one of the two Man 6-P receptors [100]. Only lysosomal enzymes become phosphorylated because the *N*-acetylglucosamine phosphotransferase specifically recognizes a peptide sequence that is uniquely present in these enzymes. The second class comprises the pituitary glycohormones which are rich in complex-type glycans based on GalNAc(β1–4)GlcNAc (lacdiNAc) rather than Gal(β1–4)GlcNAc (lacNAc). It has been proposed that the pathway is directed toward the synthesis of such oligosaccharide chains by a hormone-specific *N*-acetylgalactosaminyltransferase that recognizes a peptide determinant present in the polypeptide of these hormones [101].

23.14 Synthesis of Type 1 (Gal(β1–3)GlcNAcβ–R) versus Type 2 (Gal(β1–4)GlcNAcβ–R) Chains

Most complex-type chains are based on lacNAc and are referred to as type 2 chains. In some instances, however, protein-linked chains are based on lacto-N-biose (Gal(β1–3)GlcNAc) and are referred to as type-1 chains. Notable examples of glycoproteins carrying such chains are bovine fetuin [102] and rat $α_1$-acid glycoprotein [103]. Although glycoproteins with this type of chain are rather rare, a major fraction of human milk oligosaccharides is based on the type-1 chain compound lacto-N-tetraose (Gal(β1–3)GlcNAc(α1–3)Gal(β1–4)Glc) [104]. For a long time the only β3-GalT involved in the synthesis of type-1 chains was that described in porcine trachea [105]; this activity was not reported in tissues such as mammary gland. Recently the cloning of four human GlcNAcβ–R-specific β3-GalTs [106–109] and the mouse orthologs of three of these [110] has been reported. Indeed these β3-GalTs seem to have a rather restricted tissue expression.

Like lacNAc termini, type-1 structures may be fucosylated by α2-FucT and α4-FucT (Lewis enzyme, FucT III) to yield the H-type-1 and the Lewis[a] structures, respectively [24, 81]. In turn the H-type-1 structure can be converted into a Lewis[b], or a blood group A or B active compound, by the action of α4-FucT, and blood group A- (α3-GalNAcT) or B-enzyme (α3-GalT), respectively [24, 81]. Alternatively, type-1 structures can be disialylated by the concerted action of a α3-NeuAcT and a α6-NeuAcT to yield the terminal NeuAc(α2–3)Gal(β1–3)[NeuAc(α2–6)]-GlcNAc structure [111, 112].

23.15 The LacdiNAc Pathway of Complex-Type Oligosaccharide Synthesis

As another alternative to type-2 chain complex-type oligosaccharide formation a GalNAc can be introduced to accepting GlcNAc residues yielding a GalNAc(β1–4)GlcNAc (lacdiNAc) unit [40]. In particular bovine milk glycoproteins carry oligosaccharide chains which are rich in such units, but they also occur on many other glycoproteins including several of human origin (reviewed in Ref. [40]). As mentioned above it has been proposed that a glycoprotein hormone-specific β4-GalNAcT exists which specifically recognizes a peptide determinant present in the polypeptide chain of these glycoproteins [101]. In schistosomes [113, 114], snail [115], and bovine mammary gland [116], however, a β4-GalNAcT has been described which is hormone-non-specific and strongly resembles β4-GalT in acceptor properties. It has been proposed that this β4-GalNAcT controls the lacdiNAc pathway of complex-type oligosaccharide synthesis [40].

Several enzymes that act on lacNAc can also act on lacdiNAc, in conformity with their essential requirements for activity described above. α6-NeuAcT and α3-

FucT can catalyze the formation of NeuAc(α2–6)GalNAc(β1–4)GlcNAc [66] and GalNAc(β1–4)[Fuc(α1–3)]GlcNAc (lacdiNAc variant of Lewis^x) [67], respectively. By way of contrast α3-NeuAcT and α3-GalT do not act on GalNAc(β1–4)GlcNAc [117] and β3-GlcNAcT only acts with very low activity [40]. Furthermore a sulfotransferase catalyzing the transfer of sulfate to OH-4 of GalNAc has been identified [118], as has a β3-GalT catalyzing the formation of Galβ1–GalNAc(β1–4)GlcNAc [119].

23.16 Protein *O*-Glycosylation

In contrast with N-glycosylation, formation of O-linked chains as occurs on mucins is initiated by the addition of a single α-GalNAc residue (rather than of a preformed oligosaccharide) to the OH-group of a Ser or Thr in the polypeptide chain. It is believed that this step occurs in the *cis*-Golgi [22]. Additions to this GalNAc residue of Gal in (β1–3)-, GlcNAc in (β1–3)- and (β1–6)-, and GalNAc in (α1–3)-linkage might subsequently occur in the *medial*-Golgi which leads to different core structures (Figure 11).

These reactions are all catalyzed by glycosyltransferases specific for O-linked chains. Also, direct addition of NeuAc to GalNAc and to the core 1 disaccharide (Gal(β1–3)GalNAc) can occur as catalyzed by NeuAcTs that are highly specific for these O-linked sugars (Figure 11, right). In the *trans*-Golgi terminal GlcNAc residues on the different core structures can be further acted upon by β4-GalT yielding O-linked complex-type chains. Polylactosaminoglycans can also be formed on these chains by the concerted action of β3-GlcNAcT and β4-GalT. The chains can be completed by addition of NeuAc, Fuc and sulfate residues as catalyzed by the same sialyltransferases, fucosyltransferases, and sulfotransferases that act in the synthesis of N-linked chains. Although not all glycosyltransferases acting on lacNAc in N-linked chains (see Figure 6) act with the same efficiency on lacNAc in O-linked chains (*e.g.* α6-NeuAcT does not act well on Gal(β1–4)GlcNAc(β1–6)[Gal(β1–3)]-GalNAcα–O–R) elongation and termination of the complex-type O-linked structures follows in principle the same pathways as for N-linked chains and are catalyzed by the same enzymes that act in the terminal stages of N-glycosylation. Thus on O-linked chains blood group determinants such as (sialyl-)Lewis^x, H(O), A, B, Sd^a, i, I, and type-1 chain-based H, Lewis^a and Lewis^b can be formed in an analogous way [22].

23.17 Glycosyltransferase Families

Before the days of molecular cloning of glycosyltransferases the classification of these enzymes was based on common nucleotide–sugar donor substrate usage with sub-classification based on the linkage formed and the acceptor sugar used.

23.17 Glycosyltransferase Families

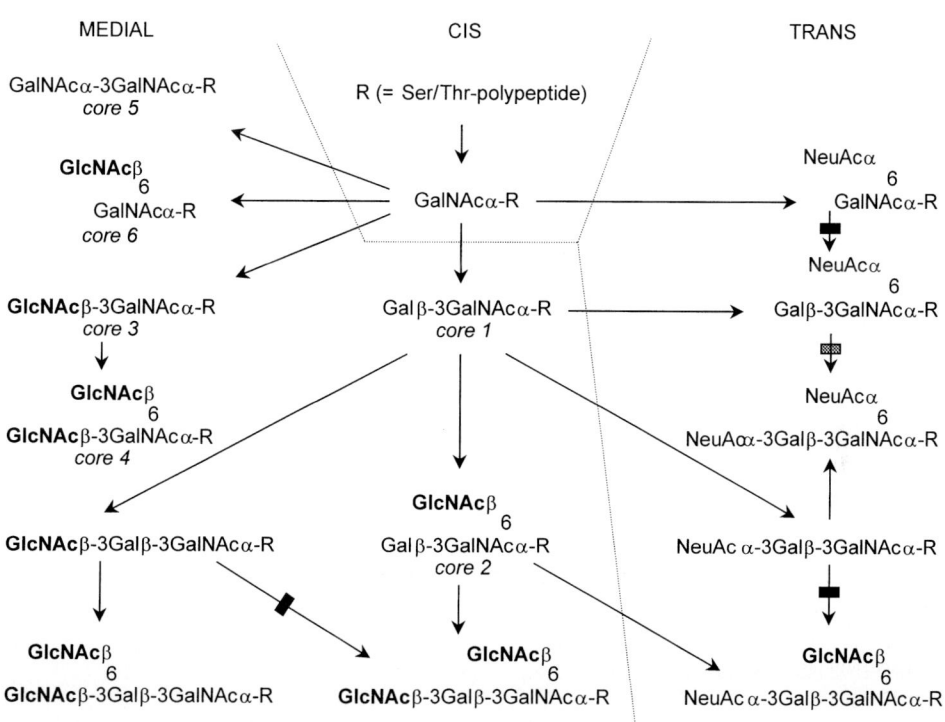

Figure 11. Major pathways in O-linked oligosaccharide synthesis. The reactions leading to the sialylated and the different core structures are depicted. Once a GlcNAc residue is attached it can be acted upon by β4-GalT to form a lacNAc unit, which can be further modified as outlined in Figures 6 and 10, and Table 1.

Thus for a long time families of sialyltransferases, fucosyltransferases, galactosyltransferases, *etc.* have been distinguished. During the past 14 years the cloning of the cDNAs of over 50 glycosyltransferases in eukaryotes has been achieved, (A glance at the web site www.vei.co.uk/tgn/gt_guide.htm. gives an impression of the vast number of glycosyltransferases that have been cloned. Note, however, that the list has not been updated recently.) The structural information obtained not only led to the insight that all of these enzymes share a common domain structure (Figure 5), but also has provided a rationale for the aforementioned classification system. On the basis of sequence similarities at least 11 different glycosyltransferase families can now be identified (Table 2).

Members within a family have extensive sequence identity at the amino acid level, which is often confined to specific boxes, and have one or more enzymatic properties in common. In the following paragraphs these glycosyltransferase families are discussed in more detail.

Table 2. Glycosyltransferase families.

Transferase family	Linkage-type established	Number of enzymes with sequence similarity	Notes
Sialyl	α3-/α6-/α8-NeuAc to Gal(NAc)/NeuAc	17	All contain sialyl motif L and S
α2-Fucosyl	α2-Fuc to Gal	2	H- and Se-enzyme
α3/4-Fucosyl	α3-/α4-Fuc to Glc(NAc)	6	FucT III–VII and IX; α3-fucosyl motif
α3-Galactosyl/N-Acetylgalactosaminyl	α3-Gal(NAc) to Gal(NAc)	4	A- and (pseudo) B-enzyme, Forssman glycolipid synthase
β4-N-Acetylglucosaminyl	β4-GlcNAc to Man	2	GlcNAcT IV is the only N-glycan branching GlcNAcT known so far that is not a "stand alone"
β6-N-Acetylglucosaminyl	β6-GlcNAc to Gal(NAc)	3	O-linked core 2 and 4 synthases and I-enzyme (GlcNAcT-V is NOT a member)
α-N-Acetylgalactosaminyl	α-GalNAc to Ser/Thr	7	Enzymes have different peptide specificities
β4-N-Acetylgalactosaminyl	β4-GalNAc to [NeuAcα3]Gal	2	Sd^a- and GM_2-synthase
β4-Galactosyl	β4-Gal/β4-GlcNAc to Glc(NAc)/Xyl	8	α-lactalbumin-sensitive and insensitive β4-GalTs and chitobio-synthase; all share several β4-galactosyl motifs
β3-Galactosyl	β3-Gal to GlcNAc or GalNAc	5	Type 1 chain synthases and ganglioside GM_1-synthase (Core 1 synthase is NOT a member)
β3-Glucuronyl	β3-GlcA to Gal	3	Involved in HNK-1 epitope and proteoglycan core synthesis

23.18 Sialyltransferase Family

All members of the sialyltransferase family share a highly conserved so called 'sialyl motif' (designated L for large) in their primary amino acid sequence; this motif has been found to participate in the binding of the common donor substrate CMP–NeuAc [120]. In addition to this sialyl motif L a smaller more C-terminally located common sequence motif has been found; this is designated sialyl motif S (S for small) [121]. Once these motifs had been identified they were exploited to clone novel sialyltransferases by homology and today the cDNAs of seventeen of such enzymes have been identified (reviewed in Refs. [122] and [123]). A unique feature of this family is that it comprises enzymes that establish different linkages ((α2–3)-, (α2–6)- or (α2–8)-) and use different acceptor sugars (Gal, GalNAc, or NeuAc). This is quite different from, for instance, the galactosyltransferases where enzymes with different linkage and acceptor specificity have very little sequence similarity and consequently are grouped in separate families (see below).

Members of the sialyltransferase family are now commonly sub-classified on the basis of the linkage they establish and the accepting sugar [123]. There are six α3-NeuAcTs that act on terminal Gal residues (ST3Gal I–VI). Best substrate structures for these enzymes are ST3Gal I, Gal(β1–3)GalNAcα (protein linked); ST3Gal II, Gal(β1–3)GalNAcβ (lipid linked); ST3Gal III, Gal(β1–3)GlcNAcβ; ST3Gal IV, Gal(β1–4)GlcNAcβ (protein linked); ST3Gal V, Gal(β1–4)Glcβ1–Cer, and ST3Gal VI, Gal(β1–4)GlcNAcβ (lipid or protein linked), respectively [122, 124–126]. Only one α6-NeuAcT occurs (ST6Gal I); this specifically acts on Galβ–4GlcNAcβ (Fig. 6, Table 1) and GalNAcβ–4GlcNAcβ [66] but not on type 1 chains or O-linked α-GalNAc and core 1. Five α6-NeuAcTs (ST6GalNAc I–V) have been cloned that use a GalNAc residue in different structures as accepting sugar (see also Fig. 11). ST6GalNAc I acts on O-linked GalNAcα-, Gal(β1–3)GalNAcα, and NeuAc(α2–3)Gal(β1–3)GalNAcα- but requires a protein aglycone; ST6GalNAc II acts on Galβ1–3GalNAcα, and NeuAcα(2–3)Galβ1–3GalNAcα- but not on GalNAcα; ST6GalNAc III, -IV and -V all three act on O-linked as well as lipid linked NeuAcα(2–3)Gal(β1–3)GalNAcα/β [122, 127–129, 189]. The latter three enzymes tolerate aromatic aglycons, but do not act on non-sialylated substrate structures. ST6GalNAc III has no pronounced preference for either acceptor structure, but ST6GalNAc IV prefers O-linked acceptor chains, while ST6GalNAc V acts with preference on lipid linked substrates. In contrast to ST6GalNAc I–IV, which seem to be involved in the sialylation of O-glycans, ST6GalNAc V has been suggested to function in the synthesis of ganglioside $GD_{1\alpha}$. Five α8-NeuAcTs (ST8Sia I–V) are known of which ST8Sia II and IV are involved in the synthesis of protein-linked polysialic acid chains, whereas ST8Sia I, III and V seem to function in the synthesis of NeuAc(α2–8)NeuAcα and NeuAc(α2–8)NeuAc(α2–8)NeuAcα sequences occurring on gangliosides such as GD_3, GT_{1a}, GT_3, and GQ_{1b} [122].

23.19 α2-Fucosyltransferase Family

Two human α2-FucTs have been cloned [130, 131]. One enzyme, occurring in serum and plasma, is the product of the *H* gene and controls the synthesis of the blood group H-antigen (Fuc(α1–2)Galβ), which is the precursor of the A- and B-antigens, on erythrocytes. The other α2-FucT occurs in milk, saliva, and other secretions, and is the product of the *Secretor (Se)* gene. It determines the expression of soluble H- (and A- and B-) antigen in secretions and the Lewis[b] (Fuc(α1–2)Gal(β1–3)[Fuc(α1–4)]GlcNAc) blood group antigens on red cells [24, 81]. The enzymes have 68%

identity at the amino acid level. They both catalyze the formation of a Fuc(α1–2)Galβ linkage, but have different acceptor specificities. Whereas the H enzyme highly prefers type-2 chain acceptors and efficiently acts on Galβ-phenyl, the secretor enzyme has a preference for type-1 chain-based acceptors and for Gal(β1–3)GalNAc [131–133]. The latter enzyme also has much lower affinity for the donor substrate GDP-Fuc than the H enzyme [131, 133]. In rabbit three α2-FucTs occur; one resembles the human H enzyme in molecular and enzymatic properties, the others are molecularly similar to, and have the enzymatic characteristics of, the secretor enzyme [134].

23.20 α3/4-Fucosyltransferase Family

Within the α3/4-fucosyltransferase family six different members have been identified. They are numbered in the order they were cloned FucT III–VII and IX. Human FucT III (Lewis enzyme), V and VI (liver/plasma enzyme) are highly related enzymes with >90% identity at the amino acid level in their catalytic domain; their genes are all located on human chromosome 19 [135]. By contrast, FucT IV (myeloid enzyme, chromosome 11) and FucT VII (leukocyte enzyme, chromosome 9) have a more distal relationship to each other and to the other members of this family [135, 136]. Recently a novel member, FucT IX, which has highest homology to FucT IV, was cloned from a mouse library [137]. Recently the human ortholog of FucT IX was described [138].

A highly conserved common motif consisting of some 10–15 amino acids in the primary structures of these enzymes could be identified upon the cloning of a bacterial (*Helicobacter pylori*) α3-FucT [139, 140]. The members of this transferase family all utilize GDP-Fuc as a common donor but it remains to be established whether this 'α3-fucosyl motif' is somehow involved in the binding of this nucleotide–sugar. All members of this family use a substituted GlcNAc or Glc as accepting sugar and all but one establish a Fuc(α1–3)-linkage. The exception is formed by FucT III which preferentially acts on type 1 acceptors yielding the Lewis[a] structure in which the Fuc is linked (α1–4) (Gal(β1–3)[Fuc(α1–4)]GlcNAc), whereas the other α3-FucTs exclusively act on type-2 chain-based acceptors yielding Lewis[x] blood group determinant-based structures. FucT III can, however, also support the synthesis of Lewis[x] by establishing a Fuc(α1–3)-linkage both in vitro and in vivo [2–4].

This enzyme is still the only exception to the one linkage–one enzyme concept mentioned in the introduction. Recently it was reported that a single amino acid in the hypervariable stem domain of FucT III determines the preference of this enzyme for type-1 structures [141]. Preferred substrate structures for the members of this family are: FucT III, Gal(β1–3)GlcNAcβ–R; FucT IV, Gal(β1–4)GlcNAcβ–R; FucT V and VI, both Gal(β1–4)GlcNAcβ–R and NeuAc(α2–3)Gal(β1–4)-GlcNAcβ–R; FucT VII, NeuAc(α2–3)Gal(β1–4)GlcNAcβ–R, and FucT IX, Gal(β1–4)GlcNAcβ–R, respectively [69, 92, 93, 137, 142]. FucT III–VI also can act with varying efficiencies on lactose (Gal(β1–4)Glc) and H-type-2 acceptors such as Fuc(α1–2)Gal(β1–4)Glc(NAcβ–R). With oligomers of lacNAc (polylactosaminoglycans), wherein Fuc may be introduced to penultimate and to internal GlcNAc residues, α3-FucTs show "site specificity". Whereas FucT IV highly prefers internal GlcNAc residues, FucT IX acts preferentially on the penultimate GlcNAc [92, 190]. FucT VII acts specifically on penultimate GlcNAc residues too, but requires that the lacNAc unit comprising this GlcNAc carries a terminal (α2–3)-linked NeuAc residue [92, 93] (Fig. 10). By contrast FucT VI (from human milk) can introduce an α3-Fuc to any GlcNAc in such acceptors whether the chain is α3-sialylated or not [143, 144].

23.21 α3-Galactosyl/N-Acetylgalactosaminyltransferase (Histo-Blood Group ABO) Family

This family, consisting of four members, is heterogeneous in that two member enzymes preferentially utilize UDP–Gal whereas the two others use UDP–GalNAc. Furthermore, one GalNAcT (blood group A transferase) and one GalT (blood group B transferase) act exclusively on substrates with Fuc(α1–2)Galβ- termini (blood group H structure) to yield the blood group A and B epitopes, respectively [7, 145] (Table 1). By contrast the other GalT ('pseudo B-enzyme', for which the enzymatic properties of α3-GalT have been described in detail above) [146, 147] acts exclusively on non-fucosylated structures, preferentially on Gal(β1–4)GlcNAc, whereas the other GalNAcT acts on β-linked GalNAc in globoside (GalNAc(β1–3)Gal(α1–4)Gal(β1–4)Glcβ1–Cer) to yield the Forssman glycolipid (GalNAc(α1–3)GalNAc(β1–3)Gal(α1–4)Gal(β1–4)Glcβ1–Cer) [148].

The single common enzymatic property of these enzymes is that they all establish a (α1–3)-linkage. Nevertheless the primary structures of these enzymes have extensive sequence similarities. This is particularly true for the human blood group A and B transferases which differ in four amino acids only [7]. It has long been known that for these enzymes sugar–donor promiscuity is relatively high [149]. By mutating each of the four amino acids in these blood group transferases it has been shown that two of these amino acids are crucial in determining the nucleotide–sugar specificity [7].

23.22 β4-N-Acetylglucosaminyltransferase (IV) Family

The individual GlcNAcTs involved in the branching of N-linked glycans (Fig. 4) do not generally seem to belong to a glycosyltransferase family (see *Glycosyltransferases standing alone* below). An exception is human GlcNAcT IV that has been described to occur in two isoforms, which show 62% identity at the amino acid level [191]. So far it is not clear whether these two forms have different specificities.

23.23 β6-N-Acetylglucosaminyltransferase Family

This family comprises three members; all occur in man. The enzyme that was cloned first was identified as an O-linked core 2 synthase [150] (*c.f.* Figure 11). It acts on Gal(β1–3)GalNAcα–pNP, but not on GlcNAc(β1–3)GalNAcα–pNP or structures that form part of linear polylactosaminoglycans [91, 151]. A structurally related glycosyltransferase has been shown to control the synthesis of blood group I active branching points in polylactosaminoglycans [152]. In vitro this enzyme has been reported to act on acceptors that form part of linear polylactosaminoglycans, either as a centrally acting β6-GlcNAcT [90] or a centrally **and** predistally acting enzyme [91], but not on Gal(β1–3)GalNAcα–pNP or GlcNAc(β1–3)GalNAcα–pNP. A third member of this family was found to act preferentially on Gal(β1–3)-GalNAcα–pNP, but also on GlcNAc(β1–3)GalNAcα–pNP and thus also has O-linked core 4 synthase activity [91, 151].

A mouse homolog of the human core 2 synthase seems to be involved in the synthesis of a globo-type sphingoglycolipid core structure (GlcNAc(β1–6)[Gal(β1–3)]GalNAc(β1–3)Gal(α1–4)Gal(β1–4)Glcβ1–Cer) [153], suggesting that core 2 synthase might have activity with acceptors having Gal(β1–3)GalNAcα or Gal(β1–3)GalNAcβ termini. This family might contain additional, not yet cloned enzymes which act with preference on penultimate (predistal) or terminal Gal residues (see also Section 23.12, *Specific Modifications of Polylactosaminoglycans*).

23.24 Polypeptide *N*-Acetylgalactosaminyltransferase Family

This family contains seven members cloned from human, rat, mouse, or bovine libraries [149–155, 192–194]. These enzymes (ppGalNAcTs), which catalyze the initiating step in O-glycosylation (*c.f.* Figure 11), have overlapping yet distinctly different peptide-acceptor specificities [11, 156–158].

With most peptide substrates ppGalNAcT 1 seems to be the most potent enzyme in terms of rate. Different relative rates, however, have been found for these enzyme with various peptides and some peptides are acted upon by one enzyme but not by others, and *vice versa*. In general all enzymes seem to prefer Thr residues to Ser in most peptide substrates. ppGalNAcT 1 and 2 introduce GalNAc residues to two Thr and one Ser position in a peptide (DTRPAPG*S*TAPPAHGV*T*SAP, points of attachment in italics) which is a repeat of MUC1. Rates for ppGalNAcT 1–3 are, however, different for each of these attachment sites [11]. With another peptide substrate (PTTTPLK, repeat of MUC2) it seemed that the number of GalNAc residues that can be incorporated by ppGalNAcT 1–3 varies from one to three and that the order of attachment is also different for each enzyme [159]. ppGalNAcT 4 and ppGalNAcT 5 seem to act on a more limited range of peptides than ppGalNAcT 1–3 [157]. On the other hand ppGalNAcT 4 seems to complement ppGalNAcT 1–3 in that it acts on a Thr and a Ser site in TAPPAHGVT-*S*APD*T*RPAPGSTAPPA (repeat of MUC1) [158]. ppGalNAcT 6 resembles ppGalNAcT 3 in acceptor preferences [192]. Interestingly, ppGalNAcT 7 shows exclusive specificity for partially GalNAc glycosylated substrate peptides resulting from prior action of other ppGalNAcTs [193, 194]. This shows that the initiating steps in O-glycosylation occur in a coordinate fashion by different ppGalNAcTs.

So far all ppGalNAcTs but one are specific for UDP–GalNAc as a donor substrate. Only ppGalNAcT 2 can also efficiently utilize UDP–Gal [11]. Although the overall sequence similarity of the enzymes is *ca* 45% [155], all contain a segment of 41 amino acids that is highly conserved in all ppGalNAcTs [160]. Interestingly, a box within this segment shows strong similarity with a conserved box found in the β4-galactosyltransferase family. It has been suggested that this segment is involved in UDP–sugar binding [160].

23.25 β4-*N*-Acetylgalactosaminyltransferase Family

Of this family one human and one murine member have been identified [83, 161]. Little information is available on the specificities of the recombinant forms of these GalNAcTs, but it has been demonstrated that the murine enzyme acts on 3′-sialyllacNAc (but not on lacNAc) to yield the Sd^a blood group-active structure GalNAc(β1–4)[NeuAc(α2–3)]Gal(β1–4)GlcNAc [83]. Transfection studies using the cDNA of the other enzyme showed that the human GalNAcT can support the synthesis of ganglioside GM$_2$ from GM$_3$, and GD$_2$ from GD$_3$ [161]. Whether these enzymes have overlapping acceptor specificities is not known, but they share the common property of acting on a Gal residue bearing a sialic acid at its OH-3 and establishing a GalNAc(β1–4)[NeuAc(α2–3)]Gal linkage.

23.26 β4-Galactosyltransferase Family

This family consists of eight enzymes, seven vertebrate and one molluscan. All establish a (β1–4) linkage and although the sequence similarity among them might be lower than 40%, so far all vertebrate members are true β4-GalTs acting on GlcNAc, Glc, or Xyl. The numbering of the β4-GalTs has been confusing. Originally they were numbered in order of their cloning, the system commonly used in other glycosyltransferase families (*e.g.* sialyltransferases and α3-fucosyltransferases). Nowadays they are numbered according to their similarity to β4-GalT I [162], the first mammalian glycosyltransferase ever cloned [163–165].

Of all members β4-GalT I has by far the highest intrinsic activity. It can act on virtually any terminally exposed β-GlcNAc to form a Gal(β1–4)GlcNAc unit (see also Section 23.5, *Committed Steps in the Formation of Complex-Type Oligosaccharide Chains and Branching*). Free GlcNAc and β-GlcNAc linked to an aromatic or fatty aglycone are excellent acceptors [166, 167]. Also GlcNAc in oligosaccharides, whether related to N-protein (GlcNAc linked to Man) or O-protein (GlcNAc linked to GalNAc) linked glycans, or contained in a polylactosaminoglycan chain (GlcNAc linked to Gal), is an efficient acceptor site for β4–GalT I [38, 39, 85, 167–169]. β4-GalT I can, furthermore, act on GlcNAc(β1–3)Gal(β1–4)Glcβ1-Cer to yield Gal(β1–4)GlcNAc(β1–3)Gal(β1–4)Glcβ1-Cer (paragloboside) [166, 167]. β4-GalT I thus can function in N-protein, O-protein, and lipid glycosylation.

For over 30 years it has been known that in the lactating mammary gland this enzyme can combine with the milk protein α-lactalbumin (α-LA) to form the lactose synthase complex [170, 171]. α-LA stimulates β4-GalT I to act on Glc yielding lactose, at the same time inhibiting the action on GlcNAc. This enzyme, therefore, also serves a specific function in lactose synthesis. β4-GalT II and β4-GalT III have acceptor properties which resemble those of β4-GalT I but act much less efficiently [166]. Whereas β4-GalT II is similarly responsive to α-LA as β4-GalT I, β4-GalT III is not, however, induced by α-LA to act on Glc, and activity toward GlcNAc is not inhibited by this modifier protein [166].

β4-GalT IV is a poorly acting enzyme and no activity could be detected with glycoprotein acceptors [172]. It has, however, notable activity with glycolipid acceptors and can support the synthesis of paragloboside. α-LA does not significantly induce the enzyme to act on Glc, but rather stimulates activity toward free GlcNAc [172].

β4-GalT V, like β4-GalT I, can act on a variety of acceptors albeit at a relatively low rate [167, 173]. The enzyme is not induced by α-LA to act on Glc, but α-LA inhibits its action on free GlcNAc. The highest activity is shown toward acceptors containing a GlcNAc(β1–6)GalNAcα structural element, suggesting that this enzyme predominantly functions in the galactosylation of O-linked core 2- and core 6-based structures [167]. A similar function has been ascribed to β4-GalT IV [174], but this contradictory finding is probably a consequence of erroneous numbering of the clones resulting from the confusion in the numbering system mentioned above.

β4-GalT VI has been identified as a lactosylceramide synthase [175]. Whether it acts on substrates other than glucosylceramide has not been determined. Finally, the β4-GalT involved in the first step of the synthesis of the GlcA(β1–3)Gal(β1–3)Gal(β1–4)Xyl core region of proteoglycans has recently been cloned and seems to be another member of this family [176, 195]. It is proposed to refer to this enzyme as β4-GalT VII.

The only member of this family found in invertebrates has acceptor specificity resembling that of β4-GalT I. In stead of Gal, however, it transfers GlcNAc from the corresponding nucleotide sugar to form a GlcNAc(β1–4)GlcNAc (N,N'-diacetylchitobiose) unit and therefore is a β4-GlcNAcT [177, 178]. It has no structural relationship to any other GlcNAcT yet cloned. The enzyme is insensitive to α-LA. A mutant of this β4-GlcNAcT, in which two out of three amino acid repeats are removed, shows acquired donor promiscuity toward UDP–GalNAc coupled to an overall higher kinetic efficiency [9]. Donor promiscuity toward UDP–GalNAc is also shown by bovine milk β4-GalT (β4-GalT I) [8]. An interesting common property of β4-GalT I, β4-GalT V, and β4-GlcNAcT is that their activity is highest with acceptors containing a (β1–6)-linked GlcNAc [85, 167, 173, 178]. With branched acceptors such as GlcNAc(β1–6)[GlcNAc(β1–3)]Gal (β4-GalT I [85], β4-GlcNAcT [178]) and GlcNAc(β1–6)[GlcNAc(β1–2)]Man(α1–6)Manβ–R (β4-GalT I [169]) they act with preference on the (β1–6)-linked GlcNAc.

Despite the small amount of overall sequence similarity, several stretches of amino acids have been highly conserved in all members ('β4-galactosyl motifs') [172, 175, 179]. Interestingly one conserved stretch is very similar to a box conserved in the sequences of the pp–GalNAcTs [160]. The amino acids in this box might, therefore, be involved in UDP–sugar binding, but apparently do not determine donor sugar specificity. It should, furthermore, be noted that there is very little sequence similarity between members of the α3-galactosyltransferase, β4-

galactosyltransferase, and β3-galactosyltransferase families and hence it seems that UDP–Gal is recognized in different ways within each of these families.

Potential other members of the β4-galactosyltransferase family are the β4-GalNAcTs that occur in snail [180] and bovine mammary gland [116]. These enzymes not only strikingly resemble β4-GalT I in acceptor requirements, but also are both induced by α-LA to act on Glc yielding the lactose analog GalNAc(β1–4)Glc, suggesting that they are molecularly related to the β4-GalTs. Attempts to clone the β4-GalNAcT have, however, failed so far.

23.27 β3-Galactosyltransferase Family

This family comprises five different enzymes; all utilize UDP–Gal as donor substrate and all establish a Gal(β1–3)-linkage [106–110]. Whereas four enzymes (β3-GalT I–III and V) act on β-linked GlcNAc yielding a type-1 disaccharide unit, β3-GalT IV acts on β-linked GalNAc as in ganglioside GM_2 and thus rather functions in ganglioside biosynthesis. Only slight differences in acceptor specificities have so far been observed for the type-1 chain synthases, but it has been suggested that β3-GalT V is specifically involved in elongation of the O-linked core 3 [109]. It has been anticipated that the β3-GalT responsible for the synthesis of O-linked core 2 (Gal(β1–3)GalNAcα–Ser/Thr) (Fig. 11) was also a member of this family. The recent cloning of this enzyme, however, revealed that it shows no homology with any other galactosyltransferase in the databases [196].

23.28 β3-Glucuronyltransferase Family

This family comprises enzymes which transfer GlcA from UDP–GlcA in β3-linkage to β-galactosides. They are involved in the synthesis of either the precursor of the HNK-1 epitope (SO_4^-–3GlcA(β1–3)Gal(β1–4)GlcNAcβ–R) occurring on glycoproteins and glycolipids, or the GlcA(β1–3)Gal(β1–3)Gal(β1–4)Xyl core structure of proteoglycans. Two rat GlcATs have been cloned, GlcAT–P [181] and GlcAT–D [182]; these are 50% identical with each other at the amino acid level. Whereas GlcAT–P seems to act preferentially on protein-linked type-2 chains [183], GlcAT–D has a broader specificity and acts on both type-1 and type-2 chains whether protein- or lipid-linked [182]. The third member of this family (GlcAT-I) has been cloned from a human library [184]. This enzyme specifically acts on Gal(β1–3)Gal(β1–4)Xyl to form the proteoglycan core tetrasaccharide sequence, but does not act on Gal(β1–4)GlcNAcβ–R.

23.29 Glycosyltransferases Standing Alone

Although most glycosyltransferases can be grouped in a family, several important enzymes so far cannot. Well known examples are the GlcNAcTs involved in N-glycan branching (with the exception of GlcNAcT IV, see β4-N-*acetylglucosaminyltransferase (IV) family*). The cloning of these enzymes has been described: GlcNAcT I [43, 44], GlcNAcT II [45], GlcNAcT III [46], and GlcNAcT V [48, 49]. These enzymes are molecularly unrelated, either to each other or to any other cloned GlcNAcT. Also the α6-FucT is involved in N-glycan core fucosylation is an enzyme that does not belong to a family [185]. Finally the cloning of two β3-GlcNAcTs, both of which might be involved in polylactosaminoglycan synthesis, has revealed that these enzymes are not related to each other [86, 186]. It can, however, not be excluded that in the future molecular relatives of each of these enzymes will be discovered.

23.29 Concluding Remarks

In this chapter the pathways leading to the various oligosaccharide structures occurring on soluble and cell-surface-bound glycoconjugates have been described, and the specificities of the glycosyltransferases which control the glycosylation reactions have been discussed. As a result of the molecular cloning of a vast number of these enzymes it has become evident that many can be grouped into families. It also has become clear that the same reaction can be catalyzed by related, yet different enzymes. Tissue-specific expression of these enzymes is, therefore, a way in which oligosaccharide biosynthesis might be fine-tuned.

In addition to glycosyltransferases many other enzymes are involved in the process of glycosylation (processing glycosidases involved in the early stages of N-glycosylation; transporters carrying the nucleotide sugars across the intracellular membranes; pyrophosphorylases catalyzing the synthesis of these nucleotide sugars; enzymes of the intermediary carbohydrate metabolism, and epimerases, mutases, transaminases, acetylases, dehydrogenases, decarboxylases, dehydratases, *etc.*, yielding the various precursor sugars) and the number of these 'glycosylation enzymes' might easily exceed 500.

Within a few years the entire human genome will have been sequenced and the functional genes will become known. Some 3–5% of these genes can be estimated to be somehow involved in the process of glycosylation [187]. It will be a challenge for the future to identify these 'glycosylation genes' and to find the function of their individual protein products.

Note added in proof

Very recently two additional glycosyltransferases were cloned. The first one is a sialyltransferase (ST6GalNAc VI) that acts on gangliosides but not on glycoproteins. It seems to function in the synthesis of ganglioside GD1α, GQ1bα and GT1aα (T. Okajima, H.-H. Chen, H. Ito, M. Kiso, T. Tai, K. Furukawa, T. Urano and K. Furukawa, *J. Biol. Chem.*, **2000**, *275*, 6717–6723). The other enzyme is an additional β6-GlcNAcT that seems to function in the synthesis of the O-linked core

2 structure (T. Schwientek, J.-C. Yeh, S. B. Levery, B. Keck, G. Merkx, A. Geurts van Kessel, M. Fukuda and H. Clausen, *J. Biol. Chem.*, **2000**, *275*, 11106–11113). Consequently the number of cloned sialyltransferases counts 18 and of cloned β6-GlcNAcTs 4.

References

1. A. Hagopian and E. H. Eylar, *Arch. Biochem. Biophys.*, **1968**, *128*, 422–433.
2. J. F. Kukowska Latallo, R. D. Larsen, R. P. Nair and J. B. Lowe, *Genes & Dev.*, **1990**, *4*, 1288–1303.
3. J. B. Lowe, J. F. Kukowska Latallo, R. P. Nair, R. D. Larsen, R. M. Marks, B. A. Macher, R. J. Kelly and L. K. Ernst, *J. Biol. Chem.*, **1991**, *266*, 17467–17477.
4. E. Grabenhorst, M. Nimtz, J. Costa and H. S. Conradt, *J. Biol. Chem.*, **1998**, *273*, 30985–30994.
5. C. H. Hokke, J. G. Van der Ven, J. P. Kamerling and J. F. G. Vliegenthart, *Glycoconjugate J.*, **1993**, *10*, 82–90.
6. A. D. Yates, P. Greenwell and W. M. Watkins, *Biochem. Soc. Trans.*, **1983**, *11*, 300–301.
7. F. Yamamoto and S. Hakomori, *J. Biol. Chem.*, **1990**, *265*, 19257–19262.
8. M. M. Palcic and O. Hindsgaul, *Glycobiology*, **1991**, *1*, 205–209.
9. H. Bakker, A. van Tetering, M. Agterberg, A. B. Smit, D. H. Van den Eijnden and I. van Die, *J. Biol. Chem.*, **1997**, *272*, 18580–18585.
10. H. Kitagawa, H. Shimakawa and K. Sugahara, *J. Biol. Chem.*, **1999**, *274*, 13933–13937.
11. H. H. Wandall, H. Hassan, E. Mirgorodskaya, A. K. Kristensen, P. Roepstorff, E. P. Bennett, P. A. Nielsen, M. A. Hollingsworth, J. Burchell, J. Taylorpapadimitriou and H. Clausen, *J. Biol. Chem.*, **1997**, *272*, 23503–23514.
12. R. Kornfeld and S. Kornfeld, *Annu. Rev. Biochem.*, **1985**, *54*, 631–664.
13. H. Schachter, *Glycobiology*, **1991**, *1*, 453–461.
14. S. C. Basu, *Glycobiology*, **1991**, *1*, 469–475.
15. C. Abeijon and C. B. Hirschberg, *TIBS*, **1992**, *17*, 32–36.
16. G. van Echten and K. Sandhoff, *J. Biol. Chem.*, **1993**, *268*, 5341–5344.
17. Schachter H., in: Molecular Glycobiology, eds. M. Fukuda and O. Hindsgaul (IRL Press, Oxford, 1994) p. 88.
18. F. W. Hemming in *Glycoproteins* (J. Montreuil, H. Schachter and J. F. G. Vliegenthart, eds.), Elsevier, Amsterdam, **1995**, pp. 127–143.
19. A. Verbert in *Glycoproteins* (J. Montreuil, H. Schachter and J. F. G. Vliegenthart, eds.), Elsevier, Amsterdam, **1995**, pp. 145–152.
20. H. Schachter in *Glycoproteins* (J. Montreuil, H. Schachter and J. F. G. Vliegenthart, eds.), Elsevier, Amsterdam, **1995**, pp. 153–199.
21. H. Schachter in *Glycoproteins* (J. Montreuil, H. Schachter and J. F. G. Vliegenthart, eds.), Elsevier, Amsterdam, **1995**, pp. 281–286.
22. I. Brockhausen, in *Glycoproteins* (J. Montreuil, H. Schachter and J. F. G. Vliegenthart, eds.), Elsevier, Amsterdam, **1995**, pp. 201–259.
23. J. Roth, in *Glycoproteins* (J. Montreuil, H. Schachter and J. F. G. Vliegenthart, eds.), Elsevier, Amsterdam, **1995**, pp. 287–312.
24. W. M. Watkins in *Glycoproteins* (J. Montreuil, H. Schachter and J. F. G. Vliegenthart, eds.), Elsevier, Amsterdam, **1995**, pp. 313–339.
25. C. B. Hirschberg, P. W. Robbins and C. Abeijon, *Annu. Rev. Biochem.*, **1998**, *67*, 49–69.
26. D. J. Kelleher, G. Greibich and R. Gilmore, *Cell*, **1992**, *69*, 55–65.
27. D. J. Kelleher and R. Gilmore, *J. Biol. Chem.*, **1994**, *269*, 12908–12917.
28. A. Lal, P. Pang, S. Kalelkar, P. A. Romero, A. Herscovics and K. W. Moreman, *Glycobiology*, **1998**, *8*, R 4.
29. S. Weng and R. G. Spiro, *Glycobiology*, **1996**, *6*, 861–868.
30. W. J. Ou, P. H. Cameron, D. Y. Thomas and J. J. Bergeron, *Nature*, **1993**, *364*, 771–776.
31. C. Hammond, I. Braakman and A. Helenius, *Proc. Natl. Acad. Sci. USA*, **1994**, *91*, 913–917.
32. C. Hammond and A. Helenius, *Curr. Opin. Cell Biol.*, **1995**, *7*, 523–529.

33. C. Labriola, J. J. Cazzulo and A. J. Parodi, *J. Cell Biol.*, **1995**, *130*, 771–779.
34. M. Sousa and A. J. Parodi, *EMBO J.*, **1995**, *14*, 4196–4203.
35. E. Staudacher, F. Altmann, J. Glössl, L. März, H. Schachter, J. P. Kamerling, K. Hård and J. F. G. Vliegenthart, *Eur. J. Biochem.*, **1991**, *199*, 745–751.
36. H. Leiter, J. Mucha, E. Staudacher, R. Grimm, J. Glössl and F. Altmann, *J. Biol. Chem.*, **1999**, *274*, 21830–21839.
37. Y. C. Zeng, G. Bannon, V. H. Thomas, K. Rice, R. Drake and A. Elbein, *J. Biol. Chem.*, **1997**, *272*, 31340–31347.
38. M. R. Paquet, S. Narasimhan, H. Schachter and M. A. Moscarello, *J. Biol. Chem.*, **1984**, *259*, 4716–4721.
39. W. M. Blanken, A. van Vliet and D. H. Van den Eijnden, *J. Biol. Chem.*, **1984**, *259*, 15131–15135.
40. D. H. Van den Eijnden, H. Bakker, A. P. Neeleman, I. M. Van den Nieuwenhof and I. van Die, *Biochem. Soc. Trans.*, **1997**, *25*, 887–893.
41. J. C. Paulson and K. J. Colley, *J. Biol. Chem.*, **1989**, *264*, 17615–17618.
42. D. H. Joziasse, *Glycobiology*, **1992**, *2*, 271–277.
43. R. Kumar, J. Yang, R. D. Larsen and P. Stanley, *Proc. Natl. Acad. Sci. USA*, **1990**, *87*, 9948–9952.
44. M. Sarkar, E. Hull, Y. Nishikawa, R. J. Simpson, R. L. Moritz, R. Dunn and H. Schachter, *Proc. Natl. Acad. Sci. USA*, **1991**, *88*, 234–238.
45. G. A. F. Dagostaro, A. Zingoni, R. L. Moritz, R. J. Simpson, H. Schachter and B. Bendiak, *J. Biol. Chem.*, **1995**, *270*, 15211–15221.
46. A. Nishikawa, Y. Ihara, M. Hatakeyama, K. Kangawa and N. Taniguchi, *J. Biol. Chem.*, **1992**, *267*, 18199–18204.
47. M. T. Minowa, S. Oguri, A. Yoshida, T. Hara, A. Iwamatsu, H. Ikenaga and M. Takeuchi, *J. Biol. Chem.*, **1998**, *273*, 11556–11562.
48. M. G. Shoreibah, G. S. Perng, B. Adler, J. Weinstein, R. Basu, R. Cupples, D. Wen, J. K. Browne, P. Buckhaults, N. Fregien and M. Pierce, *J. Biol. Chem.*, **1993**, *268*, 15381–15385.
49. H. Saito, A. Nishikawa, J. Gu, Y. Ihara, H. Soejima, Y. Wada, C. Sekiya, N. Niikawa and N. Taniguchi, *Biochem. Biophys. Res. Commun.*, **1994**, *198*, 318–327.
50. L. Puglielli, E. C. Mandon, D. M. Rancour, A. K. Menon and C. B. Hirschberg, *J. Biol. Chem.*, **1999**, *274*, 4474–4479.
51. D. H. Joziasse, W. E. C. M. Schiphorst, D. H. Van den Eijnden, J. A. Van Kuik, H. van Halbeek and J. F. G. Vliegenthart, *J. Biol. Chem.*, **1987**, *262*, 2025–2033.
52. M. Nemansky and D. H. Van den Eijnden, *Glycoconjugate J.*, **1993**, *10*, 99–108.
53. W. M. Blanken and D. H. Van den Eijnden, *J. Biol. Chem.*, **1985**, *260*, 12927–12934.
54. D. H. Van den Eijnden, A. H. L. Koenderman and W. E. C. M. Schiphorst, *J. Biol. Chem.*, **1988**, *263*, 12461–12471.
55. M. Nemansky, W. E. C. M. Schiphorst and D. H. Van den Eijnden, *FEBS Lett.*, **1995**, *363*, 280–284.
56. K. Yamashita, Y. Tachibana, T. Ohkura and A. Kobata, *J. Biol. Chem.*, **1985**, *260*, 3963–3969.
57. S. Narasimhan, H. Schachter and S. Rajalakshmi, *J. Biol. Chem.*, **1988**, *263*, 1273–1281.
58. E. W. Easton, J. G. Bolscher and D. H. Van den Eijnden, *J. Biol. Chem.*, **1991**, *266*, 21674–21680.
59. E. W. Easton, I. Blokland, A. A. Geldof, B. R. Rao and D. H. Van den Eijnden, *FEBS Lett.*, **1992**, *308*, 46–49.
60. M. Pierce, P. Buckhaults, L. Chen and N. Fregien, *Glycoconjugate J.*, **1997**, *14*, 623–630.
61. A. Kobata, *Glycoconjugate J.*, **1998**, *15*, 323–331.
62. D. H. Joziasse, W. E. C. M. Schiphorst, D. H. Van den Eijnden, J. A. Van Kuik, H. van Halbeek and J. F. G. Vliegenthart, *J. Biol. Chem.*, **1985**, *260*, 714–719.
63. M. Nemansky, H. T. Edzes, R. A. Wijnands and D. H. Van den Eijnden, *Glycobiology*, **1992**, *2*, 109–117.
64. D. H. Van den Eijnden, D. H. Joziasse, A. H. L. Koenderman, W. M. Blanken, W. E. C. M. Schiphorst and P. L. Koppen, in Proc. VIIIth Int. Symp. Glycoconjugates. (E. A. Davidson, J. C. Williams and N. M. Di Ferrante, eds.), Praeger, New York, **1985**, Vol. I, pp. 285–286.
65. M. J. Elices and I. J. Goldstein, *J. Biol. Chem.*, **1989**, *264*, 1375–1380.
66. M. Nemansky and D. H. Van den Eijnden, *Biochem. J.*, **1992**, *287*, 311–316.

67. A. A. Bergwerff, J. A. Van Kuik, W. E. C. M. Schiphorst, C. A. M. Koeleman, D. H. Van den Eijnden, J. P. Kamerling and J. F. G. Vliegenthart, *FEBS Lett.*, **1993**, *334*, 133–138.
68. K. B. Wlasichuk, M. A. Kashem, P. V. Nikrad, P. Bird, C. Jiang and A. P. Venot, *J. Biol. Chem.*, **1993**, *268*, 13971–13977.
69. T. De Vries, C. A. Srnka, M. M. Palcic, S. J. Swiedler, D. H. Van den Eijnden and B. A. Macher, *J. Biol. Chem.*, **1995**, *270*, 8712–8722.
70. K. Sujino, C. Malet, O. Hindsgaul and M. M. Palcic, *Carbohydr. Res.*, **1997**, *305*, 483–489.
71. J. Rabina, J. Natunen, R. Niemela, H. Salminen, K. Ilves, O. Aitio, H. Maaheimo, J. Helin and O. Renkonen, *Carbohydr. Res.*, **1997**, *305*, 491–499.
72. C. L. M. Stults, B. A. Macher, R. Bhatti, O. P. Srivastava and O. Hindsgaul, *Glycobiology*, **1999**, *9*, 661–668.
73. S. Oka, K. Terayama, C. Kawashima and T. Kawasaki, *J. Biol. Chem.*, **1992**, *267*, 22711–22714.
74. Y. Kato and R. G. Spiro, *J. Biol. Chem.*, **1989**, *264*, 3364–3371.
75. K. Uchimura, H. Muramatsu, T. Kaname, H. Ogawa, T. Yamakawa, Q. W. Fan, C. Mitsuoka, R. Kannagi, O. Habuchi, I. Yokoyama, K. Yamamura, T. Ozaki, A. Nakagawara, K. Kadomatsu and T. Muramatsu, *J. Biochem. Tokyo*, **1998**, *124*, 670–678.
76. D. H. Van den Eijnden, H. Winterwerp, P. Smeeman and W. E. C. M. Schiphorst, *J. Biol. Chem.*, **1983**, *258*, 3435–3437.
77. A. Takeya, O. Hosomi, N. Shimoda and S. Yazawa, *J. Biochem. Tokyo*, **1992**, *112*, 389–395.
78. P. Bailly, F. Piller, B. Gillard, A. Veyrieres, D. Marcus and J. P. Cartron, *Carbohydr. Res.*, **1992**, *228*, 277–287.
79. J. Nakayama, J.-C. Yeh, A. K. Misra, S. Ito, T. Katsuyama and M. Fukuda, *Proc. Natl. Acad. Sci. USA*, **1999**, *96*, 8991–8996.
80. C. H. Hokke, A. P. Neeleman, C. A. M. Koeleman and D. H. Van den Eijnden, *Glycobiology*, **1998**, *8*, 393–406.
81. W. M. Watkins, P. Greenwell, A. D. Yates and P. H. Johnson, *Biochimie*, **1988**, *70*, 1597–1611.
82. K. Angata, M. Suzuki and M. Fukuda, *J. Biol. Chem.*, **1998**, *273*, 28524–28532.
83. P. L. Smith and J. B. Lowe, *J. Biol. Chem.*, **1994**, *269*, 15162–15171.
84. H. Bakker, I. Friedmann, S. Oka, T. Kawasaki, N. Nifantev, M. Schachner and N. Mantei, *J. Biol. Chem.*, **1997**, *272*, 29942–29946.
85. W. M. Blanken, G. J. M. Hooghwinkel and D. H. Van den Eijnden, *Eur. J. Biochem.*, **1982**, *127*, 547–552.
86. K. Sasaki, K. Kuratamiura, M. Ujita, K. Angata, S. Nakagawa, S. Sekine, T. Nishi and M. Fukuda, *Proc. Natl. Acad. Sci. USA*, **1997**, *94*, 14294–14299.
87. F. Piller, J. P. Cartron, A. Maranduba, A. Veyrieres, Y. Leroy and B. Fournet, *J. Biol. Chem.*, **1984**, *259*, 13385–13390.
88. A. H. Koenderman, P. L. Koppen and D. H. Van den Eijnden, *Eur. J. Biochem.*, **1987**, *166*, 199–208.
89. Y. Sakamoto, T. Taguchi, Y. Tano, T. Ogawa, A. Leppanen, M. Kinnunen, O. Aitio, P. Parmanne, O. Renkonen and N. Taniguchi, *J. Biol. Chem.*, **1998**, *273*, 27625–27632.
90. P. Mattila, H. Salminen, L. Hirvas, J. Niittymaki, H. Salo, R. Niemela, M. Fukuda, O. Renkonen and R. Renkonen, *J. Biol. Chem.*, **1998**, *273*, 27633–27639.
91. J. C. Yeh, E. Ong and M. Fukuda, *J. Biol. Chem.*, **1999**, *274*, 3215–3221.
92. R. Niemela, J. Natunen, M. L. Majuri, H. Maaheimo, J. Helin, J. B. Lowe, O. Renkonen and R. Renkonen, *J. Biol. Chem.*, **1998**, *273*, 4021–4026.
93. C. J. Britten, D. H. Van den Eijnden, W. McDowell, V. A. Kelly, S. J. Witham, M. R. Edbrooke, M. I. Bird, T. De Vries and N. Smithers, *Glycobiology*, **1998**, *8*, 321–327.
94. H. Salminen, K. Ahokas, R. Niemela, L. Penttila, H. Maaheimo, J. Helin, C. E. Costello and O. Renkonen, *FEBS Lett.*, **1997**, *419*, 220–226.
95. G. Pfeiffer, K. H. Strube, M. Schmidt and R. Geyer, *Eur. J. Biochem.*, **1994**, *219*, 331–348.
96. A. Dell, H. R. Morris, R. L. Easton, M. Panico, M. Patankar, S. Oehninger, R. Koistinen, H. Koistinen, M. Seppala and G. F. Clark, *J. Biol. Chem.*, **1995**, *270*, 24116–24126.
97. R. T. Camphausen, H.-A. Yu and D. A. Cumming, in *Glycoproteins* (J. Montreuil, H. Schachter and J. F. G. Vliegenthart, eds.), Elsevier, Amsterdam, **1995**, pp. 391–414.
98. M. J. Treuheit, C. E. Costello and H. B. Halsall, *Biochem. J.*, **1992**, *283*, 105–112.
99. D. A. Cumming, *Glycobiology*, **1991**, *1*, 115–130.

100. S. Kornfeld, *FASEB J.*, **1987**, *1*, 462–468.
101. S. M. Manzella, L. V. Hooper and J. U. Baenziger, *J. Biol. Chem.*, **1996**, *271*, 12117–12120.
102. E. D. Green, G. Adelt, J. U. Baenziger, S. Wilson and H. van Halbeek, *J. Biol. Chem.*, **1988**, *263*, 18253–18268.
103. H. Yoshima, A. Matsumoto, T. Mizuochi, T. Kawasaki and A. Kobata, *J. Biol. Chem.*, **1981**, *256*, 8476–8484.
104. A. Kobata, *Methods Enzymol.*, **1972**, *28*, 262–271.
105. B. T. Sheares and D. M. Carlson, *J. Biol. Chem.*, **1983**, *258*, 9893–9898.
106. F. Kolbinger, M. B. Streiff and A. G. Katopodis, *J. Biol. Chem.*, **1998**, *273*, 433–440.
107. M. Amado, R. Almeida, F. Carneiro, S. B. Levery, E. H. Holmes, M. Nomoto, M. A. Hollingsworth, H. Hassan, T. Schwientek, P. A. Nielsen, E. P. Bennett and H. Clausen, *J. Biol. Chem.*, **1998**, *273*, 12770–12778.
108. S. Isshiki, A. Togayachi, T. Kudo, S. Nishihara, M. Watanabe, T. Kubota, M. Kitajima, N. Shiraishi, K. Sasaki, T. Andoh and H. Narimatsu, *J. Biol. Chem.*, **1999**, *274*, 12499–12507.
109. D. Zhou, E. G. Berger and T. Hennet, *Eur. J. Biochem.*, **1999**, *263*, 571–576.
110. T. Hennet, A. Dinter, P. Kuhnert, T. S. Mattu, P. M. Rudd and E. G. Berger, *J. Biol. Chem.*, **1998**, *273*, 58–65.
111. J. C. Paulson, J. Weinstein and S. de Souza, *Eur. J. Biochem.*, **1984**, *140*, 523–530.
112. H. T. de Heij, P. L. Koppen and D. H. Van den Eijnden, *Carbohydr. Res.*, **1986**, *149*, 85–99.
113. A. P. Neeleman, W. P. W. Van der Knaap and D. H. Van den Eijnden, *Glycobiology*, **1994**, *4*, 641–651.
114. J. Srivatsan, D. F. Smith and R. D. Cummings, *J. Parasitol.*, **1994**, *80*, 884–890.
115. H. Mulder, B. A. Spronk, H. Schachter, A. P. Neeleman, D. H. Van den Eijnden, M. De Jong-Brink, J. P. Kamerling and J. F. G. Vliegenthart, *Eur. J. Biochem.*, **1995**, *227*, 175–185.
116. I. M. Van den Nieuwenhof, W. E. C. M. Schiphorst, I. van Die and D. H. Van den Eijnden, *Glycobiology*, **1999**, *9*, 115–123.
117. D. H. Van den Eijnden, A. P. Neeleman, W. P. W. Van der Knaap, H. Bakker, M. Agterberg and I. van Die, *Biochem. Soc. Trans.*, **1995**, *23*, 175–179.
118. T. P. Skelton, L. V. Hooper, V. Srivastava, O. Hindsgaul and J. U. Baenziger, *J. Biol. Chem.*, **1991**, *266*, 17142–17150.
119. H. Mulder, H. Schachter, M. de Jong Brink, J. G. Van der Ven, J. P. Kamerling and J. F. G. Vliegenthart, *Eur. J. Biochem.*, **1991**, *201*, 459–465.
120. A. K. Datta and J. C. Paulson, *J. Biol. Chem.*, **1995**, *270*, 1497–1500.
121. K. Drickamer, *Glycobiology*, **1993**, *3*, 2–3.
122. S. Tsuji, *J. Biochem. Tokyo*, **1996**, *120*, 1–13.
123. S. Tsuji, A. K. Datta and J. C. Paulson, *Glycobiology*, **1996**, *6*, R 5–R 7.
124. A. Ishii, M. Ohta, Y. Watanabe, K. Matsuda, K. Ishiyama, K. Sakoe, M. Nakamura, J. Inokuchi, Y. Sanai and M. Saito, *J. Biol. Chem.*, **1998**, *273*, 31652–31655.
125. M. Kono, S. Takashima, H. Liu, M. Inoue, N. Kojima, Y. C. Lee, T. Hamamoto and S. Tsuji, *Biochem. Biophys. Res. Commun.*, **1998**, *253*, 170–175.
126. T. Okajima, S. Fukumoto, H. Miyazaki, H. Ishida, M. Kiso, K. Furukawa and T. Urano, *J. Biol. Chem.*, **1999**, *274*, 11479–11486.
127. M. L. E. Bergh and D. H. Van den Eijnden, *Eur. J. Biochem.*, **1983**, *136*, 113–118.
128. E. R. Sjoberg, H. Kitagawa, J. Glushka, H. van Halbeek and J. C. Paulson, *J. Biol. Chem.*, **1996**, *271*, 7450–7459.
129. Y. C. Lee, M. Kaufmannn, S. Kitazume-Kawaguchi, M. Kono, S. Takashima, N. Kurosawa, H. Liu, H. Pircher and S. Tsujii, *J. Biol. Chem.*, **1999**, *274*, 11958–11967.
130. V. P. Rajan, R. D. Larsen, S. Ajmera, L. K. Ernst and J. B. Lowe, *J. Biol. Chem.*, **1989**, *264*, 11158–11167.
131. R. J. Kelly, S. Rouquier, D. Giorgi, G. G. Lennon and J. B. Lowe, *J. Biol. Chem.*, **1995**, *270*, 4640–4649.
132. T. Kumazaki and A. Yoshida, *Proc. Natl. Acad. Sci. USA*, **1984**, *81*, 4193–4197.
133. A. Sarnesto, T. Kohlin, O. Hindsgaul, J. Thurin and M. Blaszczyk-Thurin, *J. Biol. Chem.*, **1992**, *267*, 2737–2744.
134. S. Hitoshi, S. Kusunoki, I. Kanazawa and S. Tsuji, *J. Biol. Chem.*, **1996**, *271*, 16975–16981.
135. B. W. Weston, P. L. Smith, R. J. Kelly and J. B. Lowe, *J. Biol. Chem.*, **1992**, *267*, 24575–24584.

136. K. Sasaki, K. Kurata, K. Funayama, M. Nagata, E. Watanabe, S. Ohta, N. Hanai and T. Nishi, *J. Biol. Chem.*, **1994**, *269*, 14730–14737.
137. T. Kudo, Y. Ikehara, A. Togayachi, M. Kaneko, T. Hiraga, K. Sasaki and H. Narimatsu, *J. Biol. Chem.*, **1998**, *273*, 26729–26738.
138. M. Kaneko, T. Kudo, H. Iwasaki, Y. Ikehara, S. Nishihara, S. Nakagawa, K. Sasaki, T. Shiina, H. Inoko, N. Saitou and H. Narimatsu, *FEBS Lett.*, **1999**, *452*, 237–242.
139. S. L. Martin, M. R. Edbrooke, T. C. Hodgman, D. H. Van den Eijnden and M. I. Bird, *J. Biol. Chem.*, **1997**, *272*, 21349–21356.
140. Z. M. Ge, N. W. C. Chan, M. M. Palcic and D. E. Taylor, *J. Biol. Chem.*, **1997**, *272*, 21357–21363.
141. F. Dupuy, J. M. Petit, R. Mollicone, R. Oriol, R. Julien and A. Maftah, *J. Biol. Chem.*, **1999**, *274*, 12257–12262.
142. T. De Vries, M. P. Palcic, P. S. Schoenmakers, D. H. Van den Eijnden and D. H. Joziasse, *Glycobiology*, **1997**, *7*, 921–927.
143. T. De Vries, T. Norberg, H. Lönn and D. H. Van den Eijnden, *Eur. J. Biochem.*, **1993**, *216*, 769–777.
144. T. De Vries and D. H. Van den Eijnden, *Biochemistry*, **1994**, *33*, 9937–9944.
145. F. Yamamoto, J. Marken, T. Tsuji, T. White, H. Clausen and S. Hakomori, *J. Biol. Chem.*, **1990**, *265*, 1146–1151.
146. D. H. Joziasse, J. H. Shaper, D. H. Van den Eijnden, A. J. Van Tunen and N. L. Shaper, *J. Biol. Chem.*, **1989**, *264*, 14290–14297.
147. R. D. Larsen, V. P. Rajan, M. M. Ruff, J. Kukowska Latallo, R. D. Cummings and J. B. Lowe, *Proc. Natl. Acad. Sci. USA*, **1989**, *86*, 8227–8231.
148. D. B. Haslam and J. U. Baenziger, *Proc. Natl. Acad. Sci. USA*, **1996**, *93*, 10697–10702.
149. A. D. Yates, P. Greenwell and W. M. Watkins, *Biochem. Soc. Trans.*, **1983**, *11*, 300–301.
150. M. F. A. Bierhuizen and M. Fukuda, *Proc. Natl. Acad. Sci. USA*, **1992**, *89*, 9326–9330.
151. T. Schwientek, M. Nomoto, S. B. Levery, G. Merkx, A. G. Van Kessel, E. P. Bennett, M. A. Hollingsworth and H. Clausen, *J. Biol. Chem.*, **1999**, *274*, 4504–4512.
152. M. F. A. Bierhuizen, M. G. Mattei and M. Fukuda, *Genes & Dev.*, **1993**, *7*, 468–478.
153. M. Sekine, K. Nara and A. Suzuki, *J. Biol. Chem.*, **1997**, *272*, 27246–27252.
154. F. K. Hagen, B. Van Wuyckhuyse and L. A. Tabak, *J. Biol. Chem.*, **1993**, *268*, 18960–18965.
155. E. P. Bennett, H. Hassan and H. Clausen, *J. Biol. Chem.*, **1996**, *271*, 17006–17012.
156. F. K. Hagen, K. G. Tenhagen, T. M. Beres, M. M. Balys, B. C. Vanwuyckhuyse and L. A. Tabak, *J. Biol. Chem.*, **1997**, *272*, 13843–13848.
157. B. G. Tenhagen, F. K. Hagen, M. M. Balys, T. M. Beres, B. Vanwuyckhuyse and L. A. Tabak, *J. Biol. Chem.*, **1998**, *273*, 27749–27754.
158. E. P. Bennett, H. Hassan, U. Mandel, E. Mirgorodskaya, P. Roepstorff, J. Burchell, J. Taylorpapadimitriou, M. A. Hollingsworth, G. Merkx, A. G. Van Kessel, H. Eiberg, R. Steffensen and H. Clausen, *J. Biol. Chem.*, **1998**, *273*, 30472–30481.
159. S. Iida, H. Takeuchi, H. Hassan, H. Clausen and T. Irimura, *FEBS Lett.*, **1999**, *449*, 230–234.
160. F. K. Hagen, B. Hazes, R. Raffo, D. deSa and L. A. Tabak, *J. Biol. Chem.*, **1999**, *274*, 6797–6803.
161. Y. Nagata, S. Yamashiro, J. Yodoi, K. O. Lloyd, H. Shiku and K. Furukawa, *J. Biol. Chem.*, **1992**, *267*, 12082–12089.
162. N. W. Lo, J. H. Shaper, J. Pevsner and N. L. Shaper, *Glycobiology*, **1998**, *8*, 517–526.
163. N. L. Shaper, J. H. Shaper, J. L. Meuth, J. L. Fox, H. Chang, I. R. Kirsch and G. F. Hollis, *Proc. Natl. Acad. Sci. USA*, **1986**, *83*, 1573–1577.
164. H. Narimatsu, S. Sinha, K. Brew, H. Okayama and P. K. Qasba, *Proc. Natl. Acad. Sci. USA*, **1986**, *83*, 4720–4724.
165. H. E. Appert, T. J. Rutherford, G. E. Tarr, J. S. Wiest, N. R. Thomford and D. J. McCorquodale, *Biochem. Biophys. Res. Commun.*, **1986**, *139*, 163–168.
166. R. Almeida, M. Amado, L. David, S. B. Levery, E. H. Holmes, G. Merkx, A. G. Van Kessel, E. Rygaard, H. Hassan, E. Bennett and H. Clausen, *J. Biol. Chem.*, **1997**, *272*, 31979–31991.
167. I. van Die, A. van Tetering, W. E. C. M. Schiphorst, T. Sato, K. Furukawa and D. H. Van den Eijnden, *FEBS Lett.*, **1999**, *450*, 52–56.
168. D. H. Van den Eijnden, W. E. C. M. Schiphorst and E. G. Berger, *Biochim. Biophys. Acta*, **1983**, *755*, 32–39.

169. M. Ujita, J. McAuliffe, O. Hindsgaul, K. Sasaki, M. N. Fukuda and M. Fukuda, *J. Biol. Chem.*, **1999**, *274*, 16717–16726.
170. U. Brodbeck, W. L. Denton, N. Tanahashi and K. E. Ebner, *J. Biol. Chem.*, **1967**, *242*, 1391–1397.
171. K. Brew, T. C. Vanaman and R. L. Hill, *Proc. Natl. Acad. Sci. USA*, **1968**, *59*, 491–497.
172. T. Schwientek, R. Almeida, S. B. Levery, E. H. Holmes, E. Bennett and H. Clausen, *J. Biol. Chem.*, **1998**, *273*, 29331–29340.
173. T. Sato, K. Furukawa, H. Bakker, D. H. Van den Eijnden and I. van Die, *Proc. Natl. Acad. Sci. USA*, **1998**, *95*, 472–477.
174. M. Ujita, J. McAuliffe, T. Schwientek, R. Almeida, O. Hindsgaul, H. Clausen and M. Fukuda, *J. Biol. Chem.*, **1998**, *273*, 34843–34849.
175. T. Nomura, M. Takizawa, J. Aoki, H. Arai, K. Inoue, E. Wakisaka, N. Yoshizuka, G. Imokawa, N. Dohmae, K. Takio, M. Hattori and N. Matsuo, *J. Biol. Chem.*, **1998**, *273*, 13570–13577.
176. T. Okajima, K. Yoshida, T. Kondo and K. Furukawa, *J. Biol. Chem.*, **1999**, *274*, 22915–22918.
177. H. Bakker, M. Agterberg, A. van Tetering, C. A. M. Koeleman, D. H. Van den Eijnden and I. van Die, *J. Biol. Chem.*, **1994**, *269*, 30326–30333.
178. H. Bakker, P. S. Schoenmakers, C. A. M. Koeleman, D. H. Joziasse, I. van Die and D. H. Van den Eijnden, *Glycobiology*, **1997**, *7*, 539–548.
179. I. van Die, H. Bakker and D. H. Van den Eijnden, *Glycobiology*, **1997**, *7*, R 5–R 8.
180. A. P. Neeleman and D. H. Van den Eijnden, *Proc. Natl. Acad. Sci. USA*, **1996**, *93*, 10111–10116.
181. K. Terayama, S. Oka, T. Seiki, Y. Miki, A. Nakamura, Y. Kozutsumi, K. Takio and T. Kawasaki, *Proc. Natl. Acad. Sci. USA*, **1997**, *94*, 6093–6098.
182. Y. Shimoda, Y. Tajima, T. Nagase, K. Harii, N. Osumi and Y. Sanai, *J. Biol. Chem.*, **1999**, *274*, 17115–17122.
183. K. Terayama, T. Seiki, A. Nakamura, K. Matsumori, S. Ohta, S. Oka, M. Sugita and T. Kawasaki, *J. Biol. Chem.*, **1998**, *273*, 30295–30300.
184. H. Kitagawa, Y. Tone, J. Tamura, K. W. Neumann, T. Ogawa, S. Oka, T. Kawasaki and K. Sugahara, *J. Biol. Chem.*, **1998**, *273*, 6615–6618.
185. S. Yanagidani, N. Uozumi, Y. Ihara, E. Miyoshi, N. Yamaguchi and N. Taniguchi, *J. Biochem. Tokyo*, **1997**, *121*, 626–632.
186. D. P. Zhou, A. Dinter, R. G. Gallego, J. P. Kamerling, J. F. G. Vliegenthart, E. G. Berger and T. Hennet, *Proc. Natl. Acad. Sci. USA*, **1999**, *96*, 406–411.
187. D. H. Van den Eijnden and D. H. Joziasse, *Curr. Opin. Struct. Biol.*, **1993**, *3*, 711–721.
188. A. Van Tetering, W. E. C. M. Schiphorst, D. H. Van den Eijnden and I. Van Die, *FEBS Lett.*, **1999**, *461*, 311–314.
189. Y. Ikehara, N. Shimizu, M. Kono, S. Nishihara, H. Nakanishi, T. Kitamura, H. Narimatsa, S. Tsuji and M. Tatematsu, *FEBS Lett.*, **1999**, *463*, 92–96.
190. S. Nishihara, H. Iwasaki, M. Kaneko, A. Tawada, M. Ito and H. Narimatsu, *FEBS Lett.*, **1999**, *462*, 289–294.
191. A. Yoshida, M. T. Minowa, S. Takamatsu, T. Hara, H. Ikenaga and M. Takeuchi, *Glycoconjugate J.*, **1998**, *15*, 1115–1123.
192. E. P. Bennet, H. Hassan, U. Mandel, M. A. Hollingsworth, N. Akisawa, Y. Ikematsu, G. Merkx, A. Geurts van Kessel, S. Olofsson and H. Clausen, *J. Biol. Chem.*, **1999**, *274*, 25362–25370.
193. K. G. Ten Hagen, D. Tetaert, F. K. Hagen, C. Richet, T. M. Beres, J. Gagnon, M. M. Balys, B. VanWuyckhuyse, G. S. Bedi, P. Degand and L. A. Tabak, *J. Biol. Chem.*, **1999**, *274*, 27867–27874.
194. E. P. Bennet, H. Hassan, M. A. Hollingsworth and H. Clausen, *FEBS Lett.*, **1999**, *460*, 226–230.
195. R. Almeida, S. B. Levery, U. Mandel, H. Kresse, T. Schwientek, E. P. Bennett and H. Clausen, *J. Biol. Chem.*, **1999**, *274*, 26165–26171.
196. J. Tongzhong, K. W. Bremer, A. D'Souza, R. D. Cummings and W. M. Canfield, *Glycobiology*, **1999**, *9*, 1123.

24 Synthesis of Sugar Nucleotides

Reinhold Öhrlein

24.1 Introduction

Recent progress in molecular glycobiology has revealed the important biological roles of numerous glycoconjugates [1]. In particular cell-surface carbohydrates play key roles in cellular communication processes *via* selective adhesion phenomena [2, 3]. They guide *e.g.* leukocyte extravasation [4], the homing of lymphocytes and adhesion of myeloid cells to activated platelets and the endothelium [5–7]. Besides these interactions carbohydrates are involved in fertilization events, cell growth, and parasitic infections of viruses and bacteria [8]. A very recent finding is the involvement of certain cell surface α-galactosides in hyperacute rejection reactions; this is of pharmaceutical interest in organ transplantation [9].

This makes sugars worthwhile targets for pharmaceutical applications [10, 11]. Although chemical synthesis of oligosaccharides has reached a mature stage [12–14] the overall synthetic sequences are still lengthy and cumbersome [15]. Biologically active carbohydrates are, moreover, composed of various different monosaccharide units and are often only active if presented in a polyvalent or multiantennary array to their respective receptors [16, 17]. These multiantennary heteropolymers, in particular, have resisted efficient chemical synthesis. Recently biocatalysts have been found to be valuable tools for complementing the classical synthetic approach [18].

Nature has invented several ways of assembling complex oligosaccharides [1]. The Leloir pathway is one of the preferred routes of mammalian systems. The Leloir enzymes—the glycosyl transferases—transfer a monosaccharide unit from a nucleotide-activated donor substrate to a growing oligosaccharide chain with rigorous regio- and stereoselectivity [19] (see Scheme 1). Exclusively one of the many OH groups of the acceptor is glycosylated, in either an α or β mode! Neither the donor nor the acceptor sugar needs protection from the highly homofunctional reactants—often OH groups only are present on the sugars. Thus, selective protecting group manipulations, a preponderant task in the classical, chemical oligosaccharide assemblage can be reduced to an absolute minimum.

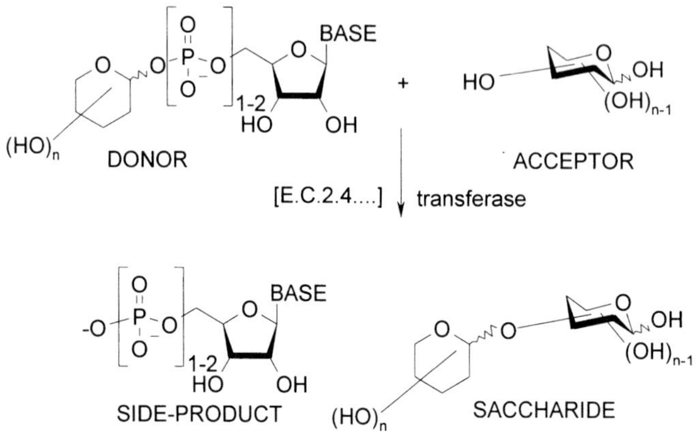

Scheme 1. Leloir glycosylation pathway.

The availability of the glycosyltransferases and the activated donor substrates is, however, a prerequisite for the application of the biocatalytic methodology. Initially the transferases were isolated from body fluids such as human and bovine milk, *e.g.* β(1–4)galactosyl transferase [20, 21], or blood, *e.g.* α(1–3)fucosyl transferase [22], or a variety of animal tissues such as mouse kidney, *e.g.* the core-II N-acetylglucosaminyl transferase [23], and rat liver, *e.g.* α(2–3)- and α(2–6)sialyl transferases [24, 25], to mention just a few. Modern cloning and expression techniques make an ever growing number of glycosyl transferases available to the glycochemist [26–28].

Astonishingly, only eight nucleotide-activated donors are used by mammalian cells (see Scheme 2) [29]. In principal all are commercially available, although on a small scale only. The high prices of the natural donor substrates and the unavailability of non-natural derivatives of those donors is the reason why several chemical and biomimetic pathways have been probed to increase access to these crucial compounds [29, 30]. The ensuing sections will summarize the most recent attempts to synthesize these donors, with emphasis on non-natural derivatives.

24.2 Synthesis of Sugar Nucleotides

24.2.1 Chemical Synthesis

UDP-Activated Donors

All the UDP-activated donors (Scheme 2) are α-pyrophosphate sugars and thus are inherently prone to degradation in acidic media, leading to monosaccharide units and uridine diphosphate [31]. This lability must always be taken into account at the outset of an efficient synthetic strategy. In addition the separation of the highly polar target compounds from side-products in the reaction mixture necessitates the

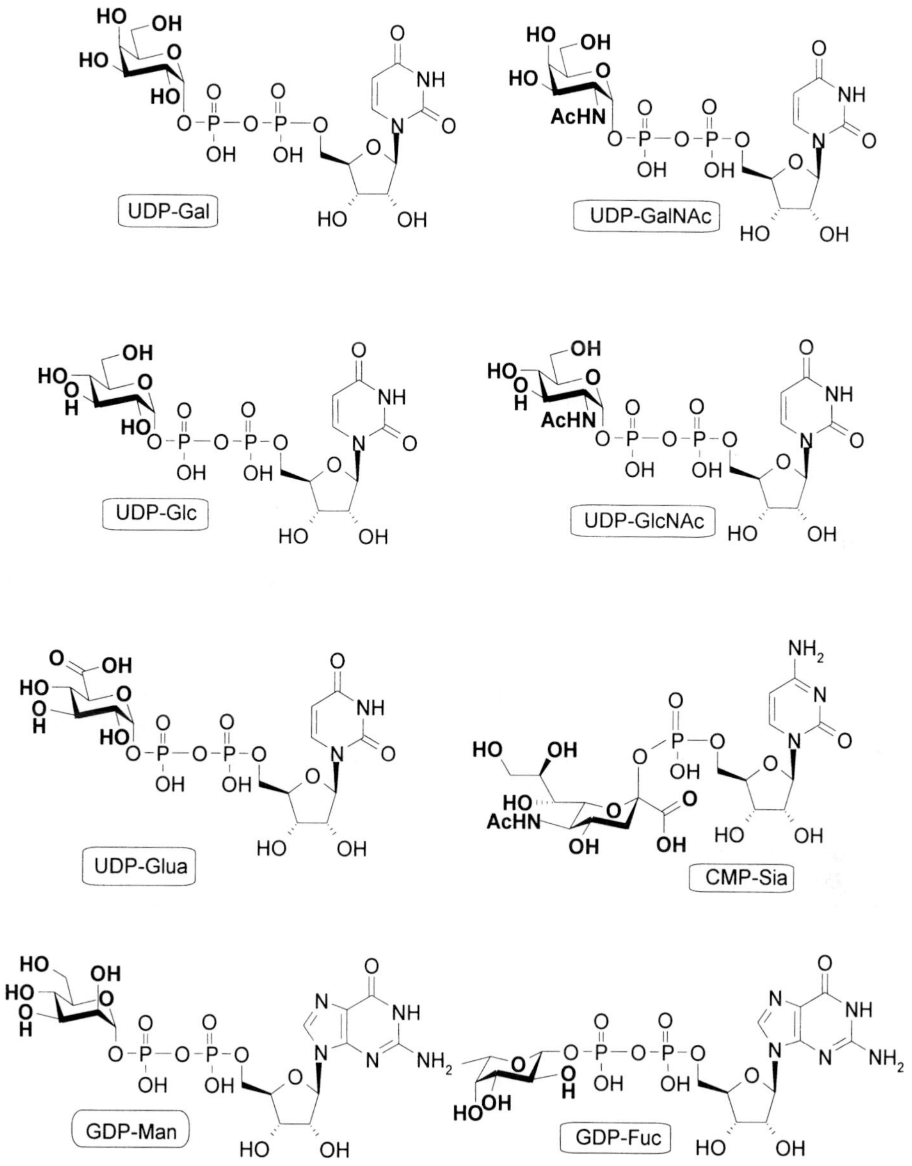

24.2 Synthesis of Sugar Nucleotides

Scheme 2. Nucleotide-activated donor sugars.

strict adherence to sophisticated purification protocols. Generally, overall yields are poor. General approaches towards UDP-activated sugars are outlined in Scheme 3.

The key intermediates in the synthesis of nucleotide α-diphosphates are the α-sugar phosphates. The MacDonald procedure [32] proved to be a good protocol for generation of numerous α-glycosyl phosphates (see Scheme 3). They are obtained in

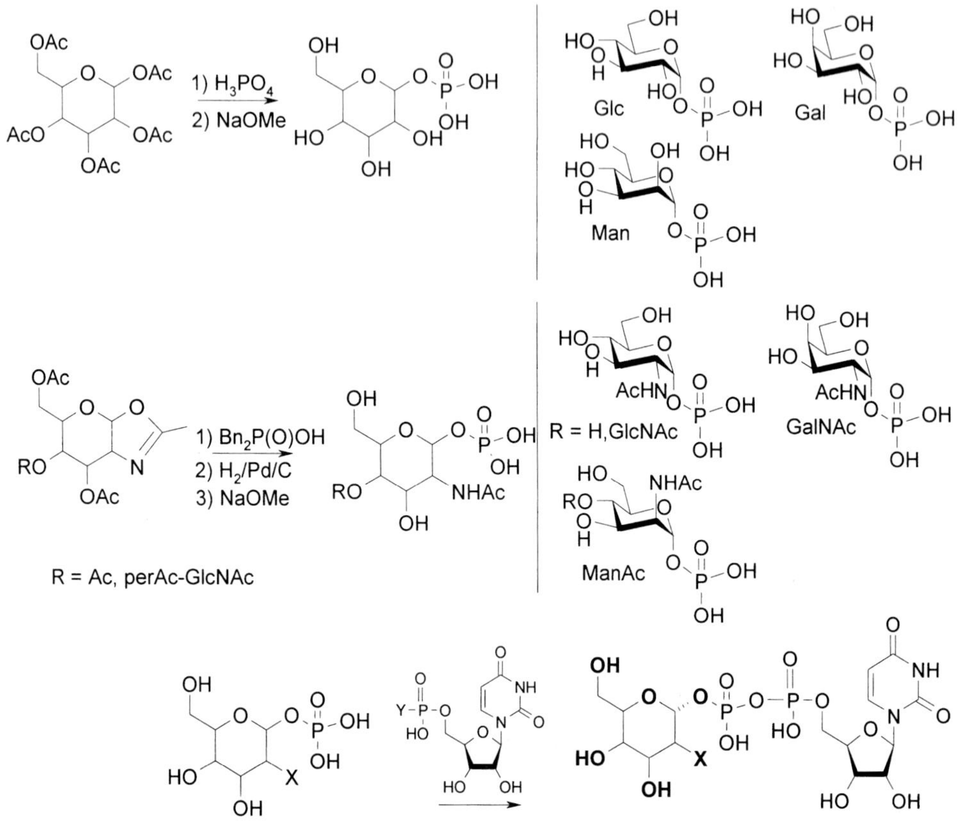

Scheme 3. Formation of nucleotide α-diphosphate sugars; Y: leaving group.

ca 30–40% yield by fusion of an α/β mixture of the peracetylated hexopyranosides with phosphoric acid, then base-catalyzed deacetylation. The low yields give rise to the assumption of the concomitant formation of the undesired β isomers. It has, however, been claimed that these hydrolyze more readily in the deacetylation step and can thus be removed completely [32, 33]. α-Glucose-1-phosphate has also been prepared in 87% yield directly from glucose and phosphoric acid in the presence of propylene oxide [34].

Alternative routes for the synthesis of α-sugar-1-phosphates started from protected sugars with a free anomeric OH group which is deprotonated in a first step and subsequently reacted with diphenyl chlorophosphate [35]. The authors found variable ratios of α/β mixtures depending, e.g., on the base used, the reaction temperature, and the structure of the reacted hexoses. Hydrogenation over PtO$_2$ and treatment with base gave the unprotected sugar phosphates. Although the α-phosphates of N-acetylglucosamine and also chitobiose could be obtained from the corresponding oxazoline precursors (Scheme 3), the α-sugar-1-dibenzylphosphate intermediates proved quite unstable [36]. Accordingly, the even more unstable α-

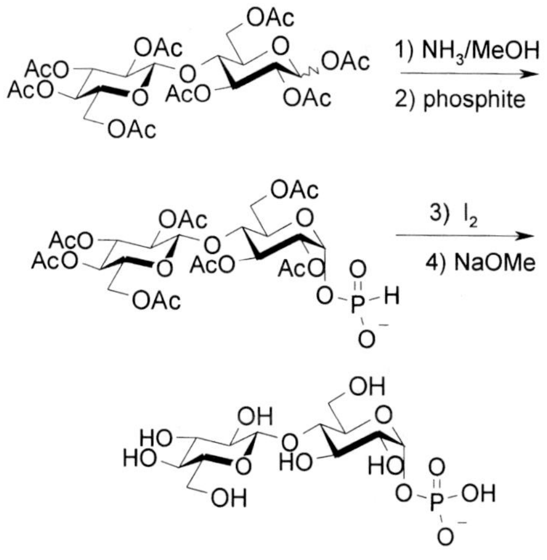

Scheme 4. Synthesis of α-cellobiosyl phosphate.

mannosyl-1-phosphate was obtained in 61% yield [37]. Yet another approach towards α-sugar-1-phosphates of mannose, galactose, glucose, and cellobiose was reported by Luu and coworkers; this is exemplified in Scheme 4 for α-cellobiosyl phosphate [38].

The peracetylated sugars were first treated with ammonia in methanol to remove the anomeric acetate and the resulting 1-OH sugars were subsequently phosphitylated. Only α-phosphites were obtained; in the cellobiose synthesis illustrated the final α-phosphate could be isolated in ca 50% yield after iodine oxidation and deacetylation. Besides the α-phosphates of the naturally occurring sugars (Scheme 2) a number of α-sugar-1-phosphates of non-natural derivatives have been prepared by the above phosphorylation procedures. These sugar phosphates are reacted with activated nucleotide phosphates as is exemplified in Scheme 3 to form the desired UDP sugars. Usually the commercial uridine 5'-monophosphomorpholidate has been used (Y = morpholine). We found the *in situ* activation of the uridine monophosphate by carbonyldiimidazole to be superior [39] (Y = imidazole).

Some recent examples of the preparation of non-natural UDP- and dTDP-activated sugars by use of this general pathway (Scheme 3) are compiled in Table 1. The literature listed should be consulted because the individual experimental details can vary.

The simultaneous replacement of uridine by deoxythymidine and galactose by a number of deoxy sugars to give non-natural donor substrates has also been reported [50]. Partially protected α-sugar-1-phosphates were coupled to morpholidate-activated deoxythymidine-5-phosphate and subsequently deprotected in the presence of lithium hydroxide to give the corresponding dTDP sugars (see Table 1).

CMP–Activated Sugars

The stereochemical preparation of CMP–sialic acid is quite a synthetic challenge. This is because of the absence of neighboring group participation to assist the

Table 1. Yields and scales for the coupling of sugar phosphates to nucleotide phosphates.

Sugar	Couple %	Scale (mg)	Ref.	Sugar	Couple %	Scale (mg)	Ref.
2-F-Glc	17*	14	[40]	2-F-Man	60*	69	[41]
3-F-Glc	10*	15	[40]	3-F-Glc (OH variant)	82*	82	[41]
3-OMe-Glc	14*	12	[42]	4-OMe-Glc	10*	17	[42]
4-OMe-Glc	11*	17	[42]	6-OMe-Glc	22*	72	[42]
2-deoxy-Glc	19*	85	[43]	2-F-Glc analog	13*	12	[43]
Glc	—*	—	[46]	2-deoxy-Glc analog	25*	18	[44]
2-S-Glc	32*	33	[45]	GlcNAc-2-F	67*	90	[47]
GlcNAc-F	71*	70	[47]	GlcNAc-S	18*	—	[48]
GlcNAc	66*	33	[49]	GalNAc	71*	35	[49]
GalNAc variant	80*	40	[49]	2-deoxy-Gal	44**	42	[50]
deoxy sugar	34**	33	[50]	GalNAc variant	27**	36	[50]
2-amino-Glc	37**	33	[50]				

* UDP sugar
** dTDP sugar.

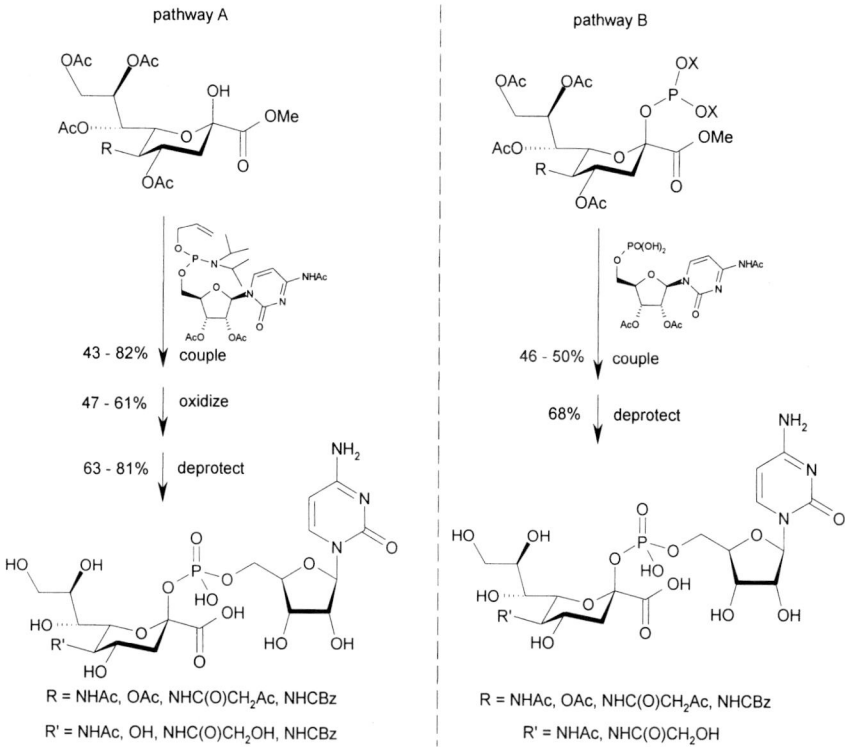

Scheme 5. Synthesis of CMP–sialic acid and derivatives.

stereoselective phosphorylation of the tertiary anomeric center. Two strategies have been followed recently to prepare the parent CMP–sialic acid and some derivatives thereof.

Pathway A relies on the coupling of a protected cytidine phosphoramidite with protected sialic acid [51, 52]. The resulting phosphite is subsequently oxidized with a *t*-butyl hydroperoxide solution and finally deprotected (see Scheme 5). Next to the natural CMP–sialic acid, CMP–ketodeoxyoctulonic acid and the CMP–*N*-glycolylsialic acid were obtained by this protocol, albeit in the low mg-range!

Pathway B starts from dialkyl sialyl phosphites [53, 54] (see Scheme 5). In this case the sialyl phosphite group is directly replaced by the N/O-acetylated cytidine phosphate with retention of configuration.

Application of strategy B also made accessible nucleotide monophosphosialic acid derivatives with the cytidine replaced by other pyrimidine and purine bases, although in modest overall yields only [55]. By following pathway B a CMP–'disialyl' derivative could be synthesized. This compound has a second sialic acid unit α-linked to the 8-OH group of the CMP–activated sialic acid [56].

GDP-Activated Donors

D-Mannose and L-fucose are related biogenetically and are transferred from their corresponding guanosine diphosphate-activated donors [57] by the corresponding transferases. GDP–Man constitutes an α-phosphate whereas in GDP–Fuc the diphosphate bridge is β-linked to the fucose unit.

The synthesis of GDP–Man was accomplished analogously as described for UDP–Gal (see Scheme 3). In this case an α-mannosyl phosphate [58] is coupled with a GMP–morpholidate in the presence of 1H-tetrazole [59] to give GDP–Man. Coupling yields vary from 40–76% (scale: 13–147 mg). Also GDP-3-N-acetyl-3-deoxy-α-D-mannose was obtained in 35% (46 mg) yield according the morpholidate procedure [60].

Alternatively, the coupling of a phosphinothioate-activated GMP with α-mannosyl phosphates has also been successful [61] (see Scheme 6).

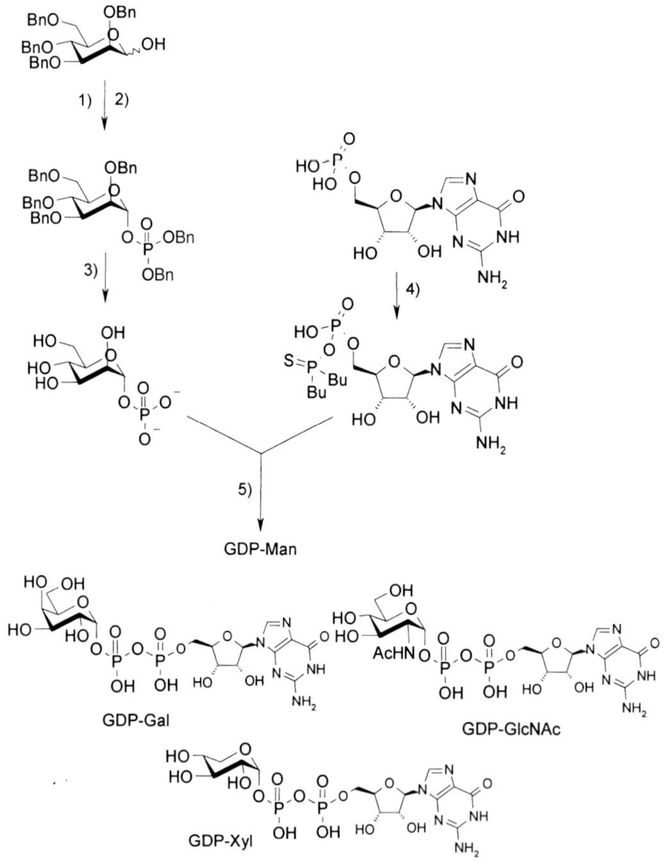

Scheme 6. Synthesis of GDP–man and congeners. 1) NaH, CCl$_3$CN, 81%; 2) HOP(O)(OBn)$_2$, 99%; 3) Na, liquid NH$_3$ then BaCl$_2$, 60%; 4) Bu$_2$P(S)Br, Bu$_3$N, THF, 61%; 5) AgOAc, pyr, 40%.

2,3,4,6-Tetra-O-benzyl-1-hydroxymannose is first deprotonated and then treated with trichloroacetonitrile to give the trichloroacetamidate intermediate which is subsequently reacted with dibenzylphosphoric acid to give the α-perbenzylated phosphate ester. All benzyl groups could be removed by sodium in liquid ammonia. The resulting α-mannosyl phosphate is subsequently treated with the guanosine 5′-monophosphate dibutyl phosphinothioic anhydride in the presence of silver acetate to form the desired GDP–Man. The necessary thiophosphoryl-activated anhydride is obtained from GMP and commercial dibutylthiophosphoryl bromide. The thiophosphoryl approach was also employed for the synthesis of GDP–Gal, GDP–GlcNAc and GDP–Xyl (see Scheme 6).

GDP–Fuc is a particularly delicate compound. This is reflected by the numerous papers describing various, albeit low-yielding synthetic approaches. The early investigations [62–64] were plagued by the inefficient and cumbersome access to the β-fucosyl-1-phosphate (Scheme 7). The most reliable, fully chemical preparations have been reported by Hindsgaul [65] and Whitesides [66]. The full synthetic pathway is outlined in Scheme 7. Commercial L-fucose is first peracetylated and subsequently treated with conc. hydrobromic acid to give the α-fucosyl bromide in quantitative yield. The bromide can be stored at −20 °C for several months as a solid. In the presence of silver carbonate the bromide is reacted with dibenzyl phosphate to give the dibenzyl 2,3,4-tri-O-acetyl-α-L-fucosyl phosphate. This compound is completely deprotected to form the labile β-fucosyl phosphate which is stirred with morpholidate-activated guanosine-5′-monophosphate at room temperature for a prolonged time. The pure GDP–Fuc is obtained from this reaction mixture by repeated passages over Dowex and Sephadex columns. Thus gram-amounts of GDP–Fuc were prepared.

According the synthetic pathway depicted in Scheme 7, a series of non-natural GDP–deoxyfucoses have been prepared successfully (compare Table 2). Hindsgaul and coworkers [67, 68] also reported the synthesis of GDP–Fuc derivatives which bear alkyl chains, aminoalkyl residues capable of being linked to fluorescent labels, and oligosaccharide moieties on the fucose C-6. Surprisingly, those GDP–'fucose' derivatives are accepted by certain fucosyl transferases [67, 69].

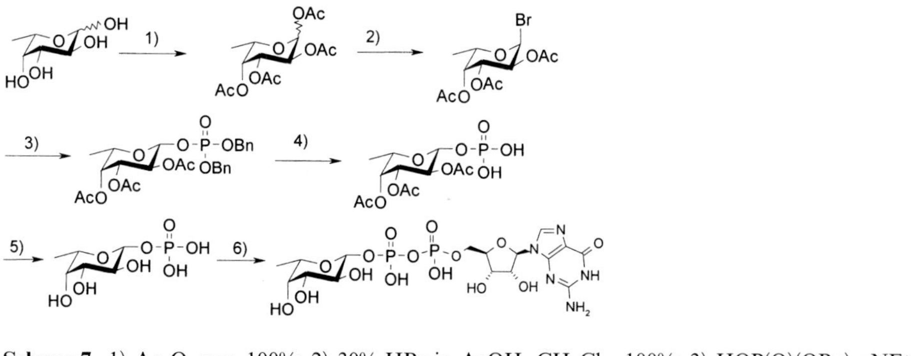

Scheme 7. 1) Ac$_2$O, pyr, 100%; 2) 30% HBr in AcOH, CH$_2$Cl$_2$, 100%; 3) HOP(O)(OBn)$_2$xNEt$_3$, Ag$_2$CO$_3$, −78 °C, CH$_2$Cl$_2$, 72%; 4) H$_2$/Pd/C, MeOH, 100%; 5) NaOMe, MeOH, 92%; 6) GMP–morpholidate, pyr, 27–43%.

Table 2. Successful preparations of non-natural GDP–deoxyfucoses.

Position	Residue	Ref.
R_b	H	[37]
R_d	H	[37]
R_c	H	[39]
X	CH_2	[40]
R_a	F	[41]
R_d	Spacer	[42, 43]

X = O, unless otherwise stated.
Always only one residue is altered, the remaining residues are either OH for R_a, R_b, R_c or CH_3 for R_d.
Spacer = fluorescent label added to spacer $H_2N(CH_2)_2S(CH_2)_2OCH_2-$, $C_5H_{11}-$, $C_7H_{15}-$.

The critical intermediate in the outlined synthesis is the β-fucosyl phosphate, which can hardly be separated cleanly from the concomitant α isomer [65, 70]. The ensuing coupling to the morpholidate-activated GMP proceeds sluggishly and is always accompanied by epimerization and hydrolysis. This makes the synthesis a low-yielding process and impedes the isolation of the pure GDP–fucose from this reaction mixture.

Comments

The first hurdle which must be overcome in the chemical synthesis of sugar nucleotides is the preparation of anomerically pure sugar phosphates. In sialic acid there is no neighboring group which can control the stereoselective phosphorylation of the C-2 center to form the desired β-phosphates selectively. Optimization of protecting groups and phosphorylation protocols led to modest improvements only of the α/β ratios. This problem is not yet solved satisfactorily.

In GDP–Fuc and derivatives the anomeric phosphate group is β-linked to the C-1 atom. Peracetate protection of the fucose enables the quantitative synthesis of the α-bromide intermediate which is converted to the desired β-phosphate almost exclusively (see Scheme 7).

The UDP-activated sugars and GDP–Man are α-phosphates and their chemical synthesis poses greater problems. Although various protecting group combinations and phosphorylation protocols were investigated, none led to the exclusive formation of the desired α-phosphates (see Schemes 3 and 6). The purification of those compounds is further hampered by their notorious instability.

The ensuing step is pyrophosphate formation, generally a very sluggish reaction, even if one of the two phosphate groups—usually that of the nucleotide—is activated by a good leaving group like morpholine. The diphosphate coupling reaction often takes up to 3–5 days at room temperature and is still not complete. An increase in temperature to promote the reaction is not advisable because of the lability of the reaction products. The introduction of *in situ* activation of one phosphate group by tetrazoles or imidazoles has only led to exceptional improvements in the coupling yields (see Scheme 7). The long reaction time is generally accompanied by decomposition of the sugar phosphates. Thus the reaction mixture often contains a number of polar compounds which are difficult to separate from the desired sugar nucleotides. This cuts down overall yields. A few sugars only could be obtained in a satisfactory yield (see Table 1).

For these reasons the full chemical preparation of nucleotide donor sugars is restricted to the laboratory scale only. Large-scale production still needs much developmental effort.

24.2.2 Chemo–Enzymatic Synthesis

The delicate nature of the activated donor sugars and the problems encountered with their chemical synthesis prompted glycochemists to investigate preparative alternatives mimicking the *in vivo* synthesis [30, 71, 72]. The activated donors are obtained either by a combined chemo–enzymatic or a fully enzymatic approach. The individual approach depends on the availability of the required biocatalysts.

Although some donor sugars could even be generated *in situ* via multi-enzyme systems before enzymatic transfer [73], this route seems limited to natural donors only; it will be discussed in more detail in Section 24.3.

Uridine Diphosphate-Activated Donor Sugars

Whitesides and coworkers extensively studied various enzymatic routes offered by nature to prepare UDP–activated donors on a gram scale (see Scheme 8) [74, 75].

It is apparent that the key intermediates are again the α-sugar-1-phosphates. Galactose and 2-aminogalactose were phosphorylated directly at the anomeric centers in the presence of galactokinase and ATP [75]. The resulting α-sugar phosphates were only crudely purified and converted to the desired UDP–sugars in a multi-enzyme cycle in the presence of UTP, UDP–Glc, UDP–glucose pyrophosphorylase (E.C. 2.7.7.9), pyrophosphatase (E.C. 3.6.1.1), and galactose-1-phosphate uridyl transferase (E.C. 2.7.7.12). This last enzyme transfers a UMP moiety from UDP–Glc on to the phosphate group of galactose and 2-amino-2-deoxygalactose to produce the corresponding UDP–sugar (see Scheme 8) and glucose-1-phosphate, which is recycled by the other enzymes to form again the UDP–Glc cosubstrate with UTP as the phosphate source.

With glucose or glucosamine [75] a phosphate group is first transferred from ATP on to the 6-OH group by a hexokinase. The 6-phosphate sugars are then rearranged with either a phosphoglucomutase or a *N*-acetylglucosamine phosphomutase, respectively, to give the α-phosphates of glucose and *N*-acetylglucosamine. Glucose-1-phosphate is subsequently incubated with UTP and UDP–glucose pyrophosphorylase, and a pyrophosphatase to decompose the pyrophosphate side-product to give

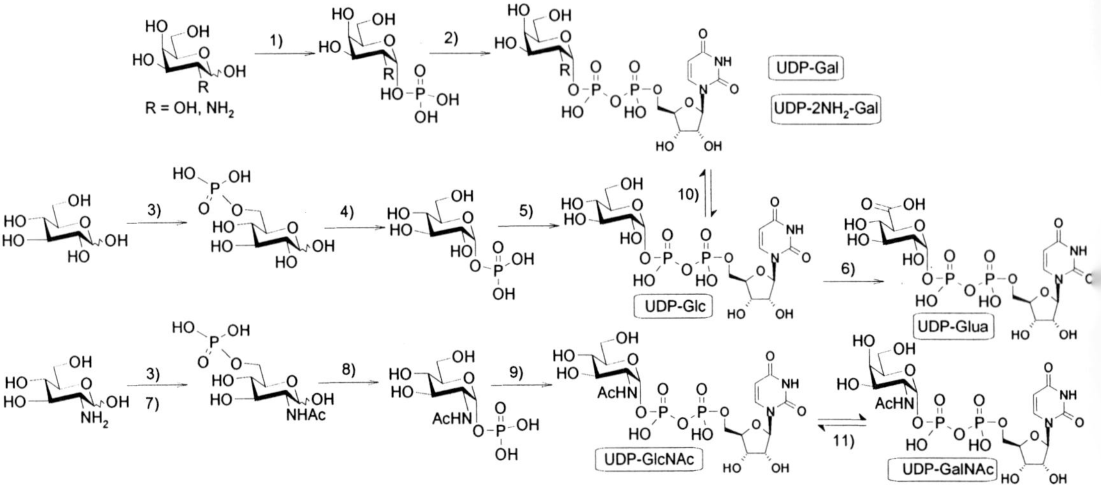

Scheme 8. 1) galactokinase, ATP; 2) galactose-1-phosphate uridyltransferase, UDP–glucose, UTP, pyrophosphatase; 3) hexokinase, ATP, pyrophosphatase; 4) phosphoglucomutase; 5) UDP–glucose pyrophosphorylase, UTP, pyrophosphatase; 6) UDP–glucose dehydrogenase; 7) N-acetoxysuccinimide; 8) N-acetylglucosamine phosphomutase; 9) UDP–N-acetylglucosamine pyrophosphorylase, UTP, pyrophosphatase; 10) if R = OH then UDP–glucose-4-epimerase; 11) UDP–N-acetylglucose-4-epimerase.

UDP–Glc. UDP–GlcNAc is obtained via an analogous incubation with UDP–N-acetylglucosamine pyrophosphorylase and UTP, and a pyrophosphatase.

UDP–glucuronic acid is synthesized by oxidation of the 6-OH group of UDP–glucose with UDP–glucose dehydrogenase and NAD^+ as a hydrogen acceptor. The process works most efficiently if the generated $NADH^+$ is immediately re-oxidized e.g. with L-lactic acid dehydrogenase and pyruvic acid [74].

UDP–glucose and UDP–N-acetylglucosamine can also be converted *in situ* into UDP–galactose or UDP–N-acetylgalactosamine, and *vice versa*, by treatment with commercial UDP–glucose epimerase (E.C. 5.1.3.2) or UDP–N-acetylglucosaminyl epimerase (E.C. 5.1.3.7), respectively. These enzymes convert the equatorial OH groups of glucose and N-acetylglucosamine, into axial 4-OH groups by an oxidation–reduction sequence. The equilibria, however, lie far on the side of the starting *gluco* isomers. The application of the epimerization reaction is, therefore, useful only when the generated UDP–Gal and UDP–GalNAc are consumed immediately, e.g. by a successive transferase reaction.

The enzymatic and semi-enzymatic routes towards UDP–sugars have been nicely compiled by Heidlas et al. [29]. Many other enzymes and enzyme sources are listed by Elling [71]. Although a vast variety of mammalian and non-mammalian sugar nucleotides might be accessible in this manner, most of the enzymes or enzyme systems are not generally available and have not been probed on a preparative scale. There are, however, a limited number of reports concerning the enzymatic synthesis of non-natural UDP–sugars such as UDP–2F-galactose [76]. Also some uridine analogs, *e.g.* 5-fluorouridine 5′-monophosphate, have also been coupled to give the corresponding 'UDP'–galactose derivatives. This reaction was catalyzed by the microorganism *C. sauitoana* in 74% yield on a 250 mg scale [77]. The synthetic routes exactly follow the route outlined for the preparation of UDP–Gal (Scheme 8).

A direct microbial conversion of glucose and UMP to UDP–Glc has also been explored [78] as has the production of α-galactosyl phosphate from bacteria [79].

Hollow fiber reactor technology has also been investigated for the production of UDP–glucosamine and UDP–GlcNAc [80]. The principles of this preparation are similar to those outlined in Scheme 8. In a first step the sugars are phosphorylated at the 6-OH group by a hexokinase and UTP, which is regenerated *in situ* from phosphoenol pyruvate in the presence of a phosphokinase. The 6-phosphate is subsequently rearranged to the sugar-1-phosphate in the same reactor. In a second step the α-anomeric phosphate is treated with UTP and UDP–glucose pyrophosphorylase to give UDP–glucosamine. The amino group can be acetylated with acetic anhydride to give UDP–GlcNAc.

CMP–Activated Sugars

Following the biosynthetic pathway has proved to be the most efficient means of preparing CMP–sialic acid and some derivatives thereof [29, 30, 71]. Glc-NAc is epimerized in a first step to give Man-NAc which is condensed with pyruvic acid in the presence of sialic acid aldolase to produce sialic acid (Scheme 9). The sialic acid is subsequently converted enzymatically to CMP–sialic acid by CMP–sialic acid synthetase. The required CTP can be produced *in situ* from CMP and phosphoenolpyruvate as the ultimate phosphate source via several phosphate-transfer reactions catalyzed by phosphokinases (see box in Scheme 9). During the coupling of sialic acid to CTP a pyrophosphate is again released. This side-product must be decomposed by a phosphatase to shift the equilibrium toward the product. The whole synthesis has been scaled up to a multi-gram production of pure CMP–sialic acid [81] by use of a synthetase cloned and overexpressed in *E. coli*. Recently, the sequence has been further improved by a Japanese group who employed a deacetylase to remove selectively any non-epimerized glcNAc which inhibits the ensuing aldolase-reaction [82].

The sialic acid synthetase could also be employed successfully to prepare a series of non-natural CMP–Sia analogs [83] (see Scheme10).

The 9-amino derivative was used to attach a fluorescenyl photolabel [83]. All the donor substrates listed in the table have been probed on various α(2–6) and α(2–3) sialyl transferases [84].

An important activated donor sugar of Gram-positive bacteria is CMP–**k**eto**d**eoxy**o**ctulosonic acid (3-deoxy-D-mannooctulosonic acid). This compound and its

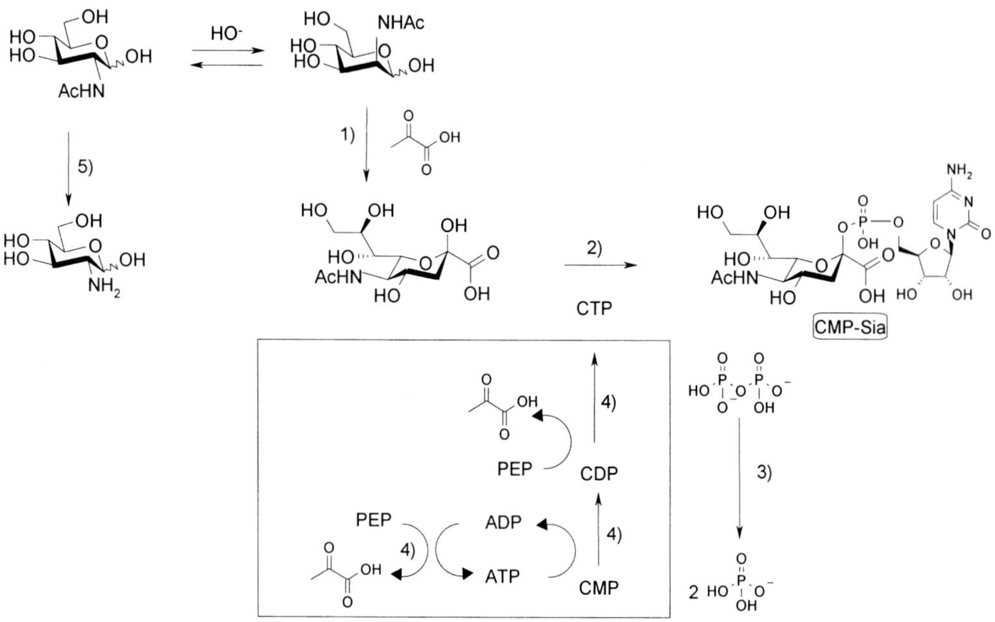

Scheme 9. 1) aldolase; 2) sialic acid synthetase; 3) pyrophosphatase; 4) phosphokinase; 5) deacylase.

Scheme 10. Synthesis of non-natural CMP–'Sia' derivatives with sialic acid synthetase.

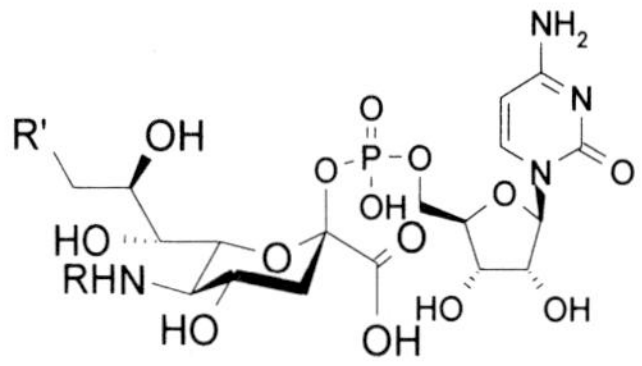

R	R'	R	R'
CH$_3$C(O)	OH	CH$_3$C(O)	$^+$NH$_3$
CH$_3$C(O)	N$_3$	CH$_3$C(O)	CH$_3$C(O)NH$^-$
CH$_3$C(O)	CH$_3$C(S)NH$^-$	HC(O)	OH
CH$_3$C(O)	CH$_3$(CH$_2$)$_4$C(O)NH$^-$	H$_3$NCH$_2$C(O)	H$_3$N$^+$CH$_2$C(O)
CH$_3$C(O)	PhC(O)NH$^-$	CH$_3$C(S)	CH$_3$C(S)

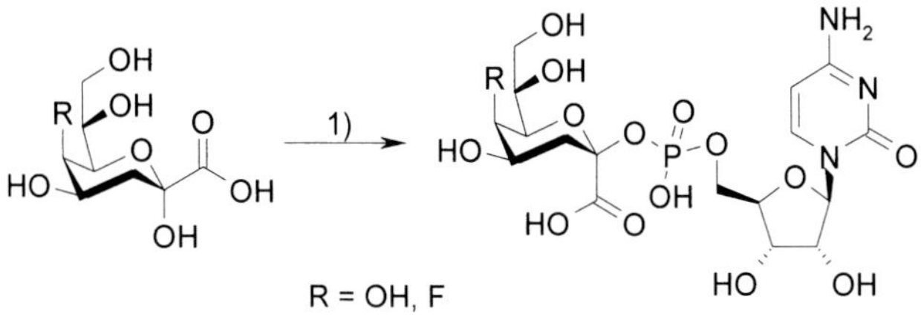

Scheme 11. 1) CMP–Kdo-synthetase, CTP.

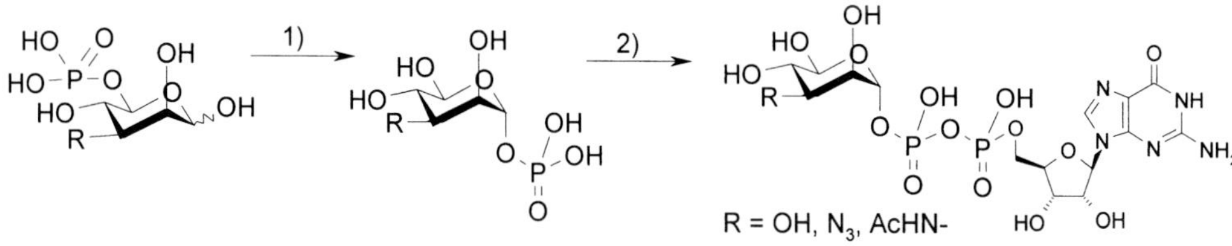

Scheme 12. 1) Mannose phosphomutase; 2) GDP–mannose pyrophosphorylase, GTP, pyrophosphatase.

5-fluoro congener (see Scheme 11) have been prepared recently on a preparative scale by use of cloned and overexpressed microbial CMP–Kdo-synthetase (E.C. 2.7.7.38) [85].

GDP-Activated sugars

The in vivo synthesis of GDP–mannose is equivalent to that of the UDP–sugars [29, 30] (see Scheme 12). Mannose-1-phosphate, prepared from mannose-6-phosphate either by a phosphomutase reaction or chemically [61], is converted to GDP–mannose by a GDP–mannose pyrophosphorylase in the presence of GTP [86]. Also in this case a pyrophosphate side-product is generated which must be decomposed by a pyrophosphatase to shift the equilibrium to the desired GDP–Man. GDP–3-azido- and GDP–3-N-acetylmannose have also been obtained successfully by following this pathway [60].

Highly efficient access to GDP–fucose and several non-natural analogs is gained by the combined application of chemical and enzymological methods (Scheme 13).

A series of peracetylated fucose analogs have been prepared by the chemical route described for the peracetylated β-fucosyl phosphate (Scheme 7) [87]. Two decisive improvements have been introduced with regard to the fully chemical synthesis of GDP–Fuc (Scheme 13). Firstly, the acetylated 'fucosyl' phosphates used for the coupling reaction are more stable than the unprotected variety and they are far more soluble in the coupling solvent than the unprotected ones. This speeds up the diphosphate formation without any detectable, concomitant anomerization of

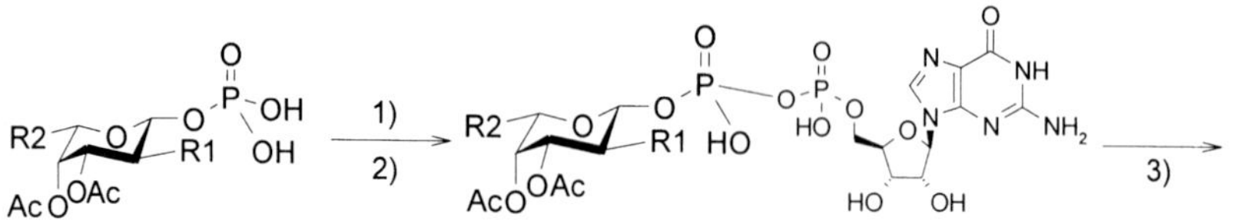

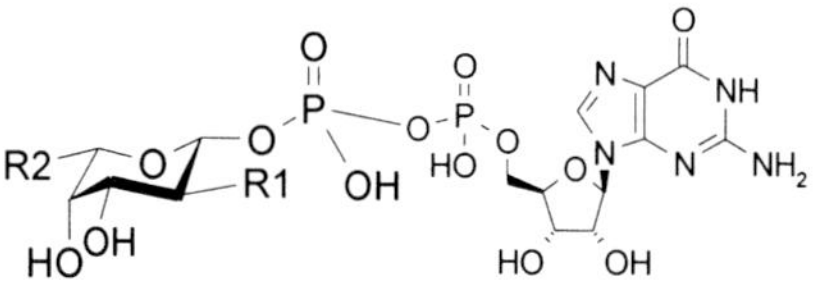

R1	R2	% overall
OH	CH₃	89
NH₂	CH₃	50
F	CH₃	80
OH	H	77
OH	CH₂OH	70
OH	CH₂OH	46*

Scheme 13. 1) a: DMF, GMPxNBu₃ and carbonyldiimidazole, b: DOWEX–H⁺; 2) H₂O, pH = 7, calf intestine alkaline phosphatase; 3) H₂O, pH = 6.8, acetylesterase, 0.1 M NaOH; *4-epi-OH.

the 'fucosyl' phosphates. In addition the resulting product mixture is subsequently treated with calf intestine alkaline phosphatase to remove any unreacted monophosphate selectively. This facilitates the isolation of the protected GDP–sugars. They are simply precipitated from this mixture by the addition of ethanol. In a subsequent step the desired unprotected donor-sugars are obtained under very mild conditions via incubation with commercial acetylesterase and final ethanol precipitation. The overall sequence benefits from the absence of any chromatographic purification step, high overall yields, and broad versatility. Thus the fucose moiety has been replaced by D-arabinose, L-glucose and L-galactose or 2-amino- and 2-fluorofucose (Scheme 13). The procedure could be applied likewise to prepare adenosine diphosphate-, xanthosine diphosphate-, and inosine diphosphate-fucose [87].

Investigation of the human FX protein has recently revealed its remarkable involvement in the *in vivo* synthesis of GDP–Fuc from GDP–Man. The purified enzyme had combined epimerase and NADPH-reductase activity, converting GDP–4-keto-6-deoxymannose into GDP–Fuc [88]. The very same reaction sequence is also achieved by use of a crude extract from *A. aerogenes*, and has been used to prepare ¹⁴C-labeled GDP–Fuc [89].

Another microbial production of GDP–Fuc has been reported by a Japanese group [90]. GDP–Man obtained by fermentation was converted to GDP–Fuc by *A. radiobacter*. A crude extract from hog submaxillary glands was also found to convert fucose directly into GDP–Fuc by a two-step enzymatic process (see Scheme 14) [91].

The protein extract probably contained both a fucokinase to generate fucose-1-phosphate and GDP–fucose pyrophosphorylase. This same analytical protocol was later on applied by chemists to synthesize small amounts of GDP–Fuc [92].

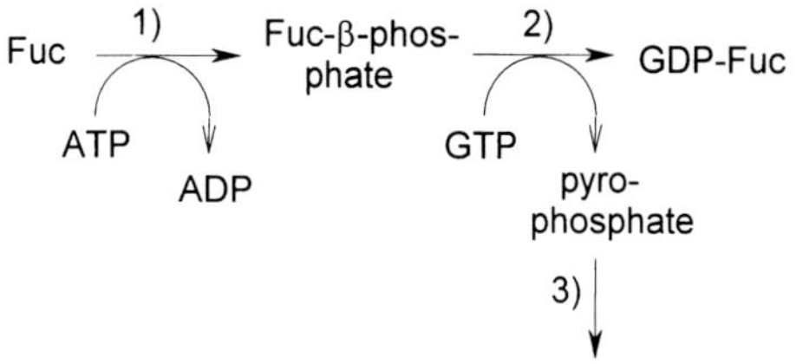

$$\text{Fuc} \xrightarrow[\text{ATP} \quad \text{ADP}]{1)} \text{Fuc-}\beta\text{-phosphate} \xrightarrow[\text{GTP} \quad \text{pyrophosphate}]{2)} \text{GDP-Fuc}$$

$$3) \downarrow$$

Scheme 14. 1) fucokinase; 2) GDP–fucose pyrophosphorylase; 3) pyrophosphatase.

Comments

In comparison with the fully chemical synthesis of nucleotide-activated donor sugars the chemo–enzymatic or enzymatic approaches are distinguished by fewer protecting group manipulations and high chemo- and stereoselectivity. This simplifies purification protocols significantly. These features reduce the number of synthetic steps and increase overall yields.

The fully enzymatic preparation of sugar nucleotides, however, requires the availability of a series of enzymes. In addition, their mutual interactions presupposes exact knowledge of their physiological properties by the user; this, in particular, makes chemists hesitate to use biocatalysts. Not all the involved enzymes may tolerate altered, non-natural substrates. A combined chemo–enzymatic approach therefore offers the possibility of combining the advantages of the chemical synthesis and those of the enzymatic preparation, *e.g.* to prepare non-natural nucleotide activated substrates.

24.3 *In situ* Generation of Sugar Nucleotides

Several papers describe enzymatic glycosylations with *in situ* generation of the required activated donor sugars; these are impressively compiled in Refs. [30] and [73]. Especially impressive is the synthesis of the trisaccharide Siaα(2–6)Galβ(1–4)GlcNAcOH with the *in situ* generation of CMP–sialic acid from GlcNAc (compare Scheme 9) and UDP–galactose from glucose (compare Scheme 8) coupled with the enzymatic transfer of those sugars. In this one-pot reaction a total of nine enzymes has been applied successfully [93]!

The synthesis of the linear B-trisaccharide Galα(1–3)Galα(1–4)GlcNAcβOR, an important human xenoantigen [94], has also been tackled by a multi-enzyme one-pot procedure, albeit only on a microscale (see Scheme 15) [95].

Especially noteworthy is the use of sucrose, Glcα(1–2)Fru, as the ultimate source of UDP–Gal. In a first step sucrose synthase from rice cleaves the disaccharide and forms UDP–Glc in the presence of UDP. UDP–Glc is subsequently epimerized at the 4-position by UDP–galactose epimerase to give the required UDP–Gal. This donor substrate is first used by β(1–4)galactosyl transferase to produce the *N*-acetyllactosamine intermediate and then by α(1–3)galactosyl transferase to gal-

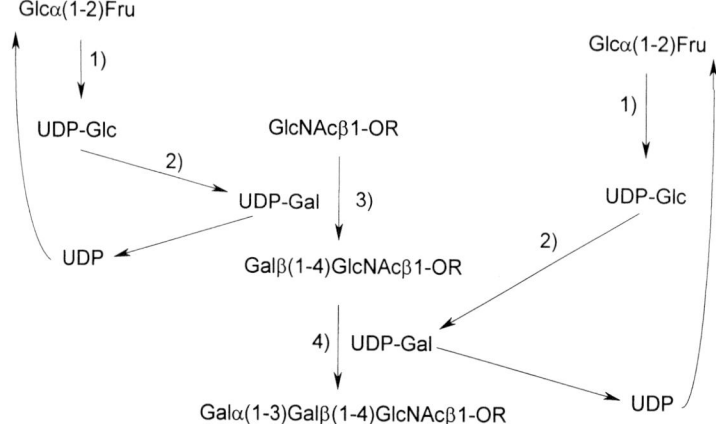

Scheme 15. 1) sucrose synthase, UTP; 2) UDP–glucose-4′-epimerase; 3) β(1–4)galactosyl transferase; 4) α(1–3)galactosyl transferase.

actosylate the intermediate disaccharide at the 3-OH group of the previously introduced galactose unit. The released UDP is recycled again. Neither the β(1–4)galactosyl transferase nor the α(1–3)galactosyl transferase attacks the sucrose.

The *in situ* generation of GDP–Fuc from β-fucose-1-phosphate coupled with the transfer of the fucose moiety on to *N*-acetyllactosamine has also been probed on an analytical scale [96]. In the presence of GTP β-fucose-1-phosphate is converted to GDP–Fuc by GDP–fucose pyrophosphorylase. GDP–Fuc is then used by an α(1–3)fucosyl transferase to transfer the fucose moiety on to *N*-acetyllactosamine. The released GDP is recycled and phosphorylated to give GTP by a phosphokinase with phosphoenol pyruvate as phosphate source. The pyrophosphate produced in the complete cycle is decomposed by pyrophosphatase (see Scheme 16).

There is only one report of the transfer of a non-natural galactose derivative (see Scheme 17) [97]. 2-Deoxyglucose-6-phosphate, which is itself produced by a hexokinase reaction, is first equilibrated to α-2-deoxyglucose-1-phosphate by a phosphoglucomutase. In the presence of UTP and UDP–glucose pyrophosphatase this sugar is converted to UDP–2-deoxyglucose which is subsequently epimerized to UDP–2-deoxygalactose by galactose epimerase. The resulting UDP–2-deoxygalactose is then transferred by a β(1–4)galactosyl transferase on to the 4-OH group of *N*-acetylglucosamine to give the deoxylactosamine derivative. The released UDP is re-phosphorylated to UTP in the presence of phosphoenolpyruvate by pyruvate kinase to close the reaction cycle. Thus milligram amounts of 2′-deoxy-*N*-acetyllactosamine were synthesized in ca 40% yield.

Comments

It goes without saying that the most elegant oligosaccharide synthesis is achieved by enzymatic assemblage of sugars starting from simple, unprotected monosaccharide building blocks and harvesting the desired oligosaccharide. This can be achieved by use of cyclic enzyme systems which activate and transfer the corresponding mono-

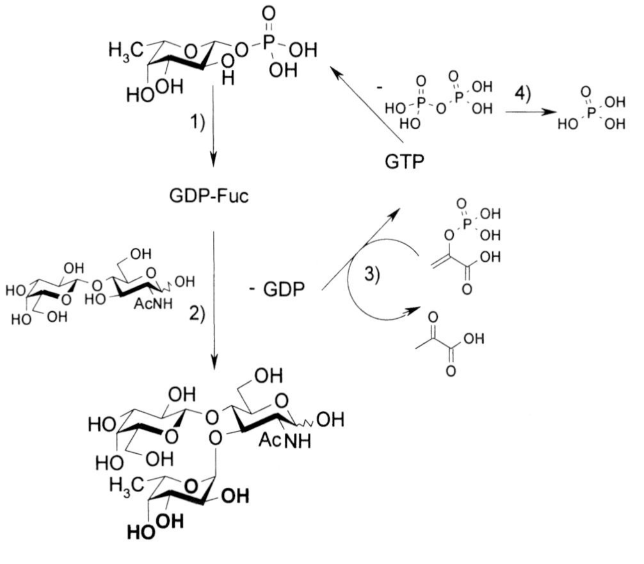

24.3 *In situ Generation of Sugar Nucleotides* 643

Scheme 16. 1) GDP–fucose pyrophosphorylase; 2) α(1–3)fucosyl transferase; 3) phosphokinase; 4) pyrophosphatase.

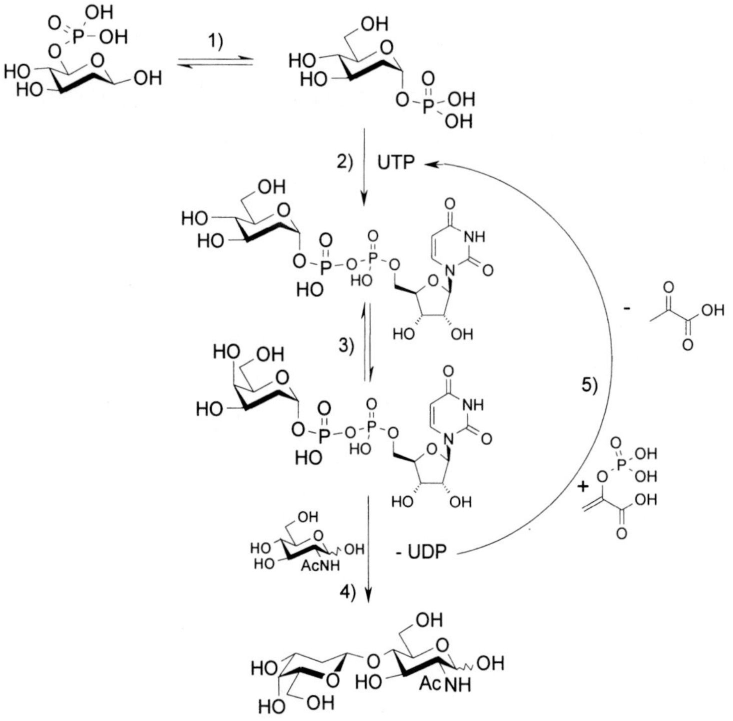

Scheme 17. 1) phosphoglucomutase; 2) UDP–glucose pyrophosphorylase; 3) UDP–galactose-4′-epimerase; 4) β(1–4)galactosyl transferase; 5) phosphokinase.

saccharides to form the desired product. No sugar phosphates and sugar nucleotides have to be isolated. In contrast, the released nucleotide diphosphate is re-used by the enzymes to start a new activation cycle (see Scheme 17).

This task can also be achieved by use of microorganisms—the so called wholecell synthesis—provided the microorganisms do not metabolize the product oligosaccharide.

The practicability of such cyclic multi-enzyme preparations has been confirmed, e.g. for the synthesis of N-acetyllactosamine or the tetrasaccharide, sialyl Lewis^x, containing four different monosaccharide units.

There have also been a limited number of reports showing that slightly altered natural sugars can be introduced into such cyclic multi-enzyme systems to generate non-natural oligosaccharides. Carbohydrates with major non-natural elements will, however, still have to be prepared chemically in the near future.

24.4 Outlook

The steady progress in cloning and expression techniques on the one hand and improvements in cell cultivation and fermentation procedures on the other hand will make an increasing number of enzymes available to the carbohydrate chemist. Tailor-made biocatalysts which will convert non-natural substrates also will be provided by these biotechnological methods. Additionally, screenings of microorganisms will increase access to rare and unknown sugars. Thus the preparation of natural and non-natural oligosaccharides will become cheaper. This will lead to the identification of a growing number of therapeutically active glycoconjugates and consequently an increased demand for sugars by the pharmaceutical industry.

There are already suppliers who offer kilogram amounts of activated natural donor sugars for enzymatic synthesis on request (Boehringer, Wandrey, Yamasa Corp. Japan). Provided the required sugar transferases are available, combinatorial techniques will be probed in enzymatic oligosaccharide synthesis. This will give easy and rapid access to oligosaccharides and will promote studies of oligosaccharide–protein interactions. Nonetheless, chemistry will not be totally and abruptly banned from the carbohydrate field, although the era of lengthy and sophisticated assemblage of oligosaccharides by purely chemical means will wane.

References

1. O. Hindsgaul and M. Fukuda, *Molecular Glycobiology*, Oxford University Press **1994**.
2. A. Varki, *Glycobiology* **1993**, *3*, 97.
3. R. A. Dwek, *Chem. Rev.* **1996**, *96*, 683.
4. P. Sears, C.-H. Wong, *Chem. Commun.* **1998**, 1161.
5. F. R. Carbone, P. A. Gleesen, *Glycobiology* **1997**, *7*, 725.
6. C. R. Bertozzi, *Chem. & Biol.* **1995**, *2*, 703.
7. L. A. Lasky, *Ann. Rev. Biochem.* **1995**, *64*, 113.

8. A series of papers on parasites in *Science* **1994**, *264*, 1857–1886.
9. K. Gustafsson, K. Strahan, A. Preece, *Immunol. Rev.* **1994**, *141*, 59.
10. J. M. McAuliffe, O. Hindsgaul, *Chem. & Ind.* **1997**, *3*, 70.
11. Z. J. Witczak, *Curr. Med. Chem.* **1995**, *1*, 392.
12. Compare preceding chapters of this volume.
13. S. Hanessian, *Preparative Carbohydrate Chemistry*, Marcel Dekker **1997**.
14. F. Barresi, O. Hindsgaul, *Modern Synth. Methods* **1995**, 283.
15. F. Barresi, O. Hindsgaul, *Carbohydr. Chem.* **1995**, *14*, 1043.
16. T. Feizi, *Curr. Opin. Struct. Biol.* **1993**, *3*, 701.
17. Y. Nagai, *Pure & Appl. Chem.* **1997**, *69*, 1893.
18. H. J. M. Gijsen, L. Qiao, W. Fitz, C.-H. Wong, *Chem. Rev.* **1996**, *96*, 443.
19. S. C. Crawley, M. M. Palcic, *Modern Methods Carbohydr. Synth.* **1996**, 492.
20. P. H. Johnson, W. M. Watkins, *Glycocojugate J.* **1992**, *9*, 241.
21. H. E. Appert, T. J. Rutherford, G. E. Tarr, N. R. Thomford, J. McCorquindale, *Biochem. Biophys. Res. Commun.* **1986**, *138*, 224.
22. A. Sarnesto, T. Köhlin, O. Hindsgaul, K. Vogele, M. Blaszczyk-Thurin, J. Thurin, *J. Biol. Chem.* **1992**, *267*, 2745.
23. R. Öhrlein, O. Hindsgaul, M. M. Palcic, *Carbohydr. Res.* **1993**, *244*, 149.
24. L. J. Melkerson-Watson, L. C. Sweeley, *J. Biol. Chem.* **1991**, *266*, 4448.
25. J. C. Paulson, J. L. Rearick, R. L. Hill, *J. Biol. Chem.* **1977**, *252*, 2363.
26. M. C. Field, L. J. Wainwright, *Glycobiology* **1995**, *5*, 463.
27. M. Fukuda, M. F. A. Bierhuizen, J. Nakayama, *Glycobiology* **1996**, *6*, 683.
28. J. B. Lowe, *Sem. Cell Biol.* **1991**, *2*, 289.
29. J. E. Heidlas, K. W. Williams G. M. Whitesides, *Acc. Chem. Res.* **1992**, *25*, 307.
30. C.-H. Wong, R. L. Halcomb, Y. Ichikawa, T. Kajimoto, *Angew. Chem. Int. Ed. Engl.* **1995**, *34*, 412, 521.
31. N. Kochetkov, V. N. Shibaev, *Adv. Carbohydr. Chem. Biochem.* **1973**, *28*, 307.
32. D. L. MacDonald, *Methods Enzymol.* **1966**, *8*, 121.
33. C. D. Warren, R. W. Jeanloz, *Biochemistry* **1973**, *12*, 5031.
34. S. B. Tzokov, I. T. Devedjiev, E. K. Bratovanova, D. Petkov, *Angew. Chem. Int. Ed. Engl.* **1994**, *33*, 2302.
35. S. Sabesan, S. Neira, *Carbohydr. Res.* **1992**, *223*, 169.
36. J. Lee, J. K. Coward, *J. Org. Chem.* **1992**, *57*, 4126.
37. T. Yamazaki, C. D. Warren, A. Herscovics, R. W. Jeanloz, *Carbohydr. Res.* **1980**, *79*, C9–C12.
38. L. Knerr, X. Pannecoucke, B. Luu, *Tetrahedron Lett.* **1998**, *39*, 273.
39. F. Cramer, H. Neunhoeffer, K. H. Scheit, G. Schneider, J. Tennigkeit, *Angew. Chem.* **1962**, *74*, 387.
40. M.-C. Chapeau, P. A. Frey, *J. Org. Chem.* **1994**, *59*, 6994.
41. A. Burton, P. Wyatt, G. J. Boons, *J. Chem. Soc. Perkin Trans. 1* **1997**, *16*, 2375.
42. T. Endo, Y. Kajihara, H. Kodama, H. Hashimoto, *Bioorg. Med. Chem.* **1996**, *4*, 1939.
43. Y. Kajihara, T. Endo, H. Ogasawara, H. Kodama, H. Hashimoto, *Carbohydr. Res.* **1995**, *269*, 273.
44. G. Srivastava, O. Hindsgaul, M. M. Palcic, *Carbohydr. Res.* **1993**, *245*, 137.
45. H. Yuasa, O. Hindsgaul, M. M. Palcic, *J. Am. Chem. Soc.* **1992**, *114*, 5891.
46. H. Yuasa, O. Hindsgaul, M. M. Palcic, *Can. J. Chem.* **1995**, *73*, 2190.
47. L. Thomas, S. A. Abbas, K. L. Abbas, *Carbohydr. Res.* **1988**, *184*, 77.
48. O. Tsuruta, G. Shinohara, H. Yuasa, H. Hashimoto, *Bioorg. Med. Chem. Lett.* **1997**, *7*, 2523.
49. T. Yamazaki, C. D. Warren, A. Herscovic, R. W. Jeanloz, *Can. J. Chem.* **1981**, *59*, 2247.
50. S. Wiemann, W. Klaffke, *Liebigs Ann. Chem.* **1995**, 1779.
51. M. D. Chappell, R. L. Halcomb, *Tetrahedron* **1997**, *53*, 11109.
52. S. Makino, Y. Ueno, M. Ichikawa, Y. Hayakawa, T. Hata, *Tetrahedron Lett.* **1993**, *34*, 2775.
53. T. J. Martin, R. R. Schmidt, *Tetrahedron Lett.* **1992**, *33*, 6123.
54. T. Yoshino, R. R. Schmidt, *Carbohydr. Lett.* **1995**, *1*, 329.
55. T. J. Martin, H. Braun, R. R. Schmidt, *Bioorg. Med. Chem.* **1994**, *2*, 1203.
56. Y. Kajihara, T. Ebata, K. Koseki, H. Kodama, H. Matsushita, H. Hashimoto, *J. Org. Chem.* **1995**, *60*, 5732.

57. V. Ginsburg, *J. Biol. Chem.* **1961**, *236*, 2389.
58. com. available from SIGMA, CH-9741 Buchs, Switzerland, P.O. 260.
59. V. Wittmann, C.-H. Wong, *J. Org. Chem.* **1997**, *62*, 2144.
60. W. Klaffke, *Carbohydr. Res.* **1995**, *266*, 285.
61. J. E. Pallanca, N. J. Turner, *Chem. Soc. Perkin Trans. I* **1993**, 3017.
62. H. A. Nunez, J. O'Connor, P. L. Rosevar, R. Baker, *Can. J. Chem.* **1981**, *59*, 2086.
63. R. R. Schmidt, B. Wegmann, K.-H. Jung, *Liebigs Ann. Chem.* **1991**, 121.
64. B. M. Heskamp, H. J. G. Broxtermann, G. A. van der Marel, H. van Boom, *J. Carbohydr. Chem.* **1996**, *15*, 611.
65. U. B. Gokhale, O. Hindsgaul, M. M. Palcic, *Can. J. Chem.* **1990**, *68*, 1063.
66. K. Adelhorst, G. M. Whitesides, *Carbohydr. Res.* **1993**, *242*, 69.
67. C. Hällgren, O. Hindsgaul, *J. Carbohydr. Chem.* **1995**, *14*, 453.
68. C. Vogel, C. Bergmann, A.-J. Ott, T. K. Lindhorst, J. Thiem, W. V. Dahlhoff, C. Hällgren, M. M. Palcic, O. Hindsgaul, *Liebigs Ann. Chem.* **1997**, 601.
69. G. Srivastava, K. J. Kaur, O. Hindsgaul, M. M. Palcic, *J. Biol. Chem.* **1992**, *267*, 22356.
70. Y. Ichikawa, M. M. Sim, C.-H. Wong, *J. Org. Chem.* **1992**, *57*, 2943.
71. L. Elling, *Adv. Biochem. Eng. & Biotechnol.* **1997**, *58*, 91.
72. S. David, C. Auge, C. Gautheron, *Adv. Carbohydr. Chem. Biochem.* **1991**, *49*, 175.
73. Y. Ichikawa, R. Wang, C.-H. Wong, *Methods Enzymol.* **1994**, *247*, 107.
74. E. J. Toone, E. S. Simon, G. M. Whitesides, *J. Org. Chem.* **1991**, *56*, 5603.
75. J. E. Heidlas, W. J. Lees, G. M. Whitesides, *J. Org. Chem.* **1992**, *57*, 146, 152.
76. T. Hayashi, B. W. Murray, R. Wang, C.-H. Wong, *Bioorg. Med. Chem.* **1997**, *5*, 497.
77. K.-i. Fujita, A. Matsukawa, K. Shibata, T. Tanaka, M. Taniguchi, S. Oi, *Carbohydr. Res.* **1994**, *265*, 299.
78. J. H. Ko, H.-S. Shin, Y. S. Kim, D.-S. Lee, C.-H. Kim, *Appl. Biochem. Biotechnol.* **1996**, *60*, 41.
79. B. Nidetzky, A. Weinhäusel, R. Grießner, K. D. Kulbe, *J. Carbohydr. Chem.* **1995**, *14*, 1017.
80. P. A. Ropp, P.-W. Cheng, *Anal. Biochem.* **1990**, *187*, 104.
81. M. Kittelmann, T. Klein, U. Kragl, C. Wandrey, O. Ghisalba, *Appl. Microbiol. Biotechnol.* **1995**, *44*, 59.
82. A. Kuboki, H. Okazaki, T. Sugai, H. Ohta, *Tetrahedron* **1997**, *53*, 2387.
83. R. Brossmer, H. J. Gross, *Methods Enzymol.* **1994**, *247*, 153.
84. H. J. Gross, R. Brossmer, *Glycoconjugate J.* **1995**, 739.
85. T. Sugai, C.-H. Lin, G.-J. Shen, C.-H. Wong, *Bioorg. Med. Chem.* **1995**, *3*, 313.
86. E. S. Simon, S. Grabowski, G. M. Whitesides, *J. Org. Chem.* **1990**, *55*, 1834.
87. G. Baisch, R. Öhrlein, *Bioorg. Med. Chem.* **1997**, *5*, 383.
88. M. Tonetti, L. Sturla, A. Bisso, U. Benatti, A. De Flora, *J. Biol. Chem.* **1996**, *271*, 27274.
89. V. Ginsburg, *Methods Enzymol.* **1966**, *8*, 293.
90. K. Yamamoto, T. Maruyama, H. Kumagai, T. Tochkura, T. Seno, H. Yamaguchi, *Agric. Biol. Chem.* **1984**, *48*, 823.
91. R. Prohaska, H. Schenkel-Brunner, *Anal. Biochem.* **1975**, *69*, 536.
92. R. Stiller, J. Thiem, *Liebigs Ann. Chem.* **1992**, 467.
93. Y. Ichikawa, J. L.-C. Liu, G.-J. Shen, C.-H. Wong, *J. Am. Chem. Soc.* **1991**, *113*, 6300.
94. U. Galili, K. Swanson, *Proc. Natl. Acad. Sci. USA* **1988**, 7401.
95. C. H. Hokke, A. Zervose, L. Elling, D. H. Joziasse, D. H. van den Eijnden, *Glycoconjugate J.* **1996**, *13*, 687.
96. Y. Ichikawa, Y.-C. Lin, D. P. Dumas, G.-J. Shen, E. Garcia-Junceda, M. A. Williams, R. Bayer, C. Ketcham, L. E. Walker, J. E. Paulson, C.-H. Wong, *J. Am. Chem. Soc.* **1992**, *114*, 9283.
97. J. Thiem, T. Wiemann, *Angew. Chem. Int. Ed. Engl.* **1991**, *30*, 1163.

25 Enzymatic Glycosylations with Glycosyltransferases

Ossi Renkonen

25.1 Introduction

Samples of pure molecular species of protein- and lipid-bound oligosaccharides are needed for studies of their biological functions. These functions, in turn, are mediated by interactions between oligosaccharides and other biomolecules. Successful studies of these interactions require the use of naturally occurring compounds in the pure form and often also the use of non-natural analogs of protein- and lipid-bound oligosaccharides. The multitude of possible isomeric oligosaccharide structures has hampered the development of glycobiology [1] because isolation of pure samples from naturally occurring mixtures of oligosaccharide isomers has been a difficult task. To avoid this difficulty some researchers rely on enzyme-assisted in vitro synthesis of oligosaccharides. The major branch of this approach is based on Leloir's description of biosynthesis of glycosidic bonds [2].

In Leloir-type of reactions, a monosaccharide unit is enzymatically transferred from a sugar nucleotide to a saccharide acceptor. The reactions proceed stereoselectively with unprotected acceptors, produce one glycosidic linkage in each step, and are applicable to a number of acceptors. A particular advantage of the Leloir type synthesis lies in the availability of several glycosyltransferases in recombinant form; the number of commercially available recombinant glycosyltransferases is in rapid growth and the number of enzymes cloned in research laboratories increases even more rapidly [3]. Another booster is the rapidly improving availability of sugar nucleotides required for the enzymatic synthesis of oligosaccharides (*cf.* Chapter 24, by R. Öhrlein, in this book). Regeneration of sugar nucleotides during the transferase reactions shows promise in large scale enzymatic oligosaccharide synthesis (*cf.* the chapter by C.-H. Wong in this book).

This chapter will describe several Leloir-type transferase reactions that have been used for stepwise construction of a oligosaccharides comprising five different monosaccharides, GlcNAc, GalNAc, Gal, Fuc, and Neu5Ac. This is a very limited approach among the myriads of possible oligosaccharides that can be potentially constructed from the few hundreds of different monosaccharides expressed in Nature. The limitations are remarkable even when considering the number of reported glycosyltransferase-catalyzed reactions. The examples actually selected for discussion represent steps to synthetic oligosaccharides with particularly interesting biomedical properties. Some of the final products are potent in vitro inhibitors of murine sperm-to-egg adhesion, and antagonists of mammalian leukocyte adhesion to activated endothelium. The examples discussed will show that relatively easily synthesized glycans can have considerable contraceptive and anti-inflammatory potential. To compensate for the limitations of this material, the reader is referred to the biosynthetic chapters of this book, and to recent reviews published elsewhere on the general theme, including those of Sears and Wong [3] and Van den Steen et al. [4].

25.2 *In vitro* Synthesis of the Core Region of *O*-Glycans

Many plasma membrane proteins contain domains of mucin type with clustered O-linked glycans; the peptide cores within these clusters adopt stiff and extended conformations, representing rods from which the saccharides protrude [5].

25.2.1 Initialization of *O*-Glycan Biosynthesis

The addition of the first sugar in mucin-type *O*-glycosylation is directed by the family of UDP–GalNAc:polypeptide *N*-acetylgalactosaminyltransferases (EC 2.4.1.41) (ppGaNTases). This activity has been partially purified from several sources [6]. More recently, several isoforms of the mammalian ppGaNTase family have been cloned and functionally expressed [7–9]. The acceptor specificities of the individual enzymes are mostly broad and overlapping, yet distinct. The large number of the transferases is probably responsible for the lack of a general consensus sequence for mucin type *O*-glycosylation. An algorithm for predicting *O*-glycosylation sites in mammalian proteins is available, however [10].

25.2.2 Synthesis of Core 1

In vitro synthesis of *O*-glycan Core 1 (Gal(β1–3)GalNAc) by UDP–Gal:GalNAc–R β3-Gal-transferase (EC 2.4.1.122) from pig submaxillary glands was described almost thirty years ago by Roseman and colleagues [11]. Porcine submaxillary gland microsomes catalyzed the transfer of galactose from UDP–[^{14}C]galactose to sialidase-treated ovine submaxillary mucin, which carries single α-linked GalNAc residues in clustered arrays on serine and threonine residues. As much as 3.4 µmol [^{14}C]galactose was transferred to the protein in an experiment involving 10 µmol multivalent mucin acceptor, 20 µmol UDP–[^{14}C]Gal and 23 mg microsomal protein. Treatment with β-galactosidase from C. *perfringens*, liberated [^{14}C]galactose from the mucin product, and its treatment with 0.05 M KOH + 1 M NaBH$_4$ released a reduced oligosaccharide. The latter gave [^{14}C]galactose and galactosaminitol in a molar ratio of 1.0:0.75 upon acid hydrolysis. Accordingly, it represented [^{14}C]GalβGalNAc$_{red}$. The product was chromatographically and electrophoretically identical with authentic Gal(β1–3)GalNAc$_{red}$ obtained from gangliosides by a known procedure; other positional isomers were not available for comparison and could not be excluded, however.

This pioneering piece of work illuminates many features characteristic of enzyme-assisted synthesis of oligosaccharides. Firstly, it is possible to obtain apparently homogeneous saccharide products by using unprotected acceptors, radiolabeled sugar nucleotides and even *very* crude enzymes. Secondly, product characterization is a difficult and a crucially important task. Later work by Medicino et al. [12] and Cheng and Bona [13] on purified Core 1 galactosyltransferase from tracheal mucosa established the formation of the Gal(β1–3)GalNAc linkage firmly; permethylation and periodate oxidation were used in the product analysis.

A family of homologous β3-galactosyltransferase genes encoding enzymes that transfer to GlcNAc and to GalNAc has been identified recently [14], but the Core 1 Gal transferase has not yet been cloned.

25.2.3 Synthesis of Core 2

In vitro synthesis of *O*-glycan Core 2 [Gal(β1–3)[GlcNAc(β1–6)]GalNAc] by UDP–GlcNAc:Gal(β1–3)GalNAc–R (GlcNAc to GalNAc) β6–GlcNAc-transferase activity of dog submaxillary gland microsomes was reported by Williams and Schachter [15]. The enzyme is conveniently abbreviated as C2GnT. Preliminary identification of the mucin product of this branch-generating reaction became possible when [^{14}C]Gal(β1–3)GalNAc–[ovine submaxillary mucin] was synthesized in vitro (see above) and was used as acceptor in the C2GnT reaction. The resulting mixture of acceptor- and product-oligosaccharides was released from the protein by reductive β-elimination, was methylated, and was finally hydrolyzed. Only 2,3,4,6-tetra-*O*-methyl-[^{14}C]galactose was found in the hydrolysate, implying that the new GlcNAc unit had been transferred to the GalNAc residue of the mucin acceptor.

Later, Williams et al. performed preparative experiments using Gal(β1–3)GalNAc-α-*O*-benzyl as the acceptor, generating a few hundred nanomoles of the trisaccharide product [16]. With this amount of material they could apply methylation analysis and NMR spectroscopy for product characterization, establishing that the canine submaxillary microsomes had catalyzed the formation of the GlcNAc(β1–6)GalNAc linkage. At least two enzyme isoforms are involved in the biosynthesis of Core 2; C2GnT of L-type is restricted to Core 2 synthesis and C2GnT of M-type synthesizes Core 2, Core 4 (see below) and GlcNAc(β1–3)[GlcNAc(β1–6)]Gal–OR which is known as blood group I antigen (see below) [17].

A cDNA has been isolated that encodes the C2GnT of the L-type [18]. Also the C2GnT of the M-type has been cloned, and the recombinant forms of the two distinct C2GnTs have been used for in vitro synthesis of Core 2 *O*-benzyl glycosides

[19]. In our laboratory the analogous transformation has been performed by use of hog gastric mucosal microsomes, which catalyzed the conversion of the free Core 1 disaccharide Gal(β1–3)GalNAc into Gal(β1–3)[GlcNAc(β1–6)]GalNAc [20].

Functionally active molecules of P-selectin glycoprotein ligand-1 (PSGL-1) are not expressed in wild type chinese hamster ovary (CHO) cells, but are expressed in CHO cells transfected with the cDNA encoding L-type C2GnT [21]. This implies that the physiological counter-receptor of P-selectin has to carry O-glycans of Core 2-type in order to be recognized by P-selectin.

25.2.4 Synthesis of Core 3 and Core 4

Brockhausen et al. [22] described in vitro synthesis of O-glycan Core 3 (GlcNAc(β1–3)GalNAc) and Core 4 [GlcNAc(β1–3)[GlcNAc(β1–6)]GalNAc] by UDP-GlcNAc:GalNAc–R β3-GlcNAc-transferase and UDP-GlcNAc:GlcNAc(β1–3)GalNAc–R (GlcNAc to GalNAc) β6-GlcNAc-transferase, respectively. The enzymes are conveniently abbreviated as C3GnT and C4GnT. GalNAc-α-phenyl and GalNAc-[ovine submaxillary mucin] served as acceptors in the reactions studied, and the mucin products (0.26–0.43 μmol) were characterized by liberating the oligosaccharides by reductive β-elimination, followed by NMR spectroscopy and methylation analysis of the liberated alditols. Later experiments have shown that C3GnT activity is high in human colonic and respiratory tissues and in salivary glands from several species. This enzyme has, however, proven difficult to solubilize by detergents and is unstable in the solubilized form, which has hampered its further study [23]. Recently, C4GnT activity was demonstrated in the recombinant form of C2GnT of M-type [19].

25.2.5 *In vitro* Extension of Core 1 Glycans

Mucin type O-glycans carry often single or multiple N-acetyllactosamine (LacNAc) units. Brockhausen et al. [24] showed that pig gastric mucosal microsomes, upon incubation with UDP-[^{14}C]GlcNAc and Gal(β1–3)GalNAcα1–OR (where R is an appropriate mucin or antifreeze glycoprotein), are able to generate [^{14}C]GlcNAc(β1–3) Gal(β1–3)GalNAcα1–OR. The transfer of [^{14}C]GlcNAc amounted to 1.65 μmol. The resulting [^{14}C]glycans were released from the protein by reductive β-elimination, and HPLC was used to isolate altogether 100 nmol pure [^{14}C]GlcNAc(β1–3)Gal(β1–3)GalNAc$_{red}$. Methylation analysis and NMR spectroscopy were used to establish its structure. This β3-GlcNAc transferase is a distinct initiator of polylactosamine chain biosynthesis on Core 1, but it is unlikely to support the further growth of the chain. The data of Brockhausen et al. [25] showed that pig gastric mucosal microsomes fail to transfer GlcNAc to Gal(β1–4)GlcNAc. The trisaccharide sequence of Core 1 type is most likely β4-galactosylated at the non-reducing end in vitro by a number of Gal-transferases and further extended in the polylactosamine mode (see below).

25.2.6 *In vitro* Extension of Core 2 Glycans

The growth of Core 2 structures resembles that of Core 1 saccharides, but can occur both at the Gal and the GlcNAc residues. Brockhausen et al. showed that pig gastric mucosa contains β3-GlcNAc transferase activity, which converts Core 2 into GlcNAc(β1–3)Gal(β1–3)[GlcNAc(β1–6)]GalNAcα1–R, where R is the polypeptide backbone of either a mucin or antifreeze glycoprotein [24]. It is likely that this reaction is catalysed by the same enzyme that elongates Core 1 disaccharide.

Core 2 trisaccharide in unconjugated form was quantitatively galactosylated by incubation with UDP-Gal and β4GalT I from bovine milk (EC 2.4.1.90), which is commercially available at reasonable prices [20]. The resulting tetrasaccharide Gal(β1–4)GlcNAc(β1–6)[Gal(β1–3)]GalNAc was isolated in pure form by gel filtration on Bio-Gel P-2 and characterized by NMR. β4Galactosylation of Core 2 is catalyzed even more effectively by β4GalT-IV than by β4GalT-I [26].

The β1,3-GlcNAc transferase activity (iGnT) of human serum is known to transfer to Gal(β1–4)GlcNAc, Gal(β1–4)Glc, their simple glycosides as well as with Gal(β1–4)GlcNAc(β1–3)Gal(β1–4)Glc [27–29], but it works poorly with Gal(β1–3)GalNAc [28]. Put together, the acceptor-data on the β1,3-GlcNAc transferases of human serum and pig gastric mucosal microsomes seem to be distinct and complementary in their substrate specificities, such that the serum enzyme probably extends the 6-linked arm and the pig stomach mucosal enzyme extends the 3-linked arm of the tetrasaccharide Gal(β1–3)[Gal(β1–4)GlcNAc(β1–6)]GalNAc.

25.2.7 Extension of Core 3 and Core 4 Glycans

Extension of Core 3 and Core 4 glycans is known to occur [4, 30], but in vitro syntheses of the extended species have not been described.

25.3 Enzymatic *in vitro* Synthesis of Polylactosamine Backbones

Polylactosamines consist of *N*-acetyllactosamine (LacNAc) chains linked to sphingoglycolipids of neolacto series and to proteins; both *N*- and *O*-glycosidic forms of protein-bound polylactosamines are common. Polylactosamines have been known for a long time as the backbones of keratan sulfates and as undersulfated components in membranes of red blood cells and embryonic carcinoma cells. The polylactosamine backbones are expressed with and without distal decorations, which can consist of sulfate groups, single monosaccharides, e.g. α-Neu5Ac, α-Fuc, α-Gal, α-GalNAc, β-GalNAc, α-GlcNAc, and sulfo-3-β-GlcA (HNK-1) and remarkably numerous oligosaccharide determinants. Fucose and sulfate decorations are expressed commonly also along the polylactosamine chain. The multitude of different capping decorations on lactosamineglycans are responsible for a large number of different saccharide–lectin interactions on mammalian cell surfaces.

The polylactosamine backbones are bonded to: (i) the Man-cores of complex N-glycans, (ii) Cores 1–4 of O-glycans, and (iii) the lactose core of sphingoglycolipids. The same β3GlcNAc transferases are probably responsible for initiation and extension of some, but not all, polylactosamines. The present concept of polylactosamine biosynthesis after initialization involves: (i) the generation of primary backbones consisting of chains of LacNAc(β1–3')LacNAc type (LacNAc, Gal(β1–4)GlcNAc), (ii) generation of eventual backbone branches to yield arrays of the type LacNAc(β1–3')[LacNAc(β1–6')]LacNAc, and (iii) decoration reactions.

25.3.1 Enzymatic Synthesis of the Primary Chains of Blood Group i-Type

The β3GlcNAc transferase responsible for extension of terminal N-acetyllactosamine residues of free oligosaccharides was reported independently by three groups, in Novikoff ascites tumor cells [31] and in human serum [27, 28]. Yates and Watkins [27] succeeded in synthesizing 3.8 μmol purified [^{14}C]GlcNAc(β1–3)Gal(β1–4)-GlcNAc from 80 μmol LacNAc and 12 μmol UDP-[^{14}C]GlcNAc. They also synthesized 7.5 μmol pure [^{14}C]GlcNAc(β1–3)Gal(β1–4)Glc from 200 μmol Gal(β1–4)Glc and 30 μmol UDP-[^{14}C]GlcNAc. β-N-Acetylhexosaminidase treatment liberated the newly transferred [^{14}C]GlcNAc residues from both products, implying that β-linkages had been formed by the serum transferase. Methylation analysis revealed that the galactosyl groups of both products were 3-O-substituted with GlcNAc.

Simple glycosides of LacNAc and lactose were also good acceptors, as was Gal(β1–4)GlcNAc(β1–3)Gal(β1–4)Glc. Piller and Cartron [28] reported similar experiments. Their data show that Fuc(α1–2)Gal(β1–4)GlcNAc and Gal(β1–4)-[Fuc(α1–3)]GlcNAc are not acceptors for the serum β3-GlcNAc-transferase, and the disaccharide Gal(β1–3)GlcNAc is a poor acceptor. By contrast, the polylactosamines Gal(β1–4)GlcNAc(β1–3)Gal(β1–4)GlcNAc-O-Me and Gal(β1–4)-GlcNAc(β1–3)Gal(β1–4)GlcNAc(β1–3)Gal(β1–4)GlcNAc-O-Me were as good acceptors as Gal(β1–4)GlcNAc; the resulting products were not structurally characterized, however. Because linear polylactosamines are known as blood group i-determinants, the enzymes responsible for these reactions are called iGnT.

The serum iGnT transferred GlcNAc effectively also to desialylated fetuin and desialylated α$_1$-acid glycoprotein and to the pentasaccharides Gal(β1–4)-GlcNAc(β1–2)Man(α1–3)Man(β1–4)GlcNAc and Gal(β1–4)GlcNAc(β1–2)-Man(α1–6)Man(β1–4)GlcNAc, representing distinct branches of complex N-glycans [28] Accordingly, serum iGnT is able to transfer to terminal Gal(β1–4)GlcNAc residues of glycoprotein-linked complex N-glycans. By contrast, asialoglycophorin, which mostly contains Gal(β1–3)GalNAc bound to serine or threonine residues, was not an acceptor. The serum iGnT converts also GalNac(β1–4)GlcNAc to GlcNAc(β1–3)GalNAc(β1–4)GlcNAc [32]. The serum iGnT extends also β6-linked branches of polylactosamines, converting the hexasaccharide LacNAc(β1–3')[LacNAc(β1–6')]LacNAc into the octasaccharide GlcNAc(β1–3')LacNAc(β1–3')[GlcNAc(β1–3)' LacNAc(β1–6')]LacNAc [33].

iGnT has been purified from calf serum [34], and a human cDNA clone encoding a similar enzyme has been isolated by Sasaki et al. [35]. A transferase experiment with the crude recombinant enzyme and 500 nmol lacto-*N-neo*-tetraose gave 0.7 nmol [^{3}H]GlcNAc(β1–3)Gal(β1–4)GlcNA(β1–3)Gal(β1–4)Glc.

25.3.2 Distal Branching of i-Type Polylactosamine Backbones

The first branching reaction of polylactosamine backbones was described by Piller et al. [36] who found an appropriate β1,6-GlcNAc transferase activity in hog gastric mucosal microsomes. This activity catalyzed the reaction shown in eq. (1).

$$\text{GlcNAc}(\beta1-3)\text{Gal}(\beta1-4)\text{Glc} + \text{UDP-GlcNAc}$$
$$\rightarrow \text{GlcNAc}(\beta1-3)[\text{GlcNAc}(\beta1-6)]\text{Gal}(\beta1-4)\text{Glc} \qquad (1)$$

The product was identified by methylation, followed by methanolysis, O-acetylation and GLC–mass spectrometry of the derivatized monosaccharides. The GLC data showed that 1,2,4,6-tetra-*O*-methyl-3-*O*-acetylgalactose, which was present among the components obtained from the trisaccharide acceptor, had completely disappeared. It was replaced by α and β isomers of 1,2,4-tri-*O*-methyl-3,6-di-*O*-acetylgalactose. Compared with the acceptor, the amount of 1,3,4,6-tetra-*O*-methyl-2-deoxy-*N*-methylacetamidoglucose obtained from the tetrasaccharide product was twofold. Mass spectrometric analysis of the derivatized galactose obtained from the tetrasaccharide product confirmed that the new substituent was linked to carbon 6. The activity in hog gastric mucosal microsomes was unable to transfer to acceptors that terminated with a complete LacNAc unit at the non-reducing end. The microsomes transferred also to the glycolipid GlcNAc(β1–3)Gal(β1–4)Glcβ1-Ceramide. The primary experiment of Piller et al. (*loc. cit.*) was repeated by Seppo et al. [37], using GlcNAc(β1–3)Gal(β1–4)GlcNAc as the acceptor. Here, the tetrasaccharide GlcNAc(β1–3)[GlcNAc(β1–6)]Gal(β1–4)GlcNAc was formed, as was shown by partial acid hydrolysis of the product and chromatographic identification of all di- and trisaccharides of the hydrolysate.

More recently, Helin et al. [38] incubated the doubly labeled pentasaccharide GlcNAc(β1–3)[^{3}H]Gal(β1–4)GlcNAc(β1–3)[^{14}C]Gal(β1–4)GlcNAc with UDP-GlcNAc and the gastric mucosal microsomes, obtaining the hexasaccharide GlcNAc(β1–3)[GlcNAc(β1–6)][^{3}H]Gal(β1–4)GlcNAc(β1–3)[^{14}C]Gal(β1–4)GlcNAc. Here, product analysis was performed by partial acid hydrolysis. Paper chromatography of the hydrolysate gave [^{3}H]-labeled, but not [^{14}C]-labeled, glycans that contained the newly formed GlcNAc(β1–6)Gal bond. The diagnostic cleavage products included the disaccharide GlcNAc(β1–6)[^{3}H]Gal, the trisaccharides GlcNAc(β1–6)[^{3}H]Gal(β1–4)GlcNAc and GlcNAc(β1–3)[GlcNAc(β1–6)][^{3}H]Gal, and the tetrasaccharide GlcNAc(β1–3)[GlcNAc(β1–6)][^{3}H]Gal(β1–4)GlcNAc. In addition to the degradation data, positive-ion MALDI–TOF mass spectrometry of the original hexasaccharide product revealed a major peak of monoisotopic m/z 1177.1 that was assigned to $(M + N)^+$ of GlcNAc$_4$Gal$_2$ (calculated m/z 1177.4). The ^{1}H NMR spectrum of the hexasaccharide product also confirmed its structure. The spectrum revealed reporter group signals similar to those of the acceptor pentasaccharide, but also contained an additional, one-proton doublet at 4.592 ppm that was assigned H-1 of the β1,6-bonded GlcNAc. In summary, the data of Helin et al. show that only the penultimate galactose unit close to the non-reducing end of the pentasaccharide served as a primary acceptor site, despite the presence of an additional, internal galactose residue. We call the branching activity of the gastric mucosal microsomes as dIGnT to emphasize its distal site specificity and the formation of blood group I-type structures.

Interestingly, incubation of the hexasaccharide GlcNAc(β1–3)[GlcNAc(β1–6)]-Gal(β1–4)GlcNAc(β1–3)Gal(β1–4)GlcNAc with UDP–GlcNAc and gastric mucosal microsomes gave a small amount of the doubly branched heptasaccharide product GlcNAc(β1–3)[GlcNAc(β1–6)]Gal(β1–4)GlcNAc(β1–3)[GlcNAc(β1–6)]-Gal(β1–4)GlcNAc [38]. It was not clear whether an analogous, slow reaction at the mid-chain galactose had occurred also at the linear pentasaccharide GlcNAc(β1–3)Gal(β1–4)GlcNAc(β1–3)Gal(β1–4)GlcNAc at a rate that was below the detection limit. Alternatively, dIGnT can catalyze reactions of different rates at the mid-chain galactoses of different polylactosamine acceptors.

Quite recently, Yeh et al. [19] cloned the cDNA of C2GnT-M and expressed it in functionally active form. Besides the C2GnT activity, the recombinant enzyme had significant C4GnT activity and a small amount of dIGnT activity when tested using the acceptors GlcNAc(β1–3)GalNAcα1-para-nitrophenol and GlcNAc(β1–3)-Gal(β1–4)GlcNAc(β1–6)Man(α1–6)Manβ1-octyl, respectively. These data support the notion that the dIGnT-activity of hog gastric mucosal microsomes used in the experiments of Piller et al. [36], Seppo et al. [37], and Helin et al. [38] might have represented C2GnT of M-type.

Starting from GlcNAc, a polylactosamine consisting of seven N-acetyllactosamine units in a highly branched array was synthesized in our laboratory, by use of a multistep process involving the dIGnT branching reaction [39]. Besides reactions catalyzed by hog gastric mucosal microsomes at two distinct stages, the process consisted also of several reactions catalyzed by the chain extension enzymes iGnT and β4GalT I (see below). The tetradecameric product was synthesized in milligram amounts and subsequently decorated by: (i) four distal α1,3-bonded Gal residues, (ii) four distal β1,3-linked GlcNAc units [39], (iii) four distal α2,3-linked Neu5Ac residues, (iv) four peridistal α3-bonded Fuc units, and (v) a combination of four distal α2,3-linked Neu5Ac and four peridistal α3-bonded Fuc units [40]. The products were characterized by NMR spectroscopy and MALDI–TOF mass spectrometry, and were studied as putative inhibitors of murine sperm-to-egg adhesion and binding of lymphocytes to capillary endothelium of inflamed tissue.

A glycolipid from hog gastric mucosa contains a large, branched polylactosamine backbone [41] that is strikingly similar to the tetradecameric polylactosamine backbone of Seppo et al. [39].

25.3.3 Central Branching of i-Type Polylactosamine Backbones

Another type of polylactosamine branching was reported by Anne Leppänen et al. [42]. In their experiments, the doubly labeled linear pentasaccharide GlcNAc(β1–3)[^{3}H]Gal(β1–4)GlcNAc(β1–3)[^{14}C]Gal(β1–4)GlcNAc was converted into the branched hexasaccharide GlcNAc(β1–3)[^{3}H]Gal(β1–4)GlcNAc(β1–3)-[GlcNAc(β1–6)][^{14}C]Gal(β1–4)GlcNAc when incubated with UDP–GlcNAc and human serum. The product was identified by digestion with *endo*-β-galactosidase that cleaved the hexasaccharide completely as shown by paper chromatography. All of the ^{3}H radioactivity of the digest was found in the disaccharide fraction, co-chromatographing with the GlcNAc(β1–3)Gal marker. Nearly all of the [^{14}C] label in the digest was found in a tetrasaccharide, which was identified as GlcNAc(β1–3)[GlcNAc(β1–6)][^{14}C]Gal(β1–4)GlcNAc by partial acid hydrolysis and subsequent chromatography.

These data imply that the original hexasaccharide product was cleaved by *endo*-β-galactosidase only at the galactosidic bond of the [^{3}H]Gal located close to the non-reducing end. This was compatible with the postulated structure of the hexasaccharide product, the GlcNAc(β1–3)[GlcNAc(β1–6)]Gal(β1–4)GlcNAc determinant is known to resist the action of *endo*-β-galactosidase. Accordingly, the serum enzyme transferred solely to the [^{14}C]-labeled mid-chain Gal of the pentasaccharide acceptor. We call this activity cIGnT to emphasize the central site-specificity of its action. Gu et al. [43] have reported similar activity in rat serum.

Later experiments showed that human serum cIGnT transfers to position 6 of the internal Gal also in the tetrasaccharide Gal(β1–4)GlcNAc(β1–3)**Gal**(β1–4)GlcNAc [44]. The cIGnT of rat serum gave a similar product. Accordingly, this enzyme is not sensitive to the distal galactosylation status of the chain. It works with chains capped by many different groups (Leppänen et al., unpublished experiments), but it transfers effectively only to acceptors having at least one complete Gal(β1–4)GlcNAc unit bonded at position 3 to the acceptor Gal.

The structure of the pentasaccharide product generated by rat serum cIGnT was established firmly by 2D NMR experiments, for which a relatively large sample of the pure pentasaccharide had to be prepared [45]. The tetrasaccharide Gal(β1–4)GlcNAc(β1–3)Gal(β1–4)GlcNAc was first synthesized in amounts of a few micromoles, starting from LacNAc that was extended in a reaction catalyzed by the iGnT activity of human serum; the resulting GlcNAc(β1–3')LacNAc was purified and then β1,4-galactosylated by bovine milk β4GalT I. Finally, the branching reaction of the tetrasaccharide using rat serum cIGnT gave 2.7 μmol of the pure pentasaccharide product.

Enzymatic generation of multiple branches to a linear polylactosamine backbone was first described by Leppänen et al. [44]. In these experiments the linear hexasaccharide LacNAc(β1–3')LacNAc(β1–3')LacNAc was converted into the doubly branched octasaccharide LacNAc(β1–3')[GlcNAc(β1–6')]LacNAc(β1–3')-[GlcNAc(β1–6')]LacNAc by incubation with UDP-GlcNAc and cIGnT activity of rat serum. Analogous experiments of Salminen et al. [46] with the linear octasaccharide LacNAc(β1–3')LacNAc(β1–3')LacNAc(β1–3')LacNAc gave the triply branched undecasaccharide LacNAc(β1–3')[GlcNAc(β1–6')]LacNAc(β1–3')-[GlcNAc(β1–6')]LacNAc(β1–3')[GlcNAc(β1–6')]LacNAc, that was converted to the tetradecasaccharide LacNAc(β1–3')[LacNAc(β1–6')]LacNAc(β1–3')-[LacNAc(β1–6')]LacNAc(β1–3')[LacNAc(β1–6')]LacNAc by a treatment with UDP-Gal and β4-GalT I. Branched polylactosamine backbones of this type are expressed in human erythrocytes and embryonic carcinoma cells.

Recent experiments have led to identification of cIGnT in human embryonic carcinoma cells [47], cloning of the cDNA encoding the cIGnT of these cells [48], expression and purification of a truncated, functional form of this cIGnT [49], as well as purification of a cIGnT from hog intestinal mucosa [50]. The recombinant cIGnT transfers efficiently also to Gal(β1–4)GlcNAc(β1–3)Gal(β1–4)GlcNAc(β1–6)Man(α1–6)Manβ1-octyl, a mimic of the polylactosamine-bearing branch of N-glycans [19]. In due time, these advances might enable transfer of multiple branches to linear polylactosamines on a large scale.

25.3.4 β4-Galactosylation in Polylactosamine Backbones

The β4GalT I from bovine and human milk was recognized as the only enzyme of its kind for a while, but novel members of this family (β4GalT-II to β4GalT-VI) have been isolated recently (reviewed by Ujita et al. [26]). β4GalT-IV initiates polylactosamine synthesis in the 6-linked arm of Core 2 O-glycan more effectively than β4GalT-I [26], but the bovine milk β4GalT-I elongates primary polylactosamine chains efficiently in vitro [44, 51]. β4GalT-I initiates the growth of β1,6-linked GlcNAc branches of polylactosamines efficiently [52], and it also elongates β1,6-bonded GlcNAc(β1–3)Gal(β1–4)GlcNAc branches [53].

25.4 α3-Sialylation of N-Acetyllactosaminoglycans at the Terminal Gal

Distal α3-sialylation of branched polylactosamines was catalyzed successfully by the activity (ST3Gal IV?) present in human placental microsomes [40, 52]; both 3- and 6-linked short polylactosamine branches reacted well. More recently, large, tetravalent sialopolylactosamines have been synthesized in low micromolar amounts (L. Penttilä et al., unpublished experiments). Even long polylactosamine chains of blood group i-type are effectively sialylated by placental microsomes [54].

Several α3-sialyltransferases have been cloned, and ST3Gal III is commercially available in recombinant form. A particularly interesting report of Gilbert et al. [55] describe a fusion protein consisting of CMP-Neu5Ac synthetase and α2,3-sialyltransferase from *Neisseria meningitidis*. This polypeptide was able to catalyze the reactions shown in eqs (2) and (3).

$$CTP + Neu5Ac \rightarrow CMP\text{-}Neu5Ac + PP \tag{2}$$

$$CMP\text{-}Neu5Ac + Gal\text{–}OR \rightarrow Neu5Ac(\alpha 2\text{–}3)Gal\text{–}OR + CMP \tag{3}$$

In small-scale reactions the fusion protein worked well with glycosides of LacNAc and lactose, as well as the bi-antennary N-linked glycan, and used as donors N-acetylneuraminic acid, N-propionylneuraminic acid, and N-glycolylneuraminic acid.

The two tethered enzymes were recovered by a simple procedure in functionally pure form, free from contaminating enzyme activity that can hydrolyze sugar nucleotides or other components of the cofactor regeneration system. Accordingly, the fusion protein could be used in a sugar nucleotide cycle in which only catalytic amounts of expensive sugar nucleotides and transferase-inhibiting nucleoside phosphates are used; the former are continuously regenerated *in situ* from low cost precursors by appropriate enzyme reactions [56]. In this way the fusion protein was used to produce α2,3-sialyllactose on a 100-g scale starting from lactose, sialic acid, phosphoenolpyruvate, and catalytic amounts of ATP and CMP.

Taken together, the data of Gilbert et al. [55] show that production of the enzymes required for oligosaccharide synthesis might be simplified by fusing together carefully selected enzymes that catalyze neighboring steps in the synthetic processes.

25.5 α3-Fucosylation of Lactosamine Saccharides

Gal(β1–4)[Fuc(α1–3)]GlcNAc- and Neu5Ac(α2–3)Gal(β1–4)[Fuc(α1–3)]GlcNAc sequences, known as Lewis^x- and as sialyl Lewis^x-determinants, respectively, are common distal elements in lactosaminoglycans. Lewis^x (Lex) is an adhesive sugar that can bind to other epitopes of its kind in the presence of Ca^{2+} [57], and sialyl Lewis^x (sLex) is the prototype ligand of the E-, P- and L-selectins, participating in leukocyte–endothelium adhesion, which is a process of considerable biomedical importance [58, 59]. Also α3′-galactosylated and α2′-fucosylated LacNAc units at chain termini are acceptors for α3-fucosylation, and the products, i.e. Gal(α1–3)Gal(β1–4)-
[Fuc(α1–3)]GlcNAc and Fuc(α1–2)Gal(β1–4)[Fuc(α1–3)]GlcNAc, respectively, are also exciting cell adhesion molecules, participating in murine sperm-to-egg adhesion [60] and adhesion of *Helicobacter pylorii* to human stomach [61]. In man, at least five different transferases (FucT III to VII) catalyze α3-fucosylation. These reactions occur at distal LacNAc termini and also at mid-chain positions of polylactosamine extensions.

Enzyme-assisted in vitro synthesis of α3-fucosylated LacNAc-glycans was initially performed with naturally occurring transferases, often with enzymes from human milk. For instance, de Vries et al. [62] showed that partially purified human milk α3/4 transferases (now assumed to be a mixture of Fuc TIII and VI) reacted with all LacNAc units of LacNAc(β1–3′)LacNAc(β1–3′)LacNAcβ1–OR converting the acceptor into the triply fucosylated product Lex(β1–3′)Lex(β1–3′)Lexβ1–OR [Lex, Lewis^x, Gal(β1–4)[Fuc(α1–3)]GlcNAc]. More recently, Niemelä et al. [63] studied partial α3-fucosylation of the free hexasaccharide LacNAc(β1–3′)-LacNAc(β1–3′)LacNAc with a similar enzyme sample isolated from human milk. They found that among the three acceptor sites of the unconjugated hexasaccharide, the middle LacNAc reacted most rapidly and the non-reducing LacNAc most slowly. All individual isomers of the resulting monofucosyl- and difucosyl-derivatives were isolated in the pure form and characterized in these experiments. The ^{1}H NMR spectra of the monofucosyl hexasaccharides revealed distinct differences between the isomers.

Conversion of the trisaccharide Neu5Ac(α2–3′)LacNAc into the tetrasaccharide Neu5Ac(α2–3′)Le^x, was first performed using cloned human Fuc-TIII [64]. In a similar experiment, de Vries et al. [65] incubated 6.6 μmol Neu5Ac(α2–3′)LacNAc and 8.1 μmol GDP-[^{14}C]fucose for 96 h at 22 °C in 6 mL 50 mM sodium cacodylate buffer, pH 6.5 in the presence of 60 μmol MnCl$_2$ with 490 mU recombinant Fuc-TV immobilized via protein A on 0.6 mL IgG–Sepharose. The resulting mixture was applied to a Bio-Gel P-2 column and eluted with 500 mM ammonium acetate, pH 6.8. The sialyl Lewis^x tetrasaccharide, eluting as a sharp peak at a position corresponding to $0.53 \times V_{tot}$, contained 81% of the total [^{14}C]fucose present in the reaction mixture. It was identified as Neu5Ac(α2–3)Gal(β1–4)[Fuc(α1–3)]GlcNAc by 1D ^{1}H NMR spectroscopy.

Using Neu5Ac(α2–3′)LacNAc(β1–3′)LacNAc(β1–3′)LacNAc as the acceptor Niemelä et al. [54] showed that that the full length recombinant Fuc–TVII transfers primarily at the distal, sialylated LacNAc unit, converting the acceptor into a long chain [sialyl–Le^x]-glycan. By contrast, recombinant Fuc-TIV transferred fastest to the two other LacNAc residues of this acceptor. Using Fuc-TIV and Fuc-TVII in a concerted manner, the authors synthesized in vitro the triply fucosylated determinant Neu5Ac(α2–3′)Le^x(β1–3′)Le^x(β1–3′)Le^x, that is expressed on the P-selectin ligand PSGL-1 of HL-60 cells.

Even polylactosamines carrying α3-fucosyl units at specific sites along the chains can be synthesized with enzyme preparations isolated from human milk, although these enzymes per se transfer in a rather random fashion. Fucosylation performed early, before the polylactosamine chains have been fully developed, leads to glycans which contain the Lewis^x epitope close to the reducing end [66, 67]. Another possibility is to generate temporary GlcNAc branches to the linear i-type chains; the ensuing α3fucosylation reaction is prevented at the branch-bearing LacNAc units, but not elsewhere along the acceptor chain [68].

α3-Fucosyltransferase preparations isolated from human milk have also been used to convert two tetravalent [sialyl(α2–3′)LacNAc]-polylactosamines into the corresponding tetravalent [sialyl(α2–3′)Le^x]-glycans [40, 52]. The fucosylation transfer was restricted to the distal, sialylated LacNAc units, because all other LacNAc residues carry branches. This restriction is probably general; all human α3-fucosyltransferases seem to require the presence of a free hydroxyl group at position 6′ of the acceptor LacNAc units.

As first described, these syntheses yielded ca 50–100 nmol of the final products, sufficient for Stamper–Woodruff binding experiments between lymphocytes and capillary-wall endothelium on tissue slices, MALDI–TOF mass spectrometry, and one dimensional ^{1}H NMR spectroscopy. The tetravalent [sialyl–Le^x] glycans proved to be very potent inhibitors of L-selectin mediated adhesion of lymphocytes to capillary endothelium of rejecting organ transplants [40, 52, 69]. They also inhibited L-selectin-mediated adhesion of lymphocytes to peripheral lymph node high endothelial venules [70]. As L-selectin is expressed on lymphocyte surfaces in clustered arrays on tips of microvilli, we believe that the oligovalent [sialyl Le^x] glycans might bind particularly tightly to lymphocyte L-selectin by crosslinking several monovalent L-selectin molecules in the cellular clusters. More recently, the syntheses of the two tetravalent [sialyl-Le^x] glycans have been scaled up 20–50 fold, to generate material for experiments involving in vivo inhibition of leukocyte rolling in mouse cremaster muscle capillaries (L. Penttilä et al., unpublished work).

Multifucosylated and terminally α3-sialylated polylactosamine chains of i-type, are also effective E-selectin ligands [71, 72]. Using an enzyme sample purified from human milk, Toppila et al. [53] transferred four α3-fucose units to a bi-antennary array of five LacNAc units that was α3-sialylated at both long antennal. The resulting tetrafucosyl bisialosaccharide was a very potent inhibitor of L-selectin-mediated lymphocyte–endothelium adhesion, both at capillaries of rejecting heart and at peripheral lymph nodes. Accordingly, the determinant Neu5Ac(α2–3′)Lex(β1–3′)Lex is recognized better than Neu5Ac(α2–3′)Lex also by L-selectin.

Of particular interest was that in the experiments of Toppila et al. (*loc. cit.*), the tetrafucosyl bisialosaccharide inhibitor was much more potent in its action at the rejecting heart than at the lymph node. This specificity of action suggests that by use of saccharide inhibitors of cell adhesion it might be possible to inhibit lymphocyte recruitment at inflamed tissues without endangering the normal lymphocyte patrolling that requires extravasation at lymph nodes. The molecular mechanisms responsible for this organ selectivity of the saccharide inhibitor remain unknown at present; one possibility is that endothelium of inflamed tissues carries less adhesion-promoting saccharides, or less active saccharides, than the endothelium of lymph nodes.

References

1. Laine, R. A (1994) Glycobiology, 4, 759–67.
2. Leloir, L. F. (1971) Science 172, 1299–303.
3. Sears, P. and Wong, C.-H. (1998) Cellular and Molecular Life Sciences 54, 223–252.
4. Van den Steen, P., Rudd., P., Dwek, R. A., and Opdenakker, G. (1998) Crit. Rev. Biochem. Mol. Biol., 33, 151–208.
5. Jentoft, N (1990) Trends Biochem. Sci. 15, 291–94.
6. Elhammer, A. and Kornfeld, S. (1986) J. Biol. Chem. 261, 5249–55.
7. Wandall H. H., Hassan H., Mirgorodskaya E., Kristensen A. K., Roepstorff P., Bennett E. P., Nielsen P. A., Hollingsworth M. A., Burchell J., Taylor-Papadimitriou J., Clausen H. (1997) J. Biol. Chem. 272, 23503–14.
8. Hagen, F. K. and Nehrke, K. (1998) J. Biol. Chem. 273, 8268–77.
9. Hagen, K. B., Hagen, F. K., Balys, M. M., Beres, T. M., Van Wuyckhuyse, B., Tabak, L. A. (1998) J. Biol. Chem. 273, 27749–54.
10. Hansen, J. E., Lund, O., and Brunak, S. (1997) Nucleic Acids Res. 25, 278–82.
11. Schachter H., McGuire, E. J. and Roseman, S. (1971) J. Biol. Chem. 246, 5321–28.
12. Mendicino, J., Sivakami, S., Davila, M. and Chandrasekaran, E. V. (1982) J. Biol. Chem. 257, 3987–94.
13. Cheng, P.-W and Bona, S. J. (1982) J. Biol. Chem. 257, 6251–58.
14. Amado M., Almeida R., Carneiro F., Levery S. B., Holmes E. H., Nomoto M., Hollingsworth M. A., Hassan H., Schwientek T., Nielsen P. A., Bennett E. P., Clausen H. (1998) J. Biol. Chem. 273, 12770–08.
15. Williams D. and Schachter, H. (1980) J. Biol. Chem. 255, 11247–52.
16. Williams D., Longmore G. D., Matta, K. L. and Schachter, H. (1980) J. Biol. Chem. 255, 11253–61.
17. Kuhns, W., Ruth, V., Paulsen, H., Matta, K. L., Baker, M. A., Barner, M., Granovsky, M., Brockhausen, I. (1993) Glycoconjugate J. 10, 381–394.
18. Bierhuizen, M. F. A. and Fukuda, M. (1992) Proc. Natl. Acad. Sci. USA, 89, 9326–30.

19. Yeh, J-C., Ong, E. and Fukuda, M. (1999) J. Biol. Chem. 274, 3215–21.
20. Maaheimo, H., Penttilä L. and Renkonen, O. (1994) FEBS Lett. 349, 55–59.
21. Li, F., Wilkins, P. P., Crawley, S., Weinstein, J., Cummings, R. D. and McEver, R. P. (1996) J. Biol. Chem. 271, 3255–64.
22. Brockhausen I., Matta, K. L., Orr, J., and Schacter H. (1985) Biochemistry, 24, 1866–74.
23. Vavasseur, F., Yang, J.-M. Dole, K., Paulsen, H. and Brockhausen, I. (1995) Glycobiology, 5, 351–57.
24. Brockhausen, I., Orr, J. and Schachter, H. (1984) Can. J. Biochem. Cell Biol. 62, 1081–90.
25. Brockhausen, I., Williams, D., Matta, K. L., Orr, J. and Schachter, H. (1983) Can. J. Biochem. Cell Biol. 61, 1322–33.
26. Ujita, M., McAuliffe, J., Schwientek, T., Almeida, R., Hindsgaul, O., Clausen, H. and Fukuda, M. (1998) J. Biol. Chem. 273, 34843–49.
27. Yates, A. D. and Watkins, W. M. (1983) Carbohydrate Res. 120, 251–268.
28. Piller, F. and Cartron, J.-P. (1983) J. Biol. Chem. 258, 12993–99.
29. Hosomi, O., Takeya, A, and Kogure, T. (1984) J. Biochem. 95, 1655–59.
30. Podolsky D. K. (1985) J. Biol. Chem. 260, 8262–71.
31. Van den Eijnden, D. H. Winterwerp, H., Smeeman, P., and Schiphorst, W. E. C. M. (1983) J. Biol. Chem. 258, 3435–37.
32. Salo, H., Niemelï, R., Ilves, K., Aitio, O. and Renkonen, O (1998) XIXth International Carbohydrate Symposium, San Diego, Abstract BP 255.
33. Vilkman, A., Niemelä, R., Penttilä, L., Helin, J., Leppänen, A., Seppo, A., Maaheimo, H., Lusa, S., and Renkonen, O. (1992) Carbohydrate Res. 226, 155–74.
34. Kawashima, H., Yamamoto, K., Osawa, T. and Irimura, T. (1993) J. Biol. Chem. 268, 27118–26.
35. Sasaki, K., Kurata-Miura K., Ujita, M., Angata, K., Nakagawa, S., Sekine, S., Nishi, T. and Fukuda, M. (1997) Proc. Natl. Acad. Sci. USA 94, 14294–99
36. Piller, F., Cartron, J-P., Maranduba, A., Veyrieres, A., Leroy, Y., Fournet, B. (1984) J. Biol. Chem. 259, 13385–90.
37. Seppo, A. Penttilä, L., Makkonen, A., Leppänen, A., Niemelä, R., Jäntti, J., Helin, J., and Renkonen, O. (1990) Biochem. Cell Biol. 68, 44–53.
38. Helin, J., Penttilä, L., Leppänen, A., Maaheimo, H., Laauri, S. Costello, C. E., and Renkonen, O. (1997) FEBS Lett. 412, 637–642.
39. Seppo, A., Penttilä, L., Niemelä, R., Maaheimo, H., Renkonen, O. and Keane, A. (1995) Biochemistry 34, 4655–61.
40. Seppo A., Turunen, J.-P. Penttilä, L., Keane, A. Renkonen, O. and Renkonen, R. (1996) Glycobiology, 6, 65–71.
41. Slomiany, A. and Slomiany, B. L. (1980) Biochem. Biophys. Res. Commun. 93, 770–775
42. Leppänen A., Penttilä, L., Niemelä, R., Helin, J., Seppo, A., Lusa, S., and Renkonen, O. (1991) Biochemistry, 30, 9287–96.
43. Gu, J., Nishikawa, A., Fujii, S., Gasa, S. and Taniguchi, N. (1992) J. Biol. Chem. 267, 2994–99.
44. Leppänen, A., Salminen, H., Zhu, Y., Maaheimo, H., Helin, J., Costello, C. E. and Renkonen, O. (1997) Biochemistry, 36, 7026–36.
45. Maaheimo, H., Räbinä, J. and Renkonen, O. (1997) Carbohydr. Res. 297, 145–51.
46. Salminen, H., Ahokas, K., Niemelä, R., Penttilä, L., Maaheimo, H, Helin, J., Costello, C. E. and Renkonen, O. (1997) FEBS Lett. 419, 220–26.
47. Leppänen, A., Zhu, Y., Maaheimo, H., Helin J., Lehtonen, E., Renkonen, O. (1998) J. Biol. Chem. 273, 17399–405.
48. Bierhuizen, M. F. A. Mattei, M.-G. and Fukuda. M. (1993) Genes Dev. 7, 468–78.
49. Mattila, P., Salminen, H., Hirvas, L., Niittymäki, J., Salo, H., Niemelä, R., Fukuda, M., Renkonen, O. and Renkonen, R. (1998) J. Biol. Chem. 273, 27633–39.
50. Sakamoto, Y. Taguchi, T. Tano Y., Ogawa, T., Leppänen, A., Kinnunen, M., Aitio, O., Parmanne, P., Renkonen, O. and Taniguchi, N. (1998) J. Biol. Chem. 273, 27625–32.
51. Helin, J., Maaheimo, H., Seppo, A., Keane, A., Renkonen, O. (1995) Carbohydrate Res. 266, 191–209.
52. Renkonen, O, Toppila, S., Penttilä, L., Salminen, H., Helin, J. Maaheimo, H., Costello, C. E., Turunen, J. P. and Renkonen, R. (1997) Glycobiology 7, 453–61.

53. Toppila, S., Renkonen, R. Penttilä, L., Natunen, J., Salminen, H., Helin, J., Maaheimo, H. and Renkonen O. (1999) Eur. J. Biochem. 261, 208–215.
54. Niemelä R., Natunen, J., Majuri, M. L., Maaheimo, H., Helin, J., Lowe, J. B., Renkonen, O., and Renkonen, R (1998) J. Biol. Chem. 273, 4021–26.
55. Gilbert, M., Bayer R., Cunningham, A.-M., Defrees, S., Gao, Y., Watson, D. C., Young, N. M., and Wakarchuk, W. W. (1998) Nature Biotechnology, 16, 769–72.
56. Ichikawa, Y., Shen, G.-J., and Wong. C.-H. (1991) J. Am. Chem. Soc. 113, 4698–700.
57. Eggens, I., Fenderson, B. A. Toyokuni, T., Dean, B., Stroud, M., and Hakomori, S. (1989) J. Biol. Chem. 264, 9476–84.
58. Lasky, L. A. (1995) Annu. Rev. Biochem. 64, 113–39.
59. Springer, T. A. (1995) Annu. Rev. Physiol. 57, 827–72.
60. Johnston, D. S., Wright, W., Shaper, J. H., Hokke, C. H., van den Eijnden, D. H., Joziasse, D. H. (1998) J. Biol. Chem. 273, 1888–95.
61. Boren, T., Falk, P., Roth K. A., Larson G., and Normark, S. (1993) Science, 262, 1892–95.
62. de Vries, T., Norberg, T., Lönn, H. and van den Eijnden, D. H. (1993) Eur. J. Biochem. 216, 769–77.
63. Niemelä, R. Natunen, J. Penttilä, L., Salminen, H., Helin, J., Maaheimo, H. Costello, C. E. and Renkonen O, (1999) Glycobiology, 9, 517–526.
64. Dumas, D. P., Ichikawa, Y., Wong, C. H. Lowe, J. B. and Nair, R. P. (1991) Bioorg. Med. Chem. Lett. 8, 425–28.
65. de Vries, T. van den Eijnden, D. H., Schultz, J. and O'Neill, R. (1993) FEBS Lett., 330, 243–48.
66. Kashem., M. A. Wlasichuk, K. B., Gregson, J. M. and Venot, A. P. (1993) Carbohydrate Res. 250, 129–44.
67. Räbinä, J., Natunen, J., Niemelä, R., Salminen, H., Ilves K., Aitio, O., Maaheimo, H., Helin, J. and Renkonen, O. (1998) Carbohydrate Res. 305, 491–99.
68. Niemelä, R., Räbinä, J., Leppänen, A., Maaheimo, H Costello, C. E. and Renkonen, O. (1995) Carbohydrate Res. 279, 331–38.
69. Turunen J. P., Majuri, M. L., Seppo, A., Seppo, A., Tiisala, S., Paavonen, T., Miyasaka, M., Lemström, K., Penttilä, L., Renkonen, O. and Renkonen, R. (1995) J. Exp. Med. 182, 1133–42.
70. Toppila, S., Lauronen, J., Mattila, P., Turunen, J. P., Penttilä, L., Paavonen, T., Renkonen, O. and Renkonen, R. (1997) Eur. J. Immunol. 27, 1360–65.
71. Patel, T. P., Goelz, S. P., Lobb, R. R. and Parekh, R. B. (1994) Biochemistry, 33, 14815–24.
72. Stroud, M. R., Handa, K., Dalyan. M. E. K., Ito, K., Levery, S. B., Hakomori, S., Reinhold, B. B., and Reinhold V. N. (1996) Biochemistry, 35, 770–78.

26 Recycling of Sugar Nucleotides in Enzymatic Glycosylation

Kathryn M. Koeller and Chi-Huey Wong

26.1 Introduction

The sugar nucleotide-dependent glycosyltransferases are effective catalysts for the enzymatic synthesis of oligosaccharides, glycopeptides, and glycoproteins [1]. These enzymes catalyze the formation of glycosidic linkages in a stereo- and regiospecific manner [2]. The use of glycosyltransferases therefore largely eliminates the extensive protecting group manipulations required in the chemical synthesis of carbohydrates. Although in recent years, many of the glycosyltransferases and their sugar nucleotide substrates have become commercially available, two major drawbacks remain to utilizing glycosyltransferases for synthetic purposes. First and foremost, nucleoside diphosphates (NDPs) generated as products during transferase reactions inhibit the enzyme [3]. For micro-scale reactions this problem can be solved by addition of a phosphatase to break down the NDP product [4]. For preparative scale synthesis, however, this solution does not overcome the expense of stoichiometric quantities of sugar nucleotides.

To circumvent these barriers, methods for recycling sugar nucleotides in enzymatic glycosylation have been developed [5]. These multi-enzyme systems require that sugar nucleotides be present in only catalytic quantities, thereby eliminating product inhibition by NDPs and reducing expense. Regeneration strategies have been employed in multi-gram syntheses of complex carbohydrates in high yields. This chapter reviews the sugar nucleotide recycling systems utilized to date for carbohydrate synthesis and highlights related advances in the field.

26.2 Glycosyltransferases of the Leloir Pathway and their Sugar Nucleotide Substrates

A considerable number of eukaryotic glycosyltransferases have been cloned to date, each with characteristic linkage specificity and substrate preference [1b]. Leloir

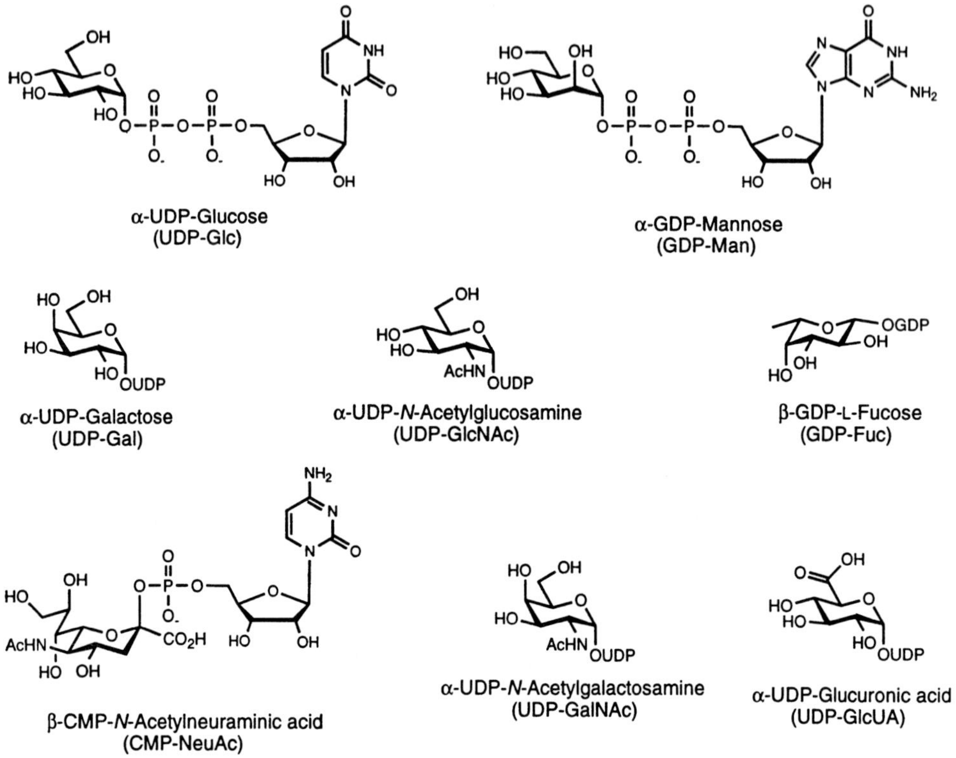

Figure 1. Sugar nucleotide substrates for glycosyltransferases of the Leloir pathway.

pathway glycosyltransferases utilize sugar nucleotides as donor substrates, and are responsible for the synthesis of most cell-surface glycoforms in mammalian systems [6]. It is remarkable that such a broad range of enzymes has converged on eight general sugar nucleotides for use as glycosyl donor substrates (Figure 1).

Glucosyl- and galactosyltransferases employ substrates activated with uridine diphosphate as the anomeric leaving group (α-UDP–Glc, α-UDP–GlcNAc, α-UDP–GlcUA, α-UDP–Gal, α-UDP–GalNAc), whereas fucosyl- and mannosyltransferases utilize guanosine diphosphate (β-GDP–Fuc, α-GDP–Man). Sialyltransferases are unique in that the glycosyl donor is activated by cytidine *mono*phosphate (β-CMP–NeuAc). Preparative-scale syntheses of relevant sugar nucleotides have been developed previously [7].

Sugar nucleotide recycling systems attempt to take advantage of the efficiency of biosynthetic pathways in enzymatic synthesis. In biological systems, simple monosaccharides are the precursors to sugar nucleotides (Figure 2) [6b].

In NDP–sugar biosynthesis an initial glycosyl phosphate is formed through kinase-mediated phosphorylation of a monosaccharide. Reaction of the glycosyl phosphate with a nucleoside triphosphate (NTP) is then catalyzed by the corre-

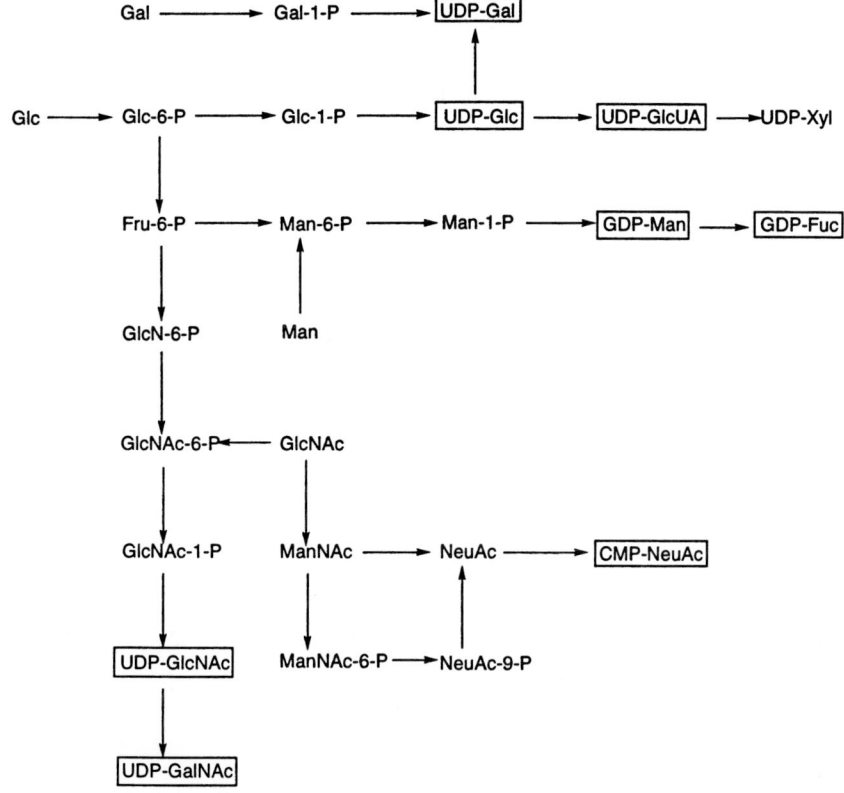

Figure 2. Biosynthetic pathways to sugar nucleotides from simple monosaccharides.

sponding NDP–sugar pyrophosphorylase (eq. (1)). This results in the production of the activated NDP–sugar. In contrast, the nucleoside monophosphate (NMP)–sugar CMP–NeuAc is formed by direct condensation of NeuAc with CTP, catalyzed by the enzyme CMP–NeuAc synthetase (eq. (2)).

$$\text{Sugar-1-phosphate} + \text{NTP} \rightarrow \text{NDP–sugar} + \text{PP}_i \qquad (1)$$

$$\text{NeuAc} + \text{CTP} \rightarrow \text{CMP–NeuAc} + \text{PP}_i \qquad (2)$$

26.3 Design of Regeneration Systems

The similarities between glycosyltransferase reactions that employ NDP–sugars as substrates enable the construction of a nearly universal NDP–sugar recycling scheme (Scheme 1). In theory, the NDP that is released by the transferase can be

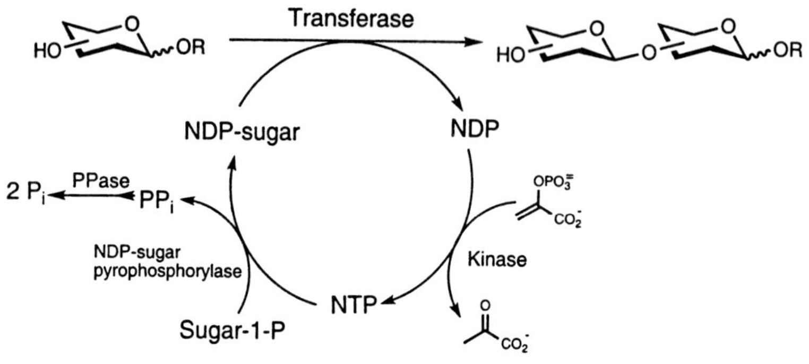

Scheme 1. A general theoretical recycling system for NDP–sugars.

converted into the corresponding NDP–sugar in a two-step enzymatic process. Such a method would involve kinase-mediated phosphorylation of the NDP to convert it to an NTP. Condensation of the NTP with a glycosyl phosphate would then afford the activated sugar nucleotide.

The regeneration of CMP–NeuAc is slightly more complicated, in that the product released by the transferase is an NMP rather than an NDP. Thus, an additional phosphorylation step and, likewise, an additional kinase are required before condensation of NeuAc with CTP can occur.

26.4 Practical Regeneration Systems

In practice, the availability of the biosynthetic enzymes and substrates is often a limiting factor in the design of a sugar nucleotide recycling system. This generally makes the overall regeneration strategy more complex and lengthier than the two-enzyme theoretical scheme. The following sections describe recycling systems for sugar nucleotides that have found application in carbohydrate synthesis.

26.4.1 UDP–Galactose

In a system where UDP–Gal regeneration is the goal, the inclusion of a one-step conversion of UTP to UDP–Gal using the enzyme UDP–Gal pyrophosphorylase would be ideal. Unfortunately, this enzyme is not commercially available, nor is it easy to prepare from biological sources [8]. Luckily, an alternative biosynthetic pathway to UDP–Gal exists. In nature, UDP–Glc is a central intermediate for the biosynthesis of UDP–Gal and UDP–GlcUA, as well as other sugar nucleotides. The enzyme UDP–Glc pyrophosphorylase (UDPGP; EC 2.7.7.9) required for the conversion of UTP to UDP–Glc can be purchased. The UDP–Gal 4-epimerase

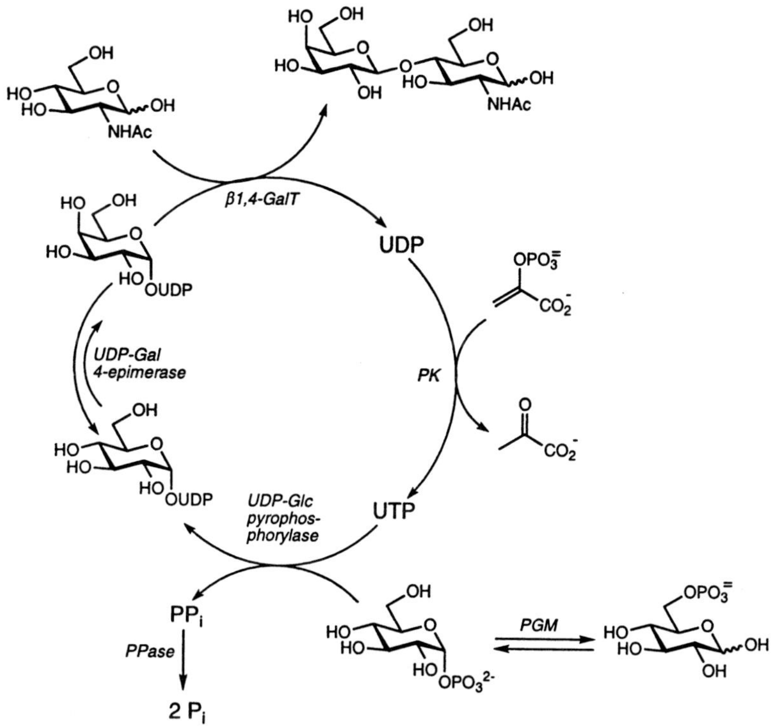

Scheme 2. Recycling of UDP–Gal employing UDP–Gal 4-epimerase.

(UDPGE; EC 5.1.3.2) needed for the subsequent conversion of UDP–Glc to UDP–Gal is also readily available.

A system employing these concepts was the first sugar nucleotide recycling method described in the literature. Wong et al. used this scheme to synthesize multi-gram quantities of *N*-acetyllactosamine (LacNAc) from *N*-acetylglucosamine (Scheme 2) [9]. This method has several salient features worth discussion that are not altogether obvious. For example, in the UDPGE-catalyzed reaction the equilibrium lies in the direction of the undesired product UDP–Glc. Continuous removal of UDP–Gal by β1,4-galactosyltransferase (β1,4-GalT; EC 2.4.1.22) drives the reaction in the direction of UDP–Gal production, however. The UDPGP substrate glucose-1-phosphate (Glc-1-P) is provided by the action of phosphoglucomutase (PGM; EC 5.4.2.2) on Glc-6-P, which is more stable and more readily available than Glc-1-P. Occasionally hexokinase (EC 2.7.1.1) can be employed to provide Glc-6-P from glucose. Finally, pyrophosphatase (PPase; EC 3.6.1.1) functions to remove the inorganic pyrophosphate product generated in the UDPGP reaction, thereby driving it forward.

Other groups have used routes based on this initial concept for the synthesis of oligosaccharides [10]. A similar recycling system was reported by Auge et al. for

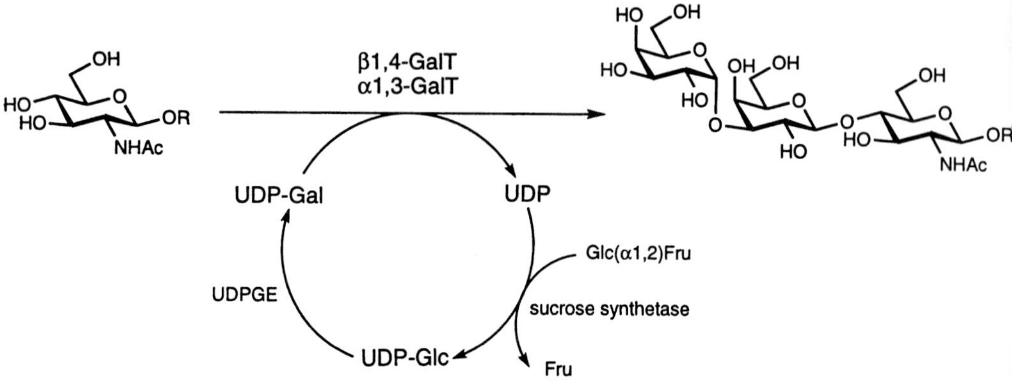

Scheme 3. Recycling of UDP–Gal employing sucrose synthetase for the one-pot synthesis of the α-gal epitope.

complex branched penta- and heptasaccharide synthesis [11]. Thiem and Wiemann have also applied this strategy to the recycling of UDP–2-deoxy-Gal for the synthesis of 2′-deoxy-LacNAc [12].

The UDP–Gal epimerase system was also one of the early methods to employ pyruvate kinase (PK; EC 2.7.1.40) for the conversion of UDP to UTP, with the simultaneous formation of pyruvate from phospho(enol)pyruvate (PEP). Previous kinase systems employed were either more expensive (nucleoside diphosphate kinase(EC 2.7.4.6)/ATP) or were thermodynamically less favorable (acetate kinase(EC 2.7.2.1)/acetyl phosphate). Another potentially useful kinase system has recently been described. Noguchi and Shiba have utilized polyphosphate kinase (PPK) for the formation of UTP from UDP in a UDP–Gal recycling system [13]. Previously, PPK was shown to catalyze the reversible transfer of phosphate from ATP to ADP. Recently, it has been discovered that PPK will also accept other NDP and NTP substrates. For application to recycling systems, poly(phosphate) is more affordable than PEP as a phosphoryl donor. Synthetically, replacement of PK with PPK in the UDPGE-based recycling system efficiently provided LacNAc from GlcNAc on a multi-gram scale.

The epimerase-based recycling system for UDP–Gal has been modified to contain the enzyme sucrose synthetase (EC 2.4.2.13) (Scheme 3) [14]. This enzyme catalyzes the formation of sucrose (Glc-α1,2-Fru) from fructose with simultaneous conversion of UDP–Glc to UDP. The readily reversible nature of the synthetase reaction enabled the authors to employ it in the production of UDP–Glc. As previously, removal of UDP–Glc by UDPGE provided the desired forward driving force in this enzymatic cascade. Initially, this strategy was utilized for the production of LacNAc from GlcNAc. Hokke et al. then expanded this methodology to synthesis of the α-gal trisaccharide epitope, employing both α1,3-GalT and β1,4-GalT in a one-pot reaction [15]. The success of this strategy relied on the fact that only the β1,4-GalT accepts the non-reducing terminal GlcNAc residue as a

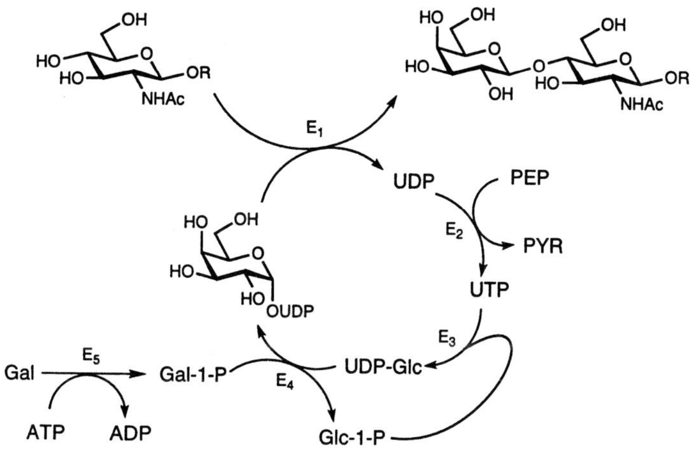

Scheme 4. Recycling of UDP–Gal employing Gal-1-P uridyltransferase for the synthesis of LacNAc.

substrate. Subsequently, the Wang group also made use of a one-pot, two-GalT system for the synthesis of the α-gal pentasaccharide epitope [16].

In an alternative strategy for the regeneration of UDP–Gal, Wong et al. employed Gal-1-P-uridyltransferase (Gal-1-P UT; EC 2.7.7.12) for the conversion of Gal-1-P to UDP–Gal (Scheme 4) [17]. This conversion is coupled to the formation of Glc-1-P from UDP–Glc by use of UDPGP. An attractive feature of this method is the utilization of galactose as starting material. This economical approach is possible through use of galactokinase (EC 2.7.1.6), an ATP-dependent enzyme that produces the galactosyl phosphate. LacNAc and several LacNAc derivatives have been synthesized using this recycling scheme.

The Gal-1-P UT regeneration system has also been used in syntheses of glycopeptides and homogeneous glycoproteins. This UDP–Gal recycling system was combined with a subtilisin-based peptide synthesis to afford several disaccharide-linked glycopeptides [18]. Furthermore, in a study of glycoprotein remodeling, the Gal-1-P UT method was employed to append galactose on to an *N*-linked GlcNAc on the surface of RNase B (Scheme 5) [19]. Further elaboration to the sialyl Lewis^x (sLe^x) tetrasaccharide was then performed with α2,3-sialyltransferase (α2,3-SiaT) and α1,3-fucosyltransferase (α1,3-FucT). Thus, recycling systems have shown their utility in simple saccharide synthesis and in the construction of complex glycopeptides and glycoproteins.

26.4.2 Other UDP–Sugars

Although UDP–Glc has great utility as an intermediate in the regeneration of other sugar nucleotides, it also occasionally functions as the desired glycosyl donor.

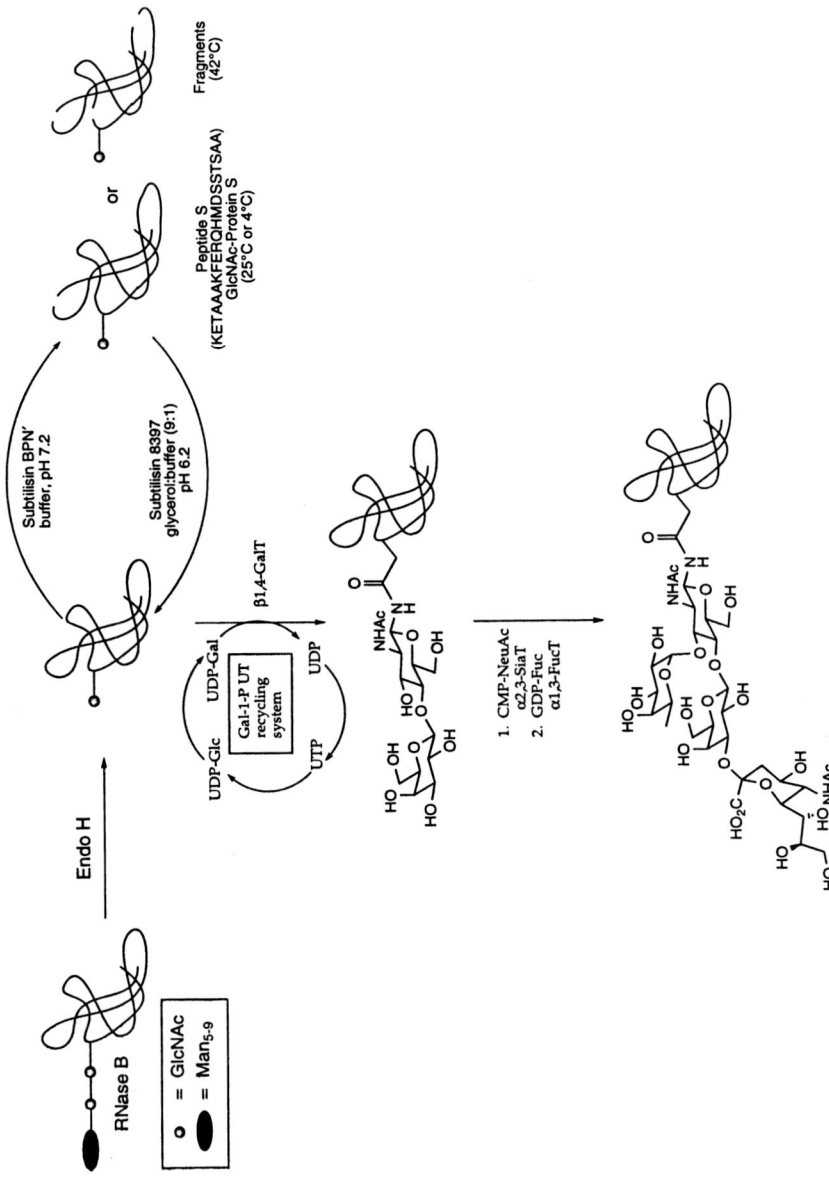

Scheme 5. Use of a UDP–Gal recycling system in glycoform remodeling of RNase B.

Haynie and Whitesides have developed a regeneration scheme for UDP–Glc in the syntheses of sucrose and trehalose [20]. As previously, the sucrose synthetase conversion of UDP to UDP–Glc is coupled to the transfer of Glc from sucrose. The authors obtained the synthetase in partially purified form from wheat germ. UDP–Glc is, furthermore, an intermediate in the biosynthesis of UDP–GlcUA, a donor substrate for β-glucuronosyltransferases (EC 2.4.1.17) [21]. For the recycling of UDP–GlcUA, the enzyme UDP–Glc dehydrogenase (EC 1.1.1.22) is required for the oxidation of the UDP–Glc 6-hydroxyl to the 6-carboxylic acid. This conversion is coupled with the reduction of 2 NAD^+ to 2 NADH. Through this recycling system it was shown that GlcUA could be transferred to various phenolic acceptors. Instead of individually immobilized enzymes, Gygax et al. utilized a crude liver homogenate containing all the enzymes necessary for the regeneration scheme.

Regenerations of UDP–GlcNAc and UDP–GlcUA have been successfully combined in the synthesis of a hyaluronic acid polymer of ~1500 sugar residues (Scheme 6) [22]. In the presence of Glc-1-P and GlcNAc-1-P, both UDP–GlcUA and UDP–GlcNAc can be recycled. The presence of UDP–Glc dehydrogenase and lactate dehydrogenase are also required for the oxidation of UDP–Glc to UDP–GlcUA. For this study, the hyaluronic acid synthase and most other enzymes were commercially available, although the UDP–GlcNAc pyrophosphorylase was obtained by overexpression in *E. coli*.

A recycling system for UDP–GlcNAc has been reported for use with β1,6-GlcNAc-transferase (EC 2.4.1.148) [23]. In this study, the authors attempted to use crude yeast cell extracts as the source of UDP–GlcNAc pyrophosphorylase (EC 2.7.7.23). Because of problems with contaminating phosphatases, however, this method was unsuccessful. Stepwise synthesis with whole yeast cells eventually provided the desired trisaccharide product.

26.4.3 CMP–NeuAc

The regeneration system for CMP–NeuAc is more complicated than that for NDP-sugars (Scheme 7) [24]. An additional phosphorylation step must be incorporated, because CMP, a nucleoside monophosphate, is released after reaction with the sialyltransferase. For recycling purposes, nucleoside monophosphate kinase (NMK; EC 2.7.4.4) or myokinase (MK; EC 2.7.4.3) is added for the conversion of CMP to CDP. In this reaction, the phosphoryl donor is ATP. Subsequently, both CDP and ADP must be re-phosphorylated to CTP and ATP, respectively. Thus, for regeneration of CMP–NeuAc, an additional kinase and two equivalents of PEP are required. The condensation of NeuAc with CTP is catalyzed by CMP–NeuAc synthetase (EC 2.7.7.43). This system was used for the large-scale synthesis of 6′-sialyl-LacNAc(6′-SLN) from LacNAc catalyzed by α2,6-SiaT (EC 2.7.7.43) in 97% yield.

A recent modification of this recycling system was in the expression of an α2,3-SiaT/CMP–NeuAc synthetase fusion protein [25]. This construct was more soluble than α2,3-SiaT alone, which is a poorly soluble transmembrane protein. The fusion

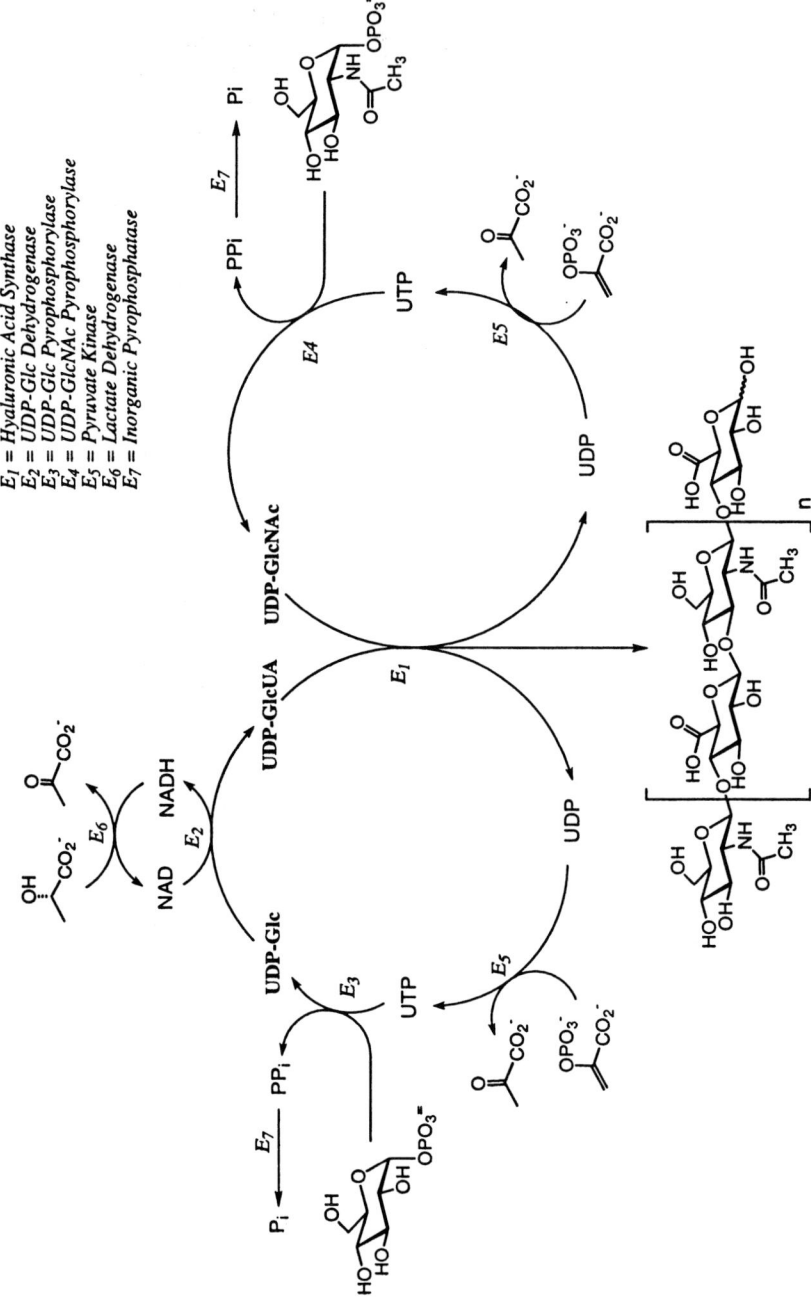

Scheme 6. Recycling of UDP–GlcUA and UDP–GlcNAc for the synthesis of a hyaluronic acid polymer.

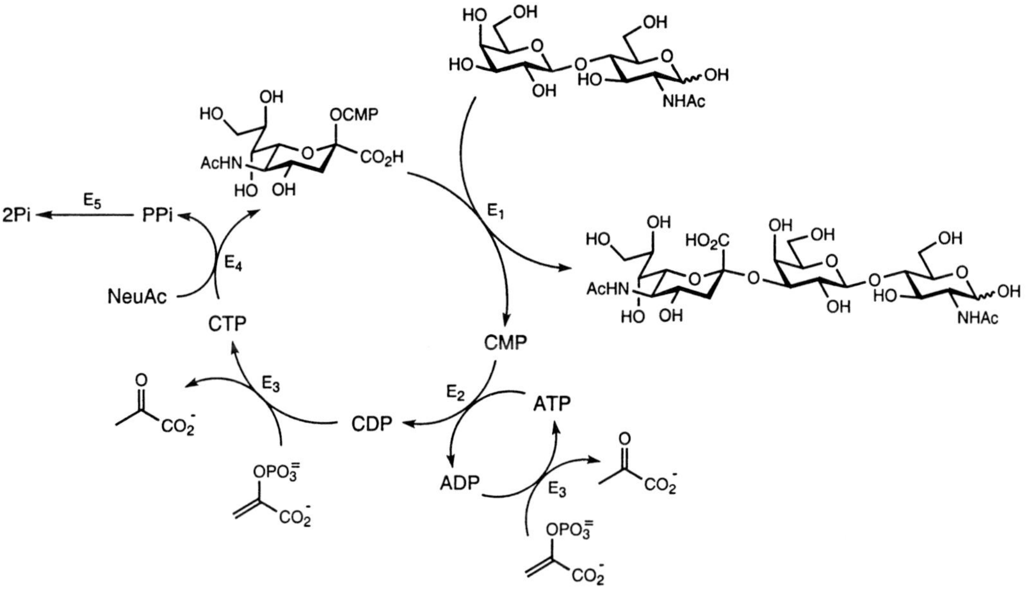

Scheme 7. Recycling of CMP–NeuAc for the synthesis of 3′-sialyl-LacNAc. E_1 = α2,3-sialyltransferase; E_2 = nucleoside monophosphate kinase; E_3 = pyruvate kinase; E_4 = CMP–NeuAc synthetase; E_5 = pyrophosphatase.

protein functioned efficiently at pH 7.5, even though the optimum pH for α2,3-SiaT (pH 6.0) and CMP–NeuAc synthetase (pH 8.5) differ significantly. This heterodimer was utilized in a 100-gram synthesis of 3′-sialyllactose.

In another study, the CMP–NeuAc recycling system was coupled to UDP–Gal regeneration for a one-pot synthesis of 6′-SLN (Scheme 8) [26]. In the strategy reported by Ichikawa et al., NeuAc is generated from the reaction of ManNAc with pyruvate, catalyzed by NeuAc aldolase (EC 4.1.3.3). In turn, the pyruvate generated by the UDP–Gal and CMP–NeuAc recycling systems serves as a substrate for NeuAc aldolase. Remarkably, the multi-enzyme systems operated efficiently in tandem, without problems of product inhibition. Thus only the monosaccharide substrates GlcNAc, ManNAc, and Glc-1-P were required for the formation of the 6′-SLN trisaccharide. In principle, this one-pot multi-system strategy could be extended to the production of other complex oligosaccharides, such as the branched tetrasaccharide sLe^x.

The monosaccharide sialic acid remains relatively costly for preparative-scale synthesis. Although it has been demonstrated that NeuAc can be generated *in situ* by reaction of ManNAc with pyruvate, catalyzed by NeuAc aldolase [27], ManNAc is also relatively expensive and difficult to prepare. Therefore, a method for the generation of NeuAc from the inexpensive monosaccharide GlcNAc in a two-enzyme system might also prove useful in regeneration schemes. It has been shown that GlcNAc can be converted to ManNAc chemically [28], or enzymatically by the

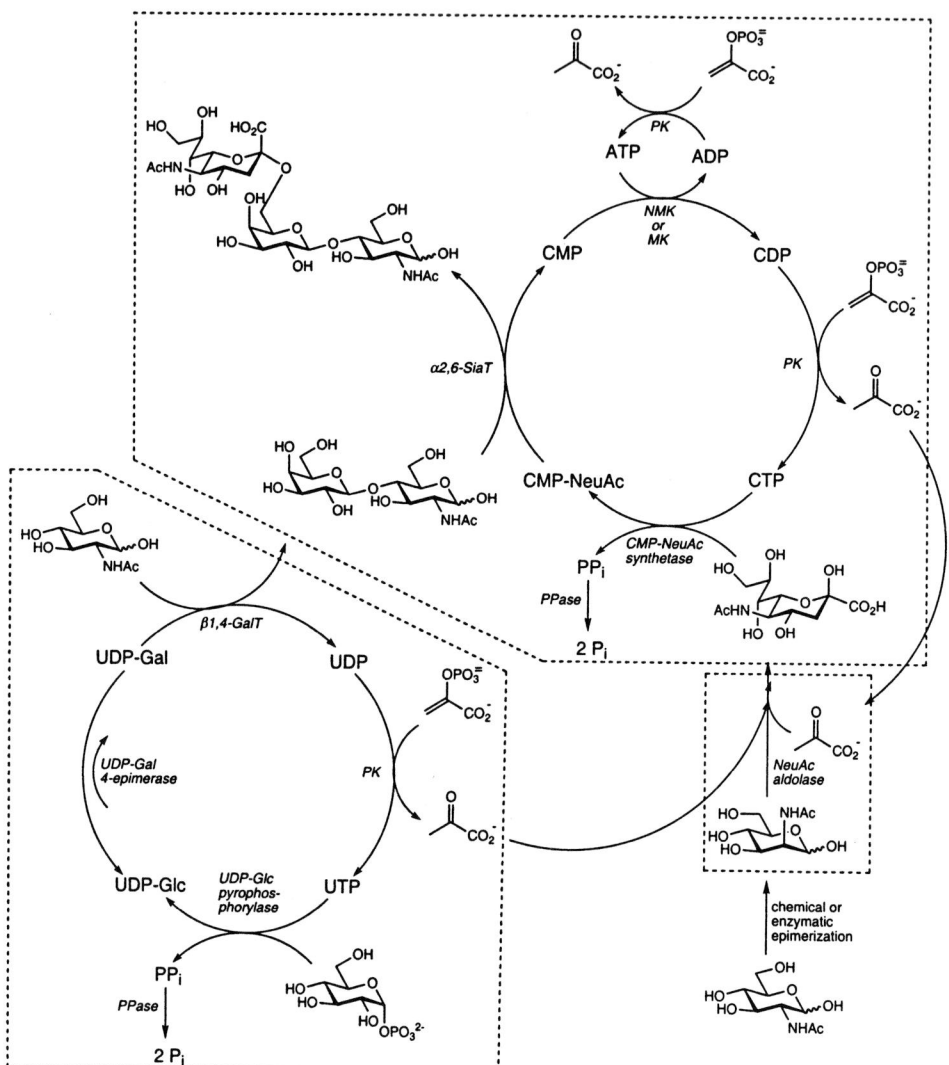

Scheme 8. Recycling of UDP–Gal and CMP–NeuAc for the one-pot synthesis of 6′-sialyl-LacNAc.

action of GlcNAc-2-epimerase [29]. Recently, an overexpression system for the epimerase was developed for industrial-scale synthesis [30]. The resulting ManNAc product provided by the epimerase could then be condensed with pyruvate to produce NeuAc.

CMP–NeuAc regeneration has also been coupled to β-galactosidase (EC 3.2.1.23) activity (Scheme 9). [31] The biological role of glycosidases is the hydro-

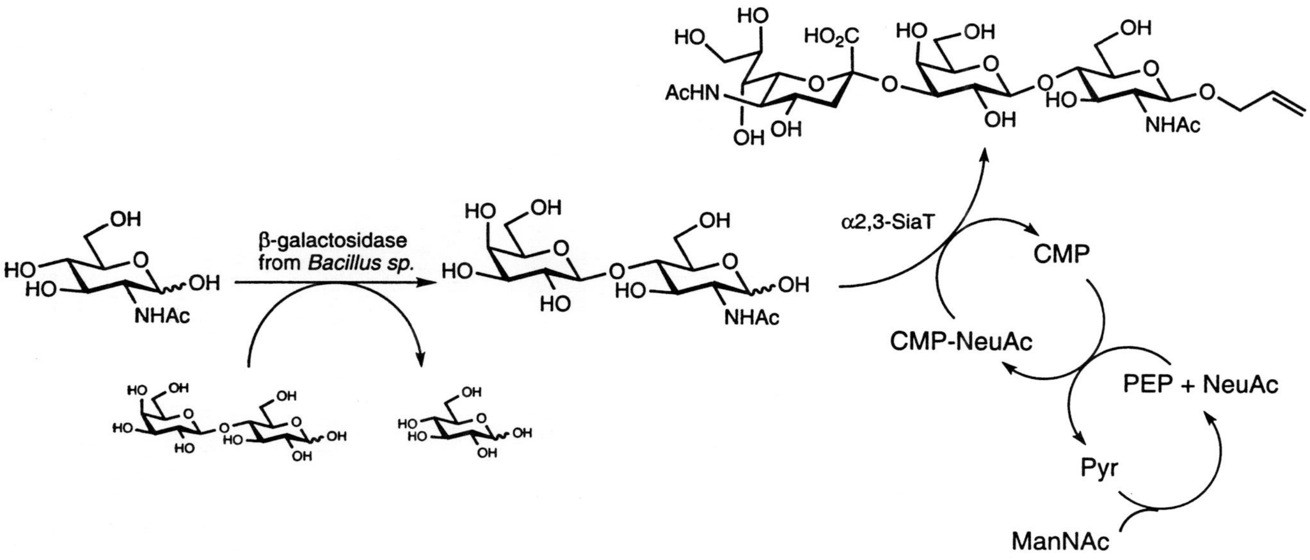

26.4 Practical Regeneration Systems

Scheme 9. Combination of β-galactosidase and α2,3-SiaT for the synthesis of 3′-sialyl-LacNAc.

lysis of glycosidic linkages, but these enzymes can be made to catalyze the reverse reaction under appropriate conditions. As such, removal of product by a transferase can shift the equilibrium to the bond-forming direction. β-Galactosidases can utilize either the lactose disaccharide or a *p*-nitrophenyl galactoside as the galactosyl donor substrate for synthetic purposes [32]. In this respect, β-galactosidase has been shown to be an effective catalyst in the synthesis of LacNAc and various derivatives thereof. In one system, LacNAc then served as a substrate for α2,6-SiaT in the production of 6′-SLN. Gambert and Thiem have, moreover, used this strategy in the synthesis of the sialylated Thomsen–Friedenreich antigen [33]. Compared with glycosyltransferases the use of glycosidases is advantageous in terms of stability, cost, and availability. As a drawback, glycosidases are also more promiscuous with regard to linkage specificity.

Another system that couples glycohydrolase activity to glycosyltransferase activity has been reported. Ito and Paulson have developed a CMP–NeuAc recycling scheme that operates in conjunction with a *trans*-sialidase activity from *Trypanosoma cruzi* (Scheme 10) [34]. This *trans*-sialidase catalyzes the reversible transfer of NeuAc from NeuAc(α2,3)Gal-OR to virtually any β-linked galactoside substrate. The CMP–NeuAc recycling system is employed to circumvent the expense of the sialylated pentasaccharide donor substrate. Synthetically, NeuAc is first transferred to Gal(β1,3)-GlcNAc(β1,3)Gal(β1,4)Glc under catalysis by α2,3-SiaT. This pentasaccharide product is then a substrate for the *trans*-sialidase. Several novel sialylated molecules were produced by this technique.

26.4.4 GDP–Sugars

The regeneration of GDP–Man has been employed in the synthesis of mannose-based oligosaccharides (Scheme 11) [35]. GDP–Man can be produced from Man-1-P and GTP by the action of GDP–Man pyrophosphorylase (GDPMP; EC 2.7.7.22) in a manner analogous to previously described for NDP–sugars. The system described employed an α1,2-mannosyltransferase (α1,2-ManT) that was overexpressed in *E. coli*. The α1,2-ManT accepted mannose, mannobiose, and *O*-mannosyl glycopeptides as substrates.

In the same fashion, GTP can be converted to GDP–Man as an intermediate in the recycling of GDP–Fuc (Scheme 12) [36]. This conversion can be accomplished by utilizing GDPMP from dried yeast cells and 'GDP–Fuc-generating enzymes' partially purified from the bacterium *Klebsiella pneumonia*. This system must be coupled to an alcohol dehydrogenase (EC 1.1.1.2), which catalyzes the oxidation of 2-propanol to acetone, along with the reduction of $NADP^+$ to NADPH. Alternatively, Fuc-1-P can be biosynthesized from fucose by the action of fucokinase (EC 2.7.1.52) from porcine liver in the presence of ATP (Scheme 13). Reaction of Fuc-1-P with GTP catalyzed by GDP–Fuc pyrophosphorylase (EC 2.7.7.30) then affords GDP–Fuc [37]. Both of these methods have been utilized in the synthesis of sLe^x from 3′-SLN.

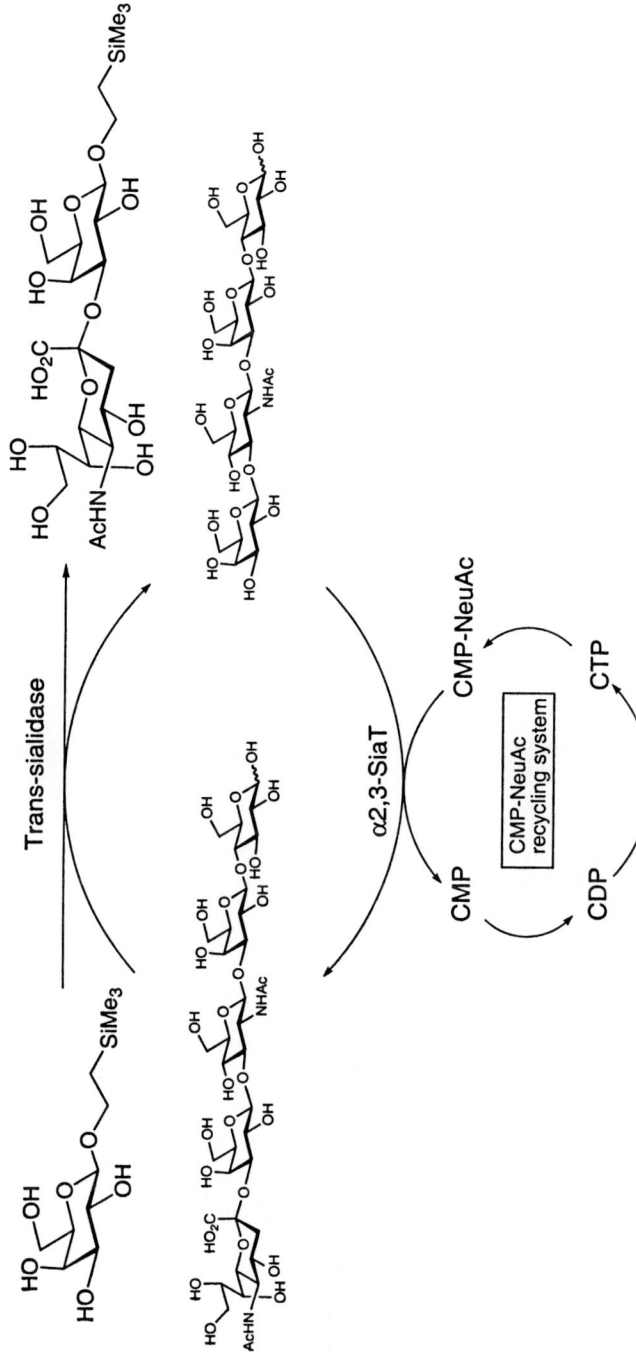

Scheme 10. Recycling of CMP–NeuAc coupled to *trans*-sialidase activity for the synthesis of novel sialosides.

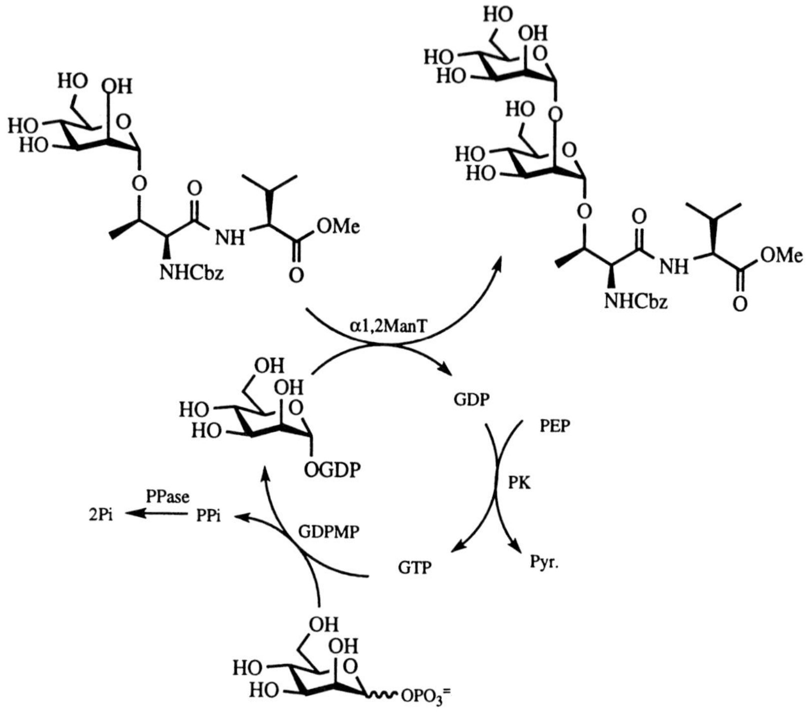

Scheme 11. Recycling of GDP–Man for the synthesis of mannosyl glycopeptides.

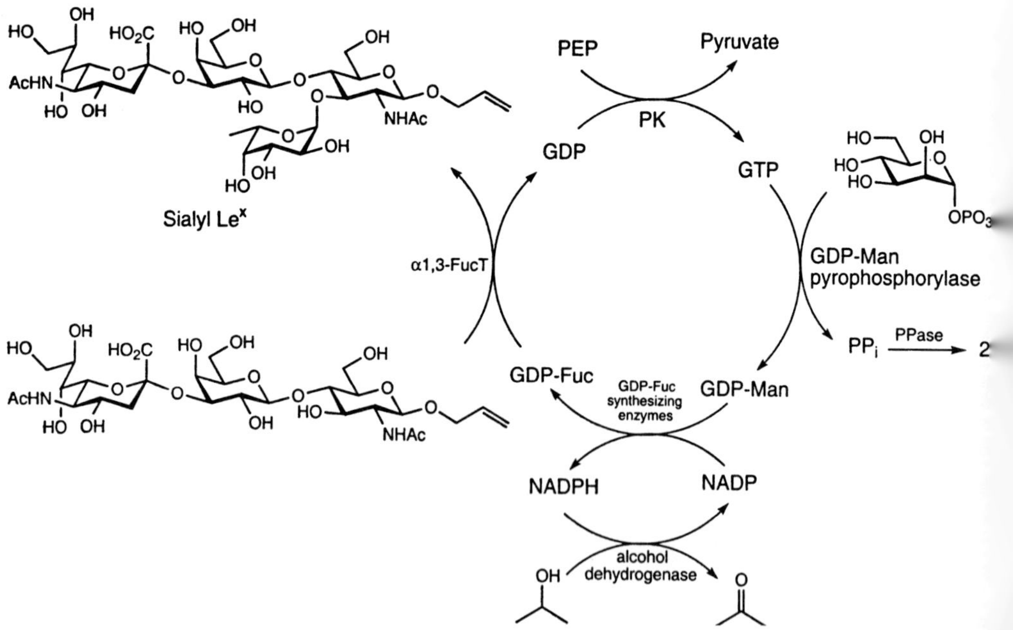

Scheme 12. Recycling of GDP–Fuc through GDP–Man for the synthesis of sLe^x.

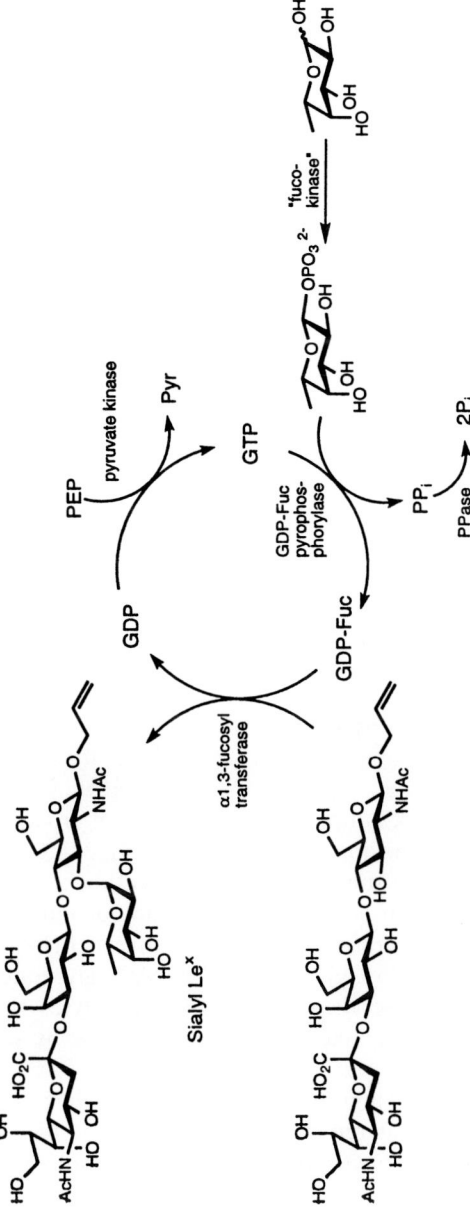

Scheme 13. Recycling of GDP–Fuc for the synthesis of sLe^x.

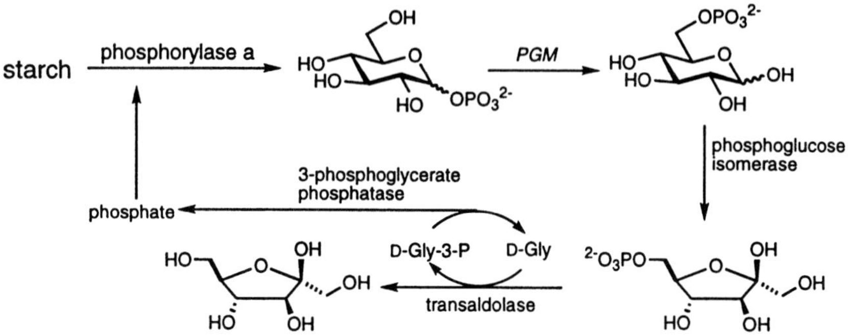

Scheme 14. Multi-enzyme synthesis of D-fructose from starch.

In situ recycling methods for GDP–Fuc are not yet optimized for preparative-scale synthesis. This is mainly because of difficulty obtaining adequate quantities of the key biosynthetic enzymes. GDP–Man pyrophosphorylase, 'GDP–synthesizing enzymes', fucokinase, and GDP–Fuc pyrophosphorylase are currently not available commercially.

26.4.5 Other Carbohydrate-Based Regeneration Systems

Other recycling systems that do not involve the synthesis of sugar nucleotides are also relevant in carbohydrate synthesis. For example, the production of fructose from starch has been accomplished in a five-enzyme system (Scheme 14) [38]. Conversion of starch to Glc-1-P is catalyzed by phosphorylase a. Isomerization to Glc-6-P, and further conversion to Fru-6-P is then achieved by phosphoglucomutase and phosphoglucose isomerase, respectively. A *trans*-aldolase then converts Fru-6-P to fructose, with the simultaneous conversion of glycerol to 3-phosphoglycerate. The action of 3-phosphoglycerate phosphatase then irreversibly hydrolyses the phosphate, providing the forward driving force, and completing the cycle by recycling inorganic phosphate.

Sulfation of carbohydrates is a topic of growing interest. The first recycling system used for the sulfation of carbohydrates was reported by Lin et al. The regeneration of 3′-phosphoadenosine 5′-phosphosulfate (PAPS), the universal biological sulfuryl donor, was accomplished by a multi-enzyme cascade (Scheme 15) [39]. In conjunction with a Nod factor sulfotransferase, N,N'-diacetylchitobiose was sulfated at the reducing terminal GlcNAc 6-hydroxyl. This sulfation system was also used to generate 6-sulfo-LacNAc, a key intermediate in the preparation of 6-sulfo-sLe^x.

A newly developed sulfation system accomplishes a similar goal, while also reducing the number of coupling enzymes necessary for the regeneration of PAPS (Scheme 16) [40]. In this protocol, PAPS is recycled from PAP through catalysis

26.4 Practical Regeneration Systems

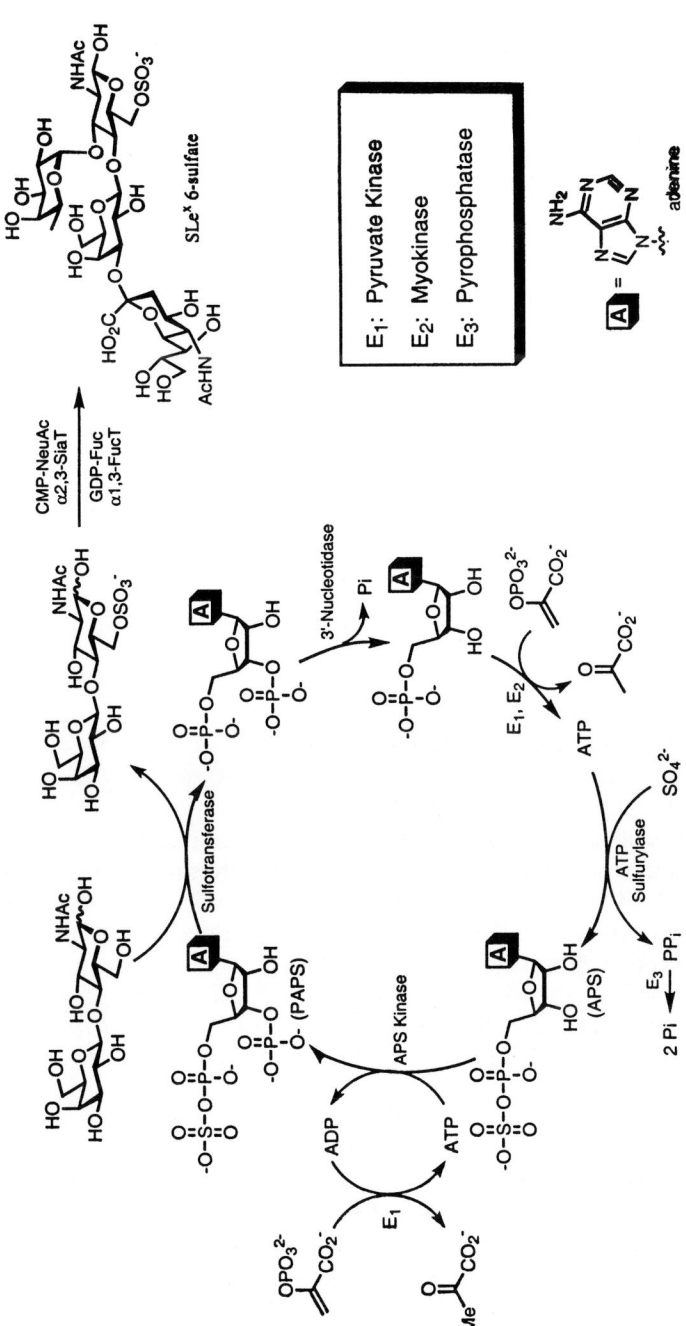

Scheme 15. Recycling of PAPS for the synthesis of sulfated carbohydrates.

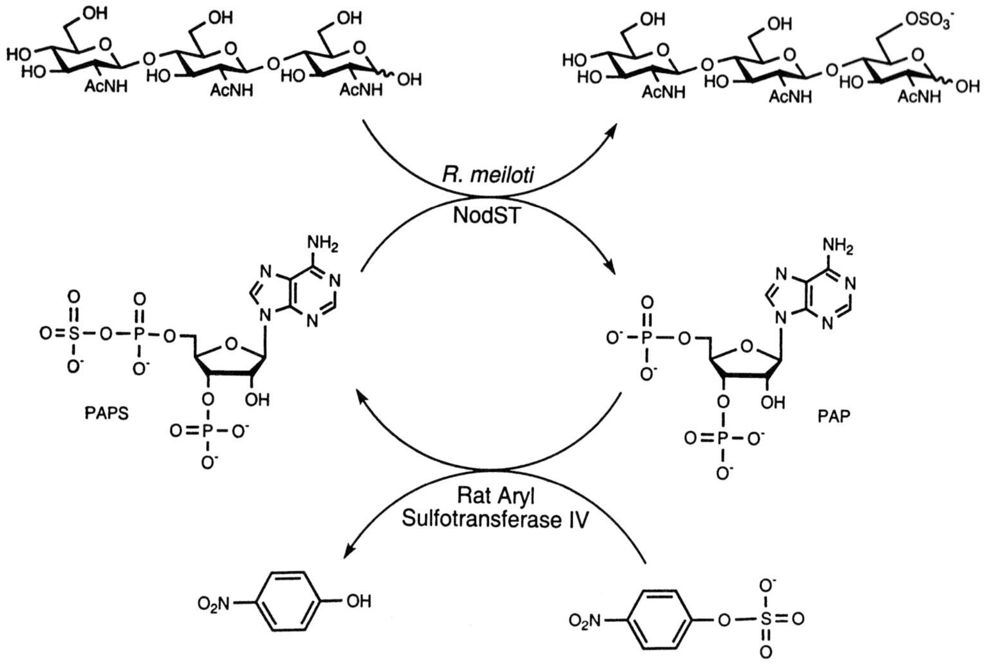

Scheme 16. Two-enzyme system for the recycling of PAPS.

by a single enzyme, aryl sulfotransferase. Because the only cofactor in this scheme is *p*-nitrophenol sulfate, the regeneration system has been greatly simplified, and is much more cost-effective. Chitotriose was found to be the best acceptor for the sulfotransferase. Notably, although LacNAc was a poor substrate, LacNAc(β1,3)-LacNAc was a good acceptor in this system.

26.5 Conclusion

Methods for the regeneration of sugar nucleotides in enzymatic glycosylation have proven extremely useful for synthetic purposes. Enzymatic synthesis can be both efficient and cost-effective in an appropriate system. Although recycling schemes for sugar nucleotides such as UDP–Gal and CMP–NeuAc are well-stocked by a commercial arsenal, others have not been as readily developed. For example, GDP–Fuc and UDP–GlcNAc regenerating schemes have yet to prove effective on a preparative-scale, because of lack of availability of the relevant biosynthetic enzymes. This field is, however, actively maturing with the continuous development of novel enzymes for use in recycling schemes. Complex substrates, such as glycopeptides, glycoproteins, and sulfated oligosaccharides, are currently being pursued to further the utility of recycling systems in carbohydrate synthesis.

References

1. a) Y. Ichikawa in *Glycopeptides and Related Compounds* (Eds. D. G. Large, C. D. Warren), Marcel Dekker, New York, **1997**, p. 79; b) P. Sears, C.-H. Wong, *Cell Mol. Life Sci.* **1998**, *54*, 223; c) C.-H. Wong, R. L. Halcomb, Y. Ichikawa, T. Kajimoto, *Angew. Chem. Int. Ed. Engl.* **1995**, *34*, 412 and 521.
2. A. T. Beyer, J. E. Sadler, J. I. Rearick, J. C. Paulson, R. L. Hill, *Adv. Enzymol.* **1981**, *52*, 24.
3. B. W. Weston, R. P. Nair, R. D. Larsen, J. B. Lowe, *J. Biol. Chem.* **1992**, *267*, 4152.
4. C. Unverzagt, H. Kunz, J. C. Paulson, *J. Am. Chem. Soc.* **1990**, *112*, 9308.
5. Y. Ichikawa, R. Wang, C.-H. Wong, *Methods Enzymol.* **1994**, *247*, 107.
6. a) L. F. Leloir, *Science* **1971**, *172*, 1299; b) R. Kornfeld, S. Kornfeld, *Ann. Rev. Biochem.* **1985**, *54*, 631.
7. C.-H. Wong, G. M. Whitesides, *Enzymes in Synthetic Organic Chemistry*, Pergamon, Oxford, **1994**, p. 252.
8. S. E. Heidlas, W. J. Lees, G. M. Whitesides, *J. Org. Chem.* **1992**, *57*, 152.
9. C.-H. Wong, S. L. Haynie, G. M. Whitesides, *J. Org. Chem.* **1982**, *47*, 5418.
10. a) M. M. Palcic, O. P. Srivastava, O. Hindsgaul, *Carbohydr. Res.* **1987**, *159*, 315; b) C. Auge, S. David, C. Mathieu, C. Gautheron, *Tetrahedron Lett.* **1984**, *25*, 1467.
11. C. Auge, C. Mathieu, C. Merienne, *Carbohydr. Res.* **1986**, *151*, 147.
12. J. Thiem, T. Wiemann, *Angew. Chem. Int. Ed. Engl.* **1991**, *30*, 1163.
13. T. Noguchi, T. Shiba, *Biosci. Biotechnol. Biochem.* **1998**, *62*, 1594.
14. L. Elling, M. Grothus, M.-R. Kula, *Glycobiology* **1993**, *3*, 349.
15. C. H. Hokke, A. Zervosen, L. Elling, D. H. Joziasse, D. H. van den Eijnden, *Glycoconjugate J.* **1996**, *13*, 687.
16. J. Fang, J. Li, X. Chen, Y. Zhang, J. Wang, Z. Guo, W. Zhang, L. Yu, K. Brew, P. G. Wang, *J. Am. Chem. Soc.* **1998**, *120*, 6635.
17. C.-H. Wong, R. Wang, Y. Ichikawa, *J. Org. Chem.* **1992**, *57*, 4343.
18. C.-H. Wong, M. Schuster, P. Wang, P. Sears, *J. Am. Chem. Soc.* **1993**, *115*, 5893.
19. K. Witte, P. Sears, R. Martin, C.-H. Wong, *J. Am. Chem. Soc.* **1997**, *119*, 2114.
20. S. L. Haynie, G. M. Whitesides, *Appl. Biochem. Biotech.* **1990**, *23*, 155.
21. D. Gygax, P. Spies, T. Winkler, U. Pfaar, *Tetrahedron* **1991**, *47*, 5119.
22. C. DeLuca, M. Lansing, I. Martini, F. Crescenzi, G.-J. Shen, M. O'Regan, C.-H. Wong, *J. Am. Chem. Soc.* **1995**, *117*, 5869.
23. G. C. Look, Y. Ichikawa, G.-J. Shen, P.-W. Cheng, C.-H. Wong, *J. Org. Chem.* **1993**, *58*, 4326.
24. Y. Ichikawa, G.-J. Shen, C.-H. Wong, *J. Am. Chem. Soc.* **1991**, *113*, 4698.
25. M. Gilbert, R. Bayer, A.-M. Cunningham, S. DeFrees, Y. Gao, D. C. Watson, N. M. Young, W. W. Wakarchuk, *Nat. Biotechnol.* **1998**, *16*, 769.
26. a) Y. Ichikawa, J. L.-C. Liu, G.-J. Shen, C.-H. Wong, *J. Am. Chem. Soc.* **1991**, *113*, 6300; b) P. Stangier, W. Treder, J. Thiem, *Glycoconjugate J.* **1993**, *10*, 26.
27. a) C. Auge, S. David, C. Gautheron, *Tetrahedron* **1984**, *25*, 4663; b) M.-J. Kim, W. J. Hennan, H. M. Sweers, C.-H. Wong, *J. Am. Chem. Soc.* **1992**, *114*, 10138.
28. E. S. Simon, M. D. Bednarski, G. M. Whitesides, *J. Am. Chem. Soc.* **1988**, *110*, 7159.
29. U. Kragl, D. Gygax, O. Ghisalba, C. Wandrey, *Angew. Chem. Int. Ed. Engl.* **1991**, *30*, 82.
30. I. Maru, J. Ohnishi, Y. Ohta, Y. Tsukada, *Carbohydr. Res.* **1998**, *306*, 575.
31. G. F. Herrmann, Y. Ichikawa, C. Wandrey, F. C. A. Gaeta, J. C. Paulson, C.-H. Wong, *Tetrahedron Lett.* **1993**, *34*, 3091.
32. S. Takayama, M. Shimazaki, L. Qiao, C.-H. Wong, *Bioorg. Med. Chem. Lett.* **1996**, *6*, 1123.
33. U. Gambert, J. Thiem, *Eur. J. Org. Chem.* **1999**, *1*, 107.

34. Y. Ito, J. C. Paulson, *J. Am. Chem. Soc.* **1993**, *115*, 7862.
35. P. Wang, G.-J. Shen, Y.-F. Wang, Y. Ichikawa, C.-H. Wong, *J. Org. Chem.* **1993**, *58*, 3985.
36. Y. Ichikawa, Y.-C. Lin, D. P. Dumas, G.-J. Shen, E. Garcia-Junceda, M. A. Williams, R. Bayer, C. Ketcham, L. E. Walker, J. C. Paulson, C.-H. Wong, *J. Am. Chem. Soc.* **1992**, *114*, 9283.
37. See also R. Stiller, J. Thiem, *Liebigs Ann. Chem.* **1992**, *5*, 467.
38. A. Moradian, S. A. Benner, *J. Am. Chem. Soc.* **1992**, *114*, 6980.
39. C.-H. Lin, G.-J. Shen, E. Garcia-Junceda, C.-H. Wong, *J. Am. Chem. Soc.* **1995**, *117*, 8031.
40. M. D. Burkart, M. Izumi, C.-H. Wong, *Angew. Chem. Int. Ed. Engl.* **1999**, *38*, 2747.

27 Enzymatic Glycosylations with Non-Natural Donors and Acceptors

Xiangping Qian, Keiko Sujino, and Monica M. Palcic

27.1 Introduction

The demonstration that oligosaccharides can serve as recognition markers in diverse biological events [1, 2] has stimulated interest in the synthesis of oligosaccharides and their analogs [3–7]. The availability of such molecules can provide further insights into their biological functions and might lead to the discovery of novel carbohydrate-based therapeutics [8]. Despite many advances in the chemical synthesis of oligosaccharides, it still remains a challenge, especially for the preparation of analogs where additional steps must be employed in the synthetic route to achieve a modification at a specified position [4]. As well, a replication system for amplifying minute amounts of carbohydrates is not currently available, nor is instrumentation commercially available for the solid-phase synthesis of oligosaccharides.

Oligosaccharides are biosynthesized in vivo by glycosyltransferases that sequentially transfer a single pyranosyl residue from a sugar nucleotide donor to a growing carbohydrate chain [9–12]. Enzymatic synthesis using glycosyltransferase has several advantages over the chemical synthesis of oligosaccharides—high regio- and stereoselectivity, no requirement for chemical protection and deprotection, and very mild reaction conditions [13–15].

Although there are more than 100 different mammalian glycosyltransferases, each biosynthesizing a unique glycosidic linkage, only nine main nucleotide donors are used as building blocks by these glycosyltransferases to construct the diverse and complex oligosaccharides found in mammals. Recognition of the donor substrates by glycosyltransferases is primarily based on the nucleotide portion. Non-natural oligosaccharide analogs can be prepared by using glycosyltransferases to transfer a modified monosaccharide from donors. Studies probing the specificity of glycosyltransferases using acceptor analogs where the hydroxyl group is replaced with H, OCH_3, NH_2 or other substituents, have helped to establish structural requirements for acceptor recognition [16]. It has been found that only a few of the hydroxyl groups of the acceptor, termed 'key polar groups' [17–19], are necessary for binding to glycosyltransferases. Chemical modifications can therefore be introduced at positions other than those bearing key polar groups to produce nonnatural acceptor precursors for enzymatic synthesis.

With advances in molecular biology and biotechnology more than 30 glycosyltransferases have been cloned and many glycosyltransferases are now readily available in quantities sufficient for use in *in vitro* synthesis [20]. The increasing availability of glycosyltransferases along with their flexibility in the recognition of both donor and acceptor substrates makes enzymatic glycosylation with non-natural donors and/or non-natural acceptors an increasingly practical alternative for the preparation of non-natural oligosaccharides [16, 21–24]. The term 'non-natural', in the current context, is used to indicate that the acceptor or donor is chemically modified and different from the natural ones that are used by glycosyltransferases. Biosynthetically glycosyltransferases act on very large and complex glycoproteins and glycolipids, therefore small molecule acceptors might be viewed as non-natural. In this contribution, however, we do not consider a change of aglycone or adding/removing a sugar unit(s) at the reducing end as creating a non-natural acceptor.

27.2 Enzymatic Glycosylations

27.2.1 Galactosylations

β1,4-Galactosyltransferase

β1,4-Galactosyltransferase (β1,4-GalT, E.C. 2.4.1.22/38/90) has been commercially available for many years in unit quantities. It is the most widely studied glycosyltransferase with regard to substrate specificity and use in preparative synthesis. It catalyzes the transfer of Gal with inversion of configuration from UDP–Gal to OH-4 of terminal β-linked GlcNAc to form *N*-acetyllactosamine. In the presence of α-lactalbumin, the enzyme prefers to use glucose as an acceptor to produce lactose.

Non-natural donors

β1,4-Galactosyltransferase has been shown to tolerate modifications on any OH group of the Gal residue of UDP–Gal donors. It transfers 2-deoxy-Gal at a rate similar to Gal [25–28]. 3-Deoxy [29], 4-deoxy [30], 6-deoxy [31, 32], and 6-deoxy-6-fluoro Gal [31, 32] are also transferred. The enzyme can also utilize UDP–Ara [30, 33], UDP–GalNAc [34], UDP–GalNH$_2$ [27], UDP–Glc [30, 35, 36], and UDP–GlcNH$_2$ [34] as donors (Scheme 1).

Interestingly, UDP–5′-thio-Gal, with the ring oxygen of Gal replaced by a sulfur atom, was active as a donor substrate for β1,4-GalT [37]. UDP–5′-thio-GalNAc was also found to be a substrate in the presence of lactalbumin [38].

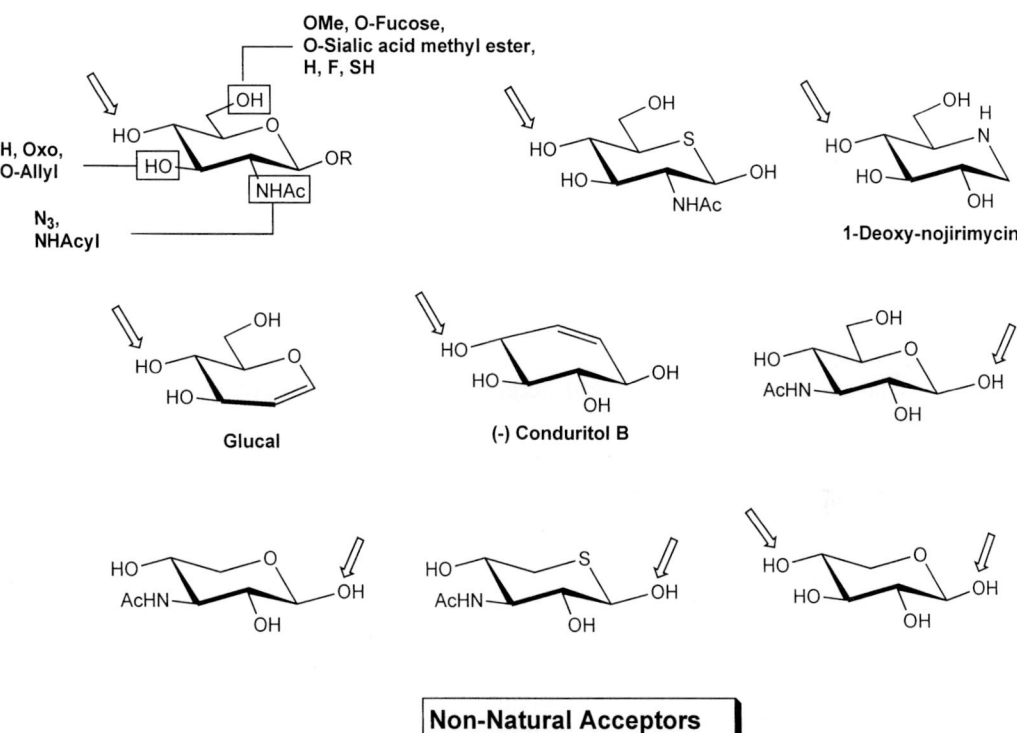

Scheme 1. Non-natural donors and acceptors for β1,4-GalT.

Non-natural acceptors

As shown in Scheme 1, the acceptor specificity of β1,4-GalT is extremely relaxed because numerous modified GlcNAc analogs are active as acceptors. Basically, β1,4-GalT tolerates modifications everywhere on the sugar ring, including the ring oxygen, as long as the 4-OH remains available for glycosylation.

The 2-NHAc group can be replaced with azido [27, 39], N-propanoyl [40, 41], N-butanoyl [40], allylcarbamate [42], and many amide derivatives [43] including charged groups, highly bulky heterocycles, and glycuronamides. 2-Ethylamino-2-deoxy, 2-N-methylacetamido-2-deoxy, and 2-O-acetyl-β-D-glucoside were, however, inactive as acceptors and inhibitors for β1,4-GalT [44].

Analogs with the 3-OH group deoxygenated [44, 45], alkylated with a methyl or allyl group, or oxidized to the ketone are active as acceptors although the relative rates of transfer are much lower than to N-acetylglucosamine [45].

The 6-OH group of GlcNAc can be methylated [41, 45], fucosylated [41], deoxygenated [44], or substituted with F or SH [44]. Although addition of α-linked sialic acid to 6-OH of GlcNAc is not tolerated [41], when the carboxylic acid of NeuAc is derivatized to the methyl ester, the resulting compound proved to be a weak acceptor for β1,4-GalT. Although the relative rate of transfer is only 4% that for GlcNAc, this is sufficient for preparative synthesis and generation of product [41].

The great tolerance of β1,4-GalT for acceptor modifications is further exemplified by its transfer of Gal from UDP–Gal to 5′-thio-Glc and 1-deoxynojirimycin, which have the ring oxygen modified, and to glucal, which has a flattened ring conformation [45]. The enzyme can even resolve racemic (±) conduritol B to give a single galactosylated product of (−)-conduritol B [46]. More interestingly, the enzyme transfers galactose to the β-anomeric position of 3-acetamido-3-deoxy-D-glucose acceptors resulting in the formation of an unusual β1–1 (trehalose type) linkage [47]. This 'frame-shifted' galactosylation [48] was also observed with N-acetylgentosamine [49], N-acetyl-5′-thiogentosamine [50], and xylose [51] acceptors. A large variety of immobilized acceptors has also been employed in preparative reactions with β1,4-GalT [52–57].

α1,3-Galactosyltransferase

α1,3-Galactosyltransferase (α1,3-GalT, E.C. 2.4.1.151) catalyzes the transfer of Gal with retention of configuration from UDP–Gal to 3-OH of the Gal residue in Gal(β1–4)GlcNAc-R to form Gal(α1–3)Gal(β1–4)GlcNAc epitopes [58–61], the major xenoactive antigens responsible for hyperacute rejection in xenotransplantation [62]. An enzyme for preparative synthesis has been isolated from porcine and bovine tissues and recombinant porcine α1,3-galactosyltransferase is now commercially available in unit quantities.

Non-natural donors

Unlike β1,4-GalT, α1,3-galactosyltransferase has a rather restricted specificity for the nucleotide donor substrate as revealed by studies with recombinant murine α1,3-GalT [63]. Although UDP–Glc, UDP–GalNAc, UDP–GlcNAc, and UDP–glucuronic acid are not substrates for the enzyme [63], UDP–2-deoxy-Gal is a better substrate than UDP–Gal. The relative rates of transfer of 3-deoxy, 4-deoxy, or 6-deoxy donors are 0.2, 0.6 and 2% that of UDP–Gal [64].

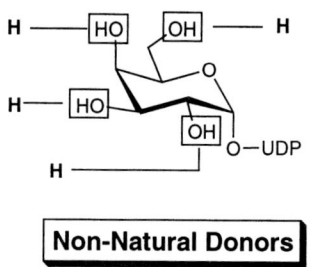

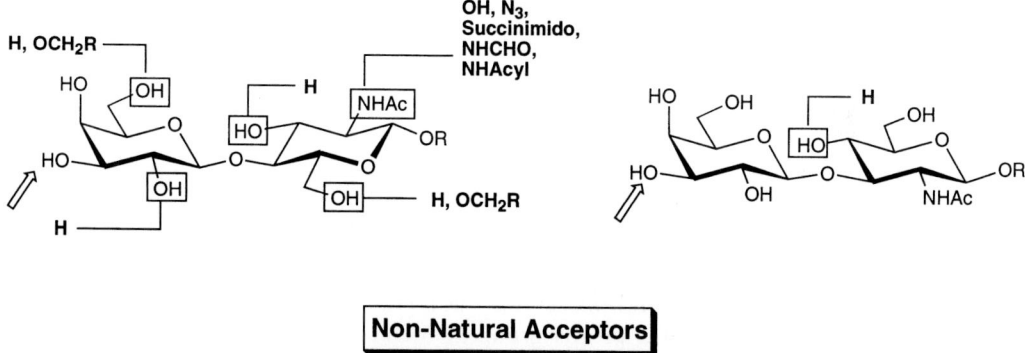

Scheme 2. Non-natural donors and acceptors for α1,3-GalT.

Non-natural acceptors

Besides using Gal(β1–4)GlcNAc as an acceptor, the enzyme can transfer Gal to Gal(β1–3)GlcNAc and Gal(β1–4)Glc [65]. Substitutions of 2-NHAc of the GlcNAc residue with azido and succinimido groups are tolerated [63]. The N-acetyl group can be replaced with many acyl groups of various sizes, hydrophilicities, or lipophilicities [66], although replacement of the 2-NHAc with an amino group abolishes activity [63]. Deoxygenation of the 3-OH of the GlcNAc residue is tolerated whereas any substitution or derivatization at this position is not. As shown in Scheme 2, analogs with modifications (deoxygenation and O-alkylation) on 6-OH of GlcNAc, or 2-OH, and 6-OH of the terminal Gal residue are substrates [67].

Modification on 4-OH of Gal is not tolerated by the enzyme, suggesting that this group is a key polar group essential for binding to α1,3-GalT [67]. The 4-deoxy analog of Gal(β1–3)GlcNAcβ-OR was found to be as active as Gal(β1–3)GlcNAcβ-OR [67].

27.2.2 Fucosylations

The fucosylated oligosaccharides on cell surfaces are involved in numerous intercellular recognition events [1]. α1,3/4-Fucosyltransferases catalyze the transfer of a fucose residue with retention of configuration from GDP–fucose to a variety of acceptors to form the blood-group related antigenic determinants Le^a and Le^x [68]. α1,3/4-Fucosyltransferases are a multigene family of enzymes with different substrate specificities toward Gal(β1–4)GlcNAc, Gal(β1–3)GlcNAc, and NeuAc(α2–3)Gal(β1–4)GlcNAc acceptors, inhibitor sensitivity, pH optima, and tissue distributions [12]. The most common source for isolation of a preparatively useful fucosyltransferase is human milk [10]. There are two different human milk fucosyltransferases, α1,3/4-fucosyltransferase and α1,3-fucosyltransferase. Five different human α1,3/4-fucosyltransferases have been cloned, FucT III to FucT VII, and both FucT V and VI are commercially available.

Human Milk α1,3/4-Fucosyltransferase

Non-natural donors

GDP-3-deoxyfucose, GDP-arabinose and GDP-L-galactose are active as donor substrates for human milk α1,3/4-FucT (E.C. 2.4.1.65) when Gal(β1–3)GlcNAcβ-O(CH$_2$)$_8$COOMe is used as an acceptor [69]. GDP-3-deoxy-Fuc and GDP–L-Gal were shown to be preparatively useful, using Gal(β1–4)GlcNAcβ-O(CH$_2$)$_8$COOMe as an acceptor for human milk α1,3/4-FucT [70]. The enzyme was also shown to tolerate the addition of a propyl group at 6-OH of L-Gal in GDP–L-Gal. Large substituents, *e.g.* tethered blood group B trisaccharide, can be introduced at C-6 of the Fuc residue in GDP–Fuc [71]. Essentially, any group can be attached to the C-6 position of Fuc, as was demonstrated by the transfer of modified fucose residues containing biotin and blood group A trisaccharide biolabels (Scheme 3) [72].

Non-natural acceptors

Human milk α1,3/4-FucT uses both Gal(β1–3)GlcNAc (type I) and Gal(β1–4)GlcNAc (type II) acceptors. Chemical mapping studies with a series of monodeoxygenated and modified acceptor substrates showed that modifications are tolerated at every hydroxyl group in the sugar rings except 6-OH of the Gal and 3- or 4-OH of the GlcNAc residue to which fucose is transferred [73]. The 2-NHAc group of the GlcNAc residue in NeuAc(α2–3)Gal(β1–4)GlcNAcβ-O(CH$_2$)$_8$COOMe or NeuAc(α2–3)Gal(β1–3)GlcNAcβ-O(CH$_2$)$_8$COOMe can be replaced with azido, amino, or propionamido groups [74]. Thio-linked *N*-acetyllactosamine, in which the inter-glycosidic oxygen is replaced by sulfur, is also a good acceptor for the enzyme [75]. Ether- and imino-linked octyl *N*-acetyl-5a′-carba-β-lactosamides were also found to be acceptors for human milk α1,3/4-FucT [76]. Surprisingly, human milk α1,3/4-FucT can even tolerate the introduction of a large methyl group directly at the site of fucosylation; it transfers fucose to the hindered tertiary alcohol acceptor 3-*C*-methyl-*N*-acetyllactosamine [77]. This unexpected finding demonstrated that glycosyltransferases can overcome inherent steric limitations and can be used to

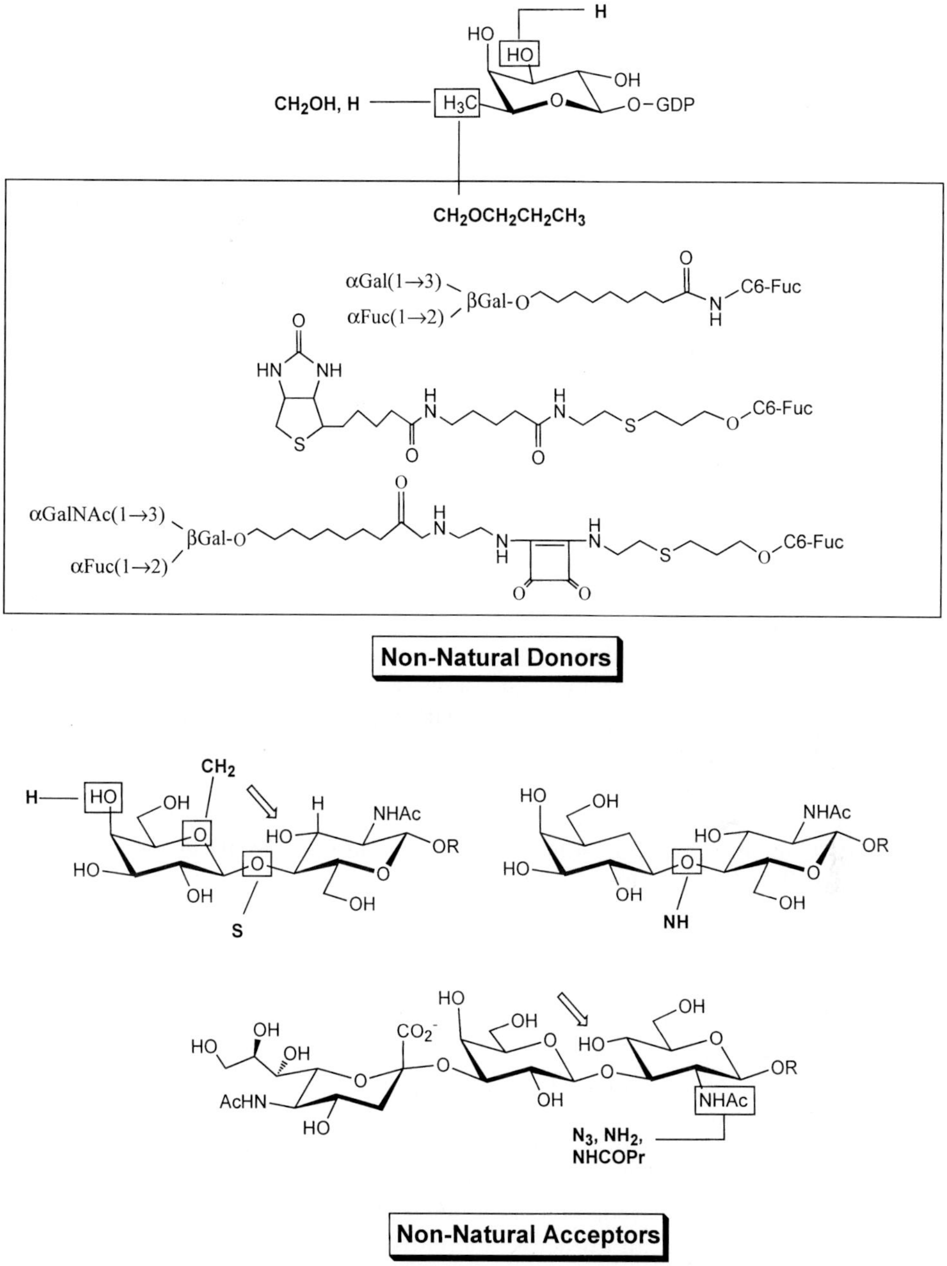

27.2 Enzymatic Glycosylations 691

Scheme 3. Non-natural donors and acceptors for human milk α1,3/4-FucT.

produce oligosaccharide analogs that cannot easily be prepared by chemical methods.

FucT III and VI

FucT III transfers fucose on to the 4-OH group of GlcNAc residues of Gal(β1–3)-GlcNAc or NeuAc(α2–3)Gal(β1–3)GlcNAc to form Le^a or sialyl Le^a respectively, whereas FucT VI uses Gal(β1–4)GlcNAc or NeuAc(α2–3)Gal(β1–4)GlcNAc to form Le^x or sialyl Le^x products. Despite their close sequence homologies, they behave differently in their recognition of non-natural donors [78]. FucT III transfers L-Gal, arabinose, L-Glc, 2-amino-2-deoxyfucose, and 2-fluoro-2-deoxyfucose whereas FucT VI cannot tolerate modifications on the 2-OH group of the fucose. FucT VI still transfers L-Gal and arabinose, however. Both enzymes tolerate diverse replacements of the N-acetyl group of the GlcNAc unit, as seen with β1,4-GalT [78–80]. FucT VI has been shown to tolerate replacement of the GlcNAc unit with glucal and even with cyclohexane diol [24]. The utility of FucT III and FucT VI has further been expanded to enable construction of libraries of sialyl Le^a or sialyl Le^x derivatives using both non-natural donors and non-natural acceptors (Schemes 4 and 5) [81–83].

FucT III and FucT VI have been probed with non-natural donors where the purine base was modified. Both enzymes can tolerate exchange of the guanine by other purine bases such as adenine [84]. These non-natural donors proved to be preparatively efficient in the enzymatic synthesis of Le^a or Le^x, suggesting that costly sugar nucleotide donors can be replaced with inexpensive ones to reduce the cost of enzymatic synthesis.

FucT V

Similar to human milk α1,3/4-FucT, FucT V requires the 6-OH group of Gal and OH-3 or OH-4 of GlcNAc for substrate binding. Gal(β1–4)Glucal [39], Gal(β1–4)(-5-S)-Glc [39], 3-sulfo-Gal(β1–4)GlcNAc [85] and Gal(β1–4)-6-sulfo-GlcNAc [85] are acceptors for FucT V.

27.2.3 Sialylations

Cell-surface sialic acid residues play important roles in diverse biological processes [86]. Sialic acids are usually found in terminal positions linked through an α-glycosidic linkage. The stereoselective synthesis of α-sialosides remains a challenge because the glycosides have the thermodynamically unfavorable equatorial orientation and the anomeric carbon is a very hindered quaternary center [4]. Enzymatic sialylation is, therefore, an attractive alternative means of preparation of α-sialosides in an efficient and stereocontrolled manner.

α2,3-Sialyltransferase and α2,6-Sialyltransferase

α2,3-Sialyltransferase and α2,6-sialyltransferase from rat liver (α2,3-SialT, E.C. 2.4.99.6; α2,6-SialT, E.C. 2.4.99.1) have been cloned and expressed and are now

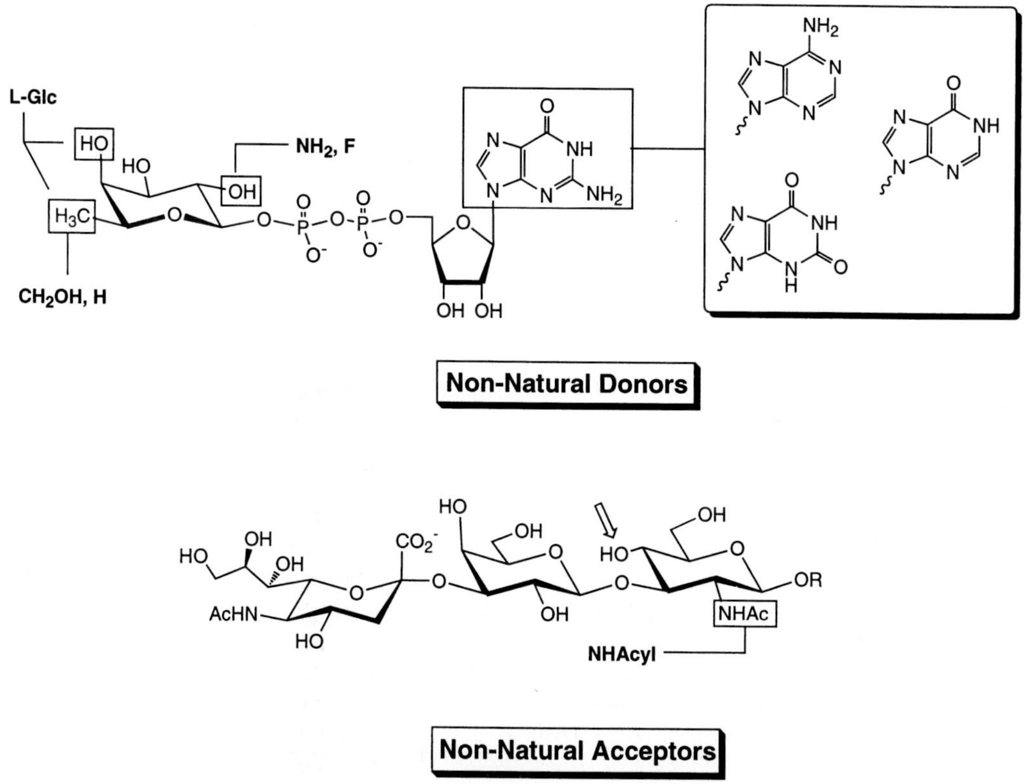

Scheme 4. Non-natural donors and acceptors for FucT III.

commercially available in quantities sufficient for use in preparative synthesis. α2,3-SialT from rat liver transfers a sialic acid unit from CMP–sialic acid to OH-3 of the terminal Gal residue in Gal(β1–4)GlcNAc or Gal(β1–3)GlcNAc sequences whereas rat liver α2,6-SialT transfers the sialic acid to OH-6 of the terminal Gal residue in Gal(β1–4)GlcNAc [87].

Non-natural donors

The finding that CMP–sialic acid synthetase can accept many sialic acid analogs has facilitated the synthesis of CMP–sialic acid analogs for studies of the donor-specificity of these enzymes [88–90]. Both α2,3-SialT and α2,6-SialT tolerate substitutions at C-9 of CMP–sialic acid (Schemes 6 and 7) [89, 91, 92].

The 9-OH group can be replaced with fluoro, azido, amino, acetamido, hexanoylamido, benzamido, and even fluorescent labels. 9-*O*-Acetyl sialic acid can also be transferred. Donors in which the 5-NHAc group of sialic acid is replaced with OH, NHC(O)CH$_2$OH, or NHCbz (Cbz = benzyloxycarbonyl) are also accepted by α2,3-SialT [93]. CMP–4-deoxysialic acid was found to be a donor substrate for α2,6-SialT [94].

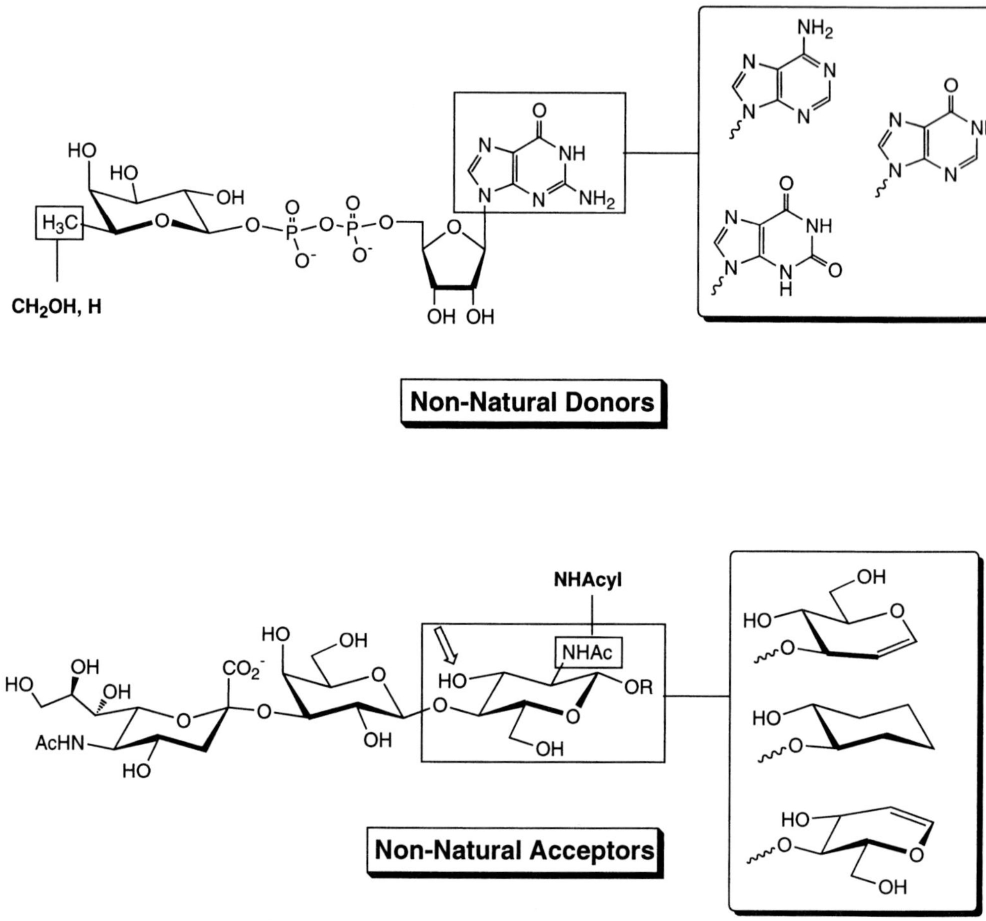

Scheme 5. Non-natural donors and acceptors for FucT VI.

Non-natural acceptors

Chemical mapping studies have indicated that OH-6 of Gal and 2-NHAc are required for binding to rat liver α2,6-SialT whereas rat liver α2,3-SialT requires an intact 3,4,6-triol system on the Gal residue [95]. Analogs of Gal(β1–4)GlcNAc(β1–2)Manα-*O*-Oct, where the 3- or 4-OH group of the Gal residue is deoxygenated or substituted with a fluoro atom, were found to be active as substrates for α2,6-SialT (Scheme 7) [96]. The 4″-*O*-methyl derivative of the trisaccharide is an acceptor for α2,3-SialT [96]. A wide range of substitutions on the *N*-acetyl group of both type I or type II acceptors were accepted by α2,3-SialT [97, 98]. α2,3-SialT also transfers sialic acid to lactal, lactose, and 2-*O*-pivaloyllactose [39].

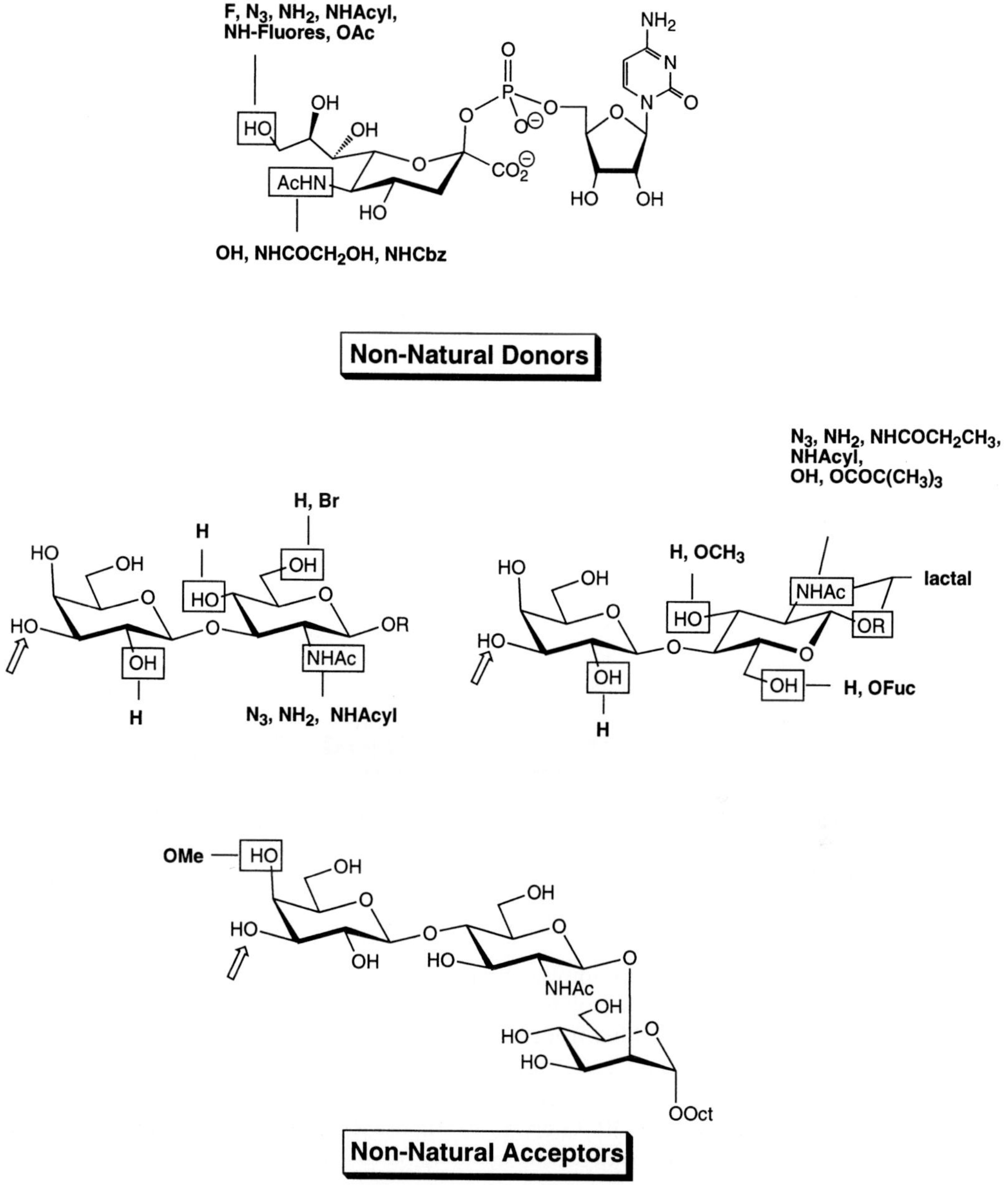

Scheme 6. Non-natural donors and acceptors for α2,3-SialT.

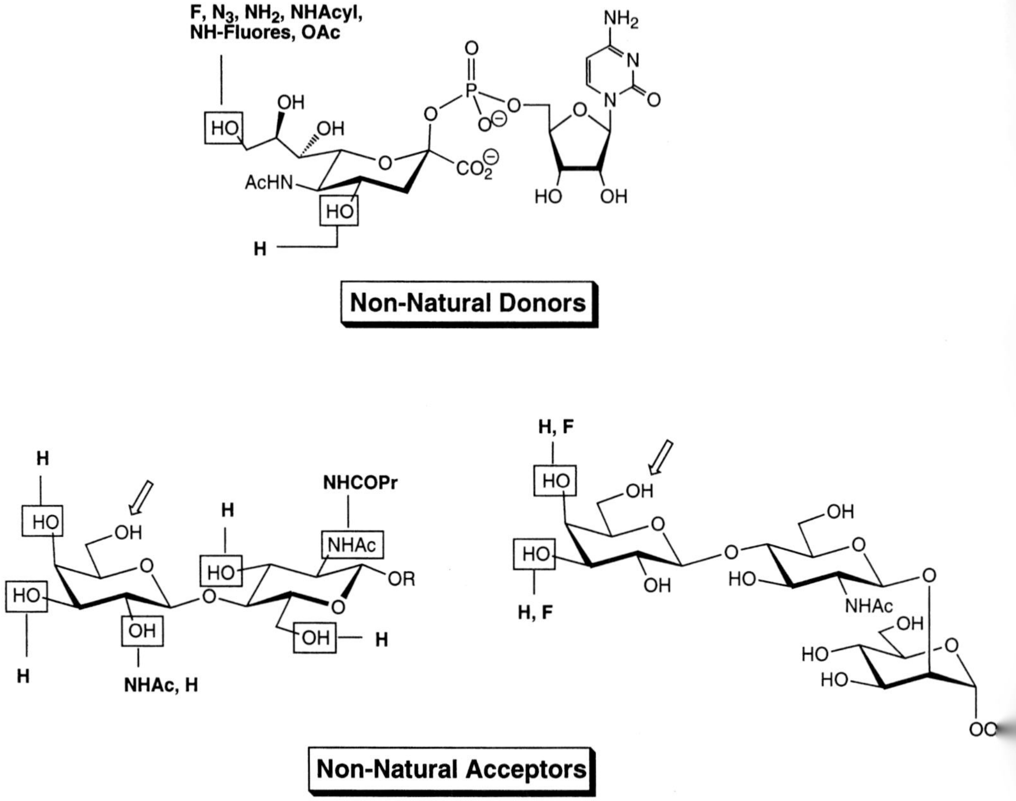

Scheme 7. Non-natural donors and acceptors for α2,6-SialT.

27.2.4 *N*-Acetylglucosaminylation

The branching pattern of complex *N*-glycans is controlled by a series of *N*-acetylglucosaminyltransferases numbered GnT I–VI [12, 99]. These enzymes transfer an *N*-acetylglucosamine residue from UDP–GlcNAc to different sites on the tri-mannose core structure Man(α1–6 [Man(α1–3)]Manβ, as shown in Scheme 8.

Biosynthetically, *N*-acetylglucosaminyltransferases catalyze the transfer of a GlcNAc residue from UDP–GlcNAc to oligosaccharide acceptors having the tri-mannose core structure attached to GlcNAc(β1–4)GlcNAcβ-Asn. The finding that GlcNAc(β1–4)GlcNAcβ-Asn can be replaced with a hydrophobic group made it feasible to study the donor and acceptor specificities of these enzymes using smaller oligosaccharide acceptors which are readily prepared [100, 101].

N-Acetylglucosaminyltransferase I, II, and III

The C-3, C-4, and C-6 deoxy analogs of UDP–GlcNAc were found to act as donor substrates for GnT I (E.C. 2.4.1.101) from human milk using Man(α1–6 [Man(α1–

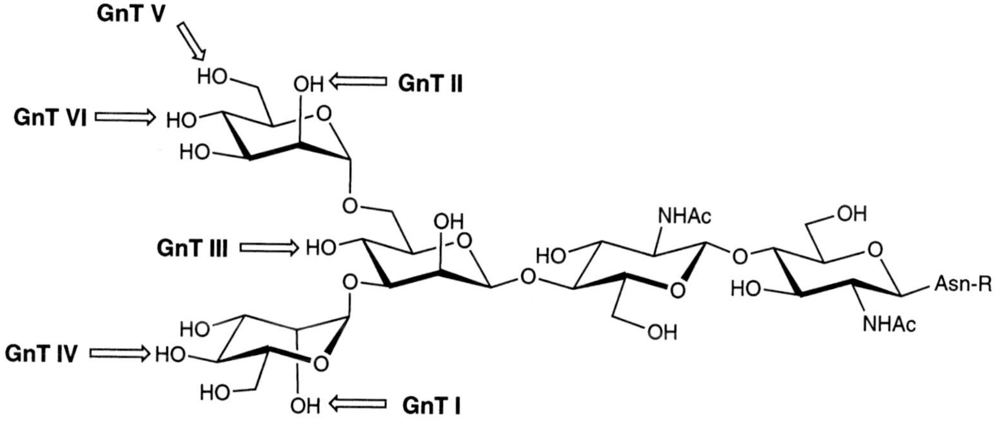

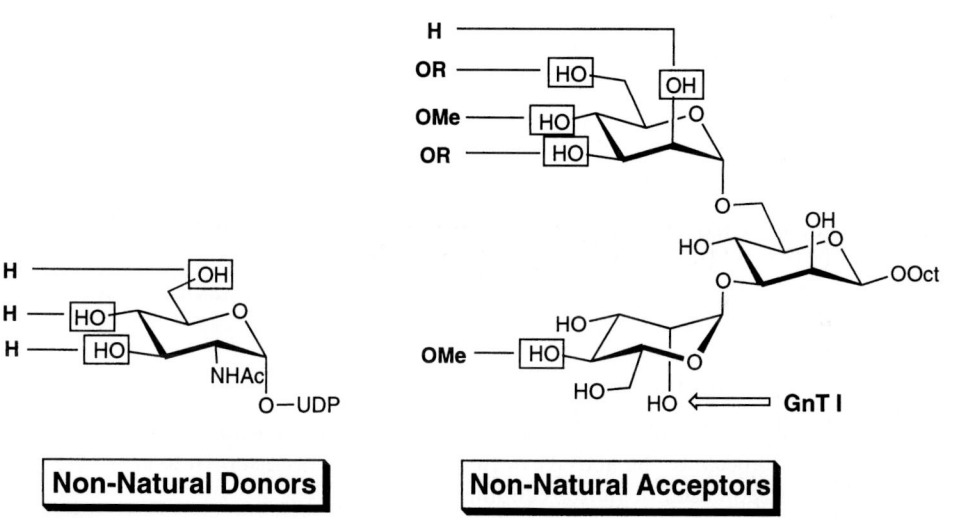

Scheme 8. Glycosylation sites for GnT I–VI.

Scheme 9. Non-natural donors and acceptors for GnT I.

3)]Manβ-O(CH$_2$)$_8$COOMe as the acceptor [102]. The tetrasaccharide products of the above reactions have been evaluated with GnT II [102]. Deoxygenation at C-3 of GlcNAc unit was not tolerated by GnT II and removal of the 4-OH or 6-OH of GlcNAc caused only modest decreases in activity [102].

As shown in Scheme 9, GnT I can use non-natural acceptors that are deoxygenated or O-alkylated on the αMan(1–6) residue [103–105].

Alkylations at OH-4 of αMan1–3, the mannose residue to which transfer occurs, are tolerated. Analogs where OH-3, 4, or 6 of the αMan1–6 residue was deoxygenated or methylated were found to be substrates of GnT II (Scheme 10) [103].

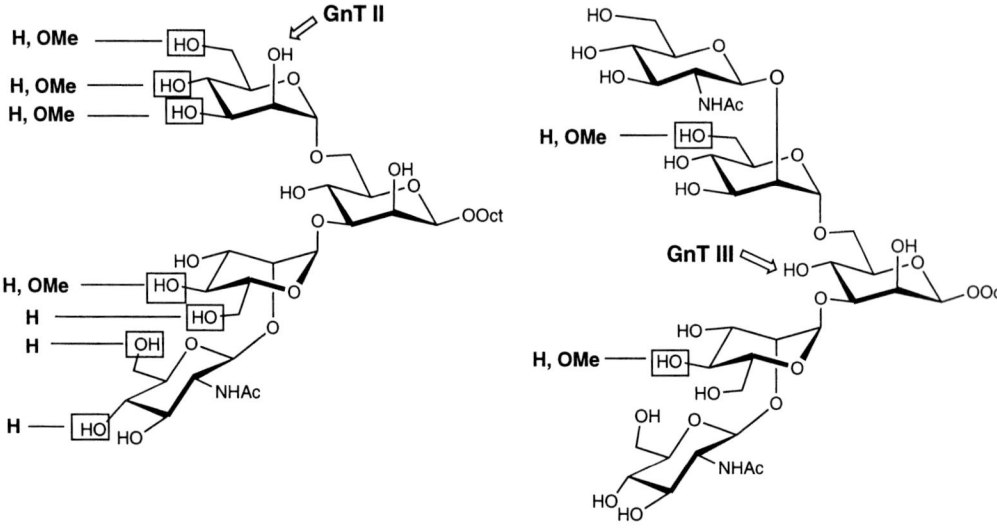

Scheme 10. Non-natural acceptors for GnT II and III.

Deoxygenation at C-4 [106], C-6 [107] or methylation at C-4 [99, 103] of αMan1–3 are also tolerated by GnT II.

A dideoxygenated pentasaccharide, which was chemo-enzymatically synthesized by use of GnT II [106], and a di-O-methylated pentasaccharide were active as substrates for GnT III (E.C. 2.4.1.144) producing the expected hexasaccharide analogs [108, 109].

N-Acetylglucosaminyltransferase V

There has been particular interest in N-acetylglucosaminyltransferase V (GnT V, E.C. 2.4.1.155) because the activity of this enzyme has been shown to correlate with the metastatic potential of tumor cell lines [110–114].

Whereas a heptasaccharide is the smallest physiological substrate, trisaccharides GlcNAc(β1–2)Man(α1–6)Manβ-OR [115, 116] and GlcNAc(β1–2)Man(α1–6)-Glcβ-OR [117] were found to be excellent acceptors for GnT V. The β-GlcNAc residue can also be replaced by a β-Glc residue [118]. Studies using analogs where OH-3, 4, or 6 of the β-GlcNAc residue of GlcNAc(β1–2)Man(α1–6)Glcβ-OR were modified indicated that an intact 3,4,6-triol system is required for recognition by GnT V [119, 120]. Modifications on the α-Man residue and β-Glc residues are tolerated by this enzyme (Scheme 11) [117, 121–123].

Kinetic studies using restricted trisaccharide analogs revealed that GnT V preferentially recognizes the gg rotamer [124]. Analogs where the oxygen atoms in the glycosidic linkages were replaced with sulfur are also good acceptors for GnT V producing the corresponding tetrasaccharides [125].

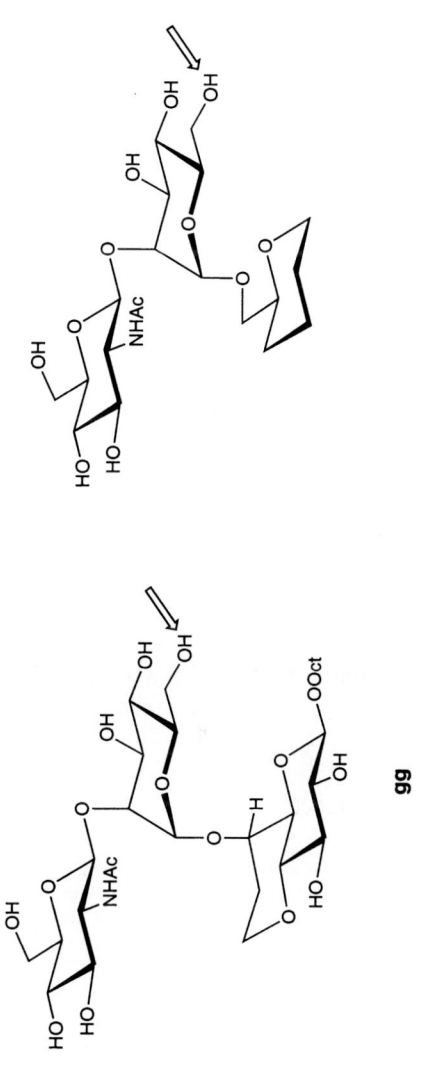

Scheme 11. Non-natural acceptors for GnT V.

27.3 Summary

Glycosyltransferases are remarkably flexible in their donor and acceptor specificities. The use of glycosyltransferases greatly simplifies the synthetic procedure for the preparation of non-natural oligosaccharides starting from non-natural donors and/or non-natural acceptors. With progress in molecular biology, even more glycosyltransferases will become available in sufficient quantities for use in chemical synthesis. As the structures of glycosyltransferases become well defined, they can be tailored using site-directed mutagenesis to have enhanced stabilities, even broader specificities and high affinities towards non-natural substrates while maintaining regio- and stereospecificity. Large-scale synthesis of oligosaccharides using a CMP–Neu5Ac synthetase/α2,3-sialyltransferase fusion has been reported [126]. Combining two or more enzymes into a fusion protein will simplify the production and purification of enzymes, reduce costs and facilitate synthesis. As improved and alternate methods for the regeneration [11] and large-scale production [127] of sugar nucleotide donors are developed, enzymatic glycosylation will become increasingly routine for the construction of large quantities of many non-natural oligosaccharides.

Acknowledgments

Research support from the Natural Sciences and Engineering Research Council of Canada (to M.M.P.) and the Alberta Research Council for a graduate scholarship in Carbohydrate Chemistry (to X.Q.) are gratefully acknowledged.

References

1. Varki, A. *Glycobiology* **1993**, *3*, 97.
2. Dwek, R. A. *Chem. Rev.* **1996**, *96*, 683.
3. Toshima, K.; Tatsuta, K. *Chem. Rev.* **1993**, *93*, 1603.
4. Khan, S. H.; Hindsgaul, O. In *Molecular Glycobiology*; Fukuda, M., Hindsgaul, O. Eds.; Oxford: New York, **1994**; pp 206.
5. Barresi, F.; Hindsgaul, O. *J. Carbohydr. Chem.* **1995**, *14*, 1043.
6. Boons, G.-J. *Tetrahedron* **1996**, *52*, 1095.
7. Schmidt, R. R. In *Glycosciences*; Gabius, H.-J., Gabius, S. Eds.; Chapman & Hall: Weinheim, **1997**; pp 31–53.
8. McAuliffe, J.; Hindsgaul, O. *Chem. Ind.* **1997**, 170.
9. Wong, C.-H.; Whitesides, G. M. *Enzymes in Synthetic Organic Chemistry*; Elsevier: Oxford, **1994**; pp. 252–311.
10. Palcic, M. M. *Methods Enzymol.* **1994**, *230*, 300.
11. Wong, C.-H.; Halcomb, R. L.; Ichikawa, Y.; Kajimoto, T. *Angew. Chem., Int. Ed. Engl.* **1995**, *34*, 521.
12. Brockhausen, I.; Schachter, H. In *Glycosciences*; Gabius, H.-J., Gabius, S. Eds.; Chapman & Hall: Weinheim, **1997**; pp 79–113.
13. Gijsen, H. J. M.; Qiao, L.; Fitz, W.; Wong, C.-H. *Chem. Rev.* **1996**, *96*, 443.
14. Takayama, S.; McGarvey, G. J.; Wong, C.-H. *Chem. Soc. Rev.* **1997**, *26*, 407.

15. Ichikawa, Y. In *Glycopeptides and Related Compounds: Synthesis, Analysis and Application*, Large, D. G., Warren, C. D. Eds.; Dekker: New York, **1997**; pp 79–205.
16. Palcic, M. M.; Hindsgaul, O. *Trends Glycosci. Glycotechnol.* **1996**, *8*, 37.
17. Lemieux, R. U. In *Proc. VIIIth Int. Symp. Med. Chem.*; Swedish Pharmaceutical Press: Stockholm, **1985**; pp. 329–351.
18. Lemieux, R. U. *Chem. Soc. Rev.* **1989**, *18*, 347.
19. Lemieux, R. U. *Acc. Chem. Res.* **1996**, *29*, 373.
20. Sears, P.; Wong, C.-H. *Cell. Mol. Life Sci.* **1998**, *54*, 223.
21. Matta, K. L. In *Modern Methods in Carbohydrate Synthesis*; Khan, S. H.; O'Neil, R. A. Eds.; Harwood Academic: Amsterdam, **1995**; pp 437–466.
22. Crawley, S. C.; Palcic, M. M. In *Modern Methods in Carbohydrate Synthesis*; Khan, S. H.; O'Neil, R. A. Eds.; Harwood Academic: Amsterdam, **1995**; pp 492–517.
23. Elhalabi, J. M.; Rice, K. G. *Curr. Med. Chem.* **1999**, *6*, 93.
24. Öhrlein, R. *Top. Curr. Chem.* **1999**, *200*, 227.
25. Thiem, J.; Wiemann, T. *Angew. Chem., Intl. Ed. Engl.* **1991**, *30*, 1163.
26. Thiem, J.; Wiemann, T. *Synthesis* **1992**, 141.
27. Wong, C.-H.; Wang, R.; Ichikawa, Y. *J. Org. Chem.* **1992**, *57*, 4343.
28. Srivastava, G.; Hindsgaul, O.; Palcic, M. M. *Carbohydr. Res.* **1993**, *245*, 137.
29. Hindsgaul, O.; Kaur, K. J.; Gokhale, U. B.; Srivastava, G.; Alton, G.; and Palcic, M. M. *ACS Symp. Ser.* **1991**, *466*, 38.
30. Berliner, L. J.; Robinson, R. D. *Biochemistry* **1982**, *21*, 6340.
31. Kodama, H.; Kajihara, Y.; Endo, T.; Hashimoto, H. *Tetrahedron Lett.* **1993**, *34*, 6419.
32. Kajihara, Y., Endo, T., Ogasawara, H., Kodama, H.; Hashimoto, H. *Carbohydr. Res.* **1995**, *269*, 273.
33. Arlt, M.; Hindsgaul, O. *J. Org. Chem.* **1995**, *60*, 14.
34. Palcic, M. M.; Hindsgaul, O. *Glycobiology* **1991**, *1*, 205–209.
35. Do, K.-Y.; Do, S.-I.; Cummings, R. D. *J. Biol. Chem.* **1995**, *270*, 18447.
36. Andree, P. J.; Berliner, L. J. *Biochim. Biophys. Acta* **1978**, *544*, 489.
37. Yuasa, H.; Hindsgaul, O.; Palcic, M. M. *J. Am. Chem. Soc.* **1992**, *114*, 5891.
38. Tsuruta, O.; Shinohara, G.; Yuasa, H.; Hashimoto, H. *Bioorg. Med. Chem. Lett.* **1997**, *7*, 2523.
39. Ichikawa, Y.; Lin, Y.-C.; Dumas, D. P.; Shen, G.-J.; Garcia-Junceda, E.; Williams, M. A.; Bayer, R.; Ketcham, K.; Walker, L. E.; Paulson, J. C.; Wong, C.-H. *J. Am. Chem. Soc.* **1992**, *114*, 9283.
40. Berliner, L. J.; Davies, M. E.; Ebner, K. E.; Beyer, T. A.; Bell, J. E. *Mol. Cell. Biochem.* **1984**, *62*, 37.
41. Palcic, M. M.; Srivastava, O. P.; Hindsgaul, O. *Carbohydr. Res.* **1987**, *159*, 315.
42. Öhrlein, R.; Ernst, B.; Berger, E. G. *Carbohydr. Res.* **1992**, *236*, 335.
43. Baisch, G.; Öhrlein, R.; Ernst, B. *Bioorg. Med. Chem. Lett.* **1996**, *6*, 749.
44. Kajihara, Y.; Kodama, H.; Endo, T.; Hashimoto, H. *Carbohydr. Res.* **1998**, *306*, 361.
45. Wong, C.-H.; Ichikawa, Y.; Krach, T.; Gautheron-Le Narvor, C.; Dumas, D. P.; Look, G. C. *J. Am. Chem. Soc.* **1991**, *113*, 8137.
46. Yu, L.; Cabrera, R.; Ramirez, J.; Malinovskii, V. A.; Brew, K.; Wang, P. G. *Tetrahedron Lett.* **1995**, *36*, 2897.
47. Nishida, Y.; Wiemann, T.; Sinnwell, V.; Thiem, J. *J. Am. Chem. Soc.* **1993**, *115*, 2536.
48. Gambert, U.; Thiem, J. *Top. Curr. Chem.* **1997**, *186*, 21.
49. Nishida, Y.; Wiemann, T.; Thiem, J. *Tetrahedron Lett.* **1992**, *33*, 8043.
50. Nishida, Y.; Wiemann, T.; Thiem, J. *Tetrahedron Lett.* **1993**, *34*, 2905.
51. Wiemann, T.; Nishida, Y.; Sinnwell, V.; Thiem, J. *J. Org. Chem.* **1994**, *59*, 6744.
52. Nishimura, S.-I.; Lee, K. B.; Matsuoka, K.; Lee, Y. C. *Biochem. Biophys. Res. Commun.* **1994**, *199*, 249.
53. Matsuoka, K.; Nishimura, S.-I. *Macromolecules* **1995**, *28*, 2961.
54. Wiemann, T.; Taubken, N.; Zehavi, U.; Thiem, J. *Carbohydr. Res.* **1994**, *257*, C1.
55. Kopper, S. *Carbohydr. Res.* **1994**, *265*, 161.
56. Schuster, M.; Wang, P.; Paulson, J. C.; Wong, C.-H. *J. Am. Chem. Soc.* **1994**, *116*, 1135.
57. Meldal, M.; Auzanneau, F.-I.; Hindsgaul, O.; Palcic, M. M. *J. Chem. Soc., Chem. Commun.* **1994**, 1849.

58. Joziasse, D. H.; Shaper, N. L.; Salyer, L. S.; van den Eijnden, D. H.; van der Spoel, A. C.; Shaper, J. H. *Eur. J. Biochem.* **1990**, *191*, 75.
59. Fang, J.; Li, J.; Chen, X.; Zhang, Y.; Wang, J.; Guo Z.; Zhang, W.; Yu, L.; Brew, K.; Wang, P. G. *J. Am. Chem. Soc.* **1998**, *120*, 6635.
60. Hokke, C. H.; Zervosen, A.: Elling, L.; Joziasse, D. H.; van den Eijnden, D. H. *Glycoconjugate Journal* **1996**, *13*, 687.
61. Galili, U. *Immunol. Today* **1993**, *14*, 480.
62. Butler, D. *Nature* **1998**, *391*, 320.
63. Stults, C. L. M.; Macher, B. A.; Bhatti, R.; Srivastava, O. P.; Hindsgaul, O. *Glycobiology* **1999**, *9*, 661.
64. Sujino, K.; Uchiyama, T.; Hindsgaul, O.; Seto, N. O. L.: Wakarchuk, W. W.; Palcic, M. M. *J. Am. Chem. Soc.* **2000**, *122*, 1261.
65. Blanken, W. M.; Van den Eijnden, D. H. *J. Biol. Chem.* **1985**, *260*, 12927.
66. Baisch, G.; Öhrlein, R.; Kolbinger, F.; Streiff, M. *Bioorg. Med. Chem. Lett.* **1998**, *8*, 1575.
67. Sujino, K.; Malet, C.; Hindsgaul, O.; Palcic, M. M. *Carbohydr. Res.* **1998**, *305*, 483.
68. Erika, S. *Trends Glycosci. Glycotechnol.* **1996**, *8*, 391.
69. Gokhale, U. B.; Hindsgaul, O.; Palcic, M. M. *Can. J. Chem.* **1990**, *68*, 1063.
70. Stangier, K.; Palcic, M. M.; Bundle, D. R.; Hindsgaul, O. Thiem, J. *Carbohydr. Res.* **1998**, *305*, 511.
71. Srivastava, G.; Kaur, K. J.; Hindsgaul, O.; Palcic, M. M. *J. Biol. Chem.* **1992**, *267*, 22356.
72. Hallgren, C.; Hindsgaul, O. *J. Carbohydr. Chem.* **1995**, *14*, 453–464.
73. Gosselin, S.; Palcic, M. M. *Bioorg. Med. Chem.* **1996**, *4*, 2023.
74. Nikrad, P. V.; Kashem, M. A.; Wlasichuk, K. B.; Alton, G.; Venot, A. P. *Carbohydr. Res.* **1993**, *250*, 145.
75. Ding, Y.; Hindsgaul, O.; Li, H.; Palcic, M. M. *Bioorg. Med. Chem. Lett.* **1998**, *8*, 3199.
76. Ogawa, S.; Matsunaga, N.; Li, H.; Palcic, M. M. *Eur. J. Org. Chem.* **1999**, 631.
77. Qian, X.; Hindsgaul, O.; Li, H.; Palcic, M. M. *J. Am. Chem. Soc.* **1998**, *120*, 2184.
78. Baisch, G.; Öhrlein, R.; Katopodis, A.; Streiff, M.; Kolbinger, F. *Bioorg. Med. Chem. Lett.* **1997**, *7*, 2447.
79. Baisch, G.; Öhrlein, R.; Katopodis, A.; Ernst, B. *Bioorg. Med. Chem. Lett.* **1996**, *6*, 759.
80. Baisch, G.; Öhrlein, R.; Streiff, M. *Bioorg. Med. Chem. Lett.* **1998**, *8*, 161.
81. Baisch, G.; Öhrlein, R.; Katopodis, A. *Bioorg. Med. Chem. Lett.* **1997**, *7*, 2431.
82. Baisch, G.; Öhrlein, R.; Streiff, M.; Kolbinger, F. *Bioorg. Med. Chem. Lett.* **1998**, *8*, 755.
83. Baisch, G.; Öhrlein, R. *Bioorg. Med. Chem.* **1998**, *6*, 1673.
84. Baisch, G.; Öhrlein, R.; Katopodis, A. *Bioorg. Med. Chem. Lett.* **1996**, *6*, 2953.
85. Chandrasekaran, E. V.; Jain, R. K.; Larsen, R. D.; Wlasichuk, K.; DiCioccio, R. A.; Matta, K. L. *Biochemistry* **1996**, *35*, 8925.
86. Reuter, W.; Stache, R.; Stehling, P., Baum, O. In *Glycosciences*; Gabius, H.-J., Gabius, S. Eds.; Chapman & Hall: Weinheim, **1997**; pp 245–259.
87. Harduin-Lepers, A.; Recchi, M.-A.; Delannoy, P.; *Glycobiology* **1995**, *5*, 741.
88. Higa, H. H.; Paulson, J. C. *J. Biol. Chem.* **1985**, *260*, 8838.
89. Gross, H. J.; Bunsch, A.; Paulson, J. C.; Brossmer, R. *Eur. J. Biochem.* **1987**, *168*, 595.
90. Brossmer, R.; Gross, H. J. *Methods Enzymol.* **1994**, *247*, 153.
91. Gross, H. J.; Rose, U.; Krause, J. M.; Paulson, J. C.; Schmid, K.; Feeney, R. E.; Brossmer, R. *Biochemistry* **1989**, *28*, 7386.
92. Gross, H. J.; Brossmer, R. *Eur. J. Biochem.* **1988**, *177*, 583.
93. Chappell, M. D.; Halcomb, R. L.; *J. Am. Chem. Soc.* **1997**, *119*, 3393.
94. Gross, H. J.; Brossmer, R. *Glycoconjugate J.* **1987**, *4*, 145–156.
95. Wlasichuk, K. B.; Kashem, M. A.; Nikrad, P. V.; Bird, P.; Jiang, C.; Venot, A. P. *J. Biol. Chem.* **1993**, *268*, 13971.
96. Van Dorst, J. A. L. M.; Tikkanen, J. M.; Krezdorn, C. H.; Streiff, M. B.; Berger, E. G.; Van Kuik, J. A.; Kamerling, J. P.; Vliegenthart, J. F. G. *Eur. J. Biochem.* **1996**, *242*, 674.
97. Baisch, G.; Öhrlein, R.; Streiff, M.; Ernst, B. *Bioorg. Med. Chem. Lett.* **1996**, *6*, 755.
98. Baisch, G.; Öhrlein, R.; Streiff, M. *Bioorg. Med. Chem. Lett.* **1998**, *8*, 157.
99. Taniguchi, N.; Ihara, Y. *Glycoconjugate J.* **1995**, *12*, 733.
100. Vella, G. J.; Paulsen, H.; Schachter, H. *Can. J. Biochem. Cell. Biol.* **1984**, *62*, 409.

101. Palcic, M. M.; Heerze, L. D.; Pierce, M.; Hindsgaul, O. *Glycoconjugate J.* **1988**, *5*, 49.
102. Srivastava, G.; Alton, G.; Hindsgaul, O. *Carbohydr. Res.* **1990**, *207*, 259.
103. Reck, F.; Springer, M.; Paulsen, H.; Brockhausen, I.; Sarkar, M.; Schachter, H. *Carbohydr. Res.* **1994**, *259*, 93.
104. Möller, G.; Reck, F.; Paulsen, H.; Kaur, K. J.; Sakar, M.; Schachter, H.; Brockhausen, I. *Glycoconjugate J.* **1992**, *9*, 180.
105. Reck, F.; Springer M.; Meinjohanns, E.; Paulsen, H.; Brockhausen, I.; Schachter, H. *Glycoconjugate J.* **1995**, *12*, 747.
106. Kaur, K. J.; Hindsgaul, O. *Carbohydr. Res.* **1992**, *226*, 219.
107. Reck, F.; Meinjohanns, E.; Springer M.; Wilkens, R.; Van Drost J. A. L. M.; Paulsen, H.; Möller, G.; Brockhausen, I.; Schachter, H. *Glycoconjugate J.* **1994**, *11*, 210.
108. Alton, G.; Kanie, Y.; Hindsgaul, O. *Carbohydr. Res.* **1993**, *238*, 339.
109. Khan, S. H.; Compston, C. A.; Palcic, M. M.; Hindsgaul, O. *Carbohydr. Res.* **1994**, *262*, 283.
110. Yamashita, K.; Tachibana, Y.; Ohkura, T.; Kobata, A. *J. Biol. Chem.* **1985**, *260*, 3963.
111. Arango, J.; Pierce, M. *J. Cell. Biochem.* **1988**, *37*, 225.
112. Dennis, J. W.; Kosh, E.; Bryce, D.-M.; Breitman, M. L. *Oncogene* **1989**, *4*, 853.
113. Dennis, J. W.; Laferte, S.; Waghorne, C.; Breitman, M. L.; Kerbel, R. S. *Science* **1987**, *236*, 582.
114. Dennis, J. W. *Cancer Surveys* **1988**, *7*, 573.
115. Tahir, S. H.; Hindsgaul, O. *Can. J. Chem.* **1986**, *64*, 1771.
116. Hindsgaul, O.; Tahir, S. H.; Srivastava, O. P.; Pierce, M. *Carbohydr. Res.* **1988**, *173*, 263.
117. Srivastava, O. P., Hindsgaul, O.; Shoreibah, M.; Pierce, M. *Carbohydr. Res.* **1988**, *179*, 137.
118. Kanie, O.; Palcic, M. M.; Hindsgaul, O. *RIKEN Rev.* **1995**, 853.
119. Kanie, O.; Crawley, S. C.; Palcic, M. M.; Hindsgaul, O. *Carbohydr. Res.* **1993**, *243*, 139.
120. Kanie, O.; Crawley, S. C.; Palcic, M. M.; Hindsgaul, O. *Bioorg. Med. Chem.* **1994**, *2*, 1231.
121. Linker, T.; Crawley, S. C.; Hindsgaul, O. *Carbohydr. Res.* **1993**, *245*, 323.
122. Khan, S. H.; Duus, J. D.; Crawley, S. C.; Palcic, M. M.; Hindsgaul, O. *Tetrahedron Asym.* **1994**, *5*, 2415.
123. Ogawa, S.; Furuya, T.; Tsunoda, H.; Hindsgaul, O.; Stangier, K.; Palcic, M. M. *Carbohydr. Res.* **1995**, *271*, 197.
124. Lindh, I.; Hindsgaul, O. *J. Am. Chem. Soc.* **1991**, *113*, 216.
125. Lu, P.-P.; Hindsgaul, O.; Li, H.; Palcic, M. M. *Can. J. Chem.* **1997**, *75*, 790.
126. Gilbert, M.; Bayer, R.; Cunningham, A.-M.; DeFrees, S.; Gao, Y.; Watson, D. C.; Young, N. M.; Wakarchuk, W. W. *Nature Biotech.* **1998**, *16*, 769.
127. Koizumi, S.; Endo, T.; Tabata, K.; Ozaki, A. *Nature Biotech.* **1998**, *16*, 847.

28 Solid-Phase Synthesis with Glycosyltransferases

Claudine Augé, Christine Le Narvor, and André Lubineau

28.1 Introduction

In recent years chemical solid-phase synthesis has gained growing interest in parallel to the development of the concept of combinatorial chemistry, a promising approach to the discovery of new biologically active compounds. In peptide and oligonucleotide chemistry this technology has been established for a long time, but not in oligosaccharide chemistry. Pioneering studies in this field suffered from the lack of effective glycosyl donors and high stereoselective coupling methods [1–3]. Recent advances in this area, and development of highly reactive sugar donors and protective group chemistry have stimulated a renewed interest in solid-phase synthesis of oligosaccharides [4–6] and glycopeptides [7, 8].

28.2 General Aspects

Solid-phase synthesis has several advantages over traditional solution-based methods. Firstly, large excesses of reagents can be used so that the coupling steps can be repeated to drive reactions to completion. Secondly, products are isolated at each step by simple filtration, an obvious simplification of the work-up procedures, which might enable future automation.

Development of polymer-supported synthesis of oligosaccharides using purely chemical glycosylation methods, a chemo–enzymatic approach based on the use of glycosyltransferases has also started to develop. Glycosyltransferases have become widely used in the past ten years as efficient tools for glycosylation; as opposed to glycosidases, another class of enzymes of interest in oligosaccharide solution phase synthesis, these enzymes catalyze sugar unit transfer from a sugar–nucleotide donor on to an unprotected sugar acceptor, with complete regio and stereoselectivity.

Thus these also seem appropriate reagents for use in solid-phase synthesis. These enzymes have, moreover, become increasingly available as a result of genetic engineering, and soluble forms of glycosyltransferases, originally membrane-bound proteins, have been generated. Solid-phase enzymatic synthesis raises extra problems, however, such as accessibility of the interior of the solid matrix to the enzyme or, for chemo–enzymatic synthesis, the biocompatibility of the support with both aqueous and organic solvents. Indeed this approach requires that the resins swell in aqueous buffers which, for example, precludes the use of the hydrophobic Merrifield type resin. Roughly, the amounts of swelling in different solvents give information on the overall polar or hydrophobic character of the support [9]. But even the microporous polyoxyethylene–polystyrene, sold under the trade name of Tentagel, which has good swelling properties in polar solvents, does not allow macromolecules such as enzymes of 20–50 kDa and more, to penetrate into the interior of the beads. A recent study on chymotrypsin-catalyzed proteolysis of peptide substrates bound to Tentagel beads has shown that only the surface of the beads, i.e. 15% of the total functional sites at most, are available to the enzyme [10].

The general scheme for solid-phase synthesis using glycosyltransferases is depicted in Scheme 1. Solid-phase synthesis requires a covalent linker group to attach the small molecule, a monosaccharide or an amino acid in the case of a glycopeptide, on to the polymeric resin. Such a linker must be stable to the reaction con-

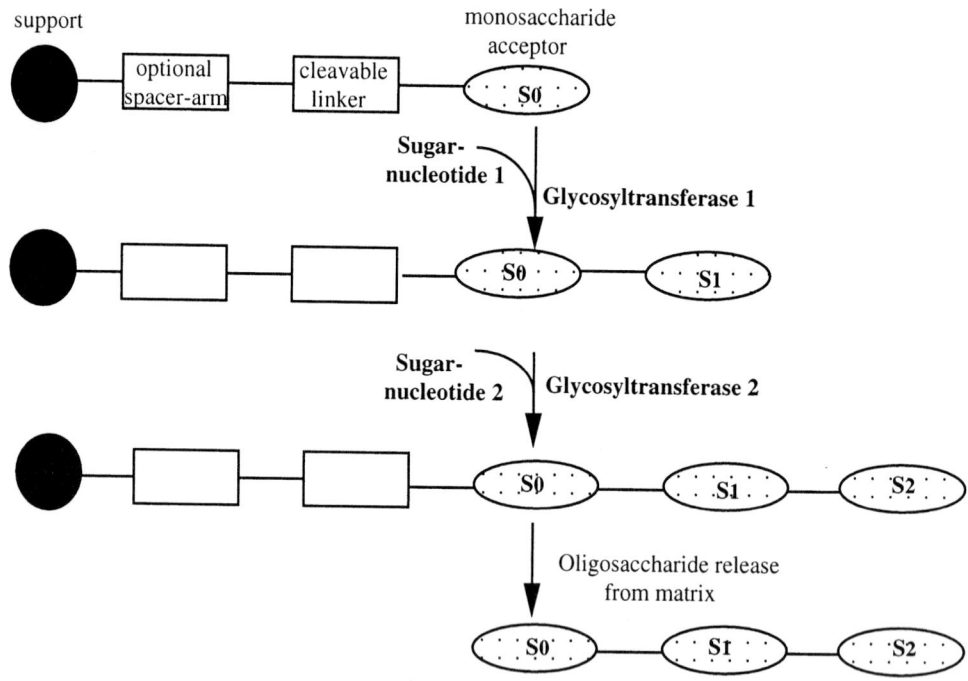

Scheme 1. General scheme for the solid-phase synthesis of oligosaccharides using glycosyltransferases.

ditions used during the synthesis but it needs to be cleaved selectively at the end of the synthesis, releasing the oligosaccharide or the glycopeptide from the resin into solution; light-sensitive or enzyme-susceptible linkers are particularly suitable for this purpose. A spacer-arm providing distance between the support and the acceptor substrate to eliminate steric problems can be optionally intercalated between the matrix and the linker. Sugar-chain elongation is the result of the sequential action of glycosyltransferases in the presence of the corresponding sugar–nucleotide donor.

We shall describe enzymatic glycosylations that have been achieved on both insoluble and soluble supports, because, irrespective of the nature of the support, polymer-supported synthesis is based on the same concept, and good results with soluble supports have been reported in the literature.

28.3 Enzymatic Synthesis on Insoluble Supports

28.3.1 Enzymatic Synthesis of Oligosaccharides

Use of an Amino-Functionalized Water-Compatible Polyacrylamide Gel

Pioneering work in the solid-phase enzymatic synthesis of oligosaccharides was accomplished by Zehavi and co-workers. The authors used an amino-functionalized water-compatible polyacrylamide gel; it is well-known that polyacrylamide gels, the preferred polymer gels for electrophoresis, allow proteins to diffuse inside. Zehavi et al. first attached one glucose unit to a photo-removable linker, methyl 4-hydroxymethyl-3-nitrobenzoate. In the second step the unprotected glycoside **1** was linked to aminoethyl-substituted polyacrylamide gel beads *via* an amide linkage (Scheme 2). The glucopolymer **2** although obtained in a low yield, served as an acceptor in $\beta(1–4)$galactosyltransferase-catalyzed reaction, in the presence of α-lactalbumin, with ^{14}C radiolabeled UDP–Gal.

The lactose disaccharide **4** was subsequently released from the lactosyl polymer **3** by irradiation at 320 nm, in quantitative yield, but nevertheless the overall yield of lactose, estimated from radiolabelling was disappointingly very low (<1%), probably because of inaccessibility of the substrate to the enzyme [11]. Such light-sensitive polymers with maltose and maltotriose side chains were also prepared for use as acceptors in the glycogen synthase reaction, but only slightly better transfers (7%) were observed [12, 13].

To improve the accessibility of saccharide acceptors in glycosylation reactions, polymers with longer spacers were prepared [14]. Thus a glucopolymer **5** with an aminooctyl spacer between the polyacrylamide support and the photolabile linker became an effective support for enzymatic glycosidation catalyzed by $\beta(1–4)$galactosyltransferase. The reaction was conducted on 0.5 mmol scale with *in situ* generation of UDP–Gal from UDP–Glc by adding UDP–Glc epimerase. Lactose **4** released from lactopolymer **6** by photolysis was isolated in an overall yield of 51% (Scheme 2) [15].

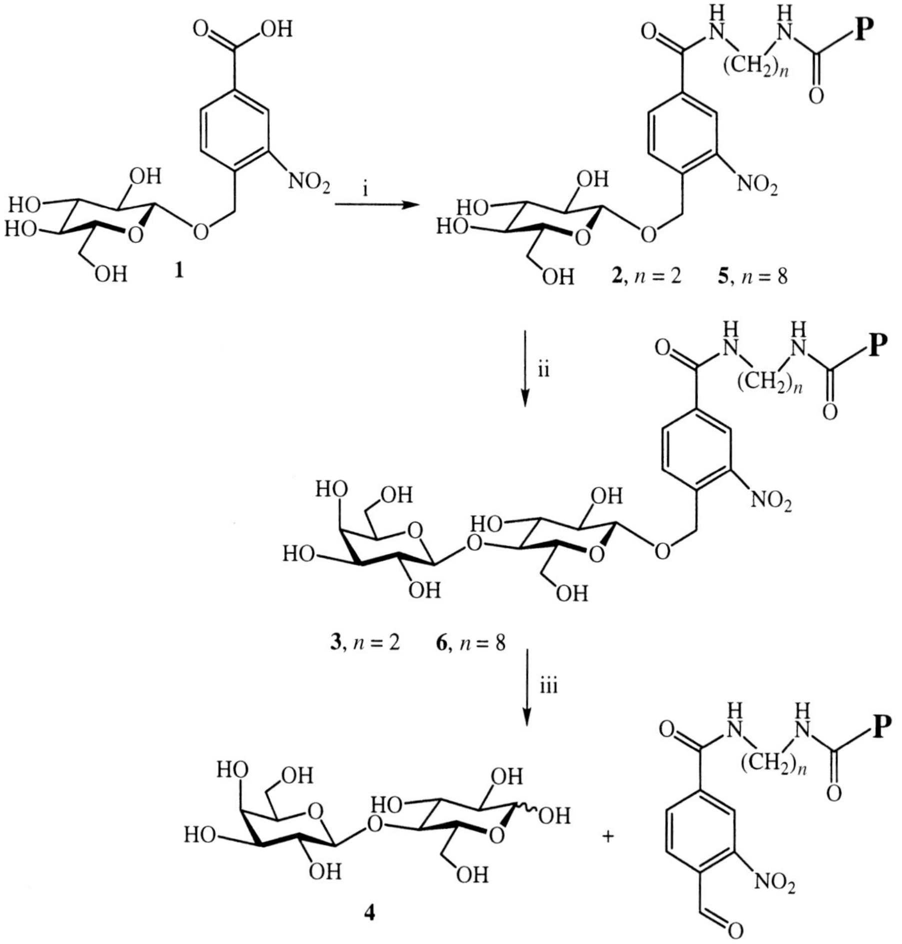

Scheme 2. Synthesis of lactose *via* enzymatic glycosylation using a photolabile solid support. Reagents: i) $NH_2-(CH_2)_n-NH-CO-P$, water–DMF, EDCD, 10% for **2**, 37% for **5**; ii) α-lactalbumin, GT, UDP–Gal (3 eq.), cacodylate buffer pH 7, 3 mM $MnCl_2$, for **3**; α-lactalbumin, GT, UDP–Glc (1.1 eq), UDPGE, cacodylate buffer pH 7, 0.3 mM $MnCl_2$, 0.02% NaN_3, for **6**; iii) hν, 1% for **4** from **2**, 51% for **4** from **5**.

Use of a Sepharose Matrix

In 1980 Barker and Nunez [16] reported the synthesis of Galβ1-4GlcNAcβ-hexanolamine by use of a solid-phase system in which the GlcNAcβ-hexanolamine glycoside was covalently linked to CNBr-activated Sepharose 4B, with 6 to 10 μmol ligand mL^{-1} packed wet gel; it is noteworthy that the same coupling method has been widely applied by David *et al.* to enzyme immobilization [17]. With partially purified β(1–4)galactosyltransferase from bovine milk, and a twofold excess of UDP–Gal, the Sepharose-bound *N*-acetylglucosamine **7** was converted into **8** in

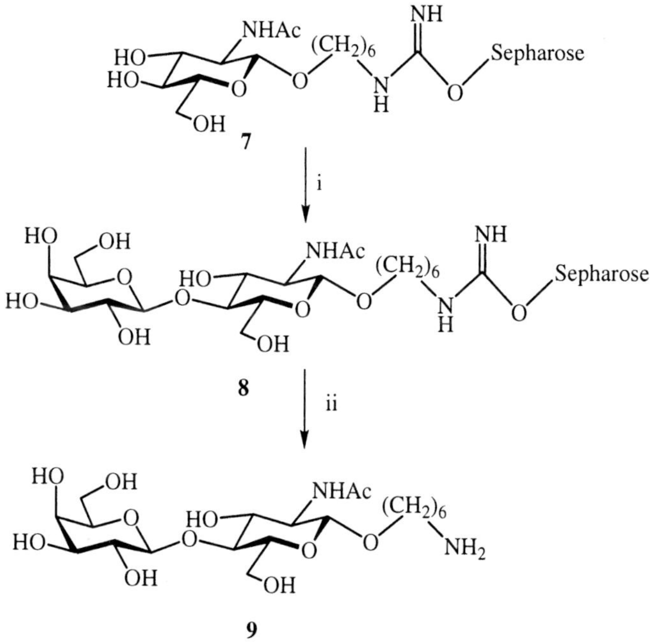

Scheme 3. Enzymatic synthesis of LacNAcβ-hexanolamine on Sepharose. Reagents: i) GT, UDP–Gal (2 eq.), cacodylate buffer pH 7, 5 mM MnCl$_2$, 0.1% mercaptoethanol, 18 h, 35 °C, 85%; ii) 2 M NaOH, 40 °C, 3 h, 80%.

85% yield; then the disaccharide glycoside **9** could be cleaved from the matrix, in 80% yield, by alkaline treatment (Scheme 3).

Very recently, in a related approach, Norberg et al. presented the solid-phase enzymatic fucosylation of a disaccharide acceptor linked to Sepharose *via* a disulfide linkage and a linker arm of 12 atoms with a loading of 5 to 10 µmol disaccharide ligand mL^{-1} wet gel [18]. The sepharose-bound disaccharide **10** incubated with human milk α(1–3/4)fucosyltransferase and GDP–Fuc in fivefold excess was converted into the Sepharose-bound trisaccharide **11** in 68% glycosylation yield (Scheme 4). The Lewis[a] trisaccharide derivative **12** could be released from the matrix, together with the starting disaccharide **13**, by treatment with 2-mercaptoethanol, in 91% yield.

Because the difficulties encountered in this enzymatic fucosylation step could not be improved, even after extensive washing of the gel and recycling with fresh enzyme and sugar-nucleotide, it was assumed that low reactivity was a result of steric factors. The same matrix was, therefore, further substituted with longer linkers of the PEG-type. Sepharose gels with linker lengths of 35, 47, 59 and 71 atoms between the matrix and the *N*-acetylglucosamine acceptor were built and subjected to enzymatic galactosylation with β(1–4)galactosyltransferase and excess UDP–Gal [19]. Aliquots were removed and treated with DTT; the ^{1}H NMR spectrum of the

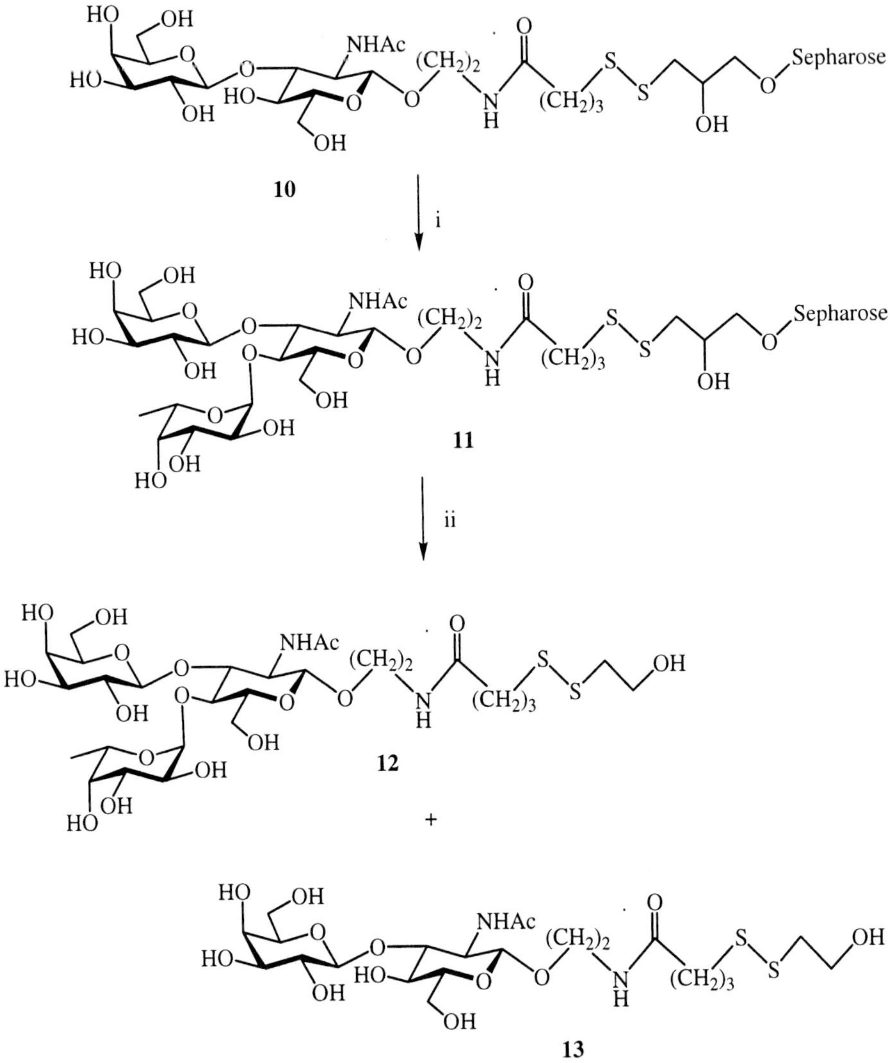

Scheme 4. Solid-phase enzymatic synthesis of a Lewis[a] trisaccharide by use of a disaccharide acceptor bound to Sepharose. Reagents: i) cacodylate buffer pH 6.8, 5 mM $MnCl_2$, 0.05% NaN_3, FT, GDP–Fuc (5 eq.) 68%; ii) 2-mercaptoethanol, 1 h, 60 °C, 91%.

filtrate enabled estimation of the yield. The longest linker turned out to give the best yield (98%). The GlcNAc Sepharose with the longest linker was then successfully converted into the sialyl Lewis[x] tetrasaccharide **16**, by successive incubation of the gel **14** with $\beta(1–4)$galactosyltransferase, recombinant $\alpha(2–3)$sialyltransferase and human milk $\alpha(1–3/4)$fucosyltransferase, in the presence of the corresponding sugar–nucleotides, used in three to fourfold excess, and alkaline phosphatase,

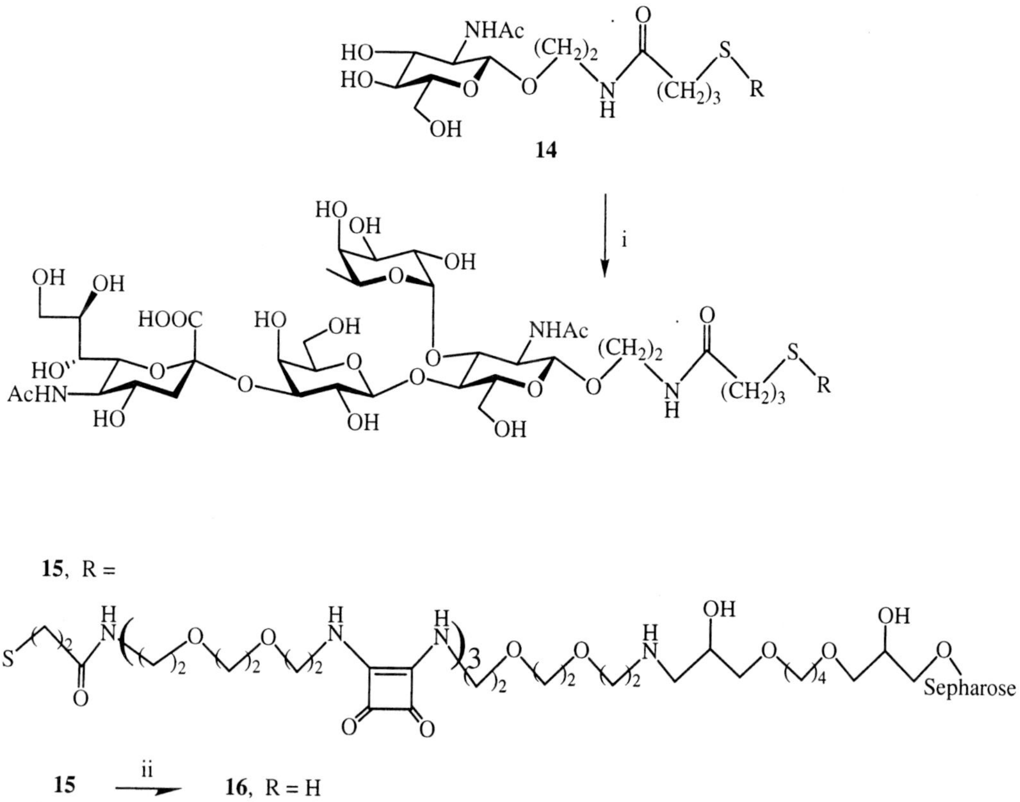

Scheme 5. Solid-phase enzymatic synthesis of a Lewis^x tetrasaccharide using a linker of 71 atoms between the monosaccharide acceptor and the Sepharose matrix. Reagents: i) 1. cacodylate buffer pH 7.5, 5 mM MnCl$_2$, α-lactalbumin, GT, UDP–Gal (3 eq.), CIAP, 37 °C, 5 days; 2. cacodylate buffer pH 7.35, 0.1% triton X-100, ST3, CMP–NeuAc (2.5 eq.), 3 mM MnCl$_2$, 0.02% NaN$_3$, 35 °C, 5 days; 3. cacodylate buffer pH 6.5, 5 mM MnCl$_2$, FT, GDP–Fuc (4 eq.), 37 °C, 5 days; ii) 1. DTT, 0.1 M phosphate buffer; 2. BioGel P-2, 57% from **14**.

followed by final treatment of the gel **15** with dithiothreitol (57% overall yield, Scheme 5).

Use of Controlled-Pore Glass

Wong and coworkers promoted the use of aminopropyl silica for solid-phase enzymatic synthesis [20]. This rigid, non-swelling support turned out to be compatible with biomolecules, but it required a spacer for rendering the substrate more accessible to the enzyme. The authors prepared a disaccharide target carrying at the reducing end a C-6 spacer-arm with a carboxylic function that was condensed *via* its cesium salt to commercially available *N*-iodoacetylaminopropyl controlled-pore glass, affording the solid-supported disaccharide acceptor **17** (Scheme 6). This ac-

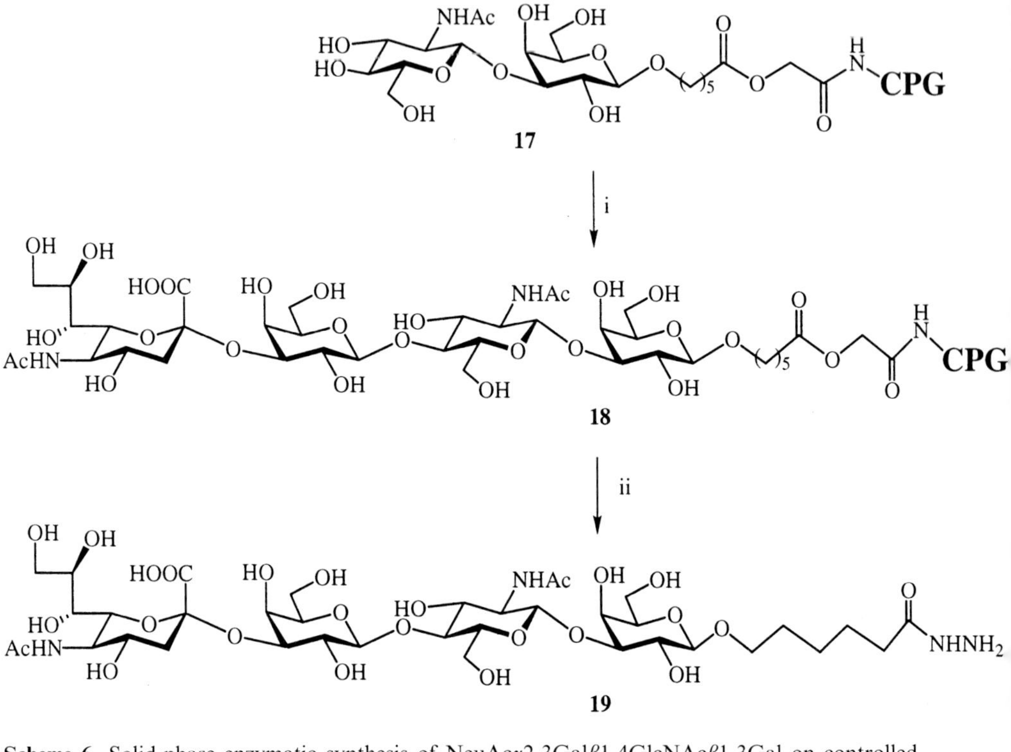

Scheme 6. Solid-phase enzymatic synthesis of NeuAcα2-3Galβ1-4GlcNAcβ1-3Gal on controlled-pore glass. Reagents: i) 1. HEPES buffer pH 7.2, 5 mM MnCl$_2$, GT, UDP–Gal (3 eq.), 0.5% DTT, 48 h; 2. cacodylate buffer pH 7.5, CMP–NeuAc (2 eq.), CIAP, 48 h, q.y.; ii) NH$_2$-NH$_2$, room temp., 24 h.

ceptor could be enzymatically galactosylated with β(1–4)galactosyltransferase and UDP–Gal, and then sialylated upon treatment with CMP–NeuAc, α(2–3)sialyl-transferase and alkaline phosphatase, commonly used to hydrolyze the inhibitor CMP, released in the reaction. Both glycosylation steps seemed quantitative, because after cleavage of the ester linkage on the conjugate **18** by treatment with hydrazine, the tetrasaccharide hydrazide **19** was the only observed product.

28.3.2 Enzymatic Synthesis of Glycopeptides

Use of Controlled-Pore Glass

The same functionalized silica support was used for glycosyltransferase-catalyzed sugar-chain elongation on a glycopeptide [21]. A *N*-linked glycodipeptide was attached to the support *via* a spacer made of six glycine units and an α-chymotrypsin-sensitive phenylalanine ester bond, giving the silica-supported *N*-Boc-Asn (GlcNAcβ)–Gly–Phe **20** with a loading of ca 0.2 mmol glycopeptide g^{-1} dry silica (Scheme 7). This supported glycopeptide was first galactosylated with β(1–4)

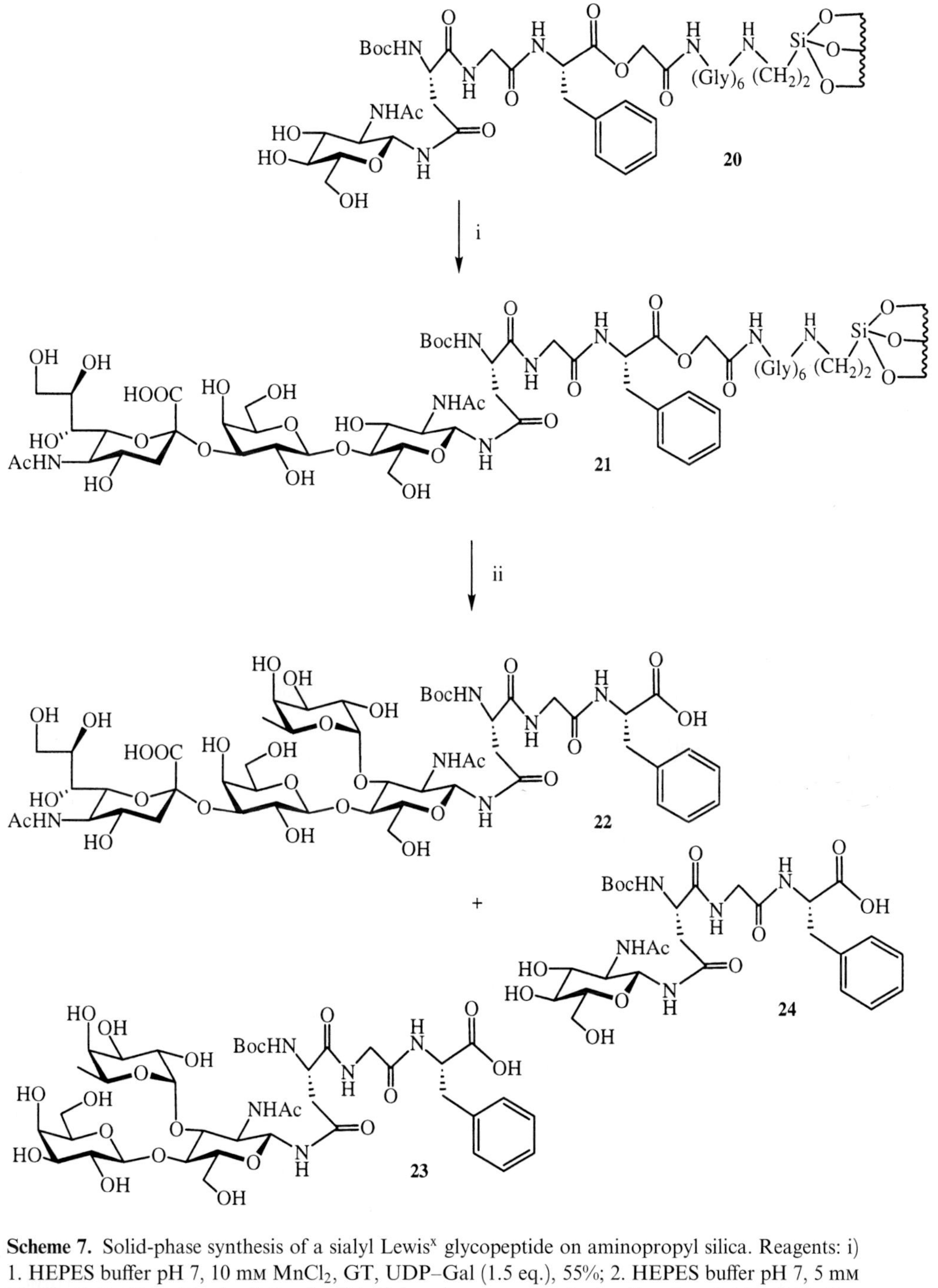

28.3 Enzymatic Synthesis on Insoluble Supports 713

Scheme 7. Solid-phase synthesis of a sialyl Lewis^x glycopeptide on aminopropyl silica. Reagents: i) 1. HEPES buffer pH 7, 10 mM MnCl$_2$, GT, UDP–Gal (1.5 eq.), 55%; 2. HEPES buffer pH 7, 5 mM MnCl$_2$, ST3, CMP–NeuAc (1.5 eq.), 65%; ii) 1. α-chymotrypsin; 2. ultrafiltration; 3. FT, GDP–Fuc (2.5 eq.), HEPES buffer pH 7, 95%.

galactosyltransferase and UDP–Gal in 55% yield, and subsequently sialylated in the presence of CMP–NeuAc and α(2–3)sialyltransferase in 65% yield affording the silica-bound sialylated glycopeptide **21**. Yields were evaluated by cleavage of the glycopeptide from an aliquot of the functionalized silica with α-chymotrypsin. Because galactosylation and sialylation reactions were incomplete, the sialylated tripeptide Boc-Asn–(NeuAcα2-3Galβ1-4GlcNAcβ)–Gly-Phe-OH was released from the solid support, together with partially glycosylated products, by cleavage with α-chymotrypsin. Then fucosylation in a traditional solution-phase reaction with α(1–3/4)fucosyltransferase and GDP–Fuc, afforded a mixture of the sialyl Lewis^x tripeptide **22** (35%), the fucosylated tripeptide **23** (20%) and unreacted glycopeptide **24** (45%).

The solid-phase synthesis of an *O*-glycopeptide by a chemo–enzymatic approach was reported by Wong and Seitz [22]. By use of the same support, aminopropyl–CPG, the starting amino acid, alanine, was attached through the flexible, acid- and base-stable Hycron anchor, first introduced by Kunz [23], which can be cleaved as an allyl ester with palladium(0) under mild conditions. The glycooctapeptide **26** incorporating the preformed glycosyl amino acid, Thr(GlcNAcβ), was first assembled according to classical solid-phase peptide synthesis in an overall yield of 20% starting from **25** (Scheme 8). According to the authors, CPG turned out not to be the

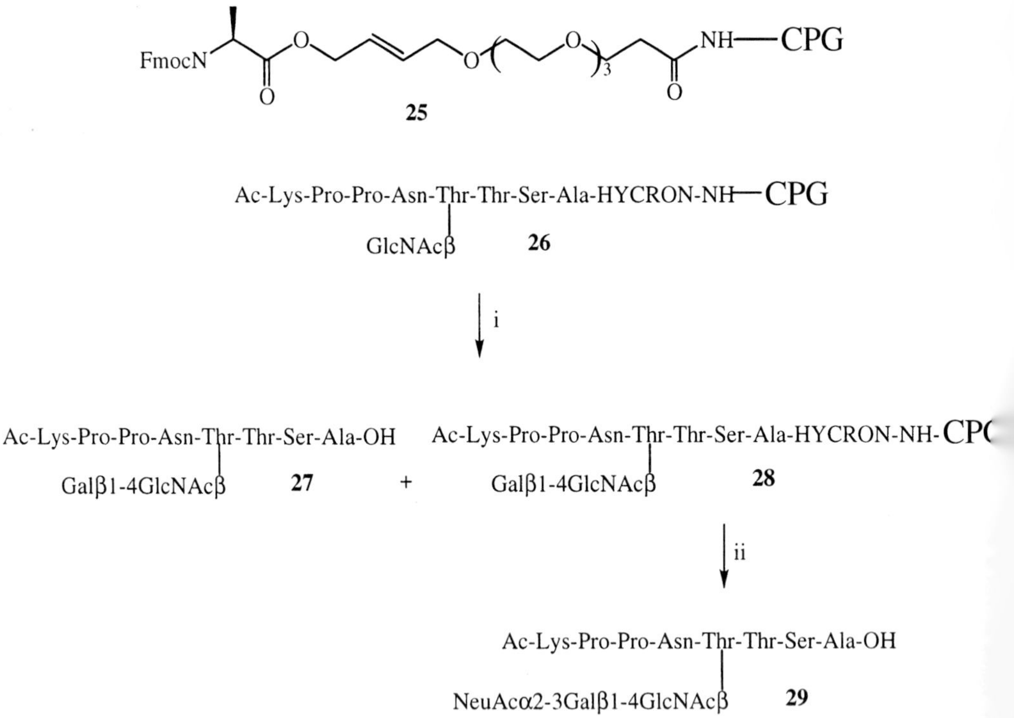

FmocN-...-O-...-O-(...)$_3$-...-NH—CPG

25

Ac-Lys-Pro-Pro-Asn-Thr-Thr-Ser-Ala-HYCRON-NH—CPG
  |
  GlcNAcβ   **26**

↓ i

Ac-Lys-Pro-Pro-Asn-Thr-Thr-Ser-Ala-OH   Ac-Lys-Pro-Pro-Asn-Thr-Thr-Ser-Ala-HYCRON-NH-CPG
  | |
  Galβ1-4GlcNAcβ   **27**   +   Galβ1-4GlcNAcβ   **28**

↓ ii

Ac-Lys-Pro-Pro-Asn-Thr-Thr-Ser-Ala-OH
  |
  NeuAcα2-3Galβ1-4GlcNAcβ   **29**

Scheme 8. Chemo–enzymatic synthesis of a *O*-sialyl-LacNAc octapeptide. Reagents: i) HEPES buffer pH 7, 5 mM MnCl$_2$, UDP–Gal (2 eq.), GT, CIAP, 37 °C, 3 days; ii) HEPES buffer pH 7, ST3, CIAP, CMP–NeuAc (1.5 eq.), 37 °C, 4 days, 76%.

optimum support for chemical peptide synthesis, but the most suitable for enzymatic reactions [24]. The loading of the support used in this preparation was also very low (60 µmol g^{-1}). The glycopeptide **26** was then submitted to enzymatic galactosylation in the presence of β(1–4)galactosyltransferase and a twofold excess of UDP–Gal under the usual conditions. Unexpectedly, during incubation at neutral pH and 37 °C, hydrolysis of the allyl ester linkage was observed in 53% yield, affording the *O*-LacNAc octapeptide **27**. The remaining supported product could be released from the silica beads by treatment with palladium(0) and morpholine. Altogether the *O*-LacNAc peptide **27** was synthesized in 15% yield on the basis of the loading of the initial support **25**, giving a 75% yield for the galactosylation step. Enzymatic sialylation was subsequently performed on the supported glycopeptide **28** but again, peptide cleavage from the support occurred during incubation and the major part (76%) of the sialylated glycopeptide **29** was finally recovered from the filtrate.

Use of Polyethylene Glycol Polyacrylamide (PEGA)

An alternative to the use of silica gel described above is a PEGA resin originally developed by Meldal for peptide solid-phase synthesis [25]. PEGA is a poly(ethylene glycol)–polyacrylamide copolymer with high swelling capacity both in organic media and in water, and has free amino groups for functionalization. The acid-labile Rink's linker (4-(α-amino-2′,4′-dimethoxybenzyl)phenoxyacetic acid) was first attached to the PEGA resin and this supported linker was then reacted with Fmoc-Gly-*O*-pentafluorophenyl ester and, after removal of the Fmoc group, the resin-bound amino acid was coupled to protected glycopeptide Asn(GlcNAcβ) by conventional peptide chemistry, to give, after deprotection, the resin-bound glycopeptide **29** with a loading capacity of 0.16 mmol g^{-1} resin (Scheme 9).

Compound **29** was then galactosylated in the presence of β(1–4)galactosyltransferase and a large excess of UDP–Gal, in quantitative yield [26]. It is worth mentioning that pre-incubation of the resin with enzyme and buffer for three days at 4 °C before adding UDP–Gal was beneficial, affording conversion of **29** into **30** in a much shorter time. The final lactosamine glycopeptide **31** was recovered by cleavage of **30** with 95% aqueous trifluoroacetic acid. Diverse PEGA resins with different amounts of cross-linking and loading capacity could be prepared and studied for their compatibility with biomolecules; this study led to the conclusion that a more cross-linked resin gave a lower yield in enzymatic galactosylation [27].

28.4 Enzymatic Synthesis of Oligosaccharides and Glycoconjugates on Soluble Supports

28.4.1 Enzymatic Synthesis of Oligosaccharides

Use of Water-Soluble Amino-Substituted Poly(vinyl alcohol)

In view of the low yields observed for enzymatic transfer of galactose on insoluble polyacrylamide polymers [11], Zehavi *et al.* developed the use of a water-soluble

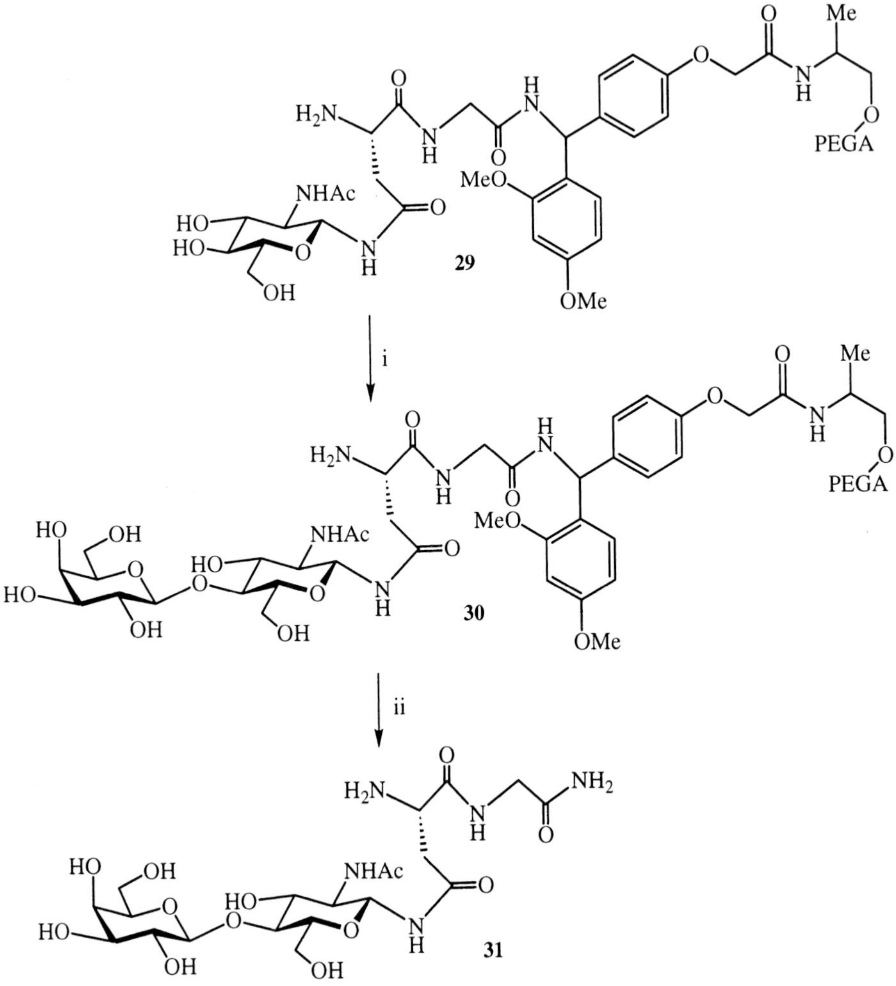

Scheme 9. Chemo–enzymatic synthesis of a *N*-glycopeptide on a polyethylene glycol polyacrylamide (PEGA) resin. Reagents: i) cacodylate buffer pH 7.4, 5 mM $MnCl_2$, UDP–Gal (3 eq.), GT, 23 °C, 48 h, 95%; ii) 95% aqueous TFA, 23 °C, 2 h.

polymer that afforded improved accessibility of substrate to enzyme. This linear polymer consisted of poly(vinyl alcohol) substituted with amino groups. The light-sensitive gluco polymer **32** carrying 0.15 mmol D-glucose g^{-1}, was glycosylated as previously described using ^{14}C-labeled UDP–Gal and β(1–4)galactosyltransferase, affording, after purification by ultrafiltration, the lactopolymer **33** in 34% yield (Scheme 10). Upon UV irradiation of **33**, radiolabeled lactose **4** was subsequently recovered in 88% yield [28].

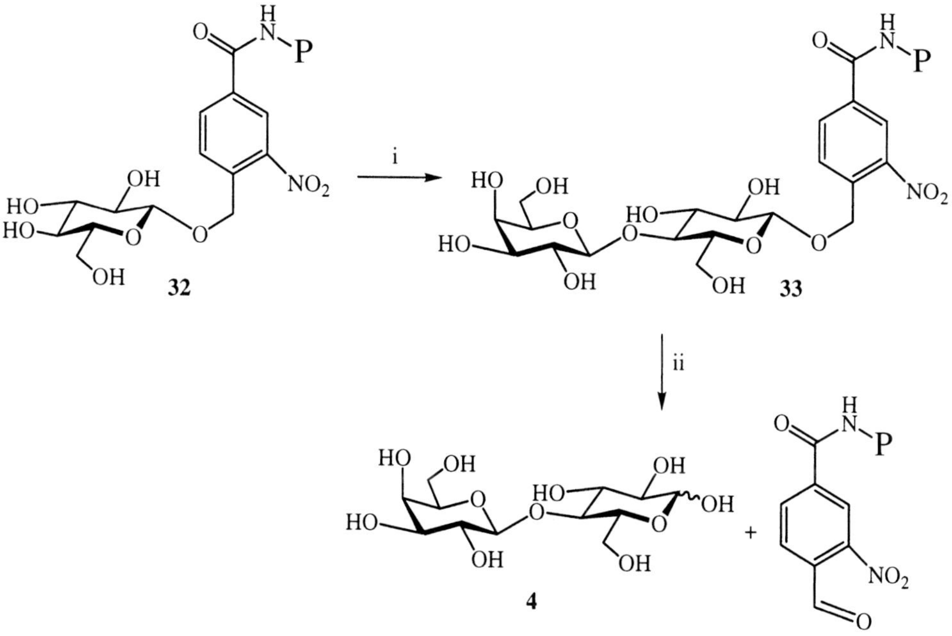

Scheme 10. Enzymatic synthesis of lactose by use of a photolabile water-soluble amino-substituted poly(vinyl alcohol). Reagents: i) cacodylate buffer pH 7, 3 mM $MnCl_2$, UDP–Gal (1 eq.), GT, α-lactalbumin, 0.1% mercaptoethanol, 37 °C, 18 h, 34%; ii) 1. hv, 20 h, 88%; 2. ultrafiltration.

Use of Water-Soluble Glycopolymer Synthesized by Polymerization

Nishimura et al. introduced the use of water-soluble glycopolymers. They synthesized the glycopolymer **34** by radical copolymerization of sugar monomer having an n-pentenyl group as a polymerizable aglycone, with acrylamide in an aqueous medium in the presence of ammonium persulfate and N,N,N',N'-tetramethylethylenediamine as initiators, according to Scheme 11 [29, 30]. Polymers of more than 40 kDa were obtained with a composition varying from 1:5 to 1:23 for the carbohydrate monomer-to-acrylamide ratio, depending on copolymerization conditions. Galactosylation of the GlcNAc polymer **34**, performed by means of UDP–Gal and β(1–4)galactosyltransferase, led to the LacNAc polymer **35** in quantitative yield, the n-pentenyl group providing the acceptor sugar with an adequately flexible spacer-arm. As opposed to insoluble polymer, such a water-soluble polymer enabled direct reaction monitoring by ^1H NMR measurement in D_2O [31].

As an extension of this work, the authors applied the same approach to the preparation of free oligosaccharides, by introducing suitable, removable linker arms between the saccharide and the polymer backbone. Thus, radical copolymerization with acrylamide of monomers **36** and **37** having spacer-arms of different lengths, under the previous conditions, yielded water-soluble GlcNAc acceptor polymers **38** and **39**, both having a linker selectively cleavable by hydrogenolysis [32, 33]. Enzy-

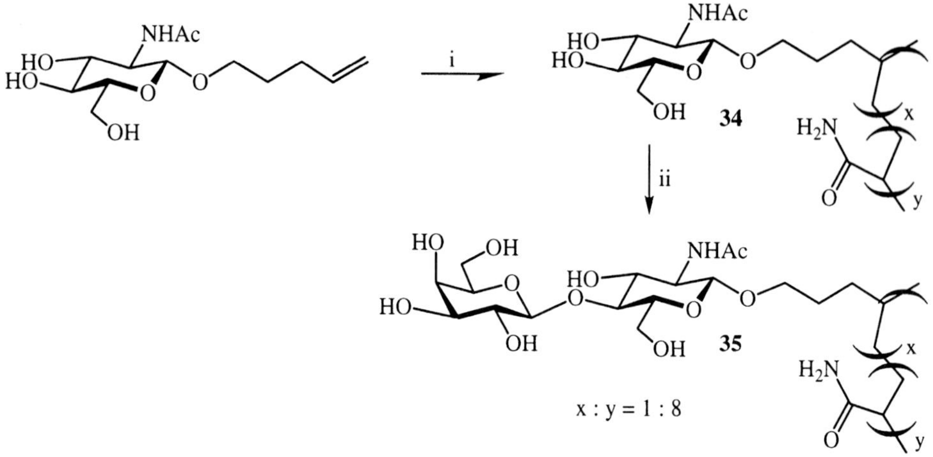

Scheme 11. Enzymatic galactosylation of a water-soluble GlcNAc-bearing polyacrylamide. Reagents: i) $CH_2=CH-CO-NH_2$, TEMED, APS, H_2O, room temp., 2 h; ii) 1. HEPES buffer pH 6, 10 mM $MnCl_2$, UDP–Gal (1.5 eq.), GT, 37 °C, 24 h; 2. gel filtration on Sephadex G-50, q.y.

matic sugar chain-elongation was performed with β(1–4)galactosyltransferase and UDP–Gal. Quantitative galactosylation could be achieved with polymer **39** bearing the long spacer-arm, whereas only 30% galactosylation was observed with polymer **38** bearing the short spacer-arm. N-acetyllactosamine **40** was finally released in high yield from polymer **41** by hydrogenolysis (Scheme 12).

Other types of GlcNAc polymer, **43** and **44**, containing an L-phenylalanine residue in the spacer-arm moiety as an α-chymotrypsin-sensitive structure, were also prepared in a similar manner [33, 34]. Galactosylation with β(1–4)galactosyltransferase and subsequent sialylation with rat liver α(2–6)sialyltransferase in the presence of the sugar–nucleotides were conducted as shown in Scheme 13. Sugar transfer reactions were achieved with high efficiency using only a small excess of UDP–Gal and CMP–NeuAc, affording, after filtration and lyophilization, the polymer **45** bearing sialyl α(2–6)N-acetyllactosamine branches, in almost quantitative yield. Finally by hydrolytic action of α-chymotrypsin, sialyl α(2–6)N-acetyllactosamine derivative **46** was recovered in 72% overall yield from **44**. It is noteworthy that whereas quantitative galactosylation was also observed with polymer **43**, it was not possible to release lactosamine from the polymer by the action of α-chymotrypsin, because of the lack of the flexible spacer-arm between phenylalanine and the polymer backbone [33].

28.4.2 Enzymatic Synthesis of Glycolipids on Water-Soluble Polyacrylamide–Poly(N-acryloxysuccinimide) (PAN)

The water-soluble polyacrylamide–poly(N-acryloxysuccinimide) (PAN) first described by Whitesides for enzyme immobilization [35], was used by Zehavi et al. in

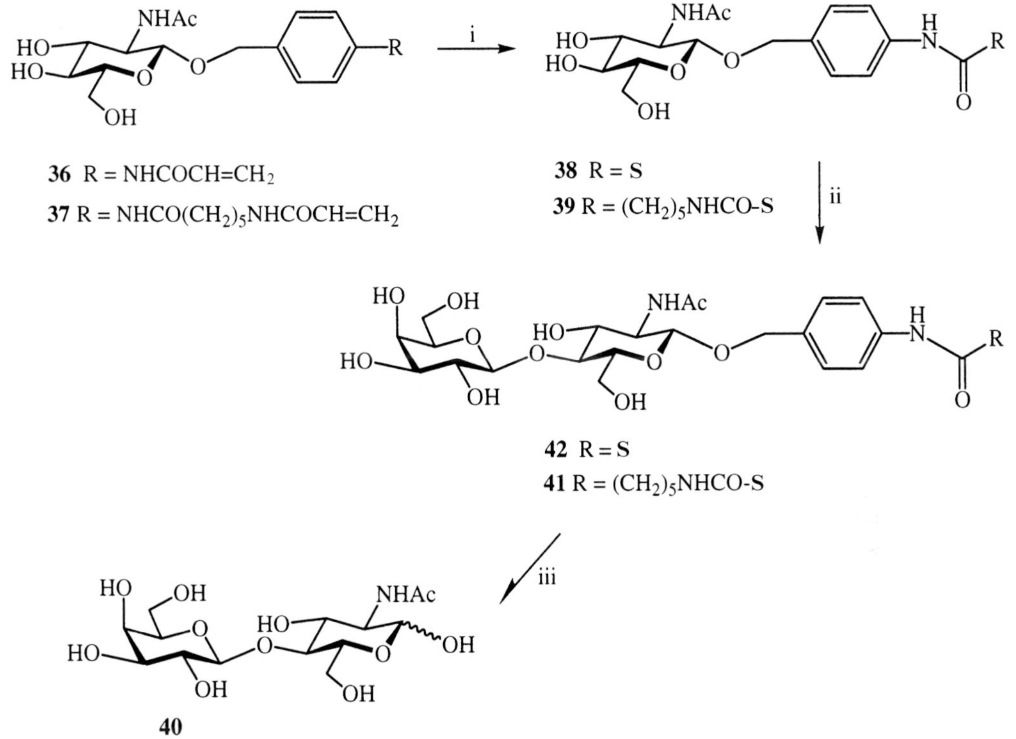

28.4 Enzymatic Synthesis of Oligosaccharides and Glycoconjugates

36 R = NHCOCH=CH$_2$
37 R = NHCO(CH$_2$)$_5$NHCOCH=CH$_2$

38 R = S
39 R = (CH$_2$)$_5$NHCO-S

42 R = S
41 R = (CH$_2$)$_5$NHCO-S

40

Scheme 12. Enzymatic synthesis of N-acetyllactosamine using a water-soluble glycopolymer with a linker cleavable by hydrogenolysis. Reagents: i) CH$_2$=CH–CO–NH$_2$, TEMED, APS, DMSO–H$_2$O, room temp., 24 h, 61% for **37**, 83% for **38**; ii) 1. HEPES buffer pH 6, 10 mM MnCl$_2$, UDP–Gal (1 eq.), GT, 37 °C, 24 h; 2. gel filtration on Sephadex G-50, q.y. for **40**, 30% for **41**; iii) H$_2$, Pd/C, H$_2$O–MeOH, 95%.

enzymatic supported synthesis of glycolipids. The hydrazide function in compound **47** was reacted with PAN yielding the glucopolymer **48** which served as an acceptor in the β(1–4)galactosyltransferase-catalyzed reaction in the presence of α-lactalbumin and ^{14}C labeled UDP–Gal [36]. The lactosyl sphingosine polymer **49** was obtained in 36% yield, after purification by extensive dialysis; further photolysis of the 2-nitrobenzylurethane group followed by acylation with stearoyl chloride, provided lactosylceramide **50** in 54% yield (Scheme 14).

Alternatively the polymer **51** carrying 0.24 meq lactosylβ(1–1)sphingosine g^{-1}, chemically prepared from PAN, was sialylated in 55% yield by use of recombinant rat liver α(2–3)sialyltransferase, ^{14}C labeled CMP–NeuAc, and calf intestinal alkaline phosphatase [37]. Again photolysis of the 2-nitrobenzylurethane group in the sialylated polymer **52** followed by acylation with stearoyl chloride afforded the GM3 ganglioside, NeuAcα(2–3)Galβ(1–4)Glcβ(1–1)Cer **53**.

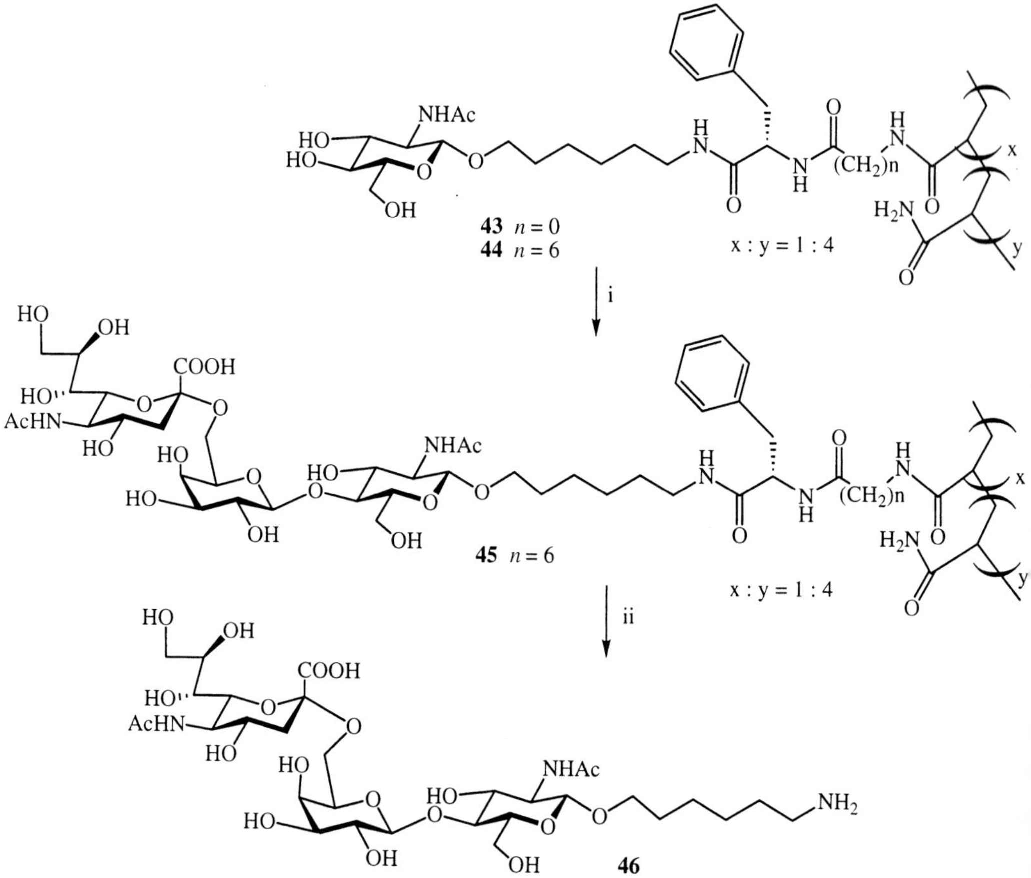

Scheme 13. Enzymatic synthesis of a sialylated trisaccharide derivative, by stepwise sugar-chain elongation, using a water-soluble chymotrypsin-sensitive glycopolymer. Reagents: i) 1. HEPES buffer pH 6, UDP–Gal (1.5 eq.), GT, 37 °C, 24 h; 2. cacodylate buffer pH 7.4, ST6, CIAP, CMP–NeuAc (1.5 eq.), 37 °C, 48 h, gel filtration on Sephadex G-25, q.y.; ii) 1. α-chymotrypsin, Tris HCl buffer pH 7.8, 40 °C, 24 h; 2. gel filtration on Sephadex G-15, 87%.

In conclusion, compared with solution-phase synthesis, solid-phase enzymatic synthesis, generally, needs higher concentrations of enzyme, longer reaction times, and a large excess of donors. Although there is still room for improvement in terms of yields and scale-up of syntheses, in solid-phase synthesis purification procedures are straightforward; so it is expected that this approach will open the route to automated oligosaccharide synthesis in the near future.

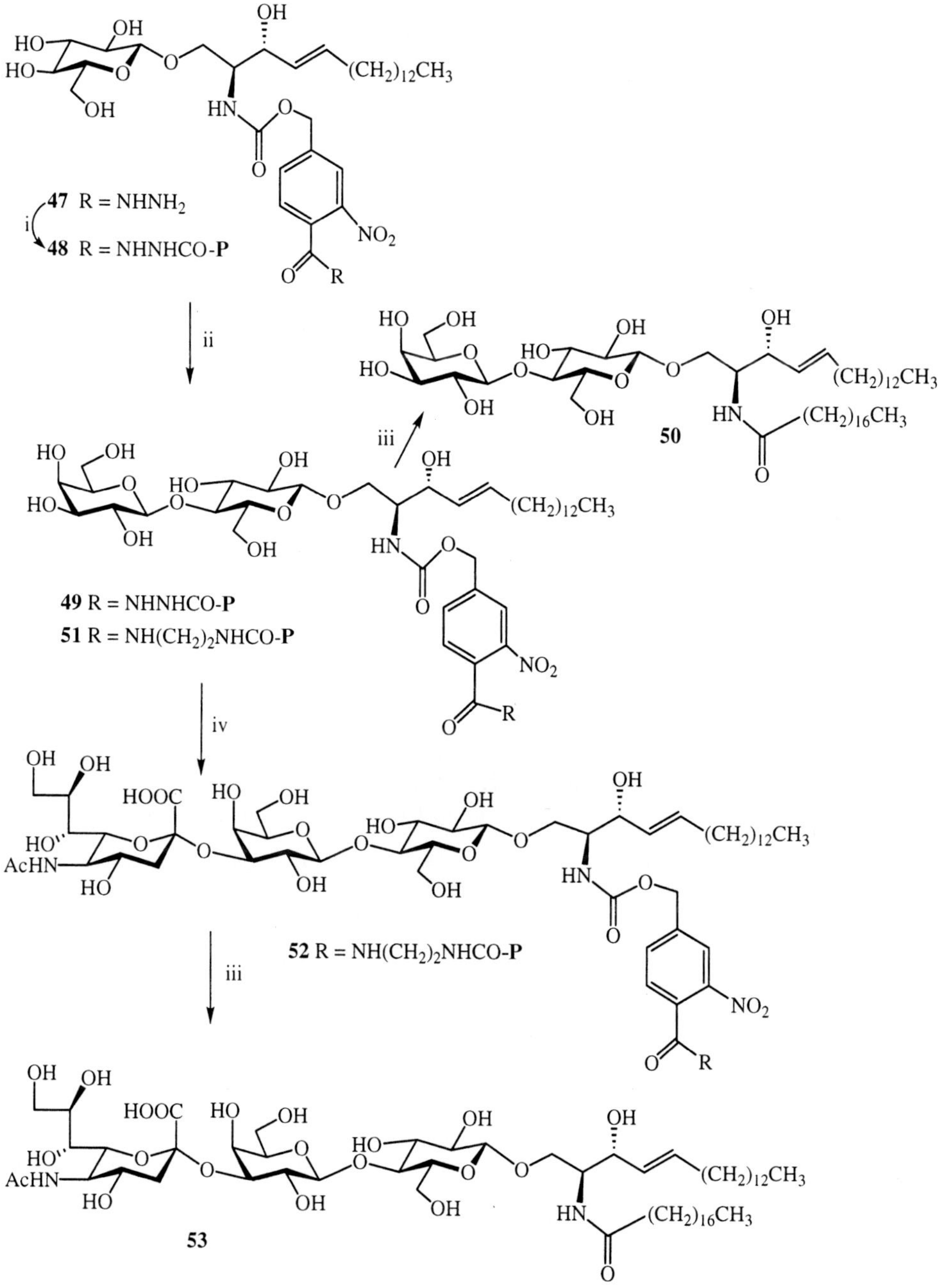

Scheme 14. Enzymatic synthesis of lactosyl ceramide and GM3 using a water-soluble polyacrylamide gel. Reagents: i) 1. PAN, Et$_3$N; 2. Sephadex G-25; ii) cacodylate buffer pH 7, 3 mM MnCl$_2$, UDP–Gal (0.3 eq.), GT, α-lactalbumin, 0.1% mercaptoethanol, 37 °C, 24 h, 36%; iii) 1. hv, THF–water, room temp., 9.5 h, 2. 50% CH$_3$COONa–THF, CH$_3$(CH$_2$)$_{16}$COCl, 54% for **50**; iv) cacodylate buffer pH 6.5, ST3, CIAP, CMP–NeuAc (2 eq.), 55%.

References

1. J. M. Fréchet, C. Schuerch, *Carbohydr. Res.*, **1972**, *22*, 399–412.
2. G. Excoffier, D. Gagnaire, J.-P. Utille, M. Vignon, *Tetrahedron*, **1975**, *31*, 549–553.
3. R. Eby, C. Schuerch, *Carbohydr. Res.*, **1975**, *39*, 151–155.
4. S. J. Danishefsky, K. F. McClure, J. T. Randolph, R. B. Ruggeri, *Science*, **1993**, *260*, 1307–1309.
5. R. Liang, L. Yan, J. Loebach, M. Ge, Y. Uozumi, K. Sekanina, N. Horan, J. Gildersleeve, C. Thompson, A. Smith, K. Biswas, W. C. Still, D. Kahne, *Science*, **1996**, *274*, 1520–1522.
6. J. Rademann, R. R. Schmidt, *Tetrahedron Lett.*, **1996**, *23*, 3989–3990.
7. Z.-W. Guo, Y. Nakahara, Y. Nakahara, T. Ogawa, *Angew. Chem. Int. Ed. Engl.*, **1997**, *36*, 1464–1466.
8. W. Kosch, J. März, H. Kunz, *Reactive Polymers*, **1994**, *22*, 181–187.
9. M. Meldal, *Methods Enzymol.*, **1997**, *289*, 83–104.
10. J. Vagner, G. Barany, K. S. Lam, V. Krchnak, N. F. Sepetov, J. A. Ostrem, P. Strop, M. Lebl, *Proc. Natl. Acad. Sci, USA*, **1996**, *93*, 8194–8199.
11. U. Zehavi, S. Sadeh, M. Herchman, *Carbohydr. Res.*, **1983**, *124*, 23–34.
12. U. Zehavi, M. Herchman, *Carbohydr. Res.*, **1986**, *151*, 371–378.
13. U. Zehavi, M. Herchman, S. Köpper, *Carbohydr. Res.*, **1992**, *228*, 255–263.
14. S. Köpper, U. Zehavi, *Reactive Polymers*, **1994**, *22*, 171–180.
15. S. Köpper, *Carbohydr. Res.*, **1994**, *265*, 161–166.
16. H. A. Nunez, R. Barker, *Biochemistry*, **1980**, *19*, 489–495.
17. S. David, C. Augé, C. Gautheron, *Adv. in Carbohydr. Chem. Biochem.*, **1991**, *49*, 175–237.
18. O. Blixt, T. Norberg, *J. Carbohydr. Chem.*, **1997**, *16*, 143–154.
19. O. Blixt, T. Norberg, *J. Org. Chem.*, **1998**, *63*, 2705–2710.
20. R. L. Halcomb, H. M. Huang, C.-H. Wong, *J. Am. Chem. Soc.*, **1994**, *116*, 11315–11322.
21. M. Schuster, P. Wang, J. C. Paulson, C.-H. Wong, *J. Am. Chem. Soc.*, **1994**, *116*, 1135–1136.
22. O. Seitz, C.-H. Wong, *J. Am. Chem. Soc.*, **1997**, *119*, 8766–8776.
23. O. Seitz, H. Kunz, *J. Org. Chem.*, **1997**, *62*, 813–826.
24. U. Slomczynska, F. Albericio, F. Cardenas, E. Giralt, *Biomed. Biochim. Acta*, **1991**, *50*, 67–73.
25. M. Meldal, *Tetrahedron Lett.*, **1992**, *33*, 3077–3080.
26. M. Meldal, F.-I. Auzanneau, O. Hindsgaul, M. Palcic, *J. Chem. Soc. Chem. Commun.*, **1994**, 1849–1850.
27. F.-I. Auzanneau, M. Meldal, K. Bock, *J. Peptide Sci.*, **1995**, *1*, 31–34.
28. U. Zehavi, M. Herchman, *Carbohydr. Res.*, **1984**, *128*, 160–164.
29. S.-I. Nishimura, K. Matsuoka, K. Kurita, *Macromolecules*, **1990**, *23*, 4182–4184.
30. K. Matsuoka, S.-I. Nishimura, *Macromolecules*, **1995**, *28*, 2961–2968.
31. S.-I. Nishimura, K. Matsuoka, Y. C. Lee, *Tetrahedron Lett.*, **1994**, *35*, 5657–5660.
32. S.-I. Nishimura, K. B. Lee, K. Matsuoka, Y. C. Lee, *Biochem. Biophys. Res. Commun.*, **1994**, *199*, 249–254.
33. K. Yamada, E. Fujita, S.-I. Nishimura, *Carbohydr. Res.*, **1997**, *305*, 443–461.
34. K. Yamada, S.-I. Nishimura, *Tetrahedron Lett.*, **1995**, *36*, 9493–9496.
35. A. Pollak, A. Blumenfeld, M. Wax, R. L. Baughn, G. M. Whitesides, *J. Am. Chem. Soc.*, **1980**, *102*, 6324–6336.
36. U. Zehavi, M. Herchman, R. R. Schmidt, T. Bär, *Glycoconjugate J.*, **1990**, *7*, 229–234.
37. U. Zehavi, A. Tuchinsky, *Glyconconjugate J.*, **1998**, *15*, 657–662.

29 Glycosidase-Catalysed Oligosaccharide Synthesis

David J. Vocadlo and Stephen G. Withers

29.1 Introduction

The use of glycosidases for the synthesis of oligosaccharides has a long history, dating back over 85 years [1]. The most difficult problem with their use as synthetic catalysts relates to their normal role being in hydrolysis. Given the 55 M concentration of water, hydrolysis is generally the preferred process in aqueous solution. A second problem can be that of control of the regiochemistry of bond formation. Several approaches have been followed in attempts to solve these problems, and these are discussed briefly within this article. More detailed accounts can be found in some useful recent reviews [2-4] and in some earlier overviews [5-7]. In this chapter we have chosen to present a brief review including some background on the enzymes themselves, the basic approaches currently used to solve the hydrolysis and regiochemistry problems, and some future perspective. The major content of this chapter is, however, a tabulation of a large number of examples of the use of glycosidases in the synthesis of specific linkages. Hopefully this will prove a useful resource for those interested in the application of such technology.

29.2 Background on Glycosidases

Glycosidic bonds occur within a fantastically broad range of contexts in natural systems, as described in Volume 2 of this series. Given the large number of glycosidic bonds and the frequent need for relatively specific cleavage thereof it is no surprise that very large numbers of glycosidases exist to carry out this function. Amino acid sequences of well over 1000 different glycoside hydrolases are now available, primarily deduced from their gene sequences, and these have been arranged into some 70 different families on the basis of sequence similarities [8]. At the time this article is being written an excellent web site is available that gives updated listings of these family members, plus some other information on the identities of important amino acid residues (http://afmb.cnrs-mrs.fr/~pedro/CAZY/db.html). The site also provides links to the original papers and three-dimensional structures of enzymes, where available. This site therefore provides an excellent starting point in the search for a specific enzyme activity because each family contains enzymes of closely related activity and the full spectrum of specificities is represented.

29.3 Basic Mechanisms

Hydrolysis occurs with one of two possible stereochemical outcomes, inversion or retention of anomeric configuration. All enzymes within a family hydrolyze their substrate with the same stereochemical outcome. These different outcomes demand different mechanisms, proposals for which were made by Koshland in 1953 [9]. Subsequent mechanistic and structural studies have largely substantiated these proposals and have identified the key active site residues as almost always a pair of carboxylic acids [10–13]. Glycosidases hydrolysing with inversion of anomeric configuration employ a mechanism involving direct displacement of the leaving group in an acid- or base-catalyzed process proceeding *via* an oxocarbenium ion-like transition state as shown in Figure 1. The two carboxylic acids are appropriately spaced (10–11 Å) to perform their role as general acidic or basic catalysts.

The vast majority of glycosidases that hydrolyze with net retention of anomeric configuration also have an active site containing a pair of carboxylic acids, but now only approximately 5 Å apart. This smaller separation of the two groups reflects their different roles in the two-step double-displacement mechanism. One residue functions as a general acid catalyst, protonating the glycosidic oxygen, whereas the other functions as a nucleophile, attacking at the anomeric center to form a covalent glycosyl-enzyme intermediate as shown in Figure 2.

In a second step, water attacks this intermediate in a general base-catalyzed process to yield the product of retained anomeric configuration. Both steps again proceed *via* oxocarbenium ion-like transition states (not shown). A variation on this mechanism is seen for some *N*-acetylhexosaminidases (at this stage apparently restricted to those from family 20 and chitinases from family 18). Such enzymes do not seem to contain a carboxylic acid that functions as a nucleophile. Rather, the *N*-acetyl group functions in that role, the reaction proceeding through an oxazoline intermediate as shown in Figure 3.

29.4 Synthesis by the 'Thermodynamic' Approach

One approach that has been employed to favor synthesis over hydrolysis involves running the reaction in the presence of very high concentrations (typically molar) of

29.4 Synthesis by the 'Thermodynamic' Approach

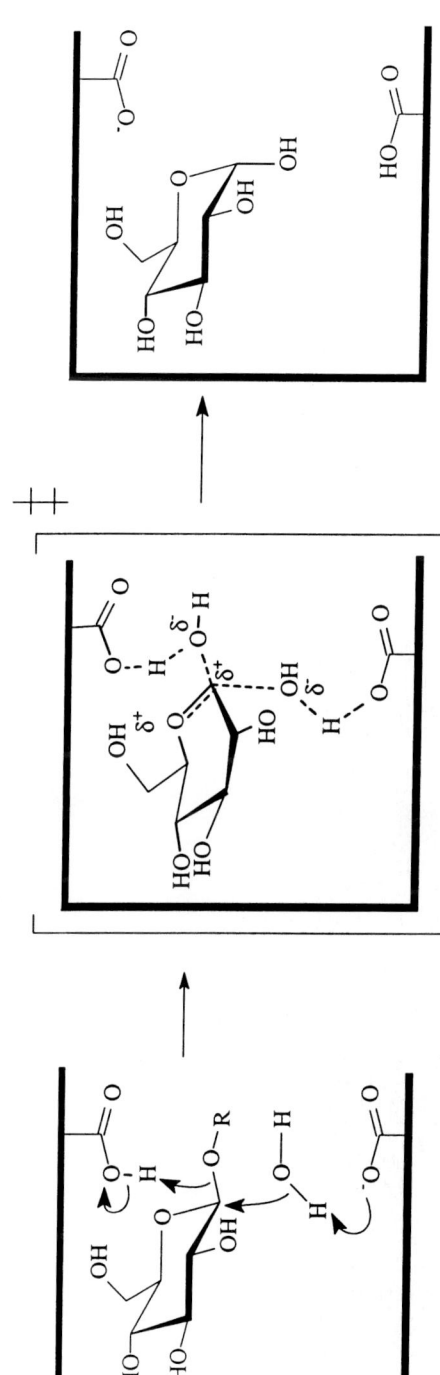

Figure 1. Proposed mechanism of β-inverting glucosidases.

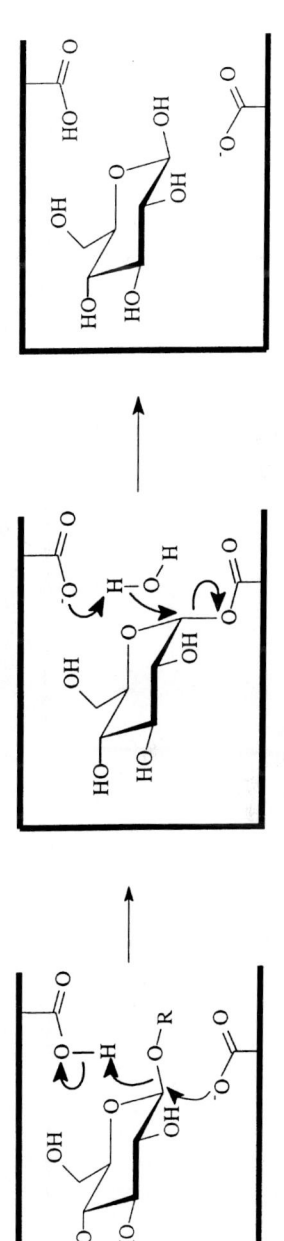

Figure 2. Proposed mechanism of β-retaining glucosidases.

726 29 *Glycosidase-Catalysed Oligosaccharide Synthesis*

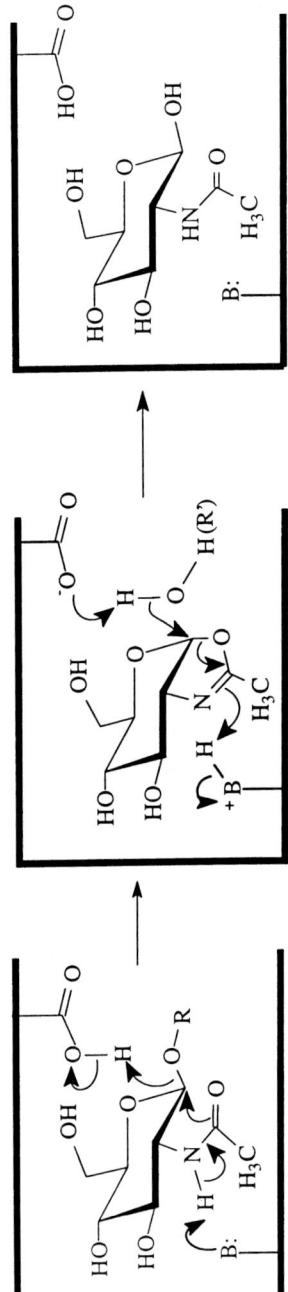

Figure 3. Proposed mechanism for family 20 *N*-acetyl β-hexosaminidases.

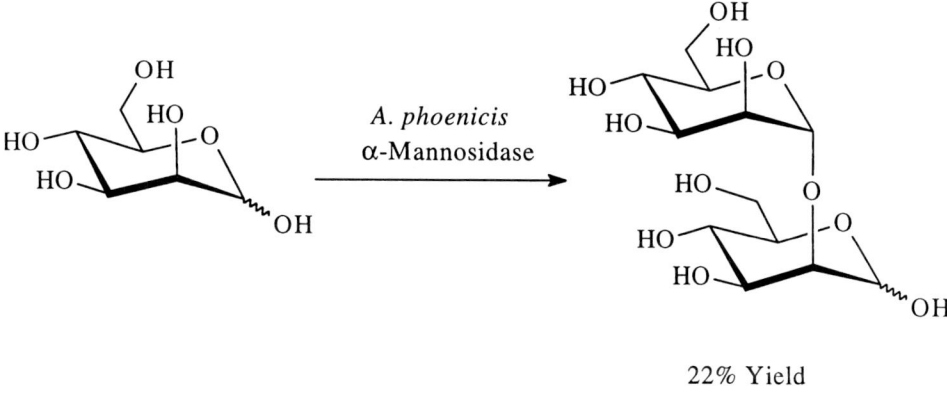

22% Yield

Figure 4. Enzymatic synthesis of α-D-Man-(1–2)-D-Man using α-mannosidase from *Aspergillus phoenicis*.

sugar under so-called 'thermodynamic' conditions such that the condensation reaction becomes favorable. Organic solvents can help in displacing this equilibrium, as long as the enzyme is compatible with, and the sugars soluble in, this medium. Similarly, high salt concentrations can be used to reduce the activity of the water and reduce product oligosaccharide solubility, thereby displacing the equilibrium [14]. The advantage of this approach is its simplicity, hence its cost-effectiveness. Yields are, however, typically poor, rarely exceeding 15%, and control of regiochemistry is essentially non-existent. The approach is therefore of limited use for the coupling of two sugars because it frequently produces intractable mixtures. There are, however, occasional exceptions involving the coupling of two units of the same sugar; useful (20%) yields of a disaccharide have occasionally been obtained [15, 16], Figure 4.

This approach can, however, provide an attractive route to the synthesis of glycosides of simple hydrophilic (ideally liquid) alcohols; reactions are performed in the presence of a very high concentration of the alcohol as shown in the example in Figure 5.

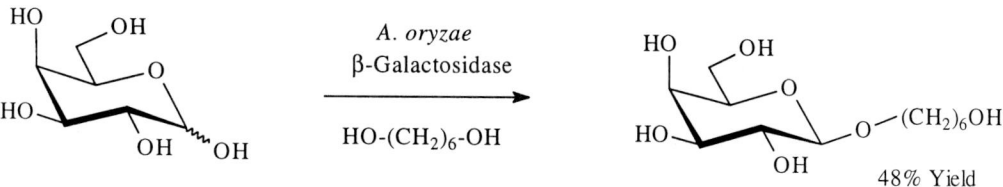

48% Yield

Figure 5. Enzymatic synthesis of 6-hydroxyhexyl β-D-galactoside using β-galactosidase from *Aspergillus oryzae*.

Crout has recently shown that this approach can provide good (up to 61%) yields if the reaction is performed at an elevated temperature (50 °C) and time is spent in determining the optimum water activity to be used for the enzyme under study [17]. This can be achieved either by directly adjusting the alcohol/water ratio or by varying the concentration of a co-solvent. This approach can therefore provide useful syntheses of simple glycosides as starting materials for chemical or enzymatic synthesis.

29.5 The Kinetic Approach

The alternative approach, which can be employed only with retaining glycosidases, involves intercepting the glycosyl–enzyme intermediate that is formed at the active site with an acceptor moiety which reacts in place of water as shown in Figure 6.

This approach relies upon the presence of an aglycone binding site that will bind a sugar in the correct orientation for one of the hydroxyl groups to react with the glycosyl–enzyme. Interactions between this second sugar and the protein will serve to stabilize the transition state for glycosyl transfer, reducing the activation barrier relative to that for hydrolysis in exactly the same way that such interactions serve to stabilize the transition state for disaccharide hydrolysis.

The process therefore involves reaction of an activated donor sugar with a suitable acceptor sugar in the presence of the appropriate retaining glycosidase as shown in the example in Figure 7.

The success of this approach depends upon two factors. Firstly, the transglycosylation process must occur in preference to hydrolysis of the glycosyl-enzyme. This can be controlled to some extent through the concentration of the acceptor sugar used, because occupancy of the acceptor site requires sugar concentrations higher than the dissociation constant for that site. The second important point is that the product disaccharide must function as a significantly poorer glycosyl donor than the activated donor substrate employed. This is best controlled by use of the most reactive (highest k_{cat}/K_m) donor possible, because the ratio of the rates of consumption of donor and product depends upon the relative k_{cat}/K_m values for the two species *and* on the concentration of each species present at that time according to the expression:

$$k_{rel} = \{(k_{cat}/K_m)_P \times [P])\}/\{(k_{cat}/K_m)_D \times [D]\}$$

Nitrophenyl glycosides and glycosyl fluorides are particularly attractive donors in this regard, not only because of their generally high k_{cat}/K_m values, but also because the departed aglycone is itself a very poor acceptor and thus will not compete as an acceptor. It is also very easily separated from the products.

Although some disaccharides, *e.g.* lactose, can serve as very inexpensive donors, complications can arise if the glucose liberated itself acts as an acceptor. A more unusual donor that has been used is the sialyl lactoside of a water-soluble polymer,

29.5 The Kinetic Approach 729

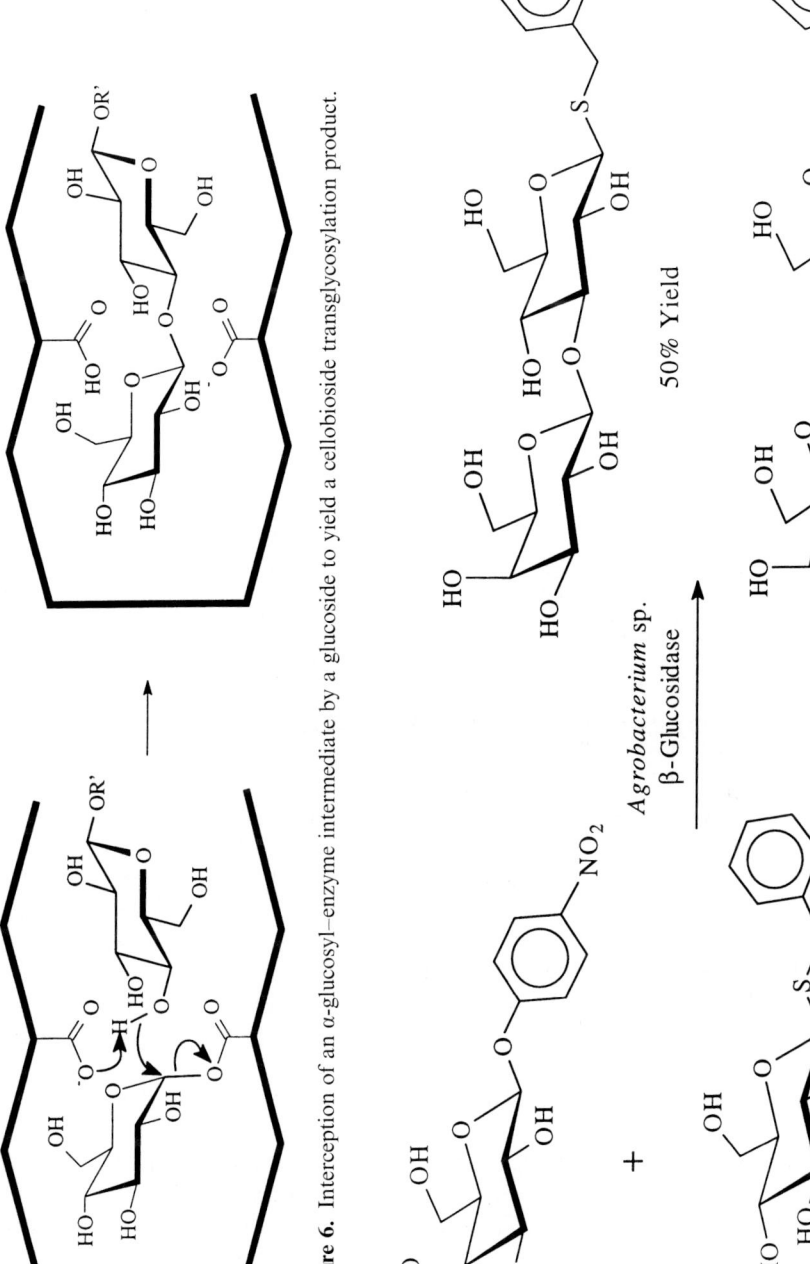

Figure 6. Interception of an α-glucosyl–enzyme intermediate by a glucoside to yield a cellobioside transglycosylation product.

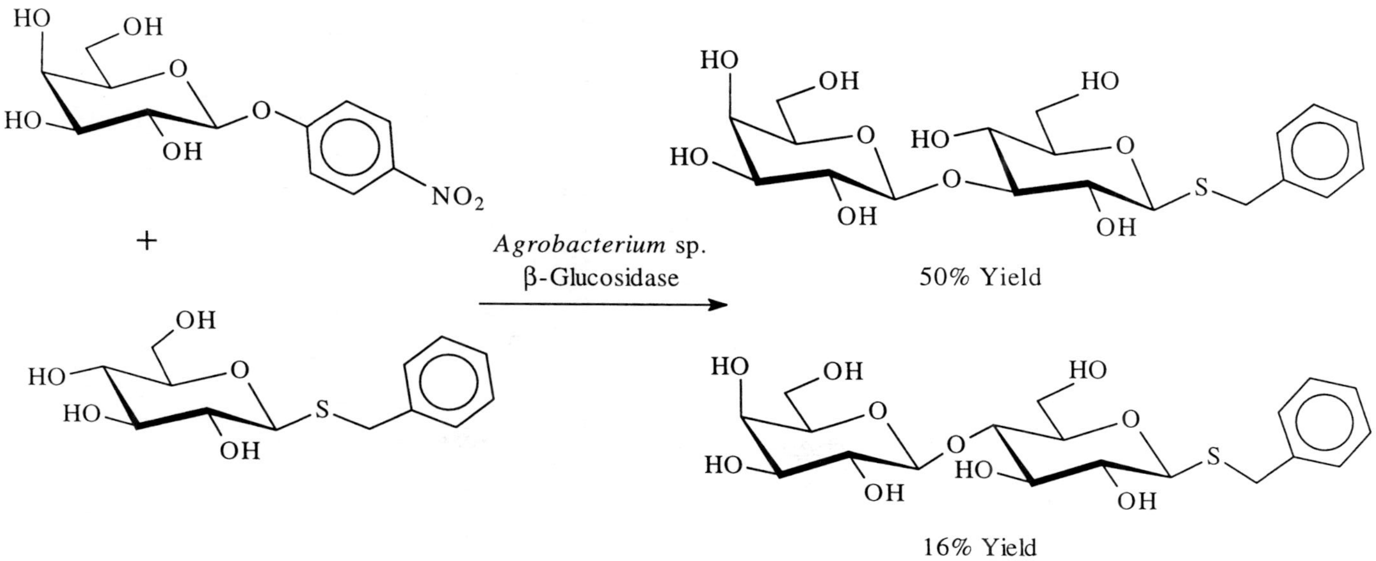

Figure 7. *Agrobacterium* sp. β-glucosidase-catalyzed synthesis of disaccharides using *p*NP-β-D-Gal as an activated donor.

which was used in conjunction with a leech ceramide glycanase for the transfer of sialyl lactose to ceramide, thereby yielding ganglioside GM_3 in high (61%) yield [18]. Sugar oxazolines have been used as donors for N-acetylhexosaminidases that react through the oxazoline mechanism. For example, incubation of *Bacillus* sp. chitinase with the oxazoline derivative of chitobiose resulted in rapid production of chito-oligosaccharides as shown in Figure 8 [19].

If these factors are controlled well, respectable yields in the range of 10–60% can be obtained. This, however, generally requires exquisite timing of product harvesting or the use of some trick to 'pull' the equilibrium of the reaction in the synthetic direction. Because no universal approach exists, these 'tricks' these must be developed separately for each enzyme. One way of achieving this end again involves the use of an organic co-solvent to reduce the activity of the water, thereby minimizing the hydrolytic process. For example, when p-nitrophenyl β-galactoside was used as donor and 3-O-methyl glucose as acceptor in reactions catalyzed by the *Kluyveromyces fragilis* β-galactosidase a 14% yield of the 6-linked disaccharide product was obtained if reaction was performed in 67% triethyl phosphate whereas *only* hydrolysis was observed when the reaction was conducted in aqueous buffer [20].

Another approach has been to remove one product continuously in some way. In a *very* few fortuitous cases this has occurred *via* crystallization of the product *in situ*; this is obviously not a general approach. Removal of disaccharide products by preferential absorption on to activated charcoal chromatography columns has also been reported [21]. A more generally useful approach has been to use a second 'coupling' enzyme, which will react with the product of interest only and, ideally, convert it directly to the desired product. This approach is particularly useful if the second enzyme is a glycosyl transferase, directly yielding the trisaccharide desired as shown below in Wong's synthesis of sialyl LacNAc (Figure 9) [22]. Other examples include Paulson's synthesis of sialyl 2,3-Gal β-OR [23] and Thiem's preparation of the sialyl T-antigen [24].

Control of the regiochemistry of bond formation is more difficult to achieve. Probably the best method at present is to screen the available glycosidases, hopefully guided by the literature, for the enzyme that gives the product desired. We hope our compilation of enzymes will be useful in this process. A logical approach to such screening, *if* the product is available, is to screen the enzymes for their ability to cleave the product of interest. Although the principle of microscopic reversibility would indicate that an enzyme capable of hydrolysing the desired product *can* form the product of interest, a positive result would not necessarily prove that it would form the desired product preferentially.

A somewhat improved approach, which does not require access to the product, is described in Section 29.6. All such screening approaches, however, are currently limited by the availability of enzymes. A second approach that has met with some success has involved manipulation of solvent conditions to change the regiochemistry of bond formation. For example, in studies on the *Bacillus circulans* β-galactosidase using p-nitrophenyl galactoside as donor and p-nitrophenyl GlcNAc as acceptor the Gal-β-1,4-GlcNAc-pNP product was obtained as the exclusive product (21% yield) when reaction was performed in the presence of 50% acetoni-

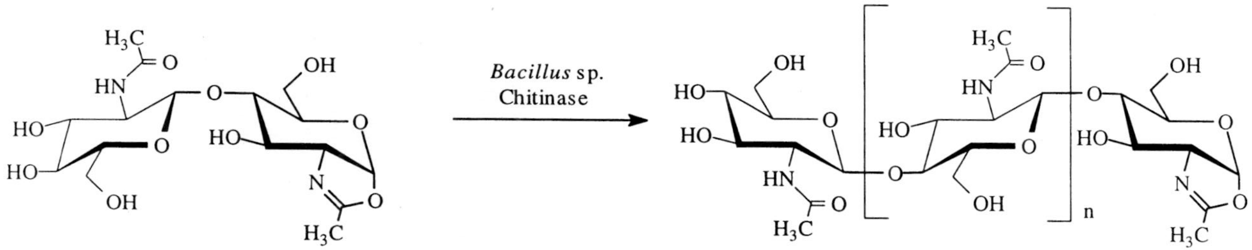

29.5 The Kinetic Approach 731

Figure 8. Enzymatic synthesis of chitin using *Bacillus* sp. chitinase with the oxazoline derivative of chitobiose as donor.

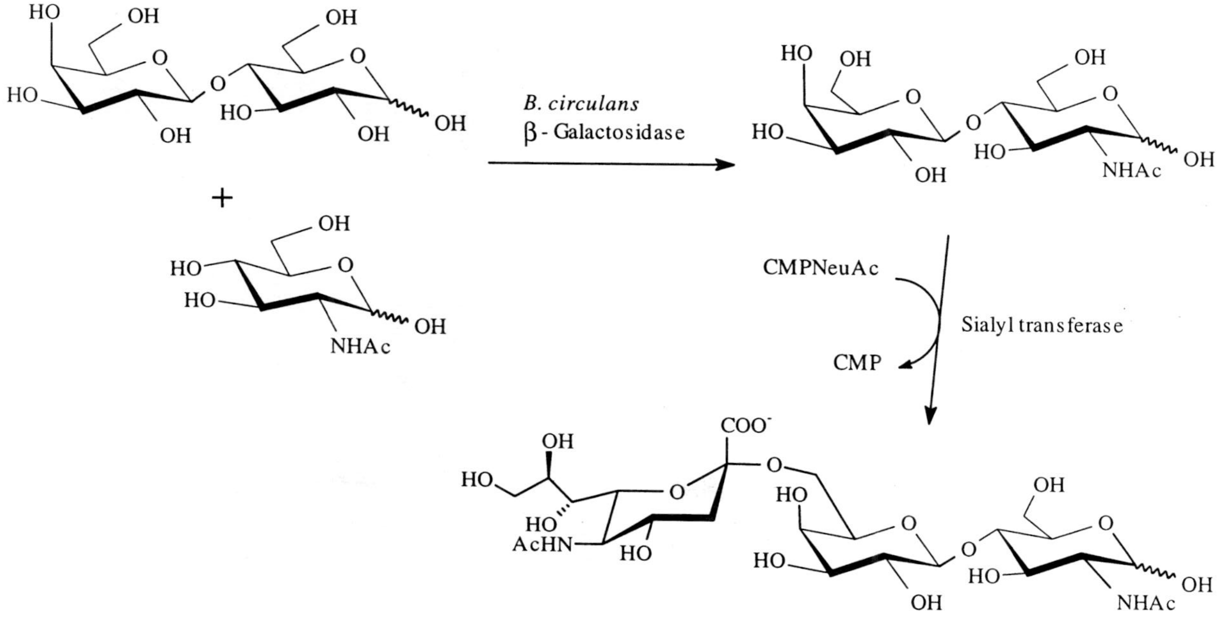

Figure 9. One-pot synthesis of a sialyl oligosaccharide by the combined use of a sialyl transferase and a β-galactosidase.

trile. When reaction was performed at lower (20%) acetonitrile concentrations, however, Gal-β-1,6-GlcNAc-*p*NP was instead obtained in 25.5% yield [25].

Finally, there have been several examples of success in manipulating regiochemical outcomes by changing the anomeric stereochemistry of the acceptor sugar. Thus, for example, Nilsson has shown that when methyl α-galactoside is used as the acceptor for the transfer of Gal from *p*-nitrophenyl α-galactoside by the coffee bean α-galactosidase, the predominant product (27%) is the 1–3 linked isomer whereas when methyl β-galactoside is the acceptor the 1–6 linked product is formed preferentially (18% compared with 9% of the 1–3 isomer). Similarly with *E. coli* β-galactosidase, using *p*-nitrophenyl β-galactoside as donor, the 1–6 isomer was dominant (14% yield) for transfer to methyl α-galactoside whereas when transfer was to methyl β-galactoside the 1–3 linked product predominated over the 1–6 (22% compared with 3%) [26].

In a further example Crout demonstrated a major difference in product distribution when the *N*-acetylgalactosaminidase from *Aspergillus oryzae* transfers GalNAc to methyl αGalNAc or to methyl βGalNAc. Transfer to methyl α-galactoside resulted in 81% yield of the 1–4 linked product and only 9% of the 1–6 isomer whereas transfer to methyl β-galactoside resulted in almost equal amounts of the two isomers, but at lower yields (20% of the 1–4 isomer and 23% of the 1–6 isomer) [27].

In evaluating the synthetic potential of glycosidases using the 'kinetic' method it is important to recognize that the yields and even the regiochemical outcomes will vary significantly with enzyme concentration and the reaction time employed. These differences in regiochemical outcome can arise because the transglycosylation products obtained are, themselves, substrates for the enzyme and can, therefore, be processed further. The net consequence will therefore tend to be a progression from an initial 'kinetically controlled' product mix towards a 'thermodynamically controlled' transglycosylation product mixture and ultimately to hydrolysis products. Consequently it is frequently observed that the product ratios obtained do not reflect expectations on the basis of cleavage specificities because the product that forms fastest will probably also be rearranged fastest. Different regiochemical outcomes arising from variation of solvent composition or acceptor structure could, therefore, be a consequence of the different relative rates of these processes rather than specific effects on enzyme structure or binding mode.

29.6 Recent Developments and New Directions

A completely new approach to the problem of controlling the hydrolytic reaction has been developed in the Withers group through the use of glycosynthases, specific mutants of retaining glycosidases that are used in conjunction with glycosyl fluoride donors of the *opposite* anomeric configuration to that hydrolyzed by the wild-type enzyme [28]. These mutants are those in which the amino acid functioning as the catalytic nucleophile has been replaced by a residue with a non-nucleophilic side-chain, alanine being the substitution of choice at the time of writing. Such mutants are completely hydrolytically inactive towards disaccharide substrates because they are unable to form the reactive glycosyl–enzyme intermediate. If, however, the mutant is presented with a glycosyl fluoride substrate of the same anomeric configuration as the intermediate that would ordinarily have formed, it is accepted by the enzyme as a surrogate for this intermediate and is transferred to a suitable acceptor as shown in Figure 10. Because the products so formed are completely hydrolytically stable, excellent yields (up to 92% isolated yields to date) can be obtained. This approach was published first with the β-glucosidase from *Agrobacterium* sp. [28]. A second example has recently appeared [29] and several others are being developed at the time of writing.

A more direct large-scale screen for enzymes of interest has very recently been developed (Blanchard & Withers, unpublished). This involves first inactivating the enzyme under investigation *via* the trapping of a 2-deoxy-2-fluoroglycosyl-enzyme intermediate by treatment of the enzyme with the appropriate 2,4-dinitrophenyl 2-deoxy-2-fluoro glycoside. Once freed of excess inactivator aliquots of this inactive, yet catalytically competent, intermediate species are incubated in the presence of a series of potential acceptors for a defined time period during which transglycosylation, releasing free enzyme, can occur, Figure 11. Each aliquot is then assayed with a standard substrate, positive results indicating that the ligand in question is capable of acting as an acceptor. The approach has been adapted to a 96-well format, enabling the facile screening of large numbers of acceptors, although the approach does not provide any insights into the regiochemistry of bond formation.

Most published work to date has been on the use of *exo*-glycosidases, which usually cleave a single sugar residue at a time from the non-reducing terminus. Approaches involving the transfer of large blocks of sugars will, however, require the use of *endo*-glycosidases. Some very impressive work in this area has been pioneered by Takagawa and coworkers, who demonstrated that an *endo*-*N*-acetyl glucosaminidase from *Arthrobacter protophormiae* (endoA) has transglycosylation activity and could be used to synthesize neoglycoproteins. They used (Man)$_6$(GlcNAc)$_2$ as donor and transferred to GlcNAc(Asn)peptide acceptors, thereby producing (Man)$_6$(GlcNAc)$_2$(Asn)-peptide products. They also showed that the approach could be used with partially deglycosylated ribonuclease B as acceptor thereby producing a modified glyco form, as shown below (Figure 12) [30–34]. This approach was further developed by Y.-C. Lee for the production of (Man)$_9$(GlcNAc)$_2$-peptide analogs by use of pre-synthesized GlcNAc–peptide analogs as acceptors [35]. Lee also showed that organic solvents such as acetone could significantly improve (up to 89%) yields of products in such block transfer reactions [33, 36]. This approach has enormous potential in the remodeling of glycoproteins.

The role of oligosaccharides in biological processes has become a field of intensive research and accordingly facile methods for their preparation are highly desirable. The following tables should hopefully provide a useful resource for the researcher interested in synthesizing such oligosaccharides *via* an enzymatic or chemoenzymatic route. Table 1 lists linkages formed by *exo*-glycosidases, Table 2 includes all products formed by *endo*-glycosidases, and Table 3 includes alkyl glycosides synthesized by *exo*-glycosidases. References included in each table were selected primarily on the basis of interest and adequate product characterization.

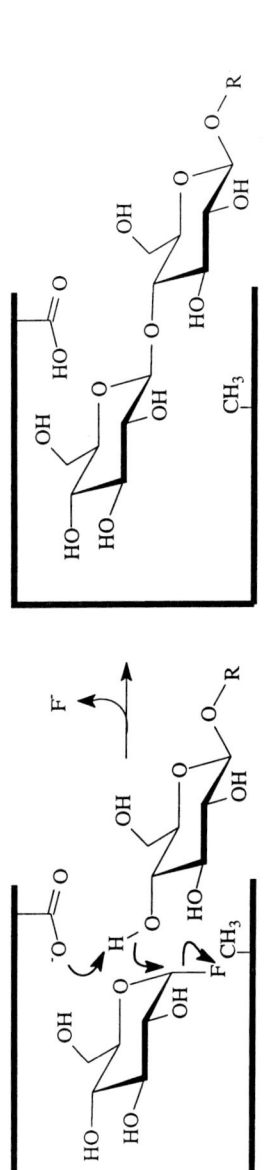

Figure 10. Glucosynthase-catalyzed synthesis of a cellobioside.

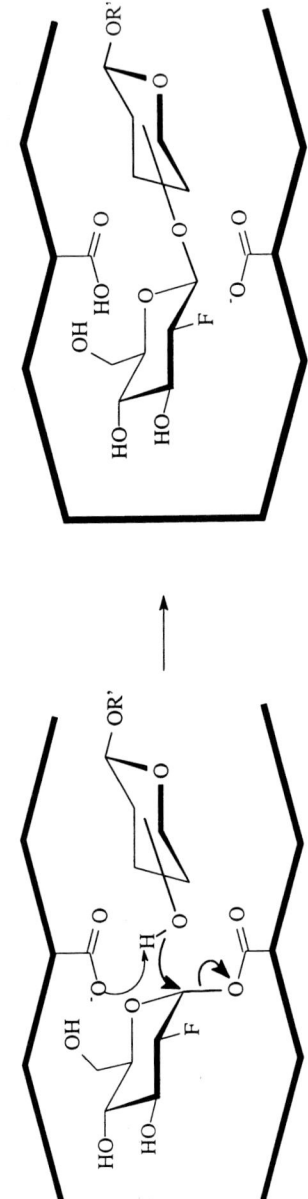

Figure 11. Reactivation of a 2-fluoro-α-glucosyl-enzyme intermediate by transglycosylation to an acceptor sugar.

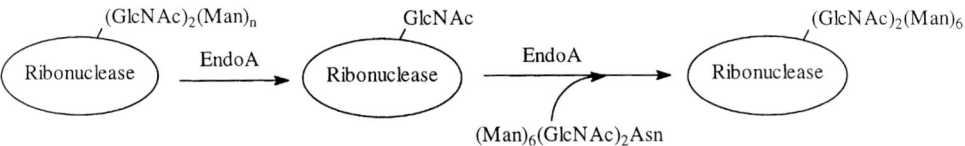

Figure 12. Remodeling of RNase B glycoforms through the use of *endo* A.

In each table the linkages formed are listed alphabetically with α-linkages listed first followed by β-linkages. In the linkage column, where the configuration of the anomeric carbon in the second saccharide unit is indicated, the acceptor molecule is either an oligosaccharide or a glycoside. Glycoside acceptors are differentiated by the inclusion of the hetero-atom involved in the glycosidic linkage after the second saccharide unit. Yields listed are calculated based on donor, acceptor, or in a few cases, transfer ratios. Identical products synthesized by different groups are listed as separate entries. When the same principal investigator has referred to the synthesis in different publications there is only one entry listing each reference.

Table 1. Enzymatic synthesis of oligosaccharides using *exo*-glycosidases.

Entry	Linkage	Enzyme Source	Donor	Acceptor	Product	Yield (%)	Reference
1	α-D-2dGlc(1-1)-α-D-2dGlc	Trehalase *Lobosphaera* sp.	2dGlc	2dGlc	α-D-2dGlc(1-1)-α-D-2dGlc	16	37
2	α-D-2dGlc(1-1)-α-D-Glc	Trehalase *Lobosphaera* sp.	2dGlc	Glc	α-D-2dGlc(1-1)-α-D-Glc	6	37
3	α-D-2dGlc(1-3)-D-2dGlc	Glucoamylase *Aspergillus niger*	2dGlc	2dGlc	α-D-2dGlc(1-3)-D-2dGlc	NA	38
4	α-D-2dGlc(1-3)-D-2dGlc	α-Glucosidase *Aspergillus niger*	2dGlc	2dGlc	α-D-2dGlc(1-3)-D-2dGlc	NA	38
5	α-D-2dGlc(1-4)-D-2dGlc	Glucoamylase *Aspergillus niger*	2dGlc	2dGlc	α-D-2dGlc(1-4)-D-2dGlc	NA	38
6	α-D-2dGlc(1-4)-D-2dGlc	α-Glucosidase *Aspergillus niger*	2dGlc	2dGlc	α-D-2dGlc(1-4)-D-2dGlc	NA	38
7	α-D-2dGlc(1-6)-D-2dGlc	Glucoamylase *Aspergillus niger*	2dGlc	2dGlc	α-D-2dGlc(1-6)-D-2dGlc	NA	38
8	α-D-2dGlc(1-6)-D-2dGlc	α-Glucosidase *Aspergillus niger*	2dGlc	2dGlc	α-D-2dGlc(1-6)-D-2dGlc	NA	38
9	α-D-Gal(1-1)-α-D-Gal	α-Galactosidase *Candida guilliermondii* H-404	Lactose	Lactose	α-D-Gal(1-1)-α-D-Gal	NA	39
10	α-D-Gal(1-1)-D-Fru*f*	β-Galactosidase *Escherichia coli*	Gal	Fru*f*	α-D-Gal(1-1)-D-Fru*f*	34	40
11	α-D-Gal(1-1)-D-Fru*f*	β-Galactosidase *Aspergillus oryzae*	Gal	Fru*f*	α-D-Gal(1-1)-D-Fru*f*	52	40
12	α-D-Gal(1-2)-α-D-Gal-O	α-Galactosidase Coffee bean	α*p*NPGal	α*p*NPGal	α-D-Gal(1-2)-α-D-Gal-O*p*NP	2	26

#	Product	Enzyme	Donor	Acceptor	Product	Yield	Ref
13	α-D-Gal(1-2)-α-D-Gal-O	α-Galactosidase Coffee bean	αpNPGal	αpNPGal	α-D-Gal(1-2)-α-D-Gal-OoNP	6	26
14	α-D-Gal(1-2)-α-D-Glc	α-Galactosidase Coffee bean	Melibiose	α-CD	α-D-Gal(1-2)-α-CD	2.2	41
15	α-D-Gal(1-2)-α-D-Glc	α-Galactosidase Coffee bean	Melibiose	β-CD	α-D-Gal(1-2)-β-CD	2.3	41
16	α-D-Gal(1-2)-α-D-Glc	α-Galactosidase Coffee bean	Melibiose	γ-CD	α-D-Gal(1-2)-γ-CD	1.8	41
17	α-D-Gal(1-2)-α-D-Glc	α-Galactosidase Coffee bean	Melibiose	CI$_8$	α-D-Gal(1-2)-CI$_8$	5	42
18	α-D-Gal(1-2)-D-Gal	α-Galactosidase *Candida guilliermondii* H-404	Lactose	Lactose	α-D-Gal(1-2)-D-Gal	NA	39
19	α-D-Gal(1-3)-α-D-(2,6-OAll)-Gal-O	α-Galactosidase Coffee bean	αpNPGal	α-D-(2,6-OAll)-Gal-OMe	α-D-Gal(1-3)-α-D-(2,6-OAll)-Gal-OMe	14	43
20	α-D-Gal(1-3)-α-D-(2-OAll)-Gal-O	α-Galactosidase Coffee bean	αpNPGal	α-D-(2-OAll)-Gal-OMe	α-D-Gal(1-3)-α-D-(2-OAll)-Gal-OMe	4	43
21	α-D-Gal(1-3)-α-D-(2-OBn)-Gal-O	α-Galactosidase Coffee bean	αpNPGal	α-D-(2-OBn)-Gal-OMe	α-D-Gal(1-3)-α-D-(2-OBn)-Gal-OMe	12	43
22	α-D-Gal(1-3)-α-D-Gal-O	α-Galactosidase Coffee bean	Raffinose	α-D-Gal-OAll	α-D-Gal(1-3)-α-D-Gal-OAll	4.9	44
23	α-D-Gal(1-3)-α-D-GalNAc-O	α-Galactosidase Coffee bean	αpNPGal	α-D-GalNAc-OEt	α-D-Gal(1-3)-α-D-GalNAc-OEt	21	45
24	α-D-Gal(1-3)-α-D-GalNAc-O	β-Galactosidase *Bacillus circulans*	βpNPGal	α-D-GalNAc-OBn	α-D-Gal(1-3)-α-D-GalNAc-OBn	62	46
25	α-D-Gal(1-3)-α-D-GalNAc-O	β-Galactosidase *Bacillus circulans*	βpNPGal	α-D-GalNAc-OMe	α-D-Gal(1-3)-α-D-GalNAc-OMe	53	46
26	α-D-Gal(1-3)-α-D-Gal-O	α-Galactosidase Coffee bean	αpNPGal	α-D-Gal-OMe	α-D-Gal(1-3)-α-D-Gal-OMe	27	26

Table 1 (*continued*)

Entry	Linkage	Enzyme Source	Donor	Acceptor	Product	Yield (%)	Reference
27	α-D-Gal(1-3)-α-D-Gal-O	α-Galactosidase Coffee bean	αpNPGal	αpNPGal	α-D-Gal(1-3)-α-D-Gal-OpNP	16	26
28	α-D-Gal(1-3)-α-D-Gal-O	α-Galactosidase Coffee bean	αpNPGal	αoNPGal	α-D-Gal(1-3)-α-D-Gal-OoNP	1	26
29	α-D-Gal(1-3)-α-D-Gal-O	α-Galactosidase Coffee bean	αpNPGal	αpNPGal	α-D-Gal(1-3)-α-D-Gal-OpNP	NA	47
30	α-D-Gal(1-3)-α-D-Glc	α-Galactosidase Coffee bean	Melibiose	Cl$_8$	α-D-Gal(1-3)-Cl$_8$	5	42
31	α-D-Gal(1-3)-β-D-(2,6-OAll)-Gal-O	α-Galactosidase Coffee bean	αpNPGal	β-D-(2,6-OAll)-Gal-OMe	α-D-Gal(1-3)-β-D-(2,6-OAll)-Gal-OMe	3	43
32	α-D-Gal(1-3)-β-D-(6-OAll)-Gal-O	α-Galactosidase Coffee bean	αpNPGal	β-D-(6-OAll)-Gal-OMe	α-D-Gal(1-3)-β-D-(6-OAll)-Gal-OMe	3	43
33	α-D-Gal(1-3)-β-D-Gal	α-Galactosidase *Aspergillus oryzae*	αpNPGal	β-D-Gal(1-4)-β-D-GlcNAc-SEt	α-D-Gal(1-3)-β-D-Gal(1-4)-β-D-GlcNAc-SEt	16	48
34	α-D-Gal(1-3)-β-D-Gal	α-Galactosidase Coffee bean	αpNPGal	β-D-Gal(1-4)-β-D-GlcNTeoc-SEt	α-D-Gal(1-3)-β-D-Gal(1-4)-β-D-GlcNTeoc-SEt	20	49
35	α-D-Gal(1-3)-β-D-Gal	α-Galactosidase Coffee bean	αpNPGal	β-Lactose-SEt	α-D-Gal(1-3)-β-D-Gal(1-4)-β-D-Glc-SEt	NA	50
36	α-D-Gal(1-3)-β-D-Gal	β-Galactosidase *Aspergillus oryzae*	αpNPGal	β-Lactose-SEt	α-D-Gal(1-3)-β-D-Gal(1-4)-β-D-Glc-SEt	15	50
37	α-D-Gal(1-3)-β-D-Gal	β-Galactosidase *Aspergillus oryzae*	βpNPGal	β-Lactose-SBu	α-D-Gal(1-3)-β-D-Gal(1-4)-β-D-Glc-SBu	18	50

29.6 Recent Developments and New Directions

No.	Disaccharide	Enzyme	Donor	Acceptor	Product	Yield (%)	Ref.
38	α-D-Gal(1-3)-β-D-Gal	β-Galactosidase *Aspergillus oryzae*	βpNPGal	β-Lactose-SPh	α-D-Gal(1-3)-β-D-Gal(1-4)-β-D-Glc-SPh	22	50
39	α-D-Gal(1-3)-β-D-Gal-O	α-Galactosidase Coffee bean	αpNPGal	β-D-Gal-OMe	α-D-Gal(1-3)-β-D-Gal-OMe	9	26
40	α-D-Gal(1-3)-β-D-Gal-O	α-Galactosidase Coffee bean	αpNPGal	β-D-Gal-OMe	α-D-Gal(1-3)-β-D-Gal-OMe	NA	51
41	α-D-Gal(1-3)-D-Gal	α-Galactosidase *Candida guilliermondii* H-404	Lactose	Lactose	α-D-Gal(1-3)-D-Gal	NA	39
42	α-D-Gal(1-4)-β-D-Glc	α-Galactosidase *Mortierella vinacea*	Raffinose/Melibiose	Rubusoside	13-α-D-Gal(1-6)-β-D-Glc-19-β-D-Glc-O-steviol	13	52
43	α-D-Gal(1-4)-β-D-Glc	α-Galactosidase *Absidia reflexa*	Raffinose/Melibiose	Rubusoside	13-β-D-Glc-19-α-D-Gal(1-6)-β-D-Glc-O-steviol	11	52
44	α-D-Gal(1-4)-β-D-Glc	α-Galactosidase *Mortierella vinacea*	Raffinose/Melibiose	Rubusoside	13-α-D-Gal(1-6)-α-D-Gal(1-6)-β-D-Glc-19-β-D-Glc-O-steviol	3.3	52
45	α-D-Gal(1-4)-β-D-Glc	α-Galactosidase *Absidia reflexa*	Raffinose/Melibiose	Rubusoside	13-α-D-Gal(1-6)-β-D-Glc-19-β-D-Glc-O-steviol	11	52
46	α-D-Gal(1-4)-β-D-GlcNAc-S	β-Galactosidase *Bacillus circulans*	αpNPGal	β-D-GlcNAc-SEt	α-D-Gal(1-4)-β-D-GlcNAc-SEt	50	48
47	α-D-Gal(1-4)-D-Fru$f$	β-Galactosidase *Escherichia coli*	Gal	Fru$f$	α-D-Gal(1-4)-D-Fru$f$	10	40
48	α-D-Gal(1-4)-D-Fru$f$	β-Galactosidase *Aspergillus oryzae*	Gal	Fru$f$	α-D-Gal(1-4)-D-Fru$f$	10	40
49	α-D-Gal(1-5)-D-Fru$f$	β-Galactosidase *Escherichia coli*	Gal	Fru$f$	α-D-Gal(1-5)-D-Fru$f$	6.6	40
50	α-D-Gal(1-5)-D-Fru$f$	β-Galactosidase *Aspergillus oryzae*	Gal	Fru$f$	α-D-Gal(1-5)-D-Fru$f$	9.2	40

Table 1 (*continued*)

Entry	Linkage	Enzyme Source	Donor	Acceptor	Product	Yield (%)	Reference
51	α-D-Gal(1-6)-α-D-Glc	α-Galactosidase Coffee bean	Melibiose	β-CD	α-D-Gal(1-6)-β-CD	NA	53
52	α-D-Gal(1-6)-α-D-Gal-O	α-Galactosidase Coffee bean	αpNPGal	α-D-Gal-OMe	α-D-Gal(1-6)-α-D-Gal-OMe	2	26
53	α-D-Gal(1-6)-α-D-Glc	α-Galactosidase Coffee bean	Melibiose	γ-CD	α-D-Gal(1-6)-γ-CD	38	53
54	α-D-Gal(1-6)-α-D-Glc	α-Galactosidase Coffee bean	Melibiose	α-CD	α-D-Gal(1-6)-α-CD	NA	54
55	α-D-Gal(1-6)-α-D-Glc	α-Galactosidase Coffee bean	Melibiose	α-CD	α-D-Gal(1-6)-α-CD	24	41
56	α-D-Gal(1-6)-α-D-Glc	α-Galactosidase Coffee bean	Melibiose	β-CD	α-D-Gal(1-6)-β-CD	26	41
57	α-D-Gal(1-6)-α-D-Glc	α-Galactosidase Coffee bean	Melibiose	γ-CD	α-D-Gal(1-6)-γ-CD	22	41
58	α-D-Gal(1-6)-α-D-Glc	α-Galactosidase *Morttierella vinacea*	Gal	Sucrose	α-D-Gal(1-6)-α-D-Glc(1-2)-β-D-Fruf	18	55
59	α-D-Gal(1-6)-β-D-(2-OAll)-Gal-O	α-Galactosidase Coffee bean	αpNPGal	β-D-(2-OAll)-Gal-OMe	α-D-Gal(1-6)-β-D-(2-OAll)-Gal-OMe	12	43
60	α-D-Gal(1-6)-β-D-(2-OBn)-Gal-O	α-Galactosidase Coffee bean	αpNPGal	β-D-(2-OBn)-Gal-OMe	α-D-Gal(1-6)-β-D-(2-OBn)-Gal-OMe	14	43
61	α-D-Gal(1-6)-β-D-Gal	α-Galactosidase *Aspergillus oryzae*	αpNPGal	β-D-Gal(1-4)-β-D-GlcNAc-SEt	α-D-Gal(1-6)-β-D-Gal(1-4)-β-D-GlcNAc-SEt	16	48
62	α-D-Gal(1-6)-β-D-Gal	α-Galactosidase Coffee bean	αpNPGal	β-D-Gal(1-4)-β-D-GlcNTeoc-SEt	α-D-Gal(1-6)-β-D-Gal(1-4)-β-D-GlcNTeoc-SEt	3.5	49

#	Product	Enzyme	Donor	Acceptor		
63	α-D-Gal(1-6)-β-D-Gal	α-Galactosidase *Aspergillus niger*	αpNPGal	β-Lactose-SEt	20	50
64	α-D-Gal(1-6)-β-D-Gal	α-Galactosidase Coffee bean	αpNPGal	β-Lactose-SEt	NA	50
65	α-D-Gal(1-6)-β-D-Gal	β-Galactosidase *Aspergillus oryzae*	βpNPGal	β-Lactose-SEt	9	50
66	α-D-Gal(1-6)-β-D-Gal	β-Galactosidase *Aspergillus oryzae*	βpNPGal	β-Lactose-SBu	13	50
67	α-D-Gal(1-6)-β-D-Gal	β-Galactosidase *Aspergillus oryzae*	βpNPGal	β-Lactose-SPh	17	50
68	α-D-Gal(1-6)-β-D-Gal-O	α-Galactosidase Coffee bean	αpNPGal	β-D-Gal-OMe	18	26
69	α-D-Gal(1-6)-β-D-Gal-O	α-Galactosidase Coffee bean	αpNPGal	β-D-Gal-OMe	NA	51
70	α-D-Gal(1-6)-β-D-Glc	β-Galactosidase *Escherichia coli*	Gal	Sucrose	11	55
71	α-D-Gal(1-6)-D-Fru$f$	β-Galactosidase *Escherichia coli*	Gal	Fru$f$	4.4	40
72	α-D-Gal(1-6)-D-Fru$f$	β-Galactosidase *Aspergillus oryzae*	Gal	Fru$f$	5.2	40
73	α-D-Gal(1-6)-D-Gal	α-Galactosidase *Candida guilliermondii* H-404	Lactose	Lactose	NA	39
74	α-D-GalNAc(1-3)-α-D-Gal-O	α-N-Acetylgalactosaminidase *Chamelea gallina*	αoNPGalNAc	α-D-Gal-OMe	6	56

Table 1 (continued)

Entry	Linkage	Enzyme Source	Donor	Acceptor	Product	Yield (%)	Reference
75	α-D-Glc(1-1)-5-N$_3$-5-deoxy-Fru	α-Glucosidase Yeast (Sigma)	Maltose	5-N$_3$-5-deoxy-Fru	α-D-Glc(1-1)-5-N$_3$-5-deoxy-Fru	6	57
76	α-D-Glc(1-1)-α-D-Glc	Trehalase Lobosphaera sp.	Glc	Glc	α-D-Glc(1-1)-α-D-Glc	5	37
77	α-D-Glc(1-1)-α-D-Xyl	Trehalase Trichoderma reesei	β-D-Glc-F	Xyl	α-D-Glc(1-1)-α-D-Xyl	NA	58
78	α-D-Glc(1-1)-D-Fruƒ	α-Glucosidase Saccharomyces sp.	Glc	Fruƒ	α-D-Glc(1-1)-D-Fruƒ	27	59
79	α-D-Glc-(1-2)-D-Glc	α-Glucosidase Saccharomyces sp.	Glc	Glc	α-D-Glc-(1-2)-D-Glc	1.8	60
80	α-D-Glc-(1-2)-D-Glc	Glucoamylase Rhizopus sp.	Glc	Glc	α-D-Glc-(1-2)-D-Glc	1.9	60
81	α-D-Glc(1-3)-β-D-Glc	β-amylase Aspergillus sp.	Maltose	Stevioside	13-α-D-Glc(1-3)-β-D-Glc(1-2)-β-D-Glc-19-β-D-Glc-O-steviol	NA	61
82	α-D-Glc-((1-3)-D-Glc	α-Glucosidase Saccharomyces sp.	Glc	Glc	α-D-Glc-(1-3)-D-Glc	2.7	60
83	α-D-Glc-((1-3)-D-Glc	Glucoamylase Rhizopus sp.	Glc	Glc	α-D-Glc-(1-3)-D-Glc	2.5	60
84	α-D-Glc(1-4)-5-N$_3$-5-deoxy-Fru	α-Glucosidase Yeast (Sigma)	Maltose	5-N$_3$-5-deoxy-Fru	α-D-Glc(1-4)-5-N$_3$-5-deoxy-Fru	6	57

No.	Acceptor		Donor	Enzyme	Product	Yield	Ref
85	α-D-Glc(1-4)-α-D-Glc	Glc	Trehalose	α-Glucosidase *Saccharomyces* sp.	α-D-Glc(1-4)-α-D-Glc(1-1)-α-D-Glc	2.1	62
86	α-D-Glc(1-4)-α-D-Glc	Glc	Trehalose	Glucoamylase *Rhizopus niveus*	α-D-Glc(1-4)-α-D-Glc(1-1)-α-D-Glc	2	62
87	α-D-Glc(1-4)-β-D-Glc-O	βpNPGlc	βpNPGlc	β-Galactosidase *Aspergillus oryzae* (Impurity?)	α-D-Glc(1-4)-β-D-Glc-O*p*NP	11	63
88	α-D-Glc(1-4)-β-D-Glc-O	βpNPGlc	βpNPGlc	β-Galactosidase *Aspergillus oryzae* (Impurity?)	α-D-Glc(1-4)-β-D-Glc-O*p*NP	11	64
89	α-D-Glc(1-4)-D-Fru*f*	Glc	Fru*f*	α-Glucosidase *Saccharomyces* sp.	α-D-Glc(1-4)-D-Fru*f*	14	59
90	α-D-Glc-(1-4)-D-Glc	Glc	Glc	α-Glucosidase *Saccharomyces* sp.	α-D-Glc-(1-4)-D-Glc	1.8	60
91	α-D-Glc-(1-4)-D-Glc	Glc	Glc	Glucoamylase *Rhizopus* sp.	α-D-Glc-(1-4)-D-Glc	2.2	60
92	α-D-Glc(1-4)-D-Glc	Glc	Glc	Glucoamylase *Rhizopus oryzae*	α-D-Glc(1-4)-D-Glc	6.4	65
93	α-D-Glc(1-4)-D-Glc	Glc	Glc	Glucoamylase *Rhizopus oryzae*	α-D-Glc(1-4)-D-Glc	3.5	66
94	α-D-Glc(1-5)-D-Fru*f*	Glc	Fru*f*	α-Glucosidase *Saccharomyces* sp.	α-D-Glc(1-5)-D-Fru*f*	4.5	59
95	α-D-Glc(1-6)-α-D-Glc	Glc	Glc	Glucoamylase *Rhizopus oryzae*	α-D-Glc(1-6)-α-D-Glc(1-4)-D-Glc	2	66
96	α-D-Glc(1-6)-α-D-Glc	Glc	Glc	Glucoamylase *Rhizopus oryzae*	α-D-Glc(1-6)-α-D-Glc(1-6)-D-Glc	6	66

Table 1 (*continued*)

Entry	Linkage	Enzyme Source	Donor	Acceptor	Product	Yield (%)	Reference
97	α-D-Glc(1-6)-α-D-Glc	α-Glucosidase *Saccharomyces* sp.	Glc	Trehalose	α-D-Glc(1-6)-α-D-Glc(1-1)-α-D-Glc	6.4	62
98	α-D-Glc(1-6)-α-D-Glc	Glucoamylase *Rhizopus niveus*	Glc	Trehalose	α-D-Glc(1-6)-α-D-Glc(1-1)-α-D-Glc	13	62
99	α-D-Glc(1-6)-β-D-Glc	β-amylase *Aspergillus* sp.	Maltose	Stevioside	13-β-D-Glc(1-2)-β-D-Glc-19-α-D-Glc(1-6)-β-D-Glc-O-steviol	NA	61
100	α-D-Glc(1-6)-β-D-Glc	β-amylase *Aspergillus* sp.	Maltose	Stevioside	13-α-D-Glc(1-6)-β-D-Glc(1-2)-β-D-Glc-19-β-D-Glc-O-steviol	NA	61
101	α-D-Glc(1-6)-D-α-Glc	Glucoamylase *Rhizopus oryzae*	Glc	Glc	α-D-Glc(1-6)-α-D-Glc(1-4)-D-Glc	3.5	65
102	α-D-Glc(1-6)-D-α-Glc	Glucoamylase *Rhizopus oryzae*	Glc	Glc	α-D-Glc(1-6)-α-D-Glc(1-6)-D-Glc	6	65
103	α-D-Glc(1-6)-D-Fruf	α-Glucosidase *Saccharomyces* sp.	Glc	Fruf	α-D-Glc(1-6)-D-Fruf	5	59
104	α-D-Glc-((1-6)-D-Glc	α-Glucosidase *Saccharomyces* sp.	Glc	Glc	α-D-Glc-(1-6)-D-Glc	24	60
105	α-D-Glc(1-6)-D-Glc	Glucoamylase *Rhizopus* sp.	Glc	Glc	α-D-Glc(1-6)-D-Glc	24	60
106	α-D-Glc(1-6)-D-Glc	Glucoamylase *Rhizopus oryzae*	Glc	Glc	α-D-Glc(1-6)-D-Glc	26	65
107	α-D-Glc(1-6)-D-Glc	Glucoamylase *Rhizopus oryzae*	Glc	Glc	α-D-Glc(1-6)-D-Glc	26	66

108	α-D-Man(1-1)-α-D-Man	α-Mannosidase Jack bean	Man	Man	α-D-Man(1-1)-α-D-Man	3.3	67
109	α-D-Man(1-2)-α-D-Man	α-Mannosidase Aspergillus phoenicis	Man	Man	α-D-Man(1-2)-α-D-Man(1-2)-D-Man	8	16
110	α-D-Man(1-2)-α-D-Man-O	α-Mannosidase Jack bean	αpNPMan	α-D-Man-OMe	α-D-Man(1-2)-α-D-Man-OMe	18	26
111	α-D-Man(1-2)-α-D-Man-O	α-Mannosidase Jack bean	αpNPMan	αpNPMan	α-D-Man(1-2)-α-D-Man-OpNP	8	26
112	α-D-Man(1-2)-D-Man	α-Mannosidase Jack bean	Man	Man	α-D-Man(1-2)-D-Man	16	68
113	α-D-Man(1-2)-D-Man	α-Mannosidase Jack bean	Man	Man	α-D-Man(1-2)-D-Man	7.2	67
114	α-D-Man(1-2)-D-Man	α-Mannosidase Jack bean	Man	Man	α-D-Man(1-2)-D-Man	NA	69
115	α-D-Man(1-2)-D-Man	α-Mannosidase Aspergillus phoenicis	Man	Man	α-D-Man(1-2)-D-Man	22	16
116	α-D-Man(1-2)-D-Man	α-Mannosidase Aspergillus niger	Man	Man	α-D-Man(1-2)-D-Man	2	15
117	α-D-Man(1-3)-D-Man	α-Mannosidase Jack bean	Man	Man	α-D-Man(1-3)-D-Man	3.6	68
118	α-D-Man(1-3)-D-Man	α-Mannosidase Jack bean	Man	Man	α-D-Man(1-3)-D-Man	1.3	67
119	α-D-Man(1-3)-D-Man	α-Mannosidase Jack bean	Man	Man	α-D-Man(1-3)-D-Man	NA	69
120	α-D-Man(1-3)-D-Man	α-Mannosidase Aspergillus niger	Man	Man	α-D-Man(1-3)-D-Man	0.2	15
121	α-D-Man(1-4)-D-Man	α-Mannosidase Jack bean	Man	Man	α-D-Man(1-4)-D-Man	4.2	67
122	α-D-Man(1-4)-D-Man	α-Mannosidase Jack bean	Man	Man	α-D-Man(1-4)-D-Man	NA	69

Table 1 (continued)

Entry	Linkage	Enzyme Source	Donor	Acceptor	Product	Yield (%)	Reference
123	α-D-Man(1-4)-D-Man	α-Mannosidase Rhizopus niveus	Man	Man	α-D-Man(1-4)-D-Man	0.3	70
124	α-D-Man(1-6)-α-D-Glc	α-Mannosidase Jack bean	Man	α-CD	α-D-Man(1-6)-α-CD	14	71
125	α-D-Man(1-6)-α-D-Glc	α-Mannosidase Jack bean	Man	β-CD	α-D-Man(1-6)-β-CD	10	71
126	α-D-Man(1-6)-α-D-Glc	α-Mannosidase Jack bean	Man	γ-CD	α-D-Man(1-6)-γ-CD	11	71
127	α-D-Man(1-6)-α-D-Glc	α-Mannosidase Jack bean	α-D-Man-OMe	α-D-Glc(1-6)-β-CD	α-D-Man(1-6)-α-D-Glc(1-6)-β-CD	5.3	72
128	α-D-Man(1-6)-α-D-Glc	α-Mannosidase Jack bean	α-D-Man-OMe	α-D-Glc(1-4)-α-D-Glc(1-6)-α-CD	α-D-Man(1-6)-α-D-Glc(1-4)-α-D-Glc(1-6)-α-CD	6.4	72
129	α-D-Man(1-6)-α-D-Glc	α-Mannosidase Jack bean	α-D-Man-OMe	α-D-Glc(1-4)-α-D-Glc(1-6)-β-CD	α-D-Man(1-6)-α-D-Glc(1-4)-α-D-Glc(1-6)-β-CD	6.1	72
130	α-D-Man(1-6)-α-D-Man-O	αpNPMan	α-D-Man-OMe	α-D-Man(1-6)-α-D-Man-OMe	4	26
131	α-D-Man(1-6)-β-D-Gal-O	α-Mannosidase Jack bean	αpNPMan	βpNPGal	α-D-Man(1-6)-β-D-Gal-OpNP	5	73
132	α-D-Man(1-6)-D-Man	α-Mannosidase Jack bean	Man	Man	α-D-Man(1-6)-D-Man	14	68
133	α-D-Man(1-6)-D-Man	α-Mannosidase Jack bean	Man	Man	α-D-Man(1-6)-D-Man	9.8	67
134	α-D-Man(1-6)-D-Man	α-Mannosidase Aspergillus niger	Man	Man	α-D-Man(1-6)-D-Man	7.6	15
135	α-D-NeuAc(2-3)-α-D-Gal	Sialidase Trypanosoma cruzi	αMuNeuAc	β-D-Gal(1-4)-β-D-Xyl-OpNP	α-D-NeuAc(2-3)-β-D-Gal(1-4)-β-D-Xyl-OpNP	33	74

#		Sialidase					Ref
136	α-D-NeuAc(2-3)-α-D-Gal	Sialidase Salmonella typhirium LT2	αpNPNeuAc	Le^x	sLe^x	9.3	75
137	α-D-NeuAc(2-3)-α-D-Gal	Sialidase Salmonella typhirium LT2	αpNPNeuAc	(2-OH)-Le^x	(2-OH)-sLe^x	13	75
138	α-D-NeuAc(2-3)-α-D-Gal	Sialidase Salmonella typhirium LT2	αpNPNeuAc	Le^a	sLe^a	12	75
139	α-D-NeuAc(2-3)-α-D-Gal	Sialidase Salmonella typhirium LT2	αpNPNeuAc	(2-OH)-Le^a	(2-OH)-sLe^a	15	75
140	α-D-NeuAc(2-3)-α-D-Gal-O	Sialidase Vibrio cholerae	αpNPNeuAc	α-D-Gal-OMe	α-D-NeuAc(2-3)-α-D-Gal-OMe	3	76
141	α-D-NeuAc(2-3)-β-D-Gal	Sialidase Vibrio cholerae	αpNPNeuAc	β-Lactose-OMe	α-D-NeuAc(2-3)-β-D-Gal(1-4)-β-D-Glc-OMe	5	76
142	α-D-NeuAc(2-3)-β-D-Gal	Sialidase Vibrio cholerae	αpNPNeuAc	β-D-Gal(1-4)-D-GlcNAc	α-D-NeuAc(2-3)-β-D-Gal(1-4)-D-GlcNAc	4	76
143	α-D-NeuAc(2-3)-β-D-Gal	Sialidase Trypanosoma cruzi	αMuNeuAc	β-D-Gal(1-4)-D-GlcNAc	α-D-NeuAc(2-3)-β-D-Gal(1-4)-D-GlcNAc	60	77
144	α-D-NeuAc(2-3)-β-D-Gal	Sialidase Clostridium perfringens	αpNPNeuAc	β-D-Gal(1-3)-D-Glc	α-D-NeuAc(2-3)-β-D-Gal(1-3)-D-Glc	3.7	78
145	α-D-NeuAc(2-3)-β-D-Gal	Sialidase Vibrio cholerae	αpNPNeuAc	β-D-Gal(1-3)-D-Glc	α-D-NeuAc(2-3)-β-D-Gal(1-3)-D-Glc	1	78
146	α-D-NeuAc(2-3)-β-D-Gal	Sialidase Newcastle disease virus	α-D-NeuAc(2-8)-D-NeuAc	β-D-Gal(1-3)-D-Glc	α-D-NeuAc(2-3)-β-D-Gal(1-3)-D-Glc	10	78
147	α-D-NeuAc(2-3)-β-D-Gal	Sialidase Newcastle disease virus	α-D-NeuAc(2-8)-D-NeuAc	β-D-Gal(1-3)-D-GlcNAc	α-D-NeuAc(2-3)-β-D-Gal(1-3)-D-GlcNAc	8	78

Table 1 (*continued*)

Entry	Linkage	Enzyme Source	Donor	Acceptor	Product	Yield (%)	Reference
148	α-D-NeuAc(2-3)-β-D-Gal	Sialidase Newcastle disease virus	α-D-NeuAc(2-8)$_n$	β-D-Gal(1-3)-D-Glc	α-D-NeuAc(2-3)-β-D-Gal(1-3)-D-Glc	2.7	78
149	α-D-NeuAc(2-3)-β-D-Gal	Sialidase Newcastle disease virus	αpNPNeuAc	β-D-Gal(1-3)-D-Glc	α-D-NeuAc(2-3)-β-D-Gal(1-3)-D-Glc	7.7	78
150	α-D-NeuAc(2-3)-β-D-Gal	Sialidase Newcastle disease virus	αpNPNeuAc	β-D-Gal(1-3)-D-GlcNAc	α-D-NeuAc(2-3)-β-D-Gal(1-3)-D-GlcNAc	12	78
151	α-D-NeuAc(2-3)-β-D-Gal-O	Sialidase Vibrio cholerae	αpNPNeuAc	β-D-Gal-OMe	α-D-NeuAc(2-3)-β-D-Gal-OMe	7	76
152	α-D-NeuAc(2-3)-β-D-Gal-O	Sialidase Trypanosoma cruzi	α-D-NeuAc(2-3)-β-D-Gal(1-3)-β-D-GalNAc-O-(CH$_2$)$_2$Si(CH$_3$)$_3$	β-D-Gal-O-(CH$_2$)$_2$Si(CH$_3$)$_3$	α-D-NeuAc(2-3)-β-D-Gal-O-(CH$_2$)$_2$Si(CH$_3$)$_3$	NA	23
153	α-D-NeuAc(2-3)-β-D-Glc-O	Sialidase Vibrio cholerae	αpNPNeuAc	β-D-Glc-OMe	α-D-NeuAc(2-3)-β-D-Glc-OMe	4	76
154	α-D-NeuAc(2-6)-α-D-Gal-O	Sialidase Vibrio cholerae	αpNPNeuAc	α-D-Gal-OMe	α-D-NeuAc(2-6)-α-D-Gal-OMe	21	76
155	α-D-NeuAc(2-6)-β-D-Gal	Sialidase Vibrio cholerae	αpNPNeuAc	β-Lactose-OMe	α-D-NeuAc(2-6)-β-D-Gal(1-4)-β-D-Glc-OMe	11	76
156	α-D-NeuAc(2-6)-β-D-Gal	Sialidase Vibrio cholerae	αpNPNeuAc	β-D-Gal(1-4)-D-GlcNAc	α-D-NeuAc(2-6)-β-D-Gal(1-4)-D-GlcNAc	10	76
157	α-D-NeuAc(2-6)-β-D-Gal	Sialidase Clostridium perfringens	α-D-NeuAc(2-8)-D-NeuAc	β-D-Gal(1-3)-D-Glc	α-D-NeuAc(2-6)-β-D-Gal(1-3)-D-Glc	8.5	78

#	Substrate	Enzyme	Donor	Acceptor	Product	Yield	Ref
158	α-D-NeuAc(2-6)-β-D-Gal	Sialidase *Clostridium perfringens*	α-D-NeuAc(2-8)-D-NeuAc	β-D-Gal(1-3)-D-GlcNAc	α-D-NeuAc(2-6)-β-D-Gal(1-3)-D-GlcNAc	5.5	78
159	α-D-NeuAc(2-6)-β-D-Gal	Sialidase *Clostridium perfringens*	{α-D-NeuAc(2-8)}$_n$-D-NeuAc	β-D-Gal(1-3)-D-Glc	α-D-NeuAc(2-6)-β-D-Gal(1-3)-D-Glc	2.4	78
160	α-D-NeuAc(2-6)-β-D-Gal	Sialidase *Clostridium perfringens*	αpNPNeuAc	β-D-Gal(1-3)-D-Glc	α-D-NeuAc(2-6)-β-D-Gal(1-3)-D-Glc	8.7	78
161	α-D-NeuAc(2-6)-β-D-Gal	Sialidase *Arthrobacter ureafaciens*	α-D-NeuAc(2-8)-D-NeuAc	β-D-Gal(1-3)-D-Glc	α-D-NeuAc(2-6)-β-D-Gal(1-3)-D-Glc	5.6	78
162	α-D-NeuAc(2-6)-β-D-Gal	Sialidase *Arthrobacter ureafaciens*	αpNPNeuAc	β-D-Gal(1-3)-D-Glc	α-D-NeuAc(2-6)-β-D-Gal(1-3)-D-Glc	7.4	78
163	α-D-NeuAc(2-6)-β-D-Gal	Sialidase *Vibrio cholerae*	α-D-NeuAc(2-8)-D-NeuAc	β-D-Gal(1-3)-D-Glc	α-D-NeuAc(2-6)-β-D-Gal(1-3)-D-Glc	4.1	78
164	α-D-NeuAc(2-6)-β-D-Gal	Sialidase *Vibrio cholerae*	αpNPNeuAc	β-D-Gal(1-3)-D-Glc	α-D-NeuAc(2-6)-β-D-Gal(1-3)-D-Glc	9.2	78
165	α-D-NeuAc(2-6)-β-D-Gal	Sialidase Newcastle disease virus	α-D-NeuAc(2-8)$_n$	β-D-Gal(1-3)-D-Glc	α-D-NeuAc(2-6)-β-D-Gal(1-3)-D-Glc	0.9	78
166	α-D-NeuAc(2-6)-β-D-Gal	Sialidase Newcastle disease virus	αpNPNeuAc	β-D-Gal(1-3)-D-Glc	α-D-NeuAc(2-6)-β-D-Gal(1-3)-D-Glc	2.5	78
167	α-D-NeuAc(2-6)-β-D-Gal-O	Sialidase *Vibrio cholerae*	αpNPNeuAc	β-D-Gal-OMe	α-D-NeuAc(2-6)-β-D-Gal-OMe	15	76
168	α-D-NeuAc(2-6)-β-D-Glc-O	Sialidase *Vibrio cholerae*	αpNPNeuAc	β-D-Glc-OMe	α-D-NeuAc(2-6)-β-D-Glc-OMe	12	76
169	α-L-Fuc(1-2)-β-D-Gal-O	α-Fucosidase Porcine liver	α-L-Fuc F	β-D-Gal-OMe	α-L-Fuc(1-2)-β-D-Gal-OMe	6.5	79

Table 1 (continued)

Entry	Linkage	Enzyme Source	Donor	Acceptor	Product	Yield (%)	Reference
170	α-L-Fuc(1-2)-β-D-Gal-O	α-Fucosidase Corynebacterium sp.	αpNPFuc	β-D-Gal-OMe	α-L-Fuc(1-2)-β-D-Gal-OMe	25	80
171	α-L-Fuc(1-2)-D-Gal	α-Fucosidase Corynebacterium sp.	αpNPFuc	Gal	α-L-Fuc(1-2)-D-Gal	18	80
172	α-L-Fuc(1-3)-D-Glc	α-L-Fucosidase Penicillium multicolor	αpNPFuc	Glc	α-L-Fuc(1-3)-D-Glc	28	81
173	α-L-Fuc(1-3)-D-Glc	α-Fucosidase Aspergillus niger	αpNPFuc	Glc	α-L-Fuc(1-3)-D-Glc	61	80
174	α-L-Fuc(1-3)-D-GlcNAc	α-L-Fucosidase Penicillium multicolor	αpNPFuc	GlcNAc	α-L-Fuc(1-3)-D-GlcNAc	49	81
175	α-L-Fuc(1-3)-D-GlcNAc	α-Fucosidase Aspergillus niger	αpNPFuc	GlcNAc	α-L-Fuc(1-3)-D-GlcNAc	58	80
176	α-L-Fuc(1-3)-D-GlcNAc	α-Fucosidase Aspergillus niger	αpNPFuc	GlcNAc	α-L-Fuc(1-3)-D-GlcNAc	24	82
177	α-L-Fuc(1-4)-β-D-(6-OBn)-GlcN-S	α-Fucosidase Bovine testes	αpNPFuc	β-D-(6-OBn)-GlcN-SEt	α-L-Fuc(1-4)-β-D-(6-OBn)-GlcN-SEt	33	83
178	α-L-Fuc(1-4)-β-D-(6-OBn)-GlcN-S	β-Fucosidase Bovine kidney	αpNPFuc	β-D-(6-OBn)-GlcN-SEt	α-L-Fuc(1-4)-β-D-(6-OBn)-GlcN-SEt	50	84
179	α-L-Fuc(1-6)-β-D-Gal-O	α-Fucosidase Porcine liver	α-L-Fuc F	β-D-Gal-OMe	α-L-Fuc(1-6)-β-D-Gal-OMe	10	79

#	Donor	Enzyme		Acceptor	Product		
180	α-L-Fuc(1-6)-β-D-Gal-O	α-Fucosidase Ampullaria	αpNPFuc	β-D-Gal-OMe	α-L-Fuc(1-6)-β-D-Gal-OMe	14	80
181	β-D-2dGlc(1-3)-β-D-Glc-S	β-Glucosidase Agrobacterium sp.	Glucal	β-D-Glc-SPh	2-deoxy-β-D-Glc(1-3)-β-D-Glc-SPh	15	85
182	β-D-2dGlc(1-3)-β-D-Xyl-S	β-Glucosidase Agrobacterium sp.	Glucal	β-D-Xyl-SBn	2-deoxy-β-D-Glc(1-3)-β-D-Xyl-SBn	16	85
183	β-D-2dGlc(1-4)-β-D-Glc-S	β-Glucosidase Agrobacterium sp.	Glucal	β-D-Glc-SPh	2-deoxy-β-D-Glc(1-4)-β-D-Glc-SPh	12	85
184	β-D-2dGlc(1-4)-β-D-Xyl-S	β-Glucosidase Agrobacterium sp.	Glucal	β-D-Xyl-SBn	2-deoxy-β-D-Glc(1-4)-β-D-Xyl-SBn	16	85
185	β-D-Fruf(2-1)-α-D-Glc	β-Fructofuranosidase Arthrobacter sp. K-1	Sucrose	Maltose	β-D-Fruf(2-1)-α-D-Glc(1-4)-D-Glc	55	86
186	β-D-Fruf(2-1)-α-D-Glc	β-Fructofuranosidase Arthrobacter sp. K-1	Sucrose	α-D-Glc(1-6)-D-Glc	β-D-Fruf(2-1)-α-D-Glc(1-6)-D-Glc	51	86
187	β-D-Fruf(2-1)-β-D-Fruf	β-Fructofuranosidase Aspergillus sydowi	Sucrose	Trehalose	β-D-Fruf(2-1)-β-D-Fruf-β-D(2-6)-α-D-Glc(1-1)-α-D-Glc	NA	87
188	β-D-Fruf(2-1)-β-D-Fruf	β-Fructofuranosidase Aspergillus sydowi	Sucrose	Trehalose	β-D-Fruf(2-1)-β-D-Fruf-β-D(2-1)-β-D-Fruf-(2-6)-α-D-Glc(1-1)-α-D-Glc	NA	87
189	β-D-Fruf(2-1)-β-D-Fruf	β-Fructofuranosidase Aspergillus oryzae	Sucrose	Sucrose	β-D-Fruf(2-1)-β-D-Fruf(2-1)-α-D-Glc	27	88
190	β-D-Fruf(2-1)-β-D-Fruf	β-Fructofuranosidase Aspergillus oryzae	Sucrose	Sucrose	β-D-Fruf(2-1)-β-D-Fruf(2-1)-β-D-Fruf(2-1)-α-D-Glc	17	88

Table 1 (continued)

Entry	Linkage	Enzyme Source	Donor	Acceptor	Product	Yield (%)	Reference
191	β-D-Fruf-(2-1)-β-D-Gal	β-Fructofuranosidase Arthrobacter sp. K-1	Sucrose	Lactose	β-D-Fruf-(2-1)-β-D-Gal(1-4)-D-Glc	44	86
192	β-D-Fruf-(2-1)-D-Ara	β-Fructofuranosidase Arthrobacter sp. K-1	Sucrose	D-Ara	β-D-Fruf-(2-1)-D-Ara	40	86
193	β-D-Fruf-(2-1)-D-Gal	β-Fructofuranosidase Arthrobacter sp. K-1	Sucrose	Gal	β-D-Fruf-(2-1)-D-Gal	52	86
194	β-D-Fruf-(2-1)-D-Xyl	β-Fructofuranosidase Arthrobacter sp. K-1	Sucrose	Xyl	β-D-Fruf-(2-1)-D-Xyl	62	86
195	β-D-Fruf-(2-1)-L-Ara	β-Fructofuranosidase Arthrobacter sp. K-1	Sucrose	L-Ara	β-D-Fruf-(2-1)-L-Ara	33	86
196	β-D-Fruf-(2-1)-L-Fuc	β-Fructofuranosidase Arthrobacter sp. K-1	Sucrose	L-Fuc	β-D-Fruf-(2-1)-L-Fuc	70	86
197	β-D-Fruf-(2-2)-L-Sor	β-Fructofuranosidase Arthrobacter sp. K-1	Sucrose	L-Sor	β-D-Fruf-(2-2)-L-Sor	55	86
198	β-D-Fruf-(2-3)-D-Gal	β-Fructofuranosidase Arthrobacter sp. K-1	Sucrose	Gal	β-D-Fruf-(2-3)-D-Gal	7.5	86
199	β-D-Fruf-(2-4)-L-Ara	β-Fructofuranosidase Arthrobacter sp. K-1	Sucrose	L-Ara	β-D-Fruf-(2-4)-L-Ara	27	86

No.	Product	Enzyme	Donor	Acceptor	Product		
200	β-D-Fruf(2-6)-α-D-Glc	β-Fructofuranosidase Aspergillus sydowi	Sucrose	Trehalose	β-D-Fruf(2-6)-α-D-Glc(1-1)-α-D-Glc	NA	87
201	β-D-Fruf(2-6)-β-D-Fruf	β-Fructofuranosidase Saccharomyces cerevisiae	Sucrose	Sucrose	β-D-Fruf(2-6)-β-D-Fruf(2-1)-α-D-Glc	7.1	89
202	β-D-Fruf(2-6)-β-D-Glc	β-Fructofuranosidase Microbacterium sp. H-1	Sucrose	13-β-D-Glc(1-3)-[β-D-Glc(1-2)]-β-D-Glc-19-β-D-Fruf(2-6)-β-D-Glc-O-steviol	13-β-D-Glc(1-3)-[β-D-Glc(1-2)]-β-D-Glc-19-β-D-Fruf(2-6)-β-D-Glc-O-steviol	82	90
203	β-D-Fruf(2-6)-β-D-Glc-O	β-Fructofuranosidase Arthrobacter sp. K-1	Sucrose	13-β-D-Glc(1-3)-[β-D-Glc(1-2)]-β-D-Glc-19-β-D-Fruf(2-6)-β-D-Glc-O-steviol	13-β-D-Glc(1-3)-[β-D-Glc(1-2)]-β-D-Glc-19-β-D-Fruf(2-6)-β-D-Glc-O-steviol	88	91
204	β-D-Fruf(2-6)-β-D-Glc-O	β-Fructofuranosidase Arthrobacter sp. K-1	Sucrose	β-D-Glc-OMe	β-D-Fruf(2-6)-β-D-Glc-OMe	68	86
205	β-D-Fruf(2-6)-D-Fruf	β-Fructofuranosidase Saccharomyces cerevisiae	Sucrose	Sucrose	β-D-Fruf(2-1)-D-Fruf	2.6	89
206	β-D-Gal(1-1)-D-Glc3NAc	β-Galactosidase Bacillus circulans	Lactose	Glc3NAc	β-D-Gal(1-1)-D-Glc3NAc	4	92
207	β-D-Gal(1-2)-α-D-Gal-O	β-Galactosidase Aspergillus oryzae	βoNPGal	α-D-Gal-OMe	β-D-Gal(1-2)-α-D-Gal-OMe	12	93
208	β-D-Gal(1-2)-α-D-Gal-O	β-Galactosidase Aspergillus oryzae	βVGal	α-D-Gal-OMe	β-D-Gal(1-2)-α-D-Gal-OMe	18	93
209	β-D-Gal(1-2)-α-D-Glc	β-Galactosidase Aspergillus oryzae	Gal	Trehalose	β-D-Gal(1-2)-α-D-Glc(1-1)-α-D-Glc	1.6	62

Table 1 (*continued*)

Entry	Linkage	Enzyme Source	Donor	Acceptor	Product	Yield (%)	Reference
210	β-D-Gal(1-2)-α-D-Glc	β-Galactosidase *Penicillium multicolor*	Lactose	CI$_8$	β-D-Gal(1-2)-CI$_8$	6.7	42
211	β-D-Gal(1-2)-α-D-Glc	β-Galactosidase *Bacillus circulans*	Lactose	CI$_8$	β-D-Gal(1-2)-CI$_8$	17	42
212	β-D-Gal(1-2)-D-Glc	β-Galactosidase *Escherichia coli*	Gal	Glc	β-D-Gal(1-2)-D-Glc	0.5	40
213	β-D-Gal(1-2)-D-Glc	β-Galactosidase *Bacillus circulans*	Lactose	Lactose	β-D-Gal(1-2)-D-Glc	NA	94
214	β-D-Gal(1-2)-D-Glc	β-Galactosidase *Bacillus circulans*	Lactose	Lactose	β-D-Gal(1-4)-[β-D-Gal(1-2)]-D-Glc	NA	94
215	β-D-Gal(1-2)-D-Xyl	β-Galactosidase Bovine testes	βoNPGal	Xyl	β-D-Gal(1-2)-D-Xyl	4	95
216	β-D-Gal(1-2)-D-Xyl	β-Galactosidase Intestinal lactase Lamb	βoNPGal	Xyl	β-D-Gal(1-2)-D-Xyl	9	95
217	β-D-Gal(1-2)-D-Xyl	β-Galactosidase *Aspergillus oryzae*	βoNPGal	Xyl	β-D-Gal(1-2)-D-Xyl	18	95
218	β-D-Gal(1-2)-D-Xyl	β-Galactosidase *Saccharomyces fragilis*	βoNPGal	Xyl	β-D-Gal(1-2)-D-Xyl	1	95
219	β-D-Gal(1-2)-D-Xyl	β-Galactosidase *Escherichia coli*	βoNPGal	Xyl	β-D-Gal(1-2)-D-Xyl	3	95

No.	Product	Enzyme	Donor	Acceptor	Product	Yield	Ref.
220	β-D-Gal(1-2)-D-Xyl	β-Galactosidase *Escherichia coli*	βpNPGal	Xyl	β-D-Gal(1-2)-D-Xyl	4.5	96
221	β-D-Gal(1-2)-L-Xyl	β-Galactosidase Bovine testes	βoNPGal	L-Xyl	β-D-Gal(1-2)-L-Xyl	2	95
222	β-D-Gal(1-2)-L-Xyl	β-Galactosidase Intestinal lactase Lamb	βoNPGal	L-Xyl	β-D-Gal(1-2)-L-Xyl	5	95
223	β-D-Gal(1-2)-L-Xyl	β-Galactosidase *Aspergillus oryzae*	βoNPGal	L-Xyl	β-D-Gal(1-2)-L-Xyl	3	95
224	β-D-Gal(1-2)-L-Xyl	β-Galactosidase *Saccharomyces fragilis*	βoNPGal	L-Xyl	β-D-Gal(1-2)-L-Xyl	11	95
225	β-D-Gal(1-2)-L-Xyl	β-Galactosidase *Escherichia coli*	βoNPGal	L-Xyl	β-D-Gal(1-2)-L-Xyl	11	95
226	β-D-Gal(1-3)-(6-OAc)-Glucal	β-Galactosidase *Escherichia coli*	βpNPGal	(6-O-Ac)-Glucal	β-D-Gal(1-3)-(6-O-Ac)-Glucal	42	97
227	β-D-Gal(1-3)-α-D-GalNAc-O	β-Galactosidase Bovine testes	βpNPGal	α-D-GalNAc-OEt	β-D-Gal(1-3)-α-D-GalNAc-OEt	11	45
228	β-D-Gal(1-3)-α-D-GalNAc-O	β-Galactosidase *Bacillus circulans*	βpNPGal	α-D-GalNAc-O-Ser-(N-Z)-OAll	β-D-Gal(1-3)-α-D-GalNAc-O-Ser-(N-Z)-OAll	68	98
229	β-D-Gal(1-3)-α-D-GalNAc-O	β-Galactosidase Bovine testes	βpNPGal	α-D-GalNAc-O-Ser	β-D-Gal(1-3)-α-D-GalNAc-O-Ser	22	99
230	β-D-Gal(1-3)-α-D-GalNAc-O	β-Galactosidase Bovine testes	βpNPGal	α-D-GalNAc-O-Thr	β-D-Gal(1-3)-α-D-GalNAc-O-Thr	28	99
231	β-D-Gal(1-3)-α-D-GalNAc-O	β-Galactosidase Porcine testes	Lactose	αpNPGalNAc	β-D-Gal(1-3)-α-D-GalNAc-OpNP	16	100
232	β-D-Gal(1-3)-α-D-GalNAc-O	β-Galactosidase Bovine Testes	βpNPGal	α-D-GalNAc-O-(CH$_2$)$_2$O(CH$_2$)$_2$Cl	β-D-Gal(1-3)-α-D-GalNAc-O-(CH$_2$)$_2$O(CH$_2$)$_2$Cl	40	101
233	β-D-Gal(1-3)-α-D-GalNAc-O	β-Galactosidase Bovine Testes	βpNPGal	α-D-GalNAc-O-(CH$_2$)$_2$O(CH$_2$)$_2$I	β-D-Gal(1-3)-α-D-GalNAc-O-(CH$_2$)$_2$O(CH$_2$)$_2$I	35	101

Table 1 (*continued*)

Entry	Linkage	Enzyme Source	Donor	Acceptor	Product	Yield (%)	Reference
234	β-D-Gal(1-3)-α-D-GalNAc-O	β-Galactosidase Bovine Testes	βpNPGal	α-D-GalNAc-O-(CH$_2$)$_2$O(CH$_2$)$_2$N$_3$	β-D-Gal(1-3)-α-D-GalNAc-O-(CH$_2$)$_2$O(CH$_2$)$_2$N$_3$	35	101
235	β-D-Gal(1-3)-α-D-GalNAc-O	β-Galactosidase Bovine Testes	βpNPGal	α-D-GalNAc-O-CH$_2$CH(NH$_2$)COOH	β-D-Gal(1-3)-α-D-GalNAc-O-CH$_2$CH(NH$_2$)COOH	22	101
236	β-D-Gal(1-3)-α-D-GalNAc-O	β-Galactosidase Bovine Testes	βpNPGal	αpNPGalNAc	β-D-Gal(1-3)-α-D-GalNAc-OpNP	21	101
237	β-D-Gal(1-3)-α-D-GalNAc-O	β-Galactosidase Bovine Testes	βpNPGal	α-D-GalNAc-OAll	β-D-Gal(1-3)-α-D-GalNAc-OAll	22	101
238	β-D-Gal(1-3)-α-D-GalNAc-O	β-Galactosidase Bovine Testes	βpNPGal	α-D-GalNAc-O-CH(CH$_3$)CH(NH$_2$)COOH	β-D-Gal(1-3)-α-D-GalNAc-O-CH(CH$_3$)CH(NH$_2$)COOH	28	101
239	β-D-Gal(1-3)-α-D-GalNAc-O	β-Galactosidase Bovine testes	Lactose	α-D-GalNAc-O-Ser-(N-Aloc)-OMe	β-D-Gal(1-3)-α-D-GalNAc-O-Ser-(N-Aloc)-OMe	20	102
240	β-D-Gal(1-3)-α-D-Gal-O	β-Galactosidase *Aspergillus oryzae*	βoNPGal	α-D-Gal-OMe	β-D-Gal(1-3)-α-D-Gal-OMe	6	93
241	β-D-Gal(1-3)-α-D-Gal-O	β-Galactosidase *Aspergillus oryzae*	βVGal	α-D-Gal-OMe	β-D-Gal(1-3)-α-D-Gal-OMe	6	93
242	β-D-Gal(1-3)-α-D-Gal-O	β-Galactosidase Bovine Testes	βpNPGal	α-D-Gal-OAll	β-D-Gal(1-3)-α-D-Gal-OAll	15	101
243	β-D-Gal(1-3)-α-D-Glc	β-Galactosidase *Aspergillus oryzae*	Gal	Trehalose	β-D-Gal(1-3)-α-D-Glc(1-1)-α-D-Glc	1.8	62
244	β-D-Gal(1-3)-α-D-Glc	β-Galactosidase *Penicillium multicolor*	Lactose	Cl$_8$	β-D-Gal(1-3)-Cl$_8$	9.7	42

No.	Target	Enzyme/Source	Donor	Acceptor	Product	Yield	Ref
245	β-D-Gal(1-3)-α-D-GlcNAc-O	β-Galactosidase Bovine Testes	βpNPGal	α-D-GlcNAc-OAll	β-D-Gal(1-3)-α-D-GlcNAc-OAll	13	101
246	β-D-Gal(1-3)-α-D-GlcNAc-O	β-Galactosidase Bovine Testes	βpNPGal	α-D-GlcNAc-OBn	β-D-Gal(1-3)-α-D-GlcNAc-OBn	32	101
247	β-D-Gal(1-3)-α-L-Fuc-O	β-Galactosidase Barley	βpNPGal	α-L-Fuc-OMe	β-D-Gal(1-3)-α-L-Fuc-OMe	28	51
248	β-D-Gal(1-3)-β-D-(6-O-Ac)-Gal	β-Galactosidase Escherichia coli	βpNPGal	β-D-(6-O-Ac)-Gal(1-4)-β-D-Xyl-OMu	β-D-Gal(1-3)-β-D-(6-O-Ac)-Gal(1-4)-β-D-Xyl-OMu	16	103
249	β-D-Gal(1-3)-β-D-(6-OAc)-Gal-O	β-Galactosidase Barley	βpNPGal	β-D-(6-OAc)-Gal-OMe	β-D-Gal(1-3)-β-D-(6-OAc)-Gal-OMe	4	51
250	β-D-Gal(1-3)-β-D-(6-OBn)-GlcN-S	β-Galactosidase Bovine testes	βpNPGal	β-D-(6-OBn)-GlcN-SEt	β-D-Gal(1-3)-β-D-(6-OBn)-GlcN-SEt	30	84
251	β-D-Gal(1-3)-β-D-Gal	β-Galactosidase Bifidobacterium bifidum	Lactose	Lactose	β-D-Gal(1-3)-β-D-Gal(1-4)-D-Glc	NA	104
252	β-D-Gal(1-3)-β-D-Gal	β-Galactosidase Bacillus circulans	Lactose	NeuAc	β-D-Gal(1-3)-β-D-Gal(1-8)-D-NeuAc	2.5	105
253	β-D-Gal(1-3)-β-D-GalNAc-O	β-Galactosidase Bovine testes	βoNPGal	β-D-GalNAc-O-EtBr	β-D-Gal(1-3)-β-D-GalNAc-O-EtBr	2.5	45
254	β-D-Gal(1-3)-β-D-GalNAc-O	β-Galactosidase Porcine testes	Lactose	α-D-GalNAcOpNP	β-D-Gal(1-3)-β-D-GalNAc-OpNP	13	100
255	β-D-Gal(1-3)-β-D-Gal-O	β-Galactosidase Aspergillus oryzae	βoNPGal	βoNPGal	β-D-Gal(1-3)-β-D-Gal-OoNP	11	63
256	β-D-Gal(1-3)-β-D-Gal-O	β-Galactosidase Escherichia coli	βoNPGal	βoNPGal	β-D-Gal(1-3)-β-D-Gal-OoNP	14	63
257	β-D-Gal(1-3)-β-D-Gal-O	β-Galactosidase Escherichia coli	βoNPGal	β-D-Gal-OMe	β-D-Gal(1-3)-β-D-Gal-OMe	22	26
258	β-D-Gal(1-3)-β-D-Gal-O	β-Galactosidase Escherichia coli	Lactose	β-D-Gal-OAll	β-D-Gal(1-3)-β-D-Gal-OAll	7.3	44

Table 1 (continued)

Entry	Linkage	Enzyme Source	Donor	Acceptor	Product	Yield (%)	Reference
259	β-D-Gal(1-3)-β-D-Gal-O	β-Galactosidase Escherichia coli	Lactose	β-D-Gal-OBn	β-D-Gal(1-3)-β-D-Gal-OBn	10	44
260	β-D-Gal(1-3)-β-D-Gal-O	β-Galactosidase Escherichia coli	Lactose	β-D-Gal-O-EtSi(Me)$_3$	β-D-Gal(1-3)-β-D-Gal-O-EtSi(Me)$_3$	5.8	44
261	β-D-Gal(1-3)-β-D-Gal-O	β-Galactosidase Escherichia coli	Lactose	β-D-Gal-O-Ser-(N-Aloc)-OMe	β-D-Gal(1-3)-β-D-Gal-O-Ser-(N-Aloc)-OMe	13	106
262	β-D-Gal(1-3)-β-D-Gal-O	β-Galactosidase Bovine liver	βoNPGal	monoallyl ether (β-D-Glc-O)-1,3,5-benzene-trimethanol	monoallyl ether (β-D-Gal(1-3)-β-D-Glc-O)-1,3,5-benzene-trimethanol	2	107
263	β-D-Gal(1-3)-β-D-Gal-O	β-Galactosidase Bacillus circulans	Lactose	β-D-Gal-OMe	β-D-Gal(1-3)-β-D-Gal-OMe	5.5	92
264	β-D-Gal(1-3)-β-D-Gal-S	β-Glucosidase Agrobacterium sp.	βpNPGal	β-D-Gal-SPh	β-D-Gal(1-3)-β-D-Gal-SPh	9.5	85
265	β-D-Gal(1-3)-β-D-Gal-S	β-Glucosidase Agrobacterium sp.	βpNPGal	β-D-Gal-SBn	β-D-Gal(1-3)-β-D-Gal-SBn	8.8	85
266	β-D-Gal(1-3)-β-D-GlcNAc-O	β-Galactosidase Bovine testes	βoNPGal	β-D-GlcNAc-O-EtSiMe$_3$	β-D-Gal(1-3)-β-D-GlcNAc-O-EtSiMe$_3$	7	45
267	β-D-Gal(1-3)-β-D-GlcNAc-O	β-Galactosidase Bovine testes	βpNPGal	β-D-GlcNAc-OMe	β-D-Gal(1-3)-β-D-GlcNAc-OMe	about 10	45
268	β-D-Gal(1-3)-β-D-GlcNAc-S	β-Galactosidase Bovine Testes	Lactose	β-D-GlcNAc-SEt	β-D-Gal(1-3)-β-D-GlcNAc-SEt	17	108
269	β-D-Gal(1-3)-β-D-Glc-O	β-Galactosidase Bacillus circulans	Lactose	β-D-Glc-OMe	β-D-Gal(1-3)-β-D-Glc-OMe	19	92

#	Donor	Acceptor	Enzyme	Product	Yield (%)	Ref.
270	β-D-Gal(1-3)-β-D-Glc-O	Lactose	β-Galactosidase *Bacillus circulans*	βpNPGlc	β-D-Gal(1-3)-β-D-Glc-OpNP	3.7	109
271	β-D-Gal(1-3)-β-D-Glc-S	βpNPGal	β-Glucosidase *Agrobacterium sp.*	β-D-Glc-SPh	β-D-Gal(1-3)-β-D-Glc-SPh	38	85
272	β-D-Gal(1-3)-β-D-Glc-S	βpNPGal	β-Glucosidase *Agrobacterium sp.*	β-D-Glc-SBn	β-D-Gal(1-3)-β-D-Glc-SBn	50	85
273	β-D-Gal(1-3)-β-D-Xyl-O	βpNPGal	β-Glucosidase *Agrobacterium sp.*	β-D-Xyl-OBn	β-D-Gal(1-3)-β-D-Xyl-OBn	39	85
274	β-D-Gal(1-3)-β-D-Xyl-O	αGalF	β-Glucosynthase *Agrobacterium sp.*	βpNPXyl	β-D-Gal(1-3)-β-D-Xyl-OpNP	81	28
275	β-D-Gal(1-3)-β-D-Xyl-O	βpNPGal	β-Galactosidase *Escherichia coli*	βMuXyl	β-D-Gal(1-3)-β-D-Xyl-OMu	4.6	103, 110
276	β-D-Gal(1-3)-β-D-Xyl-S	βpNPGal	β-Glucosidase *Agrobacterium sp.*	β-D-Xyl-SPh	β-D-Gal(1-3)-β-D-Xyl-SPh	37	85
277	β-D-Gal(1-3)-β-D-Xyl-S	βpNPGal	β-Glucosidase *Agrobacterium sp.*	β-D-Xyl-SBn	β-D-Gal(1-3)-β-D-Xyl-SBn	38	85
278	β-D-Gal(1-3)-β-D-Xyl-S	βpNPGal	β-Galactosidase *Bacillus circulans*	β-D-Xyl-SEt	β-D-Gal(1-3)-β-D-Xyl-SEt	25	111
279	β-D-Gal(1-3)-D-GalNAc	Lactose	β-Galactosidase Bovine testes	GalNAc	β-D-Gal(1-3)-D-GalNAc	21	112
280	β-D-Gal(1-3)-D-GalNAc	βpNPGal	β-Galactosidase *Bifidobacterium bifidum*	GalNAc	β-D-Gal(1-3)-D-GalNAc	trace	113

Table 1 (continued)

Entry	Linkage	Enzyme Source	Donor	Acceptor	Product	Yield (%)	Reference
281	β-D-Gal(1-3)-D-GalNAc	β-Galactosidase Penicillium multicolor	βpNPGal	GalNAc	β-D-Gal(1-3)-D-GalNAc	trace	113
282	β-D-Gal(1-3)-D-GalNAc	β-Galactosidase Streptococcus 6646K	βpNPGal	GalNAc	β-D-Gal(1-3)-D-GalNAc	3.7	113
283	β-D-Gal(1-3)-D-GalNAc	β-Galactosidase Aspergillus oryzae	βpNPGal	GalNAc	β-D-Gal(1-3)-D-GalNAc	1.4	113
284	β-D-Gal(1-3)-D-Glc	β-Galactosidase Escherichia coli	Gal	Glc	β-D-Gal(1-3)-D-Glc	0.5	40
285	β-D-Gal(1-3)-D-Glc	β-Galactosidase Bifidobacterium bifidum	Lactose	Lactose	β-D-Gal(1-3)-D-Glc	NA	104
286	β-D-Gal(1-3)-D-Glc	β-Galactosidase Bacillus circulans	Lactose	Lactose	β-D-Gal(1-3)-D-Glc	NA	94
287	β-D-Gal(1-3)-D-GlcA	β-Galactosidase Bacillus circulans	Lactose	GlcA	β-D-Gal(1-3)-D-GlcA	23	105
288	β-D-Gal(1-3)-D-GlcNAc	β-Galactosidase Streptococcus 6646K	βpNPGal	GlcNAc	β-D-Gal(1-3)-D-GlcNAc	4.4	113
289	β-D-Gal(1-3)-D-GlcNAc	β-Galactosidase Bovine Testes	Lactose	GlcNAc	β-D-Gal(1-3)-D-GlcNAc	12	108

29.6 Recent Developments and New Directions

290	β-D-Gal(1-3)-D-GlcNAc	β-Galactosidase Diplococcus pneumoniae	βpNPGal	GlcNAc	β-D-Gal(1-3)-D-GlcNAc	12	78
291	β-D-Gal(1-3)-D-Xyl	β-Galactosidase Bovine testes	βoNPGal	Xyl	β-D-Gal(1-3)-D-Xyl	39	95
292	β-D-Gal(1-3)-D-Xyl	β-Galactosidase Intestinal lactase Lamb	βoNPGal	Xyl	β-D-Gal(1-3)-D-Xyl	26	95
293	β-D-Gal(1-3)-D-Xyl	β-Galactosidase Aspergillus oryzae	βoNPGal	Xyl	β-D-Gal(1-3)-D-Xyl	17	95
294	β-D-Gal(1-3)-D-Xyl	β-Galactosidase Saccharomyces fragilis	βoNPGal	Xyl	β-D-Gal(1-3)-D-Xyl	4	95
295	β-D-Gal(1-3)-D-Xyl	β-Galactosidase Escherichia coli	βoNPGal	Xyl	β-D-Gal(1-3)-D-Xyl	7	95
296	β-D-Gal(1-3)-D-Xyl	β-Galactosidase Escherichia coli	βpNPGal	Xyl	β-D-Gal(1-3)-D-Xyl	6.4	96
297	β-D-Gal(1-3)-Glucal	β-Galactosidase Escherichia coli	βpNPGal	Glucal	β-D-Gal(1-3)-Glucal	35	97
298	β-D-Gal(1-3)-L-Xyl	β-Galactosidase Bovine testes	βoNPGal	L-Xyl	β-D-Gal(1-3)-L-Xyl	17	95
299	β-D-Gal(1-3)-L-Xyl	β-Galactosidase Intestinal lactase Lamb	βoNPGal	L-Xyl	β-D-Gal(1-3)-L-Xyl	25	95
300	β-D-Gal(1-3)-L-Xyl	β-Galactosidase Aspergillus oryzae	βoNPGal	L-Xyl	β-D-Gal(1-3)-L-Xyl	13	95
301	β-D-Gal(1-3)-L-Xyl	β-Galactosidase Saccharomyces fragilis	βoNPGal	L-Xyl	β-D-Gal(1-3)-L-Xyl	6	95

Table 1 (*continued*)

Entry	Linkage	Enzyme Source	Donor	Acceptor	Product	Yield (%)	Reference
302	β-D-Gal(1-3)-L-Xyl	β-Galactosidase *Escherichia coli*	βoNPGal	L-Xyl	β-D-Gal(1-3)-L-Xyl	15	95
303	β-D-Gal(1-4)-α-D-Gal-O	β-Galactosidase Barley	βpNPGal	α-D-Gal-OMe	β-D-Gal(1-4)-α-D-Gal-OMe	NA	51
304	β-D-Gal(1-4)-α-D-Gal-O	β-Galactosidase Snail	βpNPGal	α-D-Gal-OMe	β-D-Gal(1-4)-α-D-Gal-OMe	NA	51
305	β-D-Gal(1-4)-α-D-Glc	β-Galactosidase *Bacillus circulans*	Lactose	{α-D-Glc(1-4)}₄-α-D-Glc-O*p*NP	β-D-Gal(1-4)-{α-D-Glc(1-4)}₄-α-D-Glc-O*p*NP	12	114
306	β-D-Gal(1-4)-α-D-Glc	β-Galactosidase *Bacillus circulans*	Lactose	α-D-Glc(1-6)-α-CD	β-D-Gal(1-4)-α-D-Glc(1-6)-α-CD	20	115
307	β-D-Gal(1-4)-α-D-Glc	β-Galactosidase *Bacillus circulans*	Lactose	α-D-Glc(1-4)-α-D-Glc(1-6)-α-CD	β-D-Gal(1-4)-α-D-Glc(1-4)-α-D-Glc(1-6)-α-CD	22	115
308	β-D-Gal(1-4)-α-D-Glc	β-Galactosidase *Bacillus circulans*	Lactose	α-D-Glc(1-6)-β-CD	β-D-Gal(1-4)-α-D-Glc(1-6)-β-CD	20	115
309	β-D-Gal(1-4)-α-D-Glc	β-Galactosidase *Bacillus circulans*	Lactose	α-D-Glc(1-4)-α-D-Glc(1-6)-β-CD	β-D-Gal(1-4)-α-D-Glc(1-4)-α-D-Glc(1-6)-β-CD	22	115
310	β-D-Gal(1-4)-α-D-Glc	β-Galactosidase *Bacillus circulans*	Lactose	α-D-Glc(1-6)-γ-CD	β-D-Gal(1-4)-α-D-Glc(1-6)-γ-CD	17	115
311	β-D-Gal(1-4)-α-D-Glc	β-Galactosidase *Bacillus circulans*	Lactose	α-D-Glc(1-4)-α-D-Glc(1-6)-γ-CD	β-D-Gal(1-4)-α-D-Glc(1-4)-α-D-Glc(1-6)-γ-CD	21	115

#	Product	Enzyme	Donor	Acceptor	Product	Yield	Ref.
312	β-D-Gal(1-4)-α-D-Glc	β-Galactosidase *Aspergillus oryzae*	Gal	Trehalose	β-D-Gal(1-4)-α-D-Glc(1-1)-α-D-Glc	1.2	62
313	β-D-Gal(1-4)-α-D-GlcNAc-O	β-Galactosidase *Bacillus circulans*	βpNPGal	α-D-GlcNAc-OBn	β-D-Gal(1-4)-α-D-GlcNAc-OBn	14	116
314	β-D-Gal(1-4)-α-D-Glc-S	β-Galactosidase *Bacillus circulans*	βpNPGal	α-D-Glc-SEt	β-D-Gal(1-4)-α-D-Glc-SEt	30	111
315	β-D-Gal(1-4)-β-D-2-deoxy-2-Cl-Glc-O	β-Glucosynthase *Agrobacterium* sp.	αGalF	β-D-2-deoxy-2-Cl-Glc-O-2,4-DNP	β-D-Gal(1-4)-β-D-2-deoxy-2-Cl-Glc-O-2,4-DNP	64	28
316	β-D-Gal(1-4)-β-D-Gal	β-Galactosidase *Bacillus circulans*	Lactose	GlcA	β-D-Gal(1-4)-β-D-Gal(1-3)-D-GlcA	5	105
317	β-D-Gal(1-4)-β-D-Gal	β-Galactosidase *Bacillus circulans*	Lactose	α-D-Glc(1-6)-α-CD	β-D-Gal(1-4)-β-D-Gal(1-4)-α-D-Glc(1-6)-α-CD	5.7	115
318	β-D-Gal(1-4)-β-D-Gal	β-Galactosidase *Bacillus circulans*	Lactose	α-D-Glc(1-4)-α-D-Glc(1-6)-α-CD	β-D-Gal(1-4)-β-D-Gal(1-4)-α-D-Glc(1-4)-α-D-Glc(1-6)-α-CD	5.3	115
319	β-D-Gal(1-4)-β-D-Gal	β-Galactosidase *Bacillus circulans*	Lactose	α-D-Glc(1-6)-β-CD	β-D-Gal(1-4)-β-D-Gal(1-4)-α-D-Glc(1-6)-β-CD	4.2	115
320	β-D-Gal(1-4)-β-D-Gal	β-Galactosidase *Bacillus circulans*	Lactose	α-D-Glc(1-4)-α-D-Glc(1-6)-β-CD	β-D-Gal(1-4)-β-D-Gal(1-4)-α-D-Glc(1-4)-α-D-Glc(1-6)-β-CD	5.3	115
321	β-D-Gal(1-4)-β-D-Gal	β-Galactosidase *Bacillus circulans*	Lactose	α-D-Glc(1-6)-γ-CD	β-D-Gal(1-4)-β-D-Gal(1-4)-α-D-Glc(1-6)-γ-CD	3.2	115

Table 1 (continued)

Entry	Linkage	Enzyme	Source	Donor	Acceptor	Product	Yield (%)	Reference
322	β-D-Gal(1-4)-β-D-Gal	β-Galactosidase	Bacillus circulans	Lactose	α-D-Glc(1-4)-α-D-Glc(1-6)-γ-CD	β-D-Gal(1-4)-β-D-Gal(1-4)-α-D-Glc(1-4)-α-D-Glc(1-6)-γ-CD	8.2	115
323	β-D-Gal(1-4)-β-D-Gal	β-Galactosidase	Bacillus circulans	βpNPGal	Lactose	β-D-Gal(1-4)-β-D-Gal(1-4)-D-Glu	52	117
324	β-D-Gal(1-4)-β-D-Gal	β-Galactosidase	Bacillus circulans	Lactose	Lactose	β-D-Gal(1-4)-β-D-Gal(1-4)-D-Glc	NA	94
325	β-D-Gal(1-4)-β-D-GalNAc	β-Galactosidase	Bacillus circulans	Lactose	βpNPGalNAc	β-D-Gal(1-4)-β-D-GalNAc-OpNP	17	25
326	β-D-Gal(1-4)-β-D-Gal-O	β-Galactosidase	Barley	βpNPGal	β-D-Gal-OMe	β-D-Gal(1-4)-β-D-Gal-OMe	3	51
327	β-D-Gal(1-4)-β-D-Gal-O	β-Galactosidase	Bacillus circulans	Lactose	β-D-Gal-OMe	β-D-Gal(1-4)-β-D-Gal-OMe	32	92
328	β-D-Gal(1-4)-β-D-Gal-S	β-Galactosidase	Bacillus circulans	βpNPGal	β-D-Gal-SEt	β-D-Gal(1-4)-β-D-Gal-SEt	60	111
329	β-D-Gal(1-4)-β-D-Gal	exo-β-galactanase	Aspergillus niger	Galactan	β-D-Gal-{β-D-Gal(1-4)}$_n$-D-Gal	β-D-Gal-{β-D-Gal(1-4)}$_{(n+1)}$-D-Gal	31	118
330	β-D-Gal(1-4)-β-D-Glc	β-Galactosidase	Bacillus circulans	Lactose	Rubusoside	13-β-D-Gal(1-4)-β-D-Glc-19-β-D-Glc-O-steviol	7.3	52

Entry		Enzyme	Donor	Acceptor	Product		
331	β-D-Gal(1-4)-β-D-Glc	β-Galactosidase *Bacillus circulans*	Lactose	Rubusoside	13-β-D-Glc-19-β-D-Gal(1-4)-β-D-Glc-O-steviol	17	52
332	β-D-Gal(1-4)-β-D-Glc	β-Glucosynthase *Agrobacterium* sp.	αGalF	β-D-Glc(1-4)-β-D-Glc-O*p*NP	β-D-Gal(1-4)-β-D-Glc(1-4)-β-D-Glc-O*p*NP	92	28
333	β-D-Gal(1-4)-β-D-Glc	β-Glucosynthase *Agrobacterium* sp.	αGalF	β-D-Glc(1-4)-β-D-Glc-O*p*(OMe)Ph	β-D-Gal(1-4)-β-D-Glc(1-4)-β-D-Glc-O*p*OMePh	88	28
334	β-D-Gal(1-4)-β-D-GlcN₃-O	β-Galactosidase *Bullera singularis*	Lactose	β-D-GlcN₃-OMe	β-D-Gal(1-4)-β-D-GlcN₃-OMe	13	83
335	β-D-Gal(1-4)-β-D-GlcN₃-S	β-Galactosidase *Bacillus circulans*	β*p*NPGal	β-D-GlcN₃-SEt	β-D-Gal(1-4)-β-D-GlcN₃-SEt	4	116
336	β-D-Gal(1-4)-β-D-GlcNAc	β-Galactosidase *Bacillus circulans*	β*p*NPGal	β-D-GlcNAc(1-6)-D-GalNAc	β-D-Gal(1-4)-β-D-GlcNAc(1-6)-D-GalNAc	48	117
337	β-D-Gal(1-4)-β-D-GlcNAc	β-Galactosidase *Bacillus circulans*	β*p*NPGal	β-D-GlcNAc(1-2)-D-Man	β-D-Gal(1-4)-β-D-GlcNAc(1-2)-D-Man	4.2	119
338	β-D-Gal(1-4)-β-D-GlcNAc	β-Galactosidase *Bacillus circulans*	β*p*NPGal	β-D-GlcNAc(1-6)-D-Man	β-D-Gal(1-4)-β-D-GlcNAc(1-6)-D-Man	17	119
339	β-D-Gal(1-4)-β-D-GlcNAc-O	β-Galactosidase *Diplococcus pneumoniae*	Lactose	β-D-GlcNAc-O-Ser-(N-Z)-OEt	β-D-Gal(1-4)-β-D-GlcNAc-O-Ser-(N-Z)-OEt	20	102
340	β-D-Gal(1-4)-β-D-GlcNAc-O	β-Galactosidase *Bacillus circulans*	Lactose	β*p*NPGlcNAc	β-D-Gal(1-4)-β-D-GlcNAc-O*p*NP	21	25

Table 1 (continued)

Entry	Linkage	Enzyme Source	Donor	Acceptor	Product	Yield (%)	Reference
341	β-D-Gal(1-4)-β-D-GlcNAc-O	β-Galactosidase *Bacillus circulans*	β*p*NPGal	β-D-GlcNAc-O-6-hydroxy-methyl-hexanoate	β-D-Gal(1-4)-β-D-GlcNAc-O-6-hydroxy-methyl-hexanoate	57	116
342	β-D-Gal(1-4)-β-D-GlcNAc-O	β-Galactosidase *Bacillus circulans*	β*p*NPGal	β-D-GlcNAc-OBn	β-D-Gal(1-4)-β-D-GlcNAc-OBn	14	116
343	β-D-Gal(1-4)-β-D-GlcNAc-O	β-Galactosidase *Bacillus circulans*	β*p*NPGal	β-D-GlcNAc-O-$(CH_2)_7CH_3$	β-D-Gal(1-4)-β-D-GlcNAc-O-$(CH_2)_7CH_3$	10	116
344	β-D-Gal(1-4)-β-D-GlcNAc-O	β-Galactosidase Bovine testes	β*p*NPGal	β-D-GlcNAc-OMe	β-D-Gal(1-4)-β-D-GlcNAc-OMe	about 10	45
345	β-D-Gal(1-4)-β-D-GlcNAc-O	β-Galactosidase *Bacillus circulans*	6'-oxo-β-D-Gal-O-*p*NP	β-D-GlcNAc-OMe	β-D-Gal(1-4)-β-D-GlcNAc-OMe	60	120
346	β-D-Gal(1-4)-β-D-GlcNAc-S	β-Galactosidase *Bacillus circulans*	β*p*NPGal	β-D-GlcNAc-SEt	β-D-Gal(1-4)-β-D-GlcNAc-SEt	14	116
347	β-D-Gal(1-4)-β-D-GlcNAc-S	β-Galactosidase *Bullera singularis*	Lactose	β-D-GlcNAc-SEt	β-D-Gal(1-4)-β-D-GlcNAc-SEt	11	83
348	β-D-Gal(1-4)-β-D-GlcNAc-S	β-Galactosidase *Bacillus circulans*	β*p*NPGal	β-D-GlcNAc-SEt	β-D-Gal(1-4)-β-D-GlcNAc-SEt	49	111
349	β-D-Gal(1-4)-β-D-GlcNPht-S	β-Galactosidase *Bullera singularis*	Lactose	β-D-GlcNPht-SEt	β-D-Gal(1-4)-β-D-GlcNPht-SEt	20	84

#	Product	Enzyme	Donor	Acceptor		
350	β-D-Gal(1-4)-β-D-GlcNPth-S	β-Galactosidase *Bullera singularis*	Lactose	β-D-GlcNPth-SEt	β-D-Gal(1-4)-β-D-GlcNPth-SEt	10	83
351	β-D-Gal(1-4)-β-D-GlcNPth	β-Galactosidase *Bullera singularis*	Lactose	β-D-GlcNPth	β-D-Gal(1-4)-β-D-GlcNPth	10	83
352	β-D-Gal(1-4)-β-D-GlcNTeoc-S	β-Galactosidase *Bullera singularis*	Lactose	β-D-GlcNTeoc-SEt	β-D-Gal(1-4)-β-D-GlcNTeoc-SEt	19	83
353	β-D-Gal(1-4)-β-D-Glc-O	β-Glucosynthase *Agrobacterium* sp.	αGalF	βpNPGlc	β-D-Gal(1-4)-β-D-Glc-OpNP	84	28
354	β-D-Gal(1-4)-β-D-Glc-O	β-Galactosidase *Bacillus circulans*	Lactose	β-D-Glc-OMe	β-D-Gal(1-3)-β-D-Glc-OMe	23	92
355	β-D-Gal(1-4)-β-D-Glc-O	β-Galactosidase *Bacillus circulans*	Lactose	βpNPGlc	β-D-Gal(1-4)-β-D-Glc-OpNP	17	109
356	β-D-Gal(1-4)-β-D-Glc-S	β-Glucosidase *Agrobacterium* sp.	βpNPGal	β-D-Glc-SPh	β-D-Gal(1-4)-β-D-Glc-SPh	20	85
357	β-D-Gal(1-4)-β-D-Glc-S	β-Glucosidase *Agrobacterium* sp.	βpNPGal	β-D-Glc-SBn	β-D-Gal(1-4)-β-D-Glc-SBn	16	85
358	β-D-Gal(1-4)-β-D-Glc-S	β-Galactosidase *Bacillus circulans*	βpNPGal	β-D-Glc-SEt	β-D-Gal(1-4)-β-D-Glc-SEt	36	111
359	β-D-Gal(1-4)-β-D-Man-O	β-Glucosynthase *Agrobacterium* sp.	αGalF	βpNPMan	β-D-Gal(1-4)-β-D-Man-OpNP	66	28

Table 1 (continued)

Entry	Linkage	Enzyme Source	Donor	Acceptor	Product	Yield (%)	Reference
360	β-D-Gal(1-4)-β-D-Man-S	β-Glucosidase *Agrobacterium sp.*	βpNPGal	β-D-Man-SBn	β-D-Gal(1-4)-β-D-Man-SBn	24	85
361	β-D-Gal(1-4)-β-D-Xyl-O	β-Glucosidase *Agrobacterium sp.*	βpNPGal	β-D-Xyl-OBn	β-D-Gal(1-4)-β-D-Xyl-OBn	11	85
362	β-D-Gal(1-4)-β-D-Xyl-O	β-Galactosidase *Escherichia coli*	βpNPGal	βMuXyl	β-D-Gal(1-4)-β-D-Xyl-OMu	17	103, 110
363	β-D-Gal(1-4)-β-D-Xyl-O	β-Galactosidase *Escherichia coli*	Lactose	βpNPX	β-D-Gal(1-4)-β-D-Xyl-OpNP	14	74
364	β-D-Gal(1-4)-β-D-Xyl-O	β-Glucosidase Guinea pig liver	βpNPGal	βpNPXyl	β-D-Gal(1-4)-β-D-Xyl-OpNP	NA	121
365	β-D-Gal(1-4)-β-D-Xyl-S	β-Glucosidase *Agrobacterium sp.*	βpNPGal	β-D-Xyl-SPh	β-D-Gal(1-4)-β-D-Xyl-SPh	9	85
366	β-D-Gal(1-4)-β-D-Xyl-S	β-Glucosidase *Agrobacterium sp.*	βpNPGal	β-D-Xyl-SBn	β-D-Gal(1-4)-β-D-Xyl-SBn	11	85
367	β-D-Gal(1-4)-D-GalNAc	β-Galactosidase *Bacillus circulans*	Lactose	GalNAc	β-D-Gal(1-4)-D-GalNAc	16	92
368	β-D-Gal(1-4)-D-GalNAc	β-Galactosidase *Bifidobacterium bifidum*	βpNPGal	GalNAc	β-D-Gal(1-4)-D-GalNAc	33	113
369	β-D-Gal(1-4)-D-GalNAc	β-Galactosidase *Penicillium multicolor*	βpNPGal	GalNAc	β-D-Gal(1-4)-D-GalNAc	1	113

#	Product	Enzyme	Donor	Acceptor	Yield	Ref
370	β-D-Gal(1-4)-D-GalNAc	β-Galactosidase Streptococcus 6646K	βpNPGal	GalNAc	1.2	113
371	β-D-Gal(1-4)-D-GalNAc	β-Galactosidase Aspergillus oryzae	βpNPGal	GalNAc	0.7	113
372	β-D-Gal(1-4)-D-GalNAc	β-Galactosidase Bacillus circulans	Lactose	GalNAc	13	25
373	β-D-Gal(1-4)-D-Glc	β-Galactosidase Escherichia coli	Gal	Glc	1.7	40
374	β-D-Gal(1-4)-D-GlcN	β-Galactosidase Clonezyme library™	Lactose	GlcN	23	122
375	β-D-Gal(1-4)-D-GlcNAc	β-Galactosidase Bacillus circulans	Lactose	GlcNAc	NA	22
376	β-D-Gal(1-4)-D-GlcNAc	β-Galactosidase Bacillus circulans	Lactose	GlcNAc	36	92
377	β-D-Gal(1-4)-D-GlcNAc	β-Galactosidase Diplococcus pneumoniae	βoNPGal	GlcNAc	31	20
378	β-D-Gal(1-4)-D-GlcNAc	β-Galactosidase Bacillus circulans	βoNPGal	GlcNAc	37	20
379	β-D-Gal(1-4)-D-GlcNAc	β-Galactosidase Streptococcus 6646K	βpNPGal	GlcNAc	31	113
380	β-D-Gal(1-4)-D-GlcNAc	β-Galactosidase Bifidobacterium bifidum	βpNPGal	GlcNAc	38	113

Table 1 (continued)

Entry	Linkage	Enzyme Source	Donor	Acceptor	Product	Yield (%)	Reference
381	β-D-Gal(1-4)-D-GlcNAc	β-Galactosidase Escherichia coli	Gal	GlcNAc	β-D-Gal(1-4)-D-GlcNAc	0.9	21
382	β-D-Gal(1-4)-D-GlcNAc	β-Galactosidase Aspergillus oryzae	Gal	GlcNAc	β-D-Gal(1-4)-D-GlcNAc	2.5	40
383	β-D-Gal(1-4)-D-GlcNAc	β-Galactosidase Clonezyme library^{TM}	βoNPGal	GlcNAc	β-D-Gal(1-4)-D-GlcNAc	61	123
384	β-D-Gal(1-4)-D-GlcNAc	β-Galactosidase Bacillus circulans	Lactose	GlcNAc	β-D-Gal(1-4)-D-GlcNAc	42	124, 125
385	β-D-Gal(1-4)-D-GlcNAc	β-Galactosidase Bacillus circulans	Lactose	GlcNAc	β-D-Gal(1-4)-D-GlcNAc	23	126
386	β-D-Gal(1-4)-D-Xyl	β-Galactosidase Bovine testes	βoNPGal	Xyl	β-D-Gal(1-4)-D-Xyl	3	95
387	β-D-Gal(1-4)-D-Xyl	β-Galactosidase Intestinal lactase Lamb	βoNPGal	Xyl	β-D-Gal(1-4)-D-Xyl	1	95
388	β-D-Gal(1-4)-D-Xyl	β-Galactosidase Aspergillus oryzae	βoNPGal	Xyl	β-D-Gal(1-4)-D-Xyl	2	95
389	β-D-Gal(1-4)-D-Xyl	β-Galactosidase Saccharomyces fragilis	βoNPGal	Xyl	β-D-Gal(1-4)-D-Xyl	12	95
390	β-D-Gal(1-4)-D-Xyl	β-Galactosidase Escherichia coli	βoNPGal	Xyl	β-D-Gal(1-4)-D-Xyl	43	95

#	Product	Enzyme	Donor	Acceptor		
391	β-D-Gal(1-4)-D-Xyl	β-Galactosidase *Escherichia coli*	βpNPGal	Xyl	39	96
392	β-D-Gal(1-4)-L-Xyl	β-Galactosidase Bovine testes	βoNPGal	L-Xyl	12	95
393	β-D-Gal(1-4)-L-Xyl	β-Galactosidase Intestinal lactase Lamb	βoNPGal	L-Xyl	9	95
394	β-D-Gal(1-4)-L-Xyl	β-Galactosidase *Aspergillus oryzae*	βoNPGal	L-Xyl	29	95
395	β-D-Gal(1-4)-L-Xyl	β-Galactosidase *Saccharomyces fragilis*	βoNPGal	L-Xyl	2	95
396	β-D-Gal(1-4)-L-Xyl	β-Galactosidase *Escherichia coli*	βoNPGal	L-Xyl	20	95
397	β-D-Gal(1-5)-α-L-Araf-O	β-Glucosidase Guinea pig liver	βpNPGal	α-L-Araf-OpNP	NA	121
398	β-D-Gal(1-6)-α-D-GalNAc-O	β-Galactosidase Porcine testes	Lactose	αpNPGalNAc	5.6	100
399	β-D-Gal(1-6)-α-D-Gal-O	β-Galactosidase *Escherichia coli*	βoNPGal	α-D-Gal-OMe	14	26
400	β-D-Gal(1-6)-α-D-Gal-O	β-Galactosidase Barley	βpNPGal	α-D-Gal-OMe	NA	51
401	β-D-Gal(1-6)-α-D-Gal-O	β-Galactosidase Snail	βpNPGal	α-D-Gal-OMe	NA	51
402	β-D-Gal(1-6)-α-D-Gal-O	β-Galactosidase *Aspergillus oryzae*	βoNPGal	α-D-Gal-OMe	57	93
403	β-D-Gal(1-6)-α-D-Gal-O	β-Galactosidase *Aspergillus oryzae*	βVGal	α-D-Gal-OMe	70	93

Table 1 (*continued*)

Entry	Linkage	Enzyme Source	Donor	Acceptor	Product	Yield (%)	Reference
404	β-D-Gal(1-6)-α-D-Glc	β-Galactosidase *Bacillus circulans*	Lactose	α-D-Glc(1-6)-α-CD	β-D-Gal(1-6)-α-D-Glc(1-6)-α-CD	1.1	115
405	β-D-Gal(1-6)-α-D-Glc	β-Galactosidase *Bacillus circulans*	Lactose	α-D-Glc(1-4)-α-D-Glc(1-6)-α-CD	β-D-Gal(1-6)-α-D-Glc(1-4)-α-D-Glc(1-6)-α-CD	2.6	115
406	β-D-Gal(1-6)-α-D-Glc	β-Galactosidase *Bacillus circulans*	Lactose	α-D-Glc(1-6)-β-CD	β-D-Gal(1-6)-α-D-Glc(1-6)-β-CD	1.2	115
407	β-D-Gal(1-6)-α-D-Glc	β-Galactosidase *Bacillus circulans*	Lactose	α-D-Glc(1-4)-α-D-Glc(1-6)-β-CD	β-D-Gal(1-6)-α-D-Glc(1-4)-α-D-Glc(1-6)-β-CD	2.3	115
408	β-D-Gal(1-6)-α-D-Glc	β-Galactosidase *Bacillus circulans*	Lactose	α-D-Glc(1-4)-α-D-Glc(1-6)-γ-CD	β-D-Gal(1-6)-α-D-Glc(1-4)-α-D-Glc(1-6)-γ-CD	1.6	115
409	β-D-Gal(1-6)-α-D-Glc	β-Galactosidase *Penicillium multicolor*	Lactose	α-D-Glc(1-6)-α-CD	β-D-Gal(1-6)-α-D-Glc(1-6)-α-CD	6.8	115
410	β-D-Gal(1-6)-α-D-Glc	β-Galactosidase *Penicillium multicolor*	Lactose	α-D-Glc(1-4)-α-D-Glc(1-6)-α-CD	β-D-Gal(1-6)-α-D-Glc(1-4)-α-D-Glc(1-6)-α-CD	10	115
411	β-D-Gal(1-6)-α-D-Glc	β-Galactosidase *Penicillium multicolor*	Lactose	α-D-Glc(1-6)-β-CD	β-D-Gal(1-6)-α-D-Glc(1-6)-β-CD	14	115
412	β-D-Gal(1-6)-α-D-Glc	β-Galactosidase *Penicillium multicolor*	Lactose	α-D-Glc(1-4)-α-D-Glc(1-6)-β-CD	β-D-Gal(1-6)-α-D-Glc(1-4)-α-D-Glc(1-6)-β-CD	13	115

29.6 Recent Developments and New Directions

413	β-D-Gal(1-6)-α-D-Glc	β-Galactosidase *Penicillium multicolor*	Lactose	α-D-Glc(1-6)-γ-CD	β-D-Gal(1-6)-α-D-Glc(1-6)-γ-CD	9.5	115
414	β-D-Gal(1-6)-α-D-Glc	β-Galactosidase *Penicillium multicolor*	Lactose	α-D-Glc(1-4)-α-D-Glc(1-6)-γ-CD	β-D-Gal(1-6)-α-D-Glc(1-4)-α-D-Glc(1-6)-γ-CD	8	115
415	β-D-Gal(1-6)-α-D-Glc	β-Galactosidase *Escherichia coli*	βoNPGal	Sucrose	β-D-Gal(1-6)-α-D-Glc(1-2)-β-D-Fru*f*	NA	127
416	β-D-Gal(1-6)-α-D-Glc	β-Galactosidase *Aspergillus oryzae*	Gal	Trehalose	β-D-Gal(1-6)-α-D-Glc(1-1)-α-D-Glc	13	62
417	β-D-Gal(1-6)-β-D-GalNAc-O	β-Galactosidase Porcine testes	Lactose	β*p*NPGalNAc	β-D-Gal(1-6)-β-D-GalNAc-O*p*NP	2.6	100
418	β-D-Gal(1-6)-β-D-Gal-O	β-Galactosidase *Aspergillus oryzae*	βoNPGal	βoNPGal	β-D-Gal(1-6)-β-D-Gal-OoNP	21	63
419	β-D-Gal(1-6)-β-D-Gal-O	β-Galactosidase *Escherichia coli*	βoNPGal	βoNPGal	β-D-Gal(1-6)-β-D-Gal-OoNP	13	63
420	β-D-Gal(1-6)-β-D-Gal-O	β-Galactosidase *Escherichia coli*	βoNPGal	β-D-Gal-OMe	β-D-Gal(1-6)-β-D-Gal-OMe	3	26
421	β-D-Gal(1-6)-β-D-Gal-O	β-Galactosidase *Escherichia coli*	Lactose	β-D-Gal-OAll	β-D-Gal(1-6)-β-D-Gal-OAll	2.4	44
422	β-D-Gal(1-6)-β-D-Gal-O	β-Galactosidase *Escherichia coli*	Lactose	β-D-Gal-OBn	β-D-Gal(1-6)-β-D-Gal-OBn	3	44
423	β-D-Gal(1-6)-β-D-Gal-O	β-Galactosidase *Escherichia coli*	Lactose	β-D-Gal-O-Ser-(N-Aloc)-OMe	β-D-Gal(1-6)-β-D-Gal-O-Ser-(N-Aloc)-OMe	6	106
424	β-D-Gal(1-6)-β-D-Gal-O	β-Galactosidase Bovine liver	βoNPGal	monoallyl ether (β-D-Glc-O)-1,3,5-benzene-trimethanol	monoallyl ether (β-D-Gal(1-6)-β-D-Gal-O)-(β-D-Glc-O)-1,3,5-benzene-trimethanol	10	107

Table 1 (*continued*)

Entry	Linkage	Enzyme Source	Donor	Acceptor	Product	Yield (%)	Reference
425	β-D-Gal(1-6)-β-D-Gal-O	β-Galactosidase *Bacillus circulans*	Lactose	β-D-Gal-OMe	β-D-Gal(1-6)-β-D-Gal-OMe	1.1	92
426	β-D-Gal(1-6)-β-D-Gal-O	β-Glucosidase Guinea pig liver	βpNPGal	βpNPGal	β-D-Gal(1-6)-β-D-Gal-OpNP	NA	121
427	β-D-Gal(1-6)-β-D-Gal-S	β-Glucosidase *Agrobacterium* sp.	βpNPGal	β-D-Gal-SPh	β-D-Gal(1-6)-β-D-Gal-SPh	8.8	85
428	β-D-Gal(1-6)-β-D-Gal-S	β-Glucosidase *Agrobacterium* sp.	βpNPGal	β-D-Gal-SBn	β-D-Gal(1-6)-β-D-Gal-SBn	5.2	85
429	β-D-Gal(1-6)-β-D-GlcNAc-O	β-Galactosidase *Bacillus circulans*	Lactose	βpNPGlcNAc	β-D-Gal(1-6)-β-D-GlcNAc-OpNP	26	25
430	β-D-Gal(1-6)-β-D-GlcNAc-S	β-Galactosidase *Escherichia coli*	Lactose	β-D-GlcNAc-SEt	β-D-Gal(1-6)-β-D-GlcNAc-SEt	28	108
431	β-D-Gal(1-6)-β-D-GlcNPth-S	β-Galactosidase Bovine testes	Lactose	β-D-GlcNPth-SEt	β-D-Gal(1-6)-β-D-GlcNPth-SEt	10	83
432	β-D-Gal(1-6)-β-D-Glc-O	β-Galactosidase *Bacillus circulans*	Lactose	Rubusoside	13-β-D-Gal(1-6)-β-D-Glc-19-β-D-Glc-O-steviol	2.6	52
433	β-D-Gal(1-6)-β-D-Glc-O	β-Galactosidase *Bacillus circulans*	Lactose	Rubusoside	13-β-D-Glc-19-β-D-Gal(1-6)-β-D-Glc-O-steviol	12	52
434	β-D-Gal(1-6)-β-D-Glc-O	β-Galactosidase *Escherichia coli*	Lactose	Rubusoside	13-β-D-Gal(1-6)-β-D-Glc-19-β-D-Glc-O-steviol	0.5	52

29.6 Recent Developments and New Directions

435	β-D-Gal(1-6)-β-D-Glc-O	β-Galactosidase *Escherichia coli*	Lactose	Rubusoside	13-β-D-Glc-19-β-D-Gal(1-6)-β-D-Glc-O-steviol	12	52
436	β-D-Gal(1-6)-β-D-Glc-O	β-Galactosidase *Aspergillus oryzae*	Lactose	Rubusoside	13-β-D-Gal(1-6)-β-D-Glc-19-β-D-Glc-O-steviol	2.2	52
437	β-D-Gal(1-6)-β-D-Glc-O	β-Galactosidase *Aspergillus oryzae*	Lactose	Rubusoside	13-β-D-Glc-19-β-D-Gal(1-6)-β-D-Glc-O-steviol	3.2	52
438	β-D-Gal(1-6)-β-D-Glc-O	β-Galactosidase *Penicillium multicolor*	Lactose	Rubusoside	13-β-D-Gal(1-6)-β-D-Glc-19-β-D-Glc-O-steviol	2.3	52
439	β-D-Gal(1-6)-β-D-Glc-O	β-Galactosidase *Penicillium multicolor*	Lactose	Rubusoside	13-β-D-Glc-19-β-D-Gal(1-6)-β-D-Glc-O-steviol	3.8	52
440	β-D-Gal(1-6)-β-D-Glc-O	β-Galactosidase Bovine liver	βoNPGal	monoallyl ether (β-D-Glc-O)-1,3,5-benzene-trimethanol	monoallyl ether (β-D-Gal(1-6)-β-D-Glc-O)-1,3,5-benzene-trimethanol	6	107
441	β-D-Gal(1-6)-β-D-Glc-O	β-Galactosidase *Bacillus circulans*	Lactose	β-D-Glc-OMe	β-D-Gal(1-6)-β-D-Glc-OMe	9.1	92
442	β-D-Gal(1-6)-β-D-Glc-O	β-Galactosidase *Bacillus circulans*	Lactose	βpNPGlc	β-D-Gal(1-6)-β-D-Glc-O*p*NP	0.9	109
443	β-D-Gal(1-6)-β-D-Glc-S	β-Galactosidase *Bacillus circulans*	βpNPGal	β-D-Glc-SEt	β-D-Gal(1-6)-β-D-Glc-SEt	3	111
444	β-D-Gal(1-6)-D-(3-OMe)-Glc	β-Galactosidase *Escherichia coli*	βpNPGal	(3-OMe)-D-Glc	β-D-Gal(1-6)-D-(3-OMe)-Glc	47	20
445	β-D-Gal(1-6)-D-(3-OMe)-Glc	β-Galactosidase *Kluyveromyces fragilis*	βoNPGal	(3-OMe)-D-Glc	β-D-Gal(1-6)-D-(3-OMe)-Glc	14	20

Table 1 (continued)

Entry	Linkage	Enzyme Source	Donor	Acceptor	Product	Yield (%)	Reference
446	β-D-Gal(1-6)-D-(3-OMe)-Glc	β-Galactosidase *Aspergillus oryzae*	βoNPGal	(3-OMe)-D-Glc	β-D-Gal(1-6)-D-(3-OMe)-Glc	11	20
447	β-D-Gal(1-6)-D-(3-OMe)-Glc	β-Galactosidase *Diplococcus pneumoniae*	βoNPGal	(3-OMe)-D-Glc	β-D-Gal(1-6)-D-(3-OMe)-Glc	33	20
448	β-D-Gal(1-6)-D-(3-OMe)-Glc	β-Galactosidase *Bacillus circulans*	βoNPGal	(3-OMe)-D-Glc	β-D-Gal(1-6)-D-(3-OMe)-Glc	40	20
449	β-D-Gal(1-6)-D-Gal	β-Galactosidase *Bifidobacterium bifidum*	Lactose	Lactose	β-D-Gal(1-6)-D-Gal	NA	104
450	β-D-Gal(1-6)-D-GalNAc	β-Galactosidase *Bacillus circulans*	Lactose	GalNAc	β-D-Gal(1-6)-D-GalNAc	7.6	92
451	β-D-Gal(1-6)-D-GalNAc	β-Galactosidase *Bifidobacterium bifidum*	βpNPGal	GalNAc	β-D-Gal(1-6)-D-GalNAc	19	113
452	β-D-Gal(1-6)-D-GalNAc	β-Galactosidase *Penicillium multicolor*	βpNPGal	GalNAc	β-D-Gal(1-6)-D-GalNAc	17	113
453	β-D-Gal(1-6)-D-GalNAc	β-Galactosidase *Aspergillus oryzae*	βpNPGal	GalNAc	β-D-Gal(1-6)-D-GalNAc	26	113
454	β-D-Gal(1-6)-D-GalNAc	β-Galactosidase *Escherichia coli*	Lactose	GalNAc	β-D-Gal(1-6)-D-GalNAc	20	128

455	β-D-Gal(1-6)-D-Glc	β-Galactosidase *Escherichia coli*	Gal	Glc	β-D-Gal(1-6)-D-Glc	3.2	40
456	β-D-Gal(1-6)-D-Glc	β-Galactosidase *Bifidobacterium bifidum*	Lactose	Lactose	β-D-Gal(1-6)-D-Glc	NA	104
457	β-D-Gal(1-6)-D-Glc	β-Galactosidase *Bifidobacterium bifidum*	Lactose	Lactose	β-D-Gal(1-6)-[β-D-Gal(1-4)]-D-Glc	NA	104
458	β-D-Gal(1-6)-D-Glc	β-Galactosidase *Bacillus circulans*	Lactose	Lactose	β-D-Gal(1-6)-D-Glc	NA	94
459	β-D-Gal(1-6)-D-Glc	β-Galactosidase *Bacillus circulans*	Lactose	Lactose	β-D-Gal(1-6)-[β-D-Gal(1-4)]-D-Glc	NA	94
460	β-D-Gal(1-6)-D-Glc3NAc	β-Galactosidase *Bacillus circulans*	Lactose	Glc3NAc	β-D-Gal(1-6)-D-Glc3NAc	9.3	92
461	β-D-Gal(1-6)-D-GlcN	β-Galactosidase Clonezyme™ library	Lactose	GlcN	β-D-Gal(1-6)-D-GlcN	10	122
462	β-D-Gal(1-6)-D-GlcNAc	β-Galactosidase *Escherichia coli*	βoNPGal	GlcNAc	β-D-Gal(1-6)-D-GlcNAc	30	20
463	β-D-Gal(1-6)-D-GlcNAc	β-Galactosidase *Kluyveromyces fragilis*	βoNPGal	GlcNAc	β-D-Gal(1-6)-D-GlcNAc	12	20
464	β-D-Gal(1-6)-D-GlcNAc	β-Galactosidase *Bifidobacterium bifidum*	βpNPGal	GlcNAc	β-D-Gal(1-6)-D-GlcNAc	2.5	113
465	β-D-Gal(1-6)-D-GlcNAc	β-Galactosidase *Escherichia coli*	Gal	GlcNAc	β-D-Gal(1-6)-D-GlcNAc	9.1	21

Table 1 (*continued*)

Entry	Linkage	Enzyme Source	Donor	Acceptor	Product	Yield (%)	Reference
466	β-D-Gal(1-6)-D-GlcNAc	β-Galactosidase *Aspergillus oryzae*	Gal	GlcNAc	β-D-Gal(1-6)-D-GlcNAc	12	40
467	β-D-Gal(1-6)-D-GlcNAc	β-Galactosidase *Bacillus circulans*	Lactose	GlcNAc	β-D-Gal(1-6)-D-GlcNAc	2.3	126
468	β-D-Gal(1-6)-D-GlcNAc	β-Galactosidase *Kluyveromyces lactis*	Lactose	GlcNAc	β-D-Gal(1-6)-D-GlcNAc	18	126
469	β-D-Gal(1-6)-D-GlcNAc	β-Galactosidase *Bacillus circulans*	Lactose	GlcNAc	β-D-Gal(1-6)-D-GlcNAc	1.9	92
470	β-D-Gal(1-6)-D-Glucal	β-Galactosidase *Escherichia coli*	βpNPGal	D-Glucal	β-D-Gal(1-6)-D-Glucal	15	97
471	β-D-Gal(1-8)-D-NeuAc	β-Galactosidase *Bacillus circulans*	Lactose	NeuAc	β-D-Gal(1-8)-D-NeuAc	8.5	105
472	β-D-Gal(1-9)-D-NeuAc	β-Galactosidase *Bacillus circulans*	Lactose	NeuAc	β-D-Gal(1-9)-D-NeuAc	8.5	105
473	β-D-GalNAc(1-3)-β-D-Gal-O	β-Hexosaminidase *Chamelea gallina*	βpNPGalNAc	β-D-Gal-OMe	β-D-GalNAc(1-3)-β-D-Gal-OMe	6	56
474	β-D-GalNAc(1-3)-β-D-Glc	β-Hexosaminidase *Aspergillus oryzae*	βpNPGalNAc	β-D-Glc(1-6)-β-D-Glc-SEt	β-D-GalNAc(1-3)-β-D-Glc(1-6)-β-D-Glc-SEt	17	129

#							
475	β-D-GalNAc(1-3)-β-D-Glc-O	β-Hexosaminidase *Aspergillus oryzae*	βpNPGalNAc	β-D-Glc(1-4)-β-D-Glc-SEt	β-D-GalNAc(1-3)-β-D-Glc(1-6)-β-D-Glc-SEt	41	129
476	β-D-GalNAc(1-3)-β-D-Glc-O	β-Hexosaminidase *Aspergillus oryzae*	βpNPGalNAc	β-D-Glc-OMe	β-D-GalNAc(1-3)-β-D-Glc-OMe	29	130
477	β-D-GalNAc(1-4)-α-D-Glc	β-Hexosaminidase *Aspergillus oryzae*	βpNPGalNAc	α-D-Glc(1-4)-β-D-Glc-SEt	β-D-GalNAc(1-4)-α-D-Glc(1-4)-β-D-Glc-SEt	29	129
478	β-D-GalNAc(1-4)-α-D-GlcNAc-O	β-Hexosaminidase *Aspergillus oryzae*	βpNPGalNAc	α-D-GlcNAc-OMe	β-D-GalNAc(1-4)-α-D-GlcNAc-OMe	81	27
479	β-D-GalNAc(1-4)-α-D-Glc-O	β-Hexosaminidase *Aspergillus oryzae*	βpNPGalNAc	α-D-Glc-OMe	β-D-GalNAc(1-4)-α-D-Glc-OMe	30	130
480	β-D-GalNAc(1-4)-β-D-Glc	β-Hexosaminidase *Aspergillus oryzae*	βpNPGalNAc	β-D-Glc(1-4)-β-D-Glc-SEt	β-D-GalNAc(1-4)-β-D-Glc(1-4)-β-D-Glc-SEt	31	129
481	β-D-GalNAc(1-4)-β-D-Glc	β-Hexosaminidase *Aspergillus oryzae*	βpNPGalNAc	β-D-Glc(1-6)-β-D-Glc-SEt	β-D-GalNAc(1-4)-β-D-Glc(1-6)-β-D-Glc-SEt	20	129
482	β-D-GalNAc(1-4)-β-D-GlcNAc-O	β-Hexosaminidase *Aspergillus oryzae*	βpNPGalNAc	β-D-GlcNAc-OMe	β-D-GalNAc(1-4)-β-D-GlcNAc-OMe	20	27
483	β-D-GalNAc(1-4)-β-D-Glc-O	β-Hexosaminidase *Aspergillus oryzae*	βpNPGalNAc	β-D-Glc-OMe	β-D-GalNAc(1-4)-β-D-Glc-OMe	10	130
484	β-D-GalNAc(1-4)-D-GlcNAc	β-Hexosaminidase *Aspergillus oryzae*	βpNPGalNAc	GlcNAc	β-D-GalNAc(1-4)-D-GlcNAc	72	131

Table 1 (*continued*)

Entry	Linkage	Enzyme Source	Donor	Acceptor	Product	Yield (%)	Reference
485	β-D-GalNAc(1-6)-α-D-GlcNAc-O	β-Hexosaminidase *Aspergillus oryzae*	βpNPGalNAc	α-D-GlcNAc-OMe	β-D-GalNAc(1-6)-α-D-GlcNAc-OMe	8	27
486	β-D-GalNAc(1-6)-α-D-Glc-O	β-Hexosaminidase *Aspergillus oryzae*	βpNPGalNAc	α-D-Glc-OMe	β-D-GalNAc(1-6)-α-D-Glc-OMe	6	130
487	β-D-GalNAc(1-6)-D-GalNAc	β-Hexosaminidase *Aspergillus oryzae*	βpNPGalNAc	GalNAc	β-D-GalNAc(1-6)-D-GalNAc	38	117
488	β-D-GalNAc(1-6)-D-GlcNAc	β-Hexosaminidase *Aspergillus oryzae*	βpNPGalNAc	GlcNAc	β-D-GalNAc(1-6)-D-GlcNAc	33	131
489	β-D-Glc(1-2)-β-D-Glc-O	β-Galactosidase *Aspergillus oryzae*	βpNPGlc	βpNPGlc	β-D-Glc(1-2)-β-D-Glc-O*p*NP	19	63
490	β-D-Glc(1-2)-β-D-Glc-O	β-Galactosidase *Aspergillus oryzae*	βpNPGlc	βpNPGlc	β-D-Glc(1-2)-β-D-Glc-O*p*NP	19	64
491	β-D-Glc(1-2)-D-Glc	β-Glucosidase Almond	Glc	Glc	β-D-Glc-(1-2)-D-Glc	3.1	60, 132
492	β-D-Glc(1-2)-D-Glc	β-Glucosidase *Penicillium funiculosum*	Glc	Glc	β-D-Glc-(1-2)-D-Glc	3.3	60
493	β-D-Glc(1-2)-D-Glc	β-Glucosidase Almond	Glc	Glc	β-D-Glc(1-2)-D-Glc	5.7	133
494	β-D-Glc(1-3)-β-D-Xyl-O	β-Glucosynthase *Agrobacterium sp.*	αGlcF	βpNPXyl	β-D-Glc(1-3)-β-D-Xyl-O*p*NP	12	28

#	Product	Enzyme	Donor	Acceptor	Product	Yield	Ref
495	β-D-Glc-(1-3)-D-Glc	β-Glucosidase Almond	Glc		β-D-Glc-(1-3)-D-Glc	3.4	60, 132
496	β-D-Glc-(1-3)-D-Glc	β-Glucosidase *Penicillium funiculosum*	Glc		β-D-Glc-(1-3)-D-Glc	3.7	60
497	β-D-Glc-(1-3)-D-Glc	β-Glucosidase Almond	Glc		β-D-Glc-(1-3)-D-Glc	4.2	133
498	β-D-Glc(1-4)-α-D-Glc	β-Glucosidase Almond	Glc	Trehalose	β-D-Glc(1-4)-α-D-Glc(1-1)-α-D-Glc	0.9	62
499	β-D-Glc(1-4)-β-D-2-deoxy-2-F-Glc-O	β-Glucosynthase *Agrobacterium* sp.	αGlcF	β-D-2-deoxy-2-F-Glc-O-2,4-DNP	β-D-Glc(1-4)-β-D-2-deoxy-2-F-Glc-O-2,4-DNP	38	28
500	β-D-Glc(1-4)-β-D-Glc	β-Glucosynthase *Agrobacterium* sp.	αGlcF	βpNPXyl	β-D-Glc(1-4)-β-D-Glc(1-3)-β-D-Xyl-OpNP	51	28
501	β-D-Glc(1-4)-β-D-Glc	β-Glucosynthase *Agrobacterium* sp.	αGlcF	βpNPXyl	β-D-Glc(1-4)-β-D-Glc(1-4)-β-D-Glc(1-3)-β-D-Xyl-OpNP	3	28
502	β-D-Glc(1-4)-β-D-Glc	β-Glucosynthase *Agrobacterium* sp.	αGlcF	β-D-2-deoxy-2-F-Glc-O-2,4-DNP	β-D-Glc(1-4)-β-D-Glc(1-4)-β-D-2-deoxy-2-F-Glc-O-2,4-DNP	42	28
503	β-D-Glc(1-4)-β-D-Glc	β-Glucosynthase *Agrobacterium* sp.	αGlcF	β-D-2-deoxy-2-F-Glc-O-2,4-DNP	β-D-Glc(1-4)-β-D-Glc(1-4)-β-D-Glc(1-4)-β-D-2-deoxy-2-F-Glc-O-2,4-DNP	4	28
504	β-D-Glc(1-4)-β-D-Glc	β-Glucosidase Almond	Glc	Cellobiose	Glc₃	NA	134
505	β-D-Glc(1-4)-β-D-Glc	β-Glucosynthase *Agrobacterium* sp.	αGlcF	βPhGlc	β-D-Glc(1-4)-β-D-Glc(1-4)-β-D-Glc-OPh	34	28
506	β-D-Glc(1-4)-β-D-Glc	β-Glucosynthase *Agrobacterium* sp.	αGlcF	βpNPGlc	β-D-Glc(1-4)-β-D-Glc(1-4)-β-D-Glc-OpNP	24	28

Table 1 (continued)

Entry	Linkage	Enzyme Source	Donor	Acceptor	Product	Yield (%)	Reference
507	β-D-Glc(1-4)-β-D-Glc	β-Glucosynthase *Agrobacterium* sp.	αGlcF	βpNPGlc	β-D-Glc(1-4)-β-D-Glc(1-4)-β-D-Glc(1-4)-β-D-Glc-OpNP	10	28
508	β-D-Glc(1-4)-β-D-Glc	β-Glucosynthase *Agrobacterium* sp.	αGlcF	βoNPGlc	β-D-Glc(1-4)-β-D-Glc(1-4)-β-D-Glc-OoNP	29	28
509	β-D-Glc(1-4)-β-D-Glc	β-Glucosynthase *Agrobacterium* sp.	αGlcF	βoNPGlc	β-D-Glc(1-4)-β-D-Glc(1-4)-β-D-Glc(1-4)-β-D-Glc-OoNP	6	28
510	β-D-Glc(1-4)-β-D-Glc	β-Glucosynthase *Agrobacterium* sp.	αGlcF	βMuGlc	β-D-Glc(1-4)-β-D-Glc(1-4)-β-D-Glc-OMu	75	28
511	β-D-Glc(1-4)-β-D-Glc	β-Glucosynthase *Agrobacterium* sp.	αGlcF	βMuGlc	β-D-Glc(1-4)-β-D-Glc(1-4)-β-D-Glc(1-4)-β-D-Glc-OMu	54	28
512	β-D-Glc(1-4)-β-D-Glc	β-Glucosynthase *Agrobacterium* sp.	αGlcF	βpOMePhGlc	β-D-Glc(1-4)-β-D-Glc(1-4)-β-D-Glc-OpOMePh	66	28
513	β-D-Glc(1-4)-β-D-Glc	β-Glucosynthase *Agrobacterium* sp.	αGlcF	βpOMePhGlc	β-D-Glc(1-4)-β-D-Glc(1-4)-β-D-Glc(1-4)-β-D-Glc-OpOMePh	8	28
514	β-D-Glc(1-4)-β-D-Glc	β-Glucosynthase *Agrobacterium* sp.	αGlcF	β-D-Glc(1-4)-β-D-Glc-OpNP	β-D-Glc(1-4)-β-D-Glc(1-4)-β-D-Glc-OpNP	79	28
515	β-D-Glc(1-4)-β-D-Glc	β-Glucosynthase *Agrobacterium* sp.	αGlcF	β-D-Glc(1-4)-β-D-Glc-OpNP	β-D-Glc(1-4)-β-D-Glc(1-4)-β-D-Glc(1-4)-β-D-Glc-OpNP	64	28

#		Enzyme	Donor	Acceptor	Product	Yield	Ref.
516	β-D-Glc(1-4)-β-D-Glc	β-Glucosynthase Agrobacterium sp.	αGlcF	βpNPMan	β-D-Glc(1-4)-β-D-Glc(1-4)-β-D-Man-OpNP	42	28
517	β-D-Glc(1-4)-β-D-Glc	β-Glucosynthase Agrobacterium sp.	αGlcF	βpNPMan	β-D-Glc(1-4)-β-D-Glc(1-4)-β-D-Glc(1-4)-β-D-Man-OpNP	6	28
518	β-D-Glc(1-4)-β-D-GlcN$_3$	β-Galactosidase Bullera singularis	Cellobiose	β-D-GlcN$_3$-OMe	β-D-Glc(1-4)-β-D-GlcN$_3$-OMe	8	83
519	β-D-Glc(1-4)-β-D-GlcNPth	β-Galactosidase Bullera singularis	Cellobiose	β-D-GlcNPth-SEt	β-D-Glc(1-4)-β-D-GlcNPth-SEt	8	83
520	β-D-Glc(1-4)-β-D-GlcNPth	β-Galactosidase Bullera singularis	Cellobiose	β-D-GlcNPth	β-D-Glc(1-4)-β-D-GlcNPth	6	83
521	β-D-Glc(1-4)-β-D-Glc-O	β-Glucosynthase Agrobacterium sp.	αGlcF	βPhGlc	β-D-Glc(1-4)-β-D-Glc-OPh	48	28
522	β-D-Glc(1-4)-β-D-Glc-O	β-Glucosynthase Agrobacterium sp.	αGlcF	βpNPGlc	β-D-Glc(1-4)-β-D-Glc-OpNP	38	28
523	β-D-Glc(1-4)-β-D-Glc-O	β-Glucosynthase Agrobacterium sp.	αGlcF	βoNPGlc	β-D-Glc(1-4)-β-D-Glc-OoNP	41	28
524	β-D-Glc(1-4)-β-D-Glc-O	β-Glucosynthase Agrobacterium sp.	αGlcF	βpOMePhGlc	β-D-Glc(1-4)-β-D-Glc-OpOMePh	44	28
525	β-D-Glc(1-4)-β-D-Man-O	β-Glucosynthase Agrobacterium sp.	αGlcF	βpNPMan	β-D-Glc(1-4)-β-D-Man-OpNP	31	28
526	β-D-Glc-(1-4)-D-Glc	β-Glucosidase Almond	Glc	Glc	β-D-Glc-(1-4)-D-Glc	5.9	60, 132

Table 1 (continued)

Entry	Linkage	Enzyme Source	Donor	Acceptor	Product	Yield (%)	Reference
527	β-D-Glc-(1-4)-D-Glc	β-Glucosidase Penicillium funiculosum	Glc	Glc	β-D-Glc-(1-4)-D-Glc	6.7	60
528	β-D-Glc(1-4)-D-Glc	β-Glucosidase Almond	Glc	Glc	β-D-Glc(1-4)-D-Glc	5.7	133
529	β-D-Glc(1-6)-α-D-Glc	β-Glucosidase Almond	Glc	Trehalose	β-D-Glc(1-6)-α-D-Glc(1-1)-α-D-Glc	5	62
530	β-D-Glc(1-6)-β-D-Glc	β-Glucosidase Sesame	Cellobiose	Cellobiose	β-D-Glc(1-6)-β-D-Glc(1-4)-β-D-Glc	<25	135
531	β-D-Glc(1-6)-D-Glc	β-Glucosidase Almond	Glc	Glc	β-D-Glc(1-6)-D-Glc	19	60, 132
532	β-D-Glc(1-6)-D-Glc	β-Glucosidase Penicillium funiculosum	Glc	Glc	β-D-Glc(1-6)-D-Glc	17	60
533	β-D-Glc(1-6)-D-Glc	β-Glucosidase Almond	Glc	Glc	β-D-Glc(1-6)-D-Glc	26	133
534	β-D-GlcA(1-3)-β-D-Gal	β-Glucuronidase Bovine liver	βpNPGlcA	β-D-Gal(1-4)-β-D-Xyl-OMu	β-D-GlcA(1-3)-β-D-Gal(1-4)-β-D-Xyl-OMu	12	103
535	β-D-GlcA(1-3)-β-D-Gal	β-Glucuronidase Bovine liver	βpNPGlcA	β-D-Gal(1-4)-β-D-Gal(1-4)-β-D-Xyl-OMu	β-D-GlcA(1-3)-β-D-Gal(1-4)-β-D-Gal(1-4)-β-D-Xyl-OMu	20	103
536	β-D-GlcNAc(1-2)-D-Man	β-Hexosaminidase Bacillus circulans	GlcNAc	Man	β-D-GlcNAc(1-2)-D-Man	0.2	119
537	β-D-GlcNAc(1-3)-β-D-Gal	β-Hexosaminidase Nocardia orientalis	GlcNAc₂	β-D-Gal(1-4)-β-D-GlcNAc-OpNP	β-D-GlcNAc(1-3)-β-D-Gal(1-4)-β-D-GlcNAc-OpNP	1.5	136

#	Donor	Enzyme		Acceptor	Product	Yield	Ref
538	β-D-GlcNAc(1-3)-β-D-Gal	β-Hexosaminidase	Nocardia orientalis	GlcNAc$_2$	β-D-Gal(1-4)-β-D-Glc-OMe	β-D-GlcNAc(1-3)-β-D-Gal(1-4)-β-D-Glc-OMe	3.4	137
539	β-D-GlcNAc(1-3)-β-D-Gal	β-Hexosaminidase	Nocardia orientalis	GlcNAc$_2$	β-D-Gal(1-4)-β-D-Glc-OpNP	β-D-GlcNAc(1-3)-β-D-Gal(1-4)-β-D-Glc-OpNP	1.9	137
540	β-D-GlcNAc(1-3)-β-D-Gal	β-Hexosaminidase	Nocardia orientalis	GlcNAc$_2$	β-D-Gal(1-3)-α-D-GalNAc-OpNP	β-D-GlcNAc(1-3)-β-D-Gal(1-3)-α-D-GalNAc-OpNP	3.3	138
541	β-D-GlcNAc(1-3)-β-D-Gal	β-Hexosaminidase	Nocardia orientalis	GlcNAc$_2$	β-D-Gal(1-3)-β-D-GalNAc-OpNP	β-D-GlcNAc(1-3)-β-D-Gal(1-3)-β-D-GalNAc-OpNP	1.2	138
542	β-D-GlcNAc(1-3)-β-D-Gal-O	β-Hexosaminidase	Chamelea gallina	βpNPGlcNAc	β-D-Gal-OMe	β-D-GlcNAc(1-3)-β-D-Gal-OMe	4	56
543	β-D-GlcNAc(1-3)-β-D-Glc	β-Hexosaminidase	Aspergillus oryzae	βpNPGlcNAc	β-D-Glc-OMe	β-D-GlcNAc(1-3)-β-D-Glc-OMe	13	130
544	β-D-GlcNAc(1-4)-α-D-GlcNAc-O	β-Hexosaminidase	Aspergillus oryzae	βpNPGlcNAc	α-D-GlcNAc-OMe	β-D-GlcNAc(1-4)-α-D-GlcNAc-OMe	55	139
545	β-D-GlcNAc(1-4)-α-D-Glc-O	β-Hexosaminidase	Aspergillus oryzae	βpNPGlcNAc	α-D-Glc-OMe	β-D-GlcNAc(1-4)-α-D-Glc-OMe	12	130
546	β-D-GlcNAc(1-4)-β-D-Glc	β-Hexosaminidase	Aspergillus oryzae	βpNPGlcNAc	β-D-Glc-OMe	β-D-GlcNAc(1-4)-β-D-Glc-OMe	10	130
547	β-D-GlcNAc(1-4)-β-D-GlcNAc	β-Hexosaminidase	Aspergillus oryzae	βpNPGlcNAc	GlcNAc	β-D-GlcNAc(1-4)-β-D-GlcNAc	55	140
548	β-D-GlcNAc(1-4)-β-D-GlcNAc	β-Hexosaminidase	Aspergillus oryzae	GlcNAc$_3$	GlcNAc$_3$	GlcNAc$_5$	10	141

Table 1 (continued)

Entry	Linkage	Enzyme Source	Donor	Acceptor	Product	Yield (%)	Reference
549	β-D-GlcNAc(1-4)-β-D-GlcNAc	β-Hexosaminidase *Aspergillus oryzae*	GlcNAc$_3$	GlcNAc$_3$	GlcNAc$_6$	4	141
550	β-D-GlcNAc(1-4)-β-D-GlcNAc	β-Hexosaminidase *Aspergillus oryzae*	GlcNAc$_4$	GlcNAc$_4$	GlcNAc$_5$	20	141
551	β-D-GlcNAc(1-4)-β-D-GlcNAc	β-Hexosaminidase *Aspergillus oryzae*	GlcNAc$_4$	GlcNAc$_4$	GlcNAc$_6$	7	141
552	β-D-GlcNAc(1-4)-β-D-GlcNAc	Chitinase *Nocardia orientalis*	GlcNAc$_4$	GlcNAc$_4$	GlcNAc$_6$	34	142
553	β-D-GlcNAc(1-4)-β-D-GlcNAc	β-Hexosaminidase *Nocardia orientalis*	GlcNAc$_2$	GlcNAc$_2$	GlcNAc$_3$	13	143
554	β-D-GlcNAc(1-4)-β-D-GlcNAc-O	β-Hexosaminidase *Aspergillus oryzae*	β*p*NPGlcNAc	β-D-GlcNAc-OMe	β-D-GlcNAc(1-4)-β-D-GlcNAc-OMe	24	139
555	β-D-GlcNAc(1-4)-D-GlcNAc	Chitobiase *Bacillus* sp.	GlcNAc-oxazoline	GlcNAc	β-D-GlcNAc(1-4)-D-GlcNAc	43	144
556	β-D-GlcNAc(1-4)-D-GlcNAc	β-Hexosaminidase *Aspergillus oryzae*	GlcNAc$_2$	GlcNAc$_2$	GlcNAc$_3$	23	145
557	β-D-GlcNAc(1-4)-D-GlcNAc	β-Hexosaminidase *Aspergillus oryzae*	GlcNAc$_2$	GlcNAc$_2$	GlcNAc$_4$	7.5	145
558	β-D-GlcNAc(1-6)-α-D-GalNAc-O	β-Hexosaminidase *Nocardia orientalis*	GlcNAc$_2$	β-D-Gal(1-3)-α-D-GalNAc-O*p*NP	β-D-GlcNAc(1-6)-[β-D-Gal(1-3)]-α-D-GalNAc-O*p*NP	6.2	138

#	Product	Enzyme	Donor	Acceptor	Yield	Ref.
559	β-D-GlcNAc(1-6)-α-D-Gal-O	β-Hexosaminidase *Chamelea gallina*	βpNPGlcNAc	α-D-Gal-OMe	β-D-GlcNAc(1-6)-α-D-Gal-OMe	8	56
560	β-D-GlcNAc(1-6)-α-D-Glc-O	β-Hexosaminidase *Aspergillus oryzae*	βpNPGlcNAc	α-D-Glc-OMe	β-D-GlcNAc(1-6)-α-D-Glc-OMe	5	130
561	β-D-GlcNAc(1-6)-α-D-Man-O	β-Hexosaminidase *Chamelea gallina*	βpNPGlcNAc	α-D-Man-OMe	β-D-GlcNAc(1-6)-α-D-Man-OMe	6	56
562	β-D-GlcNAc(1-6)-α-D-Man-O	β-Hexosaminidase Jack bean	βpNPGlcNAc	α-D-Man-OMe	β-D-GlcNAc(1-6)-α-D-Man-OMe	9	45
563	β-D-GlcNAc(1-6)-β-D-Gal	β-Hexosaminidase *Nocardia orientalis*	GlcNAc$_2$	β-D-Gal(1-4)-β-D-GlcNAc-OpNP	β-D-GlcNAc(1-6)-β-D-Gal(1-4)-β-D-GlcNAc-OpNP	3.9	136
564	β-D-GlcNAc(1-6)-β-D-Gal	β-Hexosaminidase *Nocardia orientalis*	GlcNAc$_2$	β-D-Gal(1-4)-β-D-Glc-OMe	β-D-GlcNAc(1-6)-β-D-Gal(1-4)-β-D-Glc-OMe	3.6	137
565	β-D-GlcNAc(1-6)-β-D-Gal	β-Hexosaminidase *Nocardia orientalis*	GlcNAc$_2$	β-D-Gal(1-4)-β-D-Glc-OpNP	β-D-GlcNAc(1-6)-β-D-Gal(1-4)-β-D-Glc-OpNP	3	137
566	β-D-GlcNAc(1-6)-β-D-Gal	β-Hexosaminidase *Nocardia orientalis*	GlcNAc$_2$	β-D-Gal(1-3)-α-D-GalNAc-OpNP	β-D-GlcNAc(1-6)-β-D-Gal(1-3)-α-D-GalNAc-OpNP	4.5	138
567	β-D-GlcNAc(1-6)-β-D-Gal	β-Hexosaminidase *Nocardia orientalis*	GlcNAc$_2$	β-D-Gal(1-3)-β-D-GalNAc-OpNP	β-D-GlcNAc(1-6)-β-D-Gal(1-3)-β-D-GalNAc-OpNP	2.5	138
568	β-D-GlcNAc(1-6)-β-D-GalNAc-O	β-Hexosaminidase *Nocardia orientalis*	GlcNAc$_2$	β-D-Gal(1-3)-β-D-GalNAc-OpNP	β-D-GlcNAc(1-6)-[β-D-Gal(1-3)]-α-D-GalNAc-OpNP	4.3	138
569	β-D-GlcNAc(1-6)-β-D-Gal-O	β-Hexosaminidase *Chamelea gallina*	βpNPGlcNAc	β-D-Gal-OMe	β-D-GlcNAc(1-6)-β-D-Gal-OMe	6	56

Table 1 (*continued*)

Entry	Linkage	Enzyme Source	Donor	Acceptor	Product	Yield (%)	Reference
570	β-D-GlcNAc(1-6)-β-D-Gal-O	β-Hexosaminidase Jack bean	βpNPGlcNAc	β-D-Gal-OMe	β-D-GlcNAc(1-6)-β-D-Gal-OMe	14	45
571	β-D-GlcNAc(1-6)-β-D-GlcNAc-O	β-Hexosaminidase *Nocardia orientalis*	GlcNAc₂	β-D-Gal(1-4)-β-D-GlcNAc-O*p*NP	β-D-GlcNAc(1-6)-[β-D-Gal(1-4)]-β-D-GlcNAc-O*p*NP	0.8	136
572	β-D-GlcNAc(1-6)-β-D-GlcNAc-O	β-Galactosidase *Aspergillus oryzae* (Impurity?)	βpNPGlcNAc	βpNPGlcNAc	β-D-GlcNAc(1-6)-β-D-GlcNAc-O*p*NP	4	64
573	β-D-GlcNAc(1-6)-β-Glc-O	β-Hexosaminidase *Nocardia orientalis*	GlcNAc₂	β-D-Gal(1-4)-β-D-Glc-OMe	β-D-GlcNAc(1-6)-[β-D-Gal(1-4)]-β-D-Glc-OMe	10	137
574	β-D-GlcNAc(1-6)-β-Glc-O	β-Hexosaminidase *Nocardia orientalis*	GlcNAc₂	β-D-Gal(1-4)-β-D-Glc-O*p*NP	β-D-GlcNAc(1-6)-[β-D-Gal(1-4)]-β-D-Glc-O*p*NP	2.1	137
575	β-D-GlcNAc(1-6)-D-GalNAc	β-Hexosaminidase *Aspergillus oryzae*	βpNPGlcNAc	GalNAc	β-D-GlcNAc(1-6)-D-GalNAc	26	117
576	β-D-GlcNAc(1-6)-D-GlcNAc	β-Hexosaminidase *Aspergillus oryzae*	βpNPGlcNAc	GlcNAc	β-D-GlcNAc(1-6)-β-D-GlcNAc	22	140
577	β-D-GlcNAc(1-6)-D-GlcNAc	β-Hexosaminidase *Aspergillus oryzae*	GlcNAc	GlcNAc	β-D-GlcNAc(1-6)-D-GlcNAc	15	146
578	β-D-GlcNAc(1-6)-D-GlcNAc	β-Hexosaminidase *Nocardia orientalis*	GlcNAc₂	GlcNAc₂	β-D-GlcNAc(1-6)-D-GlcNAc	25	143
579	β-D-GlcNAc(1-6)-D-Man	β-Hexosaminidase *Bacillus circulans*	GlcNAc	Man	β-D-GlcNAc(1-6)-D-Man	1.7	119

No.		Enzyme	Donor	Acceptor	Product	Yield (%)	Ref.
580	β-D-Man(1-2)-β-D-Man-O	β-Mannosidase Guinea pig liver	βpClPMan	βpClPMan	β-D-Man(1-2)-β-D-Man-OpClP	0.9	147
581	β-D-Man(1-2)-β-D-Man-O	β-Mannosidase Guinea pig liver	βpNPMan	βpNPMan	β-D-Man(1-2)-β-D-Man-OpNP	1.3	147
582	β-D-Man(1-3)-α-D-Man-O	β-Galactosidase Aspergillus oryzae	α/βpNPMan (mixture)	α/βpNPMan (mixture)	β-D-Man(1-3)-α-D-Man-OpNP	7	64
583	β-D-Man(1-3)-β-D-Glc-S	β-Glucosidase Agrobacterium sp.	βManF	β-D-Glc-SPh	β-D-Man(1-3)-β-D-Glc-SPh	7	85
584	β-D-Man(1-3)-β-D-Xyl-S	β-Glucosidase Agrobacterium sp.	βManF	β-D-Xyl-SPh	β-D-Man(1-3)-β-D-Xyl-SPh	9	85
585	β-D-Man(1-4)-β-D-GlcNAc	β-Mannanase Aspergillus niger	Man$_3$	GlcNAc$_2$	β-D-Man(1-4)-β-D-GlcNAc(1-4)-D-GlcNAc	3.7	148
586	β-D-Man(1-4)-β-D-GlcNAc	β-Mannosidase Aspergillus oryzae	βpNPMan	GlcNAc$_2$	β-D-Man(1-4)-β-D-GlcNAc(1-4)-D-GlcNAc	26	149
587	β-D-Man(1-4)-β-D-Man-O	β-Galactosidase Aspergillus oryzae	α/βpNPMan (mixture)	α/βpNPMan (mixture)	β-D-Man(1-4)-β-D-Man-OpNP	13	64
588	β-D-Man(1-6)-β-D-Man-O	β-Mannosidase Guinea pig liver	βpClPMan	βpClPMan	β-D-Man(1-6)-β-D-Man-OpClP	0.7	147
589	β-D-Man(1-6)-β-D-Man-O	β-Mannosidase Guinea pig liver	βpNPMan	βpNPMan	β-D-Man(1-6)-β-D-Man-OpNP	3.7	147
590	β-D-Xyl(1-1)-β-D-Xyl	β-Xylosidase Aspergillus niger	Xyl$_2$	Xyl	β-D-Xyl(1-1)-β-D-Xyl	NA	150
591	β-D-Xyl(1-4)-D-Man	β-Xylosidase Aspergillus niger	Xyl$_2$	Man	β-D-Xyl(1-4)-D-Man	NA	150
592	β-D-Xyl(1-6)-α-D-Glc	β-Xylosidase Aspergillus awamori K4	Xyl$_2$	Trehalose	β-D-Xyl(1-6)-α-D-Glc(1-1)-α-D-Glc	10	151
593	β-D-Xyl(1-6)-D-GlcNAc	β-Xylosidase Aspergillus niger	βpNPXyl	GlcNAc	β-D-Xyl(1-6)-D-GlcNAc	36	152

Table 1 (*continued*)

Entry	Linkage	Enzyme Source	Donor	Acceptor	Product	Yield (%)	Reference
594	β-D-Xyl(1-6)-D-Man	β-Xylosidase *Aspergillus niger*	Xyl$_2$	Man	β-D-Xyl(1-6)-D-Man	NA	150
595	β-D-Xyl(1-6)-Mannitol	β-Xylosidase *Aspergillus awamori* K4	Xyl$_2$	Mannitol	β-D-Xyl(1-6)-Mannitol	<10	151
596	β-D-Xyl(1-6)-Sorbitol	β-Xylosidase *Aspergillus awamori* K4	Xyl$_2$	Sorbitol	β-D-Xyl(1-6)-Sorbitol	<10	151

Table 2. Enzymatically synthesized oligosaccharides using *endo*-glycosidases.

Entry	Linkage	Enzyme/Source	Donor	Acceptor	Product	Yield (%)	Reference
1	α-D-GalNAc-(1-?)-D-(6-OMe)-Gal	Endo-α-N-acetyl-galactosaminidase *Diplococcus pneumoniae*	β-D-Gal(1-3)-α-D-GalNAc-asialoglycoprotein	(6-OMe)-D-Gal	β-D-Gal(1-3)-α-D-GalNAc(1-?)-D-(6-OMe)-Gal	17	153
2	α-D-GalNAc-(1-?)-D-Fuc	Endo-α-N-acetyl-galactosaminidase *Diplococcus pneumoniae*	β-D-Gal(1-3)-α-D-GalNAc-asialoglycoprotein	D-Fuc	β-D-Gal(1-3)-α-D-GalNAc(1-?)-D-Fuc	20	153
3	α-D-GalNAc-(1-?)-D-Gal	Endo-α-N-acetyl-galactosaminidase *Diplococcus pneumoniae*	β-D-Gal(1-3)-α-D-GalNAc-asialoglycoprotein	Gal	β-D-Gal(1-3)-α-D-GalNAc(1-?)-D-Gal	26	153
4	α-D-GalNAc-(1-?)-D-Glc	Endo-α-N-acetyl-galactosaminidase *Diplococcus pneumoniae*	β-D-Gal(1-3)-α-D-GalNAc-asialoglycoprotein	Glc	β-D-Gal(1-3)-α-D-GalNAc(1-?)-D-Glc	22	153
5	α-D-GalNAc-(1-1)-O	Endo-α-N-acetyl-galactosaminidase *Diplococcus pneumoniae*	β-D-Gal(1-3)-α-D-GalNAc-asialoglycoprotein	Glycerol	β-D-Gal(1-3)-α-D-GalNAc-(1-1)-Glyceryl	69	153
6	α-D-GalNAc-O	Endo-α-N-acetyl-galactosaminidase *Diplococcus pneumoniae*	β-D-Gal(1-3)-α-D-GalNAc-asialoglycoprotein	Ser	β-D-Gal(1-3)-α-D-GalNAc-O-Ser	17	153
7	α-D-GalNAc-O	Endo-α-N-acetyl-galactosaminidase *Diplococcus pneumoniae*	β-D-Gal(1-3)-α-D-GalNAc-asialoglycoprotein	Thr	β-D-Gal(1-3)-α-D-GalNAc-O-Thr	12	153

Table 2 (continued)

Entry	Linkage	Enzyme/Source	Donor	Acceptor	Product	Yield (%)	Reference
8	α-D-Glc(1-3)-α-D-Glc-O	α-Amylase Streptomyces griseus	{α-D-Glc(1-4)}$_3$-D-Glc	αpNPGlc	α-D-Glc(1-4)-α-D-Glc(1-4)-α-D-Glc(1-3)-α-D-Glc-OpNP	1.5	154
9	α-D-Glc(1-3)-β-D-Glc-O	α-Amylase Streptomyces griseus	{α-D-Glc(1-4)}$_3$-D-Glc	βpNPGlc	α-D-Glc(1-4)-α-D-Glc(1-4)-α-D-Glc(1-3)-β-D-Glc-OpNP	16	154
10	α-D-Glc(1-3)-β-D-Glc-O	α-Amylase Streptomyces griseus	{α-D-Glc(1-4)}$_3$-D-Glc	βClNPGlc	α-D-Glc(1-4)-α-D-Glc(1-4)-α-D-Glc(1-3)-β-D-Glc-OClNP	18	154
11	α-D-Glc(1-4)-α-D-Glc	Cyclomaltodextrinase Bacillus sphaericus E-244	γ-CD	Glc	{α-D-Glc(1-4)}$_8$-D-Glc	20	155
12	α-D-Glc(1-4)-α-D-Glc-O	Amylase Klebsiella pneumoniae	{α-D-Glc(1-4)}$_6$-D-Glc	αpNPGlc	{α-D-Glc(1-4)}$_6$-α-D-Glc-OpNP	13	156
13	α-D-Glc(1-4)-α-D-Glc	α-Amylase Aspergillus oryzae	α-Maltosyl-F	{α-D-Glc(1-3)}$_2$-β-D-Glc-O-indoyl ethanol	{α-D-Glc(1-4)}$_2$-{α-D-Glc(1-3)}$_2$-β-D-Glc-O-indoyl ethanol	59	157
14	α-D-Glc(1-4)-α-D-Glc	Cyclomaltodextrinase Bacillus sphaericus E-244	β-CD	Glc	{α-D-Glc(1-4)}$_7$-D-Glc	39	155
15	α-D-Glc(1-4)-α-D-Glc	Pullulanase Klebsiella sp.	Pullulan	Stevioside	13-{α-D-Glc(1-4)}$_2$-β-D-Glc(1-2)-β-D-Glc-19-β-D-Glc-O-steviol	NA	61
16	α-D-Glc(1-4)-α-D-Glc	Pullulanase Klebsiella sp.	Pullulan	Stevioside	13-α-D-Glc(1-4)-β-D-Glc(1-2)-β-D-Glc-19-β-D-Glc-O-steviol	NA	61
17	α-D-Glc(1-4)-α-D-Glc	Pullulanase Klebsiella sp.	Pullulan	Stevioside	13-β-D-Glc(1-2)-β-D-Glc-19-{α-D-Glc(1-4)}$_2$-β-D-Glc-O-steviol	NA	61

#						
18	α-D-Glc(1-4)-α-D-Glc-O	Amylase Pseudomonas stutzeri	{α-D-Glc(1-4)}₄-D-Glc	αpNPGlc	{α-D-Glc(1-4)}₄-α-D-Glc-OpNP	12	158
19	α-D-Glc(1-4)-α-D-Glc-O	Amylase Pseudomonas stutzeri	{α-D-Glc(1-4)}₄-D-Glc	αpNPGlc	{α-D-Glc(1-4)}₄-α-D-Glc-OpNP	54	159
20	α-D-Glc(1-4)-α-D-Glc-O	α-Amylase Streptomyces griseus	{α-D-Glc(1-4)}₃-D-Glc	αpNPGlc	{α-D-Glc(1-4)}₃-α-D-Glc-OpNP	15	154
21	α-D-Glc(1-4)-α-D-Glc-O	Amylase Bacillus licheniformis	{α-D-Glc(1-4)}₅-D-Glc	αpNPGlc	{α-D-Glc(1-4)}₅-α-D-Glc-OpNP	13	160
22	α-D-Glc(1-4)-β-D-Glc-O	α-Amylase Streptomyces griseus	{α-D-Glc(1-4)}₃-D-Glc	βClNPGlc	α-D-Glc(1-4)-α-D-Glc(1-4)-α-D-Glc(1-4)-β-D-Glc-OClNP	13	154
23	α-D-Glc(1-4)-β-D-Glc-O	α-Amylase Streptomyces griseus	{α-D-Glc(1-4)}₃-D-Glc	βpNPGlc	α-D-Glc(1-4)-α-D-Glc(1-4)-α-D-Glc(1-4)-β-D-Glc-OpNP	11	154
24	α-D-Glc(1-6)-α-D-Glc	Isoamylase Pseudomonas amyloderamosa	α-Maltosyl-F	β-CD	α-D-Glc(1-4)-α-D-Glc(1-6)-β-CD	60	161
25	α-D-Glc(1-6)-α-D-Glc	Pullulanase Klebsiella aerogenes	Maltose	α-D-Glc(1-4)-α-D-Glc(1-6)-γ-CD	{α-D-Glc(1-4)-α-D-Glc(1-6)}₂-γ-CD	NA	162
26	α-D-Glc(1-6)-α-D-Glc	Pullulanase Aerobacter aerogenes	α-Maltosyl-F	α-CD	α-D-Glc(1-4)-α-D-Glc(1-6)-α-CD	62	161
27	α-D-Glc(1-6)-α-D-Glc	Pullulanase Aerobacter aerogenes	α-Maltosyl-F	β-CD	α-D-Glc(1-4)-α-D-Glc(1-6)-β-CD	NA	161
28	α-D-Glc(1-6)-α-D-Glc	Pullulanase Bacillus acidopullulyticus	α-Maltosyl-F	β-CD	α-D-Glc(1-4)-α-D-Glc(1-6)-β-CD	NA	161
29	α-D-Glc(1-6)-α-D-Glc	Pullulanase Bacillus acidopullulyticus	α-Maltosyl-F	α-CD	α-D-Glc(1-4)-α-D-Glc(1-6)-α-CD	25	161
30	α-D-Glc(1-6)-α-D-Glc	Isoamylase Pseudomonas amyloderamosa	α-Maltosyl-F	γ-CD	α-D-Glc(1-4)-α-D-Glc(1-6)-γ-CD	60	161
31	α-D-Glc(1-6)-α-D-Glc	Pullulanase Bacillus acidopullulyticus	α-Maltosyl-F	γ-CD	α-D-Glc(1-4)-α-D-Glc(1-6)-γ-CD	43	161
32	α-D-Glc(1-6)-α-D-Glc	Isoamylase Pseudomonas amyloderamosa	α-Maltosyl-F	α-CD	α-D-Glc(1-4)-α-D-Glc(1-6)-α-CD	60	161

Table 2 (*continued*)

Entry	Linkage	Enzyme/Source	Donor	Acceptor	Product	Yield (%)	Reference
33	α-D-Glc(1-6)-α-D-Glc	Pullulanase Aerobacter aerogenes	α-Maltosyl-F	γ-CD	α-D-Glc(1-4)-α-D-Glc(1-6)-γ-CD	NA	161
34	α-D-Glc(1-6)-α-D-Glc	Pullulanase Klebsiella aerogenes	Maltose	α-D-Glc(1-4)-α-D-Glc(1-6)₂-β-CD	{α-D-Glc(1-4)-α-D-Glc(1-6)}₂-β-CD	NA	162
35	α-D-Glc(1-6)-α-D-Glc	Pullulanase Klebsiella aerogenes	Maltose	α-CD	α-D-Glc(1-4)-α-D-Glc(1-6)-α-CD	NA	162
36	α-D-Glc(1-6)-α-D-Glc	Pullulanase Klebsiella aerogenes	Maltose	γ-CD	α-D-Glc(1-4)-α-D-Glc(1-6)-γ-CD	39	162
37	α-D-Glc(1-6)-α-D-Glc	Pullulanase Klebsiella aerogenes	Maltose	α-D-Glc(1-4)-α-D-Glc(1-6)₂-α-CD	{α-D-Glc(1-4)-α-D-Glc(1-6)}₂-α-CD	NA	162
38	α-D-Glc(1-6)-α-D-Glc	Pullulanase Klebsiella aerogenes	Maltose	β-CD	α-D-Glc(1-4)-α-D-Glc(1-6)-β-CD	NA	162
39	α-D-Glc(1-6)-α-D-Glc	Isoamylase Pseudomonas	{α-D-Glc(1-4)}₂-D-Glc	{α-D-Glc(1-4)}₂-D-Glc	α-D-Glc(1-4)-α-D-Glc(1-4)-[{α-D-Glc(1-4)}₂-α-D-Glc(1-6)]-D-Glc	NA	163
40	α-D-Glc(1-6)-α-D-Glc	Isoamylase Pseudomonas	{α-D-Glc(1-4)}₂-D-Glc	{α-D-Glc(1-4)}₂-D-Glc	{α-D-Glc(1-4)}₂-α-D-Glc(1-6)-{α-D-Glc(1-4)}₂-D-Glc	NA	163
41	α-D-Glc(1-6)-α-D-Glc	Isoamylase Pseudomonas	{α-D-Glc(1-4)}₂-D-Glc	{α-D-Glc(1-4)}₂-D-Glc	α-D-Glc(1-4)-α-D-Glc(1-4)-α-D-Glc(1-6)-[α-D-Glc(1-4)]-α-D-Glc(1-4)-D-Glc	NA	163
42	α-D-Glc(1-6)-α-D-Glc	Isoamylase Pseudomonas	Maltose	α-CD	α-D-Glc(1-4)-α-D-Glc(1-6)-α-CD	NA	163
43	α-D-Glc(1-6)-α-D-Glc	Isoamylase Pseudomonas	Maltose	β-CD	α-D-Glc(1-4)-α-D-Glc(1-6)-β-CD	NA	163
44	α-D-Glc(1-6)-α-D-Glc	Isoamylase Pseudomonas	Maltose	γ-CD	α-D-Glc(1-4)-α-D-Glc(1-6)-γ-CD	NA	163
45	α-D-Glc(1-6)-α-D-Glc	Isoamylase Pseudomonas	{α-D-Glc(1-4)}₂-D-Glc	α-CD	{α-D-Glc(1-4)}₃-α-D-Glc(1-6)-α-CD	NA	163
46	α-D-Glc(1-6)-α-D-Glc	Isoamylase Pseudomonas	{α-D-Glc(1-4)}₂-D-Glc	β-CD	{α-D-Glc(1-4)}₃-α-D-Glc(1-6)-β-CD	NA	163

29.6 Recent Developments and New Directions

#	Donor	Enzyme / Source	Acceptor	Product	Yield	Ref.
47	α-D-Glc(1-6)-α-D-Glc	Isoamylase *Pseudomonas*	{α-D-Glc(1-4)}₂-D-Glc	γ-CD	NA	163
48	α-D-Glc(1-6)-D-Glc	Neopullulanase *Bacillus subtilis*	{α-D-Glc(1-4)}₂-D-Glc	{α-D-Glc(1-4)}₂-α-D-Glc(1-6)-γ-CD	NA	163
49	α-D-Glc(1-6)-D-Glc	Neopullulanase *Bacillus subtilis*	{α-D-Glc(1-4)}₂-D-Glc	α-D-Glc(1-6)-D-Glc	10	164
50	α-D-Glc(1-6)-D-Glc	Neopullulanase *Bacillus subtilis*	{α-D-Glc(1-4)}₂-D-Glc	α-D-Glc(1-4)-α-D-Glc(1-6)-α-D-Glc(1-4)-D-Glc	5.7	164
51	α-D-Glc(1-6)-D-Glc	Neopullulanase *Bacillus subtilis*	{α-D-Glc(1-4)}₂-D-Glc	α-D-Glc(1-4)-α-D-Glc(1-6)-D-Glc	NA	164
52	α-D-Glc-O	Rice debranching enzyme	Pullulan	α-D-Glc(1-6)-α-D-Glc(1-4)-α-D-Glc	NA	164
53	β-D-3-deoxy-3-F-Glc	Cellulase *Trichoderma viridae*	MeOH	{α-D-Glc(1-4)}₂-α-D-Glc-OMe	NA	165
54	β-D-Gal(1-3)-β-D-Gal-O	β-Galactanase *Penicillium citrinum*	β-D-3-deoxy-3-F-Glc-OMe	β-D-3-deoxy-3-F-Glc-OMe	18	166, 167
55	β-D-Gal(1-4)-α-D-Gal-O	β-Galactanase *Penicillium citrinum*	β-D-Gal-OMe	β-D-Gal(1-4)-β-D-Gal-OMe	7	168
56	β-D-Gal(1-4)-α-D-Gal-O	β-Galactanase *Penicillium citrinum*	α-D-Gal-OMe	β-D-Gal(1-4)-β-D-Gal(1-4)-α-D-Gal-OMe	3.5	168
57	β-D-Gal(1-4)-β-D-Gal	β-Galactanase *Penicillium citrinum*	α-D-Gal-OMe	β-D-Gal(1-4)-α-D-Gal-OMe	19	168
58	β-D-Gal(1-4)-β-D-Gal	β-Galactanase *Penicillium citrinum*	β-D-Gal(1-4)-D-GlcNAc	β-D-Gal(1-4)-β-D-Gal(1-4)-D-GlcNAc	16	168
59	β-D-Gal(1-4)-β-D-Gal	β-Galactanase *Penicillium citrinum*	β-D-Gal(1-3)-D-GalNAc	β-D-Gal(1-4)-β-D-Gal(1-3)-D-GalNAc	17	168
	β-D-Gal(1-4)-β-D-Gal	β-Galactanase *Penicillium citrinum*	β-D-Gal(1-6)-D-GlcNAc	β-D-Gal(1-4)-β-D-Gal(1-6)-D-GlcNAc	18	168

Table 2 (continued)

Entry	Linkage	Enzyme/Source	Donor	Acceptor	Product	Yield (%)	Reference
60	β-D-Gal(1-4)-β-D-Gal	β-Galactanase *Penicillium citrinum*	Arabinogalactan	β-D-Gal(1-4)-D-Glc	β-D-Gal(1-4)-β-D-Gal(1-4)-D-Glc	11	168
61	β-D-Gal(1-4)-β-D-Gal-O	β-Galactanase *Penicillium citrinum*	Arabinogalactan	β-D-Gal-OMe	β-D-Gal(1-4)-β-D-Gal(1-4)-β-D-Gal-OMe	2.5	168
62	β-D-Gal(1-4)-β-D-Gal-O	β-Galactanase *Penicillium citrinum*	Arabinogalactan	β-D-Gal(1-3)-D-GlcNAc	β-D-Gal(1-4)-β-D-Gal(1-3)-D-GlcNAc	11	168
63	β-D-Gal(1-4)-β-D-Gal-O	β-Galactanase *Penicillium citrinum*	Arabinogalactan	βpNPGal	β-D-Gal(1-4)-β-D-Gal(1-4)-β-D-Gal-OpNP	7.5	168
64	β-D-Gal(1-4)-β-D-Gal-O	β-Galactanase *Penicillium citrinum*	Arabinogalactan	βpNPGal	β-D-Gal(1-4)-β-D-Gal-OpNP	14	168
65	β-D-Gal(1-4)-β-D-Gal-O	β-Galactanase *Penicillium citrinum*	Arabinogalactan	β-D-Gal-OMe	β-D-Gal(1-4)-β-D-Gal-OMe	16	168
66	β-D-Gal(1-4)-β-D-Gal-O	β-Galactanase *Bacillus subtilus*	{β-D-Gal(1-4)}$_2$-D-Gal	Glycerol	β-D-Gal(1-4)-β-D-Gal-O-Glyceryl	40 to 75	169
67	β-D-Gal(1-4)-β-D-Gal-O	β-Galactanase *Bacillus subtilus*	Arabinogalactan	Glycerol	β-D-Gal(1-4)-β-D-Gal-O-Glyceryl	40 to 75	169
68	β-D-Glc(1-3)-β-D-Glc	Endo-β-(1-3)-glucanase L-IV *Spisula sachalinensis*	Laminarin	[1-^{14}C]-Glc	{β-D-Glc(1-3)}$_{(1-5)}$-D-[1-^{14}C]-Glc	26	170
69	β-D-Glc(1-6)-DNJ	Cellulase *Trichoderma* sp.	Cellooligosaccharides	DNJ	β-D-Glc(1-4)-β-D-Glc(1-6)-DNJ	NA	171

	Acceptor	Enzyme	Cellooligosaccharides	DNJ	Product	% Yield	Ref
70	β-D-Glc(1-4)-DNJ	Cellulase *Trichoderma* sp.			β-D-Glc(1-4)-β-D-Glc(1-4)-DNJ	NA	171
71	β-D-Glc(1-4)-β-D-(3-OMe)-Glc-O	Cellulase *Trichoderma viridae*	β-Lac-F	β-D-(3-OMe)-Glc-OMe	β-D-Gal(1-4)-β-D-Glc(1-4)-β-D-(3-OMe)-Glc-OMe	8	166, 167
72	β-D-Glc(1-4)-β-D-Glc	(1-3),(1-4)-β-Glucanase *Bacillus licheniformis*	β-D-Glc(1-4)-β-D-Glc(1-3)-β-D-Glc-F	β-D-Glc(1-4)-β-D-Glc(1-3)-β-D-Glc-F	β-D-Glc(1-4)-β-D-Glc(1-3)-{β-D-Glc(1-4)}$_2$-β-D-Glc(1-3)-D-Glc	20	172
73	β-D-Glc(1-4)-β-D-Glc	(1-3),(1-4)-β-Glucanase *Bacillus licheniformis*	β-D-Glc(1-3)-β-D-Glc-F	β-D-Glc(1-3)-β-D-Glc-F	β-D-Glc(1-3)-β-D-Glc(1-4)-β-D-Glc(1-3)-D-Glc	10	173
74	β-D-Glc(1-4)-β-D-Glc	(1-3),(1-4)-β-Glucanase *Bacillus licheniformis*	β-D-Glc(1-3)-β-D-Glc-F	β-D-Glc(1-3)-β-D-Glc-F	{β-D-Glc(1-3)-β-D-Glc(1-4)}$_2$-β-D-Glc(1-3)-D-Glc	5	173
75	β-D-Glc(1-4)-β-D-Glc	(1-3),(1-4)-β-Glucanase *Bacillus licheniformis*	β-D-Glc(1-3)-β-D-Glc-F	β-D-Glc(1-3)-β-D-Glc-OMe	β-D-Glc(1-3)-β-D-Glc(1-4)-β-D-Glc(1-3)-β-D-Glc-OMe	40	173
76	β-D-Glc(1-4)-β-D-Glc	(1-3),(1-4)-β-Glucanosynthase *Bacillus licheniformis*	β-D-Glc(1-3)-α-D-Glc-F	β-D-Glc(1-3)-β-D-Glc-OMu	β-D-Glc(1-3)-β-D-Glc(1-4)-β-D-Glc(1-3)-β-D-Glc-OMu	NA	29
77	β-D-Glc(1-4)-β-D-Glc	Endoglucanase I *Humicola insolens*	β-Lac-F	β-D-Glc(1-4)-β-D-Glc-O-indoyl ethanol	β-D-Gal(1-4)-{β-D-Glc(1-4)}$_2$-β-D-Glc-O-indoyl ethanol	60	174
78	β-D-Glc(1-4)-β-D-Glc	(1-3),(1-4)-β-Glucanosynthase *Bacillus licheniformis*	β-D-Glc(1-3)-α-D-Glc-F	β-D-Glc(1-4)-β-D-Glc-OMu	β-D-Glc(1-3)-{β-D-Glc(1-4)}$_2$-β-D-Glc-OMu	NA	29
79	β-D-Glc(1-4)-β-D-Glc	Cellulase *Trichoderma viridae*	β-Lac-F	β-D-Glc(1-4)-β-D-Glc-OMe	β-D-Gal(1-4)-{β-D-Glc(1-4)}$_2$-β-D-Glc-OMe	36	166, 175, 167, 176
80	β-D-Glc(1-4)-β-D-Glc	Cellulase *Trichoderma viridae*	β-Lac-F	β-D-Glc(1-4)-α-D-Glc-OMe	β-D-Gal(1-4)-{β-D-Glc(1-4)}$_2$-α-D-Glc-OMe	60	166, 167

Table 2 (continued)

Entry	Linkage	Enzyme/Source	Donor	Acceptor	Product	Yield (%)	Reference
81	β-D-Glc(1-4)-β-D-Glc	Cellulase Trichoderma viridae	β-Lac-F	β-D-Glc(1-3)-β-D-Glc-OMe	β-D-Gal(1-4)-β-D-Glc(1-4)-β-D-Glc(1-3)-β-D-Glc-OMe	23	166, 167
82	β-D-Glc(1-4)-β-D-Glc	Cellulase Trichoderma viridae	β-Lac-F	β-D-Glc(1-6)-β-D-Glc-OMe	β-D-Gal(1-4)-β-D-Glc(1-4)-β-D-Glc(1-6)-β-D-Glc-OMe	26	166, 167
83	β-D-Glc(1-4)-β-D-Glc	Cellulase Trichoderma viridae	β-Lac-F	β-D-Glc(1-4)-β-D-Glc(1-4)-β-D-Glc-OMe	β-D-Gal(1-4)-{β-D-Glc(1-4)}$_2$-β-D-Glc(1-4)-β-D-Glc-OMe	27	166, 167, 176
84	β-D-Glc(1-4)-β-D-Glc	Cellulase Trichoderma viridae	β-Cellobiosyl-F	β-Cellobiosyl-F	Cellulose	54	177, 167, 178
85	β-D-Glc(1-4)-β-D-Glc	Cellulase Aspergillus niger	β-Cellobiosyl-F	β-Cellobiosyl-F	Cellulose	8	177, 167
86	β-D-Glc(1-4)-β-D-Glc	Cellulase Polyporus tulipiferae	β-Cellobiosyl-F	β-Cellobiosyl-F	Cellulose	25	177, 167
87	β-D-Glc(1-4)-β-D-Glc	Cellulase Trichoderma viridae	β-Lac-F	β-D-Glc(1-4)-β-D-Glc-SMe	β-D-Gal(1-4)-{β-D-Glc(1-4)}$_2$-β-D-Glc-SMe	23	167
88	β-D-Glc(1-4)-β-D-Glc	Cellulase Trichoderma viridae	β-Lac-F	β-D-Glc(1-4)-β-D-Glc-OAll	β-D-Gal(1-4)-{β-D-Glc(1-4)}$_2$-β-D-Glc-OAll	13	176
89	β-D-Glc(1-4)-β-D-Glc	(1-3),(1-4)-β-Glucanase Bacillus licheniformis	β-D-Glc(1-3)-β-D-Glc-F	β-D-Glc(1-3)-β-D-Glc-OMe	{β-D-Glc(1-4)-β-D-Glc(1-4)}$_2$-β-D-Glc(1-3)-β-D-Glc-OMe	0.6	173
90	β-D-Glc(1-4)-β-D-Glc-O	(1-3),(1-4)-β-Glucanosynthase Bacillus licheniformis	β-D-Glc(1-3)-α-D-Glc-F	β-D-Glc-OMu	β-D-Glc(1-3)-β-D-Glc(1-4)-β-D-Glc-OMu	88	29
91	β-D-Glc(1-4)-β-D-Glc-O	Cellulase Trichoderma viridae	β-Lac-F	β-D-Glc-OMe	β-D-Gal(1-4)-β-D-Glc(1-4)-β-D-Glc-OMe	51	166, 167, 176

29.6 Recent Developments and New Directions

#	Donor	Enzyme	Acceptor	Product	Yield	Ref
92	β-D-Glc(1-4)-β-D-Glc-S	Cellulase Trichoderma viridae	β-Lac-F	β-D-Glc-S-CH$_2$CONHPr	β-D-Gal(1-4)-β-D-Glc(1-4)-β-D-Glc-S-CH$_2$CONHPr	43	167
93	β-D-Glc(1-4)-β-D-Glc-S	Cellulase Trichoderma viridae	β-Lac-F	β-D-Glc-S-C$_{12}$H$_{25}$	β-D-Gal(1-4)-β-D-Glc(1-4)-β-D-Glc-S-C$_{12}$H$_{25}$	30	167
94	β-D-Glc(1-4)-β-D-Glc-S	Cellulase Trichoderma viridae	β-Lac-F	β-D-Glc-S-Ph	β-D-Gal(1-4)-β-D-Glc(1-4)-β-D-Glc-S-Ph	36	167
95	β-D-Glc(1-4)-β-D-Glc-S	Cellulase Trichoderma viridae	β-D-Glc-S(1-4)-β-D-Glc-F	β-D-Glc-S(1-4)-β-D-Glc-F	{β-D-Glc-S(1-4)-β-D-Glc}$_{(1-6)}$-β-D-Glc-S(1-4)-D-Glc	41	179
96	β-D-Glc(1-4)-β-D-Man-O	Cellulase Trichoderma viridae	β-Lac-F	β-D-Man-OMe	β-D-Gal(1-4)-β-D-Glc(1-4)-β-D-Man-OMe	52	166, 167
97	β-D-Glc(1-4)-β-D-Xyl	Xylanase Trichoderma viridae	β-D-Xyl(1-4)-β-D-Glc-F	β-D-Xyl(1-4)-β-D-Glc-F	{β-D-Xyl(1-4)-β-D-Glc(1-4)}$_{1-11}$-β-D-Xyl(1-4)-β-D-Glc	58	180
98	β-D-Glc(1-4)-β-D-Xyl-O	Cellulase Trichoderma viridae	β-Lac-F	β-D-Xyl-OMe	β-D-Gal(1-4)-β-D-Glc(1-4)-β-D-Xyl-OMe	25	166, 167
99	β-D-Glc-O	Endoglycoceramidase Corynebacterium sp.	GM$_1$	HO-(CH$_2$)$_4$CH$_3$	^{3}NeuAcCgOse$_4$-O-(CH$_2$)$_4$CH$_3$	22	181
100	β-D-Glc-O	Endoglycoceramidase Corynebacterium sp.	GM$_1$	HO-(CH$_2$)$_9$CH$_3$	^{3}NeuAcCgOse$_4$-O-(CH$_2$)$_9$CH$_3$	2.7	181
101	β-D-Glc-O	Endoglycoceramidase Corynebacterium sp.	GM$_2$	HO-(CH$_2$)$_5$CH$_3$	^{3}NeuAcCgOse$_3$-O-(CH$_2$)$_5$CH$_3$	22	181
102	β-D-Glc-O	Endoglycoceramidase Corynebacterium sp.	GD$_{1a}$	HO-(CH$_2$)$_5$CH$_3$	IV^{3}NeuAcII^{3}NeuAc-CgOse$_4$-O-(CH$_2$)$_5$CH$_3$	25	181
103	β-D-Glc-O	Endoglycoceramidase Corynebacterium sp.	GM$_1$	HO-(CH$_2$)$_6$CH$_3$	^{3}NeuAcCgOse$_4$-O-(CH$_2$)$_6$CH$_3$	30	181
104	β-D-Glc-O	Endoglycoceramidase Corynebacterium sp.	GM$_1$	HO-(CH$_2$)$_7$CH$_3$	^{3}NeuAcCgOse$_4$-O-(CH$_2$)$_7$CH$_3$	18	181
105	β-D-Glc-O	Endoglycoceramidase Macrobdella decora	GM$_1$	HO-(CH$_2$)$_7$CH$_3$	^{3}NeuAcCgOse$_4$-O-(CH$_2$)$_7$CH$_3$	NA	182
106	β-D-Glc-O	Endoglycoceramidase Corynebacterium sp.	GM$_1$	HO-(CH$_2$)$_5$CH$_3$	^{3}NeuAcCgOse$_4$-O-(CH$_2$)$_5$CH$_3$	35	181

Table 2 (*continued*)

Entry	Linkage	Enzyme/Source	Donor	Acceptor	Product	Yield (%)	Reference
107	β-D-Glc-O	Endoglycoceramidase *Cornybacterium* sp.	GM$_1$	MeOH	^{3}NeuAcCgOse$_4$-OMe	17	181
108	β-D-Glc-O	Endoglycoceramidase *Cornybacterium* sp.	Asialo-GM$_1$	HO-(CH$_2$)$_5$CH$_3$	CgOse$_4$-O-(CH$_2$)$_5$CH$_3$	33	181
109	β-D-Glc-O	Endoglycoceramidase *Cornybacterium* sp.	IV^{3}NeuAcβGalNAc-GbOse$_4$-O-Cer	HO-(CH$_2$)$_5$CH$_3$	IV^{3}NeuAcβGalNAc-GbOse$_4$-O-(CH$_2$)$_5$CH$_3$	9.6	181
110	β-D-Glc-O	Endoglycoceramidase Leech	LacSer-support	Ceramide	GM$_3$	61	18, 183
111	β-D-Glc-O	Endoglycoceramidase *Macrobdella decora*	GM$_1$	(HO-CH$_2$)$_3$CNH-CO(CH$_2$)$_4$COOMe	^{3}NeuAcCgOse$_4$-O-CH$_2$(HO-CH$_2$)$_2$CNHCO(CH$_2$)$_4$-COOMe	NA	182
112	β-D-Glc-O	Endoglycoceramidase *Macrobdella decora*	GM$_1$	HO-(CH$_2$)$_6$NHCOCF$_3$	^{3}NeuAcCgOse$_4$-O-(CH$_2$)$_6$NHCOCF$_3$	NA	182
113	β-D-Glc-O	Endoglycoceramidase *Macrobdella decora*	GM$_1$	HO-(CH$_2$)$_6$NH-COCO(CH$_3$)$_3$	^{3}NeuAcCgOse$_4$-O-(CH$_2$)$_6$NHCOCO(CH$_3$)$_3$	NA	182
114	β-D-Glc-O	Endoglycoceramidase *Macrobdella decora*	GM$_1$	HO-(CH$_2$)$_8$-OH	^{3}NeuAcCgOse$_4$-O-(CH$_2$)$_8$-OH	NA	182
115	β-D-Glc-O	Endoglycoceramidase *Macrobdella decora*	GM$_1$	HO-(CH$_2$)$_8$-CH=CH$_2$	^{3}NeuAcCgOse$_4$-O-(CH$_2$)$_8$-CH=CH$_2$	NA	182
116	β-D-GlcN(1-4)-β-D-GlcN	Lysozyme Hen egg white	(2-deoxy-2-monochloroacetamido-β-D-Glc)$_3$	(2-deoxy-2-chloroacetamido-β-D-Glc)$_3$	{(β-D-GlcN(1-4)}$_{(3-8)}$-D-GlcN	33	184

#	Product	Enzyme	Donor	Acceptor	Yield	Ref.
117	β-D-GlcNAc(1-?)-(3-OMe)-D-Glc	Endo-β-N-acetyl-glucosaminidase *Arthrobacter protophormiae*	Man₉GlcNAc₂Asn	(3-OMe)-D-Glc	Man₉-β-D-GlcNAc(1-?)-(3-OMe)-D-Glc	30	36
118	β-D-GlcNAc(1-?)-α-D-GlcNAc-O	Endo-β-N-acetyl-glucosaminidase *Arthrobacter protophormiae*	Man₉GlcNAc₂Asn	α-D-GlcNAc-OMe	Man₉-β-D-GlcNAc(1-?)-α-D-GlcNAc-OMe	66	36
119	β-D-GlcNAc(1-?)-D-2dGlc	Endo-β-N-acetyl-glucosaminidase *Arthrobacter protophormiae*	Man₉GlcNAc₂Asn	2dGlc	Man₉-β-D-GlcNAc(1-?)-D-2dGlc	61	36
120	β-D-GlcNAc(1-?)-D-6-deoxy-Glc	Endo-β-N-acetyl-glucosaminidase *Arthrobacter protophormiae*	Man₉GlcNAc₂Asn	6-deoxy-Glc	Man₉-β-D-GlcNAc(1-?)-D-6-deoxy-Glc	67	36
121	β-D-GlcNAc(1-?)-D-Fru	Endo-β-N-acetyl-glucosaminidase *Arthrobacter protophormiae*	Man₉GlcNAc₂Asn	Fru	Man₉-β-D-GlcNAc(1-?)-D-Fru	4.7	36
122	β-D-GlcNAc(1-?)-D-Glc	Endo-β-N-acetyl-glucosaminidase *Arthrobacter protophormiae*	Man₉GlcNAc₂Asn	Glc	Man₉-β-D-GlcNAc(1-?)-D-Glc	61	36
123	β-D-GlcNAc(1-?)-D-Xyl	Endo-β-N-acetyl-glucosaminidase *Arthrobacter protophormiae*	Man₉GlcNAc₂Asn	Xyl	Man₉-β-D-GlcNAc(1-?)-D-Xyl	16	36
124	β-D-GlcNAc(1-2)-β-L-Fuc-OMe	Endo-β-N-acetyl-glucosaminidase *Arthrobacter protophormiae*	Man₉GlcNAc₂Asn	β-L-Fuc-OMe	Man₉-β-D-GlcNAc(1-2)-β-L-Fuc-OMe	34	185
125	β-D-GlcNAc(1-3)-α-D-Glc	Lysozyme Hen egg white	GlcNAc₂	{α-D-Glc(1-4)}₄-α-D-Glc-O*p*NP	β-D-GlcNAc(1-3)-{α-D-Glc(1-4)}₃-α-D-Glc-O*p*NP	5	186

Table 2 (continued)

Entry	Linkage	Enzyme/Source	Donor	Acceptor	Product	Yield (%)	Reference
126	β-D-GlcNAc(1-3)-α-D-Glc	Lysozyme Hen egg-white	GlcNAc$_2$	α-D-Glc(1-4)-α-D-Glc-OpNP	β-D-GlcNAc(1-3)-α-D-Glc(1-4)-α-D-Glc-OpNP	8.5	187
127	β-D-GlcNAc(1-3)-α-D-GlcNAc-O	Lysozyme Hen egg-white	GlcNAc$_2$	αpNPGlcNAc	β-D-GlcNAc(1-3)-α-D-GlcNAc-OpNP	1.9	187
128	β-D-GlcNAc(1-4)-α-D-Glc	Lysozyme Hen egg-white	GlcNAc$_2$	α-D-Glc(1-4)-α-D-Glc-OpNP	β-D-GlcNAc(1-4)-α-D-Glc(1-4)-α-D-Glc-OpNP	4.2	187
129	β-D-GlcNAc(1-4)-α-D-GlcNAc-O	Lysozyme Hen egg-white	GlcNAc$_2$	αpNPGlcNAc	β-D-GlcNAc(1-3)-α-D-GlcNAc-OpNP	0.6	187
130	β-D-GlcNAc(1-4)-α-D-Glc-O	Endo-β-N-acetyl-glucosaminidase *Arthrobacter protophormiae*	Man$_9$GlcNAc$_2$Asn	αpNPGlc	Man$_9$-β-D-GlcNAc(1-4)-α-D-Glc-OpNP	75	188
131	β-D-GlcNAc(1-4)-β-D-2dGlc-O	Lysozyme Hen egg-white	GlcNAc$_2$	βpNP2dGlc	β-D-GlcNAc(1-4)-β-D-2dGlc-OpNP	3.4	189
132	β-D-GlcNAc(1-4)-β-D-Glc	Endo-β-N-acetyl-glucosaminidase *Arthrobacter protophormiae*	Man$_9$GlcNAc$_2$Asn	TIN(Glc)AS	TIN[Man$_9$GlcNAcGlc]AS	39	190
133	β-D-GlcNAc(1-4)-β-D-Glc	Endo-β-N-acetyl-glucosaminidase *Arthrobacter protophormiae*	Man$_6$GlcNAc$_2$Asn	β-D-Glc(1-6)-D-Glc	Man$_6$-β-D-GlcNAc(1-4)-β-D-Glc(1-6)-D-Glc	NA	30, 31
134	β-D-GlcNAc(1-4)-β-D-Glc-O	Lysozyme Hen egg-white	GlcNAc$_4$	βpNPGlc	β-D-GlcNAc(1-4)-β-D-Glc-OpNP	49	189
135	β-D-GlcNAc(1-4)-β-D-GlcA	Hyaluronidase Bovine testes	Hyaluronic acid	{β-D-GlcA(1-3)-β-D-GlcNAc(1-4)}$_3$-PA	{β-D-GlcA(1-3)-β-D-GlcNAc(1-4)}$_{(4-11)}$-PA	72	191

#	Donor	Enzyme	Acceptor	Product	Yield	Ref.
136	β-D-GlcNAc(1-4)-β-D-GlcA	Hyaluronidase Bovine testes	Chondroitin 6-O-SO$_3$H	{β-D-GlcA(1-3)-β-D-GlcNAc(1-4)}$_3$-PA	β-D-GlcA(1-3)-β-D-(6-O-SO$_3$H)-GalNAc(1-4)-{β-D-GlcA(1-3)-β-D-GlcNAc}$_3$-PA	2.7	191
137	β-D-GlcNAc(1-4)-β-D-GlcA	Hyaluronidase Bovine testes	Hyaluronic acid	{β-D-GlcA(1-3)-β-D-GlcNAc(1-4)}$_3$-PA	{β-D-GlcA(1-3)-β-D-GlcNAc(1-4)}$_{(1-4)}$-{β-D-GlcA(1-3)-β-D-GlcNAc}$_3$-PA	45	191
138	β-D-GlcNAc(1-4)-β-D-GlcA	Hyaluronidase Bovine testes	Hyaluronic acid	{β-D-GlcA(1-3)-β-D-(6-O-SO$_3$H)-GalNAc(1-4)}$_3$-PA	{β-D-GlcA(1-3)-β-D-GlcNAc(1-4)}$_{(1-3)}$-{β-D-GlcA(1-3)-β-D-GlcNAc}$_3$-PA	35	191
139	β-D-GlcNAc(1-4)-β-D-GlcA	Hyaluronidase Bovine testes	Chondroitin 4-O-SO$_3$H	{β-D-GlcA(1-3)-β-D-GlcNAc(1-4)}$_3$-PA	{β-D-GlcA(1-3)-β-D-(4-O-SO$_3$H)-GalNAc(1-4)}$_{(1-2)}$-{β-D-GlcA(1-3)-β-D-GlcNAc}$_3$-PA	10	191
140	β-D-GlcNAc(1-4)-β-D-GlcA	Hyaluronidase Bovine testes	Chondroitin	{β-D-GlcA(1-3)-β-D-GlcNAc(1-4)}$_3$-PA	{β-D-GlcA(1-3)-β-D-GalNAc(1-4)}$_{(1-5)}$-{β-D-GlcA(1-3)-β-D-GlcNAc}$_3$-PA	20	191
141	β-D-GlcNAc(1-4)-β-D-GlcNAc	Endo-β-N-acetyl-glucosaminidase *Arthrobacter protophormiae*	Man$_6$GlcNAc$_2$Asn	EEKYN[GlcNAc]LTSVL	EEKYN[Man$_6$GlcNAc$_2$]LTSVL	NA	32
142	β-D-GlcNAc(1-4)-β-D-GlcNAc	Chitinase *Nocardia orientalis*	GlcNAc$_5$	GlcNAc$_5$	GlcNAc$_7$	23	192
143	β-D-GlcNAc(1-4)-β-D-GlcNAc	Endo-β-N-acetyl-glucosaminidase *Mucor hiemalis*	STF-GP	CSN[GlcNAc]LST	CSN[(NeuAcGal-GlcNAcMan)$_2$-ManGlcNAc$_2$]LST	8.5	193
144	β-D-GlcNAc(1-4)-β-D-GlcNAc	Chitinase *Bacillus* sp.	GlcNAc$_2$-oxazoline	GlcNAc$_2$-oxazoline	Chitin	100	19

Table 2 (continued)

Entry	Linkage	Enzyme/Source	Donor	Acceptor	Product	Yield (%)	Reference
145	β-D-GlcNAc(1-4)-β-D-GlcNAc	Endo-β-N-acetyl-glucosaminidase Arthrobacter protophormiae	Man$_9$GlcNAc$_2$Asn	TIN[GlcNAc]AS	TIN[Man$_9$GlcNAc$_2$]AS	25	194
146	β-D-GlcNAc(1-4)-β-D-GlcNAc	Endo-β-N-acetyl-glucosaminidase Arthrobacter protophormiae	Man$_9$GlcNAc$_2$Asn	TIN[GlcNAc-CH$_2$-]AS	TIN[Man$_9$GlcNAc$_2$-CH$_2$-]AS	26	194
147	β-D-GlcNAc(1-4)-β-D-GlcNAc	Endo-β-N-acetyl-glucosaminidase Arthrobacter protophormiae	Man$_6$GlcNAc$_2$Asn	GlcNAc-RNAse B	Man$_6$GlcNAc$_2$-RNAse B	NA	32
148	β-D-GlcNAc(1-4)-β-D-GlcNAc	Endo-β-N-acetyl-glucosaminidase Arthrobacter protophormiae	Man$_6$GlcNAc$_2$	β-D-GlcNAc-Asn	Man$_6$GlcNAc$_2$Asn	NA	32
149	β-D-GlcNAc(1-4)-β-D-GlcNAc	Endo-β-N-acetyl-glucosaminidase Arthrobacter protophormiae	Man$_6$GlcNAc$_2$Asn	GlcNAc$_2$	Man$_6$GlcNAc-β-D-GlcNAc(1-4)-β-D-GlcNAc(1-4)-D-GlcNAc	NA	30, 31
150	β-D-GlcNAc(1-4)-β-D-GlcNAc	Endo-β-N-acetyl-glucosaminidase Mucor hiemalis	Asn-GlcNAc$_2$Man$_6$	CSN[GlcNAc]LST	CSN[Man$_6$GlcNAc$_2$]LST	NA	193
151	β-D-GlcNAc(1-4)-β-D-GlcNAc	Endo-β-N-acetyl-glucosaminidase Mucor hiemalis	ASTF-GP	CSN[GlcNAc]LST	CSN[(GalGlcNAcMan)$_2$-ManGlcNAc$_2$]LST	NA	193
152	β-D-GlcNAc(1-4)-β-D-GlcNAc	Chitinase Nocardia orientalis	GlcNAc$_4$	GlcNAc$_4$	GlcNAc$_6$	21	192

#	Donor	Enzyme	Acceptor	Product	Yield	Ref.
153	β-D-GlcNAc(1-4)-β-D-GlcNAc	Lysozyme Hen egg white	GlcNAc$_2$	GlcNAc$_7$	12	14
154	β-D-GlcNAc(1-4)-β-D-GlcNAc	Chitinase *Trichoderma reesei* KDR-11	GlcNAc$_4$	GlcNAc$_6$	40	14
155	β-D-GlcNAc(1-4)-β-D-GlcNAc	Lysozyme Hen egg white	GlcNAc$_2$	GlcNAc$_6$	17	14
156	β-D-GlcNAc(1-4)-β-D-GlcNAc	Lysozyme Hen egg white	GlcNAc$_2$	GlcNAc$_5$	3.3	14
157	β-D-GlcNAc(1-4)-β-D-GlcNAc-CH$_2$-	Endo-β-N-acetyl-glucosaminidase *Arthrobacter protophormiae*	Man$_9$GlcNAc$_2$Asn	TIN[Man$_9$GlcNAc$_2$-CH$_2$-]AS	26	35
158	β-D-GlcNAc(1-4)-β-D-GlcNAc-O	Lysozyme Hen egg white	GlcNAc$_5$	{β-D-GlcNAc(1-4)}$_3$-β-D-GlcNAc-O*p*NP	18	195, 196
159	β-D-GlcNAc(1-4)-β-D-GlcNAc-O	Lysozyme Hen egg white	GlcNAc$_5$	{β-D-GlcNAc(1-4)}$_4$-β-D-GlcNAc-O*p*NP	26	195, 196
160	β-D-GlcNAc(1-4)-β-D-GlcNAc-O	Lysozyme Hen egg-white	GlcNAc$_4$	β-D-GlcNAc(1-4)-β-D-GlcNAc-O*p*NP	56	189
161	β-D-GlcNAc(1-4)-β-D-GlcNAc-O	Lysozyme Hen egg-white	GlcNAc$_2$	β-D-GlcNAc(1-4)-β-D-GlcNAc-O*p*NP	6.5	187
162	β-D-GlcNAc(1-4)-β-D-GlcNAc-O	Endo-β-N-acetyl-glucosaminidase *Arthrobacter protophormiae*	Man$_9$GlcNAc$_2$Asn	Man$_9$-β-D-GlcNAc(1-4)-β-D-GlcNAc-O-(CH$_2$)$_3$CH=CH$_2$	84	33

29 Glycosidase-Catalysed Oligosaccharide Synthesis

Table 2 (continued)

Entry	Linkage	Enzyme/Source	Donor	Acceptor	Product	Yield (%)	Reference
163	β-D-GlcNAc(1-4)-β-D-GlcNAc-O	Endo-β-N-acetyl-glucosaminidase *Arthrobacter protophormiae*	$Man_9GlcNAc_2Asn$	β-D-GlcNAc-O-$(CH_2)_3$-NHCOCH=CH_2	Man_9-β-D-GlcNAc(1-4)-β-D-GlcNAc-O-$(CH_2)_3$NHCOCH=CH_2	70	33
164	β-D-GlcNAc(1-4)-β-D-GlcNAc-O	Endo-β-N-acetyl-glucosaminidase *Arthrobacter protophormiae*	$Man_9GlcNAc_2Asn$	β-D-GlcNAc-OBn	Man_9-β-D-GlcNAc(1-4)-β-D-GlcNAc-OBn	67	33
165	β-D-GlcNAc(1-4)-β-D-GlcNAc-O	Endo-β-N-acetyl-glucosaminidase *Arthrobacter protophormiae*	$Man_9GlcNAc_2Asn$	βMuGlcNAc	Man_9-β-D-GlcNAc(1-4)-β-D-GlcNAc-OMu	66	33
166	β-D-GlcNAc(1-4)-β-D-GlcNAc-O	Endo-β-N-acetyl-glucosaminidase *Arthrobacter protophormiae*	$Man_9GlcNAc_2Asn$	β*p*NPGlcNAc	Man_9-β-D-GlcNAc(1-4)-β-D-GlcNAc-O*p*NP	33	33
167	β-D-GlcNAc(1-4)-β-D-GlcNAc-O	Endo-β-N-acetyl-glucosaminidase *Arthrobacter protophormiae*	$Man_9GlcNAc_2Asn$	β-D-GlcNAc-O-$(CH_2)_6NH_2$	Man_9-β-D-GlcNAc(1-4)-β-D-GlcNAc-O-$(CH_2)_6NH_2$	81	33
168	β-D-GlcNAc(1-4)-β-D-GlcNAc-O	Endo-β-N-acetyl-glucosaminidase *Arthrobacter protophormiae*	$Man_9GlcNAc_2Asn$	β-D-GlcNAc-O-$CH_2CH=CH_2$	Man_9-β-D-GlcNAc(1-4)-β-D-GlcNAc-O-$CH_2CH=CH_2$	81	33
169	β-D-GlcNAc(1-4)-β-D-GlcNAc-S	Endo-β-N-acetyl-glucosaminidase *Arthrobacter protophormiae*	$Man_9GlcNAc_2Asn$	β-D-GlcNAc-S-CH_2-CONHCH$_2$CH(OMe)$_2$	Man_9-β-D-GlcNAc(1-4)-β-D-GlcNAc-S-$CH_2CONHCH_2$-CH(OMe)$_2$	81	33

#		Enzyme				Yield	Ref.
170	β-D-GlcNAc(1-4)-β-D-GlcNAc-S	Endo-β-N-acetyl-glucosaminidase *Arthrobacter protophormiae*	Man$_9$GlcNAc$_2$Asn	β-D-GlcNAc-S-(CH$_2$)$_3$CH$_3$	Man$_9$-β-D-GlcNAc(1-4)-β-D-GlcNAc-S-(CH$_2$)$_3$CH$_3$	78	33
171	β-D-GlcNAc(1-4)-β-D-GlcNAc-S	Endo-β-N-acetyl-glucosaminidase *Arthrobacter protophormiae*	Man$_9$GlcNAc$_2$Asn	β-D-GlcNAc-S-CH$_2$CN	Man$_9$-β-D-GlcNAc(1-4)-β-D-GlcNAc-S-CH$_2$CN	83	33
172	β-D-GlcNAc(1-4)-β-D-Man-O	Lysozyme Hen egg-white	GlcNAc$_2$	βpNPMan	β-D-GlcNAc(1-4)-β-D-Man-OpNP	10	187
173	β-D-GlcNAc(1-4)-D-Glc	Endo-β-N-acetyl-glucosaminidase *Arthrobacter protophormiae*	Man$_6$GlcNAc$_2$Asn	Glc	Man$_6$-β-D-GlcNAc(1-4)-D-Glc	NA	30, 31
174	β-D-GlcNAc(1-4)-D-GlcNAc	Endo-β-N-acetyl-glucosaminidase *Mucor hiemalis*	STF-GP	GlcNAc-Asn-FMOC	(NeuAcGalGlcNAcMan)$_2$-ManGlcNAc$_2$-Asn-FMOC	20	197
175	β-D-GlcNAc(1-4)-D-GlcNAc	Endo-β-N-acetyl-glucosaminidase *Mucor hiemalis*	ASTF-GP	GlcNAc	(GalGlcNAcMan)$_2$-ManGlcNAc$_2$	NA	198
176	β-D-GlcNAc(1-4)-D-GlcNAc	Endo-β-N-acetyl-glucosaminidase *Mucor hiemalis*	ASTP-GP	β-D-GlcNAc(1-4)-D-GlcNAc-PA	(GalGlcNAcMan)$_2$-ManGlcNAc$_3$-PA	NA	198
177	β-D-GlcNAc(1-4)-D-GlcNAc	Endo-β-N-acetyl-glucosaminidase *Mucor hiemalis*	STF-GP	ASTTTN[GlcNAc]YT	ASTTTN[(NeuAcGal-GlcNAcMan)$_2$-ManGlcNAc$_2$]YT	9	199
178	β-D-GlcNAc(1-4)-D-GlcNAc	Endo-β-N-acetyl-glucosaminidase *Mucor hiemalis*	Man$_6$GlcNAc$_2$Asn	β-D-GlcNAc(1-4)-D-GlcNAc-PA	Man$_6$GlcNAc$_3$-PA	NA	198
179	β-D-GlcNAc(1-4)-D-GlcNAc	Endo-β-N-acetyl-glucosaminidase *Arthrobacter protophormiae*	Man$_9$GlcNAc$_2$Asn	GlcNAc	Man$_9$GlcNAc$_2$	91	185

Table 2 (continued)

Entry	Linkage	Enzyme/Source	Donor	Acceptor	Product	Yield (%)	Reference
180	β-D-GlcNAc(1-4)-D-GlcNAc	Endo-β-N-acetyl-glucosaminidase *Mucor hiemalis*	STF-GP	CSN[GlcNAc]LSTCVL GKSNELH-KLNTYPRT DVGAGTP	CSN[(NeuAcGal-GlcNAcMan)$_2$-ManGlcNAc$_2$]LS-TCVLGKS-NELHKLN-TYPRTD-VGAGTP	8.5	200
181	β-D-GlcNAc(1-4)-D-GlcNAc	Endo-β-N-acetyl-glucosaminidase *Arthrobacter protophormiae*	Man$_9$GlcNAc$_2$Asn	D-GlcNAc	Man$_9$-β-D-GlcNAc(1-4)-D-GlcNAc	85	33
182	β-D-GlcNAc(1-4)-D-GlcNAc	Endo-β-N-acetyl-glucosaminidase *Mucor hiemalis*	ASTF-GP	GlcNAc-Asn-FMOC	(GalGlcNAcMan)$_2$-ManGlcNAc$_2$-Asn-FMOC	15	197
183	β-D-GlcNAc(1-4)-D-GlcNAc	Endo-β-N-acetyl-glucosaminidase *Mucor hiemalis*	STF-GP	Dansyl-ASTTTN[GlcNAc]YT	Dansyl-ASTTTN[(NeuAcGal-GlcNAcMan)$_2$-ManGlcNAc$_2$]YT	16	199
184	β-D-GlcNAc(1-4)-D-GlcNAc	Endo-β-N-acetyl-glucosaminidase *Mucor hiemalis*	Man$_6$-GP	GlcNAc-Asn-FMOC	Man$_6$GlcNAc$_2$-Asn-FMOC	8	197
185	β-D-GlcNAc(1-4)-D-GlcNAc	Endo-β-N-acetyl-glucosaminidase *Mucor hiemalis*	ASTF-GP	Dansyl-ASTTTN[GlcNAc]YT	Dansyl-ASTTTN[(GalGlcNAcMan)$_2$-ManGlcNAc$_2$]YT	15	199
186	β-D-GlcNAc(1-4)-D-GlcNAc	Endo-β-N-acetyl-glucosaminidase *Mucor hiemalis*	Man$_6$-GP	IN[GlcNAc]ATL	IN[Man$_6$GlcNAc$_2$]ATL	4	197
187	β-D-GlcNAc(1-4)-D-GlcNAc	Endo-β-N-acetyl-glucosaminidase *Mucor hiemalis*	STF-GP	IN[GlcNAc]ATL	IN[(NeuAcGal-GlcNAcMan)$_2$-ManGlcNAc$_2$]ATL	10	197

29.6 Recent Developments and New Directions

#	Linkage	Enzyme	Donor	Acceptor	Product	Yield (%)	Ref
188	β-D-GlcNAc(1-4)-D-GlcNAc	Endo-β-N-acetyl-glucosaminidase *Mucor hiemalis*	ASTF-GP	IN[GlcNAc]ATL	IN[(GalGlcNAcMan)$_2$-ManGlcNAc$_2$]ATL	7	197
189	β-D-GlcNAc(1-4)-D-Man	Endo-β-N-acetyl-glucosaminidase *Arthrobacter protophormiae*	Man$_6$GlcNAc$_2$Asn	Man	Man$_6$-β-D-GlcNAc(1-4)-D-Man	NA	30, 31
190	β-D-GlcNAc(1-4)-D-Man	Endo-β-N-acetyl-glucosaminidase *Arthrobacter protophormiae*	Man$_9$GlcNAc$_2$Asn	Man	Man$_9$-β-D-GlcNAc(1-4)-D-Man	83	185
191	β-D-GlcNAc(1-4)-D-Man	Lysozyme Hen egg-white	GlcNAc$_2$	Man	β-D-GlcNAc(1-4)-D-Man	21	187
192	β-D-GlcNAc(1-4)-L-Fuc	Endo-β-N-acetyl-glucosaminidase *Arthrobacter protophormiae*	Man$_9$GlcNAc$_2$Asn	L-Fuc	Man$_9$-β-D-GlcNAc(1-2)-L-Fuc	26	185
193	β-D-GlcNAc(1-4)-L-Gal	Endo-β-N-acetyl-glucosaminidase *Arthrobacter protophormiae*	Man$_9$GlcNAc$_2$Asn	L-Gal	Man$_9$-β-D-GlcNAc(1-2)-L-Gal	19	185
194	β-D-GlcNAc-O	Endo-β-N-acetyl-glucosaminidase *Arthrobacter protophormiae*	Man$_9$GlcNAc$_2$Asn	MeOH	Man$_9$GlcNAc-OMe	64	33
195	β-D-GlcNAc-O	Endo-β-N-acetyl-glucosaminidase *Arthrobacter protophormiae*	Man$_9$GlcNAc$_2$Asn	PrOH	Man$_9$GlcNAc-OPr	8	33
196	β-D-GlcNAc-O	Endo-β-N-acetyl-glucosaminidase *Arthrobacter protophormiae*	Man$_9$GlcNAc$_2$Asn	EtOH	Man$_9$GlcNAc-OEt	47	33

Table 2 (*continued*)

Entry	Linkage	Enzyme/Source	Donor	Acceptor	Product	Yield (%)	Reference
197	β-D-GlcNAc-O	Endo-β-N-acetyl-glucosaminidase *Arthrobacter protophormiae*	$Man_9GlcNAc_2Asn$	iPrOH	$Man_9GlcNAc$-OiPr	9.6	33
198	β-D-GlcNAc-O	Endo-β-N-acetyl-glucosaminidase *Arthrobacter protophormiae*	$Man_9GlcNAc_2Asn$	Glycerol	$Man_9GlcNAc$-(1-?)-Glyceryl	56	33
199	β-D-Xyl(1-4)-β-D-Xyl	Xylanase *Trichoderma viridae*	β-D-Xyl(1-4)-β-D-Xyl-F	β-D-Xyl(1-4)-β-D-Xyl-F	{β-D-Xyl(1-4)}$_{(2-12)}$-D-Xyl	72	201

Table 3. Enzymatically synthesised alkyl glycosides.

Entry	Linkage	Enzyme/Source	Donor	Acceptor	Product	Yield (%)	Reference
1	α-D-Gal-O	α-Galactosidase Aspergillus oryzae	αpNPGal	Chanoclavine	α-D-Gal-O-chanoclavine	24	202
2	α-D-Gal-O	α-Galactosidase Aspergillus oryzae	αpNPGal	Elymoclavine	α-D-Gal-O-elymoclavine	23	202
3	α-D-Gal-O	α-Galactosidase Coffee bean	Raffinose	HOAll	α-D-Gal-OAll	NA	44
4	α-D-Gal-O	α-Galactosidase Trichoderma reesei	αpNPGal	MeOH	α-D-Gal-OMe	NA	203
5	α-D-Gal-O	α-Galactosidase Trichoderma reesei	αpNPGal	EtOH	α-D-Gal-OEt	NA	203
6	α-D-Gal-O	α-Galactosidase Trichoderma reesei	αpNPGal	PrOH	α-D-Gal-OPr	NA	203
7	α-D-Gal-O	α-Galactosidase Trichoderma reesei	αpNPGal	iPrOH	α-D-Gal-OiPr	NA	203
8	α-D-Gal-O	α-Galactosidase Trichoderma reesei	αpNPGal	HO-(CH$_2$)$_3$CH$_3$	α-D-Gal-O-(CH$_2$)$_3$CH$_3$	NA	203
9	α-D-Gal-O	α-Galactosidase Trichoderma reesei	αpNPGal	HO-CH(CH$_3$)CH$_2$CH$_3$	α-D-Gal-O-CH(CH$_3$)CH$_2$CH$_3$	NA	203
10	α-D-Gal-O	α-Galactosidase Aspergillus niger	Gal	HO-(CH$_2$)$_6$OH	α-D-Gal-O(CH$_2$)$_6$OH	47	204

Table 3 (continued)

Entry	Linkage	Enzyme/Source	Donor	Acceptor	Product	Yield (%)	Reference
11	α-D-Gal-O	α-Galactosidase *Aspergillus niger*	Gal	HO-(CH$_2$)$_3$CH=CH$_2$	α-D-Gal-O(CH$_2$)$_3$CH=CH$_2$	37	204
12	α-D-Gal-O	α-Galactosidase *Aspergillus oryzae*	α*p*NPGal	Lysergol	α-D-Gal-O-lysergol	6	205
13	α-D-Gal-O	α-Galactosidase *Aspergillus oryzae*	α*p*NPGal	9,10-Dihydrolysergol	α-D-Gal-O-9,10-dihydrolysergol	11	205
14	α-D-Gal-O	α-Galactosidase Coffee bean	Raffinose	Ser-(N-Boc)-OMe	α-D-Gal-O-Ser-(N-Boc)-OMe	3	206
15	α-D-Gal-O	α-*Galactosidase* Coffee bean	Raffinose	Ser-(N-Aloc)-OMe	α-D-Gal-O-Ser-(N-Aloc)-OMe	8	206
16	α-D-GalNAc-O	α-N-acetyl-galactosaminidase Bovine liver	GalNac	L-Ser	α-D-GalNAc-O-L-Ser	10	207
17	α-D-GalNAc-O	α-N-acetyl-galactosaminidase Bovine liver	GalNac	D-Ser	α-D-GalNAc-O-D-Ser	NA	207
18	α-D-GalNAc-O	α-N-acetyl-galactosaminidase Bovine liver	GalNac	L-Thr	α-D-GalNAc-O-L-Thr	NA	207
19	α-D-GalNAc-O	α-N-acetyl-galactosaminidase Bovine liver	GalNac	D-Thr	α-D-GalNAc-O-D-Thr	NA	207

#		Enzyme		Acceptor	Product	Yield	Ref.
20	α-D-GalNAc-O	α-Galactosaminidase Aspergillus oryzae	αpNPGalNAc	Thr-(N-Ac)-OMe	α-D-GalNAc-O-Thr-(N-Ac)-OMe	29	208
21	α-D-GalNAc-O	α-Galactosaminidase Aspergillus oryzae	αpNPGalNAc	Ser-(N-COOMe)-OMe	α-D-GalNAc-O-Ser-(N-COOMe)-OMe	50	208
22	α-D-GalNAc-O	α-Galactosaminidase Aspergillus oryzae	αpNPGalNAc	Ser-(N-COCH$_2$CH=CH$_2$)-OMe	α-D-GalNAc-O-Ser-(N-COCH$_2$CH=CH$_2$)-OMe	10–50	208
23	α-D-GalNAc-O	α-Galactosaminidase Aspergillus oryzae	αpNPGalNAc	Ser-(N-COCH$_3$)-OMe	α-D-GalNAc-O-Ser-(N-COCH$_3$)-OMe	10–50	208
24	α-D-GalNAc-O	α-Galactosaminidase Aspergillus oryzae	αpNPGalNAc	Ser-(N-COCH$_2$C(Cl)$_3$)-OMe	α-D-GalNAc-O-Ser-(N-COCH$_2$C(Cl)$_3$)-OMe	10–50	208
25	α-D-GalNAc-O	α-Galactosaminidase Aspergillus oryzae	αpNPGalNAc	Ser-(N-COC(CH$_3$)$_3$)-OMe	α-D-GalNAc-O-Ser-(N-CO(CH$_3$)$_3$)-OMe	10–50	208
26	α-D-Glc-O	Glucoamylase Rhizopus oryzae	Glc	HO-(CH$_2$)$_4$OH	α-D-Glc-O-(CH$_2$)$_4$OH	25	209
27	α-D-Glc-O	Glucoamylase Rhizopus oryzae	Glc	CH$_3$CHOHCHOHCH$_3$	α-D-Glc-O-CH(CH$_3$)CHOHCH$_3$	2	209
28	α-D-Glc-O	α-Glucosidase Talaromyces duponti	Maltodextrins	HO-(CH$_2$)$_3$CH$_3$	α-D-Glc-O-(CH$_2$)$_3$CH$_3$	NA	210
29	α-D-Glc-O	α-Glucosidase Talaromyces duponti	Maltose	HO-(CH$_2$)$_3$CH$_3$	α-D-Glc-O-(CH$_2$)$_3$CH$_3$	23	211
30	α-D-Glc-O	α-Glucosidase Talaromyces duponti	Maltose	HO-(CH$_2$)$_4$CH$_3$	α-D-Glc-O-(CH$_2$)$_4$CH$_3$	18	211

Table 3 (*continued*)

Entry	Linkage	Enzyme/Source	Donor	Acceptor	Product	Yield (%)	Reference
31	α-D-Glc-O	α-Glucosidase *Talaromyces duponti*	Maltose	HO-(CH$_2$)$_5$CH$_3$	α-D-Glc-O-(CH$_2$)$_5$CH$_3$	11	211
32	α-D-Glc-O	α-Glucosidase *Talaromyces duponti*	Maltose	HO-(CH$_2$)$_6$CH$_3$	α-D-Glc-O-(CH$_2$)$_6$CH$_3$	6.3	211
33	α-D-Glc-O	α-Glucosidase *Talaromyces duponti*	Maltose	HO-(CH$_2$)$_7$CH$_3$	α-D-Glc-O-(CH$_2$)$_7$CH$_3$	1.3	211
34	α-D-Glc-O	α-Glucosidase Rice, type V	Maltose	9,10-Dihydrolysergol	α-D-Glc-O-9,10-dihydrolysergol	3	205
35	α-D-Man-O	α-Mannosidase Jack bean	Man	L-Ser	α-D-Man-O-L-Ser	10	207
36	α-D-Man-O	α-Mannosidase Jack bean	Man	D-Ser	α-D-Man-O-D-Ser	NA	207
37	α-D-Man-O	α-Mannosidase Jack bean	Man	L-Thr	α-D-Man-O-L-Thr	NA	207
38	α-D-Man-O	α-Mannosidase Jack bean	Man	D-Thr	α-D-Man-O-D-Thr	NA	207
39	α-D-Man-O	α-Mannosidase Jack bean	Man	Cyanuricacid	α-D-Man-O-cyanuricacid	4	212
40	α-D-Man-O	α-Mannosidase Jack bean	Man	Pentaerythritol	α-D-Man-O-pentaerythritol	8	212
41	α-D-Man-O	α-Mannosidase Jack bean	α*p*NPMan	Chanoclavine	α-D-Man-O-chanoclavine	28	202
42	α-D-Man-O	α-Mannosidase Jack bean	Man	Chanoclavine	α-D-Man-O-chanoclavine	11	202
43	α-D-Man-O	α-Mannosidase Jack bean	Man	Elymoclavine	α-D-Man-O-elymoclavine	18	202

44	α-D-Man-O	α-Mannosidase Jack bean	Man	Ergometrine	α-D-Man-O-ergometrine	13	202
45	α-D-Man-O	α-Mannosidase Almond	αpNPMan	Chanoclavine	α-D-Man-O-chanoclavine	18	202
46	α-D-Man-O	α-Mannosidase Jack bean	αpNPMan	5-phenyl-1-pentanol	α-D-Man-O-5-phenyl-1-pentyl	68	213
47	α-D-Man-O	α-Mannosidase Almond	αpNPMan	Elymoclavine	α-D-Man-O-elymoclavine	3	205
48	α-D-Man-O	α-Mannosidase Almond	αpNPMan	Lysergol	α-D-Man-O-lysergol	3	205
49	α-D-Man-O	α-Mannosidase Almond	αpNPMan	9,10-Dihydrolysergol	α-D-Man-O-9,10-dihydrolysergol	2	205
50	α-L-Ara-O	β-Galactosidase Bovine liver	α-L-pNPAra	Monobenzyl ether 1,3,5-benzene-trimethanol	monobenzyl ether 1,3-(α-L-Ara-O)-1,3,5-benzene-trimethanol	3	107
51	α-L-Fuc-O	α-L-Fucosidase Limpet	αpNPFuc	Chanoclavine	α-L-Fuc-O-chanoclavine	2	202
52	β-D-Fruf-O	Invertase Bakers yeast	Fru	HO-$(CH_2)_3CH_3$	β-D-Fruf-O-$(CH_2)_3CH_3$	3	214
53	β-D-Fruf-O	Invertase Bakers yeast	Sucrose	HO-$(CH_2)_3CH_3$	β-D-Fruf-O-$(CH_2)_3CH_3$	1	214
54	β-D-Fruf-O	Invertase Saccharomyces cervisiae	Sucrose	HO-$(CH_2)_3CH_3$	β-D-Fruf-O-$(CH_2)_3CH_3$	40	215
55	β-D-Gal(1-2)-DNJ	β-Galactosidase Bacillus circulans	Lactose	DNJ	β-D-Gal(1-2)-DNJ	6	216
56	β-D-Gal(1-3)-DNJ	β-Galactosidase Bacillus circulans	Lactose	DNJ	β-D-Gal(1-3)-DNJ	20	216
57	β-D-Gal(1-4)-DNJ	β-Galactosidase Bacillus circulans	Lactose	DNJ	β-D-Gal(1-4)-DNJ	26	216
58	β-D-Gal(1-4)-O	Rhodotorula lactosa cells	Lactose	CalysteginB2	β-D-Gal(1-4)-O-calysteginB2	10	217

Table 3 (*continued*)

Entry	Linkage	Enzyme/Source	Donor	Acceptor	Product	Yield (%)	Reference
59	β-D-Gal(1-6)-DNJ	β-Galactosidase *Bacillus circulans*	Lactose	DNJ	β-D-Gal(1-6)-DNJ	7.2	216
60	β-D-Gal-O	β-Galactosidase *Aspergillus oryzae*	βpNPGal	Cyanuricacid	β-D-Gal-O-cyanuricacid	20	212
61	β-D-Gal-O	β-Galactosidase *Aspergillus oryzae*	βpNPGal	Pentaerythritol	β-D-Gal-O-pentaerythritol	12	212
62	β-D-Gal-O	β-Galactosidase *Aspergillus oryzae*	βpNPGal	p-bis(hydroxymethyl)benzene	β-D-Gal-O-p-bis(hydroxymethyl)benzene	13	212
63	β-D-Gal-O	β-Galactosidase *Aspergillus oryzae*	βoNPGal	Serine	β-D-Gal-O-Ser	10	63
64	β-D-Gal-O	β-Hexosaminidase *Aspergillus oryzae*	βpNPGlcNAc	p-COOCH$_3$-m-NO$_2$-Bn-OH	β-D-GlcNAc-O-p-COOCH$_3$-m-NO$_2$	23	218
65	β-D-Gal-O	β-Galactosidase *Escherichia coli*	Lactose	Butan-2-ol	β-D-Gal-O-CH(CH$_3$)CH$_2$CH$_3$	32 (R: 6 ee)	219
66	β-D-Gal-O	β-Galactosidase *Escherichia coli*	Lactose	(R)-(−)-Butan-2-ol	(R)-β-D-Gal-O-CH(CH$_3$)CH$_2$CH$_3$	32	219
67	β-D-Gal-O	β-Galactosidase *Escherichia coli*	Lactose	Propane-1,2-diol	β-D-Gal-O-CH$_2$CHOHCH$_3$	26 (R: 8 ee)	219
68	β-D-Gal-O	β-Galactosidase *Escherichia coli*	Lactose	Propane-1,2-diol	β-D-Gal-O-CH(CH$_3$)CH$_2$OH	20 (R: 13 ee)	219

69	β-D-Gal-O	β-Galactosidase *Escherichia coli*	Lactose	(S)-(+)-Propane-1,2-diol	β-D-Gal-O-CH₂CHOHCH₃	38	219
70	β-D-Gal-O	β-Galactosidase *Escherichia coli*	Lactose	(S)-(+)-Propane-1,2-diol	β-D-Gal-O-CH(CH₃)CH₂OH	13	219
71	β-D-Gal-O	β-Galactosidase *Escherichia coli*	Lactose	Butane-1,3-diol	β-D-Gal-O-(CH₂)₂CHOHCH₃	53 (0 ee)	219
72	β-D-Gal-O	β-Galactosidase *Escherichia coli*	Lactose	Butane-1,3-diol	β-D-Gal-O-CH(CH₃)(CH₂)₂OH	8 (R: 33 ee)	219
73	β-D-Gal-O	β-Galactosidase *Escherichia coli*	Lactose	(S)-(+)-Butane-1,3-diol	β-D-Gal-O-(CH₂)₂CHOHCH₃	54	219
74	β-D-Gal-O	β-Galactosidase *Escherichia coli*	Lactose	(S)-(+)-Butane-1,3-diol	β-D-Gal-O-CH(CH₃)(CH₂)₂OH	7	219
75	β-D-Gal-O	β-Galactosidase *Escherichia coli*	Lactose	Propane-1,3-diol	β-D-Gal-O-(CH₂)₃OH	42	219
76	β-D-Gal-O	β-Galactosidase *Escherichia coli*	Lactose	HOAll	β-D-Gal-OAll	NA	44
77	β-D-Gal-O	β-Galactosidase *Escherichia coli*	Lactose	HOBn	β-D-Gal-OBn	NA	44
78	β-D-Gal-O	β-Galactosidase *Escherichia coli*	Lactose	HO-(CH₂)₂Si(CH₃)₃	β-D-Gal-O-(CH₂)₂Si(CH₃)₃	NA	44
79	β-D-Gal-O	β-Galactosidase *Aspergillus oryzae*	βPhGal	HO-(CH₂)₂C₆H₅	β-D-Gal-O-(CH₂)₂C₆H₅	20	220

Table 3 (continued)

Entry	Linkage	Enzyme/Source	Donor	Acceptor	Product	Yield (%)	Reference
80	β-D-Gal-O	β-Galactosidase *Escherichia coli*	Lactose	cis-1,2-Cyclopentanediol	β-D-Gal-O-cis-1,2-cyclopentanediol	30 (89 de)	221
81	β-D-Gal-O	β-Galactosidase *Escherichia coli*	Lactose	cis-1,2-Cyclohexanediol	β-D-Gal-O-cis-1,2-cyclohexanediol	10 (90 de)	221
82	β-D-Gal-O	β-Galactosidase *Escherichia coli*	Lactose	cis-Norbornenediol	β-D-Gal-O-cis-norbornenediol	24 (75 de)	221
83	β-D-Gal-O	β-Galactosidase *Escherichia coli*	Lactose	cis-Norbornanediol	β-D-Gal-O-cis-norbornanediol	28 (63 de)	221
84	β-D-Gal-O	β-Galactosidase *Escherichia coli*	Lactose	4-Cyclopentane-1,4-diol	β-D-Gal-O-cis-1,2-cyclopentanediol	36 (50 de)	221
85	β-D-Gal-O	β-Galactosidase *Aspergillus oryzae*	Lactose	cis-1,2-Cyclopentanediol	β-D-Gal-O-cis-1,2-cyclopentanediol	13 (70 de)	221
86	β-D-Gal-O	β-Galactosidase *Aspergillus oryzae*	Lactose	cis-1,2-Cyclohexanediol	β-D-Gal-O-cis-1,2-cyclohexanediol	20 (38 de)	221
87	β-D-Gal-O	β-Galactosidase *Aspergillus oryzae*	βPhGal	Gitoxigenin	β-D-Gal-O-3-gitoxigenin	26	222, 223
88	β-D-Gal-O	β-Galactosidase *Aspergillus oryzae*	βPhGal	Digitoxigenin	β-D-Gal-O-3-digitoxigenin	38	222, 223
89	β-D-Gal-O	β-Galactosidase *Aspergillus oryzae*	βPhGal	16B,17β-Epoxy-17α-digitoxigenin	β-D-Gal-O-3-16B,17β-epoxy-17α-digitoxigenin	64	222, 223

90	β-D-Gal-O	β-Galactosidase *Aspergillus oryzae*	βPhGal	Strophanthidin	β-D-Gal-O-3-strophanthidin	40	222, 223
91	β-D-Gal-O	β-Galactosidase *Aspergillus oryzae*	βPhGal	MeOH	β-D-Gal-OMe	83	222, 223
92	β-D-Gal-O	β-Galactosidase *Aspergillus oryzae*	βPhGal	nBuOH	β-D-Gal-O-$(CH_2)_3CH_3$	67	222, 223
93	β-D-Gal-O	β-Galactosidase *Aspergillus oryzae*	βPhGal	iPrOH	β-D-Gal-OiPr	74	222, 223
94	β-D-Gal-O	β-Galactosidase *Aspergillus oryzae*	βPhGal	Cyclohexanol	β-D-Gal-O-cyclohexanol	25	222, 223
95	β-D-Gal-O	β-Galactosidase *Escherichia coli*	βoNPGal	$CH_3CHOHCH_2CH_3$	β-D-Gal-O-$CH(CH_3)CH_2CH_3$	73 (R: 64 ee)	224
96	β-D-Gal-O	β-Galactosidase *Escherichia coli*	βoNPGal	$CH_3CHOH(CH_2)_5CH_3$	β-D-Gal-O-$CH(CH_3)(CH_2)_5CH_3$	17 (R: 63 ee)	224
97	β-D-Gal-O	β-Galactosidase *Escherichia coli*	βoNPGal	$CH_3CHOHCH(CH_3)_2$	β-D-Gal-O-$CH(CH_3)CH(CH_3)_2$	36 (R: 80 ee)	224
98	β-D-Gal-O	β-Galactosidase *Escherichia coli*	βoNPGal	$CH_3CHOHC_6H_5$	β-D-Gal-O-$CH(CH_3)C_6H_5$	39 (R: 98 ee)	224
99	β-D-Gal-O	β-Galactosidase *Escherichia coli*	βoNPGal	$CH_3CHOHCH_2OH$	β-D-Gal-O-$CH(CH_3)CH_2OH$	20 (R: 49 ee)	224
100	β-D-Gal-O	β-Galactosidase *Escherichia coli*	βoNPGal	$CH_3CHOHCH_2OCH_3$	β-D-Gal-O-$CH(CH_3)CH_2OCH_3$	28 (R: 59 ee)	224

Table 3 (*continued*)

Entry	Linkage	Enzyme/Source	Donor	Acceptor	Product	Yield (%)	Reference
101	β-D-Gal-O	β-Galactosidase *Escherichia coli*	βoNPGal	CH$_3$CHOH(CH$_2$)$_2$-COOCH$_3$	β-D-Gal-O-CH(CH$_3$)(CH$_2$)$_2$COOCH$_3$	46 (R: 65 ee)	224
102	β-D-Gal-O	β-Galactosidase *Escherichia coli*	βoNPGal	CH$_3$CHOH(CH$_2$)$_2$-COOCH$_2$CH$_3$	β-D-Gal-O-CH(CH$_3$)(CH$_2$)$_2$COOCH$_2$CH$_3$	37 (R: 55 ee)	224
103	β-D-Gal-O	β-Galactosidase *Escherichia coli*	βoNPGal	CH$_3$CHOH(CH$_2$)$_2$-COO(CH$_2$)$_3$CH$_3$	β-D-Gal-O-CH(CH$_3$)(CH$_2$)$_2$COO(CH$_2$)$_3$CH$_3$	8 (R: 11 ee)	224
104	β-D-Gal-O	β-Galactosidase *Kluyveromyces lactis*	Lactose	HO-(CH$_2$)$_6$OH	β-D-Gal-O-(CH$_2$)$_6$OH	49	225
105	β-D-Gal-O	β-Galactosidase *Kluyveromyces lactis*	Lactose	HO-(CH$_2$)$_2$OH	β-D-Gal-O(CH$_2$)$_2$OH	40	225
106	β-D-Gal-O	β-Galactosidase *Kluyveromyces lactis*	Lactose	Glycerol	β-D-Gal-O-1-Glyceryl	60	225
107	β-D-Gal-O	β-Galactosidase *Kluyveromyces lactis*	Lactose	HOBn	β-D-Gal-OBn	21	225
108	β-D-Gal-O	β-Galactosidase *Kluyveromyces lactis*	Lactose	EtOH	β-D-Gal-OEt	36	225
109	β-D-Gal-O	β-Galactosidase *Escherichia coli*	βoNPGal	HO-(CH$_2$)$_8$OH	β-D-Gal-O-(CH$_2$)$_8$OH	92	226
110	β-D-Gal-O	β-Galactosidase *Escherichia coli*	Lactose	HO-(CH$_2$)$_8$OH	β-D-Gal-O-(CH$_2$)$_8$OH	50	226

111	β-D-Gal-O	β-Galactosidase *Escherichia coli*	βoNPGal	HO-(CH$_2$)$_{10}$OH	β-D-Gal-O-(CH$_2$)$_{10}$OH	78	226
112	β-D-Gal-O	β-Galactosidase *Escherichia coli*	βoNPGal	HO-(CH$_2$)$_7$CH$_3$	β-D-Gal-O-(CH$_2$)$_7$CH$_3$	50	226
113	β-D-Gal-O	β-Galactosidase *Escherichia coli*	βoNPGal	HO-(CH$_2$)$_9$CH$_3$	β-D-Gal-O-(CH$_2$)$_9$CH$_3$	27	226
114	β-D-Gal-O	β-Galactosidase *Kluyveromyces lactis*	Gal	HO-(CH$_2$)$_3$CH$_3$	β-D-Gal-O-(CH$_2$)$_3$CH$_3$	18	214
115	β-D-Gal-O	β-Galactosidase *Kluyveromyces lactis*	Lactose	HO-(CH$_2$)$_3$CH$_3$	β-D-Gal-O-(CH$_2$)$_3$CH$_3$	23	214
116	β-D-Gal-O	β-Galactosidase *Aspergillus oryzae*	Lactose	HO-(CH$_2$)$_3$CH$_3$	β-D-Gal-O-(CH$_2$)$_3$CH$_3$	23	214
117	β-D-Gal-O	β-Galactosidase *Aspergillus oryzae*	Lactose	Phenylethanol	β-D-Gal-O-phenylethyl	14	227
118	β-D-Gal-O	β-Galactosidase *Escherichia coli*	β*p*NPGal	HO-CH(CH$_3$)CH$_2$CH$_3$	β-D-Gal-O-CH(CH$_3$)CH$_2$CH$_3$	52	228, 229
119	β-D-Gal-O	β-Galactosidase *Escherichia coli*	β*p*NPGal	HO-C(CH$_3$)$_3$	β-D-Gal-O-C(CH$_3$)$_3$	23	228, 229
120	β-D-Gal-O	β-Galactosidase *Escherichia coli*	β*p*NPGal	5-Phenylpentanol	β-D-Gal-O-5-phenylpentyl	66	228, 230
121	β-D-Gal-O	β-Galactosidase *Escherichia coli*	β*p*NPGal	HO-(CH$_2$)$_7$CH$_3$	β-D-Gal-O-(CH$_2$)$_7$CH$_3$	82	228, 230

Table 3 (*continued*)

Entry	Linkage	Enzyme/Source	Donor	Acceptor	Product	Yield (%)	Reference
122	β-D-Gal-O	β-Galactosidase *Escherichia coli*	βpNPGal	HO-(CH$_2$)$_9$CH$_3$	β-D-Gal-O-(CH$_2$)$_9$CH$_3$	13	228, 230
123	β-D-Gal-O	β-Galactosidase *Escherichia coli*	βpNPGal	HO-(CH$_2$)$_{11}$CH$_3$	β-D-Gal-O-(CH$_2$)$_{11}$CH$_3$	15	228, 230
124	β-D-Gal-O	β-Galactosidase *Escherichia coli*	βpNPGal	1,3-di-O-Dodecylglycerol	β-D-Gal-O-1,3-di-O-dodecylglyceryl	9	228
125	β-D-Gal-O	β-Galactosidase *Escherichia coli*	βpNPGal	1,2-di-O-Dodecylglycerol	β-D-Gal-O-1,2-di-O-dodecylglyceryl	52	228
126	β-D-Gal-O	β-Galactosidase *Escherichia coli*	βpNPGal	Cholesterol	β-D-Gal-O-cholesteryl	7.2	228, 231
127	β-D-Gal-O	β-Galactosidase *Bacillus circulans*	βpNPGal	(R)-1-Phenylethanol	β-D-Gal-O-(R)-1-phenylethyl	41	228
128	β-D-Gal-O	β-Galactosidase *Bacillus circulans*	βpNPGal	(S)-1-Phenylethanol	β-D-Gal-O-(S)-1-phenylethyl	16	228
129	β-D-Gal-O	β-Galactosidase *Bacillus circulans*	βpNPGal	(R)-1-Methyloctanol	β-D-Gal-O-(R)-1-methyloctyl	36	228
130	β-D-Gal-O	β-Galactosidase *Bacillus circulans*	βpNPGal	(S)-1-Methyloctanol	β-D-Gal-O-(S)-1-methyloctyl	<5	228, 230
131	β-D-Gal-O	β-Galactosidase *Escherichia coli*	βpNPGal	HO-(CH$_2$)$_3$CH$_3$	β-D-Gal-O-(CH$_2$)$_3$CH$_3$	67	228, 230

29.6 Recent Developments and New Directions

#		Enzyme		Acceptor	Product	Yield	Refs
132	β-D-Gal-O	β-Galactosidase Aspergillus oryzae	βpNPGal	HO-(CH₂)₃CH₃	β-D-Gal-O-(CH₂)₃CH₃	75	228, 230
133	β-D-Gal-O	β-Galactosidase Bacillus circulans	βpNPGal	HO-(CH₂)₃CH₃	β-D-Gal-O-(CH₂)₃CH₃	29	228, 230
134	β-D-Gal-O	β-Galactosidase Bovine liver	βpNPGal	HO-(CH₂)₃CH₃	β-D-Gal-O-(CH₂)₃CH₃	8	228
135	β-D-Gal-O	β-Galactosidase Escherichia coli	βpNPGal	HO-CH(CH₃)CH₂CH₃	β-D-Gal-O-CH(CH₃)CH₂CH₃	53	228, 230
136	β-D-Gal-O	β-Galactosidase Aspergillus oryzae	βpNPGal	HO-CH(CH₃)CH₂CH₃	β-D-Gal-O-CH(CH₃)CH₂CH₃	31	228, 230
137	β-D-Gal-O	β-Galactosidase Bacillus circulans	βpNPGal	HO-CH(CH₃)CH₂CH₃	β-D-Gal-O-CH(CH₃)CH₂CH₃	54	228, 230
138	β-D-Gal-O	β-Galactosidase Escherichia coli	βpNPGal	HO-C(CH₃)₃	β-D-Gal-O-C(CH₃)₃	23	228, 230
139	β-D-Gal-O	β-Galactosidase Bacillus circulans	βpNPGal	5-Phenylpentanol	β-D-Gal-O-5-phenylpentyl	72	232
140	β-D-Gal-O	β-Galactosidase Escherichia coli	βpNPGal	1,2,3,4-Tetra-O-acetyl-D-glucopyranose	β-D-Gal-O-1,2,3,4-tetra-O-acetyl-D-glucopyranosyl	66	230, 229
141	β-D-Gal-O	β-Galactosidase Escherichia coli	βpNPGal	2,3,4,6-Tetra-O-acetyl-D-glucopyranose	β-D-Gal-O-2,3,4,6-tetra-O-acetyl-D-glucopyranose	40	230, 229
142	β-D-Gal-O	β-Galactosidase Bacillus circulans	βpNPGal	2,3,4,6-Tetra-O-acetyl-D-glucopyranose	β-D-Gal-O-2,3,4,6-tetra-O-acetyl-D-glucopyranose	45	230, 231
143	β-D-Gal-O	β-Galactosidase Bacillus circulans	βpNPGal	Methacryloyl-oxyethanol	β-D-Gal-O-methacryloyl-oxyethyl	42	230, 231

Table 3 (*continued*)

Entry	Linkage	Enzyme/Source	Donor	Acceptor	Product	Yield (%)	Reference
144	β-D-Gal-O	β-Galactosidase *Bacillus circulans*	βpNPGal	pNP	β-D-Gal-O-pNP	35	230, 231
145	β-D-Gal-O	β-Galactosidase *Escherichia coli*	βpNPGal	5-Phenyl-1-pentanol	β-D-Gal-O-5-phenyl-1-pentyl	62	213
146	β-D-Gal-O	β-Galactosidase *Aspergillus oryzae*	Gal	But-3-en-2-ol	β-D-Gal-O-but-3-en-2-yl	21 (R: 40 ee)	233
147	β-D-Gal-O	β-Galactosidase *Escherichia coli*	Lactose	Ser-(N-Boc)-Gly-OMe	β-D-Gal-O-Ser-(N-Boc)-Gly-OMe	13	106
148	β-D-Gal-O	β-Galactosidase *Escherichia coli*	Lactose	Ser-(N-Boc)-Ala-OMe	β-D-Gal-O-Ser-(N-Boc)-Ala-OMe	9	106
149	β-D-Gal-O	β-Galactosidase *Escherichia coli*	Lactose	Ser-(N-Boc)-Ser-OMe	β-D-Gal-O-Ser-(N-Boc)-Ser-OMe	11	106
150	β-D-Gal-O	β-Galactosidase *Escherichia coli*	Lactose	Gly-(N-Boc)-Ser-OMe	Gly-(N-Boc)-[β-D-Gal]-O-Ser-OMe	7	106
151	β-D-Gal-O	β-Galactosidase *Escherichia coli*	Lactose	Ala-(N-Boc)-Ser-OMe	Ala-(N-Boc)-[β-D-Gal]-O-Ser-OMe	5	106
152	β-D-Gal-O	β-Galactosidase *Aspergillus oryzae*	βoNPGal	HOAll	β-D-Gal-OAll	62	93
153	β-D-Gal-O	β-Galactosidase *Aspergillus oryzae*	βVGal	HOAll	β-D-Gal-OAll	50	93

#		Enzyme		Acceptor	Product	Yield	Ref.
154	β-D-Gal-O	β-Galactosidase Bovine liver	βoNPGal	Monoallyl ether (4,6-iso-propylidine-β-D-Glc-O)-1,3,5-benzene-trimethanol	monoallyl ether (β-D-Gal-O)-(4,6-isopropylidine-β-D-Glc-O)-1,3,5-benzene-trimethanol	17	107
155	β-D-Gal-O	β-Galactosidase Bovine liver	βoNPGal	Monoallyl ether (β-D-Glc-O)-1,3,5-benzene-trimethanol	monoallyl ether (β-D-Gal-O)-(β-D-Glc-O)-1,3,5-benzene-trimethanol	10	107
156	β-D-Gal-O	β-Galactosidase *Achatina achatina*	Lactose	Ser-(N-Z)-OMe	β-D-Gal-O-Ser-(N-Z)-OMe	35	234
157	β-D-Gal-O	β-Galactosidase *Achatina achatina*	Lactose	Thr-(N-Z)-OMe	β-D-Gal-O-Thr-(N-Z)-OMe	10	234
158	β-D-Gal-O	β-Galactosidase *Achatina achatina*	Lactose	Hyp-(N-Z)-OMe	β-D-Gal-O-Hyp-(N-Z)-OMe	28	234
159	β-D-Gal-O	β-Galactosidase *Achatina achatina*	Lactose	Ser-(N-Boc)	β-D-Gal-O-Ser-(N-Boc)	4	235
160	β-D-Gal-O	β-Galactosidase *Aspergillus oryzae*	Lactose	Thr-(N-Boc)	β-D-Gal-O-Thr-(N-Boc)	4	235
161	β-D-Gal-O	β-Glucosidase Almond	Gal	HOAll	β-D-Gal-OAll	15	17
162	β-D-Gal-O	β-Galactosidase *Aspergillus oryzae*	Gal	HO-(CH$_2$)$_6$OH	β-D-Gal-O(CH$_2$)$_6$OH	48	204
163	β-D-Gal-O	β-Galactosidase *Aspergillus oryzae*	Gal	HO-(CH$_2$)$_3$CH=CH$_2$	β-D-Gal-O(CH$_2$)$_3$CH=CH$_2$	22	204
164	β-D-Gal-O	β-Galactosidase *Streptococcus thermophilus*	Lactose	HOAll	β-D-Gal-OAll	38	236

Table 3 (continued)

Entry	Linkage	Enzyme/Source	Donor	Acceptor	Product	Yield (%)	Reference
165	β-D-Gal-O	β-Galactosidase *Escherichia coli*	Lactose	Ser-(N-Boc)-OMe	β-D-Gal-O-Ser-(N-Boc)-OMe	15	206
166	β-D-Gal-O	β-Galactosidase *Escherichia coli*	Lactose	Ser-(N-Aloc)-OMe	β-D-Gal-O-Ser-(N-Aloc)-OMe	15	206
167	β-D-Gal-O	β-Galactosidase *Escherichia coli*	Lactose	Ser-(N-Z)-OMe	β-D-Gal-O-Ser-(N-Z)-OMe	9	206
168	β-D-Gal-O	β-Galactosidase *Achatina achatina*	Lactose	Ser-(N-Z)-Gly-OEt	β-D-Gal-O-Ser-(N-Z)-Gly-OEt	16	237
169	β-D-Gal-O	β-Galactosidase *Achatina achatina*	Lactose	Ser-(N-Z)-Ala-OEt	β-D-Gal-O-Ser-(N-Z)-Ala-OEt	15	237
170	β-D-Gal-O	β-Galactosidase *Achatina achatina*	Lactose	Gly-(N-Z)-Ser-OEt	Gly-(N-Z)-[β-D-Gal-O]-Ser-OEt	29	237
171	β-D-Gal-O	β-Galactosidase *Achatina achatina*	Lactose	Ala-(N-Z)-Ser-OEt	Ala-(N-Z)-[β-D-Gal-O]-Ser-OEt	8	237
172	β-D-Gal-O	β-Galactosidase *Achatina achatina*	Lactose	Ala-(N-Z)-Ser-OMe	Ala-(N-Z)-[β-D-Gal-O]-Ser-OMe	7	237
173	β-D-Gal-O	β-Galactosidase *Achatina achatina*	Lactose	Thr-(N-Z)-Gly-OEt	β-D-Gal-O-Thr-(N-Z)-Gly-OEt	9	237
174	β-D-Glc(1-3)-O	*Rhodotorula lactosa* cells	Cellobiose	CalysteginB1	β-D-Glc(1-3)-O-calysteginB1	0.9	217

#	Donor	Enzyme	Acceptor	Product	Yield %	Ref.
175	β-D-Glc(1-4)-DNJ	Cellulase *Trichoderma* sp.	Cellooligosaccharides	DNJ	β-D-Glc(1-4)-DNJ	NA	171
176	β-D-Glc(1-4)-O	Cellulase *Rhodotorula lactosa* cells	Cellobiose	CalysteginB2	β-D-Glc(1-4)-O-calysteginB2	11	217
177	β-D-Glc-O	β-Galactosidase *Aspergillus oryzae*	βPhGal	HO-(CH$_2$)$_2$C$_6$H$_5$	β-D-Glc-O-(CH$_2$)$_2$C$_6$H$_5$	39	220
178	β-D-Glc-O	β-Galactosidase *Aspergillus oryzae*	βPhGal	HO-CH$_2$CHOHCH$_3$	β-D-Glc-O-CH$_2$CHOHCH$_3$	48	220
179	β-D-Glc-O	β-Galactosidase *Aspergillus oryzae*	βPhGal	HO-CH$_2$CHOHCH$_2$Cl	β-D-Glc-O-CH$_2$CHOHCH$_2$Cl	43	220
180	β-D-Glc-O	β-Glucosidase Almond	βoNPGlc	Ser-(N-Ac)-OMe	β-D-Glc-O-Ser-(N-Ac)-OMe	25	238, 239
181	β-D-Glc-O	β-Glucosidase Almond	βpNPGlc	Ser-(N-Ac)-OMe	β-D-Glc-O-Ser-(N-Ac)-OMe	4	238, 239
182	β-D-Glc-O	β-Glucosidase Almond	Cellobiose	Ser-(N-Ac)-OMe	β-D-Glc-O-Ser-(N-Ac)-OMe	3	238, 239
183	β-D-Glc-O	β-Glucosidase Almond	βoNPGlc	Thr-(N-Ac)-OMe	β-D-Glc-O-Thr-(N-Ac)-OMe	11	238, 239
184	β-D-Glc-O	β-Glucosidase Almond	βoNPGlc	4-OH-Pro-(N-Ac)-OMe	β-D-Glc-O-4-Pro-(N-Ac)-OMe	9	238, 239
185	β-D-Glc-O	β-Glucosidase Almond	βoNPGlc	HO-CH$_2$CH(CH$_3$)-COOCH$_3$ (S)	β-D-Glc-O-CH$_2$CH(CH$_3$)COOCH$_3$ (S)	27	238, 239
186	β-D-Glc-O	β-Glucosidase Almond	βoNPGlc	HO-CH$_2$CH(CH$_3$)NHAc (S)	β-D-Glc-O-CH$_2$CH(CH$_3$)NHAc (S)	11	238, 239
187	β-D-Glc-O	β-Glucosidase Almond	βoNPGlc	Ser-(N$_3$)-OMe	β-D-Glc-O-Ser-(N$_3$)-OMe	25	239
188	β-D-Glc-O	β-Glucosidase Almond	βoNPGlc	D-Ser-(N$_3$)-OMe	β-D-Glc-O-D-Ser-(N$_3$)-OMe	12	239

Table 3 (continued)

Entry	Linkage	Enzyme/Source	Donor	Acceptor	Product	Yield (%)	Reference
189	β-D-Glc-O	β-Glucosidase Almond	βoNPGlc	HO-CH$_2$CH(CH$_3$)NHAc (R)	β-D-Glc-O-CH$_2$CH(CH$_3$)NHAc (R)	39	239
190	β-D-Glc-O	β-Glucosidase Almond	Cellobiose	oHOBnOH	β-D-Glc-O-oHOBn	3	240
191	β-D-Glc-O	β-Glucosidase Almond	Cellobiose	mHOBnOH	β-D-Glc-O-mHOBn	2	240
192	β-D-Glc-O	β-Glucosidase Almond	Cellobiose	pHOBnOH	β-D-Glc-O-pHOBn	1.9	240
193	β-D-Glc-O	β-Glucosidase Almond	Glc	oHOBnOH	β-D-Glc-O-oHOBn	1.6	240
194	β-D-Glc-O	β-Glucosidase Almond	Glc	mHOBnOH	β-D-Glc-O-mHOBn	1.1	240
195	β-D-Glc-O	β-Glucosidase Almond	Glc	pHOBnOH	β-D-Glc-O-pHOBn	1	240
196	β-D-Glc-O	β-Glucosidase Fusarium oxysporum	Cellobiose	MeOH	β-D-Glc-OMe	22	241
197	β-D-Glc-O	β-Glucosidase Fusarium oxysporum	Cellobiose	EtOH	β-D-Glc-OEt	10	241
198	β-D-Glc-O	β-Glucosidase Fusarium oxysporum	Cellobiose	PrOH	β-D-Glc-OPr	33	241
199	β-D-Glc-O	β-Galactosidase Aspergillus oryzae	βPhGlc	Gitoxigenin	β-D-Glc-O-3-gitoxigenin	43	222, 223
200	β-D-Glc-O	β-Galactosidase Aspergillus oryzae	βPhGlc	Digitoxigenin	β-D-Glc-O-3-digitoxigenin	74	222, 223

#	Donor	Enzyme		Acceptor	Product	Yield	Ref.
201	β-D-Glc-O	β-Galactosidase *Aspergillus oryzae*	βPhGal	MeOH	β-D-Glc-OMe	49	222, 223
202	β-D-Glc-O	β-Galactosidase *Aspergillus oryzae*	βPhGal	Cyclohexanol	β-D-Glc-O-cyclohexanol	26	222, 223
203	β-D-Glc-O	β-Glucosidase Almond	Glc	HO-(CH$_2$)$_4$OH	β-D-Glc-O-(CH$_2$)$_4$OH	55	242
204	β-D-Glc-O	β-Glucosidase Almond	Glc	EtOH	β-D-Glc-OEt	60	209
205	β-D-Glc-O	β-Glucosidase Almond	Glc	PrOH	β-D-Glc-OPr	39	209
206	β-D-Glc-O	β-Glucosidase Almond	Glc	iPrOH	β-D-Glc-OiPr	31	209
207	β-D-Glc-O	β-Glucosidase Almond	Glc	HO-CH$_2$CH(CH$_3$)$_2$	β-D-Glc-O-CH$_2$CH(CH$_3$)$_2$	42	209
208	β-D-Glc-O	β-Glucosidase Almond	Glc	HO-C(CH$_3$)$_3$	β-D-Glc-O-C(CH$_3$)$_3$	12	209
209	β-D-Glc-O	β-Glucosidase Almond	Glc	BuOH	β-D-Glc-O-(CH$_2$)$_3$CH$_3$	38	209
210	β-D-Glc-O	β-Glucosidase Almond	Glc	CH$_3$CHOHCH$_2$CH$_3$	β-D-Glc-O-CH(CH$_3$)CH$_2$CH$_3$	27	209
211	β-D-Glc-O	β-Glucosidase Almond	Glc	HO-(CH$_2$)$_4$CH$_3$	β-D-Glc-O-(CH$_2$)$_4$CH$_3$	12	209
212	β-D-Glc-O	β-Glucosidase Almond	Glc	HO-(CH$_2$)$_5$CH$_3$	β-D-Glc-O-(CH$_2$)$_5$CH$_3$	6	209
213	β-D-Glc-O	β-Glucosidase Almond	Glc	HO-(CH$_2$)$_4$OH	β-D-Glc-O-(CH$_2$)$_4$OH	72	209
214	β-D-Glc-O	β-Glucosidase Almond	Glc	CH$_3$CHOHCHOHCH$_3$	β-D-Glc-2-O-CH(CH$_3$)CHOHCH$_3$	40	209
215	β-D-Glc-O	β-Glucosidase *Kluyveromyces lactis*	β*p*NPGlc	HO-*p*[CO(CH$_2$)$_2$OH]C$_6$H$_4$	β-D-Glc-O-*p*[CO(CH$_2$)$_2$OH]C$_6$H$_4$	23	243

Table 3 (continued)

Entry	Linkage	Enzyme/Source	Donor	Acceptor	Product	Yield (%)	Reference
216	β-D-Glc-O	β-Glucosidase *Sulfolobus solfataricus*	βPhGlc	HO-p[CO(CH$_2$)$_2$OH]C$_6$H$_4$	β-D-Glc-O-p[CO(CH$_2$)$_2$OH]C$_6$H$_4$	42	243
217	β-D-Glc-O	β-Glucosidase Almond	Glc	HO-p[CO(CH$_2$)$_2$OH]C$_6$H$_4$	β-D-Glc-O-p[CO(CH$_2$)$_2$OH]C$_6$H$_4$	3.2	243
218	β-D-Glc-O	β-Glucosidase *Sulfolobus solfataricus*	Glc	HO-p[CO(CH$_2$)$_2$OH]C$_6$H$_4$	β-D-Glc-O-p[CO(CH$_2$)$_2$OH]C$_6$H$_4$	2.1	243
219	β-D-Glc-O	β-Glucosidase *Trichoderma viridae*	Cellobiose	HO-(CH$_2$)$_6$CH$_3$	β-D-Glc-O-(CH$_2$)$_6$CH$_3$	12	244
220	β-D-Glc-O	β-Glucosidase *Trichoderma viridae*	Cellobiose	HO-(CH$_2$)$_7$CH$_3$	β-D-Glc-O-(CH$_2$)$_7$CH$_3$	7.4	244
221	β-D-Glc-O	β-Glucosidase Almond	Glc	HO-(CH$_2$)$_3$OH	β-D-Glc-O-(CH$_2$)$_3$OH	55	245
222	β-D-Glc-O	β-Glucosidase Almond	Glc	HO-(CH$_2$)$_4$OH	β-D-Glc-O-(CH$_2$)$_4$OH	47	245
223	β-D-Glc-O	β-Glucosidase Almond	Glc	HO-(CH$_2$)$_5$OH	β-D-Glc-O-(CH$_2$)$_5$OH	40	245
224	β-D-Glc-O	β-Glucosidase Almond	Glc	HO-(CH$_2$)$_6$OH	β-D-Glc-O-(CH$_2$)$_6$OH	41	245
225	β-D-Glc-O	β-Glucosidase Almond	Glc	HO-(CH$_2$)$_3$CH$_3$	β-D-Glc-O-(CH$_2$)$_3$CH$_3$	44	214
226	β-D-Glc-O	β-Glucosidase *Trichoderma reesei*	Cellobiose	Geraniol	β-D-Glc-O-geranyl	1.2	246
227	β-D-Glc-O	β-Glucosidase *Aspergillus niger*	Cellobiose	Geraniol	β-D-Glc-O-geranyl	3.4	246

29.6 Recent Developments and New Directions

228	β-D-Glc-O	β-Glucosidase Candida molischiana	Cellobiose	Geraniol	β-D-Glc-O-geranyl	0.1	246
229	β-D-Glc-O	β-Glucosidase Almond	Cellobiose	Geraniol	β-D-Glc-O-geranyl	0.6	246
230	β-D-Glc-O	β-Glucosidase Trichoderma reesei	Cellobiose	Nerol	β-D-Glc-O-neryl	0.8	246
231	β-D-Glc-O	β-Glucosidase Aspergillus niger	Cellobiose	Nerol	β-D-Glc-O-neryl	3.3	246
232	β-D-Glc-O	β-Glucosidase Candida molischiana	Cellobiose	Nerol	β-D-Glc-O-neryl	0.1	246
233	β-D-Glc-O	β-Glucosidase Almond	Cellobiose	Nerol	β-D-Glc-O-neryl	0.3	246
234	β-D-Glc-O	β-Glucosidase Trichoderma reesei	Cellobiose	Citronellol	β-D-Glc-O-citronellyl	0.3	246
235	β-D-Glc-O	β-Glucosidase Aspergillus niger	Cellobiose	Citronellol	β-D-Glc-O-citronellyl	0.6	246
236	β-D-Glc-O	β-Glucosidase Almond	Cellobiose	Citronellol	β-D-Glc-O-citronellyl	0.1	246
237	β-D-Glc-O	β-Glucosidase Almond	Glc	2-Hydroxybenzylalcohol	β-D-Glc-O-2-hydroxybenzyl	17	247
238	β-D-Glc-O	β-Glucosidase Almond	Glc	HO-(CH$_2$)$_7$CH$_3$	β-D-Glc-O-(CH$_2$)$_7$CH$_3$	8	247
239	β-D-Glc-O	β-Glucosidase Aspergillus niger	βpNPGlc	5-Phenyl-1-pentanol	β-D-Glc-O-5-phenyl-1-pentyl	23	213
240	β-D-Glc-O	β-Glucosidase Agrobacterium tumefaciens	βpNPGlc	HO-(CH$_2$)$_7$CH$_3$	β-D-Glc-O-(CH$_2$)$_7$CH$_3$	80	248

Table 3 (*continued*)

Entry	Linkage	Enzyme/Source	Donor	Acceptor	Product	Yield (%)	Reference
241	β-D-Glc-O	β-Glucosidase Almond	Glc	HO-(CH$_2$)$_6$OH	β-D-Glc-O-(CH$_2$)$_6$OH	61	249
242	β-D-Glc-O	β-Glucosidase Almond	Glc	HO-(CH$_2$)$_6$NHCOCF$_3$	β-D-Glc-O-(CH$_2$)$_6$NHCOCF$_3$	7	249
243	β-D-Glc-O	β-Glucosidase Almond	Glc	HS-(CH$_2$)$_3$SH	β-D-Glc-S-(CH$_2$)$_3$SH	17	249
244	β-D-Glc-O	β-Glucosidase Almond	Glc	HOAll	β-D-Glc-OAll	62	249
245	β-D-Glc-O	β-Glucosidase Almond	1,2-3,5-Di-phenyl-boronate-Glc$f$	HO-(CH$_2$)$_2$OH	β-D-Glc-O-(CH$_2$)$_6$OH	7	249
246	β-D-Glc-O	β-Glucosidase Almond	1,2-3,5-Di-phenyl-boronate-Glc$f$	HS-(CH$_2$)$_3$SH	β-D-Glc-S-(CH$_2$)$_3$SH	10	249
247	β-D-Glc-O	β-Glucosidase Almond	Glc	But-3-en-2-ol	β-D-Glc-O-but-3-en-2-yl	27 (R: 20 ee)	233
248	β-D-Glc-O	β-Glucosidase Almond	Glc	Pent-1-en-3-ol	β-D-Glc-O-pent-1-en-3-yl	12 (R: 40 ee)	233
249	β-D-Glc-O	β-Glucosidase Almond	Glc	Hex-1-en-3-ol	β-D-Glc-O-hex-1-en-3-yl	5 (R: 80 ee)	233
250	β-D-Glc-O	β-Glucosidase Almond	Glc	Pent-3-en-2-ol	β-D-Glc-O-pent-3-en-2-yl	17 (R: 37 ee)	233
251	β-D-Glc-O	β-Glucosidase Almond	Glc	HO-(CH$_2$)$_6$OH	β-D-Glc-O-(CH$_2$)$_6$OH	17	250
252	β-D-Glc-O	β-Glucosidase *Sulfolobus solfataricus*	βPhGlc	HO-CH$_2$CHOHCH$_3$	β-D-Glc-O-CH$_2$CHOHCH$_3$	77	250

#		Enzyme		Acceptor	Product	Yield	Ref.
253	β-D-Glc-O	β-Glucosidase Sulfolobus solfataricus	βPhGlc	HO-CH$_2$CHOHCH$_3$	β-D-Glc-O-CH(CH$_3$)CH$_2$OH	20	250
254	β-D-Glc-O	β-Glucosidase Pyrococcus furiosus	Cellobiose	HO-(CH$_2$)$_2$OHCH$_3$	β-D-Glc-O-CH$_2$CHOHCH$_3$	63	250
255	β-D-Glc-O	β-Glucosidase Pyrococcus furiosus	Cellobiose	HO-(CH$_2$)$_2$OHCH$_3$	β-D-Glc-O-CH(CH$_3$)CH$_2$OH	15	250
256	β-D-Glc-O	β-Glucosidase Almond	Glc	HO-(CH$_2$)$_7$CH$_3$	β-D-Glc-HO-(CH$_2$)$_7$CH$_3$	NA	251
257	β-D-Glc-O	β-Glucosidase Almond	Glc	HOAll	β-D-Glc-OAll	65	17
258	β-D-Glc-O	β-Glucosidase Almond	Glc	HOBn	β-D-Glc-OBn	40	17
259	β-D-Glc-O	β-Glucosidase Almond	Glc	HO-(CH$_2$)$_6$OH	β-D-Glc-O-(CH$_2$)$_6$OH	61	204
260	β-D-Glc-O	β-Glucosidase Almond	Glc	HO-(CH$_2$)$_3$CH=CH$_2$	β-D-Glc-O-(CH$_2$)$_3$CH=CH$_2$	50	204
261	β-D-Glc-O	β-Glucosidase Almond	Glc	HO-(CH$_2$)$_2$Si(CH$_3$)$_3$	β-D-Glc-O-(CH$_2$)$_2$Si(CH$_3$)$_3$	11	204
262	β-D-GlcNAc-O	β-Hexosaminidase Aspergillus oryzae	βpNPGlcNAc	Cyanuricacid	β-D-GlcNAc-O-cyanuricacid	29	212
263	β-D-GlcNAc-O	β-Hexosaminidase Aspergillus oryzae	βpNPGlcNAc	p-bis(hydroxymethyl)benzene	β-D-GlcNAc-O-p-bis(hydroxymethyl)benzene	12	212
264	β-D-GlcNAc-O	β-Hexosaminidase Aspergillus oryzae	βpNPGlcNAc	Ser-(N-COOMe)-OMe	β-D-GlcNAc-O-Ser-(N-COOMe)-OMe	10	208
265	β-D-GlcNAc-O	β-Hexosaminidase Aspergillus oryzae	βpNPGlcNAc	Ser-(N-COCH$_2$CH=CH$_2$)-OMe	β-D-GlcNAc-O-Ser-(N-COCH$_2$CH=CH$_2$)-OMe	10–50	208

Table 3 (*continued*)

Entry	Linkage	Enzyme/Source	Donor	Acceptor	Product	Yield (%)	Reference
266	β-D-GlcNAc-O	β-Hexosaminidase Bovine kidney	βpNPGlcNAc	5-Phenyl-1-pentanol	β-D-GlcNAc-O-5-phenyl-1-pentyl	33	213
267	β-D-Glc-O	β-Glucosidase Almond	Cellobiose	Ser-(N-Aloc)-OMe	β-D-Glc-Ser-(N-Aloc)-OMe	8	206
268	β-D-Man-O	β-Mannosidase *Rhizopus niveus*	β-D-Man(1-4)-D-Man	HO-(CH$_2$)$_3$CH$_3$	β-D-Man-O-(CH$_2$)$_3$CH$_3$	1	70
269	β-D-Man-O	β-Mannosidase *Rhizopus niveus*	β-D-Man(1-4)-D-Man	HO-(CH$_2$)$_2$OH	β-D-Man-O-(CH$_2$)$_2$OH	4	70
270	β-D-Man-O	β-Mannosidase *Rhizopus niveus*	β-D-Man(1-4)-D-Man	HOBn	β-D-Man-OBn	15	70
271	β-D-Man-O	β-Mannosidase Snail	βpNPMan	MeOH	β-D-Man-OMe	67	252
272	β-D-Man-O	β-Mannosidase Snail	βpNPMan	EtOH	β-D-Man-OEt	75	252
273	β-D-Man-O	β-Mannosidase Snail	βpNPMan	HO-CH(CH$_3$)$_2$	β-D-Man-O-CH(CH$_3$)$_2$	9	252
274	β-D-Man-O	β-Mannosidase Snail	βpNPMan	HO-(CH$_2$)$_3$CH$_3$	β-D-Man-O-(CH$_2$)$_3$CH$_3$	19	252
275	β-D-Man-O	β-Mannosidase Snail	βpNPMan	HO-(CH$_2$)$_2$OH	β-D-Man-O-(CH$_2$)$_2$OH	76	252
276	β-D-Man-O	β-Mannosidase Snail	βpNPMan	HO-(CH$_2$)$_4$OH	β-D-Man-O-(CH$_2$)$_4$OH	45	252
277	β-D-Man-O	β-Mannosidase Snail	βpNPMan	HO-(CH$_2$)$_6$OH	β-D-Man-O-(CH$_2$)$_6$OH	36	252

278	β-D-Man-O-	β-Mannosidase Snail	βpNPMan	HO-(CH₂)₂CH(CH₃)OH	β-D-Man-O-CH₂)₂CH(CH₃)OH	26	252
279	β-D-Man-O-	β-Mannosidase Snail	βpNPMan	MeOH	β-D-Man-OMe	65	64
280	β-D-Man-O-	β-Mannosidase Snail	βpNPMan	EtOH	β-D-Man-OEt	75	64
281	β-D-Man-O-	β-Mannosidase Snail	βpNPMan	PrOH	β-D-Man-OPr	50	64
282	β-D-Man-O-	β-Mannosidase Snail	βpNPMan	HO-(CH₂)₃CH₃	β-D-Man-O-(CH₂)₃CH₃	20	64
283	β-D-Man-O-	β-Mannosidase Snail	βpNPMan	HO-(CH₂)₅CH₃	β-D-Man-O-(CH₂)₅CH₃	5	64
284	β-D-Man-O-	β-Mannosidase Snail	βpNPMan	HO-(CH₂)₇CH₃	β-D-Man-O-(CH₂)₇CH₃	1	64
285	β-D-Man-O-	β-Mannosidase Snail	βpNPMan	iPrOH	β-D-Man-OiPr	15	64
286	β-D-Man-O-	β-Mannosidase Snail	βpNPMan	HO-CH(CH₃)(CH₂)₂CH₃	β-D-Man-O-CH(CH₃)(CH₂)₂CH₃	2	64
287	β-D-Man-O-	β-Mannosidase Snail	βpNPMan	Cyclohexanol	β-D-Man-O-cyclohexanol	3	64
288	β-D-Man-O-	β-Galactosidase *Aspergillus oryzae* (Impurity?)	Man	Ser-(N-Boc)	β-D-Man-O-Ser-(N-Boc)	14	235
289	β-D-Man-O-	β-Galactosidase *Aspergillus oryzae* (Impurity?)	Man	Thr-(N-Boc)	β-D-Man-O-Thr-(N-Boc)	8	235
290	β-D-Man-O-	β-Mannosidase *Helix pomatia*	Man	HO-(CH₂)₆OH	β-D-Man-O-(CH₂)₆OH	12	204
291	β-D-Xyl-O-	β-Xylosidase *Aspergillus niger*	Xyl₂	MeOH	β-D-Xyl-OMe	84	253, 254

Table 3 (continued)

Entry	Linkage	Enzyme/Source	Donor	Acceptor	Product	Yield (%)	Reference
292	β-D-Xyl-O	β-Xylosidase *Aspergillus niger*	Xyl$_2$	EtOH	β-D-Xyl-OEt	84	253, 254
293	β-D-Xyl-O	β-Xylosidase *Aspergillus niger*	Xyl$_2$	PrOH	β-D-Xyl-OPr	90	253, 254
294	β-D-Xyl-O	β-Xylosidase *Aspergillus niger*	Xyl$_2$	HO-(CH$_2$)$_3$CH$_3$	β-D-Xyl-O-(CH$_2$)$_3$CH$_3$	82	253, 254
295	β-D-Xyl-O	β-Xylosidase *Aspergillus niger*	Xyl$_2$	HO-(CH$_2$)$_4$CH$_3$	β-D-Xyl-O-(CH$_2$)$_4$CH$_3$	82	253, 254
296	β-D-Xyl-O	β-Xylosidase *Aspergillus niger*	Xyl$_2$	HO-(CH$_2$)$_5$CH$_3$	β-D-Xyl-O-(CH$_2$)$_5$CH$_3$	58	253, 254
297	β-D-Xyl-O	β-Xylosidase *Aspergillus niger*	Xyl$_2$	HO-(CH$_2$)$_7$CH$_3$	β-D-Xyl-O-(CH$_2$)$_7$CH$_3$	44	253, 254
298	β-D-Xyl-O	β-Xylosidase *Aspergillus niger*	Xyl$_2$	HOBn	β-D-Xyl-OBn	20	253, 254
299	β-D-Xyl-O	β-Xylosidase *Aspergillus niger*	Xyl$_2$	iPrOH	β-D-Xyl-OiPr	94	253, 254
300	β-D-Xyl-O	β-Xylosidase *Aspergillus niger*	Xyl$_2$	HO-CH(CH$_3$)CH$_2$CH$_3$	β-D-Xyl-O-CH(CH$_3$)CH$_2$CH$_3$	88	253, 254
301	β-D-Xyl-O	β-Xylosidase *Aspergillus niger*	Xyl$_2$	HO-CH(CH$_3$)(CH$_2$)$_3$CH$_3$	β-D-Xyl-O-CH(CH$_3$)(CH$_2$)$_3$CH$_3$	16	253, 254

302	β-D-Xyl-O	β-Xylosidase *Aspergillus niger*	Xyl₂	Cyclohexanol	β-D-Xyl-O-cyclohexyl	42	253, 254
303	β-D-Xyl-O	β-Xylosidase *Aspergillus niger*	Xyl₂	HO-C(CH₃)₃	β-D-Xyl-O-C(CH₃)₃	24	253, 254
304	β-D-Xyl-O	β-Xylosidase *Trichoderma reesei*	β-D-Xyl-OMe	EtOH	β-D-Xyl-OEt	NA	255
305	β-D-Xyl-O	β-Xylosidase *Trichoderma reesei*	β-D-Xyl-OMe	PrOH	β-D-Xyl-OPr	53	255
306	β-D-Xyl-O	β-Xylosidase *Trichoderma reesei*	β-D-Xyl-OMe	HO-(CH₂)₃CH₃	β-D-Xyl-O-(CH₂)₃CH₃	NA	255
307	β-D-Xyl-O	β-Xylosidase *Trichoderma reesei*	β-D-Xyl-OMe	HO-CH(CH₃)CH₂CH₃	β-D-Xyl-O-CH(CH₃)CH₂CH₃	NA	255
308	β-D-Xyl-O	β-Xylosidase *Aspergillus pulverulentus*	Xyl₂	Hydroquinone	β-D-Xyl-O-hydroquinone	25	256

References

1. E. Bourquelot, *Ann. Chim.*, **1913**, *29*, 145.
2. G. M. Watt, P. A. S. Lowden, and S. L. Flitsch, *Curr. Opin. Struct. Biol.*, **1997**, *7*, 652–660.
3. D. H. G. Crout and G. Vic, *Curr. Opin. Struct. Biol.*, **1998**, *2*, 98–111.
4. V. Kren and J. Thiem, *Chem. Soc. Rev.*, **1997**, *26*, 463–473.
5. C. H. Wong, R. L. Halcomb, Y. Ichikawa, and T. Kajimoto, *Angew. Chem. Int. Ed. Engl.*, **1995**, *34*, 521–546.
6. J. Thiem, *FEMS Microbiol. Rev.*, **1995**, *16*, 193–211.
7. K. G. I. Nilsson, *Tibtech*, **1988**, *6*, 256–264.
8. B. Henrissat and G. Davies, *Curr. Opin. Struct. Biol.*, **1997**, *7*, 637–644.
9. D. E. Koshland, *Biol. Rev.*, **1953**, *28*, 416–436.
10. M. L. Sinnott, *Chem. Rev.*, **1990**, *90*, 1171–1202.
11. J. D. McCarter and S. G. Withers, *Curr. Opin. Struct. Biol.*, **1994**, *4*, 885–892.
12. G. Davies and B. Henrissat, *Structure*, **1995**, *3*, 853–859.
13. G. Davies, M. L. Sinnott, and S. G. Withers, in *Comprehensive Biological Catalysis*, (Ed. M. L. Sinnott), Academic Press, London, Vol. 1, 1997, pp. 119–208.
14. T. Usui, H. Matsui, and K. Isobe, *Carbohydr. Res.*, **1990**, *203*, 65–77.
15. K. Ajisaka, I. Matsuo, M. Isomura, H. Fujimoto, M. Shirakabe, and M. Okawa, *Carbohydr. Res.*, **1995**, *270*, 123–130.
16. S. Suwasono and R. A. Rastall, *Biotechnol. Lett.*, **1996**, *18*, 851–856.
17. G. Vic and D. H. G. Crout, *Carbohydr. Res.*, **1995**, *279*, 315–319.
18. S. Nishimura and K. Yamada, *J. Am. Chem. Soc.*, **1997**, *119*, 10555–10556.
19. S. Kobayashi, T. Kiyosada, and S. Shoda, *J. Am. Chem. Soc.*, **1996**, *118*, 13113–13114.
20. P. Finch and J. H. Yoon, *Carbohydr. Res.*, **1997**, *303*, 339–345.
21. K. Ajisaka, H. Nishida, and H. Fujimoto, *Biotechnol. Lett.*, **1987**, *9*, 387–392.
22. G. F. Herrmann, Y. Ichikawa, C. Wandrey, F. C. A. Gaeta, J. C. Paulson, and C.-H. Wong, *Tetrahedron Lett.*, **1993**, *34*, 3091–3094.
23. Y. Ito and J. C. Paulson, *J. Am. Chem. Soc.*, **1993**, *115*, 7862–7863.
24. V. Kren and J. Thiem, *Angew. Chem. Int. Ed. Engl.*, **1995**, *34*, 893–895.
25. T. Usui, S. Kubota, and H. Ohi, *Carbohydr. Res.*, **1993**, *244*, 315–323.
26. K. G. I. Nilsson, *Carbohydr. Res.*, **1987**, *167*, 95–103.
27. D. H. G. Crout, S. Singh, B. E. P. Swoboda, P. Critchley, and W. T. Gibson, *J. Chem. Soc., Chem. Commun.*, **1992**, 704–705.
28. L. F. Mackenzie, Q. P. Wang, R. A. J. Warren, and S. G. Withers, *J. Am. Chem. Soc.*, **1998**, *120*, 5583–5584.
29. C. Malet and A. Planas, *FEBS Lett.*, **1998**, *440*, 208–212.
30. K. Takegawa, S. Yamaguchi, A. Kondo, I. Kato, and S. Iwahara, *Biochem. Int.*, **1991**, *25*, 829–835.
31. K. Takegawa, S. Yamaguchi, A. Kondo, H. Iwamoto, M. Nakoshi, I. Kato, and S. Iwahara, *Biochem. Int.*, **1991**, *24*, 849–855.
32. K. Takegawa, M. Tabuchi, S. Yamaguchi, A. Kondo, I. Kato, and S. Iwahara, *J. Biol. Chem.*, **1995**, *270*, 3094–3099.
33. J. Q. Fan, M. S. Quesenberry, K. Takegawa, S. Iwahara, A. Kondo, I. Kato, and Y. C. Lee, *J. Biol. Chem.*, **1995**, *270*, 17730–17735.
34. K. Yamamoto and K. Takegawa, *Trends Glycoscience Glycotechnol.*, **1997**, *9*, 339–354.
35. L. X. Wang, J. Q. Fan, and Y. C. Lee, *Tetrahedron Lett.*, **1996**, *37*, 1975–1978.
36. J. Q. Fan, K. Takegawa, S. Iwahara, A. Kondo, I. Kato, C. Abeygunawardana, and Y. C. Lee, *J. Biol. Chem.*, **1995**, *270*, 17723–17729.
37. H. Nakano, M. Moriwaki, T. Washino, and S. Kitahata, *Biosci. Biotechnol. Biochem.*, **1994**, *58*, 1435–1438.
38. H. Nakano, K.-I. Hamayasu, K. Fujita, K. Hara, M. Ohi, H. Yoshizumi, and S. Kitahata, *Biosci. Biotechnol. Biochem.*, **1995**, *59*, 1732–1736.
39. H. Hashimoto, C. Katayama, M. Goto, T. Okinaga, and S. Kitahata, *Biosci. Biotechnol. Biochem.*, **1995**, *59*, 179–183.

40. K. Ajisaka, H. Fujimoto, and H. Nishida, *Carbohydr. Res.*, **1988**, *180*, 35–42.
41. K. Koizumi, T. Tanimoto, Y. Okada, K. Hara, K. Fujita, H. Hashimoto, and S. Kitahata, *Carbohydr. Res.*, **1995**, *278*, 129–142.
42. K. Koizumi, T. Tanimoto, Y. Kubota, and S. Kitahata, *Carbohydr. Res.*, **1998**, *305*, 393–400.
43. K. G. I. Nilsson and A. Fernandez-Mayoralas, *Biotechnol. Lett.*, **1991**, *13*, 715–720.
44. K. G. I. Nilsson, *Carbohydr. Res.*, **1988**, *180*, 53–59.
45. K. G. I. Nilsson, *Carbohydr. Res.*, **1989**, *188*, 9–17.
46. A. Naundorf, M. Caussette, and K. Ajisaka, *Biosci. Biotechnol. Biochem.*, **1998**, *62*, 1313–1317.
47. I. Matsuo, H. Fujimoto, M. Isomura, and K. Ajisaka, *Bioorg. Med. Chem. Lett.*, **1997**, *7*, 255–258.
48. G. Vic, C. H. Tran, M. Scigelova, and D. H. G. Crout, *J. Chem. Soc., Chem. Commun.*, **1997**, 169–170.
49. K. G. I. Nilsson, *Tetrahedron Lett.*, **1997**, *38*, 4527–4527.
50. G. Vic, M. Scigelova, J. J. Hastings, O. W. Howarth, and D. H. G. Crout, *J. Chem. Soc., Chem. Commun.*, **1996**, 1473–1474.
51. A. Millqvist-Fureby, D. A. MacManus, S. Davies, and E. N. Vulfson, *Biotechnol. Bioeng.*, **1998**, *60*, 197–203.
52. S. Kitahata, H. Ishikawa, T. Miyata, and O. Tanaka, *Agric. Biol. Chem.*, **1989**, *53*, 2929–2934.
53. K. Hara, K. Fujita, N. Kuwahara, T. Tanimoto, H. Hashimoto, K. Koizumi, and S. Kitahata, *Biosci. Biotechnol. Biochem.*, **1994**, *58*, 652–659.
54. S. Kitahata, K. Hara, K. Fujita, N. Kuwahara, and K. Koizumi, *Biosci. Biotechnol. Biochem.*, **1992**, *56*, 1518–1519.
55. K. Ajisaka and H. Fujimoto, *Carbohydr. Res.*, **1989**, *185*, 139–146.
56. K. G. I. Nilsson, *Carbohydr. Res.*, **1990**, *204*, 79–83.
57. K. Dax, M. Ebner, R. Peinsipp, and A. E. Stutz, *Tetrahedron Lett.*, **1997**, *38*, 225–226.
58. T. Kasumi, C. F. Brewer, E. T. Reese, and E. J. Hehre, *Carbohydr. Res.*, **1986**, *146*, 39–49.
59. H. Fujimoto and K. Ajisaka, *Biotechnol. Lett.*, **1988**, *10*, 107–112.
60. H. Fujimoto, H. Nishida, and K. Ajisaka, *Agric. Biol. Chem.*, **1988**, *52*, 1345–1351.
61. S. V. Lobov, R. Kasai, K. Ohtani, O. Tanaka, and K. Yamasaki, *Agric. Biol. Chem.*, **1991**, *55*, 2959–2965.
62. K. Ajisaka and H. Fujimoto, *Carbohydr. Res.*, **1990**, *199*, 227–234.
63. B. Sauerbrei and J. Thiem, *Tetrahedron Lett.*, **1992**, *33*, 201–204.
64. N. Taubken, B. Sauerbrei, and J. Thiem, *J. Carbohydr. Chem.*, **1993**, *12*, 651–667.
65. V. Laroute and R. M. Willemot, *Enzyme Microb. Technol.*, **1992**, *14*, 528–534.
66. V. Laroute and R.-M. Willemot, *Biotechnol. Lett.*, **1989**, *11*, 249–254.
67. R. A. Rastall, N. H. Rees, R. Wait, M. W. Adlard, and C. Bucke, *Enzyme Microb. Technol.*, **1992**, *14*, 53–57.
68. S. Suwasono and R. A. Rastall, *Biotechnol. Lett.*, **1998**, *20*, 15–17.
69. E. Johansson, L. Hedbys, and P.-O. Larsson, *Biotechnol. Lett.*, **1986**, *8*, 421–424.
70. H. Fujimoto, M. Isomura, and K. Ajisaka, *Biosci. Biotechnol. Biochem.*, **1997**, *61*, 164–165.
71. K. Hamayasu, K. Hara, K. Fujita, Y. Kondo, H. Hashimoto, T. Tanimoto, K. Koizumi, H. Nakano, and S. Kitahata, *Biosci. Biotechnol. Biochem.*, **1997**, *61*, 825–829.
72. K. Hara, K. Fujita, H. Nakano, N. Kuwahara, T. Tanimoto, H. Hashimoto, K. Koizumi, and S. Kitahata, *Biosci. Biotechnol. Biochem.*, **1994**, *58*, 60–63.
73. T. J. Bartlett, R. A. Rastall, N. H. Rees, M. W. Adlard, and C. Bucke, *J. Chem. Technol. Biotechnol.*, **1992**, *55*, 73–78.
74. A. Vetere, S. Ferro, M. Bosco, P. Cescutti, and S. Paoletti, *Eur. J. Biochem.*, **1997**, *247*, 1083–1090.
75. Y. Makimura, H. Ishida, A. Kondo, A. Hasegawa, and M. Kiso, *J. Carbohydr. Chem.*, **1998**, *17*, 975–979.
76. J. Thiem and B. Sauerbrei, *Angew. Chem. Int. Ed. Engl.*, **1991**, *30*, 1503–1505.
77. A. Vetere and S. Paoletti, *FEBS Lett.*, **1996**, *399*, 203–206.
78. K. Ajisaka, H. Fujimoto, and M. Isomura, *Carbohydr. Res.*, **1994**, *259*, 103–115.
79. S. C. T. Svensson and J. Thiem, *Carbohydr. Res.*, **1990**, *200*, 391–402.

80. K. Ajisaka and M. Shirakabe, *Carbohydr. Res.*, **1992**, *224*, 291–299.
81. K. Ajisaka, H. Fujimoto, and M. Miyasato, *Carbohydr. Res.*, **1998**, *309*, 125–129.
82. A. Vetere, C. Galateo, and S. Paoletti, *Biochem. Biophys. Res. Commun.*, **1997**, *234*, 358–361.
83. K. G. I. Nilsson, H. F. Pan, and U. Larsson-Lorek, *J. Carbohydr. Chem.*, **1997**, *16*, 459–477.
84. K. G. I. Nilsson, A. Eliasson, and U. Larsson-Lorek, *Biotechnol. Lett.*, **1995**, *17*, 717–722.
85. H. Prade, L. F. MacKenzie, and S. G. Withers, *Carbohydr. Res.*, **1998**, *305*, 371–381.
86. K. Fujita, N. Kuwahara, T. Tanimoto, K. Koizumi, M. Iizuka, N. Minamiura, K. Furuichi, and S. Kitahata, *Biosci. Biotechnol. Biochem.*, **1994**, *58*, 239–243.
87. M. Muramatsu and T. Nakakuki, *Biosci. Biotechnol. Biochem.*, **1995**, *59*, 208–212.
88. M. Kurakake, T. Onoue, and T. Komaki, *Appl. Microbiol. Biotechnol.*, **1996**, *45*, 236–239.
89. A. J. J. Straathof, A. P. G. Kieboom, and H. van Bekkum, *Carbohydr. Res.*, **1986**, *146*, 154–159.
90. H. Ishikawa, S. Kitahata, K. Ohtani, and O. Tanaka, *Chem. Pharm. Bull.*, **1991**, *39*, 2043–2045.
91. H. Ishikawa, S. Kitahata, K. Ohtani, C. Ikuhara, and O. Tanaka, *Agric. Biol. Chem.*, **1990**, *54*, 3137–3143.
92. T. Usui, S. Morimoto, Y. Hayakawa, M. Kawaguchi, T. Murata, Y. Matahira, and Y. Nishida, *Carbohydr. Res.*, **1996**, *285*, 29–39.
93. V. Chiffoleau-Giraud, P. Spangenberg, and C. Rabiller, *Tetrahedron: Asymmetry*, **1997**, *8*, 2017–2023.
94. S. Yanahira, T. Kobayashi, T. Suguri, M. Nakakoshi, S. Miura, H. Ishikawa, and I. Nakajima, *Biosci. Biotechnol. Biochem.*, **1995**, *59*, 1021–1026.
95. E. Montero, J. Alonso, F. J. Canada, A. Fernandez-Mayorales, and M. Martin-Loma, *Carbohydr. Res.*, **1998**, *305*, 383–391.
96. J. J. Aragón, F. J. Canada, A. Fernandez-Mayoralas, R. Lopez, M. Martinlomas, and D. Villanueva, *Carbohydr. Res.*, **1996**, *290*, 209–216.
97. G. C. Look and C. H. Wong, *Tetrahedron Lett.*, **1992**, ?, 4253–4256.
98. K. Suzuki, H. Fujimoto, Y. Ito, T. Sasaki, and K. Ajisaka, *Tetrahedron Lett.*, **1997**, *38*, 1211–1214.
99. U. Gambert and J. Thiem, *Carbohydr. Res.*, **1997**, *299*, 85–89.
100. T. Murata, T. Itoh, Y. Hayakawa, and T. Usui, *J. Biochem.*, **1996**, *120*, 851–855.
101. U. Gambert, R. G. Lio, E. Farkas, J. Thiem, V. V. Bencomo, and A. Liptak, *Bioorg. Med. Chem.*, **1997**, *5*, 1285–1291.
102. K. G. I. Nilsson, *Biotechnol. Lett.*, **1996**, *18*, 791–794.
103. T. Yasukochi, K. Fukase, Y. Suda, K. Takagaki, M. Endo, and S. Kusumoto, *Bull. Chem. Soc. Jpn.*, **1997**, *70*, 2719–2725.
104. V. Dumortier, J. Montreuil, and S. Bouquelet, *Carbohydr. Res.*, **1990**, *201*, 115–123.
105. S. Yanahira, Y. Yabe, M. Nakakoshi, S. Miura, N. Matsubara, and I. Nakajima, *Biosci. Biotechnol. Biochem.*, **1998**, *62*, 1791–1794.
106. S. Attal, S. Bay, and D. Cantacuzene, *Tetrahedron*, **1992**, *48*, 9251–9260.
107. S. Menzler, H. Seker, M. Gschrey, and M. Wiessler, *Biotechnol. Lett.*, **1997**, *19*, 269–272.
108. L. Hedbys, E. Johansson, K. Mosbach, P.-O. Larsson, A. Gunnarson, S. Swensson, and H. Lohn, *Glycoconjugate J.*, **1989**, *6*, 161–168.
109. T. Murata, S. Akimoto, M. Horimoto, and T. Usui, *Biosci. Biotechnol. Biochem.*, **1997**, *61*, 1118–1120.
110. K. Fukase, T. Yasukochi, Y. Suda, M. Yoshida, and S. Kusumoto, *Tetrahedron Lett.*, **1996**, *37*, 6763–6766.
111. G. Vic, J. J. Hastings, O. W. Howarth, and D. H. Crout, *Tetrahedron: Asymmetry*, **1996**, *7*, 709–720.
112. L. Hedbys, E. Johansson, K. Mosbach, and P.-O. Larsson, *Carbohydr. Res.*, **1989**, *186*, 217–223.
113. J. H. Yoon and K. Ajisaka, *Carbohydr. Res.*, **1996**, *292*, 153–163.
114. T. Usui, K. Ogawa, H. Nagai, and H. Matsui, *Anal. Biochem.*, **1992**, *202*, 61–67.
115. K. Koizumi, T. Tanimoto, K. Fujita, K. Hara, N. Kuwahara, and S. Kitahata, *Carbohydr. Res.*, **1993**, *238*, 75–91.

116. S. Takayama, M. Shimazaki, L. Qiao, and C. H. Wong, *Bioorg. Med. Chem. Lett.*, **1996**, *6*, 1123–1126.
117. S. Singh, M. Scigelova, G. Vic, and D. Crout, *J. Chem. Soc., Perkin Trans. 1*, **1996**, *21*, 1921–1926.
118. E. Bonnin and J. F. Thibault, *Enzyme Microb. Technol.*, **1996**, *19*, 99–106.
119. H. Fujimoto, M. Isomura, T. Miyazaki, I. Matsuo, R. Walton, T. Sakakibara, and K. Ajisaka, *Glycoconjugate J.*, **1997**, *14*, 75–80.
120. T. Kimura, S. Takayama, H. M. Huang, and C. H. Wong, *Angew. Chem. Int. Ed. Engl.*, **1996**, *35*, 2348–2350.
121. W. S. Hays, D. J. VanderJagt, B. Bose, A. S. Serianni, and R. H. Glew, *J. Biol. Chem.*, **1998**, *273*, 34941–34948.
122. J. W. Fang, W. H. Xie, J. Li, and P. G. Wang, *Tetrahedron Lett.*, **1998**, *39*, 919–922.
123. J. Li and P. G. Wang, *Tetrahedron Lett.*, **1997**, *38*, 7967–7970.
124. A. Vetere and S. Paoletti, *Biochem. Biophys. Res. Commun.*, **1996**, *219*, 6–13.
125. G. F. Herrmann, U. Kragl, and C. Wandrey, *Angew. Chem. Int. Ed. Engl.*, **1993**, *32*, 1342–1343.
126. K. Sakai, R. Katsumi, H. Ohi, T. Usui, and Y. Ishido, *J. Carbohydr. Chem.*, **1992**, *11*, 553–565.
127. K. Suyama, S. Adachi, T. Toba, T. Sohma, C.-J. Hwang, and T. Itoh, *Agric. Biol. Chem.*, **1986**, *50*, 2069–2075.
128. L. Hedbys, P.-O. Larsson, and K. Mosbach, *Biochem. Biophys. Res. Commun.*, **1984**, *123*, 8–15.
129. S. Singh, M. Scigelova, P. Critchley, and D. H. G. Crout, *Carbohydr. Res.*, **1998**, *305*, 363–370.
130. D. H. Crout, O. W. Howarth, S. Singh, B. E. P. Swoboda, P. Critchley, and W. T. Gibson, *J. Chem. Soc., Chem. Commun.*, **1991**, 1550–1551.
131. S. Singh, D. H. G. Crout, and J. Packwood, *Carbohydr. Res.*, **1995**, *279*, 321–325.
132. C. Ravet, D. Thomas, and M. D. Legoy, *Biotechnol. Bioeng.*, **1993**, *42*, 303–308.
133. K. Ajisaka, H. Nishida, and H. Fujimoto, *Biotechnol. Lett.*, **1987**, *9*, 243–248.
134. T. Tanaka and S. Oi, *Agric. Biol. Chem.*, **1985**, *49*, 1267–1273.
135. K. Kuriyama, K. Tsuchiya, and T. Murui, *Biosci. Biotechnol. Biochem.*, **1995**, *59*, 1142–1143.
136. T. Murata, A. Tashiro, T. Itoh, and T. Usui, *Biochim. Biophys. Acta G.*, **1997**, *1335*, 326–334.
137. Y. Matahira, A. Tashiro, T. Sato, H. Kawagishi, and T. Usui, *Glycoconjugate J.*, **1995**, *12*, 664–671.
138. T. Murata, T. Itoh, and T. Usui, *Glycoconjugate J.*, **1998**, *15*, 575–582.
139. S. Singh, J. Packwood, and D. H. G. Crout, *J. Chem. Soc., Chem. Commun.*, **1994**, 2227–2228.
140. S. Singh, J. Packwood, C. J. Samuel, P. Critchley, and D. H. G. Crout, *Carbohydr. Res.*, **1995**, *279*, 293–305.
141. S. Singh, R. Gallagher, P. J. Derrick, and D. H. G. Crout, *Tetrahedron: Asymmetry*, **1995**, *6*, 2803–2810.
142. F. Nanjo, K. Sakai, M. Ishikawa, K. Isobe, and T. Usui, *Agric. Biol. Chem.*, **1989**, *53*, 2189–2195.
143. F. Nanjo, M. Ishikawa, R. Katsumi, and K. Sakai, *Agric. Biol. Chem.*, **1990**, *54*, 899–906.
144. S. Kobayashi, T. Kiyosada, and S. Shoda, *Tetrahedron Lett.*, **1997**, *38*, 2111–2112.
145. S. Singh, J. Packwood, C. J. Samuel, P. Critchley, and D. H. G. Crout, *Carbohydr. Res.*, **1995**, *279*, 293–305.
146. E. Rajnochova, J. Dvorakova, Z. Hunkova, and V. Kren, *Biotechnol. Lett.*, **1997**, *19*, 869–872.
147. S. Kyosaka, S. Murata, Y. Tsuda, and M. Tanaka, *Chem. Pharm. Bull.*, **1986**, *34*, 5140–5143.
148. T. Usui, M. Suzuki, T. Sato, H. Kawagishi, K. Adachi, and H. Sano, *Glycoconjugate J.*, **1994**, *11*, 105–110.
149. S. Singh, M. Scigelova, and D. H. G. Crout, *J. Chem. Soc., Chem. Commun.*, **1996**, 993–994.
150. H. Kizawa, H. Shinoyama, and T. Yasui, *Agric. Biol. Chem.*, **1991**, *55*, 671–678.
151. M. Kurakake, S. Osada, and T. Komaki, *Biosci. Biotechnol. Biochem.*, **1997**, *61*, 2010–2014.
152. A. Vetere, M. Bosco, and S. Paoletti, *Carbohydr. Res.*, **1998**, *311*, 79–83.

153. R. M. Bardales and V. P. Bhavanandan, *J. Biol. Chem.*, **1989**, *264*, 19893–19897.
154. T. Usui, T. Murata, Y. Yabuuchi, and K. Ogawa, *Carbohydr. Res.*, **1993**, *250*, 57–66.
155. R. Uchida, A. Nasu, K. Tobe, T. Oguma, and N. Yamaji, *Carbohydr. Res.*, **1996**, *287*, 271–274.
156. K. Ogawa, O. Uejima, T. Nakakuki, T. Usui, and K. Kainuma, *Agric. Biol. Chem.*, **1990**, *54*, 581–586.
157. N. Payre, S. Cottaz, and H. Driguez, *Angew. Chem. Int. Ed. Engl.*, **1995**, *34*, 1239–1241.
158. T. Usui and T. Murata, *J. Biochem.*, **1988**, *103*, 969–972.
159. K. Ogawa, T. Murata, and T. Usui, *Carbohydr. Res.*, **1991**, *212*, 289–294.
160. K. Ogawa, T. Nakakuki, and T. Usui, *J. Carbohydr. Chem.*, **1991**, *10*, 877–886.
161. Y. Yoshimura, S. Kitahata, and S. Okada, *Agric. Biol. Chem.*, **1988**, *52*, 1655–1659.
162. S. Hizukuri, S. Kawano, J.-I. Abe, K. Koizumi, and T. Tanimoto, *Biotechnol. Appl. Biochem.*, **1989**, *11*, 60–73.
163. J.-I. Abe, N. Mizowaki, S. Hizukuri, K. Koizumi, and T. Utamura, *Carbohydr. Res.*, **1986**, *154*, 81–92.
164. T. Kuriki, M. Yanase, H. Takata, Y. Takesada, T. Imanaka, and S. Okada, *Appl. Environ. Microbiol.*, **1993**, *59*, 953–959.
165. N. Shiomi and J. Yamada, *Agric. Biol. Chem.*, **1988**, *52*, 1861–1862.
166. S. I. Shoda, K. Obata, O. Karthaus, and S. Kobyashi, *J. Chem. Soc., Chem. Commun.*, **1993**, *18*, 1402–1404.
167. O. Karthus, S. Shoda, H. Takano, K. Obata, and S. Kobayashi, *J. Chem. Soc., Perkin Trans. 1*, **1994**, 1851–1857.
168. H. Fujimoto, H. Nakano, M. Isomura, S. Kitahata, and K. Ajisaka, *Biosci. Biotechnol. Biochem.*, **1997**, *61*, 1258–1261.
169. H. Nakano, S. Kitahata, H. Kinugasa, Y. Watanabe, H. Fujimoto, K. Ajisaka, and S. Takenishi, *Agric. Biol. Chem.*, **1991**, *55*, 2075–2082.
170. P. W. Bezukladnikov and L. A. Elyakova, *Carbohydr. Res.*, **1988**, *184*, 268–270.
171. T. Kawaguchi, K. Sugimoto, H. Hayashi, and M. Arai, *Biosci. Biotechnol. Biochem.*, **1996**, *60*, 344–346.
172. J. L. Viladot, B. Stone, H. Driguez, and A. Planas, *Carbohydr. Res.*, **1998**, *311*, 95–99.
173. J. L. Viladot, V. Moreau, A. Planas, and H. Driguez, *J. Chem. Soc., Perkin Trans. 1*, **1997**, 2383–2387.
174. S. Armand, S. Drouillard, M. Schulein, B. Henrissat, and H. Driguez, *J. Biol. Chem.*, **1997**, *272*, 2709–2713.
175. S. Kobayashi, T. Kawasaki, K. Obata, and S. I. Shoda, *Chem. Lett.*, **1993**, 685–686.
176. S. Shoda, T. Kawasaki, K. Obata, and S. Kobayashi, *Carbohydr. Res.*, **1993**, *249*, 127–137.
177. S. Kobayashi, K. Kashiwa, T. Kawasaki, and S. Shoda, *J. Am. Chem. Soc.*, **1991**, *113*, 3079–3084.
178. J. H. Lee, R. M. Brown, S. Kuga, S. Shoda, and S. Kobayashi, *Proc. Nat. Acad. Sci. USA*, **1994**, *91*, 7425–7429.
179. V. Moreau and H. Driguez, *J. Chem. Soc., Perkin Trans. 1*, **1996**, 525–527.
180. M. Fujita, S. Shoda, and S. Kobayashi, *J. Am. Chem. Soc.*, **1998**, *120*, 6411–6412.
181. H. Ashida, Y. Tsuji, K. Yamamoto, H. Kumagai, and T. Tochikura, *Arch. Biochem. Biophys.*, **1993**, *305*, 559–562.
182. Y. T. Li, B. Z. Carter, B. N. N. Rao, H. Schweingruber, and S. C. Li, *J. Biol. Chem.*, **1991**, *266*, 10723–10726.
183. K. Yamada, E. Fujita, and S. I. Nishimura, *Carbohydr. Res.*, **1998**, *305*, 443–461.
184. K. Akiyama, K. Kawazu, and A. Kobayashi, *Carbohydr. Res.*, **1995**, *279*, 151–160.
185. J. Q. Fan, L. H. Huynh, B. B. Reinhold, V. N. Reinhold, K. Takegawa, S. Iwahara, A. Kondo, I. Kato, and Y. C. Lee, *Glycoconjugate J.*, **1996**, *13*, 643–652.
186. H. Matsui, H. Kawagishi, and T. Usui, *Biochim. Biophys. Acta*, **1990**, *1035*, 90–96.
187. Y. Matahira, K. Ohno, M. Kawaguchi, H. Kawagishi, and T. Usui, *J. Carbohydr. Chem.*, **1995**, *14*, 213–225.
188. K. Takegawa, K. Fujita, J. Q. Fan, M. Tabuchi, N. Tanaka, A. Kondo, H. Iwamoto, I. Kato, Y. C. Lee, and S. Iwahara, *Anal. Biochem.*, **1998**, *257*, 218–223.
189. T. Rand-Meir, F. W. Dahlquist, and M. A. Raftery, *Biochemistry*, **1969**, *8*, 4206–4214.

190. I. L. Deras, K. Takegawa, A. Kondo, I. Kato, and Y. C. Lee, *Bioorg. Med. Chem. Lett.*, **1998**, *8*, 1763–1766.
191. H. Saitoh, K. Takagaki, M. Majima, T. Nakamura, A. Matsuki, M. Kasai, H. Narita, and M. Endo, *J. Biol. Chem.*, **1995**, *270*, 3741–3747.
192. T. Usui, Y. Hayashi, F. Nanjo, K. Sakai, and Y. Ishido, *Biochim. Biophys. Acta*, **1987**, *923*, 302–309.
193. K. Haneda, T. Inazu, M. Mizuno, R. Iguchi, K. Yamamoto, H. Kumagai, S. Aimoto, H. Suzuki, and T. Noda, *Bioorg. Med. Chem. Lett.*, **1998**, *8*, 1303–1306.
194. L. X. Wang, M. Tang, T. Suzuki, K. Kitajima, Y. Inoue, S. Inoue, J. Q. Fan, and Y. C. Lee, *J. Am. Chem. Soc.*, **1997**, *119*, 11137–11146.
195. T. Usui, Y. Hayashi, F. Nanjo, and Y. Ishido, *Biochim. Biophys. Acta*, **1988**, *953*, 179–184.
196. T. Usui and H. Matsui, *Agric. Biol. Chem.*, **1989**, *53*, 383–388.
197. K. Haneda, T. Inazu, K. Yamamoto, H. Kumagai, Y. Nakahara, and A. Kobata, *Carbohydr. Res.*, **1996**, *292*, 61–70.
198. K. J. Yamamoto, S. Kadowaki, J. Watanabe, and H. Kumagai, *Biochem. Biophys. Res. Commun.*, **1994**, *203*, 244–252.
199. K. Yamamoto, K. Fujimori, K. Haneda, M. Mizuno, T. Inazu, and H. Kumagai, *Carbohydr. Res.*, **1998**, *305*, 415–422.
200. M. Mizuno, K. Haneda, R. Iguchi, I. Muramoto, T. Kawakami, S. Aimoto, K. Yamamoto, and T. Inazu, *J. Am. Chem. Soc.*, **1999**, *121*, 284–290.
201. S. Kobayashi, X. Wen, and S.-I. Shoda, *Macromolecules*, **1996**, *29*, 2698–2700.
202. M. Scigelova, V. Kren, and K. G. I. Nilsson, *Biotechnol. Lett.*, **1994**, *16*, 683–688.
203. E. V. Eneyskaya, A. M. Golubev, A. M. Kachurin, A. N. Savel'ev, and K. N. Neustroev, *Carbohydr. Res.*, **1998**, *305*, 83–91.
204. G. Vic, J. J. Hastings, and D. H. G. Crout, *Tetrahedron: Asymmetry*, **1996**, *7*, 1973–1984.
205. M. Scigelova, P. Sedmera, V. Havlicek, V. Prikrylova, and V. Kren, *J. Carbohydr. Chem.*, **1998**, *17*, 981–986.
206. D. Cantacuzene, S. Attal, and S. Bay, *Biomed. Biochim. Acta*, **1991**, *50*, S231–S236.
207. E. Johansson, L. Hedbys, and P. O. Larsson, *Enzyme Microb. Technol.*, **1991**, *13*, 781–787.
208. K. G. I. Nilsson, G. Ljunger, and P. M. Melin, *Biotechnol. Lett.*, **1997**, *19*, 889–892.
209. V. Laroute and R.-M. Willemot, *Biotechnol. Lett.*, **1992**, *14*, 169–174.
210. M.-P. Bousquet, R.-M. Willemot, P. Monsan, and E. Boures, *Enzyme Microb. Technol.*, **1998**, *23*, 83–90.
211. M.-P. Bousquet, R.-M. Willemot, P. Monsan, and E. Boures, *Appl. Microbiol. Biotechnol.*, **1998**, *50*, 167–173.
212. C. Kieburg, T. K. Lindhorst, and V. Kren, *J. Carbohydr. Chem.*, **1998**, *17*, 1239–1247.
213. T. Mori and Y. Okahata, *Tetrahedron Lett.*, **1997**, *38*, 1971–1974.
214. A. Ismail and M. Ghoul, *Biotechnol. Lett.*, **1996**, *18*, 1199–1204.
215. A. J. J. Straathof, J. P. Vrijenhoef, E. P. A. T. Sprangers, H. van Bekkum, and A. P. G. Kieboom, *J. Carbohydr. Chem.*, **1988**, *7*, 223–238.
216. M. Kojima, T. Seto, Y. Kyotani, H. Ogawa, S. Kitazawa, K. Mori, S. Maruo, T. Ohgi, and Y. Ezure, *Biosci. Biotechnol. Biochem.*, **1996**, *60*, 694–696.
217. N. Asano, A. Kato, H. Kizu, K. Matsui, R. C. Griffiths, M. G. Jones, A. A. Watson, and R. J. Nash, *Carbohydr. Res.*, **1997**, *304*, 173–178.
218. T. Wiemann, N. Taubken, U. Zehavi, and J. Thiem, *Carbohydr. Res.*, **1994**, *257*, C1–C6.
219. D. H. G. Crout, D. A. Macmanus, and P. Critchley, *J. Chem. Soc., Perkin Trans. 1*, **1990**, 1865–1868.
220. Y. Ooi, N. Mitsuo, and T. Satoh, *Chem. Pharm. Bull.*, **1985**, *33*, 5547–5550.
221. H.-J. Gais, A. Zeissler, and P. Maidonis, *Tetrahedron Lett.*, **1988**, *29*, 5743–5744.
222. Y. Ooi, T. Hashimoto, N. Mitsuo, and T. Satoh, *Chem. Pharm. Bull.*, **1985**, *33*, 1808–1814.
223. Y. Ooi, T. Hashimoto, N. Mitsuo, and T. Satoh, *Tetrahedron Lett.*, **1984**, *25*, 2241–2244.
224. S. Matsumura, H. Yamazaki, and K. Toshima, *Biotechnol. Lett.*, **1997**, *19*, 583–586.
225. D. E. Stevenson, R. A. Stanley, and R. H. Furneaux, *Biotechnol. Bioeng.*, **1993**, *42*, 657–666.
226. S. Matsumura, H. Kubokawa, and S. Yoshikawa, *Chem. Lett.*, **1991**, 945–948.
227. Y. Fortun and B. Colas, *Biotechnol. Lett.*, **1991**, *13*, 863–866.
228. Y. Okahata and T. Mori, *J. Chem. Soc., Perkin Trans. 1*, **1996**, 2861–2866.

229. T. Mori, S. Fujita, and Y. Okahata, *Chem. Lett.*, **1997**, 73–74.
230. T. Mori, S. Fujita, and Y. Okahata, *Carbohydr. Res.*, **1997**, *298*, 65–73.
231. Y. Okahata and T. Mori, *J. Mol. Cat. B Enzym.*, **1998**, *5*, 119–123.
232. T. Mori and Y. Okahata, *J. Chem. Soc., Chem. Commun.*, **1998**, 2215–2216.
233. R. R. Gibson, R. P. Dickinson, and G. J. Boons, *J. Chem. Soc., Perkin Trans. 1*, **1997**, 3357–3360.
234. S. Leparoux, Y. Fortun, and B. Colas, *Biotechnol. Lett.*, **1994**, *16*, 677–682.
235. K. G. I. Nilsson and M. Scigelova, *Biotechnol. Lett.*, **1994**, *16*, 671–676.
236. D. E. Stevenson and R. H. Furneaux, *Carbohydr. Res.*, **1996**, *284*, 279–283.
237. S. Leparoux, M. Padrines, Y. Fortun, and B. Colas, *Biotechnol. Lett.*, **1996**, *18*, 135–138.
238. N. J. Turner and M. C. Webberley, *J. Chem. Soc., Chem. Commun.*, **1991**, 1349–1350.
239. A. Baker, N. J. Turner, and M. C. Webberley, *Tetrahedron: Assymetry*, **1994**, *5*, 2517–2522.
240. G. Vic and D. Thomas, *Tetrahedron Lett.*, **1992**, *33*, 4567–4570.
241. P. Christakopoulos, M. K. Bhat, D. Kekos, and B. J. Macris, *Int. J. Biol. Macromol.*, **1994**, *16*, 331–334.
242. M. Gelo-Pujic, E. Guibe-Jampel, and A. Loupy, *Tetrahedron*, **1997**, *53*, 17247–17252.
243. A. Trincone and E. Pagnotta, *Biotechnol. Lett.*, **1995**, *17*, 45–48.
244. H. Shinoyama, K. Takel, A. Ando, T. Fujii, M. Sasaki, Y. Doi, and T. Yasui, *Agric. Biol. Chem.*, **1991**, *55*, 1679–1681.
245. A. Millqvist-Fureby, I. S. Gill, and E. N. Vulfson, *Biotechnol. Bioeng.*, **1998**, *60*, 190–196.
246. Z. Gunata, M. J. Vallier, J. C. Sapis, R. Baumes, and C. Bayonove, *Enzyme Microb. Technol.*, **1994**, *16*, 1055–1058.
247. G. Vic, D. Thomas, and D. H. G. Crout, *Enzyme Microb. Technol.*, **1997**, *20*, 597–603.
248. D. K. Watt, H. Ono, and K. Hayashi, *Biochim. Biophys.*, **1998**, *1385*, 78–88.
249. G. Vic and D. H. G. Crout, *Tetrahedron: Asymmetry*, **1994**, *5*, 2513–2516.
250. M. Gelo-Pujic, E. Guibe-Jampel, A. Loupy, and A. Trincone, *J. Chem. Soc., Perkin Trans. 1*, **1997**, 1001–1002.
251. G. Ljunger, P. Adlercreutz, and B. Mattiasson, *Enzyme Microb. Technol.*, **1994**, *16*, 751–755.
252. N. Taubken and J. Thiem, *Synthesis (C)*, **1992**, 517–518.
253. H. Shinoyama and T. Yasui, *Agric. Biol. Chem.*, **1988**, *52*, 2375–2377.
254. H. Shinoyama, Y. Kamiyama, and T. Yasui, *Agric. Biol. Chem.*, **1988**, *52*, 2197–2202.
255. P. Drouet, M. Zhang, and M. D. Legoy, *Biotechnol. Bioeng.*, **1994**, *43*, 1075–1080.
256. J. Sulistyo, Y. Kamiyama, H. Ito, and T. Yasui, *Biosci. Biotechnol. Biochem.*, **1994**, *58*, 1311–1313.

30 Production of Heterologous Oligosaccharides by Recombinant Bacteria (Recombinant Oligosaccharides)

Roberto A. Geremia and Eric Samain

30.1 Introduction

During recent years it has been firmly established that oligosaccharides, either free or conjugated, are involved in many biological recognition processes, such as those that occur during fertilization, embryogenesis, inflammation, symbiotic nitrogen fixation, and host pathogen adhesion [1]. These advances have led to increased interest in potential applications of oligosaccharides and derivatives in therapeutics and in the agricultural industry. Both chemical and enzymatic methods have been developed for production of relevant amounts of biologically active oligosaccharides. Chemical methods still need many protection and deprotection steps; as the number of steps increases with the size of the oligosaccharide, longer oligosaccharides are produced in lower yields. Enzymatic methods use purified glycosylhydrolases or glycosyltransferases to assemble the oligosaccharide. Glycosyltransferases (GTs), are enzymes that transfer a sugar from a donor to an acceptor. With Leloir GTs the sugar donor is a sugar–nucleotide (Figure 1). The most specific enzymatic methods use glycosyltransferases, consequently these methods suffer from the relatively poor availability of glycosyltransferases and from the need to regenerate *in situ* sugar–nucleotides.

Taking into account that enzymatic methods basically mimic in vitro the natural pathways of oligosaccharide biosynthesis, a reasonable short-cut is the production of the oligosaccharide in a living recombinant organism (*e.g. Escherichia coli*). Unfortunately this task is not simple to accomplish, because of the characteristics of the biosynthetic machinery involved in the synthesis of the desired products. Production in *E. coli* of active eukaryotic GTs in suitable amounts is not always possible or requires considerable effort for reasons that are not well understood. This problem can be overcome by use of homologous enzymes from prokaryotic organisms.

A second possible difficulty consists in the availability of sugar–nucleotides. Bacteria recycle these compounds, because they are necessary for the synthesis of their cell wall carbohydrate polymers (murein, lipopolysaccharide, capsular poly-

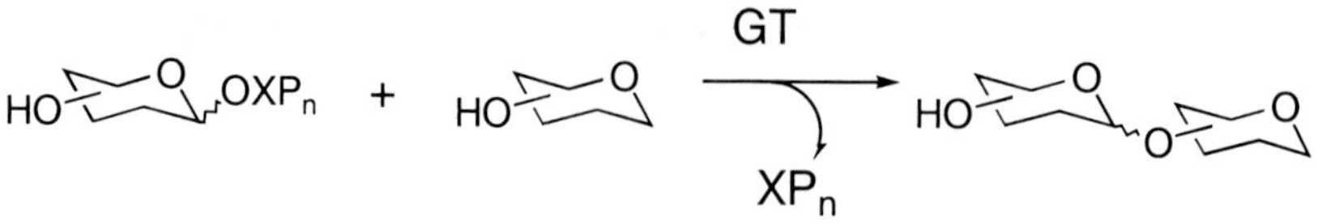

Figure 1. Reaction catalyzed by 'Leloir' glycosyltransferases.

saccharides, *etc.*) [2]. Occasionally, *e.g.* with GDP–fucose, the production of the sugar donor is probably coupled with the corresponding polysaccharide biosynthetic system, which is, in turn, induced by particular physiological or environmental conditions. For these reasons it is difficult to estimate the actual availability of the sugar donor. The solution consists either in adapting the culture conditions or genetically modifying the bacteria for the production of the desired sugar–nucleotide.

Finally, the most difficult problem to address is that of the acceptor. The physiological acceptor for glycosyltransferases involved in oligosaccharide synthesis is a mono- or oligosaccharide, which can be linked to a protein, a lipid, or another carrier molecule. The acceptor requirements are well studied for eukaryotic GTs, but little is known for the bacterial variety. Production of the oligosaccharide coupled to proteins or lipids requires engineering of the complete glycosylation pathway into the bacteria, which is an enormous task. The use of free sugars in vivo as acceptors is not feasible, because they are rapidly catabolized, thus their intracellular concentration is virtually zero. One possibility is to use lipid-linked intermediates as acceptors but they are present in small amounts [3], limiting the amount of oligosaccharide produced. The availability of an acceptor in sufficient amounts is, therefore, the limiting step for the production of oligosaccharides in bacteria.

If the desired oligosaccharide is produced, it might be necessary to add non-glycosidic substituents, requiring the expression of the appropriate modifying enzyme and the presence of donors for the modification.

Despite these drawbacks, this system remains attractive because once the genes are cloned and conditions established, the production is performed in vivo by the bacteria, without all the problems found in chemical or enzymatic synthesis. *E. coli* is commonly used for the production of heterologous DNA or proteins, and is a very well known organism. Its metabolism can be manipulated by varying the culture conditions. One of the most exciting features for our purposes is that it can be grown at high or very high cell density (OD optical density 100–200, in contrast with 2–5 under normal conditions). This means the amount of biomass obtained in a single culture can be more than 50 g L^{-1} (dry weight).

An exceptional opportunity for synthesizing oligosaccharides in *E. coli*, is provided by the rhizobial chitin oligosaccharide synthase NodC. The putative acceptor or primer for this enzyme has not yet been identified, but is present in *E. coli*, which enabled the production of chitin oligosaccharides in vivo. By use of the methodology developed by us and described below, *E. coli* strains expressing *nodC* produce 1–2 g L^{-1} chitin pentasaccharide [4]. Modification with glycosidic or non-glycosidic groups was also achieved by co-expression of the proper genes, with production yields between 0.2 and 1 g L^{-1} [4–6].

30.2 Concept and Methodology of Heterologous ('Recombinant') Oligosaccharide Production in *E. coli*

Sugar–nucleotides and glycosyltransferases are already present in *E. coli*, because it produces a series of cell-wall polysaccharides. The enzymes for the biosynthesis of sugar–nucleotides are cytoplasmic, whereas most GTs involved in the biosynthesis of outer polysaccharides are peripheral membrane proteins whose catalytic domain is thought to face the cytoplasm [7, 8]. Therefore, to use the pool of sugar–nucleotides, the recombinant GT should be targeted at the cytoplasm. An advantage of expressing recombinant bacterial glycosyltransferase in *E. coli* is that cytoplasm is their natural environment, and that having transmembrane domains, insertion into membrane in the *E. coli* host possibly occurs by similar mechanisms. Unfortunately, most eukaryotic GT are membrane-bound and present in the luminal side of endoplasmic reticulum or the Golgi apparatus, that is equivalent to the outside of cells. Because these organelles are not present in bacteria, it is necessary to remove the transmembrane domain (involved in targeting) to keep the protein in bacterial cytoplasm. It should be pointed out that the heterologous oligosaccharides are not secreted, and that they remain in the cytoplasm.

The production of heterologous oligosaccharides uses recombinant DNA technology to produce recombinant GTs, and the *E. coli* cell machinery to produce the desired oligosaccharides. We have chosen to call the heterologous oligosaccharides obtained by this approach 'recombinant oligosaccharides'.

30.2.1 Biosynthesis of Nod Factors

One of the major sources of biological N_2 fixation is the symbiosis relationship *Rhizobium*–leguminous. As part of this symbiosis a dedicated N_2 fixation organ, called nodule, is formed. The nodule is a plant structure that is invaded by rhizobia, and its formation results from a 'molecular dialog' between the plant and the bacteria [9]. Briefly, the plant secretes specific flavonoids that induce the expression of *nod* genes, resulting in the synthesis of Nod factors. Upon secretion by specific proteins, the Nod factor will induce the deformation of hairy roots, and mitosis of plant cells. Eventually, the bacteria invade the root cell, migrate towards the cell division region, and are released into these cells. Finally, bacteria differentiate into a N_2-fixing form, the bacteroid.

Nod factors consist of a chitin (tri-, tetra- or penta-) oligosaccharide, of which the non-reducing end *N*-acetyl group is substituted by a long-chain fatty acid. Several substitutions are found at the reducing- and non-reducing-end sugars, although substitutions on other residues are present in some species [10, 11]. The Nod factors

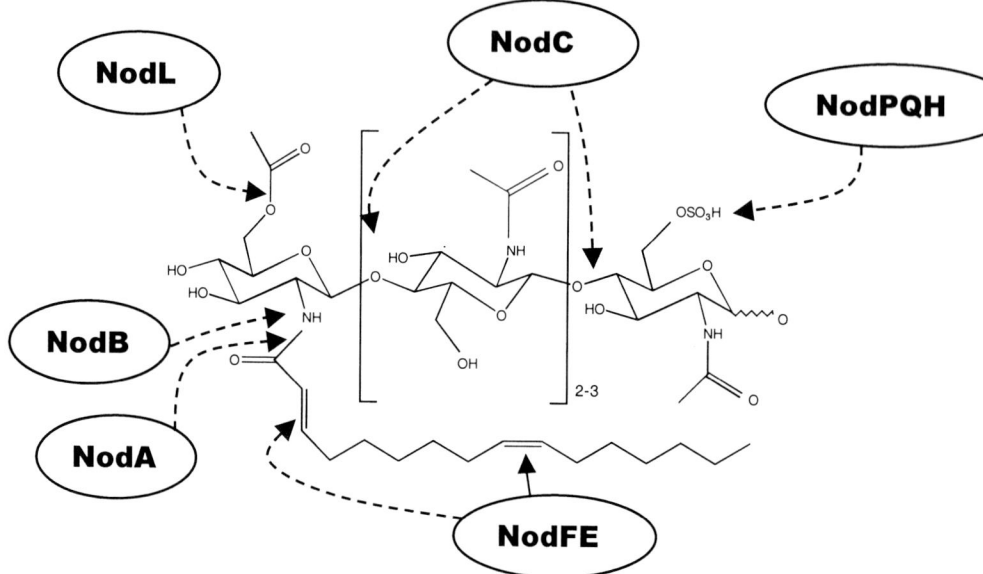

Figure 2. Structure of NodRmIV (Ac, S). Proteins involved in addition of each substituent are circled.

produced by *R. meliloti* are shown in Figure 2. The first committed step of Nod factor biosynthesis is the formation of the chitin oligomer by NodC using UDP-N-satylglucosamine (UDP–GlcNAc) as the sugar donor [11]. NodB specifically deacetylates the non-reducing terminus of the oligosaccharide. The NH_2 group is finally acylated in a step that requires NodA. Other substituents are added before or after the acylation by specific enzymes (Figure 2). The final product, is secreted by a system that requires NodI and NodJ [11].

The *nod* genes were discovered as result of the identification of rhizobial non-nodulating mutants. These genes were isolated and sequenced [12, 13] but no function was assigned to them. Discovery of Nod factors [14] brought substance to the study of Nod protein function. The first hint of the function of NodC was the sequence similarity with cellulose and chitin synthases [15]. The in vitro synthesis using rhizobial cell extracts or *E. coli* extracts containing recombinant NodC was then accomplished [16] using UDP–[^{14}C]GlcNAc as sugar donor. The oligosaccharide formed was shown to bind to wheat germ agglutinin, and to be sensitive to chitinase, but the minute amounts of the chitin oligosaccharide available precluded unambiguous structural characterization.

It was, nevertheless, discovered that the *E. coli* strain expressing NodC was able to produce, in vivo, small amounts of chitin oligosaccharides; these were characterized by mass spectrometry [17]. The product consisted of a mixture of di-, tri-, tetra- and pentasaccharides, and purification involved several steps. The use of

E. coli strains expressing several combinations of *nod* genes enabled the elucidation of the biosynthetic pathway and the function of these genes [17], by establishing the structure of the corresponding Nod metabolites. This was the first example of the production of a recombinant oligosaccharide and the starting point for the production of oligosaccharide in *E. coli*. Our further reasoning was that if a single batch culture (biomass representing OD = 2) can produce a few mg L^{-1} Nod metabolites, in a culture of high density (OD = 100) it would produce a few hundred mg L^{-1}.

It was found by sequence analysis and tagging with reporter enzymes [18], that NodC is a membrane-bound protein with four transmembrane domains (one at the NH_2 terminal end, and three at the COOH terminus) flanking a 300 amino acid cytoplasmic peptide. The cytoplasmic moiety of NodC shares sequence similarity with several glycosyltransferases and polysaccharide synthases [19–21] in two different domains (domain A towards the $-NH_2$ end and domain B towards the –COOH end). Whereas both domains are conserved in polysaccharide synthases, only domain A is present in glycosyltransferases [21]. This led to the hypothesis that domain B is involved in multiple sugar transfer. The suggested mechanism proposes the presence of one catalytic center in each domain. It was also proposed that the acceptor is formed by the enzyme [21]. To date several authors have reported the in vitro formation of chitin oligosaccharides in the absence of the putative GlcNAc acceptor [22, 23].

30.2.2 Expression Systems and Cloning Strategy

UDP–GlcNAc is a key metabolite of bacteria, because it is a precursor for the formation of murein and Lipid A, both necessaries for cell viability. UDP–GlcNAc is synthesized from fructose-6-phosphate in four steps. The two last reactions leading to UDP–GlcNAc from glucosamine-6-phosphate are catalyzed by a bifunctional enzyme encoded by *glmU* [24]. The acetyltransferase activity of this enzyme is strongly inhibited by GlcNAc-1-P and by UDP–*N*-acetylmuramic acid, suggesting that this reaction is the main point of UDP–GlcNAc synthesis regulation.

Because UDP–GlcNAc is substrate for NodC, the recombinant enzyme should be expressed at low levels, so as not to disturb cell wall synthesis. To regulate gene expression, the appropriate genes are cloned under the control of a given promoter. Two promoters were used, P_{Lac} and P_{Ara}. P_{Lac} promoter has the advantage that it allows a low-level expression in the absence of inducer, while P_{Ara} is a tight promoter [25], and can be used to tune the expression of the gene by use of an inducer (arabinose). For the production of a single oligosaccharide with glycosidic or nonglycosidic modifications, the expression of several genes is required. This can be achieved by cloning more than one gene into a given vector, or in two compatible vectors. With eukaryotic GT genes it is necessary to provide a bacterial ribosomal binding site, and to remove regions coding for transmembrane domains; one solution is to produce a fusion protein. Some of the plasmids used are shown in Figure 3; they are commonly available [4–6].

850 30 Production of Heterologous Oligosaccharides by Recombinant Bacteria

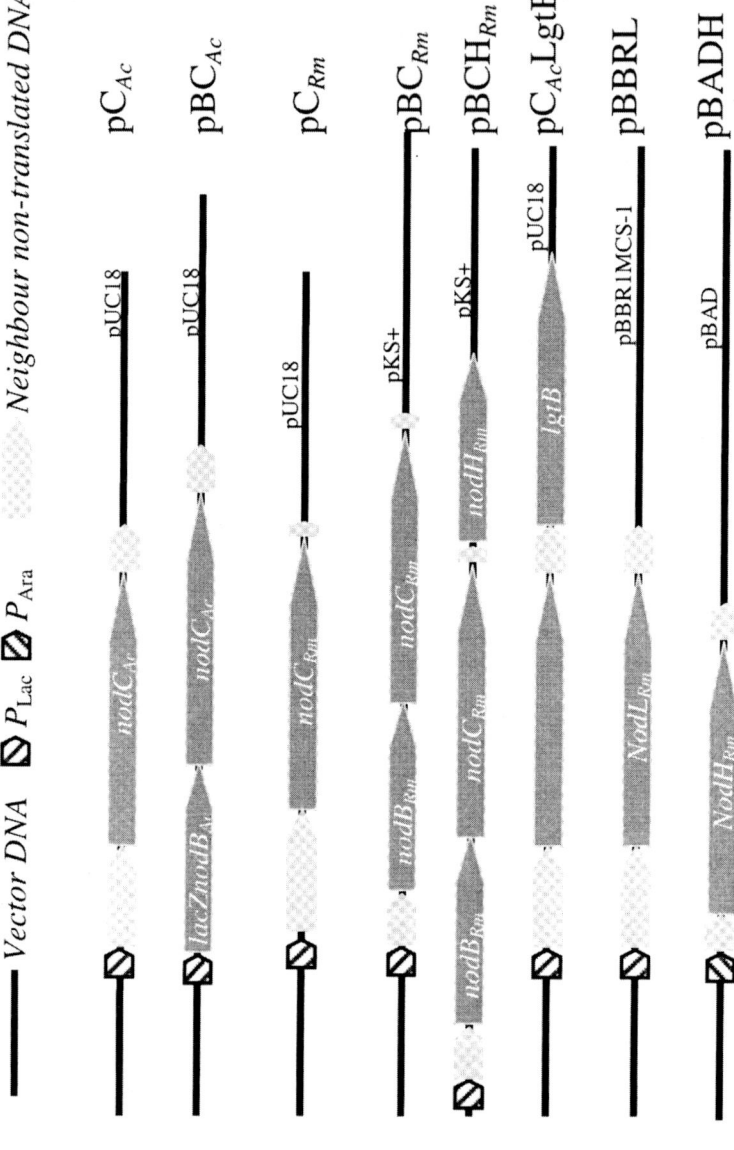

Figure 3. Plasmids containing chitin oligosaccharide synthases and different modifying enzymes. Relevant genes are in black-filled arrows, promoters in hatched arrows. Because of the cloning strategy, DNA contiguous to the desired genes are present in the constructions (gray).

30.2.3 High Cell-Density Cultivation

The cultivation of *E. coli* to high cell concentration is one way of maximizing the volumetric productivity of recombinant oligosaccharides. The main problem encountered in high cell-density culture (HCDC) is the risk of O_2 limitation, which obliges the bacteria to shift to a fermentative metabolism, resulting in the production of growth-inhibiting metabolites such as acetic acid. Because the O_2 delivery capacity of the fermenter is limited, the easiest way to prevent O_2 shortage is to limit the bacterial growth by controlled feeding of the carbon source. Different feeding strategies have been proposed [26, 27]. Although some strategies require sophisticated feedback control of substrate concentration to determine the feeding rate, reasonably high cell density can be also reached with simple fed-batch technique using pre-determined feeding rates to sustain carbon-limited growth [28]. This simple technique requires minimal fermentation equipment: a laboratory-scale fermenter with temperature, pH and O_2 controls, and a peristaltic pump for the feeding.

The protocol we routinely use to produce recombinant chitin oligosaccharides in high cell density cultures of *E. coli* includes the use of glycerol as carbon source because its catabolism does not provoke substantial accumulation of acetate, as does the use of glucose. The fed-batch strategy is outlined in Figure 4. During phase 1, cells grow exponentially until all the glycerol that was initially added in the starting media (17 g L^{-1}) had been consumed [4]. To prevent oxygen limitation,

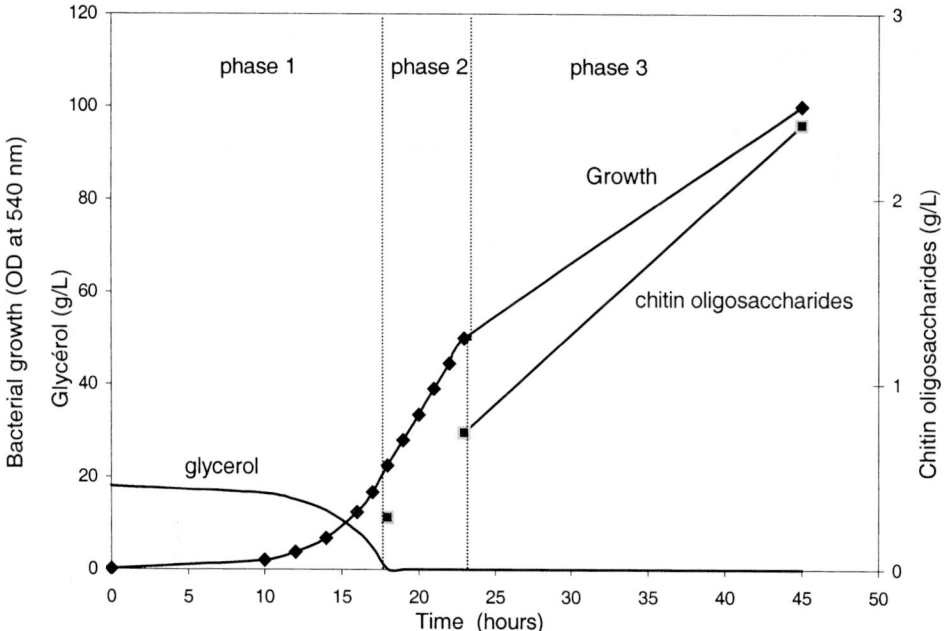

Figure 4. High cell-density culture strategy. For details see Ref. [3].

growth is then carbon limited, and it is supported by the continuous feeding of a glycerol solution. Glycerol is first supplied at a high feeding rate in phase 2 to promote a rapid increase in biomass. Finally growth is conducted at a low rate (phase 3) to increase in the intracellular concentration of recombinant oligosaccharide.

30.2.4 Purification of Recombinant Oligosaccharides

The recombinant chitin oligosaccharides are produced and accumulated in the cytoplasm and are not excreted in the culture medium. The first purification step consists of harvesting the cells by centrifugation to remove all culture supernatant. The cells are then suspended in distilled water and permeabilized either by a heat shock (30 min at 100 °C) or by a series of freeze–thaw cycles. The latter method is employed for compounds which are susceptible to damage at high temperature. These treatments enable the oligosaccharides to diffuse outside the permeabilized cell. After a second centrifugation the oligosaccharides are recovered in the supernatant and are then further purified by adsorption on charcoal–celite and selective elution with aqueous ethanol. Positively charged N-deacetylated chitin oligosaccharides and negatively charged sulfated compounds can be advantageously separated by ion-exchange chromatography. These purification procedures enable gram-scale preparation of oligosaccharides, which are found to be more than 90% pure by HPLC and NMR.

30.3 Examples of Recombinant Oligosaccharides

30.3.1 Production of Chitin Oligosaccharides in *E. coli* Expressing NodC

The production of chitin oligosaccharides is achieved in *E. coli* DH1 cells containing pC$_{Ac}$ or pC$_{Rm}$ (Figure 3). *R. meliloti* produces Nod factors in which the oligosaccharide consists of a mixture of tetra- and pentasaccharide [29], whereas *A. caulinodans* NodC produces pentasaccharide as the major product [30]. The final degree of polymerization of the oligosaccharides is determined by NodC [5, 23] but the concentration of UDP–GlcNAc influences the size of the oligosaccharide [22, 23], perhaps by acting as a chain terminator. Because the concentration of the sugar donor can change during cell growth, and under different metabolic conditions, the size of the oligosaccharides was assessed at the end of phase 1 and at the end of phase 3. On both occasions the major products were the pentasaccharide for pC$_{Ac}$, and a mixture of tetra- (70%) and pentasaccharide (30%) for pC$_{Rm}$ [5]. Whereas these oligosaccharides were almost pure at the end of phase 3, other products were found at the end of phase 1.

These products comprise mainly derivatives of the chitin oligosaccharides whose reducing end was found to be substituted with a glycerol moiety, linked in the β configuration to either the primary or secondary hydroxyl groups of glycerol [31].

Glycerol-substituted oligosaccharides are no longer detectable at the end of phase 3. During phase 1 the intracellular concentration of glycerol is high, because its uptake is mediated by the glycerol facilitator GlpF [32]. In contrast, during phases 2 and 3 the added glycerol is immediately used to sustain growth, therefore the intracellular concentration is low. These results strongly suggest that high intracellular glycerol concentration promotes its incorporation at the reducing end. The incubation in vitro of *E. coli* membranes containing NodC in the presence of glycerol and UDP[^{14}C]GlcNAc, leads to the formation of chitinase-sensitive radioactive oligosaccharides, that migrate more slowly than the corresponding chitin oligosaccharides [31]. These results suggest that NodC adds the glycerol moiety to the oligosaccharide.

30.3.2 Production of Nod Factor Precursors

As mentioned above, much work has been devoted to understanding the function of the proteins involved in Nod factor biosynthesis, but their mode of action is not fully understood. Future trends are related to Nod factor recognition and signalling [33–36]. Because chemical synthesis is long and tedious, we have proposed a new strategy involving the production of Nod factor precursors in *E. coli* expressing appropriate biosynthetic genes [37]. These oligosaccharides are provided to chemists for further modifications. In the first instance, when *E. coli* DH1 strains harboring the plasmid pBC$_{Ac}$ were employed, chitin pentasaccharides N-deacetylated at the non-reducing terminus were obtained in yields close to 1 g L^{-1} of culture [4]. This oligosaccharide was acylated by chemical means [37]. and used for characterization of Nod factor protein binding sites [38]. Further chemical modification will enable the production of Nod factor derivatives useful for affinity chromatography, fluorescence and/or photoactivatable probes. These molecules will be useful tools for isolation and biochemical characterization of the Nod factor binding sites. The substantial amounts obtained will also facilitate the study of the signal transduction pathway as well as the regulation of plant modulation genes expression.

The biological activity of *R. meliloti* Nod factors requires a tetrasaccharide and the presence of sulfate and *O*-acetate. We have used several combinations of the *nod* genes (NodB, NodC, NodH, NodL); these have enabled us to produce as many as eight different molecules (see Figures 5 and 6) in the recombinant oligosaccharide system.

In *R. meliloti*, three genes (*nodPQH*) are necessary for the sulfation of Nod factors [39]. The protein NodH catalyzes the transfer of sulfate from PAPS (3'-phosphoadenosine 5'-phosphosulfate) to the reducing terminal 6-O position of Nod factors [40]. The enzyme is also active on chitin oligosaccharides and their N-deacetylated derivatives. Kinetic analysis indicates that the affinity of NodH for chitin oligosaccharides is an order of magnitude lower than for lipo-chitin oligosaccharides, suggesting that sulfation occurs in vivo after the acylation [41]. The K_m value of ca 150–200 μM found in vitro for chitin oligosaccharides should, however, be sufficient to enable the sulfation of recombinant chitin oligosaccharides which have been shown to accumulate in the cytoplasm of *E. coli* at a concentration higher than 10 mM.

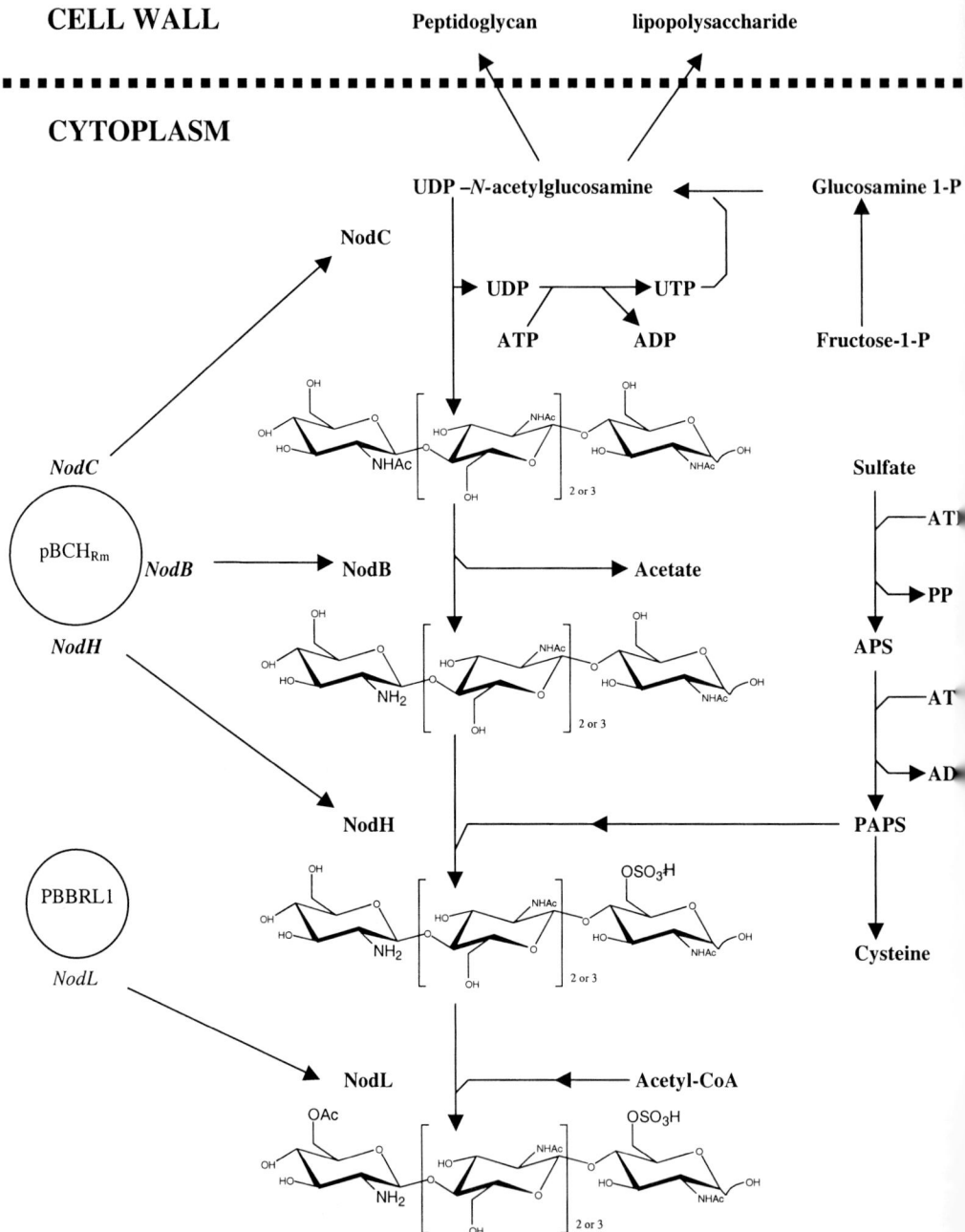

Figure 5. Metabolic pathway leading to the synthesis of Nod factor precursors in *E. coli*. (NodL and NodH can both act on unsubstituted oligosaccharides.)

The proteins NodP (ATP sulfurylase) and NodQ (adenosine 5′-phosphosulfate (APS) kinase) are associated in a sulfate-activating complex that enables the synthesis of the sulfate donor (PAPS) for Nod factor sulfation [42]. PAPS is also the sulfate donor used for the synthesis of cysteine and it is produced in *E. coli* by the normal household proteins CysC, CysD, and CysN [43, 44]. The genes *cysCDN* are part of the cysteine regulon which is repressed in the presence of cysteine and other reduced sulfur compounds [45]. This means that the *E. coli* machinery of PAPS biosynthesis can be used for chitin oligosaccharide sulfation if the bacteria are cultivated with sulfate as the only sulfur source. The total PAPS demand for the synthesis of cysteine and methionine (87 and 146 μmol g^{-1} dried cells, respectively) [46] largely exceeds the maximum demand that can be estimated for chitin oligosaccharide sulfation (around 20 μmol g^{-1} dried cells).

High cell density cultivation of strain DH1 (pBCH$_{Rm}$) leads to the production of 330 mg L^{-1} sulfated chitin oligosaccharides. This represented a sulfation yield of ca 50%. Attempts were made to improve this yield by expressing *nodH* in *trans* under the control of the P$_{BAD}$ promoter in a separate plasmid compatible with pBC$_{Rm}$. The sulfation yields were similar to that obtained with *nodH* expressed in *cis*. Moreover, this approach has another advantage, because it limits the number of plasmids that have to be constructed to obtain strains expressing different possible combinations of genes. Strain DH1 was co-transformed with plasmid pBADH and different plasmids carrying the gene *nodC* or *nodBC*. Cultivation of these strains enabled the production of sulfated N-peracetylated and non-reducing end N-monodeacetylated chitin oligosaccharides. The choice between the gene *nodC* from *A. caulinodans* or from *R. meliloti* provides the additional possibility of producing these molecules as pentamers or as tetramers.

The protein NodL is a transacetylase that utilizes acetyl-CoA as a substrate to O-acetylate lipo-chitin oligosaccharides and both N-peracetylated and non-reducing end N-monodeacetylated chitin oligosaccharides at the 6 position of the non-reducing terminal sugar [46–49]. Because acetyl-CoA is a metabolite normally present in the bacterial cytoplasm, the coexpression of *nodL* with *nodC* can result in the production of mono 6-O-acetylated chitin oligosaccharide. The gene *nodL* was sub-cloned in pBBR plasmid which is compatible with both pUC and pBAD derivatives. Analysis of chitin oligosaccharides produced by strain DH1 carrying *nodC* and *nodL* indicated that ca 70% of the total oligosaccharide was O-acetylated. Cultivation of strains expressing the four genes *nodBCHL* enables the production of the direct precursor for the preparation of NodRm-IV (Ac, S, $C_{16:2}$), *i.e.* the natural Nod factor produced by *R. meliloti*.

30.3.3 Production of Derivatives of *N*-Acetyllactosamine

The production of chitin oligosaccharides in *E. coli* opened the possibility of adding glycosyl substitutions to it. One interesting possibility is the addition of a β(1,4)-linked galactosyl residue at the non-reducing end (Figure 6), to form a terminal LacNAc, the core of several biologically important oligosaccharides. This step is catalyzed by *N*-acetyllactosamine synthases (EC 2.4.1.90). For this purpose, we choose LgtB, a bacterial β-1,4 galactosyltransferase (Swiss-Prot Acc. Number Q51116) [50]. LgtB is involved in the synthesis of the lacto-*N*-neotetraose moiety of the lipo-oligosaccharide produced by *Neisseria meningitidis* [50]. Because *E. coli* produces a β-galactosidase, namely LacZ, we have used a *lacZ*⁻ strain to prevent degradation of the terminal LacNAc residue.

Co-expression of NodC and LgtB leads to the production in vivo of the expected compound [6]. The total recombinant oligosaccharide was produced at 0.8 g L^{-1} of culture. On the basis of NMR data from the crude recombinant oligosaccharide fraction, the extent of galactosylation of the chitin oligosaccharide is close to 90%. Expression of the two genes in *cis* (pC_{Ac} LgtB) slightly increased the galactosylation rate and total amount of recombinant oligosaccharides. Analysis of the recombinant oligosaccharides produced after phase 3 showed that the major product was the galactosyl chitin pentaose (Figure 6); other minor galactosylated and non-galactosylated oligosaccharides were also detected. Spectroscopic characterization of the galactosylated pentasaccharide confirmed the proposed structure [6].

We have tested whether the Gal–(GlcNAc)$_5$ is an acceptor for the bovine α-1,3 galactosyltransferase in vitro using UDP[^{14}C] Gal. The incorporation rate was similar to that of the physiological acceptor (LacNAc) [5]. This result is promising for the production of αGal1,3-βGal-1,4(GlcNAc)$_5$ in *E. coli*.

30.4 Conclusions and Future Perspectives

In the last three years, we have produced more than 50 g oligosaccharide using two 2 L and one 10 L fermenters. Typically, purification of gram amounts of an end N-deacetylated chitin oligosaccharide takes one week. Nine different compounds have been produced (Figure 6), with production yields between 0.25 g L^{-1} and 2 g L^{-1} depending on the product (Table 1).

The work presented here represents the first few steps towards the production of tailored oligosaccharides in *E. coli*. Obviously, there are several critical points that can be improved, as well as the production of other oligosaccharides and other applications that can be envisaged.

30.4.1 Production of Labeled Chitin Oligosaccharides to Study Their Interactions with Proteins

The recognition of different sugars by specific proteins (glycosyl hydrolases, GTs, lectins) is a key factor of biological processes. Molecular probes are needed to enable better understanding of the molecular basis of sugar recognition. Chitin oligosaccharides can be used to study the interaction with chitinase, and recognition by lectins or Nod factor binding proteins. To this end we have produced ^{13}C-enriched chitin pentaose (20% enrichment) for NMR study of the mechanism of action of

30.4 Conclusions and Future Perspectives 857

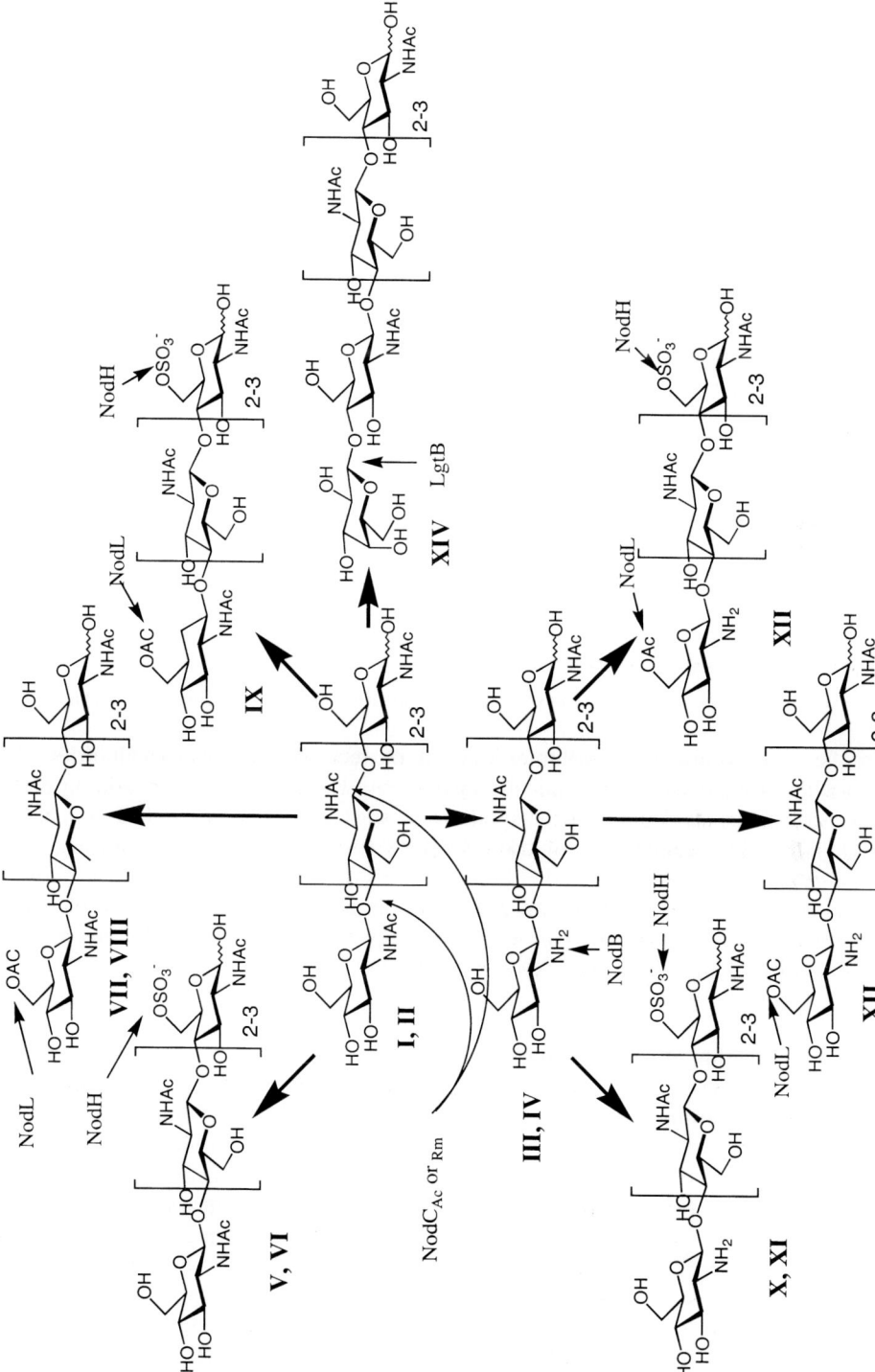

Figure 6. Recombinant oligosaccharides obtained.

Table 1. The recombinant oligosaccharides obtained.

Product	Plasmid used	Estimated production yield (g L^{-1})
I	pC$_{Rm}$	1.0
II	PC$_{Ac}$	2.0
III	PBC$_{Rm}$	0.5
IV	PBC$_{Ac}$	1.2
V	pC$_{Rm}$ + pBADH	0.4
VI	pC$_{Ac}$ + pBADH	0.6
VII	pC$_{Rm}$ + PBBRL1	0.8
VIII	PC$_{Ac}$ + pBBRL1	1.4
IX	pC$_{Rm}$ + PBBRL1 + pBADH	0.3
X	pBCH$_{Rm}$	0.25
XI	pBC$_{Ac}$ + pBADH	0.25
XII	pBC$_{Rm}$ + PBBRL1	0.4
XIII	pBCH$_{Rm}$ + PBBRL1	0.2
XIV	PC$_{Ac}$LgtB	1.0

chitinases [51]. The enrichment might be improved by using specific bacterial strains or specific growth conditions.

30.4.2 Improvement of Oligosaccharide Production, and Metabolic Engineering

In the first instance, the volumetric yield of recombinant oligosaccharides can probably be increased by genetically engineering the bacteria, either adding extra copies of the genes coding for the synthesis of sugar donors (*gmlU*: UDP–GlcNAc, *galE*: UDP–Gal epimerase), or using stronger promoters. In the second instance, sugar–nucleotides are expressed either under certain conditions (GDP–fucose) or in certain strains (CMP–Neurominic acid). The constitutive production of these sugar–nucleotides could also be engineered to enable synthesis of more diverse oligosaccharides.

30.4.3 Production of More Complex Oligosaccharides

An important limitation to this method is the availability of appropriate GTs. Although the production in *E. coli* of eukaryotic GT active in vitro was reported, the activity in vivo was not assessed [52]. Production of fully active GTs in *E. coli* will broaden the range of possible recombinant oligosaccharides. Research on the mode of action of GTs is also necessary for eventual engineering of available bacterial GTs to meet specificity requirements. Alternatively, new bacterial glycosyltransferases whose activity resembles those of eukaryotic GTs should be found. In this regard the results are encouraging, because the bacterial genes can be used for the production of oligosaccharides in *E. coli* as we did with LgtB.

Here we have presented an original strategy to produce derivatives of chitin oligosaccharides by a simple living organism. The products were obtained in amounts and purity that enable their use as substrates to perform chemical modifications or study biological properties. It is conceivable that the recombinant oligosaccharides can also be used as substrates for enzymatic methods. We think that the limits of this method have not yet been achieved, and it is possible that in the next few years the production of more complex oligosaccharides will become routine.

Acknowledgments

The authors acknowledge Drs Serge Perez and Annemarie Lellouch (Cermav, Grenoble) for helpful discussions, and Dr Hughes Driguez for reading the manuscript. This work was supported by Xenotransplantation Project BIO4CT972242 of the Biotech program from the European Union, and the program PCV 97 061 from the CNRS (France).

References

1. Varki, *Glycobiology*, **1993**, *3*, 97.
2. O. Gabriel, H. E. Umbarger, *Esherichia coli and Salmonella cellular and molecular biology*. ASM Press, **1987**, p. 504.
3. P. D. Rick, G. L. Hubbard, M. Kitaoka, H. Nagaki, T. Kinoshita, S. Dowd, V. Simplaceanu, C. Ho, *Glycobiology*, **1998**, *8*, 557.
4. E. Samain, S. Drouillard, A. Heyraud A, H. Driguez, R. A. Geremia, *Carbohydr. Res.*, **1997** *302*, 35.
5. E. Samain, V. Chazalet, R. A. Geremia, *J. Biotechnol.*, submitted.
6. E. Bettler, E. Samain, V. Chazalet, C. Bosso, A. Heyraud, D. H. Joziasse, W. W. Wakarchuk, A. Imberty, R. A. Geremia. Submitted.
7. Whitfield, M. A. Valvano, *Advances in Microbial Physiology*, **1993**, *35*, 135.
8. I. Roberts, *Annu. Rev. Microbiol.*, **1996**, *50*, 285.
9. P. van Rhijn, J. Vanderleyden, *Microbiol. Rev.*, **1995**, *59*, 124.
10. J. Denarie, F. Debelle, J. C. Prome, *Annu Rev Biochem*, **1996**, *65*, 503.
11. P. Mergaert, M. Van Montagu, M. Holsters, *Mol. Microbiol.*, **1997**, *25*, 811.
12. T. W. Jacobs, T. T. Egelhoff, S. R. Long, *J. Bacteriol.*, **1985**, *162*, 469.
13. I. Torok, E. Kondorosi, T. Stepkowski, J. Posfai, A. Kondorosi, *Nucleic Acids Res.*, **1984**, *12*, 9509.
14. P. Lerouge, P. Roche, C. Faucher, F. Maillet, G. Truchet, J. C. Prome, J. Denarie, *Nature*, **1990**, *344*, 781.
15. E. Bulawa, *Mol. Cell. Biol.*, **1992**, *12*, 1764.
16. R. A. Geremia, P. Mergaert, D. Geelen, M. Van Montagu, M. Holsters, *Proc. Natl. Acad. Sciences USA*, **1994**, *91*, 2669.
17. P. Mergaert, W. D'Haeze, D. Geelen, D. Prome, M. van Montagu, R. A. Geremia, J. C. Prome, M. Holsters, *J. Biol. Chem.*, **1995**, *270*, 29217.
18. M. A. Barny, E. Schoonejans, A. Economou, A. W. Johnston, J. A. Downie, *Mol. Microbiol.*, **1996**, *19*, 443.
19. M. Atkinson, S. R. Long, *Mol. Plant–Microbe Interact.*, **1992**, *5*, 439.
20. Debelle, C. Rosenberg, J. Denarie, *Mol. Plant–Microbe Interact.*, **1992**, *5*, 443.
21. I. M. Saxena, R. M. Brown Jr., M. Fevre, R. A. Geremia, B. Henrissat, *J. Bacteriol.*, **1995**, *177*, 1419.

22. N. I. de Iannino, S. G. Puepke, R. A. Ugalde, *Mol. Plant–Microbe Interact.*, **1995**, *8*, 292.
23. E. Kamst, J. Pilling, L. M. Raamsdonk, B. J. Lugtenberg, H. P. Spaink, *J. Bacteriol.*, **1997**, *197*, 2103.
24. D. Mengin-Lecreulx, J. J. van Heijenoort, *J. Bacteriol.*, **1994**, *176*, 5788.
25. L. M. Guzman, D. Belin, M. J. Carson, J. Beckwith, *J Bacteriol*, **1995**, *177*, 4121.
26. D. Riesenberg, *Curr. Opin. Biotechnol.*, **1991**, *2*, 380.
27. S. Y. Lee, *Trends Biotechnol.*, **1996**, *14*, 98.
28. D. J. Korz, U. Rinas, K. Hellmuth, E. A. Sanders, W. D. Deckwer. *J. Biotechnol.*, **1995**, *39*, 59.
29. M. Schultze, B. Quiclet-Sire, E. Kondorosi, H. Virelizer, J. N. Glushka, G. Endre, S. D. Gero, A. Kondorosi, *Proc. Natl. Acad. Sci. USA*, **1992**, *89*, 192.
30. P. Mergaert, M. van Montagu, J. C. Prome, M. Holsters, *Proc. Natl. Acad. Sci. USA*, **1993**, *90*, 1551.
31. R. A. Geremia, personal communication.
32. C. Maurel, J. Reizer, J. I. Schroeder, M. J. Chrispeels, M. H. Saier Jr., *J. Biol. Chem.*, **1994**, *269*, 11869.
33. J. J. Bono, J. Riond, K. C. Nicolaou, N. J. Bockovich, V. A. Estevez, J. V. Cullimore, R. Ranjeva, *Plant J.*, **1995**, *7*, 253.
34. J. C. Prome, *Curr Opin Struct Biol.*, **1996**, *6*, 671.
35. S. G. Puepke, *Crit. Rev. Biotechnol.*, **1996**, *16*, 1.
36. K. van de Sande, T. Bisseling, *Essays Biochem.*, **1997**, *32*, 127.
37. S. Drouillard, J. J. Bono, B. Henrissat, R. A. Geremia, F. Gressent, E. Samain, H. Driguez, **1997**, 9th European Carbohydrate Symposium, 6–11 July, Utrecht, Nederlands.
38. J. J. Bono, F. Gressens, personal communication. See complete reference attached.
39. P. Roche, F. Maillet, C. Plazanet, F. Debelle, M. Ferro, G. Truche, J. C. Prome, J. Denarie, *Proc. Natl. Acad. Sci. USA*, **1996**, *93*, 15305.
40. D. W. Ehrhardt, E. M. Atkinson, K. F. Faull, D. I. Freedberg, D. P. Sutherlin, R. Armstrong, S. R. Long, *J. Bacteriol.*, **1995**, *177*, 6237.
41. M. Schultze M, C. Staehelin, H. Rohrig, M. Joh, J. Schmidt, E. Kondorosi, J. Schell, A. Kondorosi, *Proc. Natl. Acad. Sci. USA*, **1995**, *92*, 2706.
42. J. S. Schwedock, C. Liu, T. S. Leyh, S. R. Long, *J. Bacteriol.*, **1994**, *176*, 7055.
43. T. S. Leyh, J. C. Taylor, G. D. Markham, *J. Biol. Chem.*, **1988**, *263*, 2409.
44. T. S. Leyh, T. F. Vogt, Y. Suo, *J. Biol. Chem.*, **1992**, *267*, 10405.
45. N. M. Kredich, *Esherichia coli and Salmonella cellular and molecular biology*. ASM press, Baltimore, **1996**, pp. 514–527.
46. F. C. Neidhardt, H. E. Umbarger, *Esherichia coli and Salmonella cellular and molecular biology*. ASM Press, **1996**, p. 15.
47. P. Spaink, D. Sheeley, A. A. van Brussel, J. Glushka, W. S. York, T. Tak, O. Geiger, E. P. Kennedy, V. N. Reinhold, B. J. Lugtenberg, *Nature*, **1991**, *354*, 125.
48. V. Bloemberg, J. E. Thomas-Oates, B. J. Lugtenberg, H. P. Spaink, *Mol. Microbiol.*, **1994**, *11*, 793.
49. G. V. Bloemberg, R. M. Lagas, S. van Leeuwen, G. A. Van der Marel, J. H. Van Boom, B. J. Lugtenberg, H. P. Spaink, *Biochemistry*, **1995**, *34*, 12712.
50. W. Wakarchuk, A. Martin, M. P. Jennings, E. R. Moxon, J. C. Richards, *J. Biol. Chem.*, **1996**, *271*, 19166.
51. E. Samain, personal communication.
52. Fang, J. Li, X. Chen, Y. Zhang, J. Wang, Z. Guo, W. Zhang, L. Yu, K. Brew, P. G. Wang, *J. Am. Chem. Soc.*, **1998**, *120*, 6635.

Part I

Volume 2

IV
Carbohydrate–Protein Interactions

31 Protein–Carbohydrate Interaction: Fundamental Considerations

Nikki F. Burkhalter, Sarah M. Dimick, and Eric J. Toone

31.1 Introduction

A major—if not the major—rationale for the study of carbohydrate chemistry derives from the roles played by glycoconjugates in biology [1]. In almost all such roles a carbohydrate ligand must bind to a protein receptor. There exists, then, tremendous potential to modulate biological activity through the creation of high affinity mimics of native saccharide receptors; such compounds have potential therapeutic value in the treatment of viral, parasitic, mycoplasmal and bacterial infections, and the treatment of a range of human cancers [2]. The study of protein–carbohydrate interaction, then, forms a key focus of this discipline of carbohydrate chemistry and biology.

The development of inhibitors of carbohydrate-mediated biological recognition is impeded by the exceptionally weak bindings that typify protein–carbohydrate interaction [3]. In this chapter we consider the energetic issues central to protein–carbohydrate interaction. We devote some space to a consideration of the general forces and interactions that provide both affinity and specificity during association in aqueous solution. A range of phenomena observed during protein–carbohydrate association appear to be dependent on the assay used to evaluate 'binding'; we therefore consider the some of the more commonly utilized assays of protein carbohydrate interaction and delineate the microscopic events each assay is designed to evaluate. Finally, the last ten years has seen tremendous activity in the development of multivalent carbohydrate ligands; we thus explore the energetic issues surrounding multivalency in ligand binding.

Throughout, we endeavor to relate energetic issues directly to protein–carbohydrate interaction. We stress, however, that a tremendous body of literature pertaining to the very issues central to protein–carbohydrate interaction exists in the context of other binding systems. Where appropriate, we draw on the lessons and experience of these adjacent disciplines in our consideration of the biophysical aspects of protein–carbohydrate interaction.

31.2 Association in Aqueous Solution

Two species combine to form a complex in water if the sum of the intermolecular forces between them more than offsets the sum of the loss of favorable interactions with solvent and any unfavorable interactions that develop between solutes during complex formation. Collectively the interactions between non-bonded species are referred to as cohesive forces, defined as those forces lost when the species are transferred to infinite separation in the gas phase. While it is common to classify chemical forces as covalent or non-covalent, the interactions are fundamentally the same; only the magnitude of the interactions varies. Cohesive, non-specific forces are weak compared to covalent interactions; typically we consider cohesive forces as those forces with strengths less than 1% of covalent bond strengths. We will see, however, that this definition is somewhat arbitrary and in fact a continuum of interaction energies exists.

That non-covalent interactions between molecules are important has long been appreciated. By the middle of the 19th century failings in the ideal gas law were apparent, and by 1873 van der Waals had postulated his equation of state. Thus the functional dependence of intermolecular interactions on internuclear distance—attractive at long separations and repulsive at very short internuclear spacings—was clear long before there existed a complete understanding of the nature of the interactions involved. Association in aqueous solution is further complicated by the highly participatory nature of the solvent; as we shall see the contribution from solvation/desolvation processes to overall thermodynamic parameters is typically greater than that provided by interactions between the two associating solutes. We begin our discussion by considering the interactions that result during the approach of two species absent the effect of solvation. We later consider the effect of water on each of these terms.

31.2.1 Gas Phase Non Covalent Interactions

Dipole–Dipole Interactions

Molecules in which the center of electron density differs from the center of mass possess a permanent dipole. The interaction of oppositely charged dipoles yields a coulombic attraction, while the analogous interaction of equivalently charged species provides a repulsive contribution to an intermolecular interaction energy. The product of the dipole separation and the partial charge of the dipoles yields the dipole moment μ. Many species containing strongly electronegative elements show no dipole moments because of a symmetrical disposition of partial charge; examples include carbon dioxide and carbon tetrachloride. While this symmetry produces a net dipole moment of zero, the molecules have higher order moments. Thus carbon dioxide has a large quadrupole moment while carbon tetrachloride shows a large octopole moment. Dipole moments of groups commonly encountered in protein–carbohydrate interaction are listed in Table 1 [4].

Table 1. Dipole moments of representative small molecules [4].

	μ^a		μ^a
H_2O	1.88	$CH_3CH_2OCH_3$	1.75
H_2S	0.98	CH_3CH_2NO	3.20
CH_3CH_2F	1.90	CH_3CH_2CN	3.50
CH_3CH_2Cl	2.00	CH_3CH_2COOH	1.25
CH_3CH_2Br	1.90	CH_3CH_2CHO	2.50
CH_3CH_2I	1.75	$CH_3CH_2CONH_2$	3.60
FCH_2CH_2OH	1.52	$CH_3CH_2NH_2$	1.30
CH_3CH_2OH	1.70	$NH_2CH_2CH_2NH_2$	1.86
CH_3CH_2OD	1.54	$CH_3CH_2NO_2$	2.25
CH_3CH_2SH	1.52	$C_6H_5NO_2$	3.95
$HOCH_2CH_2OH$	2.31	$o\text{-}NO_2C_6H_4NO_2$	5.90
$HSCH_2CH_2SH$	1.70	$m\text{-}NO_2C_6H_4NO_2$	3.79
CH_3OCH_3	1.32	$p\text{-}NO_2C_6H_4NO_2$	0.69
CH_3SCH_3	1.51		

[a] Debye

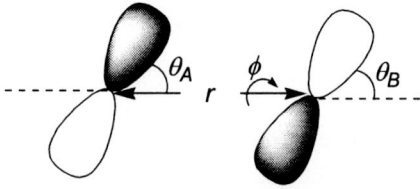

Figure 1. Orientation of dipoles.

The contribution of interactions between permanent dipoles to the overall cohesive energy between two molecules depends on both the magnitude of the dipoles and their relative orientation. A stepwise consideration of each interaction— dipole–dipole, dipole–quadrupole, quadrupole–quadrupole, etc.—provides the overall contribution of the interactions of partial charges to the total cohesive energy between two molecules. The strength of individual interactions drops off as a power function of the internuclear distance; the exact functional dependence varies with the interaction. In general, the interaction between multipoles of order n_1 and n_2 drops off as $r^{-(n_1+n_2+1)}$. If we consider the interaction of two species each with both a dipole (μ) and quadrupole (Θ) separated by an internuclear distance r and disposed relative to each other by the angles θ_A, θ_B and ϕ (Figure 1), the intermolecular force is given by [5]:

$$U_{AB} = \frac{\mu_A \mu_B}{(4\pi\varepsilon_o)r^3} f_1(\theta_A \theta_B \phi) + \frac{\mu_A \Theta_B}{(4\pi\varepsilon_o)r^4} f_2(\theta_A \theta_B \phi)$$
$$+ \frac{\mu_B \Theta_A}{(4\pi\varepsilon_o)r^4} f_2(\theta_A \theta_B \phi) + \frac{\Theta_A \Theta_B}{(4\pi\varepsilon_o)r^5} f_3(\theta_A \theta_B \phi)$$

where ε_o is the permitivity of free space. The dependence of each term on the respective dispositions of the multipoles is given by the expressions:

$$f_1(\theta_A, \theta_B, \phi) = \sin\theta_A \sin\theta_B \cos\phi - 2\cos\theta_A \cos\theta_B$$

$$f_2(\theta_A, \theta_B, \phi) = \frac{3}{2}[\cos\theta_A(3\cos^2\theta_B - 1) - 2\sin\theta_A \sin\theta_B \cos\theta_B \cos\phi]$$

$$f_3(\theta_A, \theta_B, \phi) = \frac{3}{4}[1 - 5\cos^2\theta_A - 5\cos^2\theta_B - 15\cos^2\theta_A \cos^2\theta_B$$
$$+ 2(4\cos\theta_A \cos\theta_B - \sin\theta_A \sin\theta_B \cos\phi)^2]$$

The functional form of the relationship makes intuitive sense; f_1 reaches a favorable maximum when the dipoles are aligned antiparallel to each other, an unfavorable maximum when the dipoles are aligned in a parallel fashion, and zero when the dipoles are oriented perpendicular to each other. Averaged over all orientations, the terms sum to zero. Application of a Boltzman distribution provides an energy for dipole–dipole, dipole–quadrupole and quadrupole–quadrupole interactions of:

$$\langle U \rangle_{\mu\mu} = -\frac{2\mu_A^2 \mu_B^2}{3r^6 kT (4\pi\varepsilon_o)^2}$$

$$\langle U \rangle_{\mu\Theta} = -\frac{\mu_A^2 \Theta_B^2 + \mu_B^2 \Theta_A^2}{r^8 kT (4\pi\varepsilon_o)^2}$$

$$\langle U \rangle_{\Theta\Theta} = -\frac{14\Theta_A^2 \Theta_B^2}{5r^{10} kT (4\pi\varepsilon_o)^2}$$

All interaction terms are negative (favorable) and inversely proportional to temperature. As the order of the multipole increases the distance dependence becomes increasingly severe, and interactions beyond quadrupole–quadrupole do not provide a significant contribution to overall interaction energies.

Dipole–Induced Dipole

A dipole is induced in a molecule with no net permanent dipole but with a polarizable electron cloud when it is brought into contact with a permanent dipole. The magnitude of the induced dipole and the strength of the resulting interaction with the permanent dipole depends on the magnitude of the permanent dipole and the polarizability of the second species:

$$\mu_{ind} = \alpha E$$

where E is the electric field generated by the permanent dipole and α is the susceptibility, or polarizability, of the second species. The energy of the interaction is then given by:

$$U = -\int \mu_{ind} \, dE = -\int \alpha E \, dE = -\frac{1}{2}\alpha E^2$$

The strength of the interaction again depends on the relative orientation of the two species; the effective field between a dipole and a polarizable species separated by a distance r and at an angle θ to the normal is given by:

$$E = \mu \frac{\sqrt{1 + 3\cos^2\theta}}{4\pi\varepsilon_o r^3}$$

and thus:

$$U = \frac{\alpha\mu^2(3\cos^2\theta + 1)}{2r^6(4\pi\varepsilon_o)^2}$$

As before, over all orientations the interaction sums to zero; application of a Boltzman correction yields an interaction energy given by:

$$\langle U \rangle_{ind} = \frac{\mu^2 \alpha}{(4\pi\varepsilon_o)^2 r^6}$$

Again, the magnitude of the term drops off as the sixth power of the internuclear distance, and is directional. Some values of polarizabilities of various groups commonly encountered in protein–carbohydrate interaction are shown in Table 2 [6].

Dispersive Interactions

The intuitive interactions described above—essentially Coulombic attractions between oppositely charged particles—cannot account for more than a fraction of the

Table 2. Group polarizabilities [6].

	α_g [a]		α_g [a]
–H	3.9	–CO | O	23.5
–F	4.5		
–Cl	22.1	–O–	6.7
–Br	33.1	–S–	30.0
–I	52.9	=SO	32.8
–O–H	9.6	=NH	13.7
–S–H	3.2	≡CH	13.7
–NH$_2$	16.4	–C≡C–	27.1
–CN	20.7	–C=C–C=C–	57.7
–NO$_2$	19.7	≡N	10.4
–CH$_3$	22.4	–S–S	60.8
–CH$_2$	18.4	–C=C– | |	25.6
–CO | H	22.0		
–C– |	9.8		80.5

[a] 10^{-25} cm^3

total cohesive energy between common materials in the condensed phase. By far the most important contributors to overall interaction energies are the dispersive forces first identified by Fritz London in 1930 [7, 8]. Although it is impossible to properly consider dispersive interactions in a purely classical treatment, the concept is made intuitive by considering the structures of atoms and molecules on timescales shorter than electron reorganizations (i.e. the reciprocal of the dispersion frequencies). Frozen in time, all molecules possess instantaneous dipoles; interaction of these dipoles provides the London or dispersive interaction. The interaction energy between two species arising from London forces is given by the expression:

$$\varepsilon_{\alpha_A \alpha_B} = -\frac{3N}{2} \frac{h \nu_A \nu_B}{\nu_A + \nu_B} \frac{\alpha_A \alpha_B}{r^6}$$

where ν_A and ν_B are the dispersion frequencies of A and B, and α_A and α_B the corresponding polarizabilities. Although individual dispersive interactions are weak, they have no orientational dependence. Summed over an entire molecule they are large, providing the bulk of the total cohesive energy.

Specific Forces: Hydrogen Bonding and n-σ Bonding

The forces described so far are weak, typically less than one percent the strength of a covalent bond. In many instances, interactions between non-covalently bonded species are significantly stronger than this limit. For example, the exothermic transfer of ethanol from the gas phase to triethylamine reduces the strength of the ethanol O–H covalent bond by 13.5%: clearly the interactions formed between ethanol and triethylamine are significant on the scale of covalent bonds.

These stronger interactions are referred to as specific forces, and result from the interaction of an electron lone pair with a permanent dipole (polarized σ-bond). Such forces differ from the non-specific forces described above in several important ways:

i) *The interactions are directional and stoichiometric.* Unlike other coulombic interactions that include a directional component to the force vector but that have no *a priori* defined stoichiometry, specific interactions have a discrete stoichiometry. Each lone pair of electrons forms a single interaction with an appropriately polarized n-σ bond. It is thus correct to consider true equilibrium constants for the formation of two hydrogen bonds between water and acetone:

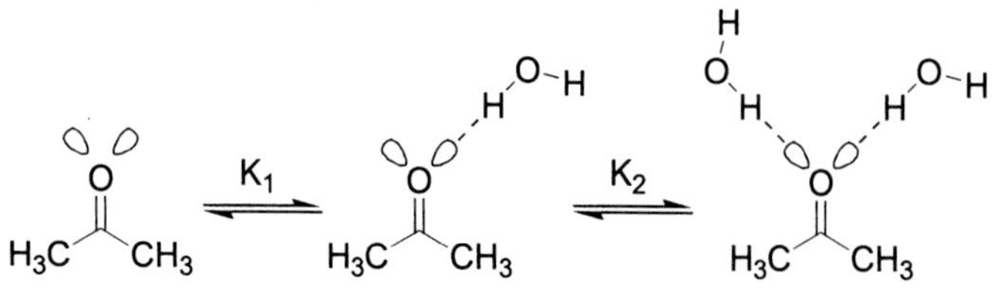

Scheme 1. Hydrogen bond formation.

H 2.2							
Li 1.0	Be 1.6	B 2.0	C 2.5	N 3.0	O 3.5	F 4.0	
Na 0.9	Mg 1.3	Al 1.6	Si 1.9	P 2.2	S 2.6	Cl 3.2	
K 0.8	Ca 1.0	Sc 1.4	Ge 2.0	As 2.2	Se 2.5	Br 3.0	
Rb 0.8	Sr 0.9	Y 1.2	Sn 1.9	Sb 2.0	Te 2.1	I 2.7	
Cs 0.8	Ba 0.9						

Figure 2. Representative Pauling electronegativities.

ii) *The formation of specific interactions significantly shortens the internuclear distance between the bound partners*—typically to less than the sum of the van der Waal radii—while significantly lengthening the σ-bond between the electropositive proton and its electronegative partner. This bond weakening is most readily observed as an increase in the IR stretch frequency.

iii) *The process is exothermic*, despite the weakening of the polarized σ-bond.

The extent of polarization in an X–H bond that is required for effective activity as a hydrogen bond donor is unclear. A useful rule of thumb is that protons bonded to elements with Pauling electronegativities greater than three act as hydrogen bond donors. Most species with a lone pair of electrons act as hydrogen bond acceptors, including oxygen (alcohols and ethers, carbonyls, phosphine and amine oxides, sulfoxides), nitrogen (amines, pyridines, nitriles), and sulfur. Stable anions also form strong hydrogen bonds.

The most reliable predictor of hydrogen bond strengths, at least in the gas phase, is the difference in proton affinity of the hydrogen bond donor and acceptor. By this criteria the FH–F$^-$ interaction is predicted as the strongest monomeric hydrogen bond; this prediction is borne out by a crystal phase heat of formation of -37 kcal mol^{-1}. The value of hydrogen bond strengths in solvent, however, is greatly influenced by the interactions of both donor and acceptor with the solvent itself, and conclusions regarding the contributions of hydrogen bonds to overall net free energies of interaction in solution cannot be drawn from proton affinity data [9].

31.2.2 The Effect of Water on Intermolecular Interactions

Having considered intermolecular interaction in the absence of solvation, we now consider the consequence of addition of solvent to a system of interacting molecules. At the outset we note that the picture changes from one that considers the

Table 3. Proton affinities of representative small molecules [9].

	PA[a]		PA[a]
$N(CH_3)_3$	226.8	H_2S	168.5
C_5H_5NH	222.3	CH_3CH_2Br	166.4
$(CH_3)_2NH$	222.2	CH_3CH_2I	173.2
$CH_3CH_2NH_2$	218.0	CH_3CH_2Cl	165.7
CH_3NH_2	214.9	CH_3CH_2F	163.3
$(CH_3)_2CO$	194.1	$HO(CH_2)_3OH$	209.4
$(CH_3)_2S$	198.6	CH_3CH_3	142.5
$(CH_3)_2O$	189.3	$CH_3CH_2CONH_2$	209.4
CH_3CH_2NC	203.5	$HOCH_2CH_2OH$	195.0
CH_3CH_2CN	189.8	CH_3CH_2SH	188.7
CF_3CF_2CN	165.4	CH_3CH_2OH	185.6
$C_6H_5CH_3$	187.4	CH_3CHCH_2	179.6
$CH_3CO_2CH_3$	196.4	$HOCH_2CH(OH)CH_2OH$	209.1
CH_3CHO	183.7	CH_3SSCH_3	194.9
CH_3OH	180.3	CH_3CH_2COOH	190.5
H_2O	165.2		

[a] kcal mol^{-1}

interaction between solutes to one that considers the *difference* in interactions between solutes and solvent *versus* solute and solute. The transfer of all species from the gas phase to water is exothermic: by a variety of specific and non-specific mechanisms water interacts with all solutes such that the free energy of the solvated species is lower than that of the isolated gas-phase species. Binding two solutes thus necessarily involves the loss of favorable interactions with solvent, a loss that is ameliorated by the favorable interactions that arise during association in the absence of solvent. Below we consider the effects of solvation on each of the classes of interaction described above. Finally, we consider the 'hydrophobic effect' and its contribution to ligand binding.

Coulombic Stabilization

We consider first the effect of aqueous solvation on all intermolecular stabilizations that derive from the interaction of charged or dipolar species. Because of its small size and significant dipole and quadrupole, water interacts strongly with all ionic and dipolar species. A binding event of charged or dipolar compounds thus proceeds with a significant loss of favorable cohesive interactions between solutes and water. The effect is most profound for ionic interactions; similar ameliorations of solute–solute interaction apply to multipole–multipole and dipole-induced dipole interactions.

Gas phase ionic interaction energies are large, typically on the order of 50–200 kcal mol^{-1}. The interaction of water with charged species is nearly as large; as a result solution-phase ionic interactions are weak, seldom more than five kcal mol^{-1}

and typically less than one kcal mol^{-1}. Examples of this diminution, or leveling, of interaction energies are myriad. The binding constant for the association of the quaternary ammonium ion choline with acetyl cholinesterase differs from that of the uncharged analogue 3,3-dimethyl propanol by a factor of only 30. Likewise, a range of monovalent anions inhibit acetoacetate decarboxylase by interacting with the positively charged active site with binding constants ranging from 10 to 10^5 M^{-1}, corresponding to binding free energies of only one to five kcal mol^{-1}. The powerful effect of solvation is also apparent in the order of effectiveness of acetoacetate decarboxylase inhibition. The halogen anions bind with decreasing affinity in the order I$^-$ > Br$^-$ > Cl$^-$ > F$^-$, the opposite order to that predicted on the basis of electronegativities. Apparently the more effective solvation of smaller, harder anions leads to reduced binding to the protein: a decreasing favorable electrostatic interaction proceeding down the periodic table is more than compensated for by a more rapidly decreasing unfavorable desolvation energy. Similar arguments are readily apparent during the consideration of multipole–multipole and dipole-induced dipole interactions. The effective strong favorable interaction of water with multipoles lost during binding precludes a significant energetic advantage from dipole-related interactions between solutes in the bound state relative to the solvated unbound state.

It is important at this point to distinguish affinity from specificity, at least as it pertains to this discussion. Dipolar interactions are directional and, in an absolute sense, strong. Improperly satisfied dipolar interactions in a complex are strongly disfavored relative to the energy of the unbound solvated system. Thus dipolar and ionic interactions play important roles in determining the specificity of solute–solute interactions and the precise structure of the bound complex, even though they likely do not contribute strongly to the overall net negative free energy of complexation.

Hydrogen Bonding

The strongest known hydrogen bond is the FH–F$^-$ interaction. In the crystal phase this interaction is favorable by some 37 kcal mol^{-1}. In aqueous solution the equilibrium constant for this hydrogen bond is 0.4 M^{-1}, corresponding to a free energy of 0.8 kcal mol^{-1}. There is no evidence for the formation of any other monofunctional hydrogen bond in aqueous solution. The situation becomes somewhat more complex for intramolecular hydrogen bonds, which form without a loss of translational and rotational entropy. Most of the data collected in this regard involve pK_a values of dibasic acids. Thus, for example, the two pK_a values of fumaric acid, for which no intramolecular hydrogen bond is possible, are 3.03 and 4.54. From this reference, both pK_a values are perturbed in the isomeric maleic acid, at 1.91 and 6.33. Presumably this larger δpK_a reflects the energetic contribution of an intramolecular hydrogen bond in the singly ionized species.

Even with a loss of net favorable interaction energy, we stress again the important roles that hydrogen bonds play in determining both the specificity of solute–solute interactions and the structure of the bound complex. Because of the strong angular dependence of hydrogen bonds on the X–H→lone pair vector, hydrogen bonds provide an even more powerful orienting force than do dipolar interactions.

Dispersive Interactions

Despite the fact that they are individually weak, dispersive forces contribute to the exothermicity of the interaction between solutes in aqueous solution. The magnitude of the dispersive component of the cohesive energy varies as the polarizability of the interacting solutes. Because oxygen polarizes very weakly, a binding cycle that replaces solute—water interactions with solute—solute interactions will be energetically favorable: the dispersive component of OH–OH self association is 470 kcal mol^{-1} Å^6, while that for CH$_2$–CH$_2$ interaction is 1160 kcal mol^{-1} Å^6 [5]. Additionally, the r^6 dependence of the interaction is a powerful orienting force during ligand binding.

31.2.3 'Hydrophobic' Interactions

It has long been appreciated that placement of most organic species into aqueous solution produces an energetically unfavorable perturbation to the structure of water. Ligand binding proceeds with the desolvation of some fraction of the interacting surfaces; to the extent that this desolvation relieves the perturbation of water structure the free energy of the bound system is lowered relative to the unbound state. This process is typically considered, at least philosophically, as a repulsion between water and 'hydrophobic' solutes. In reality, of course, the effect is the result of the strong attractive interactions between water molecules, providing a liquid with a boiling point of 100 °C and an enthalpy of vaporization of 540 cal g^{-1}, despite a molecular weight of only 18 g mol^{-1}.

The molecular origin of the hydrophobic effect remains obscure, despite nearly 100 years of intensive experimental and computational investigation. Water has a remarkably high cohesive energy density; *a priori* one might expect that the unfavorable free energy of transfer of 'hydrophobic' solutes from some non-aqueous condensed phase to water should come from the loss of water–water contacts as a cavity is formed to accommodate the solute. Such a process would lead to an unfavorable enthalpy of transfer; one would thus reasonably expect the unfavorable free energy of hydrophobic solvation to be dominated by an endothermic enthalpy of transfer. In fact, near room temperature the enthalpy of transfer of virtually all hydrophobic species is near zero or slightly exothermic. Rather, near room temperature the unfavorable free energy resulting from the transfer of 'hydrophobic' surface area to water derives from an unfavorable entropic term. Furthermore, both the enthalpy and entropy of transfer are strong functions of temperature: 'hydrophobic' dissolution proceeds with a large positive increment in the constant pressure heat capacity of the system, ΔC_p. *It is this term that most distinguishes water from all other solvents*. In a seminal work Arnett and coworkers demonstrated that the heat capacity increment during aqueous dissolution was uniquely a property of water and is not found for any other solvent, regardless of polarity [10].

During the past decades a variety of physical models designed to rationalize the observed thermodynamic behavior of nonpolar solutes in water have been proposed. It is beyond the scope of this work to describe each in detail; rather, the reader is referred to any of several reviews and monographs on the topic [11]. Here, we describe the most popular of the group—the 'clathrate' model of water—in enough detail to allow a qualitative, intuitive understanding of the energetic consequences of hydrophobic transfer to water.

The transfer of a hydrophobic solute to water can be conceptually divided into two steps: the formation of a cavity of sufficient volume to accommodate the solute, and the transfer of the solute to that cavity followed by collapse of solvent onto the solute. The first of these steps must be endothermic; that the overall process is thermoneutral or exothermic near room temperature thus requires that the second conceptual step—placing the solute in the cavity and allowing the solvent to collapse on the solute—be exothermic. Water in bulk solution can form four hydrogen bonds in a tetrahedral orientation; in a solvation shell such bonding is impossible. The clathrate model of water asserts that water in a solvation shell orients to maximize favorable hydrogen bonds. In this structure a loss in the total *number* of hydrogen bonds is compensated for by an increase in the *strength* of the hydrogen bonds that can form. This increased strength comes at the price of reduced entropy, particularly the rotational entropy of water, in the clathrate solvation shell. The enthalpically enhanced hydrogen bonds of the clathrate shell also provide a mechanism for the positive increment in constant pressure molar heat capacity. The notion of clathrate water as a physical model to explain the unusual thermodynamics of hydrophobic hydration has been parameterized by both Müller and Ben-Naim to yield a semi-quantitative description of the event [12–14]. In these constructs the enthalpy of transfer is given by the expression:

$$\Delta H = \frac{3N}{2}[(1 - f_b)\Delta H_b - (1 - f_{hs})\Delta H_{hs}]$$

where f_b and f_{hs} is the fraction of hydrogen bonds in the bulk liquid and the hydration shell, respectively, and ΔH_b and ΔH_{hs} are the enthalpies of hydrogen bond formation in bulk solution and hydration shells, respectfully. While the clathrate model of water is intuitively accessible and explains both qualitatively and quantitatively many of the phenomena associated with hydrophobic hydration, many unanswered questions remain. Specifically, physical evidence for the existence of clathrates is lacking; indeed many studies specifically designed to observe such structures fail to note any increase in the order of water structure [15–20]. The notion of clathrate water is far from universally accepted and other models of water exist, especially a series of models based on scaled particle theory that explain the peculiar phenomenology of hydrophobic hydration based on the exceptionally small size of water [21–29].

Putting aside issues of the molecular origin of 'hydrophobic effect', the more significant issue is the extent to which desolvation contributes to overall binding thermodynamics. A variety of experimental methodologies to evaluate the magnitude of this effect exist. The most commonly used tools are Hansch transfer parameters [30, 31]. Briefly, a series of compounds are partitioned between water and water-saturated octanol. The value π is then derived for a substituent according to the expression:

Table 4. Hansch parameters of representative functional groups [30, 31].

	π	
	arom[a]	aliph[b]
F	0.15	−0.73
Cl	0.70	−0.13
Br	1.02	0.04
I	1.26	
CN	−0.32	−1.47
CH3	0.52	
CF3	1.07	
CH2CH3	0.97	
n-Bu	1.90	
sec-Bu	1.82	
t-Bu	1.68	
cyclohexyl	2.51	
C6H5	1.89	
OH	−0.61	−1.80
CH2OH	−1.03	
OCH3	−0.04	−0.98
SCH3	0.62	
NH2	−1.23	−1.85
N(CH3)2	0.18	−0.95
NO2	0.24	
COOH	−0.15	−1.26
CH2COOH	−0.72	
COOCH3	−0.01	−0.91
OCOCH3		−0.91
COCH3	−0.37	−1.26
CONH2	−1.49	−2.28

[a] based on aromatic parent compound
[b] based on aliphatic parent compound

$$\pi = \log \frac{P}{P_o}$$

where P and P_o are the partition ratios of a substituted and parent compound, respectively. The value $RT \ln \pi$ is thus the incremental Gibbs free energy for the transfer of the substituent from water to water-saturated octanol, assuming at least rough group additivity. Hansch parameters for a range of commonly encountered substituents are shown in Table 4.

Another approach to issues of solvation thermodynamics utilizes the differential hydrogen bonding properties of light and heavy water [32, 33]. The use of thermodynamic solvent isotope effects, or the evaluation of the enthalpy of ligand binding in light and heavy water, allows evaluation of the extent to which desolvation aids ligand binding. Again, our goal here is not to describe the technique exhaustively; the reader is referred to recent reviews of the subject for a more complete treatment. Briefly, the D–O hydrogen bond is enthalpically favored relative to the H–O analog by roughly 10%; an offsetting decrease in the entropy of bond formation leads to a free energy difference near zero between the two. Assuming a clathrate model of water, the transfer of a solute from light to heavy water will be exothermic by an amount related to the enthalpy of clathrate formation. Straightforward extension of the Müller formalism above quantifies the enthalpy of transfer as [13, 34]:

$$\Delta\Delta H = \frac{3N}{2}[(1-f_b)\Delta\Delta H_b - (1-f_{hs})\Delta\Delta H_{hs}]$$

A schematic representation relating ligand binding in isotopic solvents can be created as:

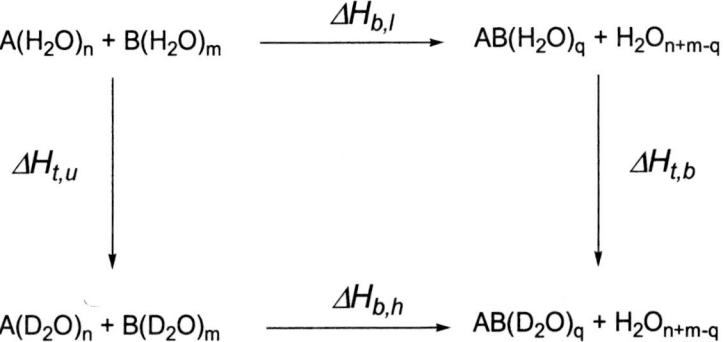

Figure 3. Cycle relating binding in isotopic waters. The subscripts represent light (l) and heavy (h) waters, and unbound (u) and bound (b) systems.

Thermodynamic parameters are state functions. Thus the difference in the two horizontal processes of Figure 3—binding in isotopic solvents—is equivalent to the difference in the two vertical processes—transfer of the free and bound systems from light to heavy water. From this formalism it is apparent that the difference in enthalpy of binding in isotopic waters is *equivalent to the enthalpy of transfer of the fraction of the ligand and protein desolvated during binding*. Because this enthalpy is related to the enthalpy of solvation, the thermodynamic solvent isotope effect is a measure of the enthalpy of binding that derives from removal of the binding site and ligand from solution.

Thermodynamic solvent isotope effects have been utilized to evaluate the extent to which desolvation aids ligand binding [35, 36]. While the assumptions implicit in the calculations likely make precise determination of ΔH_{solv} impossible, the *relative* role of desolvation for a series of bindings is clearly accessible (Table 5). This exercise facilitates two observations that are likely general. First, a significant fraction

Table 5. Solvation-associated enthalpies of ligand binding [35].

Binding pair	ΔG	ΔH	ΔH_s
Dioclea/3,6-di-(*O*-αMan)αManOMe	−8.2	−13.0	−4
Concanavalin A/αManOMe	−5.0	−7.1	−5
Concanavalin A/3,6-di-(*O*-αMan)αManOMe	−7.5	−10.7	−5
Vancomycin/diAcLys-D-Ala-D-Ala	−7.2	−11.5	−7
RNAse/CMP	−7.5	−14.1	−14
FK506/FKBP	−12.3	−17.2	−18

of overall binding enthalpy derives from solvent reorganization; this trend is likely repeated in binding free energies. Secondly, the enthalpy derived from solvent reorganization is lower for protein–carbohydrate interaction than for any other interacting system studied. This observation hardly comes as a surprise, given the hydrophilic nature and high aqueous solubility of carbohydrates. On the other hand, the relatively small enthalpy of binding available from saccharide desolvation is the most probable cause of the low binding constants that typify protein–carbohydrate interaction. *This limitation is fundamental, severe, and likely insurmountable.*

Together, the forgoing discussion leads one to a picture of binding that is largely driven by 'hydrophobic' effects. In this model, a range of polar interactions are largely responsible for the *specificity* of the interaction: if key hydrogen bonds, salt bridges and dipolar interactions are satisfied by solute–solvent interactions prior to association but lost during binding, an unfavorable contribution to the overall binding free energy results. If all interactions lost to solvent are adequately replaced in the complex, the net favorable free energy that derives from the desolvation of non-polar surface area drives formation of the complex. Lemieux described such notions two decades ago in the concept of a 'polar gate' [37]. More recently, several researchers have shown that a significant fraction of the total surface area of an oligosaccharide ligand can be replaced by non-carbohydrate surface [38–40].

31.3 The Evaluation of Protein–Carbohydrate Binding

Discussion of the effect of ligand structure on protein–carbohydrate affinity requires an evaluation of complex stability constants. A number of biophysical techniques are appropriate for the study of protein–carbohydrate interaction; many of the more enlightening strategies are the topics of separate chapters elsewhere in this volume. We describe below three techniques used extensively in glycobiology— inhibition of hemagglutination, enzyme-linked lectin assay (ELLA), and isothermal titration microcalorimetry—and we consider the types of information provided by each technique in order to facilitate appropriate interpretation of the data.

31.3.1 Precipitin Assays

By far the most common of the techniques utilized to evaluate protein–carbohydrate interaction are a group of assays based on the hemagglutination process. Anticoagulated red blood cells spontaneously separate on standing in microtitre plate wells, as the denser erythrocytes settle to the bottom of the wells. Addition of a polyvalent lectin to a suspension of erythrocytes prevents this segregation and results in the formation of a gel-phase cross-linked lattice. This cross-linking process is referred to as hemagglutination, and provides a straightforward read-out for inhibition assays. Assays of protein–carbohydrate binding function by inhibiting this hemagglutination event. Addition of a soluble saccharide sets up a competition between glycoconjugates on erythrocyte surfaces. Typically serial dilutions are performed across a microtitre plate, and the minimum concentration of soluble ligand required to inhibit 50% agglutination is reported as an IC_{50}. Because the assay is run as a serial dilution errors are typically assumed at ± one well, or a factor of two. Absolute IC_{50} values are frequently not reproducible from lab to lab, although the order of inhibitory potency through a series of ligands is generally robust.

Agglutination assays evaluate a straightforward physical process, namely the ability of a soluble ligand to prevent a cross-linking aggregation process. The danger of agglutination assays lies in the assumption that IC_{50} values are simply related to K_{eq} for a reversible protein–carbohydrate binding event. At a minimum, the activity of a soluble ligand in an agglutination assay is the inhibition of an irreversible or quasi-reversible phase transition, from a soluble lectin to an ordered cross-linked gel matrix. Ligand binding is surely requisite for the initiation of this event, but all events that follow serve to produce a complicated set of coupled equilibria, each with a distinct microscopic set of kinetic and thermodynamic constants. The situation becomes significantly more complicated when the soluble ligands are themselves multivalent, since now there exists a competitive set of cross-linking/aggregation processes. In the limit, a protein–carbohydrate binding followed by an irreversible aggregation/precipitation will appear in this assay as an exceptionally high affinity association, since the binding equilibrium is coupled with the aggregation/precipitation event.

Hemagglutination assays have been and will remain a mainstay of the study of protein–carbohydrate interaction. In many respects the assay is highly relevant to the study of biological processes; Nature doubtlessly uses the high valency of both carbohydrate ligands and binding proteins for a range of functions. On the other hand, IC_{50} values from agglutination assays cannot be interpreted in terms of protein–carbohydrate affinity. In instances where protein–carbohydrate interactions have been evaluated by agglutination assay in addition to some other biophysical technique, hemagglutination IC_{50} values do not correlate with K_a values for the equivalent process. Recently we reported a study of the binding of multivalent saccharide ligands to concanavalin A [41]. While IC_{50} values varied by a factor of 30 on a valence corrected basis, binding constants for the same ligands evaluated by titration microcalorimetry varied by less than a factor of two. Rather, hemagglutination IC_{50} values correlate with entropies of ligand binding, again demonstrating that hemagglutination evaluates an aggregation, rather than a binding, process.

31.3.2 Enzyme-Linked Lectin Assay (ELLA)

Roy and coworkers have reported a modification of the precipitin assay [42–50]. This assay evaluates competitive binding between soluble and immobilized ligands to a lectin-enzyme conjugate and is essentially a variant of quantitative ELISA. In the enzyme-linked lectin assay a high molecular weight polyvalent ligand is adsorbed to the surface of microtitre plates. The assay has been utilized most frequently for the evaluation of binding to mannose-specific proteins; in such cases yeast mannan serves as the immobilized ligand. Following blockage of non-specific binding with BSA, a lectin-horseradish peroxidase (HRP) conjugate is incubated in the microtitre plate wells with serial dilutions of a soluble ligand. Following an incubation period the microtitre plates are evacuated and the binding mixture replaced with hydrogen peroxide and a pro-dye substrate. Color formation, proportional to the amount of lectin bound to the microtitre plates, is read quantitatively with a UV plate reader.

ELLA is designed to obviate the most significant complication of agglutination assays, namely the irreversible formation of aggregates. In theory, the competition between immobilized and soluble ligand should not involve the formation of kinetically trapped aggregates. On the other hand, data from ELLA assays do show evidence of kinetic contributions. Plots of inhibitor concentration *versus* fractional inhibition frequently asymptote to values less than 100% inhibition. Often different ligands asymptote to different fractional inhibitions. The most obvious explanation for this behavior is the formation of some irreversibly bound lectin species on the surface of the microtiter plate wells. The exact nature of this species is somewhat difficult to imagine, although the use of microheterogeneous polydisperse polysaccharide as an immobilized ligand likely presents a range of binding epitopes for the lectin-HRP conjugate.

In summary, it is now clear that precipitin assays do not measure ligand binding. ELLA is a technique designed to overcome the issues that render agglutination unsuitable for evaluating ligand binding. However, until IC_{50} values obtained through this technique are verified by comparison to binding constants measured by other biophysical techniques, results from the assay should be regarded with some caution.

31.3.3 Isothermal Titration Microcalorimetry

Isothermal titration calorimetry (ITC) has long been recognized as a useful tool for the evaluation of binding constants [51]. The field of calorimetry changed markedly during the early 1990s with the introduction of commercial titration microcalorimeters [52]. These devices, available from several suppliers, operate with volumes near one to two milliliters. Virtually all the data reported to date on the thermodynamics of protein–carbohydrate interaction has been obtained with one of several generations of the Microcal ITC; these instruments have sample cell volumes near 1.3 ml (Figure 4).

The sensitivity requirements of the technique mandate a continuous power compensation design rather than the simpler and more traditional passive thermal conductivity experimental approaches. In this design, a sample and reference cell are heated at a constant rate, typically less than 100 microcalories per second. This heating rate alters the temperature of the cell contents by less than 0.1 °C during the course of a titration. The temperature between the two cells is evaluated and a second compensating voltage is applied to bring the two cells into precise thermal equilibrium. Addition of an aliquot of ligand through a syringe into the sample cell disrupts this equilibrium; in turn, the compensating voltage is adjusted to return the cells to equilibrium. The raw data is thus power as a function of time. Integration over time yields the more familiar plot of enthalpy per injection *versus* ligand concentration. The enthalpy evolved on each injection is a function of the concentrations of the two binding partners, the stoichiometry of the interaction, the binding constant (K_{eq}) and the enthalpy of binding (ΔH). A simple non-linear least squares fit to the data provides estimates of the binding stoichiometry, binding constant, and enthalpy of binding. The binding constant is related to the free energy of binding by the relationship:

$$\Delta G = -RT \ln K_{eq}$$

The entropy of binding is thus available by subtraction. The partial differential of enthalpy with respect to temperature defines the partial change in constant pressure heat capacity. Over short temperature ranges this derivative is approximated as

$$\Delta C_p = \frac{\partial \Delta H}{\partial T} \approx \frac{\Delta \Delta H}{\Delta T}$$

The required concentrations of both binding partners are determined by the binding constant. The shape of the binding curve is determined by the unitless parameter c, equivalent to the product of the binding constant and the concentration of binding sites within the calorimeter cell. Figure 5 shows curves that result from values of c ranging over 1–1000. While in principle it is possible to fit curves anywhere within this range, values of c of 10–100 give optimal results.

Practically, titration microcalorimetry operates to furnish estimates of both enthalpies and free energies of binding for systems with binding constants in the range of 10^3 to 10^7 M^{-1}. At low affinity, protein solubility becomes limiting; a binding constant of 10^3 M^{-1} requires at least millimolar protein in solution. At high affinity, sensitivity becomes limiting. A binding constant of 10^7 M^{-1} limits the concentration of binding sites to 0.1 micromolar, which in turn limits the size of individual aliquot enthalpies. Across a complete titration, peak areas (i.e. enthalpies) should average no less than 100 microcalories. The range of 10^3 to 10^7 M^{-1} thus assumes stable soluble protein and significant enthalpies of binding. The concentration of titrant in the syringe is typically 20-times the concentration of binding sites.

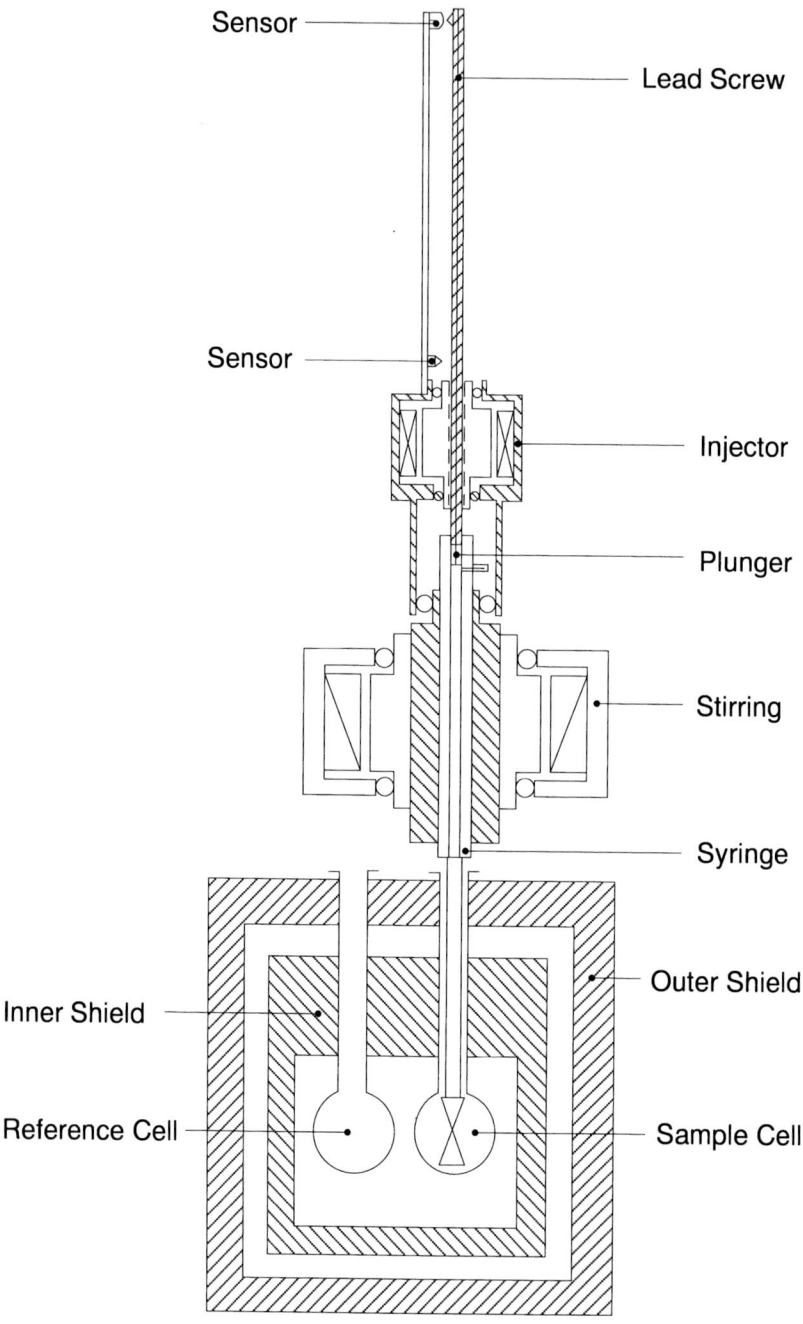

Figure 4. The Microcal Omega titration microcalorimeter.

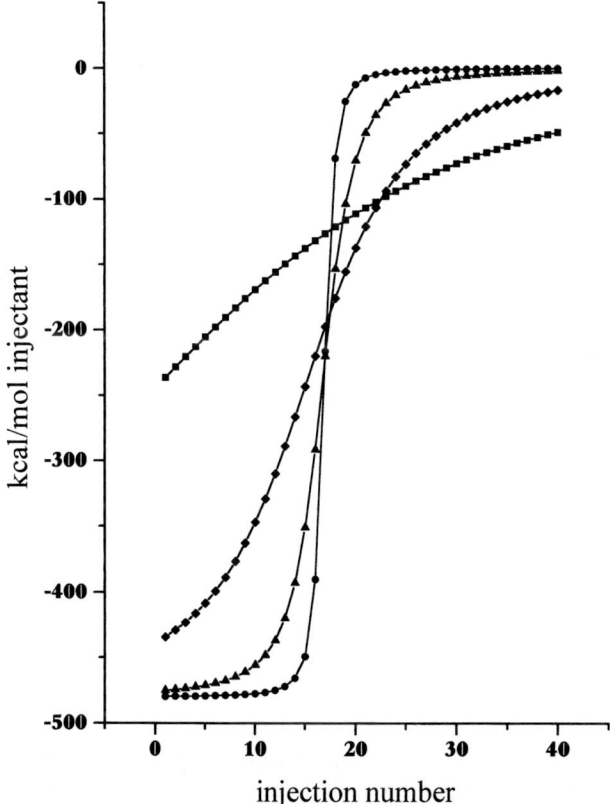

Figure 5. The effect of c on calorimetric curve shape. Data calculated for $\Delta H = -6$ kcal mol^{-1} and 1 mM binding sites. Circles $K_{eq} = 10^6$; triangles $K_{eq} = 10^5$; diamonds $K_{eq} = 10^4$; squares $K_{eq} = 10^3$.

Even beyond the high affinity limit, calorimetry remains a powerful technique for the study of intermolecular interactions. With stoichiometric binding, enthalpies of binding are still accessible in a straightforward fashion. In this instance, the total enthalpy evolved on each injection is simply divided by the quantity of titrant added. As before, evaluation of the molar enthalpy of binding as a function of temperature provides ΔC_p. Determination of the entropy of binding requires an independent determination of K_{eq}; this protocol offers the additional advantage of avoiding correlated errors.

That thermodynamic parameters are state functions provides another powerful methodology for determining both enthalpies and free energies of binding for very high affinity systems through a competitive cycle. If two species bind a single binding site with differing affinities, titration of the higher affinity ligand to binding sites loaded with a lower affinity ligand yields the difference in binding enthalpy and free energy between the two ligands. Thermodynamic parameters characterizing binding of the lower affinity ligand can be determined independently and the corresponding values for the high affinity ligand are then equivalent to the sum of the two values. An elegant demonstration of this methodology was provided by Sigurskjold and coworkers in a series of papers describing the binding of acarbose to glucoamylase (*vide infra*) [53–56].

31.4 The Interpretation of Calorimetric Data

While the measurement of thermodynamic parameters by ITC is straightforward, interpretation of the derived values is not. It is important to recall that thermodynamic properties represent the sum of all molecular processes that occur during a binding event. A range of intermolecular interactions must be considered carefully before interpreting thermodynamic properties in terms of solute–solute interaction. Below we consider some of the processes that contribute to net measured thermodynamic processes and, where possible, describe experimental techniques for the evaluation of the contribution of each set of microscopic events to overall ligand binding thermodynamics.

31.4.1 Solvation/Desolvation

It has always been fashionable to reach conclusions regarding the role of solvation during association processes based on measured enthalpies and entropies of binding. Processes with favorable entropic components are frequently referred to as 'entropy driven'—indicative of a 'hydrophobically driven' binding—while processes with large favorable enthalpies of binding are referred to as 'enthalpy driven'; such a pattern is often interpreted in terms of favorable hydrogen bonding or salt bridge formation. There is no basis for this distinction. The predicted magnitude and sign of enthalpies and entropies arising from desolvation of both ligand and binding site are unclear. Indeed, the thermodynamic parameter most uniquely associated with solvation processes in aqueous solution is ΔC_p. The compensating dependence of both enthalpy and entropy on this term requires that over a sufficient temperature range both the magnitude and sign of a measured enthalpy and entropy will change. Surely, however, the same physical processes aid and oppose binding over this range.

Two methodologies allow independent evaluation of the role of solvation in ligand binding processes. The first requires measurement of thermodynamic solvent isotope effects, the differential enthalpy of binding in light and heavy water [33]. The fraction of the enthalpy of binding arising from solvent reorganization is directly related to this quantity. In principle, an absolute solvation-associated enthalpy is obtainable through this approach; in practice, derivation of such a value requires the use of a proportionality constant not known with either precision or accuracy. Thus the *absolute* value of solvation-associated enthalpies derived in this way likely have substantial error associated with them. On the other hand, the error is a constant, deriving from a poor understanding of the precise difference in enthalpies between the H–O and D–O interaction. Comparative thermodynamic solvent isotope effects are thus free of this error. TSIE is thus most useful for evaluating the effect of ligand modification on binding thermodynamics [36]. The fraction of an overall enthalpy of binding attributable to solvation/desolvation for two ligands is readily compared by comparing solvent isotope effects. In this way assignment of *changes* in the enthalpy of binding in response to a change in ligand structure to either changes in solute–solute interaction or to changes in ligand solvation is straightforward and unambiguous.

We have previously shown that thermodynamic solvent isotope effects correlate linearly with the change in molar heat capacity that accompanies binding [35]. *This correlation demonstrates that ΔC_p is exclusively a measure of solvation*, and comparison of ΔC_p values for related ligands allows a similar interpretation of differences in binding enthalpies in terms of protein–carbohydrate interaction *versus* differences in solute–solvent interactions prior to association [57]. The relative random errors of the two methodologies are unclear; in the one instance small values must be measured accurately while in the other errors in individual determinations are propagated during curve fitting. In any event, a meaningful discussion regarding the molecular origin of measured thermodynamic parameters is impossible without some evaluation of the contribution of solvation to overall measured enthalpies of binding.

The entropy of binding arising from solvation is also available with knowledge of the change in molar heat capacity accompanying binding. Overall entropies of binding can be conceptually dissected to describe the contributions of changes in translational and rotational entropy, solvation-associated entropy, and changes in entropy from restriction of conformational degrees of freedom during binding:

$$\Delta S = \Delta S_{T+R} + \Delta S_{solv} + \Delta S_{conf}$$

Separation of overall entropies of binding into component segments in this fashion again allows relatively sophisticated conclusions to be drawn regarding the molecular origin of measured thermodynamic parameters. In this formalism, entropies of binding are divided into three contributions and knowledge of any two of the three provides the third by subtraction. Evaluation of ΔS_{solv} and ΔS_{T+R} are both feasible.

Solvation Entropy

It has long been recognized from efforts to understand protein folding that solvation associated entropies are related to changes in molar heat capacity accompanying binding. Thus,

$$\Delta S_{solv} = \Delta S^*_{solv} + \Delta C_p (\ln T/T^*)$$

where T is the temperature at which the entropy of binding was measured and T^* is the so-called entropy convergence temperature, or the temperature at which all solvation associated entropies approach zero [58, 59]. ΔS^*_{solv} includes contributions to entropy from proton transfer and electrostatic effects; for most protein–carbohydrate interactions this term is negligable. Evaluation of ΔC_p thus allows extraction of the fraction of the total entropy of binding attributable to solvent reorganization. Note that this term is temperature dependent; again, this dependence makes intuitive sense. As the temperature of the system is raised the residual order of water slowly dissipates, and at 385.15 K disappears entirely.

Translational/Rotational Entropy

A major fraction of the entropy of a particle derives from its ability to translocate in three dimensions and rotate on three axes. The magnitude of this entropy varies as the logarithm of the particle mass and the principle moment arms describing the distribution of mass. Because of the logarithmic relationship ligand binding proceeds with a loss of entropy roughly equivalent to the translational and rotational entropy of the smaller particle. The magnitude of this term has long been a contentious issue. Evaluation of ΔS_{T+R} in the gas phase is straightforward, and follows from the Sakur-Tetrode equation. Transfer to condensed phase, however, greatly complicates the matter. All solutions show greatly reduced molecular motion relative to the gas phase. Water is especially severe with regard to the restriction of molecular mobility: as a highly associated liquid with a tremendous cohesive energy density, water interacts strongly with both adjacent water molecules and with dissolved solutes. A wide range of values of the contribution of ΔS_{T+R} to complexation entropies has been suggested; these are values of 7–50 eu (Table 6). The entropy values reported in Table 6 correspond to the best estimates of the transla-

Table 6. Reported values of ΔS_{T+R}.

Investigators	ΔS_{T+R}[a]	$T\Delta S$[b]	S_{int}[a]	Ref.
Doty & Myers	122	36.4		60
Dunitz	7 (0–7)	2.1	0–7	61
Erickson	23–37	7–11		62
Horton & Lewis	21	6.2		63
Janin	100 (50)	30 (15)	50	64
Janin & Chothia	57–74	17–22		65–67
Jencks	45 (35)	13.4 (10.4)	10	68
Murphy et al.	8	2.4		58
Page & Jencks	40–50 (25–35)	12–15 (7.5–10.4)	15	69
Searle & Williams	40 (0–40)	12	0–40	70
Searle et al.	40	12		71
Spolar & Record	40–60	11.9–17.9		72
Tidor & Karplus	100 (77)	30 (23)	23	73
Williams et al.	39	11.5		74

[a] e.u.
[b] kcal mol^{-1} at 298K

tional and rotational entropy loss during association. Many of these values refer to total immobilization, i.e. total loss of translational and rotational entropy with no account for residual internal entropy. In those cases for which an estimate of the residual internal entropy was made, the corrected translational and rotational entropy loss has been placed in parentheses after the value for total immobilization, and the compensating internal energy (S_{int}) has been noted.

In order to interpret changes in solution thermodynamics in response to changes in ligand structure, we require ΔS_{solv} and ΔS_{conf}. With a methodology for evaluation of ΔS_{solv} in hand knowledge of ΔS_{T+R} would provide ΔS_{conf} by subtraction. Although accurate knowledge of ΔS_{T+R} remains elusive, it is important to recall that values of ΔS_{T+R} are relatively insensitive to molecular size and shape. Accordingly, an arbitrary value of ΔS_{T+R} can be selected for the purpose of this exercise; this value will remain constant across a set of ligands. Changes in entropies of binding arising from ligand modification can thus confidently be ascribed to changes in solvation-associated entropies, or interactions of solvent and solute prior to association, or to changes in conformational degrees of freedom in ligand and protein during binding. Again, the range of values in Table 6 precludes extraction of absolute contributions to overall measured entropies arising from the loss of conformational degrees of freedom.

31.4.2 Other Contributions to Thermodynamics of Association

Proton Transfer

Many ligand binding events proceed with the transfer of a proton, either from or to buffer; in such instances the enthalpy of buffer ionization will be included in measured values of ΔH_B. Enthalpies of buffer protonation range from near zero to greater than 10 kcal mol^{-1} (Table 7) [51]. Evaluation of enthalpies of binding as a function of buffer ionization enthalpy is a straightforward method of determining the extent of protonation contributions to overall binding enthalpies. A plot of measured enthalpies of binding *versus* buffer ionization enthalpy provides a plot with a slope equal to the number of protons transferred during binding, and a y-intercept equal to the enthalpy of binding in the absence of buffer ionization.

Salt Effects/Binding Site Reorganization

In some instances, a binding site undergoes reorganization during association. An equilibrium can be written describing the enthalpy, entropy and free energy difference between the two states. Ligand binding thus involves a coupled equilibrium, and the reported values of thermodynamic properties are sums of those values for binding site reorganization and ligand binding to the preformed site. If environmental conditions affect the position of this equilibrium, the precise conditions under which binding is measured will affect the measured enthalpy of binding. An example of such an effect is observed in the binding of mannosides by concanvalin A [36]. Here, a salt bridge between Asp16 and Arg288 in the unbound form of the protein is ruptured during ligand binding, a process that proceeds with an unfa-

Table 7. Enathalpy of ionization of common buffers [51].

Trivial Name	Formula	$\Delta H_B{}^a$
MES	O⟨⟩N$^+$HCH$_2$CH$_2$SO$_3{}^-$	12.68
bis-tris	(HOCH$_2$)$_3$CN$^+$H(CH$_2$CH$_2$OH)$_2$	29.25
ACES	H$_2$NCOCH$_2$N$^+$H$_2$CH$_2$CH$_2$SO$_3{}^-$	30.12
ADA	H$_2$NCOCH$_2$N$^+$H(CH$_2$COO$^-$)$_2$	11.51
MOPS	O⟨⟩N$^+$HCH$_2$CH$_2$CH$_2$SO$_3{}^-$	19.0
PIPES	NaO$_3$SCH$_2$CH$_2$N⟨⟩N$^+$HCH$_2$CH$_2$SO$_3{}^-$	8.70
BES	(HOCH$_2$CH$_2$)$_2$N$^+$HCH$_2$CH$_2$SO$_3{}^-$	23.10
HEPES	HOCH$_2$CH$_2$HN$^+$⟨⟩NCH$_2$CH$_2$SO$_3{}^-$	16.40
TES	(HOCH$_2$)$_3$CN$^+$H$_2$CH$_2$CH$_2$SO$_3{}^-$	29.25
Ethyl glycinate	H$_3$N$^+$CH$_2$COOCH$_2$CH$_3$	46.32
Glycinamide	H$_3$N$^+$CH$_2$CONH$_2$	44.77
HEPPS	HOCH$_2$CH$_2$H$^+$N⟨⟩NCH$_2$CH$_2$CH$_2$SO$_3{}^-$	17.95
Tricine	(HOCH$_2$)$_3$CN$^+$H$_2$CH$_2$COO$^-$	30.50
THAM	(HOCH$_2$)$_3$CN$^+$H$_3$	47.28
Glycylglycine	H$_3$N$^+$CH$_2$CONHCH$_2$COO$^-$	43.72
Bicine	(HOCH$_2$CH$_2$)$_2$N$^+$HCH$_2$COO$^-$	26.23
TAPS	(HOCH$_2$)$_3$CN$^+$H$_2$CH$_2$CH$_2$CH$_2$SO$_3{}^-$	40.12
N,N-Dimethylglycine	(CH$_3$)$_2$N$^+$HCH$_2$COO$^-$	31.51
CAPS	⟨⟩N$^+$H$_2$CH$_2$CH$_2$CH$_2$SO$_3{}^-$	48.53

^{a} kcal mol^{-1}; enthalpy of removing the H from the quaternary nitrogen atom

vorable enthalpy. Increasing ionic strength cleaves this intramolecular salt bridge prior to ligand binding, diminishing the unfavorable contribution to the overall enthalpy of binding. As a consequence, enthalpies of mannoside binding vary by a factor of two over a salt range from 5 mM to 1 M.

31.4.3 van't Hoff versus Calorimetric Enthalpies

A measured difference between van't Hoff and calorimetric enthalpies of binding has been offered as a possible test of contributions other than ligand binding to calorimetric enthalpies of binding [75–77]. van't Hoff enthalpies of binding are determined from the temperature dependence of the binding free energy according to the expression:

$$\ln K = \ln A - \frac{\Delta H_{vH}}{RT}$$

where A is a pre-exponential term related to the entropy of binding. A plot of $\ln K$

versus $1/T$ thus gives a slope equivalent to $\Delta H_{vH}/R$. This relationship has been utilized extensively by the protein folding community to observe coupled equilibria unrelated to folding. There is, however, a range of problems associated with its use in ligand binding. First, associations in aqueous solution universally proceed with a change in molar heat capacity, i.e. $\Delta C_p \neq 0$. The expression above then is predicted to be nonlinear, since ΔH is itself a function of temperature. The fit required to extract ΔH_{vH} is thus a non-linear regression for two variables using values that have themselves been extracted from a non-linear regression for three variables. Appropriate permutation of errors almost certainly precludes observation of all but extraordinarily large differences between the calorimetric and van't Hoff enthalpies with any degree of confidence. Secondly, it is somewhat difficult to devise a physical model that corresponds to a measured difference. van't Hoff enthalpies will differ from calorimetric enthalpies only to the extent that calorimetric enthalpies contain contributions from processes that do not contribute to the free energy of binding. Because thermodynamic processes are state functions, this condition can hold only for situations where perfect enthalpy-entropy compensation exists, and even then only if the compensating terms have equivalent temperature dependencies. Because ΔC_p affects both ΔH and ΔS, this condition can be true only if ΔS for the compensating process is large compared to ΔC_p. Thus while in principle differences between van't Hoff and calorimetric enthalpies contain important information regarding the precise molecular processes that occur during binding, such differences should be regarded cautiously.

31.5 The Thermodynamics of Protein–Carbohydrate Interaction

Having considered those intermolecular interactions that contribute to overall net measured thermodynamic parameters, we turn now to a review of thermodynamic measurements of protein–carbohydrate binding reported during the last five years. Values reported prior to this time can be found in earlier reviews. Calorimetrically derived changes in enthalpies, entropies, free energies and molar heat capacity that occur during protein–carbohydrate binding are shown in Table 8.

Extension of the database of known protein–carbohydrate thermodynamic parameters continues to reinforce the basic concepts identified some time ago. Virtually all of the examples in Table 8 show binding free energies of 4–8 kcal mol^{-1}, corresponding to association constants of 10^3–10^5 M^{-1}. Typically, enthalpies of binding are more negative than the corresponding free energies of binding and the entropic contribution to ΔG is unfavorable. The database of changes in molar heat capacity accompanying protein–carbohydrate binding remains small, but continues to grow. Again, in almost all cases exceptionally small values of ΔC_p are recorded, almost always less than 200 cal mol^{-1} deg^{-1}, and usually less than 100 cal mol^{-1} deg^{-1}. A small number of values in Table 8 are positive; this observation is unusual. In most instances the errors in individual values of both ΔH and the ΔH *versus* T fits are large and the validity of these data is unclear. There is at least some correlation between ΔC_p and ΔG. Thus, for example, the binding of acarbose to glu-

Table 8. Thermodynamics of protein–carbohydrate binding.

Protein	Ligand[a]	$-\Delta H$[b]	$T\Delta S$[b]	$-\Delta G$[b]	ΔC_P[c]	Ref.
Antithrombin III K[121]–A[134]	Heparin	34.5	−28.0	10.1		78
Antithrombin III K[121]ΔA	Heparin	14.9	−7.6	4.5		78
Antithrombin III K[125]ΔA	Heparin	36.0	−31.2	4.3		78
Antithrombin III R[129]ΔQ	Heparin	48.6	−45.8	4.1		78
Antithrombin III K[133]ΔA	Heparin	57.5	−53.9	4.8		78
CBDN1	βGlc(1→4)[βGlc(1→4)]$_2$GlcOH	9.7	−4.7	5.0		79
	βGlc(1→4)[βGlc(1→4)]$_3$GlcOH	12.7	−6.6	6.1	−50.0	79
	βGlc(1→4)[βGlc(1→4)]$_4$GlcOH	13.0	−6.9	6.1	−61.9	79
	Barley β-glucan	13.7	−7.4	6.3		79
	Oat β-glucan	14.4	−8.2	6.2		79
	Hydroxyethyl Cellulose	13.4	−7.3	6.1	−103	79
	Acid Swollen Cellulose	7.7	−1.8	5.9		79
Cellumonas fimi CBD	Cellulose I	1.2	9.7	10.8		80
	Cellulose II	0.2	8.2	8.4		80
Succinyl concanavalin A	1	14.5	−9.7	4.8		81
	2	13.8	−8.7	5.1		81
	3	5.0	0.2	5.2		81
fragment free concanavalin A	Manα(1→6)[Manα(1→3)]-Man	14.3	6.7	31.6		82
concanavalin A	αMeGlc	5.3	−0.7	4.6		81
		6.6	−2.1	4.5		82
		6.7	−2.1	4.6		83
	αMeMan	6.8	−1.5	5.3		81
		8.2	−2.9	5.3		82
		8.4	−2.8	5.6		84
	αMe2dMan	7.2	−1.9	5.2		83
	C-allylglucose	5.9	−0.2	5.0		81
	C-allylmannose	5.9	−0.8	5.1		81
	1	8.5	−3.5	5.0		81
	2	11.1	−5.0	6.1		81
	3	6.2	−1.0	5.2		81

31.5 The Thermodynamics of Protein–Carbohydrate Interaction

Compound				Ref
αMeGlcNAc	6.2	−2.1	4.1		82
αGlc(1→4)GlcOH	6.2	−1.9	4.3		82
αGlc(1→6)GlcOH	6.2	−1.9	4.3		83
αMe2dGlc	6.7	−1.6	5.1		83
	7.3	−2.6	4.7		82
Manα(1→2)Man	9.9	−3.6	6.3		82
Manα(1→2)ManOMe	10.5	−3.5	7.0		82
Manα(1→2)Manα(1→2)Man	10.7	−3.1	7.6		82
Manα(1→3)Man	10.2	−4.5	5.7		82
Manα(1→3)ManOMe	10.7	−4.5	6.2		82
Manα(1→6)Man	9.4	−3.8	5.6		82
Manα(1→6)ManOMe	8.4	−3.1	5.3		82
GlcNAcβ(1→2)Man	5.3	−0.1	5.2		82
Manα(1→6)[Manα(1→3)]-Man	14.1	−6.6	7.5		82
4	10.6	−2.2	8.4		84
4dManα(1→6)[Manα(1→3)]-ManOMe	10.6	−2.2	8.4		85
2dManα(1→6)[Manα(1→3)]-ManOMe	12.3	−5.5	6.8		85
	14.9	−7.1	7.8		84
2dManα(1→6)[Manα(1→3)]ManOMe	14.0	−6.2	7.1		85
4dManα(1→6)[Manα(1→3)]ManOMe	11.2	−4.9	6.3		85
6dManα(1→6)[Manα(1→3)]ManOMe	11.7	−5.7	6.0		85
Manα(1→3)[Manα(1→6)]2dManOMe	11.6	−5.5	6.1		85
Manα(1→3)[Manα(1→6)]4dManOMe	13.4	−6.5	6.9		85
3dManα(1→3)[Manα(1→6)]2dManOMe	12.1	−5.7	6.4		85
3dManα(1→3)[Manα(1→6)]4dManOMe	10.6	−4.4	6.2		85
3dManα(1→3)[Manα(1→6)]2,4dManOMe	9.7	−3.8	5.9		85
Manα(1→3)[Manα(1→6)]ManOMe	8.7	−3.2	5.5		85
2dManα(1→3)[Manα(1→6)]ManOMe	14.7	−7.1	7.6		84
3,4dManα(1→3)[Manα(1→6)]2,4dManOMe	14.1	−6.9	7.2		84
	8.9	−3.3	5.6		84
αGlcOPh	3.5	1.8	5.3	−19.4	76
1dGlc	4.2	0.7	3.5	−6.7	76
αGlcF	7.1	−4.2	2.9	41.1	76
2dGlcOH	4.8	−1.4	3.4	1.8	76

Table 8 (continued)

Protein	Ligand[a]	$-\Delta H^b$	$T\Delta S^b$	$-\Delta G^b$	ΔC_P^c	Ref.
	2FGlcOH	6.5	−3.0	3.5	40.8	76
	3FGlcOH	3.8	−0.7	3.1	−9.6	76
	αManOMe	6.8	−1.5	5.3	−50.0	36
		6.8	−1.5	5.3		86
	2-OMe αManOMe	4.8	0.4	5.2		86
	2-OEt αManOMe	2.8	3.4	6.2		86
	2-OPr αManOMe	4.2	1.2	5.4		86
	2-OBn αManOMe	6.2	−0.2	6.0		86
	3-OMe αManOMe	5.3	−0.3	5.0		86
	2,3-diOMe αManOMe	6.8	−2.0	4.8		86
	3-OMe αMan(1→3)-2-OBn αManOMe	7.5	−1.7	5.8		86
	αGlcOMe	5.3	−0.7	4.6		36
	5	5.9	−0.7	5.2		36
	αMan (1→2)αManOMe	7.0	−0.2	6.8		36
	αMan(1→3)αManOMe	7.4	−1.4	6.0	−110	36
	αMan(1→4)αManOMe	7.4	−2.3	5.0		36
	αMan(1→6)αManOMe	6.9	−1.6	5.3	−44.0	36
	αGlc(1→4)GlcOH	3.7	0.7	4.4		36
	αGlc(1→6)GlcOH	4.6	−0.2	4.4		36
	αMan(1→3)[αManα(1→6)]-αManOMe	10.2	−2.8	7.4	−93.0	36
Co-concanavalin A	αManOMe	6.4	−0.9	5.5		87
	ManOH	5.3	−1.0	4.4		87
	αGlcOMe	4.8	−0.1	4.6		87
	GlcOH	3.6	0.1	3.7		87
Ni-concanavalin A	αManOMe	6.2	−0.6	5.6		87
	ManOH	4.0	0.6	4.6		87
	αGlcOMe	4.5	0.0	4.5		87
	GlcOH	3.4	0.3	3.6		87

31.5 The Thermodynamics of Protein–Carbohydrate Interaction

Protein	Ligand					Ref
Cd-concanavalin A	αManOMe	7.5	−2.2	5.3		87
	ManOH	5.5	−1.0	4.5		87
	αGlcOMe	4.6	−0.1	4.5		87
	GlcOH	3.9	−0.1	3.8		87
concanavalin A (pH 5.2)	ManOH	5.0	−0.5	4.5		88
	αMan(1→3)ManOH	8.4	−2.7	5.6		88
	αMan(1→6)ManOH	7.2	−1.6	5.7		88
	αMan(1→3)[αMan(1→6)]αManOMe	14.2	−6.8	7.4		88
Dioclea grandiflora	αGlcOMe	4.4	−0.2	4.2		81
		5.0	−0.3	4.7		84
		4.4	−0.1	4.2		36
	αManOMe	7.8	−1.5	4.8		81
		8.2	−3.3	4.9		84
		7.8	−2.9	4.8	−56.0	36
	αGlcCallyl	2.4	2.0	4.4		81
	αManCallyl	6.3	−1.7	4.6		81
	α2dGlcOMe	7.5	−2.9	4.6		84
	αMan(1→2)ManOMe	9.9	−3.8	6.1		84
		6.4	−0.5	5.9		36
	αMan(1→3)ManOMe	10.1	−4.3	5.8		84
		11.4	−5.1	5.5	−40.0	36
	αMan(1→4)ManOMe	8.4	−3.5	5.0		84
	αMan(1→6)ManOMe	8.4	−3.3	5.1		84
		8.6	−3.6	5.1	−22.0	36
	αMan(1→3)[αMan(1→6)]αManOMe	16.2	−7.9	8.3		84
		13.0	−4.8	8.2	−96.0	36
	2dαMan(1→3)[αMan(1→6)]αManOMe	15.2	−7.2	8.0		84
	3dαMan(1→3)[αMan(1→6)]αManOMe	11.0	−4.5	6.5		84
	4dαMan(1→3)[αMan(1→6)]αManOMe	14.6	−6.8	7.8		84
	6dαMan(1→3)[αMan(1→6)]αManOMe	15.1	−7.1	8.0		84
	2dαMan(1→6)[αMan(1→3)]-αManOMe	12.8	−5.5	7.3		84
	2dαMan(1→6)[αMan(1→3)]-αManOMe	10.0	−4.0	6.0		84
	4dαMan(1→6)[αMan(1→3)]-αManOMe	10.4	−4.5	5.9		84
	6dαMan(1→6)[αMan(1→3)]-αManOMe	9.8	−4.0	5.8		84

Table 8 (continued)

Protein	Ligand[a]	$-\Delta H^b$	$T\Delta S^b$	$-\Delta G^b$	ΔC_p^c	Ref.
	αMan (1→3)[αMan (1→6)]-2dαManOMe	14.8	−7.5	7.3		84
	αMan (1→3)[αMan (1→6)]-4dαManOMe	12.8	−6.0	6.8		84
	3,4dαMan (1→3)[αMan(1→6)]2,4dαManOMe	9.9	−4.2	5.7		84
	4	4.6	1.8	6.4		84
	αGlc (1→4)GlcOH	6.1	−1.4	4.7		83
	αGlc (1→6)GlcOH	6.3	−1.6	4.7		83
	5	6.9	−1.7	5.2		36
	αGlc (1→4)GlcOH	4.4	−0.1	4.3		36
	αGlc(1→6)GlcOH	4.4	−0.2	4.2		36
Erythrina corallodendron	βGalOMe	4.4	−0.8	3.6		89
		4.9	−1.0	3.9		90
		4.4	−0.5	3.9		90
	GalNAcOH	7.1	−2.9	4.2		89
		5.6	−1.2	4.4		90
		5.5	−1.2	4.3		90
	GalOH	3.6	0.9	4.4		90
		3.3	1.1	4.4		90
	αGalOMe	5.3	−1.1	4.2		90
	βGalNAcOMe	5.2	−0.9	4.3		90
	βGal(1→4)GlcOH	6.8	−2.5	4.3		89
		6.3	−1.9	4.4		89
		10.4	−5.7	4.7		90
		9.9	−5.4	4.5		90
	βGal(1→4)GlcNAcOH	10.9	−5.9	5.0	−93.2	89
		12.6	−6.8	5.8		89
		11.0	−5.5	5.5		90
		11.3	−6.0	5.4		90
	βGal(1→4)βGlcNAcOMe	11.3	−6.3	5.0		89
	αFuc(1→2)βGal(1→4)GlcOH	3.9	1.1	5.0		90
		4.3	0.6	4.9		90

31.5 The Thermodynamics of Protein–Carbohydrate Interaction

System	Ligand					Ref
	αMe-N-dansylGalaminide	4.4	2.9	7.3		90
	FucOH	5.5	2.1	7.6		90
		1.0	2.6	3.6		90
		1.1	2.5	3.6		90
Erythrina cristagalli	βGalOMe	4.7	−0.7	4.1		89
	GalNAcOH	7.3	−2.9	4.4		89
	βGalNAcOMe	6.8	−2.5	4.3		89
	βGal(1→4)GlcOH	6.0	−1.1	4.9		89
	βGal(1→4)GlcNAcOH	10.9	−5.4	5.6	105	89
		12.8	−6.5	6.3		89
	βGal(1→4)βGlcNAcOMe	10.3	−4.8	5.5		89
Erythrina indica	βGalOMe	5.9	−2.1	3.8		89
	GalNAcOH	4.9	−0.7	4.2		89
	βGalNAcOMe	5.2	−0.8	4.4		89
	βGal(1→4)GlcOH	10.4	−5.6	4.8	−200	89
		7.4	−2.7	4.7		89
		13.7	−7.8	5.9	−244	89
		10.9	−5.1	5.9		89
	βGal(1→4)βGlcNAcOMe	10.4	−4.8	5.7		89
FGF-1 cyclic mimic	Heparan Sulfate	24.4	−16.9	7.5		91
FGF-1 Native Binding Site	Heparan Sulfate	23.2	−15.9	7.3		91
FGF-1: D-Pro136	Heparan Sulfate	22.9	−15.9	7.1		91
Gal-1 (Bovine Spleen)	βGal (1→4)Glc	5.0	0.2	5.2	66.3	75
	βGal(1→4)βGlcOMe	7.1	−2.0	5.0	6.7	75
	βGal(1→4)FrucOH	8.2	−2.9	5.4	12.5	75
	βGal(1→4)ManOH	8.4	−3.2	5.5	−48.6	75
	βGal(1→3)AraOH	8.8	−3.7	5.1	−20.0	75
	βGal(1→3)GlcNAcOH	7.1	−1.3	5.7	−35.1	75
	2-OMe βGal (1→4)GlcOH	7.3	−1.5	5.8	−29.1	75
	Galβ(1→4)GlcNAcOH	7.8	−2.3	5.5	−57.1	75
	Thio Galβ(1→1)βGal	11.1	−5.6	5.5	−82.4	75
Gal-1 (dimer)	βGal(1→4)GlcNAcOH	6.6	−1.4	5.2	−88.4	89
	Dithiogalactoside	3.8	0.9	4.8		89

Table 8 (continued)

Protein	Ligand[a]	$-\Delta H^b$	$T\Delta S^b$	$-\Delta G^b$	ΔC_p^c	Ref.
Gal-1 C2S mutant	βGal(1→4)GlcNAcOH	2.8	1.9	4.8		89
	Dithiogalactoside	2.6	2.1	4.7		89
Gal-1 triple mutant	βGal(1→4)GlcNAcOH	0.6	4.8	5.4		89
Glucoamylase G1	1-Deoxynojirimycin	1.7	4.8	6.4		53
	Acarbose[d]	7.9	8.6	16.5		53
	Methyl α,β-acarviosinide	7.4	2.1	9.5		53
	β-cyclodextrin	12.9	−6.5	6.4		55
	6	20.2	−10.2	10.0		55
	7	21.4	−11.2	10.2		55
	8	22.5	−11.8	10.7		55
	9	16.5	−5.6	11.0		55
G1: Tyr50Phe	Acarbose[d]	9.5	7.3	16.8		56
	1-Deoxynojirimycin	2.4	3.9	6.2		56
G1: Trp52Phe	Acarbose[d]	12.7	2.9	15.6		56
	1-Deoxynojirimycin	4.5	0.6	5.1		56
G1: Arg54Lys	Acarbose[d]	9.9	−0.4	9.5		56
G1: Arg54Leu	Acarbose[d]	8.6	0.4	9.0		56
G1: Ser119Tyr	Acarbose[d]	7.2	9.3	16.4		56
	1-Deoxynojirimycin	2.2	4.5	6.7		56
G1: Trp120Phe	Acarbose[d]	6.9	2.7	9.6		56
	1-Deoxynojirimycin	6.9	−0.7	6.2		56
G1: Asn171Ser	Acarbose[d]	12.3	4.3	16.7		56
	1-Deoxynojirimycin	3.6	2.9	6.5		56
G1: Gln172Asn	Acarbose[d]	11.3	5.8	17.0		56
	1-Deoxynojirimycin	3.3	2.8	6.1		56
G1: Thr173Gly	Acarbose[d]	12.8	4.0	16.8		56
	1-Deoxynojirimycin	4.1	2.4	6.5		56
G1: Gly174Cys	Acarbose[d]	8.4	8.8	17.2		56
	1-Deoxynojirimycin	2.4	4.0	6.4		56

31.5 The Thermodynamics of Protein–Carbohydrate Interaction

Protein	Ligand					Ref
G1: Asp176Asn	Acarbose[d]	9.4	6.0	15.4		56
	1-Deoxynojirimycin	3.2	2.8	6.1		56
G1: Glu180Gln	Acarbose[d]	4.7	1.4	6.1		56
	1-Deoxynojirimycin	3.4	2.4	5.8		56
G1: Ser185His	Acarbose[d]	13.0	2.2	15.2		56
	1-Deoxynojirimycin	5.4	0.4	5.0		56
G1: Arg305Lys	Acarbose[d]	7.3	−0.5	6.8		56
G1: Asp309Glu	Acarbose[d]	7.4	6.7	14.2		56
	1-Deoxynojirimycin	2.8	3.7	6.5		56
G1: Trp317Phe	Acarbose[d]	8.5	2.4	10.9		56
G1 Starch Binding Domain	GlcSGlc$_2$	9.1	3.4	4.3		54
	GlcSGlc$_3$	17.7	13.2	4.5		54
	GlcSGlc$_4$	11.1	6.4	4.7		54
	β-cyclodextrin	13.1	6.6	6.5		54
Glucoamylase G2	1-Deoxynojirimycin	2.7	3.4	6.2	−23.4	53
	Acarbose[e]	9.7	6.6	16.3	−200	53
	D-*Gluco*-dihydroacarbose	7.1	3.2	10.3		53
	L-*Ido*-dihydroacarbose	2.3	5.0	7.3		53
	αSGlc(1→4)αGlcOMe	1.4	4.8	6.2		53
	αSGlc(1→4)βGlcOMe	3.3	1.9	5.2		53
	αS,SGlc(1→4)αGlcOMe	1.4	3.4	4.8		53
	6	8.0	>3.0	>11.0		55
	7	7.6	>3.4	>11.0		55
	8	6.5	>4.3	>10.8		55
	9	9.9	>1.1	>11.0		55
G2: Tyr175Phe	Acarbose[e]	8.9	9.2	18.0		56
	1-Deoxynojirimycin	2.9	2.5	5.3		56
Gyrase B	Clorobiocin	8.8	1.0	9.8		92
	Novobiocin	9.3	0.8	10.1		92
Hevein	βGlcNAc(1→4)GlcNAcOH	4.9	−1.2	3.8	−64.5	93
	βGlcNAc(1→4)βGlcNAc(1→4)GlcNAcOH	6.0	−0.2	5.8	−83.7	93
	αGluOPh	2.6	0.7	3.3	−12.1	93
Lentil Lectin	1dGlu	3.6	−1.1	2.4	−9.0	76
	αGluF	5.9	−3.4	2.5	−44.7	76

Table 8 (*continued*)

Protein	Ligand[a]	$-\Delta H^b$	$T\Delta S^b$	$-\Delta G^b$	ΔC_P^c	Ref.
	2dGluOH	4.3	−1.7	2.6	−4.8	76
	2FGluOH	3.7	−0.6	3.1	−33.4	76
	3FGluOH	3.5	−1.3	2.2	−17.4	76
	3OMeGluOH	2.9	1.6	4.5	−53.5	76
MBP-A (rat serum)[e]	αManOMe	4.7	−0.9	3.8		94
	OMeGlcNAc	5.2	−1.4	3.8		94
	NAcYD(G-(CH$_2$)$_6$-Man)$_2$	5.2	−1.4	3.8		94
MBP-C (rat liver)[e]	αManOMe	5.1	−1.3	3.8		94
	OMeGlcNAc	0.0	0.4	0.4		94
	NAcYD(G-(CH$_2$)$_6$-Man)$_2$	4.7	−0.9	3.8		94
MBP (*E. coli*)	αGlc(1→4)GlcOH	7.4	−1.5	5.9		94
	β-cyclodextrin	2.8	10.9	8.1	−207	95
Pea Lectin	αGlcOPh	3.0	4.4	7.4	162	95
	1dGlu	3.0	0.9	3.9	−4.5	76
	αGluF	4.3	−1.4	2.9	11.0	76
	2dGluOH	4.3	−1.4	2.9	−5.0	76
	2FGluOH	5.0	−1.9	3.0	−40.3	76
	3FGluOH	3.7	0.0	3.6	−46.3	76
	3OMeGluOH	3.0	0.0	2.8	8.7	76
Peptide: YYWIGIR-NH2	sLe^x	3.4	1.2	4.6	−1.8	76
	Le^x	73.7	−67.3	6.4		96
	αNeuAc(2→3)βGal(1→4)GlcNAcOH	18.6	−12.4	6.3		96
Ricin communis Agglutinin	βGal(1→4)GlcNAcOH	3.2	0.9	4.1		96
		6.9	−0.89	6.0		97
	βGal(1→4)GlcOH	6.6	−0.6	6.1		97
		7.6	−1.7	6.0		97
		8.1	−1.6	6.0		97
		7.5	−2.1	6.0		97
	ThiodiGal	12.2	−6.8	5.4		97
		12.0	−6.5	5.5		97

31.5 The Thermodynamics of Protein–Carbohydrate Interaction

Lectin	Ligand					Ref.
Se 155.4	MumbαGal	5.8	−1.4	4.4		97
	MumbβGal	5.2	−0.8	4.5		97
		9.5	−3.6	5.9		97
	αGalOMe	10.4	−4.6	5.8		97
		5.3	−0.9	4.3		97
	βGalOMe	5.3	−1.2	4.4		97
		5.4	−0.3	5.0		97
	MumbβGalNAc	5.5	−0.8	4.8		97
	GalOH	7.2	−1.7	5.5		97
	10	5.8	−1.4	4.3		97
	11	6.8	0.3	7.1		98
	12	8.8	−1.8	7.0		98
		8.2	−1.2	7.0		98
	13	8.6	−1.5	7.0		98
	14	7.0	0.7	7.6		98
Sheep Spleen Lectin (L-14)	ThiodiGal	12.9	−7.4	5.5		99
		11.5	−5.5	6.0		99
	βGal(1→4)GlcNAcOH	10.0	−3.8	6.2		99
		10.5	−4.3	6.2		99
	MumβLac	8.6	−2.9	5.7		99
		10.3	−4.5	5.7		99
	βGal(1→4)GlcOH	8.8	−3.6	5.3		99
		8.6	−3.4	5.3		99
	MumαGal	2.4	2.4	4.8		99
		2.2	2.9	5.0		99
	GalOH	0.6	−1.7	2.4		99
	αGalOMe	0.7	−1.7	2.4		99
	βGalOMe	0.5	−1.7	2.2		99
Snowdrop Lectin	αMan(1→3)αManOMe	3.1	1.7	4.8		36
	αMan(1→6)αManOMe	2.1	2.1	4.2		36
	αMan(1→3)[αMan(1→6)]αManOMe	3.9	0.9	4.8		36
Soybean Agglutinin	βGalOMe	10.6	−6.9	3.7	−93.2	89
		8.9	−4.9	4.0		89

Table 8 (continued)

Protein	Ligand[a]	$-\Delta H$[b]	$T\Delta S$[b]	$-\Delta G$[b]	ΔC_P[c]	Ref.
	GalNAcOH	9.5	-4.1	5.4		89
	βGalNAcOMe	13.9	-8.0	6.0	-100	89
		12.1	-5.9	6.2		89
	βGal(1→4)GlcOH	5.5	-2.4	3.2		89
	βGal(1→4)GlcNAcOH	8.2	-4.3	3.9		89
	βGal(1→4)βGlcNAcOMe	7.5	-3.8	3.7		89
Urtica dioica Agglutinin[e]	βGlcNAc(1→4)GlcNAcOH	4.7	-0.8	3.9		100
	βGlcNAc(1→4)βGlcNAc(1→4)GlcNAcOH	6.3	-1.2	5.1		100
	βGlcNAc(1→4)[βGlcNAc(1→4)]₂GlcNAcOH	5.1	0.5	5.6		100
	βGlcNAc(1→4)[βGlcNAc(1→4)]₃GlcNAcOH	5.1	0.8	5.9		100
Winged Bean Agglutinin I	GalOH	6.2	-1.8	4.4	-13.1	101
	1HGal	5.9	-1.6	4.3	-3.0	101
	2HGal	3.9	-0.7	3.2	-6.0	101
	FucOH	5.0	-1.5	3.5	24.9	101
	α1FGal	6.5	-2.1	4.4		101
	2FGalOH	5.8	-1.6	4.1	-36.4	101
	6FGalOH	6.1	-1.9	4.3	8.4	101
	2OMeGalOH	7.0	-1.8	5.2	-23.2	101

[a] structures of ligands **1–14** are shown in Attachment to Table 8; d = deoxy; reducing sugars are assumed to be equilibrating mixtures of anomers
[b] kcal mol⁻¹
[c] cal mol⁻¹ deg⁻¹
[d] by displacement
[e] data obtained by fixing n

31.6 The Role of Multivalency in Protein–Carbohydrate Interaction

Attachment to Table 8: Ligands 1–14.

	R₁	R₂
10	OH	none
11	O	(CH₂)₆
12	O	(CH₂)₇
13	O	(CH₂)₉
14	S	CH₂-C₆H₄-CH₂

	m	n
6	2	0
7	2	1
8	3	2
9	3	3

Table 9. Acarbose binding by displacement method [53].

Enzyme	Ligand	K^a	$-\Delta G^b$	$-\Delta H^b$	$T\Delta S^b$
G2	1-deoxynojirimycin	3.30×10^4	6.2	2.7	3.5
	acarbose (apparent)[c]	2.70×10^7	10.1	7.0	3.1
	acarbose	8.80×10^{11}	16.3	9.7	6.6
G1	1-deoxynojirimycin	4.70×10^4	6.4	1.7	4.8
	acarbose (apparent)[c]	2.00×10^7	10.0	6.2	3.9
	acarbose	9.40×10^{11}	16.5	7.8	8.6
G1	methyl α,β-acarviosinide	7.80×10^6	9.5	7.4	2.1
	acarbose (apparent)[d]	3.80×10^4	6.3	1.7	4.6
	acarbose	2.90×10^{11}	15.8	9.1	6.6

[a] M^{-1}
[b] kcal mol^{-1}
[c] Inhibited by 1-deoxynojirimycin
[d] Inhibited by methyl α,β-acarviosinide

coamylase G2 proceeds with a free energy of binding of 16.3 kcal mol^{-1} and a ΔC_p of -200 cal mol^{-1} deg^{-1} [53]. Similarly, the binding of maltose to maltose binding protein shows a free energy of binding of 8.14 kcal mol^{-1} and an accompanying change in molar heat capacity of -207 cal mol^{-1} deg^{-1} [95]. It seems likely that the source of enhanced affinity is highly efficient removal of solvent from poorly solvated surfaces.

Sigurskjold and coworkers have utilized a displacement strategy to evaluate the binding thermodynamics of the highest affinity protein–carbohydrate complexes known, that of acarbose and *Aspergillus* glucoamylase (Table 9) [53]. The high affinity of this complex precludes direct calorimetric titration of the protein with ligand. Instead, this group takes advantage of the additivity of binding thermodynamic parameters and displaces the moderately binding ligand 1-deoxynojirimycin with the high affinity ligand acarbose. In an appropriate concentration domain, the enthalpy of acarbose binding is roughly equivalent to the sum of the enthalpy of 1-deoxynojirimycin binding and the enthalpy of 1-deoxynojirimycin displacement by acarbose. Similarly, the free energies of binding are additive, requiring that the complex stability constants are mulitiplicative.

The effect of metal ion identity on the thermodynamics of concanavalin A–oligomannoside binding has been investigated [87]. The role of metal ion in legume lectin binding has long been questioned; although metal ions are required for binding they do not appear to contact saccharide ligands. Through the series cobalt nickel and cadmium only minor alterations in enthalpies and free energies of binding were observed, suggesting a primary role for metal ion in the maintenance of protein structure rather than in ligand binding.

Previously, we and others have commented on the similarities and differences between patterns of saccharide association for lectins and antibodies [3]. Unfortunately, the database of thermodynamic parameters for antibody–carbohydrate complexation remains small. While important structural and energetic differences may exist that provide clues to the origin of both affinity and association in aqueous solution, it is impossible to reach any conclusions at this time.

31.6 The Role of Multivalency in Protein–Carbohydrate Interaction

The tremendous potential of saccharide ligands for the treatment of a wide range of human diseases continues to drive much of contemporary glycoscience. The major impediment to the use of carbohydrate or carbohydrate mimetics continues to be the low affinity of saccharide ligands for their lectin receptors. As we note above, the inherently low enthalpy and free energy available from desolvation during protein–carbohydrate complexation likely places fundamental limitations on the affinity that might be achieved through modification of monomeric ligands.

Lectins are seldom found in vivo as monomeric species; rather, they are aggregated into higher order oligomeric structures. A reasonable conclusion drawn from this observation is that nature takes advantage of the remarkable diversity of carbohydrate structures to achieve high selectivity, while overcoming the low inherent affinity through the use of multivalency. From this concept an enormous array of multivalent carbohydrate ligands has been synthesized and evaluated in a range of binding assays. Many such ligands show remarkable enhancements in activity compared to an equivalent concentration of the corresponding monovalent ligand; indeed some show affinity enhancements as large as 10^9 on a valence corrected, or per mole of saccharide, basis. In most instances the physical basis of the observed effects is unclear. Here, we review briefly the phenomonology of multivalency effects in protein–carbohydrate interaction, then consider the thermodynamic consequences of tethering recognition domains. The goal of this exercise is to provide a molecular basis for the 'cluster glycoside effect' and evaluate its potential for therapeutic application.

31.6.1 Phenomenology

The 'cluster glycoside effect', as defined by Y. C. Lee, was first observed using the hemagglutination assay [1]. Evaluating several hepatic lectins against several monosaccharide types tethered by amino acids, Lee noted the IC_{50} of these multivalent neoglycoconjugates was lower than expected based on the saccharide content of the ligand (Figure 6). Since this time an enormous variety of multivalent ligands have been prepared and bound (Table 10) [102, 103]. The magnitude of the cluster glycoside effect varies over nine orders of magnitude across the range of neoglycoconjugate structures that have been evaluated.

Some of the first multivalent ligands reported are a group of glycosylated polyacrylamides. These ligands are prepared by copolymerization of acrylamide and an acrylic acid ester [104]. Two conceptual routes have been reported, varying in the nature of the acrylic acid ester. In one instance the ester contains the saccharide recognition domain; in the other the acryloyl ester incorporates an N-hydroxysuccinamide ester that is later displaced by a saccharide recognition domain tethered through an amine spacer (Figure 6).

The molecular weight range achieved with these methods spans 10,000–450,000 Da, with larger polymers typically formed by incorporating the saccharides during

Table 10. Representative binding enhancements through multivalency.

Ligand	Lectin	Ligand valency	β^a	Assay Method	Ref.
Sialylated glycopolymer; pre-P	Viral hemagglutinin	200–2000	5.0×10^6	HA^b	104
Sialylated glycopolymer; co-P	Viral hemagglutinin	200–2000	5.0×10^5	HA	104
Sialylated glycopolymer; pre-P	Viral hemagglutinin	200–2000	1.0×10^6	ELISA^c	104
Sialylated glycopolymer; co-P	Viral hemagglutinin	200–2000	1.6×10^8	ELISA	104
Bivalent and trivalent mannosylated glycoclusters	Serum-type mannose binding protein	2–3	$1–2 \times 10^0$	precipitation	105
Bivalent glycopeptide based NAcYD(GG-ah-GlyC)	rat hepatic lectin	2	4.7×10^2	precipitation	1
Bivalent glycopeptide based YEE (ah-GlyC)	Chicken hepatic lectin	2	2.6×10^3	precipitation	1
Tris-GlcNAc glycocluster	Chicken hepatic lectin	3	5.1×10^0	precipitation	1
Sialylated PAMAM glycodendrimer	*Limax flavus* lectin	12	1.8×10^2	ELLA^d	47
Sialylated PAMAM glycodendrimer	*Limax flavus* lectin	4	2.6×10^1	ELLA	47
Mannosylated PAMAM glycodendrimer	Concanavalin A	16	6.6×10^1	ELLA	42
Mannosylated PAMAM glycodendrimer	Pea lectin	16	1.4×10^3	ELLA	42
Lysine-based mannosylated glycoclusters	Human mannose receptor	6	1.0×10^7	Displacement	106
Lysine-based mannosylated glycoclusters	Human mannose receptor	6	5.6×10^7	Displacement	106
Galabioside glycoclusters	*Strepococcus suis*	4	6.0×10^2	HA	107
Bivalent sialosides	Viral hemagglutinin	2	$1–2 \times 10^0$	HA	108
Bivalent sialosides	Intact influenza virus	2	8.3×10^2	HA	108
sLe^x dimer	Immobilized E-selectin	2	4.9×10^0	ELLA	109
Lyso-GM$_3$-PGA glycopolymer	Viral hemagglutinin	270	1.4×10^7	Fluorescence	110
oligo-GM$_1$ propylene-imine glycodendrimer	Cholera toxin B subunit	7	3.3×10^3	ELLA variation	111
oligo-GM$_1$ propylene-imine glycodendrimer	Heat labile enterotoxin	7	1.7×10^3	ELLA variation	111
C-mannosyl. Glycopolymer	Concanavalin A	3×10^5	2.6×10^5	HA	112
3′, 6′ sulfated galactosyl glycopolymer	P-selectin	25	2.0×10^1	Inhibition	113
Mannosylated glycodendrimer	Concanavalin A	4	1.2×10^2	HA	41

^{a}calculated by dividing K_a for multivalent ligand by K_a for monovalent refrence compound
^{b}Hemagglutination Assay
^{c}Enzyme Linked Immunosorbent Assay
^{d}Enzyme Linked Lectin Assay

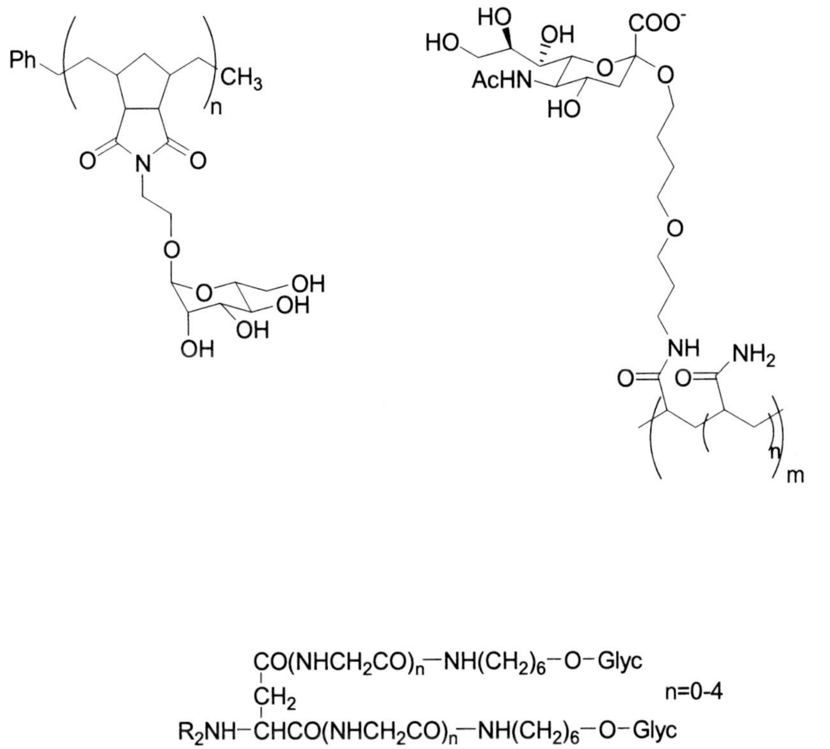

31.6 The Role of Multivalency in Protein–Carbohydrate Interaction

Figure 6. Representative multivalent ligands. From top left: mannosylated ROMP polymers; acrylamide/acryloyl ester copolymers; peptide-based dendritic ligands. See text for descriptions.

the copolymerization, rather than adding the saccharides to an activated backbone. Several other groups have reported a similar strategy [114–119].

The performance of glycopolymers in agglutination assays is remarkable; indeed, the largest cluster glycoside effects observed to date are for the acrylamide polymers. Spaltenstein and Whitesides observed a reduction of 10^5 in IC_{50} for polyacrylamide glycopolymers bearing α-sialic acid residues compared to monosaccharide assayed against influenza viral hemagglutinin, placing minimum inhibitory concentrations in the nanomolar range [120]. Remarkably the efficacy of polymeric inhibitors varied both as the content of ligand and as the method by which it was incorporated: IC_{50} values for glycopolymers prepared by the activated backbone method were 75 times lower than equivalent sized glycopolymers prepared by copolymerization of sialic acid acrylamide monomers [121]. Apparently the mechanism by which these polyvalent ligands inhibit hemagglutinin-mediated agglutination is dependent on both the content and spatial orientation of saccharide epitopes. Polyacrylamides bearing N-acetyllactosamine, prepared by Tsuchida and co-workers, showed IC_{50} values enhanced by roughly 10^3 relative to unmodified monovalent ligand in agglutination assays against plant lectins [118].

Kiessling and coworkers have prepared a range of glycopolymers using a ring-opening metathesis polymerization (ROMP) strategy (Figure 6) [112]. Again, enhancements in affinity near 10^3 on a valence-corrected basis are observed in agglutination assays. Significantly, greatly reduced enhancements in apparent affinity are observed when the ligands are assayed by other methodologies. A surface plasmon resonance study of the binding of mannosylated ROMP ligands to concanavalin A showed enhancements of five to 40-fold [122]. Kiessling and coworkers have also demonstrated a size dependence during glycopolymer inhibition [123]. Mannosylated glycopolymers showed enhancements in the range of 50–3,000 compared to monosaccharide, increasing exponentially with polymer chain lengths up to 143 units in agglutination assays against concanavalin A. Increasing chain lengths beyond this value failed to provide a continued valence corrected enhancement, although overall ligand affinities apparently continue to increase. With very large polymers (MW 10^6) enhancements near 10^5 were observed.

The cell surface presents a multivalent display of carbohydrate exquisitely suited to maximize protein–carbohydrate binding while minimizing entropic penalties. On the one hand, carbohydrate epitopes are free to orient themselves in two dimensions in a configuration that optimizes favorable contacts with the lectin. On the other hand, the bulk of the translational entropy and all of the rotational entropy has already been lost during placement of the ligand in the lipid bilayer. This situation is mimicked by a variety of synthetic glycosylated liposomes and, like glycopolymers, glycosylated liposomes show remarkably enhanced performance in agglutination assays compared to an equivalent concentration of the corresponding monovalent ligand. The general method for synthesis of glycosylated liposomes is:

i) conjugation of the glycosyl moiety to a lipid chain;
ii) incorporation of some amount of this glycolipid into a mixture of phosphatidyl choline and cholesterol; and
iii) sonication or extrusion through a membrane to form the liposome [124].

Kingery-Wood and co-workers produced a series of sialic acid containing liposomes with mole fractions of sialic acid (compared to phosphatidyl choline and cholesterol) of 2–25%. By hemagglutination assays, a maximum inhibition was observed at 5% sialic acid incorporation, with IC_{50} values in the nanomolar range. DeFrees and co-workers repeated this procedure with the sialyl Lewis[x] epitope [125]. Ketis et al. incorporated the glycoprotein glycophorin into liposomes as a 33% component for binding assays with wheat germ agglutinin [126]. Charych and co-workers have developed an innovative methodology for glycoliposome production [127–129]. Incorporation of diacetylenic lipids followed by UV (254 nm) irradiation yielded a highly colored (blue) polymerized liposome species. The incorporated chromophore provides a unique binding read-out: upon lectin binding the liposome turns red, presumably as the lipid bilayer adopts a conformation optimal for ligand binding.

A much larger group of multivalent ligands include the so-called glycoclusters and glycodendrimers. These ligands are typically smaller in size than the polymers, and require significantly greater synthetic effort. On the other hand, the resulting species are homogenous in nature and more easily characterized than polymeric ligands; additionally, greater control is achieved over the spacing and orientation of glycosyl moieties. A complete description of this group of ligands is beyond the scope of this document; rather, the reader is directed to several recent reviews of the field [102, 103].

In contrast to the remarkable performance of glycopolymers, glycoclusters and glycosylated dendrimers provide much more modest enhancements in affinity. With valences ranging to 36, typical valence-corrected enhancements in affinity are on the order of $10-10^3$, although some exceptional enhancements have been reported. Hansen and co-workers reported increases of 10^2-10^3 in IC_{50} values for several low-valency galabiose inhibitors relative to unmodified galabiose against the Gram-positive bacterium *Streptococcus suis* [107]. Nonetheless, this enhancement raised the approximate range of binding efficacy from micromolar to nanomolar, approaching ranges required for therapeutic utility. Knowles and coworkers observed a 100-fold increase in the relative potency for a series of dimeric sialosides compared to unmodified sialic acid when evaluated against the influenza virus in agglutination assays [108]. A peptide-based mannosylated glycocluster reported by Biessen and coworkers showed a cluster glycoside effect of 10^6 relative to mannose when evaluated against human mannose receptor; this dramatic increase in relative inhibitory potential is one of the largest seen with non-polymeric systems [106]. Again, the effect is also a function of the analytical methodology utilized, and apparent enhancements in binding measured by ELLA are much smaller than those seen in hemagglutination assays [42–50]. For example, many of the glycosylated PAMAM dendrimers, with valencies ranging over 2–16, show enhancements in apparent affinity relative to the unmodified saccharides ranging from 2- to 15-fold when considered on a per saccharide basis [47, 130]. Similarly, Ashton and co-workers report only modest enhancements in affinity during evaluation of glycodendrimers based on a combination aliphatic/aromatic core [131].

31.6.2 The Energetic Consequence of Ligand Linkage

Having considered the reported phenomonology of multivalency effects in protein–carbohydrate interaction, we now aim to understand the origin of the cluster glycoside effect at a molecular level. To the extent that a set of design principles relevant to high affinity can be extracted, multivalency is among the most promising strategies towards the development of therapeutically useful carbohydrates. We consider here only a brief treatment of multivalency effects in aqueous association; a more complete treatment of the topic has recently appeared [103].

A meaningful discussion of multivalency in ligand binding first requires a precise definition of the terms involved. Consider the binding of N monovalent ligands *versus* a ligand of valency N, L^N, to an N-valent lectin. Assuming the binding sites are equivalent and non-interacting, thermodynamic parameters describing the association of each monovalent ligand can be represented as ΔJ_N. The corresponding terms for association of the multivalent ligand are related to those for binding of the monovalent ligand by the expression:

$$\Delta J^N = N \Delta J_N + \Delta J_i$$

where ΔJ_i is an interaction term, describing the energetic consequences of physical linkage [68]. A discussion of multivalency effects next requires distinction be drawn between an overall affinity for the multivalent ligand and affinity on a per saccharide basis. These concepts have been variously termed *functional affinity* and *intrinsic affinity*, or *avidity* and *affinity*, respectively [103, 132]. Whitesides and coworkers provided a more precise rendering of these concepts by defining the quantities α and β as:

$$\alpha = \frac{\ln(K_{poly})}{\ln(K_{mono})^N}$$

or, in free energy terms:

$$\Delta G_{poly} = \alpha N \Delta G_{mono}$$

Defined in this way, α is a measure of what is traditionally referred to as positive cooperativity; in the analysis of Jencks, values of α greater than unity correspond to favorable (i.e. negative) interaction free energies. Determination of α values require knowledge of the number of ligands bound. Since in most instances this information is not known, the quantity β describes simple phenomenology:

$$K = \beta K_{mono}$$

At its simplest level enhanced affinity from multivalency requires only that $|\Delta G^N| > |\Delta G_N|$, where ΔG^N represents the change in Gibbs' free energy upon binding of the N-valent ligand and ΔG_N represents the change in Gibbs' free energy upon binding of each of the constituent monomeric ligands to an individual lectin binding site. Note that in this construct, the contribution of each monovalent recognition epitope to the overall binding free energy need not be greater than—or even equivalent to—that of monovalent ligand. The only requirement for the observation of enhanced affinity of the *ligand* is that $|\Delta J_i|$ be less than $|(N-1)\Delta J_N|$; in such instances β will be greater than or equal to one. At a second level, a multivalency effect requires that the interaction free energy be favorable. This case describes what is normally referred to as positive cooperativity; here $|\Delta G^N| > |N \Delta G_N|$.

The key value describing the effect of ligand multivalency on binding thermodynamics is the interaction energy, ΔG_i. A range of physical processes contribute to these interaction free energies; the concepts are intuitively more accessible if considered as separate enthalpic and entropic terms.

Enthalpic Contributions to ΔG_i

The addition of a linker region has enthalpic consequences to overall binding free energies. First, a linker must be long enough to facilitate optimal placement of ligands within binding sites. Linkers too short or too rigid to facilitate optimal placement of recognition epitopes within binding sites have the effect of contributing unfavorably to interaction enthalpies. Assuming the recognition epitopes can in fact bind successfully, the linker will almost certainly contact the surface of the protein at the periphery of the binding site, on the surface of the protein beyond the binding sites, or both. These contacts can be either favorable or unfavorable, contributing postively or negatively to interaction energies. The *a priori* prediction of favorable or unfavorable contributions to ΔH_i is virtually impossible. Combinatorial approaches using, for example, peptide spacers may be a useful method of sampling wide ranges of linker–peptide interaction. Finally, the interaction of the linker region with solvent prior to binding leads to an energetic consequence during binding if the linker region is desolvated: this desolvation could make a favorable or unfavorable contribution to ΔH_i, depending on the precise molecular details of the linker surface. Again, association in solution requires consideration both of the interactions between solutes and the interactions of both solutes with solvent prior to association.

Entropic Contributions to ΔG_i

Energetic consequences of ligand tethering are typically considered primarily in entropic terms; there is little doubt that there will be a significant contribution to binding free energies attributable to an entropic interaction term. Predicting the magnitude—or even sign—of the entropic consequence of tethering ligands, however, is not straightforward. As described above, an overall entropy of binding can be conceptually separated into a translational and rotational entropy, a conformational entropy, and a solvation-associated entropy. A similar exercise facilitates consideration of entropic contributions to ΔS_i. In this construct, ΔS_i is considered as the sum of the differences in each entropic term for binding of the N-valent ligand and N monovalent ligands; that is:

$$\Delta S_i = [\Delta S_{T+R}(M^N L^N) - N \Delta S_{T+R}(M^N L)]$$
$$+ [\Delta S_{conf}(M^N L^N) - N \Delta S_{conf}(M^N L)]$$
$$+ [\Delta S_{solv}(M^N L^N) - N \Delta S_{solv}(M^N L)]$$

where $M^N L^N$ represents multivalent and monovalent complexes, respectively. The first term describes the entropic 'savings' that derives from minimizing the loss of translational and rotational entropy—the ability of a molecule to translocate in three dimensions and rotate on three axes—by converting N-ligands into a single species. Because translational and rotational entropies scale as the logarithm of the molecular size, ΔS_{T+R} for a multivalent ligand will be approximately equivalent to that of the corresponding monovalent species. As previously noted, the magnitude of this term is largely unknown. Table 6 above shows the wide range of values currently assigned to this term.

The second term of the entropy decomposition describes the loss in conformational degrees of freedom during binding, typically the restriction of flexible dihedrals. In the context of interaction entropies, this term primarily arises from losses in degrees of freedom in a flexible linker region during binding. Although the magnitude of this term is somewhat unclear, the situation is considerably less opaque than for ΔS_{T+R}. Some agreement seems to have emerged for a value near 0.5

kcal mol^{-1} near room temperature for complete restriction of a bond previously free to equilibrate among three energetically equivalent staggered conformers [133, 134]. Obviously not all bonds restricted during binding possess such freedom prior to complexation, and not all are completely constrained following binding. Linkers designed to span binding sites separated by 20–70 Å possess considerable conformational flexibility; the loss of this flexibility will unquestionably lead to a significant unfavorable contribution to ΔS_i.

Based on an entropic analysis, the prescription for the design of linker regions is clear. Linkers must be long enough to facilitate optimal placement of the saccharide recognition domains within the binding site, but with enough rigidity to minimize a conformational entropy penalty. An alternative strategy might be to make linkers of sufficient length that residual flexibility following binding minimizes the unfavorable contribution of ΔS_{conf} to ΔS_i. A number of studies on the role of flexibility in linker regions have been reported. Glick et al. evaluated a series of bivalent sialic acid ligands for the viral hemagglutinin, varying in length and linker flexibility [108]. A linker of intermediate flexibility provided the tightest binding as evaluated by hemagglutination assay. Unfortunately, without the required thermodynamic data, molecular interpretation of the result is impossible.

Bundle and coworkers also explored the role of ligand flexibility, albeit in a somewhat different context [98, 135]. Two decades of intense study on small molecule recognition in organic solvent clearly demonstrated that preorganization and complementarity are key requirements of ligand structure for the observation of high affinity. From that observation, ligands with diminished conformational entropy, or preorganized ligands, should show enhanced affinity relative to ligands with freely rotating bonds. Monoclonal antibody Se155.4 binds the trisaccharide methyl 2-(α-D-galactopyranosyl)-3-(α-D-3,6-dideoxyxylopyranosyl)-α-D-mannopyranoside. This binding buries the dideoxyxylopyranose (abequose) moiety completely, and consequently freezes rotation about the AbeMan glycosidic bond. A series of ligands were prepared that preorganize the ligand by freezing this rotation (Figure 7). Binding of each restricted ligand proceeded with thermodynamic parameters remarkably unchanged from those of the native—presumably less ordered—ligand. The origin of this effect is unclear. On the one hand, the inherent flexibility of the localized bond is low; the exo-anomeric effect likely limits the entropic gain available. On the other hand, the tether used to preorganize the ligand may interact unfavorably in the complex.

The role of ligand flexibility/preorganization in determining affinities has also been probed through the use of C-glycosides [136, 137]. Absent the anomeric oxygen and consequent exo-anomeric effect, C-glycoside ligands should show enhanced flexibility around the glycosidic bond relative to the corresponding O-glycoside. This additional flexibility in the unbound ligand should enhance the conformational entropy loss upon binding. Again, the results to date fail to unambiguously demonstrate general design principles with regard to anomeric flexibility. A series of ligands that sequentially replace both the anomeric and pyranosidic oxygen with a methylene and thioether failed to yield a clear picture of structure–activity relationships. As with Bundle's work, however, the alteration in flexibility predicted to arise from such modification is small. Assuming an exo-anomeric effect of roughly

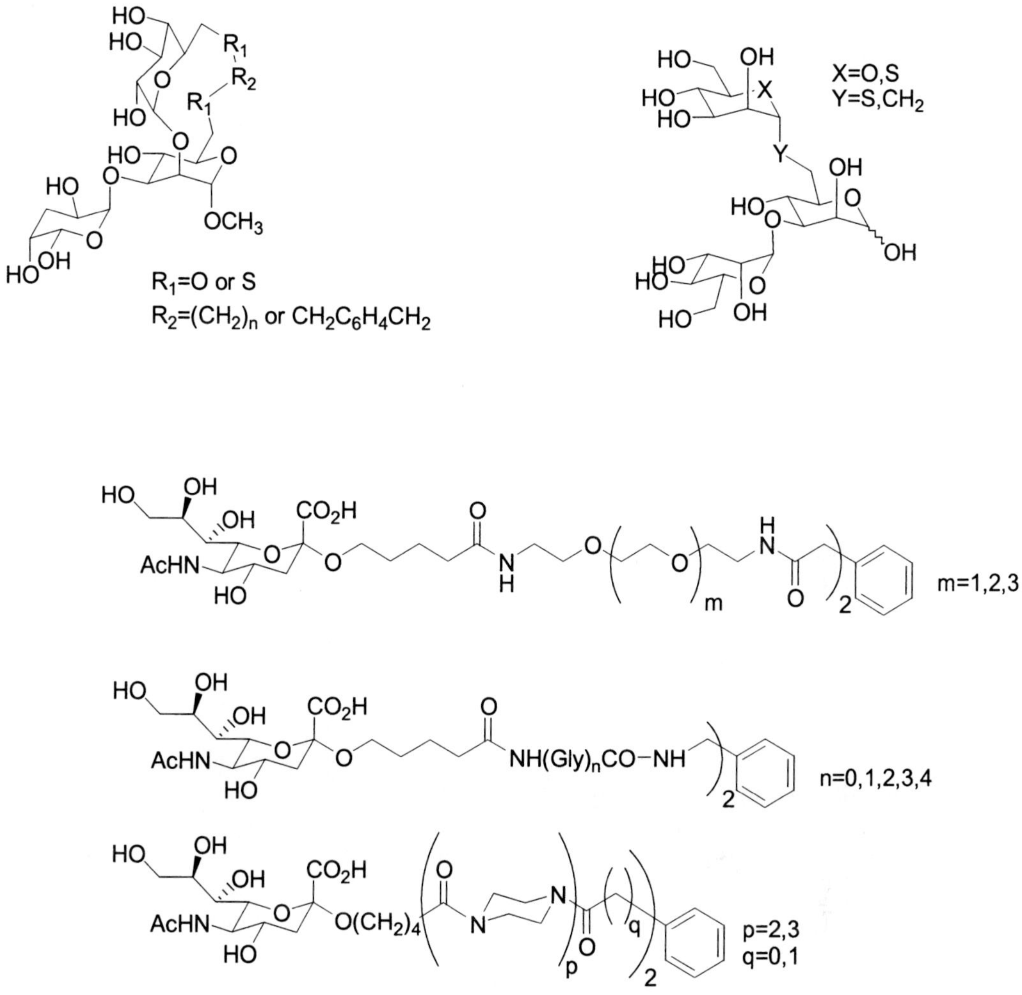

Figure 7. Ligands designed to assess the role of conformational entropy in ligand binding. See text for description.

1–2 kcal mol^{-1} and that the gg-configuration about the glycosidic linkage is precluded in both the *C*- and *O*-glycoside, deletion of the anomeric oxygen alters the predicted rotomeric distribution about the anomeric linkage by roughly a factor of two. Furthermore, the uncertain role of changes in ligand solvation renders a molecular basis for changes in affinity opaque; further calorimetric study of such ligands may illuminate the relevant issues.

Beyond concepts of linker flexibility, unpredictable values of ΔS_{solv} and ΔH_i

produce intractable design problems. The large potential energetic consequences of even small ligand modifications, coupled with the relatively weak interaction free energies that typify protein–carbohydrate interaction require a careful and stepwise exploration of the relationship between ligand structure and binding energetics. There likely do exist a simple set of rules to govern intelligent construction of multivalent ligands. Finally, it is imperative that the analytical methodology used to observe cluster glycoside effects be carefully considered before a molecular explanation of phenomenology is offered.

31.6.3 A Molecular Basis for the Cluster Glycoside Effect

Having considered the phenomenology of the cluster glycoside effect and the thermodynamic considerations surrounding aqueous association of multivalent ligands, we now turn briefly to possible molecular explanations for the affinity that apparently derives from multivalency. At this time, there are no recorded definitive examples of cooperativity in protein–carbohydrate binding deriving from favorable interaction energies. It seems most likely, then, that most cluster glycoside effects reported to date are in fact measures of the ability of multivalent saccharide ligands to drive the formation of some form of higher order aggregate. While formation of this aggregate might be favorable, providing a coupled equilibrium that would, in turn, enhance protein–carbohydrate affinities, there is no evidence to suggest that this is in fact true. Alternatively, Mammen, Choi, and Whitesides postulated a theory for the remarkable performance of glycopolymers in agglutination assays [103]. After binding several sialic acids to the viral hemagglutinin, these large, sterically demanding molecules may form a water-swollen, gel-like layer between the lectin and the erythrocyte surface. This barrier prevents additional binding from proceeding, but does not require thermodynamically enhanced saccharide-lectin binding.

In some respects, protein–carbohydrate interaction is a poor system with which to study multivalency effects in aqueous association. The low monovalent affinities, poor analytical methodology for evaluation of binding, and the propensity of lectins to aggregate in the presence of multivalent saccharide ligands provides a daunting series of complications to the interpretation of binding data. More recently the study of protein–carbohdyrate interaction using bacterial two-component toxins suggests that some enhancement in affinity may be possible through multivalency [138]. These proteins, which do not aggregate polyvalent saccharides and show relatively small distances between binding sites, are more amenable to study. The enhancements in affinity available in these systems is not yet clear.

Finally, we note that the appropriate analytical methodology and model of binding is dependent on the question asked. It may well be the case that in the context of biological systems, ligands especially capable of driving aggregation/precipitation processes might act as useful agents. The concern, rather, is a precise definition of the issue at hand. With regard to ligand binding, in the commonly accepted definition, it appears unlikely that multivalency can afford high affinity through positive cooperativity.

Acknowledgments

The authors gratefully acknowledge the assistance of Mr. J. Lundquist and Ms. Aimeé Butler for preparation of figures. Financial support for this work was provided by the National Institutes of Health (GM57179).

References

1. Lee, Y.C.; Lee, R.T. *Acc. Chem. Res.* **1995**, *28*, 321.
2. Dwek, R.A. *Chem. Rev.* **1996**, *96*, 683–720.
3. Toone, E.J. *Current Opinion in Structural Biology* **1994**, *4*, 719–728.
4. McClellan, A.L. *Tables of Experimental Dipole Moments*; Rahara Enterprises: El Cerrito, 1974.
5. Rigby, M.; Smith, E.B.; Wakeham, W.A.; Maitland, G.C. *The Forces Between Molecules*; Oxford: New York, 1986.
6. Zeegers-Huyskens, T.; Huyskens, P. *Intermolecular Forces*; Huyskens, P. L.; Luck, W. A. P.; Zeegers-Huyskens, T., Ed.; Springer-Verlag: New York, 1991, pp 1–30.
7. London, F. *Z. Physik* **1930**, *63*, 245.
8. London, F. *Z. Physik Chem. Leipzig* **1930**, *11*, 222.
9. Hunter, E.P.L.; Lias, S.G. *J. Phys. Chem. Ref. Data* **1998**, *27*, 413–656.
10. Mirejovsky, D.; Arnett, E.M. *J. Am. Chem. Soc.* **1983**, *105*, 1112.
11. *Water: A Comprehensive Treatise*, Vol. 1–7; Franks, F., Ed.; Plenum Press: New York, 1974.
12. Muller, N. *J. Solution Chem.* **1988**, *17*, 669.
13. Muller, N. *Acc. Chem. Res.* **1990**, *23*, 23–28.
14. Ben-Naim, A.M., R. *J. Phys. Chem. B* **1997**, *101*, 11221–11225.
15. Soper, A.K.; Luzar, A.J. *J. Phys. Chem.* **1996**, *100*, 1357.
16. Turner, J.; Soper, A.K. *J. Chem. Phys.* **1994**, *101*, 6116.
17. Turner, J.Z.; Soper, A.K.; Finney, J.L. *J. Chem. Phys.* **1995**, *102*, 5438.
18. Turner, J.; Soper, A.K.; Finney, J.L. *Molecular Phys.* **1990**, *70*, 679.
19. Finney, J.L.; Soper, A.K.; Turner, J.Z. *Pure Appl. Chem.* **1993**, *65*, 2521.
20. Leberman, R.; Soper, A.K. *Nature* **1995**, *378*, 364.
21. Reiss, H.; Frisch, H.L.; Lebowitz, J.L. *J. Chem. Phys.* **1959**, *31*, 369.
22. Reiss, H.; Tully-Smith, D.M. *J. Chem. Phys.* **1971**, *55*, 1674.
23. Reiss, H. *Adv. Chem. Phys.* **1966**, *9*, 1.
24. Frisch, H L. *Adv. Chem. Phys.* **1963**, *6*, 229.
25. Pierotti, R.A. *A. Chem. Rev.* **1976**, *76*, 717.
26. Stillinger, F.H. *J. Solution Chem.* **1973**, *2*, 141.
27. Ben-Naim, A.; Tenne, R. *J. Chem. Phys.* **1977**, *67*, 627.
28. Lee, B. *Biopolymers* **1991**, *31*, 993.
29. Lee, B. *Methods Enzymol.* **1995**, *259*, 555.
30. Fujita, T.; Iwasa, J.; Hansch, C. *J. Am. Chem. Soc.* **1964**, *86*, 5175.
31. Iwasa, J.; Fujita, T.; Hansch, C. *J. Med. Chem.* **1965**, *8*, 150.
32. Arnett, E.M.; McKelvey, D.R. in *Solute-Solvent Interactions* Coetzee, J.F.; Ritchie, C.D., Ed.; Marcel Dekker: New York, 1969, pp. 342.
33. Oas, T.G.; Toone, E.J. *Adv. Biophys. Chem.* **1997**, *6*, 1–52.
34. Muller, N. *J. Solution Chem.* **1991**, *20*, 669.
35. Chervenak, M.C.; Toone, E.J. *J. Am. Chem. Soc.* **1994**, *116*, 10533–10539.
36. Chervenak, M.C.; Toone, E.J. *Biochemistry* **1995**, *34*, 5685–5695.
37. Lemieux, R.U. *Chem. Soc. Rev.* **1989**, *18*, 347–374.
38. Simanek, E.E.; McGarvey, G.J.; Jablonowski, J.A.; Wong, C.H. *Chem. Rev.* **1998**, *98*, 833–862.
39. Huwe, C.M.; Woltering, T.J.; Jiricek, J.; Weitz-Schmidt, G.; Wong, C.H. *Bioorg. Med. Chem.* **1999**, *7*, 773–788.

40. Du, M.; Hindsgaul, O. *Synlett* **1997**, 395–397.
41. Dimick, S.M.; Powell, S.C.; McMahon, S.A.; Moothoo, D.N.; Naismith, J.H.; Toone, E.J. *J. Am. Chem. Soc.* **1999**, in press.
42. Pagé, D.; Zanini, D.; Roy, R. *Bioorganic & Medicinal Chemistry* **1996**, *4*, 1949–1961.
43. Pagé, D.; Roy, R. *Bioconjugate Chem.* **1997**, *8*, 714–723.
44. Pagé, D.; Roy, R. *Glycoconjugate J.* **1997**, *14*, 345–356.
45. Roy, R.; Tropper, F.D.; Romanowska, A. *Bioconjugate Chem.* **1992**, *3*, 256–261.
46. Roy, R. Pagé, D.; Perez, S.F.; Bencomo, V.V. *Glycoconjugate J.* **1998**, *15*, 251–263.
47. Zanini, D.; Roy, R. *J. Am. Chem. Soc.* **1997**, *119*, 2088–2095.
48. Tropper, F.D.; Romanowska, A.; Roy, R. *Methods in Enzymology* **1994**, *242*, 257–271.
49. Arya, P.; Kutterer, K.M.K.; Qin, H.; Roby, J.; Barnes, M.L.; Kim, J.M.; Roy, R. *Bioorg. Med. Chem. Lett.* **1998**, *8*, 1127–1132.
50. Garcia-Lopez, J.J.; Hernandez-Mateo, F.; Isac-Garcia, J.; Kim, J.M.; Roy, R.; Santoyo-Gonzalez, F.; Vargas-Berenguel, A. *J. Org. Chem.* **1999**, *64*, 522–531.
51. Grime, J.K. *Analytical Solution Chemistry*; John Wiley & Sons: New York, 1985; Vol. 79.
52. Wiseman, T.; Williston, S.; Brandts, J.F.; Lin, L.-N. *Anal. Biochem.* **1989**, *179*, 131.
53. Sigurskjold, B.W.; Berland, C.R.; Svensson, B. *Biochemistry* **1994**, *33*, 10191–10199.
54. Sigurskjold, B.W.; Svensson, B.; Williamson, G.; Driguez, H. *Eur. J. Biochem.* **1994**, *225*, 133–141.
55. Sigurskjold, B.W.; Christensen, T.; Payre, N.; Cottaz, S.; Driguez, H.; Svensson, B. *Biochemistry* **1998**, *37*, 10446–10452.
56. Berland, C.R.; Sigurskjold, B.W.; Stoffer, B.; Frnadsen, T.P.; Svensson, B. *Biochemistry* **1995**, *34*, 10153–10161.
57. Sturtevant, J.M. *Proc. Natl. Acad. Sci. USA* **1977**, *74*, 2236–2240.
58. Murphy, K.P.; Privalov, P.L.; Gill, S.J. *Science* **1990**, *247*, 559–561.
59. Murphy, K.P. *Biophys. Chem.* **1994**, *51*, 311–326.
60. Doty, P.; Meyers, G.E. *Discussions Faraday Soc.* **1953**, *13*, 51.
61. Dunitz, J.D. *Science* **1994**, *264*, 670.
62. Erickson, H.P. *J. Mol. Biol.* **1989**, *206*, 465–474.
63. Horton, N.; Lewis, M. *Protein Science* **1992**, *1*, 169–181.
64. Janin, J. *Proteins Struct. Funct. Genet.* **1995**, *21*, 30–39.
65. Janin, J.; Chothia, C. *J. Mol. Biol.* **1976**, *100*, 197–211.
66. Janin, J.; Chothia, C. *Biochemistry* **1978**, *17*, 2943–2948.
67. Janin, J.; Chothia, C. *J. Biol. Chem.* **1990**, *265*, 16027–16030.
68. Jencks, W.P. *Adv. Enzymol.* **1975**, *43*, 219–410.
69. Page, M.I.; Jencks, W.P. *Proc. Natl. Acad. Sci.* **1971**, *68*, 1678–1683.
70. Searle, M.S.;Williams, D.H. *J. Am. Chem. Soc.* **1992**, *114*, 10690–10697.
71. Searle, M.S.;Williams, D.H.; Gerhard, U. *J. Am. Chem. Soc.* **1992**, *114*, 10697–10704.
72. Spolar, R.S.; Record, M.T., Jr. *Science* **1994**, *263*, 777–784.
73. Tidor, B.; Karplus, M. *J. Mol. Biol.* **1994**, *238*, 405–414.
74. Williams, D.H.; Searle, M.S.; Mackay, J.P.; Gerhard, U.; Maplestone, R.A. *Proc. Natl. Acad. Sci. USA* **1993**, *90*, 1172–1178.
75. Schwarz, F.P.; Ahmed, H.; Bianchet, M.A.; Amzel, L.M.; Vasta, G.R. *Biochemistry* **1998**, *37*, 5867–5877.
76. Schwarz, F.P.; Misquith, S.; Surolia, A. *Biochem J.* **1996**, *316*, 123–129.
77. Schwarz, F.P. *J. Solution Chem.* **1996**, *25*, 471–484.
78. Tyler-Cross, R.; Sobel, M.; McAdory, L.E.; Harris, R.B. *Arch. Biochem. Biophys.* **1996**, *334*, 206–213.
79. Tomme, P.; Creagh, A.L.; Kilburn, D.G.; Haynes, C.A. *Biochemistry* **1996**, *35*, 13885–13894.
80. Creagh, A.L.; Ong, E.; Jervis, E.; Kilburn, D.G.; Haynes, C.A. *Proc. Natl. Acad. Sci. USA* **1996**, *93*, 12229–12234.
81. Weatherman, R.V.; Mortell, K.H.; Chervenak, M.; Kiessling, L.L.; Toone, E.J. *Biochemistry* **1996**, *35*, 3619–3624.
82. Mandal, D.; Kishore, N.; Brewer, C.F. *Biochemistry* **1994**, *33*, 1149–1156.
83. Dam, T.K.; Oscarson, S.; Sacchettini, J.C.; Brewer, C.F. *J. Biol. Chem.* **1998**, *273*, 32826–32832.

84. Dam, T.K.; Oscarson, S.; Brewer, C.F. *J. Biol. Chem.* **1998**, *273*, 32812–32817.
85. Gupta, D.; Dam, T.K.; Oscarson, S.; Brewer, C.F. *J. Biol. Chem.* **1997**, *272*, 6388–6392.
86. Chervenak, M.C.; Toone, E.J. *Bioorg. Med. Chem.* **1996**, *4*, 1963–1977.
87. Sanders, J.N.; Chenoweth, S.A.; Schwarz, F.P. *J. Inorg. Biochem.* **1998**, *70*, 71–82.
88. Swaminathan, C.P.; Surolia, N.; Surolia, A. *J. Am. Chem. Soc.* **1998**, *120*, 5153–5159.
89. Gupta, D.; Cho, M.; Cummings, R.D.; Brewer, C.F. *Biochemistry* **1996**, *35*, 15236–15243.
90. Surolia, A.; Sharon, N.; Schwarz, F.P. *J. Biol. Chem.* **1996**, *271*, 17697–17703.
91. Fromm, J.R.; Hileman, R.E.; Weiler, J.M.; Linhardt, R.J. *Arch. Biochem. Biophys.* **1997**, *346*, 252–262.
92. Tsai, F.T.F.; Singh, O.M.P.; Skarzynski, T.; Wonacott, A.J.; Weston, S.; Tucker, A.; Pauptit, R.A.; Breeze, A.L.; Poyser, J.P.; O'Brien, R.; Ladbury, J.E.; Wigley, D.B. *Proteins Struct. Funct. Genet.* **1997**, *28*, 41–52.
93. Garcia-Hernandez, E.; Zubillaga, R.A.; Rojo-Dominguez, A.; Rodriguez-Romero, A.; Hernandez-Arana, A. *Proteins: Struct., Funct., Genet.* **1997**, *29*, 467–477.
94. Quesenberry, M.S.; Lee, R.T.; Lee, Y.C. *Biochemistry* **1997**, *36*, 2724–2732.
95. Thomson, J.; Liu, Y.; Sturtevant, J.M.; Quiocho, F.A. *Biophys. Chem.* **1998**, *70*, 101–108.
96. Briggs, J.B.; Larsen, R.A.; Harris, R.B.; Sekar, K.V.S.; Macher, B.A. *Glycobiology* **1996**, *6*, 831–836.
97. Sharma, S.; Bhardwaj, S.; Surolia, A.; Podder, S.K. *Biochem. J.* **1998**, *333*, 539–542.
98. Bundle, D.R.; Alibes, R.; Nilar, S.; Otter, A.; Warwas, M.; Zhang, P. *J. Am. Chem. Soc.* **1998**, *120*, 5317–5318.
99. Ramkumar, R.; Surolia, A.; Podder, S.K. *Biochem J.* **1995**, *308*, 237–241.
100. Lee, R.T.; Gabius, H.J.; Lee, Y.C. *Glycoconjugate J.* **1998**, *15*, 649–655.
101. Swaminathan, C.P.; Gupta, D.; Sharma, V.; Surolia, A. *Biochemistry* **1997**, *36*, 13428–13434.
102. Roy, R. *Curr. Opin. Struct. Biol.* **1996**, *6*, 692–702.
103. Mammen, M.; Choi, S.K.; Whitesides, G.M. *Angew. Chem. Int. Ed.* **1998**, *37*, 2754–2794.
104. Sigal, G.B.; Mammen, M.; Dahmann, G.; Whitesides, G.M. *J. Am. Chem. Soc.* **1996**, *118*, 3789–3800.
105. Lee, R.T.; Ichikawa, Y.; Kawasaki, T.; Drickamer, K.; Lee, Y.C. *Arch. Biochem. Biophys.* **1992**, *299*, 129–136.
106. Biessen, E.A.L.; Noorman, F.; vanTeijlingen, M.E.; Kuiper, J.; BarrettBergshoeff, M.; Bijsterbosch, M.K.; Rijken, D.C.; vanBerkel, T.J.C. *J. Biol. Chem.* **1996**, *271*, 28024–23669.
107. Hansen, H.C.; Haataja, S.; Finne, J.; Magnusson, G. *J. Am. Chem. Soc.* **1997**, *119*, 6974–6979.
108. Glick, G.D.; Toogood, P.L.; Wiley, D.C.; Skehel, J.J.; Knowles, J.R. *J. Biol. Chem.* **1991**, *266*, 23660–23669.
109. Wittman, V.; Takayama, S.; Gong, K.W.; Weitz-Schmidt, G.; Wong, C.H. *J. Org. Chem.* **1998**, *63*, 5137–5143.
110. Kamitakahara, H.; Suzuki, T.; Nishigori, N.; Suzuki, Y.; Kanie, O.; Wong, C.H. *Angew. Chem. Int. Ed. Engl.* **1998**, *37*, 1524–1528.
111. Thompson, J.P.; Schengrund, C.L. *Glycoconj. J.* **1997**, *14*, 837–845.
112. Mortell, K.H.; Weatherman, R.V.; Kiessling, L.L. *J. Am. Chem. Soc.* **1996**, *118*, 2297–2298.
113. Manning, D.D.; Hu, X.; Beck, P.; Kiessling, L.L. *J. Am. Chem. Soc.* **1997**, *119*, 3161–3162.
114. Nishida, Y.; Dohi, H.; Uzawa, H.; Kobayashi, K. *Tet. Lett.* **1998**, *39*, 8681.
115. Cao, S.; Roy, R. *Tet. Let.* **1996**, *37*, 3421.
116. Bovin, N.V.; Gabius, H.J. *Chem. Soc. Rev.* **1995**, 413.
117. Bovin, N.V. *Glycoconj. J.* **1998**, *15*, 431.
118. Tsuchida, A.; Akimoto, S.; Usui, T.; Kobayashi, K. *J. Biochem.* **1998**, *123*, 715.
119. Park, K.H.; Takei, R.; Goto, M.; Maruyama, A.; Kobayashi, A.; Kobayashi, K.; Akaike, T. *J. Biochem.* **1997**, *121*, 997.
120. Spaltenstein, A.; Whitesides, G.M. *J. Am. Chem. Soc.* **1991**, *113*, 686.
121. Lees, W.J.; Spaltenstein, A.; Kingery-Wood, J. E.; Whitesides, G. M. *J. Med. Chem.* **1994**, *37*, 3419.
122. Mann, D.A.; Kanai, M.; Maly, D.J.; Kiessling, L.L. *J. Am. Chem. Soc.* **1998**, *120*, 10575–10582.
123. Kanai, M.; Mortell, K.H.; Kiessling, L.L. *J. Am. Chem. Soc.* **1997**, *119*, 9931.

124. Kingery-Wood, J.E.; Williams, K.W.; Sigal, G.B.; Whitesides, G.M. *J. Am. Chem. Soc.* **1992**, *114*, 7303.
125. DeFrees, S.A.; Phillips, L.; Guo, L.; Zalipsky, S. *J. Am. Chem. Soc.* **1996**, *118*, 6101.
126. Ketis, N.V.; Girdlestone, J.; Grant, C.W. M. *Proc. Natl. Acad. Sci. USA* **1980**, *77*, 3788.
127. Charych, D.; Cheng, Q.; Reichart, A.; Kuziemko, G.; Stroh, M.; Nagy, J.O.; Spevak, W.; Stevens, R.C. *Chem. & Biol.* **1996**, *3*, 113.
128. Spevak, W.; Foxall, C.; Charych, D.H.; Dasgupta, F.; Nagy, J.O. *J. Med. Chem.* **1996**, *39*, 1018.
129. Pan, J.J.; Charych, D. *Langmuir* **1997**, *13*, 1365.
130. Zanini, D.; Roy, R. *J. Org. Chem.* **1998**, *63*, 3486.
131. Ashton, P.R.; Hounsell, E.F.; Jayaraman, N.; Nilsen, T.M.; Spencer, N.; Stoddart, J.F.; Young, M. *J. Org. Chem.* **1998**, *63*, 3429.
132. Hornick, C.L.; Karush, F. *Immunochem.* **1972**, *9*, 341.
133. Mammen, M.; Shaknovich, E.I.; Deutch, J.M.; Whitesides, G.M. *J. Org. Chem.* **1998**, *63*, 3821–3830.
134. Mammen, M.; Shaknovich, E.I.; Whitesides, G.M. *J. Org. Chem.* **1998**, *63*, 3168–3175.
135. Alibes, R.; Bundle, D.R. *J. Org. Chem.* **1998**, *63*, 6288–6301.
136. For example: Espinosa, J.F.; Bruix, M.; Jarreton, O.; Skrydstrup, T.; Beau, J.M.; Jimenez-Barbero, J. *Chem. Eur. J.* **1999**, *5*, 442–448, and references therein.
137. Aguilera, B.; Jimenez-Barbero, J.; Fernandez-Mayoralas, A. *Carbohydr. Res.* **1998**, *308*, 19–27.
138. Toone et al., Unpublished Data.

32 Structural Analysis of Oligosaccharides: FAB-MS, ES-MS and MALDI-MS

Anne Dell, Howard R. Morris, Richard Easton, Stuart Haslam, Maria Panico, Mark Sutton-Smith, Andrew J. Reason, and Kay-Hooi Khoo

32.1 Introduction

In the glycobiology field no structural technique can match mass spectrometry for the breadth of problems that can be addressed, the complexity of samples that can be successfully analyzed, and for the amount of structural information that can be obtained from sub-nanomolar amounts of material. The introduction of fast atom bombardment-mass spectrometry (FAB-MS) at the beginning of the 1980s [1] revolutionized the structure determination of a very wide range of carbohydrate-containing biopolymers [2, 3] and this revolution has continued with the newer techniques of electrospray ionization (ES-MS) [4] and matrix assisted laser desorption ionization (MALDI-MS) [5]. A summary of each of these techniques is given below in order to facilitate understanding of their problem-solving capabilities. All three technologies permit the direct ionization/desorption of non-volatile substances and are applicable to intact glycoconjugates as well as fragments.

32.2 Fast Atom Bombardment-Mass Spectrometry (FAB-MS)

Double focusing sector mass spectrometers are employed in the majority of biopolymer FAB-MS studies. These consist of a source where the ions are generated, an analyzer comprising an electric sector for energy focusing and a magnetic sector for separating ions of different mass to charge ratios, and a detector. In the FAB experiment, an accelerated beam of atoms (usually xenon) or ions (usually cesium) is fired from an atom or ion gun towards a small metal target attached to the end of a probe (see Figure 1a). Prior to insertion of the probe into the FAB source, it is loaded with a viscous liquid called the matrix in which is dissolved the sample to be analyzed. When the atom or ion beam collides with the matrix, kinetic energy is

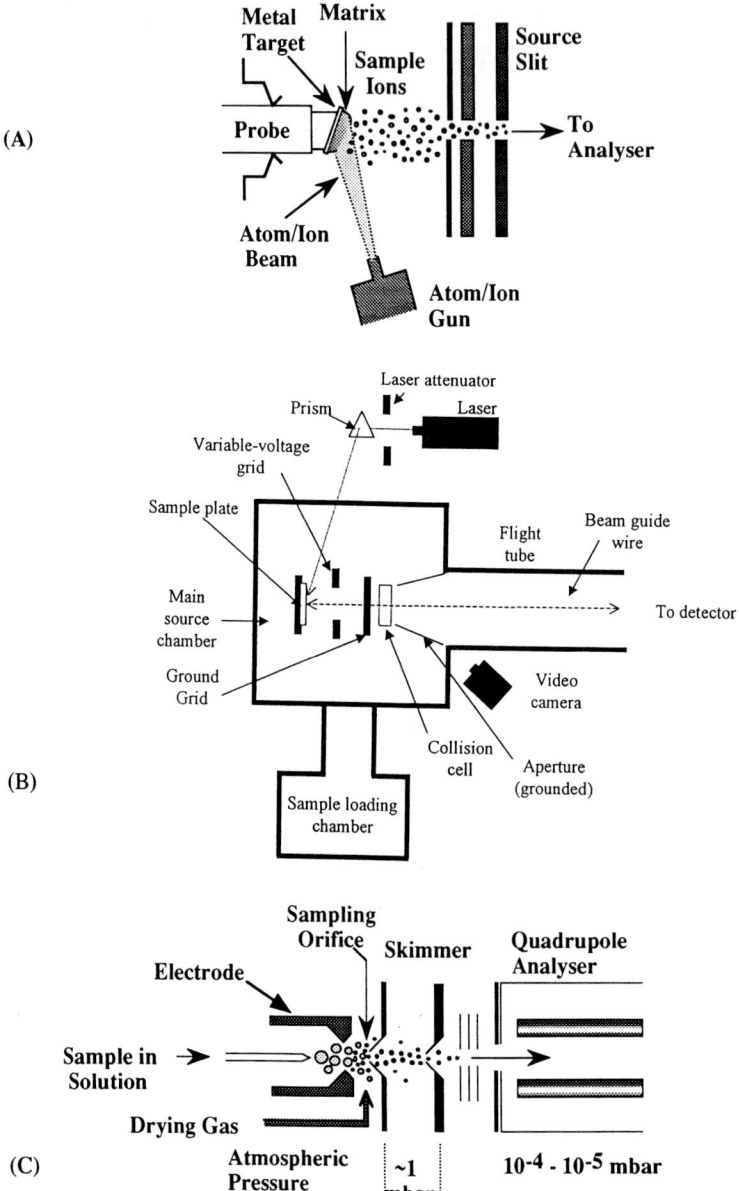

Figure 1. (**A**) Fast Atom Bombardment (FAB): the sample is dissolved in a liquid matrix and ionization/desorption is effected by a high energy beam of particles fired from an atom or ion gun. (**B**) Matrix Assisted Laser Desorption Ionisation (MALDI): the sample is dried on a metal target in the presence of a chromophoric matrix and sample ions are produced by energy transfer from matrix molecules that have absorbed energy from the laser pulse. (**C**) Electrospray (ES): a stream of liquid containing the sample of interest enters the source through a capillary interface, where the sample molecules are stripped of solvent, leaving them as multiply charged species.

transferred to the surface molecules, many of which are sputtered out of the liquid into the high vacuum of the ion source. A significant number of these molecules are ionized during the sputtering process. Thus gas-phase ions are generated without prior volatilization of the sample allowing the analysis of polar, involatile, and thermally labile compounds. Both positive and negative ions are produced during the sputtering process, and either can be recorded by an appropriate choice of instrumental parameters. Molecules are ionized by protonation or by the addition of a cation such as sodium, potassium, ammonium (positive ion formation), or by the loss of a proton, or addition of an anion such as chloride, thiocyanate (negative ion formation). During ionization, some internal energy is imparted to the molecule resulting in fragmentation of labile bonds. The sputtered molecular and fragment ions are accelerated to about 8,000 eV prior to passage through the analyzer sectors to a detector where the mass to charge ratios are recorded by a computer to give the mass spectrum.

Magnetic sector analyzers separate ions on the principle that charged molecules are deflected by strong magnetic fields. Ions of larger mass are deflected by the magnetic field less than ions of smaller mass according to the equation $m/z = B^2 r^2 / 2V$ where m/z is the mass to charge ratio of the ion, B is the strength of the magnetic field, r is the radius of the circular path through which the ion is travelling through the magnet and V is the accelerating voltage. Ions of different m/z values are brought into focus at the detector by scanning B with V and r being kept constant. Modern high field magnetic sector mass spectrometers are capable of focusing ions up to 15,000 Da at full sensitivity, but the working mass range of FAB-MS for carbohydrate polymers is limited to about 6,000 Da. Higher molecular weight samples are either too difficult to desorb from the matrix or do not produce a sufficient abundance of molecular ions to allow detection above background.

The FAB experiment often yields structurally informative fragment ions as well as molecular ions but the quantity and quality of fragmentation can be very variable depending on factors such as the purity of the sample and whether it has been derivatized. To ensure that sufficient fragment ions are produced to allow structural assignments it may be necessary to employ tandem MS techniques (MS/MS) in which ions produced in the FAB source pass through the mass analyzer into a chamber containing an inert gas. Collisions with the gas provide sufficient energy for bond cleavages and the resulting fragment ions (called daughter ions) are detected after passing through a second (tandem) analyzer. Tandem MS is especially powerful when mixtures are being analyzed because it allows unambiguous attribution of fragment ions to individual components of the mixture.

32.3 Matrix Assisted Laser Desorption Ionization-Time of Flight-Mass Spectrometry (MALDI-TOF-MS)

In the MALDI experiment (Figure 1b) the sample is embedded in a low molecular weight, UV absorbing matrix which enhances sample ionization. When the matrix absorbs the laser pulse enough energy is transferred to the sample, via mechanisms that are not well understood, to enable the formation of molecular ions. The matrix is present in a vast excess over the sample and therefore isolates individual sample molecules. This results in the volatilization of predominantly monomeric molecular ions, although dimers, trimers etc are observed in some MALDI spectra. Ions produced in MALDI-MS are usually analyzed by Time of Flight (TOF) instrumentation. TOF analyzers work on the principle that when ions are accelerated with the same potential from a fixed point and at a fixed initial time and are allowed to drift, the ions will separate according to their mass/charge ratios. Lighter ions drift more quickly to the detector and heavier ions drift more slowly. The time required for ions to reach the detector can therefore be related to their mass.

Following acceleration, ions exhibit a broad energy distribution. This energy spread can be minimized using Delayed Extraction (DE) which involves applying a high voltage pulse at a pre-determined time following ion generation in a weak electrical field leading to enhanced resolution and mass accuracy.

Most modern MALDI instruments allow both linear and reflector mass analysis. In the linear mode, ions travel without interference down the flight tube to the detector which records the m/z ratio and signal intensity. A reflector is a single-stage gridded mirror that focuses energy. In reflector mass analysis mode, a uniform electric field is applied to the mirror to reflect ions. This filters out neutral molecules, corrects dispersion of the ions occurring in the flight tube and thus provides greater mass accuracy and resolution.

In contrast to the magnetic sectors used for analysis in FAB-MS, the TOF analyzer has an almost unlimited mass range. MALDI is also considerably more efficient than FAB in ionizing large biopolymers and is the method of choice for defining the molecular weights of carbohydrate polymers, particularly at high mass.

32.4 Electrospray-Mass Spectrometry (ES-MS)

ES-MS is a method by which a stream of liquid containing the sample of interest is introduced into the atmospheric pressure ion source of a mass spectrometer (Figure 1c). This can be achieved by direct injection into buffer which is pumped at the rate of a few microliters per minute through a metal-tipped glass capillary. Alternatively the eluent of a microbore LC, after stream splitting, can be passed via the capillary into the ES source. Many newer instruments have high sensitivity nanospray sources capable of operating at flow rates of as low as a few nanoliters per minute. With this type of source, samples are introduced using a probe with an attached nanospray needle, the latter being pre-loaded with about a microliter of a solution of the sample (see 3.9.2, Protocol 2).

Irrespective of the introduction method, an aerosol of microdroplets is generated in the source which then traverses a series of skimmers, encountering a drying gas, the net effect of which is the creation of charged molecular species, devoid of solvent. These gaseous ions, whose charge depends on the number of ionizable groups in the molecule, are passed into the mass analyzer which is commonly a quadrupole mass filter. This type of analyzer is compact and relatively inexpensive. It comprises four parallel rods which act as mass filters when rf and dc fields are applied to diagonally opposed rods. By sweeping the rf and dc voltages in a fixed ratio, ions of successive mass to charge ratios follow a stable path to the detector. Ions with m/z values above about 4,000 cannot be detected with a quadrupole analyzer. However this is not too problematical for a significant portion of biopolymer ES-MS applications because of the presence of multiple functional groups capable of carrying charge. For example proteins and glycoproteins in excess of 100 kDa can be amenable to ES-MS analysis. Uncharged or poorly charged polysaccharides are not, however, amenable to ES-MS on a quadrupole instrument.

The ionization process in ES-MS is very gentle thus resulting in very few fragment ions and little or no sequence information. To overcome this problem many ES instruments have triple quadrupole analyzers for collisional activation MS/MS experiments. The first quadrupole is used to select the parent ion, the second is the collision chamber and the third separates the resulting fragment ions.

Whilst triple quadrupole instruments are very powerful, their sensitivity is limited by a number of factors including scanning ion detection and poor fragment ion resolution. These problems have been overcome in a novel mass spectrometer, called the Q-TOF, which was introduced in the mid-1990s [6, 7]. The Q-TOF has both quadrupole and TOF analyzers which are arranged orthogonally with a collision cell between them (Figure 2). For normal mass spectra, the quadrupole is used in the rf-only mode as a wide-bandpass filter to transmit a wide mass range. The collision cell is not pressurized, and ions are transmitted to the TOF for mass analysis. In the MS/MS mode the quadrupole operates in the normal resolving mode and is able to select precursor ions up to m/z 4,000 for collisional activation in the hexapole gas cell. The fragment ions are transmitted to the TOF for mass analysis. The orthogonal geometry, and parallel rather than sequential detection of the ions, leads to a significant improvement in sensitivity over scanning instruments when used to acquire full spectra. The Q-TOF is an immensely powerful instrument. The good signal/noise ratios in MS/MS on the Q-TOF correspond to low femtomole/attomole sample consumption. Definitive unambiguous sequence assignment is facilitated by daughter ion resolutions of greater than 3,000 (which allows easy assignment of the z value of m/z) and daughter ion mass accuracies of 0.05 Da.

32.5 Appearance of Mass Spectra Obtained in FAB-MS, MALDI-MS and ES-MS Experiments

FAB spectra are characterized by a high level of chemical "noise", arising from the matrix and sample, giving a peak at every mass upon which are superimposed the signals for molecular and fragment ions of the sample and matrix. The chemical noise is much lower in MALDI-MS which results in better signal to noise for sam-

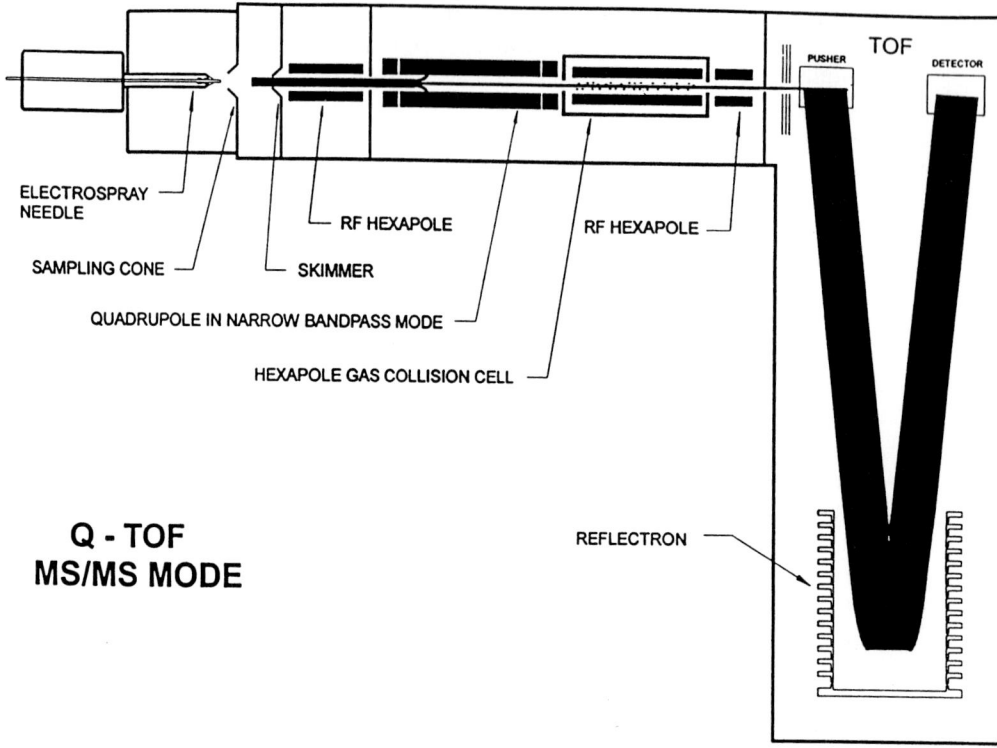

Figure 2. Schematic illustration of the Q-TOF in the MS-MS mode of operation. The quadrupole is set to transmit the parent ion of interest and the hexapole collision chamber contains the collision gas. The orthogonal TOF separates the daughter ions.

ple ions. Most of the molecular and fragment ions observed in FAB and MALDI spectra are singly charged, e.g. $[M+H]^+$, $[M-H]^-$, $[M+Na]^+$, $[M-H+2Na]^+$ etc.

In contrast to FAB-MS and MALDI-MS, electrospray mass spectra are characterized by multiply charged ions. The raw data obtained following ES-MS analyses is complex because ions can carry a range of charges due to the different ionisable sites in the molecule. Multiple m/z signals are therefore recorded for each mass value. Fortunately these data can be processed using a simple computer algorithm to produce a mass spectrum of comparable appearance to a FAB or MALDI spectrum.

Ions are observed as clusters in all types of mass spectra because of the existence of isotopes. The contribution of ^{13}C (1.1% natural abundance) means that for every carbon atom in the molecule there is a 1.1% chance that it will be a ^{13}C atom and so have an atomic mass of 13 instead of 12. The relative intensity of the different peaks in the cluster reflects this probability distribution, i.e. the height of the signal at m/z $X+1$ (where X corresponds to the molecular ion that contains only ^{12}C) reflects the probability of finding one ^{13}C atom in the molecule. The individual signals in a cluster will be observed when the instrument resolution is set to resolve nominal masses. At lower resolutions the clusters are present as an unresolved envelope. For very high molecular weight ions these envelopes may embrace more than one cluster of molecular ions, e.g. $[M+H]^+$ and $[M+Na]^+$, resulting in a very broad peak which cannot be accurately mass-assigned.

32.6 Assignment of Mass Values

There are four principal ways of denoting the mass of an ion in the mass spectrum:

i) Nominal mass—this is the sum of the integer atomic weights of the isotopes comprising the ion, giving the specific peak whose mass is being assigned.
ii) Accurate mass—this is the sum of the accurate atomic weights of the isotopes comprising the ion giving the specific peak whose mass is being assigned. Accurate masses are obtained when spectra are computer assigned by comparison with a calibration spectrum containing ions whose accurate masses are known (the calibration standard is often an alkali halide such as CsI).
iii) Average mass (chemical mass)—this is the average of the accurate isotope mass of each element, weighted by the relative abundance of the isotopes, i.e. it corresponds to the sum of the chemical atomic weights. Average masses are assigned when an isotopic cluster is recorded at such low resolution that the cluster is completely unresolved. The average mass is the centre of gravity of the unresolved cluster.
iv) Peak top mass—this is the accurate mass of the top of an unresolved cluster. At high masses, above about m/z 5,000, the peak top mass is very close to the average mass because the isotopic distribution results in a Gaussian shape for the cluster. ES-MS assignments are normally based on peak top masses.

32.7 Derivatisation

Although native samples are amenable to FAB-MS, MALDI-MS and ES-MS, it is often desirable to prepare derivatives prior to analysis. As a general rule glycans whose hydroxyl groups are protected by functional groups, such as methyl or acetyl, fragment more reliably than their native counterparts. Also derivatives are easier to obtain free from salt impurities which may prejudice the MS experiment and sensitivity is significantly improved when hydrophobic moieties are present. Derivatization methods can be broadly divided into two categories: (i) "tagging" of reducing ends, and (ii) protection of most or all of the functional groups. Commonly used tagging reagents include *p*-aminobenzoic acid ethyl ester (ABEE) and 2-aminopyridine (2-AP). This type of derivatization facilitates chromatographic purification and enhances reducing-end fragment ions in MS and MS/MS experiments. Protection of functional groups by permethylation or per(deutero)acetylation is by far the most important type of derivatization employed in carbohydrate MS. These derivatives fragment reliably to yield abundant A-type fragment ions (see below) which are extremely useful for sequencing.

32.8 Fragmentation Pathways

The following general rules apply to the fragmentation behavior of oligosaccharides and glycoconjugates [3]:

i) The most abundant fragment ions are formed by cleavages at glycosidic linkages. Glycosidic cleavage is accompanied by a hydrogen transfer to the glycosidic oxygen if the charge on the fragment ion is not specifically located at the point of cleavage.
ii) Ring cleavage, when it occurs, is best rationalized as arising from the sequential movement of electron pairs around the ring resulting in the breakage of single bonds and the formation of double bonds (see Figure 3).
iii) Fragment ions can sometimes be formed by two or more cleavage events occurring in different parts of the molecule. This phenomenon is more frequently observed in native samples than in derivatives. Unambiguous sequencing of native samples is often not possible if "double cleavage" ions are formed in abundance.
iv) The fragment ions produced by glycopeptides are derived predominantly from cleavage of the glycosidic linkages. Fragment ions resulting from cleavage of the peptide bonds are often of low abundance.
v) If a permanent charge is present in the molecule (e.g. a sulfate moiety) then fragment ions produced by cleavage of labile bonds in the vicinity of the charge dominate the spectrum.

The major fragmentation pathways relevant to carbohydrate MS are summarized below:

i) *A-type cleavage*—Glycosidic cleavage (Figure 3) yields an oxonium ion thereby locating the charge at the point of cleavage on the non-reducing fragment. The term A-type cleavage is normally applied to this type of fragmentation based on nomenclature used in electron impact mass spectrometry. A-type cleavage is the major mode of fragmentation of permethylated and per(deutero)acetylated oligosaccharides. If HexNAc residues are present in the sequence, cleavage occurs predominantly (and sometimes exclusively) at the amino sugar residues. A secondary fragmentation associated with A-type cleavage is elimination of the substituent at the 3-position of the HexNAc oxonium ion thus defining whether the 3-position is occupied by a sugar such as fucose.

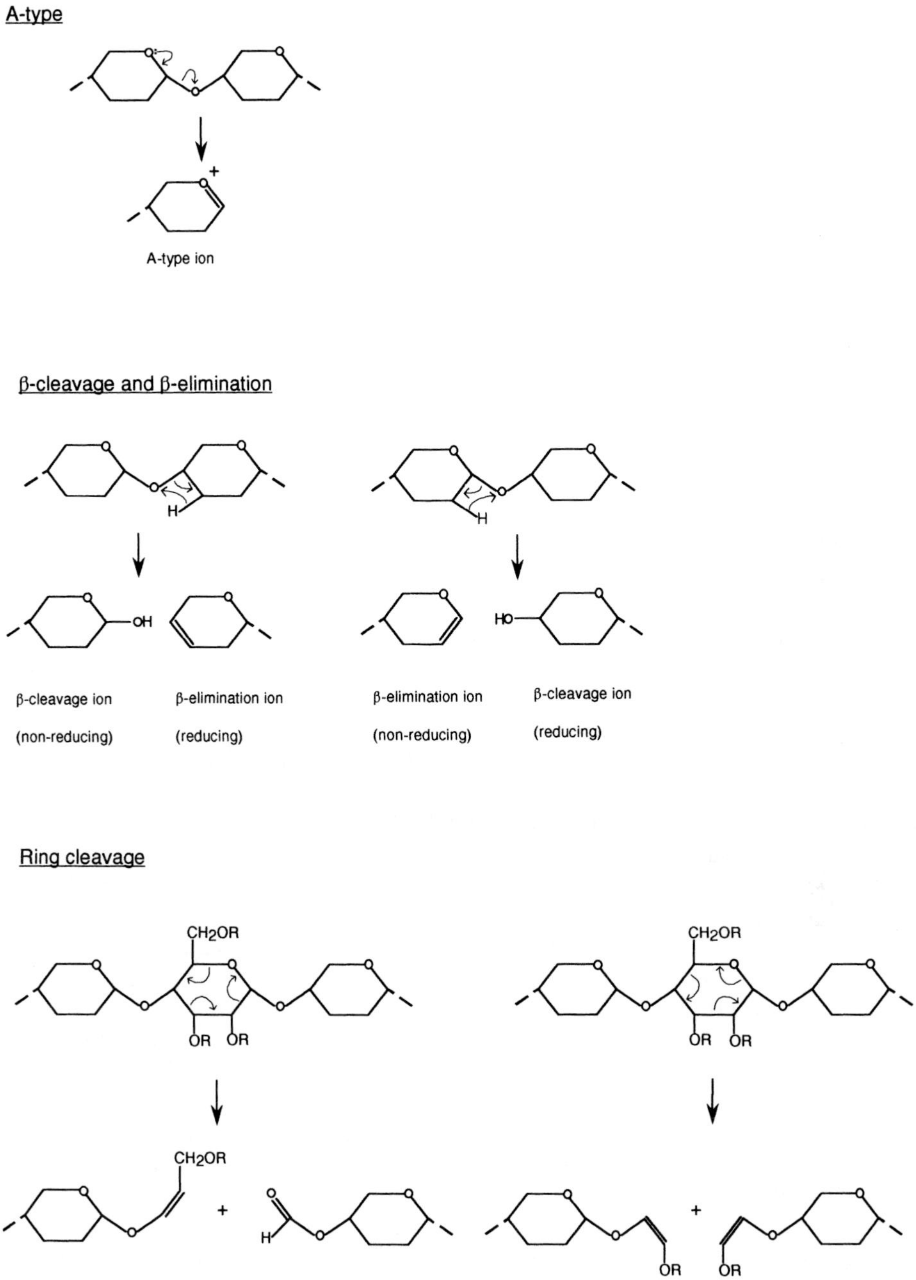
A-type
+
A-type ion
β-cleavage and β-elimination
H
OH
HO
β-cleavage ion
(non-reducing)
β-elimination ion
(reducing)
β-elimination ion
(non-reducing)
β-cleavage ion
(reducing)
Ring cleavage
CH2OR
OR OR
+
H
OR OR

Figure 3. Key fragmentation pathways in glycopolymer mass spectrometry are shown in this figure. The terms A-type cleavage, β-cleavage, ring cleavage etc are useful descriptors of these pathways. The reader should consult [19] for information on how to systematically name fragment ions arising from each of these pathways.

ii) β-*cleavage-and* β-*elimination*—This involves cleavage on either side of the glycosidic linkage with hydrogen transfer to the glycosidic oxygen resulting in one fragment having a hydroxyl group at the position of glycosidic linkage and the other having a double bond (Figure 3). These are referred to as β-cleavage and β-elimination products respectively. Note that, unlike A-type cleavage, a charged moiety is not produced as a result of cleavage. Depending on where the charge is located in the fragmenting molecule, either or both reducing and non-reducing β-cleavage and/or β-elimination ions will be observed. This type of cleavage is favored by native oligosaccharides and glycopeptides and is especially prominent in samples modified with a reducing end "tag".

iii) *Ring cleavages*—These are frequently observed in MS/MS experiments especially when cationised molecular ions are subjected to collisional activation. They can provide useful information on attachment sites of functional groups and glycosidic linkages. Examples of ring cleavages are shown in Figure 3.

32.9 Protocols for MS Analysis

In this section we describe some of the MS protocols employed in our laboratories. These are given for illustrative purposes in order to facilitate understanding of the case studies presented in later sections of this Chapter. The protocols exemplify general strategies underlying each technique and are not intended to be exclusive.

32.9.1 Protocol 1—Sample Loading for FAB-MS Analysis

i) Dissolve the sample in either 5% acetic acid (for underivatized samples) or methanol (for derivatives) to such a concentration that 1 µl aliquot of the sample contains the desired amount of sample for FAB-MS analysis.
ii) Smear about 1 µl of the matrix onto the metal target which is attached to the end of the FAB probe.
iii) Load about 1 µl of the sample on the surface of the matrix using a micropipette or syringe.
iv) Introduce the probe into the FAB source and collect data immediately.
v) When a wide mass range of scanning is required, e.g. m/z 4,000 to 200, it may be necessary depending upon instrument type to acquire scans covering the high and low mass range separately with two separate sample loadings.
vi) Use different matrices and matrix additives to improve the quality of the data or alter the molecular and/or fragment ion patterns:
   – use monothioglycerol as the matrix for general analysis
   – use *m*-nitrobenzyl alcohol matrix for salty samples.
   – add 1 µl of dilute aqueous (100–200 mM) HCl to the matrix to minimize cationization
   – other matrix additives, e.g. sodium acetate and ammonium thiocyanate can also be used in specific cases [3].

32.9.2 Protocol 2—Sample Loading for NanoES-MS and MS-MS Analysis on the Q-TOF

i) Samples should be solubilized in an ES-MS compatible solvent such as acetonitrile/0.1% aqueous TFA (1:1, v:v). It should be noted that when using nanospray, it is observed that increased beam stability is achieved sometimes when using methanol. 1–2 µl of sample can be loaded to the end of the metal coated capillary using either plastic gel loader tips or a microlitre syringe. Care must be taken in handling the capillary because the 1–2 µm tip through which the sample is sprayed is very fragile and easily broken.

ii) Place the metal coated capillary (square end first) into its holder. This consists of a knurled nut, a 5 mm length of conductive elastomer and a Swagelok union. Once secured, screw the capillary/holder onto the probe and place into the source.

iii) Having inspected the MS spectrum, and tuned the quadrupole to, for example, a double or triply charged ion of interest, the argon collision gas pressure is adjusted to approximately 10^{-4} (this will vary for optimum collision-induced fragmentation) and the collision energy is set between 10 and 60 eV relative to the size of the molecule under study, and its charge state.

iv) Collect data until summation of scans produces a good quality spectrum. Collection times may vary depending on sample concentration, its propensity for ionization and state of capillary tips. However, for a 100 femtomole per µl solution, 0.5 min should be adequate for a high quality spectrum.

v) Manipulation of the capillary tips by fracturing may be needed during ionization to maintain the spray without air bubbles and particulate blockages.

32.9.3 Protocol 3—Sample Loading for LC-ES-MS and LC-ES-MS-MS on the Q-TOF

i) Dissolve the sample in an appropriate volume by using starting condition buffer and load into a 20 µl sample loop. Elute using 0.1% aqueous trifluoroacetic acid (buffer A) and a gradient up to 100% acetonitrile in 0.1% aqueous trifluoroacetic acid (buffer B) at a flow rate of 50 µl/min. The elution is monitored by UV absorbance at 214 nm in a nanoflow cell.

ii) After the column a methoxyethanol:isopropanol (1:1, v:v) solvent mixture is added at a flow rate of 50 µl/min; this helps to counteract the signal suppressive effects of trifluoroacetic acid. The flow is then stream-split to allow collection of approximately 85 µl fractions from the column at 1 min intervals, with the remaining 15 µl/min directed on-line to the electrospray source of the Q-TOF instrument.

32.9.4 Protocol 4—Sample Loading for MALDI-MS Analysis

i) Dissolve the sample to be analyzed in pure water (for native carbohydrates) or 80:20 methanol:pure water (for derivatized carbohydrates) to produce a sample concentration of around 1–10 pmoles/µl.

ii) Prepare the appropriate matrix:
   – 2,5-Dihydroxybenzoic acid (DHB) is used as general matrix; 10 mg/ml in 90:10 water:ethanol (for native carbohydrates) or 80:20 methanol:water (for derivatized carbohydrates)
   – 2-(4-hydroxyphenylazo)-benzoic acid (HABA) is used for enhancement of polar carbohydrates; 1.3 mg/ml in 50:50 water:acetonitrile (for native carbohydrates) or 80:20 methanol:water (for derivatized carbohydrates)
iii) Pipette 1 µl of the sample solution (using a Gilson P2 pipette or similar) to the metal target followed by 1 µl of the matrix solution.
iv) Dry under vacuum.
v) Introduce into the MALDI source and operate the instrument
   – DE-MALDI-TOF-MS analysis in the reflector mode is the method of choice for carbohydrate samples in the mass range 100–10,000 Da.
   – DE-MALDI-TOF-MS analysis in the linear mode is the method of choice for carbohydrate samples in the mass range >10,000 Da.

32.10 Applications of FAB-MS, MALDI-MS and ES-MS in Glycobiology

Broadly speaking FAB-MS, MALDI-MS and ES-MS can be exploited in two general ways in the glycobiology field:

i) detailed characterization of purified individual glycopolymers or mixtures of glycopolymers which usually requires acquisition of a considerable body of structural information, and
ii) rapid screening of cell-, tissue- and whole animal-extracts where a limited number of MS experiments on relatively crude extracts are often sufficient to answer the questions addressed.

Below we have selected examples from recent glycopolymer research which exemplify the versatility of problems amenable to current technology.

32.10.1 Case Study 1—Molecular Weight Profiling of Polysaccharides by MALDI-MS

The following study of laminarin polysaccharides illustrates the potential of MALDI-MS for rapidly and sensitively defining the degree of polymerization (d.p.) of carbohydrate polymers and for revealing structural modifications which result in altered molecular weight.

Laminarins are a class of low-molecular-weight storage β-glucans of brown algae consisting of (1–3)-linked β-D-glucopyranose residues in which some 6-O-branching in the main chain and some β(1–6)-intrachain links are present. The majority of laminarins contain polymeric chains of two types, polymeric glucopyranose only (G-chains) and polymeric glucopyranose terminated with 1-O-substituted D-mannitol residues (M-chains).

Both native and permethylated laminarins give excellent MALDI data [8]. The permethylated derivatives were examined in order to facilitate discrimination between M- and G-laminarins which differ by only 2 mass units in their native form but by 16 mass units after permethylation. Typical spectra are shown in Figure 4a for native laminarin from *Laminaria cichorioides* (a mannitol-containing laminarin) and in Figure 4b for permethylated laminarin from *Chorda crinita* (a mannitol-free laminarin). The *L. cichorioides* sample shows a Gaussian-like d.p. profile peaking at about d.p. 27 and ranging over about d.p. 9–40. The *C. crinita* sample has a bimodal d.p. profile with abundant oligomers at both low and high mass. Interestingly, close examination of Figure 4b reveals two regions of the spectrum where clusters of ions suggest additional components, namely d.p. 5–10 and d.p. 16–17. The latter show abundant signals 46 mass units lower than the corresponding G-oligomer which is consistent with cyclic components of d.p. 16 and 17. The former exhibit satellite peaks 41 mass units higher than the corresponding G-oligomer. This mass interval is consistent with the presence of a HexNAc residue replacing a hexose residue in a minority of the sample. This was an unexpected discovery because aminosugars had not previously been found in laminarins and illustrates the power of the MALDI technique for revealing the presence of novel minor constituents in complex mixtures.

32.10.2 Case Study 2—Analysis of Glycoproteins by LC-ES-MS and FAB-MS

The following study of glycodelin A illustrates how the strengths of LC-ES-MS and FAB-MS can be exploited in a complementary manner to obtain structural information on low micromolar quantities of novel glycoproteins [9].

Glycodelin-A (GdA), is a human amniotic fluid-derived glycoprotein that has potent contraceptive and immunosuppressive activities. GdA has 162 amino acids and there are three potential *N*-linked glycosylation sites at Asn-28, Asn-63 and Asn-85. Together LC-ES-MS and FAB-MS gave sufficient data to define the site occupancy and the sequences of the *N*-glycans present on GdA.

The strategy for analyzing GdA is shown in Figure 5. The key experimental steps were as follows:

i) LC-ES-MS and LC-ES-MS-MS analyses of tryptic and cyanogen bromide digests of GdA confirmed the protein sequence, established the non-occupancy of Asn-85 and showed that Asn-28 and Asn-63 were glycosylated.
ii) Glycans were released from LC fractions containing the glycopeptides identified by ES-MS analysis. After permethylation these were analyzed by FAB-MS yielding abundant molecular ions and A-type fragment ions. The latter define the non-reducing sequences in the antennae of complex-type and hybrid *N*-glycans whilst the former define the overall compositions of each type of glycan including high mannose structures. FAB-MS is the method of choice for this

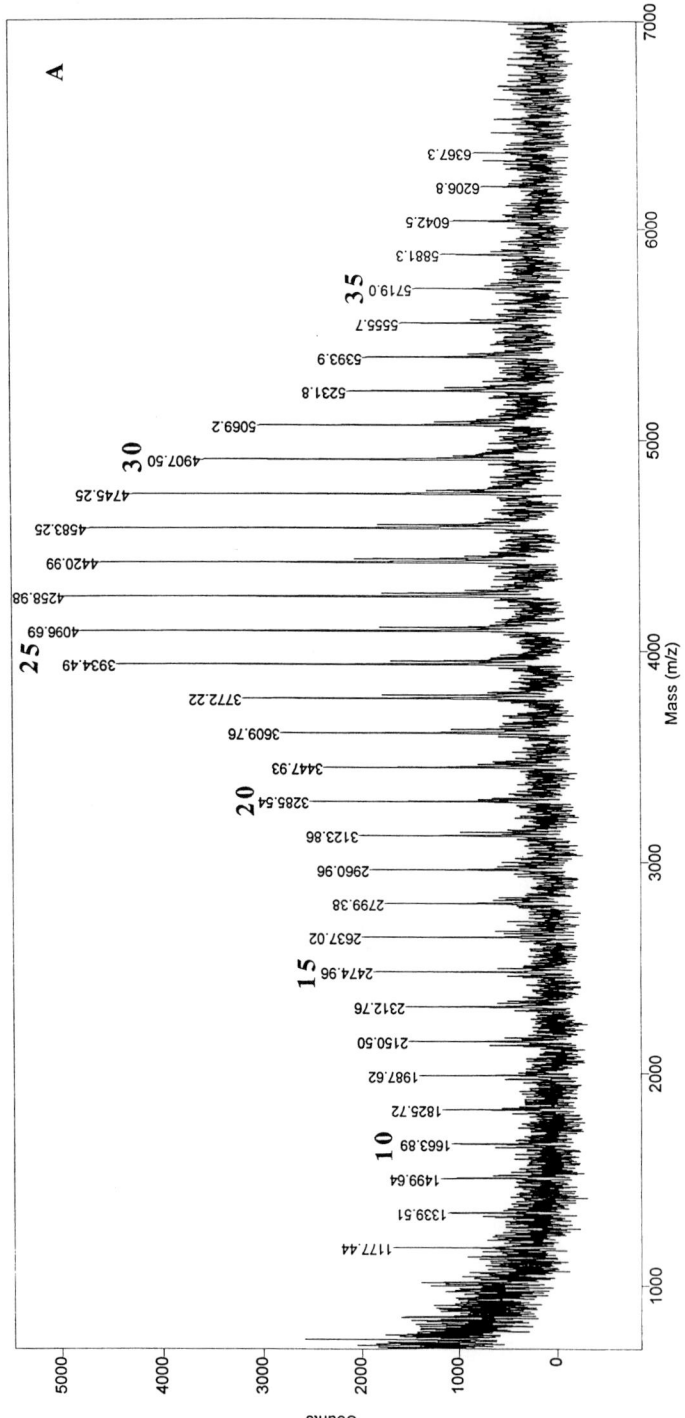

Figure 4. MALDI-TOF mass spectra of **(A)** native laminarin from *Laminura cichorioides*, and **(B)** permethylated laminarin from *Chorda crinita*.

32.10 Applications of FAB-MS, MALDI-MS and ES-MS in Glycobiology

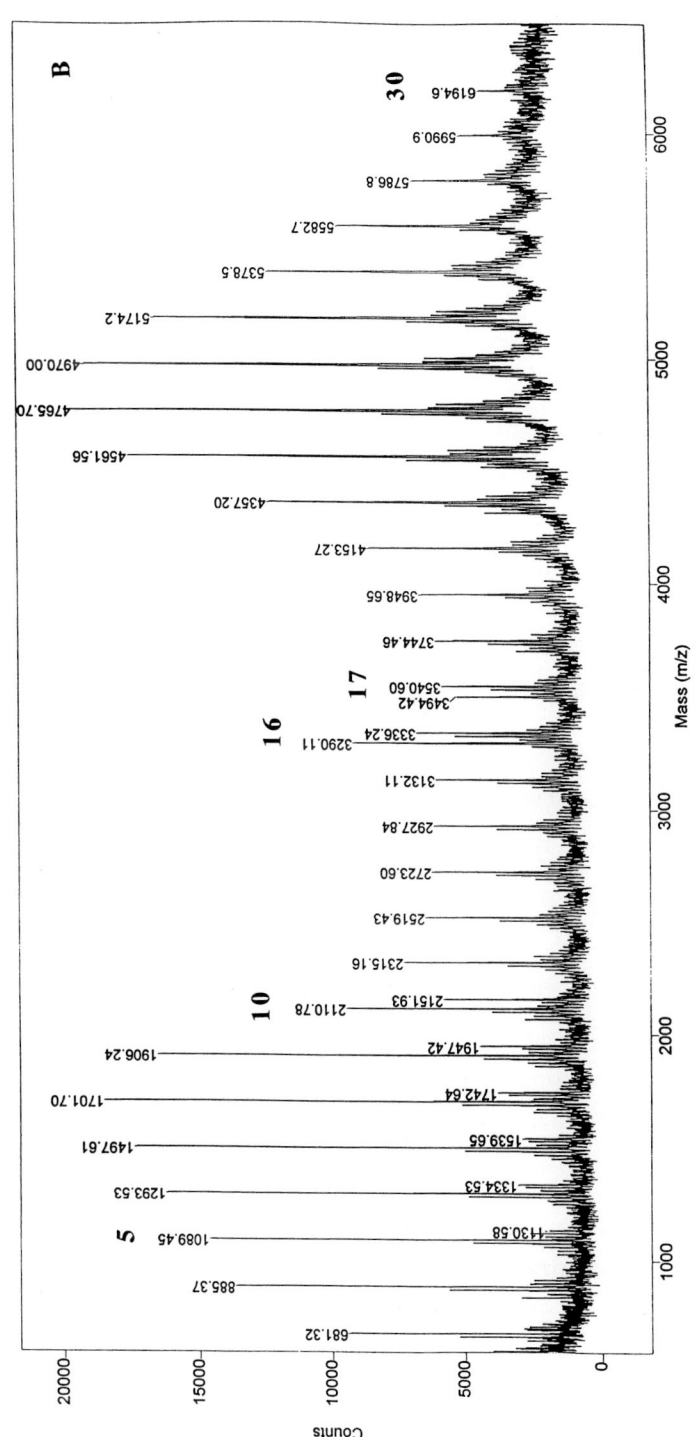

Figure 4 (*continued*)

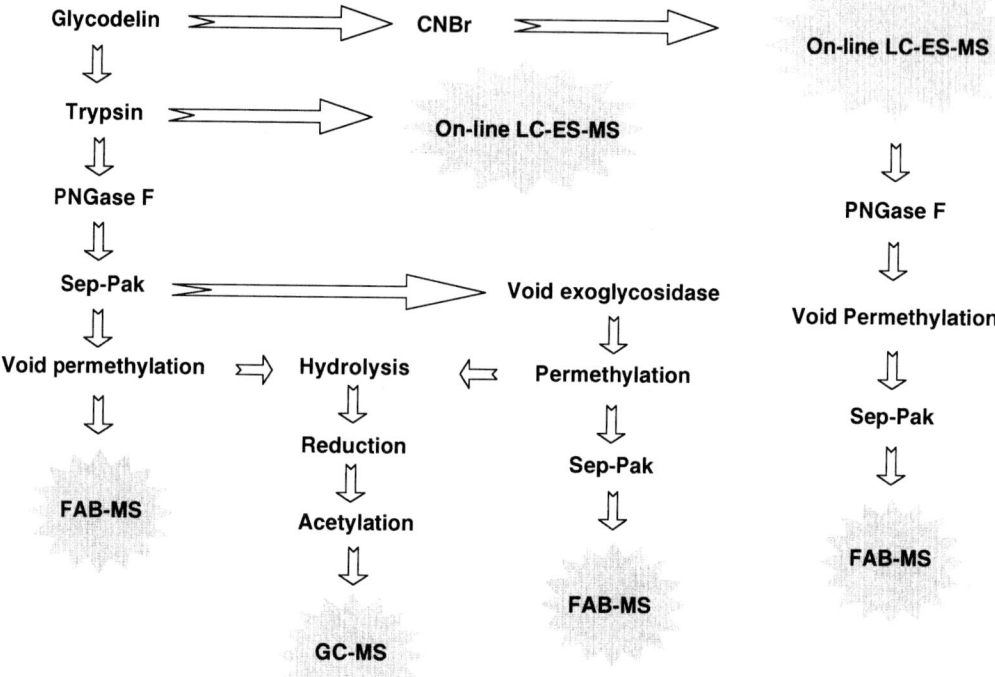

Figure 5. The experimental strategy used to characterize glycodelin A.

type of analysis because it reliably produces A-type ions without the need for MS/MS experiments.
iii) The products of various exoglycosidase digests were analyzed by FAB-MS after permethylation.
iv) Linkage analysis by GC-MS was used to define sites of glycosyl attachment in the N-glycan sequences.

From these experiments the structures of the majority of the oligosaccharides present in GdA were defined as shown in Figure 6. The Asn-28 site was shown to carry high mannose, hybrid and complex-type structures, whereas Asn-63 is exclusively occupied by complex-type glycans.

32.10.3 Case Study 3—Characterization of a Novel N-Glycan by FAB-MS and FAB-MS-MS

The following study of *Haemonchus contortus* N-glycans demonstrates how novel glycans present in complex mixtures can be rigorously identified by strategies based on FAB-MS and FAB-MS-MS analyses [10].

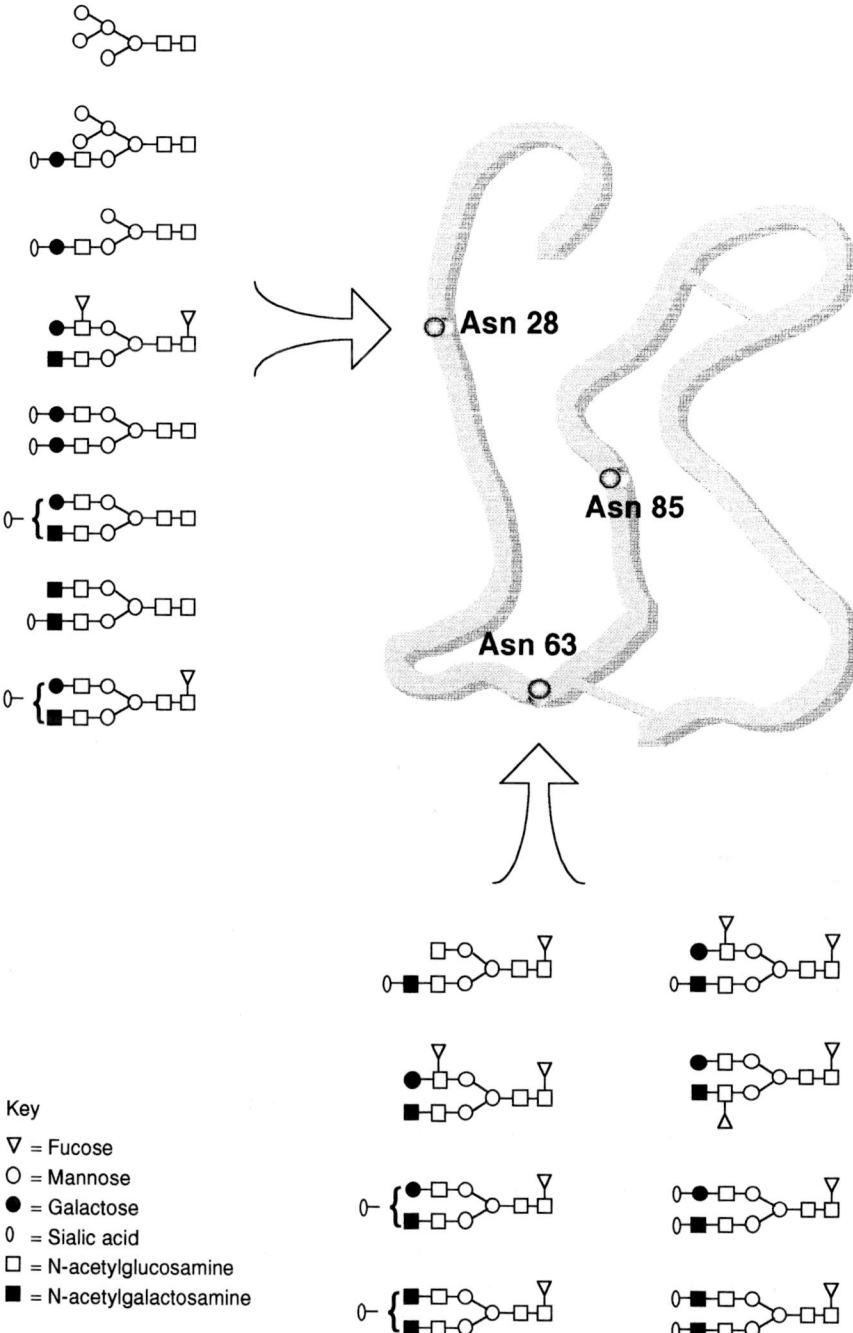

Figure 6. This cartoon structure shows the major glycoforms of glycodelin A. Further information can be found in [9].

```
              Fuc α1-6
               |
Man β1-4GlcNAc β1-4GlcNAc
      |            |
    Fuc α1-3    Fuc α1-3
```

Figure 7. Structure of the novel highly fucosylated core of some *H. contortus* N-glycans.

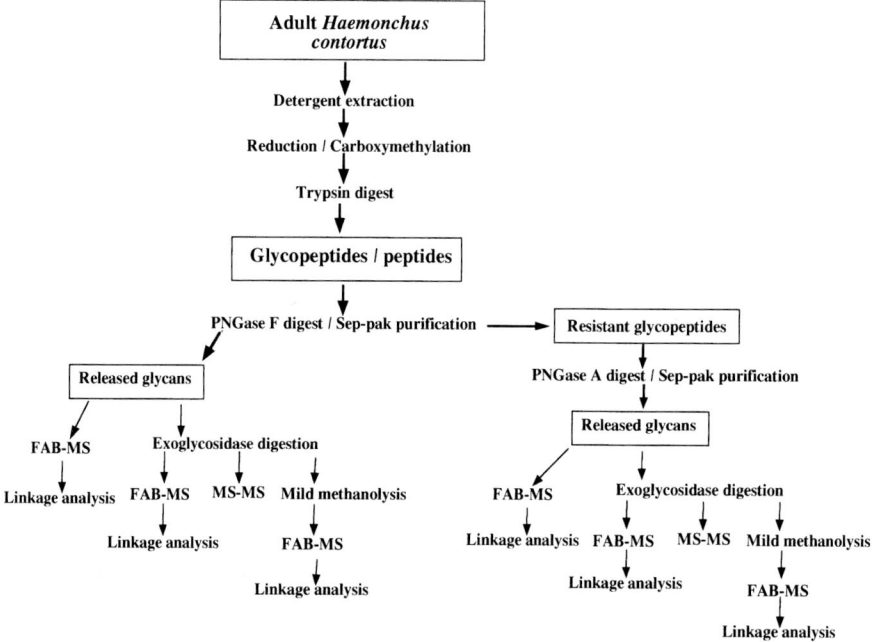

Figure 8. The experimental strategy used to characterize the *H. contortus* N-glycan shown in Figure 7.

H. contortus is an economically important nematode that parasitizes domestic ruminants. In a programme of work aimed at identifying carbohydrate antigens that could be targets for vaccine development, the structures of N-glycans present in *H. contortus* extracts have been investigated. Novel N-linked glycans with trifucosylated cores (Figure 7) have been identified in adult animals using the strategy outlined in Figure 8.

The key experimental steps were as follows:

i) FAB-MS analysis of permethylated glycans released from *H. contortus* glycopeptides by peptide N-glycosidase (PNGase F) and PNGase A (the latter releases glycans with fucose attached to the 3-position of the proximal GlcNAc of the core which are resistant to PNGase F) established the presence of high mannose structures, minor amounts of complex structures and unusual truncated glycans substituted with up to three fucose residues.

ii) Digestion of released glycans with α- and β-mannosidase greatly reduced

the complexity of the mixture of glycans facilitating subsequent MS-MS experiments.
iii) Selected molecular ions were subjected to FAB-MS-MS which provides sequence and linkage informative daughter fragment ions.
iv) Linkage analysis by GC-MS was used to confirm sites of glycosyl attachment.

An especially important element of the above structural strategy is the MS-MS component because the novel glycans could not be isolated in sufficient quantities for individual structural studies. Some of the MS-MS data which helped to establish the sites of fucosyl attachment are shown in Figure 9.

32.10.4 Case Study 4—High Sensitivity Sequencing of a Novel Glycopeptide by Q-TOF ES-MS-MS and MALDI-MS

This study of a novel glycopeptide derived from *Dictyostelium* sp. SKP1 exemplifies the power of the Q-TOF nano-electrospray mass spectrometer for high sensitivity sequencing [11]. The efficacy of MALDI-MS for screening enzyme digests is also demonstrated.

SKP1 is a cytoplasmic protein which has been identified in a variety of eukaryotes including yeast, mouse and man. It is found as part of a multiprotein complex which is involved in the ubiquitination of certain cell cycle and nutritional regulatory proteins thus condemning them to proteasomal degradation. Prior to the Q-TOF study described below, metabolic labelling and other experiments had indicated that, unusually for a cytoplasmic protein, SKP1 from *Dictyostelium* is modified by an oligosaccharide containing Fuc and Gal.

To investigate the glycosylation of SKP1, cells were metabolically labelled with ^{3}[H]Fuc, radioactive SKP1 was purified, reduced and alkylated and finally digested with endo-Lys-C. Components of the digest were separated by gel filtration and reverse phase HPLC yielding a single radioactive peak which was likely to be the expected fucoglycopeptide. Parallel experiments with non-radioactive material were performed and the fraction eluting at the position of the radioactive peak was subjected to nanospray-ES-MS-MS on the Q-TOF.

Analysis of a few picomoles in the MS-only mode gave a major $[M+3H]^{+++}$ signal at m/z 829.42 which was subjected to collisional activation yielding a spectrum which was remarkably rich in doubly and singly charged daughter ions (Figure 10). A portion of the spectrum is shown at high magnification (Figure 11) to illustrate the excellent signal to noise and resolution of even very minor fragment ions. This is an important strength of Q-TOF instrumentation especially since information such as sugar attachment sites in glycopeptides is often carried by low abundance fragment ions. Detailed interpretation of the MS-MS data allowed assignment of the carbohydrate sequence, Hex-Hex-deoxyHex-Hex-HexNAc, and the complete peptide sequence, N-D-F-T-P(OH)-E-E-E-E-Q-I-R-K and also the site of attachment of the carbohydrate, which was unexpectedly found to be hydroxyproline. Some of the key fragment ions used in the structure assignment are shown in Figure 12.

MALDI-MS of the glycopeptide gave an $[M+H]^+$ signal at m/z 2487 thus providing corroborative evidence for the structure proposed from the Q-TOF data. To

934   32 *Structural Analysis of Oligosaccharides*

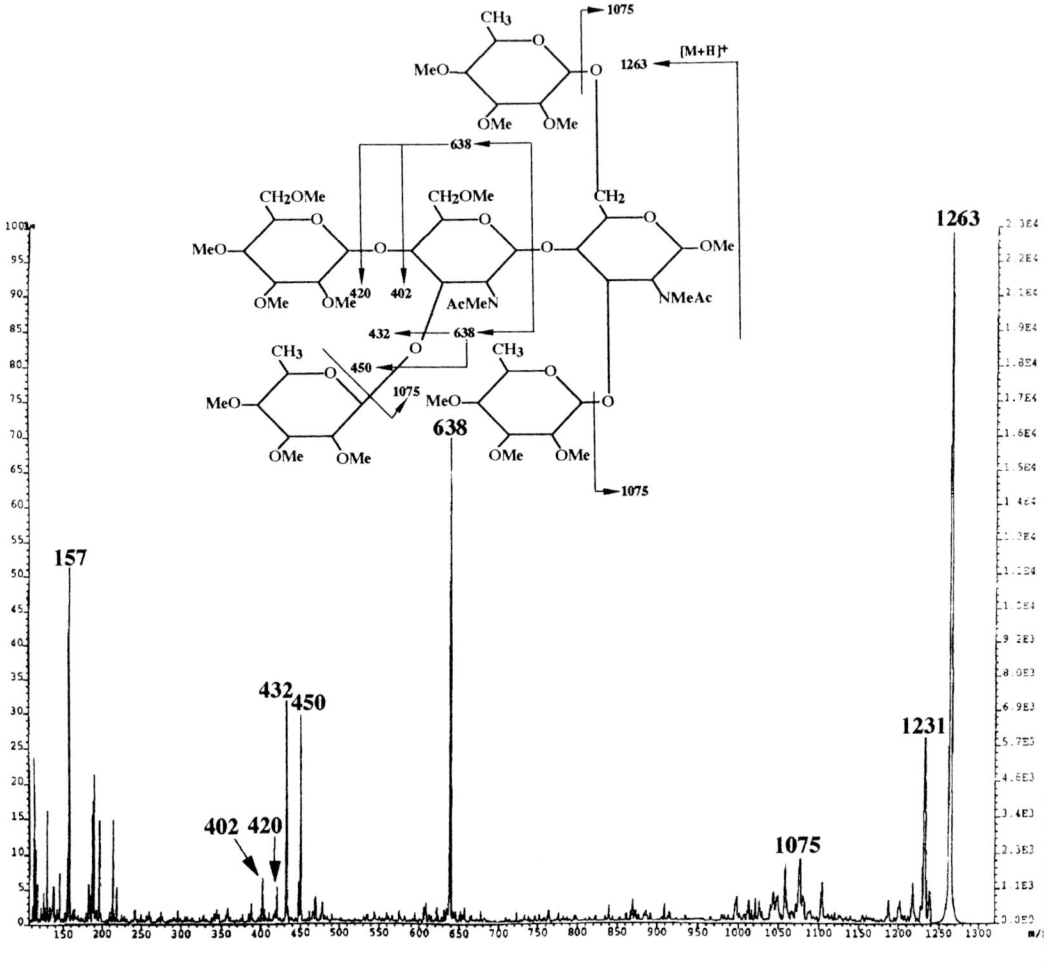

a)

Figure 9. Data from a FAB-MS-MS experiment which provided evidence for the trifucosylated structure shown in Figure 7. The [M+H]$^+$ and [M+Na]$^+$ ions at m/z 1263 **(A)** and 1285 **(B)** respectively were selected for collisional activation. Note that because of the different internal energies of these two molecular ions, the fragmentation pathways are different. Structurally useful fragment ions, produced via fragmentation pathways outlined in Figure 3, are shown on the inset. These ions provide important information on the attachment sites of the fucosyl residues.

assign linkages, and to rigorously establish the types of sugar present, the glycopeptide was treated with a variety of exoglycosidases with the products being analysed by MALDI-TOF-MS. Representative data are shown in Figure 13 which shows the evidence for the non-reducing sequence being Galα1-6Galα1-?Fucα1-2. Similar experiments showed that the proximal disaccharide in Figure 12 is Galβ1-3GlcNAc.

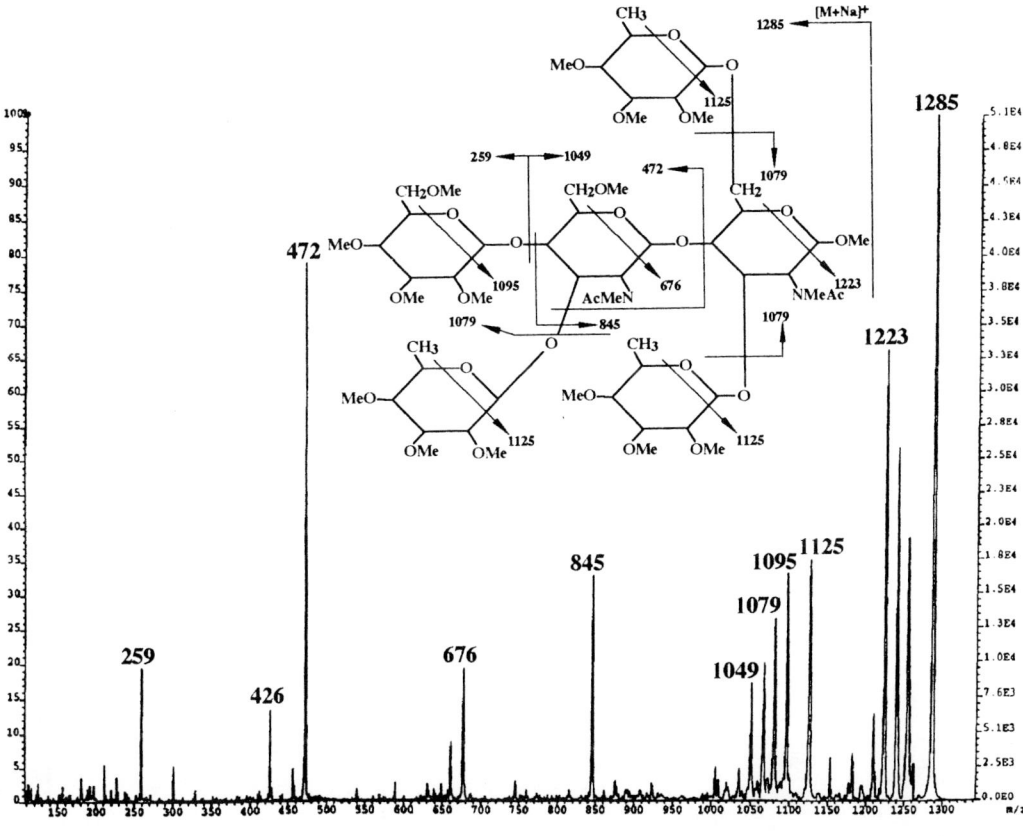

Figure 9 (*continued*)

32.10.5 Case Study 5—FAB-MS Screening of Biological Samples for Glycan Content

Although detailed structural analysis is fundamental to a full understanding of structure/function relationships, there are many situations where the rapid acquisition of partial structural information is of greater value to the program of research than a time-consuming strategy leading to complete structure assignments. For example, many months of painstaking research was required to characterise the novel *N*-glycan described in Case Study 3. Once its structure was known it was of interest to determine whether its synthesis was developmentally regulated and whether it was found in other parasites occupying a similar niche to *H. contortus*. These issues can be addressed by screening experiments [12] in which tissue extracts are analysed for their glycan content using FAB-MS or MALDI-MS to screen for molecular ions

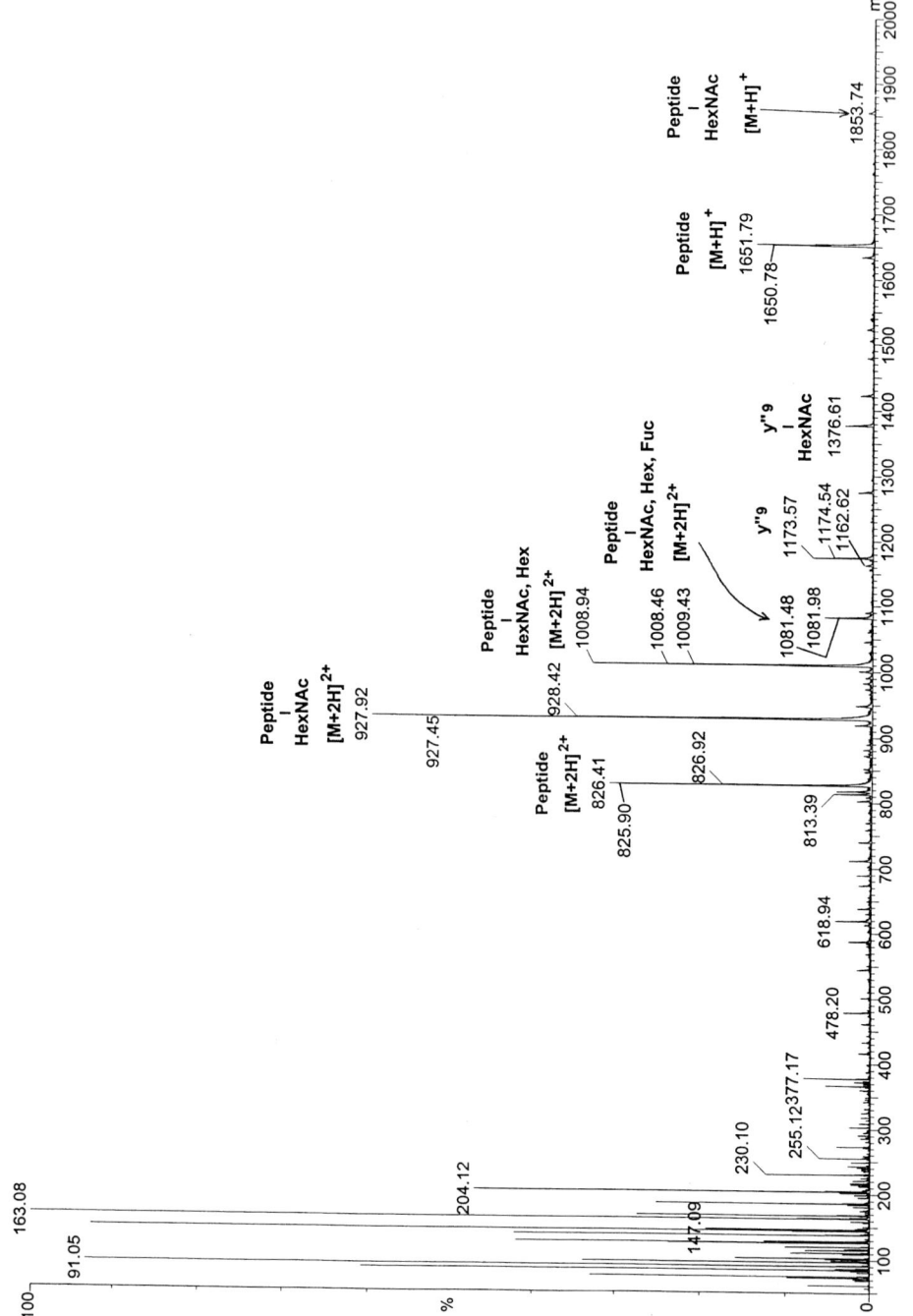

Figure 10. Q-TOF analysis of a glycopeptide from SKP1: MS-MS collision activated decomposition spectrum of m/z 829.42^{+++}

32.10 Applications of FAB-MS, MALDI-MS and ES-MS in Glycobiology 937

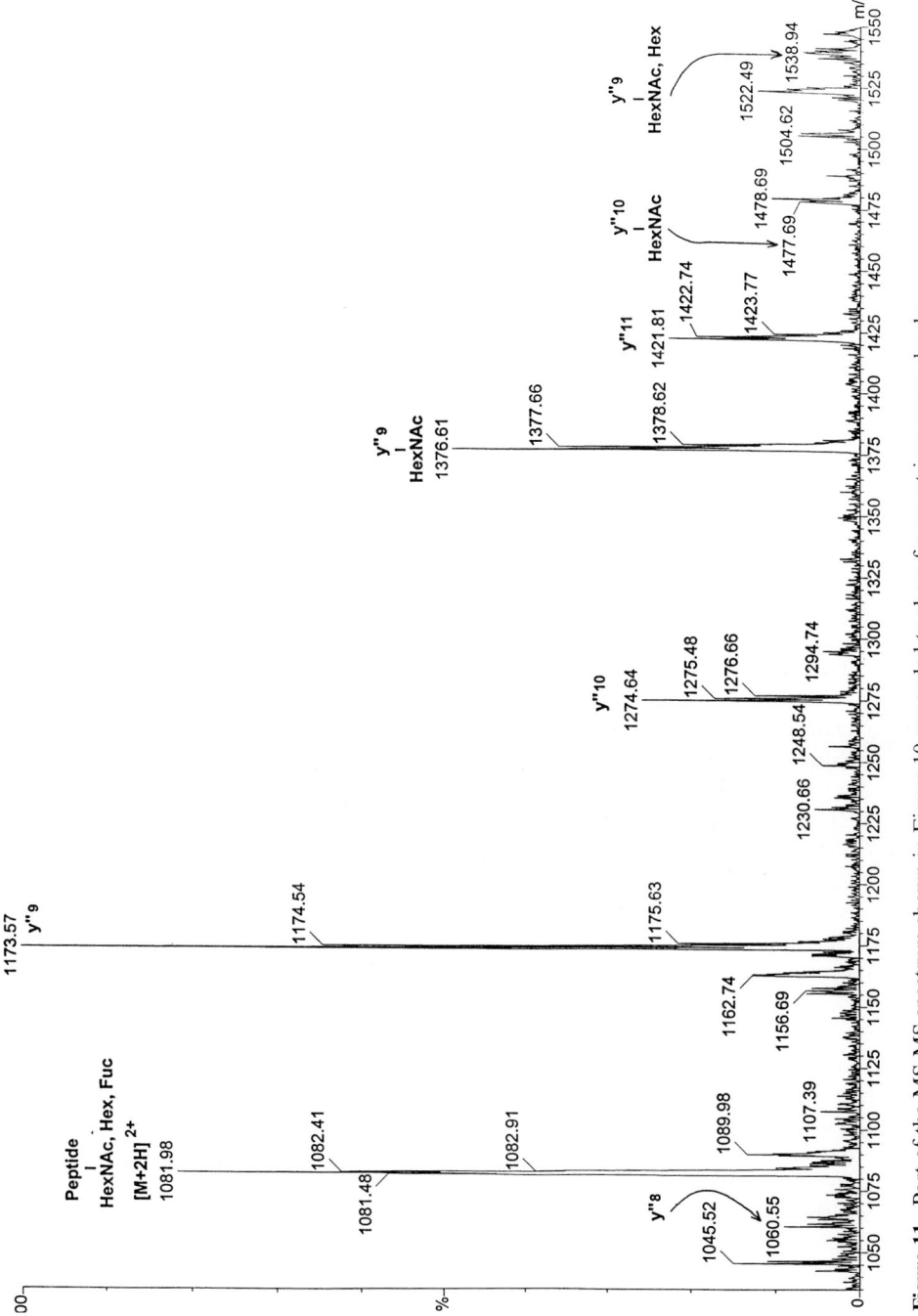

Figure 11. Part of the MS-MS spectrum shown in Figure 10 expanded to show fragment ions more clearly.

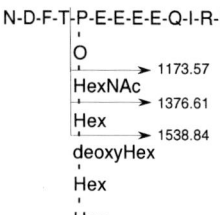

Figure 12. Structure of the SKP1 glycopeptide deduced from the Q-TOF data showing some of the diagnostic fragment ions used to assign the position of attachment of the sugar.

and FAB-MS to screen for non-reducing structures via A-type fragment ions. This type of experiment can be completed in a few weeks in contrast to the months or years required to fully characterize individual glycans in complex mixtures.

Screening methods which are applicable to a wide range of biological material, including organs, cell lines and whole parasites, are exemplified by the following profiling study of a variety of mouse organs in which the FAB-MS data are interpreted in the context of previous knowledge of murine glycosylation. This work is part of a study using knockout mice to address fundamental issues of glycan function in which FAB-MS is being used to identify changes in glycosylation occurring when particular glycosyltransferase and glycosidase genes are ablated.

The experimental strategy employs the following steps:

i) Detergent extraction of the tissue/organ followed by detergent removal
ii) Reduction/carboxymethylation, tryptic digestion and Sep-pak separation of the peptide/glycopeptide pool from salts, free sugars etc
iii) Release of *N*-glycans by peptide *N*-glycosidase F digestion
iv) Sep-pak separation of *N*-glycans from the peptide/*O*-linked glycopeptide pool
v) Permethylation of *N*-glycans and FAB-MS analysis
vi) Reductive elimination of *O*-glycans from peptide/*O*-glycopeptide pool and Dowex purification
vii) Permethylation of *O*-glycans and FAB-MS analysis

These experiments provide compositional information (via the molecular ions) on the majority of neutral and sialic acid containing glycans in the tissues/organs together with information on the types of non-reducing structures present (via the A-type ions). This is illustrated by the data in Figures 14 and 15. Figure 14 shows the molecular ion region of the *N*-glycan pools from mouse brain, liver and lung. A unique composition can be ascribed to each molecular ion as shown. Figure 15 shows the A-type fragment ions formed from these molecular ions. The A-type ions are assigned using a variety of information including their masses, the presence or absence of associated fragment ions and prior knowledge of glycosylation in the mouse. The data show that all organs are rich in both high mannose and complex-type glycans. The brain and kidney have complex-type glycans that are rich in Lewis x antennae (note the A-type ion at m/z 638 in Figures 15A and B) whilst

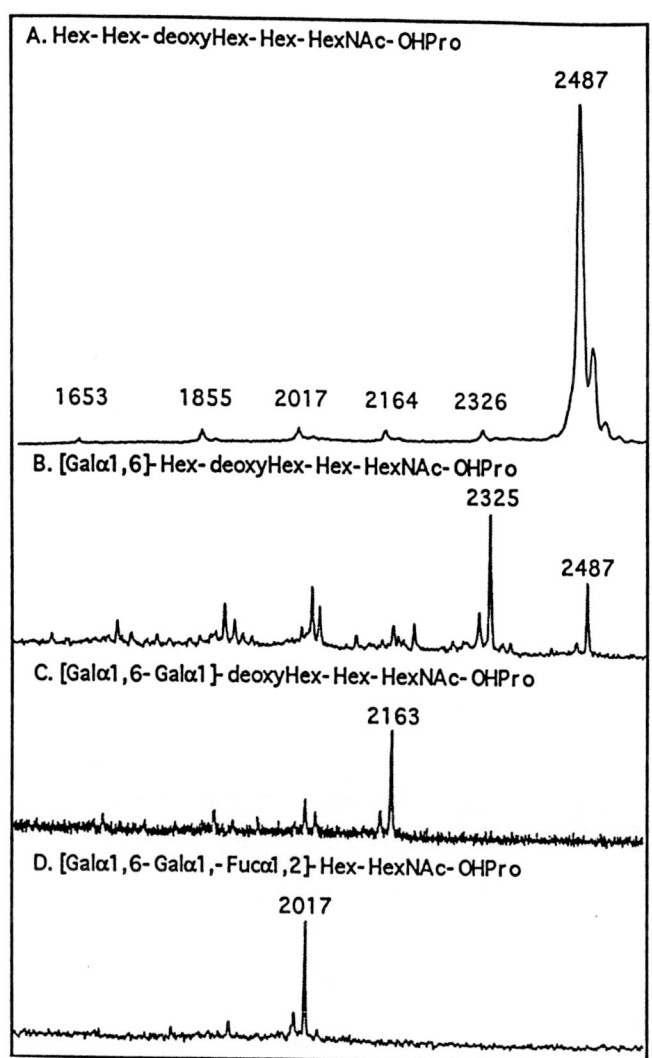

Figure 13. MALDI-TOF analysis of the SKP1 glycopeptide: **(A)** spectrum from the native glycopeptide, **(B)** spectrum after treatment with *X. manihotis* α 1–3/6 galactosidase, **(C)** spectrum after treatment with green coffee bean α-galactosidase, **(D)** spectrum after treatment with green coffee bean α-galactosidase and *X. manihotis* α 1–2 fucosidase.

those of the lung are characterised by sialylation (note the A-type ions at *m/z* 825 and 855 in Figure 15C) or alpha-Gal capping (note the A-type ion at *m/z* 668 in Figure 15C). The above screening strategy is remarkably reproducible with very little variation in relative molecular ion abundance occurring between data from different experiments on the same organ or on organs from different animals. Hence

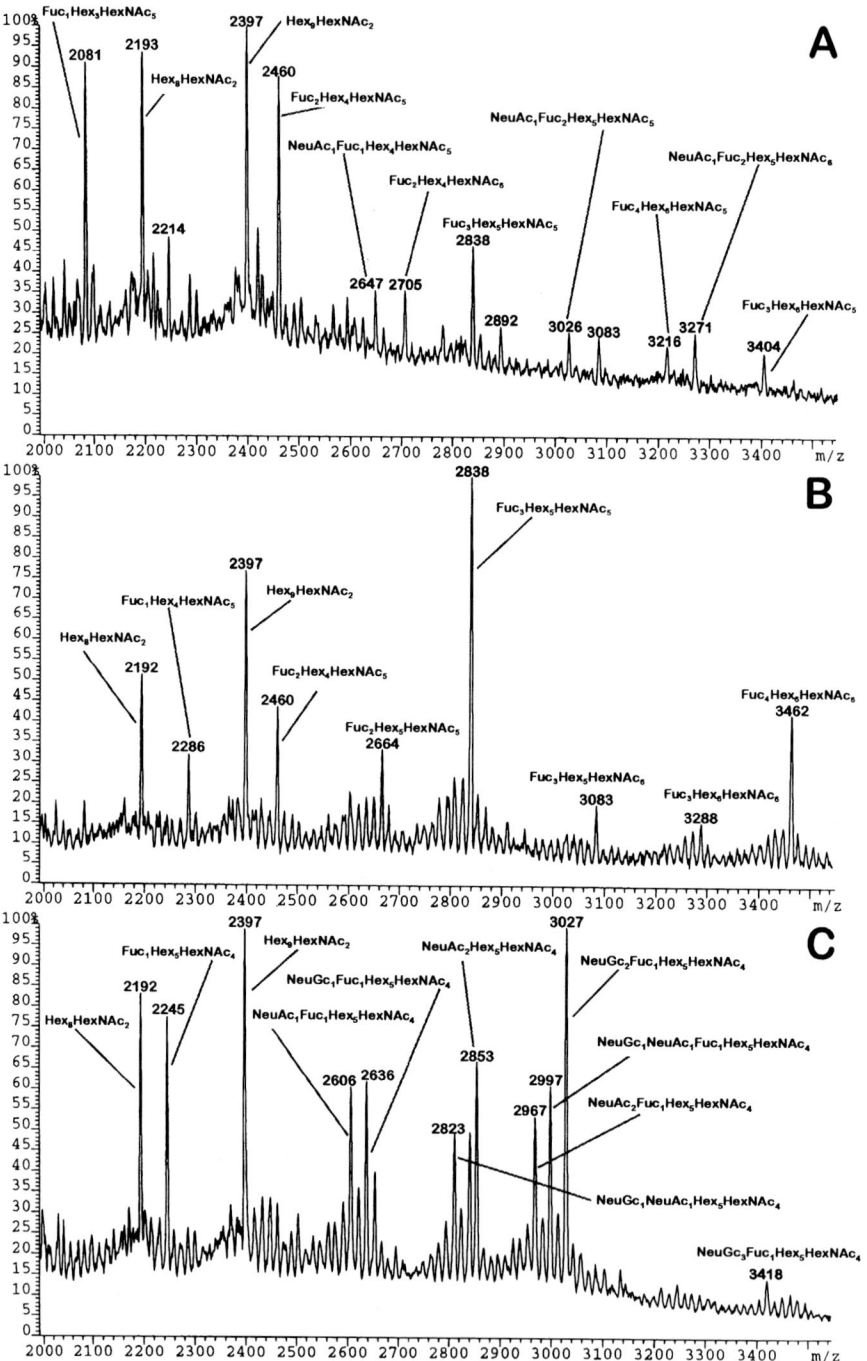

Figure 14. Molecular ion region of FAB spectra obtained from screening experiments (see text) on murine **(A)** brain, **(B)** liver and **(C)** lung. Compositions of major signals are shown.

32.10 *Applications of FAB-MS, MALDI-MS and ES-MS in Glycobiology*   941

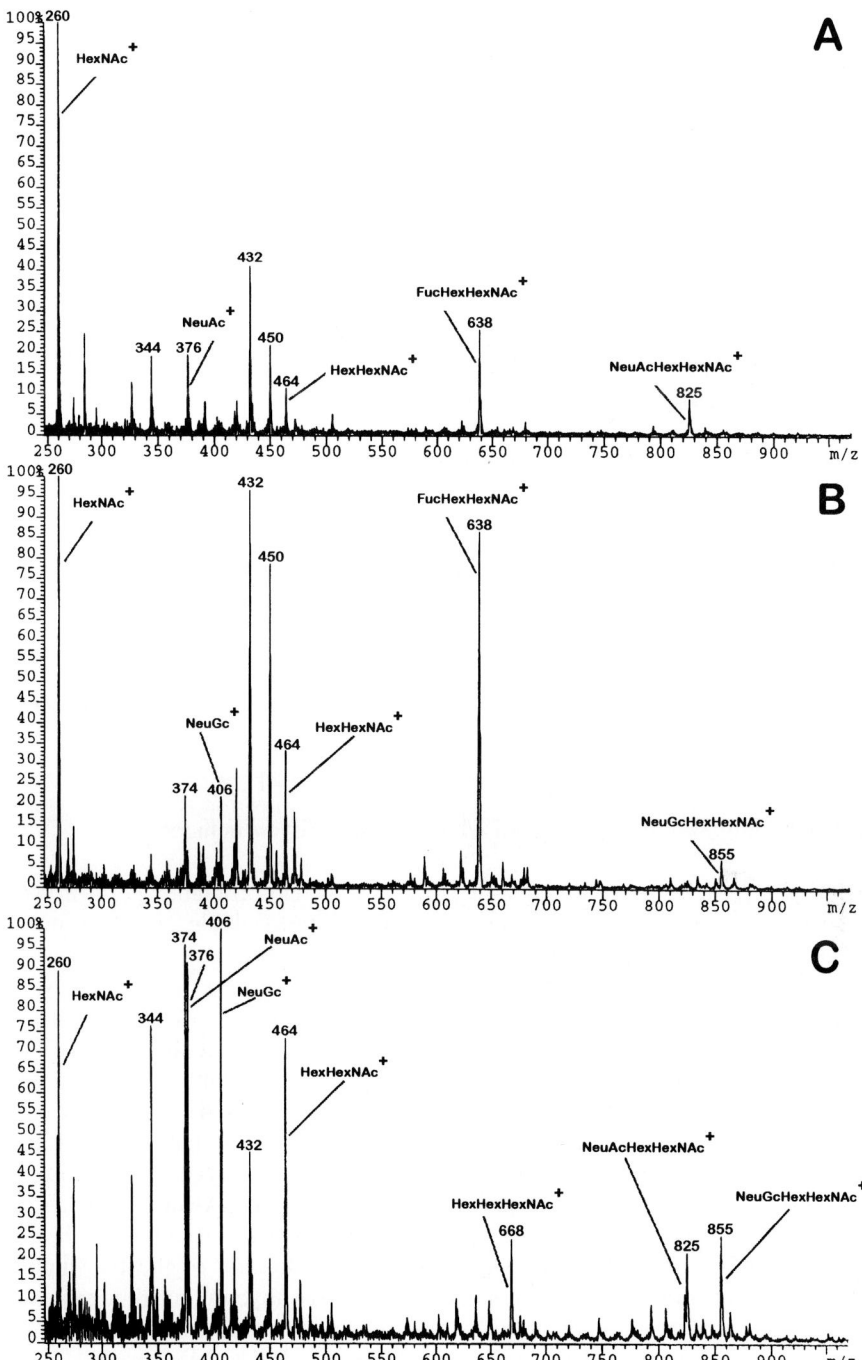

Figure 15. A-fragment ion region of the spectra giving the molecular ion data shown in Figure 14: **(A)** brain, **(B)** liver, **(C)** lung.

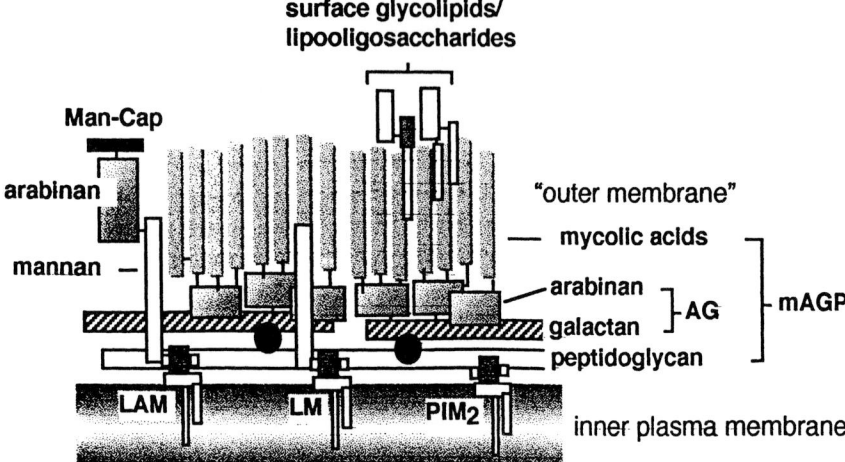

Figure 16. A chemical model of the mycobacterial cell wall. The two major components are the mycolyl arabinogalactan-peptidoglycan complex and the lipoarabinomannan/lipomannan. The amount and exact chemical nature of the surface glycolipids are species- and/or strain-specific.

quantitative as well as qualitative differences can be readily explored in experiments on knockout mice.

32.10.6 Case Study 6—MS Analysis of Mycobacterial Glycoconjugates

The carbohydrate-rich bacterial cell wall presents a structural challenge of a very different nature to the case studies above. Although the available sample amount for analysis is less of a problem in comparison to the situation with bioactive mammalian or helminth glycoproteins, the extremely diverse range of glycoconjugates, as well as novel saccharide residues and non-saccharide substituents that may be present, exerts a high premium on detailed structural characterization. Studies of mycobacterial cell wall components (Figure 16) exemplify recent trends in mass spectrometric analysis in this area of biopolymer research.

Sequencing of the species- and strains-specific surface glycolipids, including the glycopeptidolipids and the acyltrehalose-containing lipooligosaccharides, requires the kind of sequence-informative fragment ions afforded by FAB-MS analysis of permethyl and/or peracetyl derivatives, in addition to precise molecular weight determination [13, 14]. On the other hand, structural elucidation of the two highly heterogeneous homo-polymeric components, namely the mycolyl arabinogalactan (AG) and the lipoarabinomannan (LAM), requires different strategic approaches.

In the case of AG, initial efforts were directed towards GC-MS and FAB-MS analysis of the small fragments generated by partial hydrolysis. The deduced structural motifs were then pieced together to give the model shown in Figure 17 based

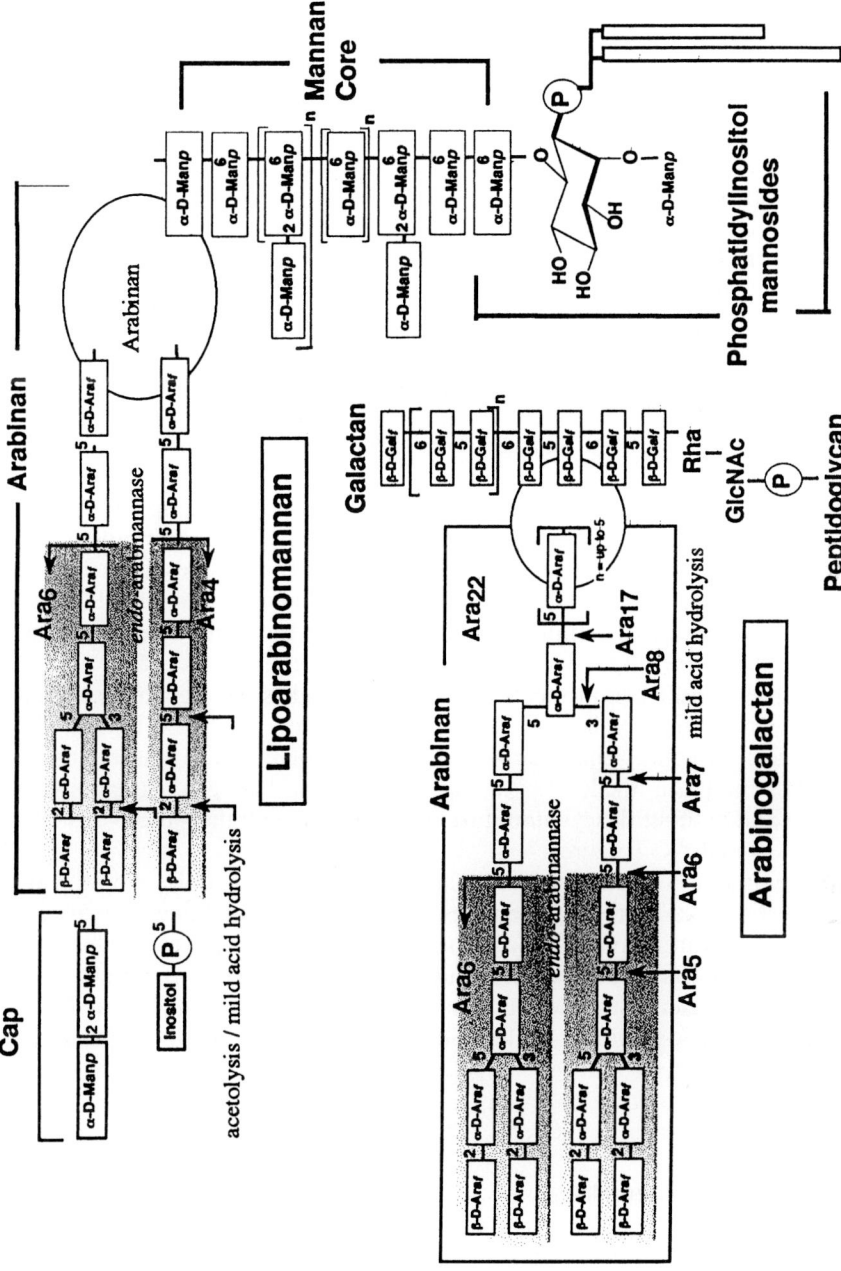

Figure 17. Structural model of LAM and AG. AG is drawn without the mycolic acids attached to the terminal Ara residues. The molecular weight of LAM has been determined by MALDI-TOF on both native and permethylated samples. The arabinan circle in both LAM and AG represents structural details not yet defined. Detailed characterizations of the various arabinan motifs were based on MS/chemical analysis of fragments obtained through acetolysis, mild acid hydrolysis and endoarabinase digestion as indicated. The branched Ara_6 motif is common to both AG and LAM whereas the linear Ara_4 terminal motif is found only on LAM. An Ara_{22} motif as drawn is deduced to be present on AG [15].

on FAB-MS analysis of larger pieces generated by very mild hydrolysis of permethylated samples [15]. Importantly, any branching position in addition to the reducing end linkage is "visualised" by the mass difference between a methyl group and exposed hydroxyl group (which can be further tagged by deuteromethyl or deuteroethyl) as a consequence of hydrolyzing off the branch(es). Thus, as illustrated, arabinan oligomers with no additional branched point were detected as Ara 5, 6, 7, 8 and then Ara 17 since Ara 9–16 cannot be generated by partial hydrolysis with only one clip at the reducing end. The success of this approach relies on the ability of MS in detecting small mass difference at high mass range, i.e. high resolution. This important information cannot at present be obtained via direct MS analysis of non-derivatized samples. Similarly, although the approximate molecular weight distribution of intact LAM can be determined by MALDI-TOF and has been used effectively to confirm the truncation in size of LAM due to ethambutol drug inhibition [16], detailed structures of LAM can only be determined through analysis of fragments derived through both chemical and enzymatic degradation [17].

The capability of MS for examining crude mixtures (see for example Case Study 6), affording precise mass values for each component present, enables rapid screening of novel modifications otherwise cryptic to conventional analysis such as the inositol cap on some LAMs [17]. Likewise, whilst the introduction of ES and MALDI has made the determination of the intact molecular weights of lipopolysaccharides (LPS), lipooligosaccharides, and other polysaccharides feasible, these techniques merely complement more detailed studies of fragments generated via partial degradation which are usually most effectively examined by FAB-MS. Thus, complete structural characterisation of LPS still involves painstaking and dedicated MS analysis at the level of the repeating units (O-antigen), core units, and lipid A with and without deacylation, just as is the case with the various structural motifs of LAM and AG (Figure 17).

Finally, it should be mentioned that the LC-ES-MS methodology now widely used for detecting eukaryotic glycopeptides in proteolytic digests (see Case Study 2) has proven successful in analyzing mycobacterial glycoproteins [18]. In this study, which provides the first firm chemical evidence for the presence of protein glycosylation in mycobacteria, the neutral loss and daughter ion scanning in ES-MS provided a very sensitive means of detection of minor glycopeptides in proteolytic digests.

32.11 Concluding Remarks

As exemplified by the case studies described above, FAB-, MALDI- and ES-MS are powerful techniques for glycopolymer structure analysis. Each has its own strengths. For example, MALDI-MS would be the method of choice for molecular weight profiling of polysaccharide mixtures (Case Study 1), FAB-MS is ideally suited to screening for non-reducing epitopes in biological samples (Case Study 5), whilst nano-ES-MS-MS on the Q-TOF provides the most sensitive means of sequencing peptides and glycopeptides (Case Study 4). However it is important to bear in mind that, with an appropriate choice of experimental strategy, many structural problems can be successfully addressed with any one of the three ionization methods. Thus, for many researchers, access to instrumentation and expertise might be more important than factors such as relative sensitivities or whether collisional activation is required to generate fragment ions.

Advances in biopolymer mass spectrometry in the past 20 years have been truly breathtaking. We hope that the information provided in this Chapter will encourage readers to delve more deeply into the vast recent literature of mass spectrometric glycopolymer analysis and to use their new-found knowledge to tackle the next generation of structural problems.

References

1. M. Barber, R.S. Bordoli, R.D. Sedgwick and A.N. Tyler, *J. Chem. Soc. Commun.*, (1981) 325–327.
2. A. Dell, H.R. Morris, H. Egge, G. Strecker and H.V. Nicolai, *Carbohydr. Res.*, 115 (1983) 41–52.
3. A. Dell *Adv. Carbohydr. Chem. Biochem.*, 45 (1987) 19–72.
4. J.B. Fenn, M. Mann, C.K. Meng, S.F. Wong, C.M. Whitehouse, *Mass Spectrom. Rev.*, 9 (1990) 37.
5. M. Karas, A. Ingendoh, U. Bahr. and F. Hillenkamp *Biomed. Environ. Mass Spectrom.*, 18 (1989) 841.
6. H.R. Morris, T. Paxton, A. Dell, J. Langhorne, M. Berg, R.S. Bordoli, J. Hoyes, R.H. Bateman. *Rapid Commun. Mass Spectrom.*, 10 (1996), 889–896.
7. H.R. Morris, T. Paxton, M. Panico, M. McDowell, A. Dell. *Mass Spectrometry of Biological Materials* Vol. 2 (eds B. Larsen & C. McEwan) Marcel Dekker, New York, (1998), pp 53–80.
8. A.O. Chizhov, A. Dell, H.R. Morris, A.J. Reason, S.M. Haslam, R.A. McDowell, O.S. Chizhov. and A.I. Usov, *Carbohydr. Res.*, 310 (1998), 203–210.
9. A. Dell, H.R. Morris, R.L. Easton, M. Panico, M. Patankar, S. Oehninger, R. Koistinen, H. Koistinen, M. Seppala, M. and G.F. Clark *J. Biol. Chem.*, 270 (1995) 24116–241266.
10. S.M. Haslam, G.C. Coles, E.A. Munn, T.S. Smith, H.F. Smith, H.R. Morris, H.R. and A. Dell *J. Biol. Chem.*, 271 (1996) 30561–30572
11. P. Teng-umnuay, H.R. Morris, A. Dell, M. Panico, T. Paxton, and C.M. West, *J. Biol. Chem.*, 273 (1998) 18242–18249.
12. S.M. Haslam, G.C. Coles, A.J. Reason, H.R. Morris and A. Dell *Mol. Biochem. Parasitol.*, 93 (1998) 143–147.
13. K.-H. Khoo, R. Suzuki, H.R. Morris, A. Dell, P.J. Brennan, and G.S. Besra, *Carbohydr. Res.*, 276 (1995) 449–455.
14. K.-H. Khoo, D. Chatterjee, A. Dell, H.R. Morris, P.J. Brennan, P.J. and P. Draper, *J. Biol. Chem.*, 271 (1996) 12333–12342.
15. G.S. Besra, M. McNeil, M., K-H. Khoo, A. Dell, H.R. Morris and P.J. Brennan, *Biochemistry* 34 (1995) 4257–4266.
16. K.-H. Khoo, E. Douglas, P. Azadi, J.M. Inamine, G. Besra, P.J. Brennan and D. Chatterjee, *J. Biol. Chem.*, 271 (1996) 28682–28690.
17. K.-H. Khoo, A. Dell, H.R. Morris, P.J. Brennan, D. Chatterjee, *J. Biol. Chem.*, 270 (1995) 12380–12389.
18. K.M. Dobos, K.-H. Khoo, K. Swiderek, P.J. Brennan, P.J. and J.T. Belisle, *J. Bacteriol.*, 178 (1996) 2498–2506.
19. B. Domon and C.E. Costello *Glycoconj. J.*, 5 (1998) 397–409.

33 Conformational Analysis in Solution by NMR

S. W. Homans

33.1 Introduction

Of the functions that have been ascribed to oligosaccharides [1], their role in molecular recognition appears to be a repetitive theme in many biological processes. Recent work on the molecular genetics of glycosylation has indicated that oligosaccharides exhibit a particularly prominent role in the development of multicellular organisms [2–9]. Morever, it has been known for many years that oligosaccharides are important for the invasion, infectivity and survival of parasites [10] and pathogens [11] in host cells. There is therefore compelling evidence to suggest that a role in cell–cell communication might be the primary function of oligosaccharides. This might explain the apparent lack of function of these moieties when examined at the level of a protein or indeed an individual cell.

In order to understand the molecular basis of oligosaccharide-mediated recognition phenomena, we need to have knowledge of the structure and dynamics of the free carbohydrate ligand in addition to the ligand–receptor complex. In this Chapter, I shall review current knowledge on the former. A discussion of ligand–receptor complexes can be found elsewhere in this volume.

33.2 Solution Conformations of Oligosaccharides

33.2.1 The NMR Technique

Since oligosaccharides in general fail to crystallize, the only technique that is able reliably to offer information on the structure of oligosaccharides in solution at atomic resolution is nuclear magnetic resonance (NMR) spectroscopy [12–15]. There are three NMR parameters that are relevant to the structural analysis of oligosaccharides. The chemical shift refers essentially to the frequency at which each discrete resonance line in the spectrum can be found. This frequency is dependent upon the precise chemical environment (hence the term chemical shift) of the nucleus that gives rise to the resonance, since this environment has the capacity to shield or deshield the nucleus from the full effects of the applied magnetic field. The chemical shift contains a wealth of information on molecular structure, but in view of the complexity of the factors that contribute to it, it is only possible to make practical use of this information in a few simple systems. Nevertheless, chemical shifts can offer useful information on carbohydrate structure when interpreted in a semi-quantitative sense [16–28]. In addition to the chemical shift, each NMR resonance line typically exhibits a multiplet structure or 'splitting'. This arises from the sensitivity of a given nuclear environment to the spin-state of nearby covalently bonded nuclei and is transmitted via the bonding electrons, becoming essentially unmeasurable over more than three bonds, at least in saturated systems. The three-bond coupling (denoted $^3J_{nn}'$) between nuclei is of considerable practical importance, since the dihedral angle formed by the outermost pair of bonds influences the magnitude of the splitting. It was shown by Karplus [29] many years ago that an empirical relationship of the form $^3J_{nn}' = A\cos^2\theta + B\cos\theta + C$ (where A, B and C are constants) can be used to derive angular information from three-bond couplings. A third parameter is known as the nuclear Overhauser effect (NOE). This parameter is not manifest in the NMR spectrum directly, but can be observed as a change in the intensity of the resonance corresponding to one nucleus when the spin-state populations of another resonance are perturbed by selective application of radio frequency energy [30]. This change in intensity is related, *inter alia*, to the inverse sixth power of the distance between the nuclei. Under the appropriate circumstances the NOE can therefore be used to measure internuclear distances up to ~0.5 nm.

33.2.2 Conformational Parameters in Oligosaccharides

Oligosaccharides are comprised of linear or branched monosaccharide units that are invariably linked from the C-1 position of one monosaccharide unit (the 'glycon') to one of several positions on the neighboring monosaccharide unit (the 'aglycon'). With few exceptions, the chemical nature of the coupling between the monosaccharide units is an ether linkage (C-O-C), which is termed the glycosidic linkage. Therefore two rotatable bonds exist about the glycosidic linkage (Figure 1). In pyranoses (the types of monosaccharide that will almost exclusively be discussed here) the geometry of each monosaccharide ring can essentially be thought of as being fixed on the timescale of the NMR experiment, usually in the 'chair' ($^4C_1$) configuration. Therefore the primary sites of conformational variation are the two glycosidic torsion angles defined by the atoms H-1–C-1–O-1–C-X and C-1–O-1–C-X–H-X, where C-X and H-X represent aglyconic atoms. These torsion angles are denoted ϕ_H and ψ_H in IUPAC convention. In addition there is a third dihedral angle about the rotatable C-5–C-6 bond in each monosaccharide residue defined by H-5–C-5–C-6–O-6. This dihedral angle is of particular importance for conforma-

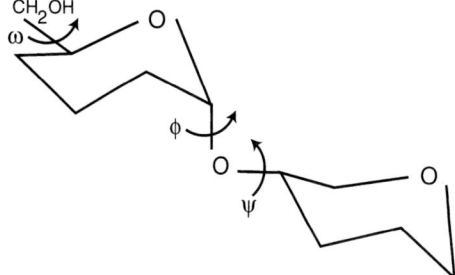

Figure 1. Conformational torsion angles in oligosaccharides. For NMR studies, these angles are usually defined as ϕ = H1-C1-O1-CX, ψ = C1-O1-CX-HX and ω = H5-C5-C6-O6, where CX and HX are aglyconic atoms.

tional studies when substitution occurs at the C-6 position and is given the symbol ω_H. Henceforth, the subscripts on ϕ_H, ψ_H and ω_H will be dropped for convenience.

33.2.3 Conformational Restraints

Since the internal geometry of each monosaccharide ring is essentially fixed, the solution conformation of an oligosaccharide can be defined by the torsion angles ϕ, ψ and ω. These angles can in principle be determined experimentally by measurement of conformational restraints from one monosaccharide across the glycosidic linkage to its neighbour. Thus, one or more NOEs can usually be measured from H-1 of the glycon to aglyconic protons proximal to the glycosidic linkage, and two ^{13}C-^{1}H three-bond coupling constants can be measured across the glycosidic linkage. Unfortunately, even when three such NOEs can be measured (usually the maximum number that can be observed) together with two coupling constants, the number of restraints is such that it can be impossible to distinguish between models involving a single conformation about the glycosidic linkage rather than a model involving substantial conformational flexibility [31]. Additional conformational restraints are required to distinguish between these two models, and much effort has been expended in recent years to achieve this goal.

33.2.4 ^{13}C Isotopic Enrichment

In order to obtain additional conformational restraints in oligosaccharides, it is useful or even essential to work with the glycan in ^{13}C-enriched form. This is by no means a trivial task. Although stable isotopic enrichment of proteins is now commonplace, this is often relatively straightforward by use of the relevant organism that over-expresses the protein in question together with isotopically enriched media. While it is possible to grow mammalian cells using such media and thus to prepare isotopically enriched glycoproteins [32], the cleavage of the glycans from the latter is not an efficient way to generate isotopically enriched material. It is usually more efficient to use a synthetic or chemoenzymatic strategy [33–36]. The latter is particularly attractive in view of its simplicity on the milligram scale which is ideal for NMR studies. In particular, once the relevant isotopically enriched building blocks have been obtained, it is straightforward to prepare oligosaccharides with a variety of different labelling regimes to suit a particular NMR technique. In general, the protocols for the chemoenzymatic synthesis of ^{13}C-enriched oligosaccharides can mirror those used for the synthesis of their natural-abundance ^{13}C counterparts. However, it is of course necessary not only to have available the relevant ^{13}C-enriched monosaccharide precursors, but also the nucleoside–diphosphate derivatives that are the correct donor substrates of the relevant glycosyltransferase. Since the only monosaccharide that is available in fully ^{13}C-enriched form at reasonable cost is glucose, a good deal of synthetic work might be necessary in order to obtain the relevant monosaccharides and their derivatives for the chemoenzymatic synthesis. As an example Figure 2 shows the protocol used in our own laboratory for the chemoenzymatic synthesis of sialyl-Lewis^x, Neu5Acα2-3Galβ1-4(Fucα1-3)GlcNAc, which required 38 steps [34].

33.2.5 Additional Conformational Restraints

Exchangeable Protons

Conventional ^{1}H NMR experiments on oligosaccharides are typically recorded in D$_2$O solvent, in order to avoid swamping proton resonances from the oligosaccharide with the much higher concentration of protons (~110 M) in pure water. Under these circumstances, the hydroxyl and amide protons exchange rapidly with the deuterated solvent and become invisible in proton NMR spectra. While highly efficient solvent suppression pulse sequences have been developed recently which enable oligosaccharide proton NMR spectra to be recorded in H$_2$O without substantial interference from the solvent resonance [37], it is still impossible to observe the hydroxyl protons at ambient temperature since they exchange too rapidly with the solvent. Pioneering work by Poppe and van Halbeek [38–40] showed that the hydroxyl proton exchange rate could be slowed sufficiently to enable the observation of discrete resonances in aqueous solution at temperatures lower than $-15\,°$C. This important development has enabled the measurement of NOEs to and from exchangeable protons, thus increasing dramatically the number of available conformational restraints in certain instances. One of the earliest applications of such NOEs concerns the study of the solution conformation of Neu5Acα2-6Galβ1-4Glc [41]. These authors also demonstrated that the observation of hydroxyl proton resonances can be very useful in demonstrating the existence of long-lived hydrogen bonds in solution. Such hydrogen bond restraints can be very powerful in the conformational analysis of oligosaccharides, since unlike the NOE, the presence of a hydrogen bond restrains the relevant atoms to a narrow range in the region of 0.25

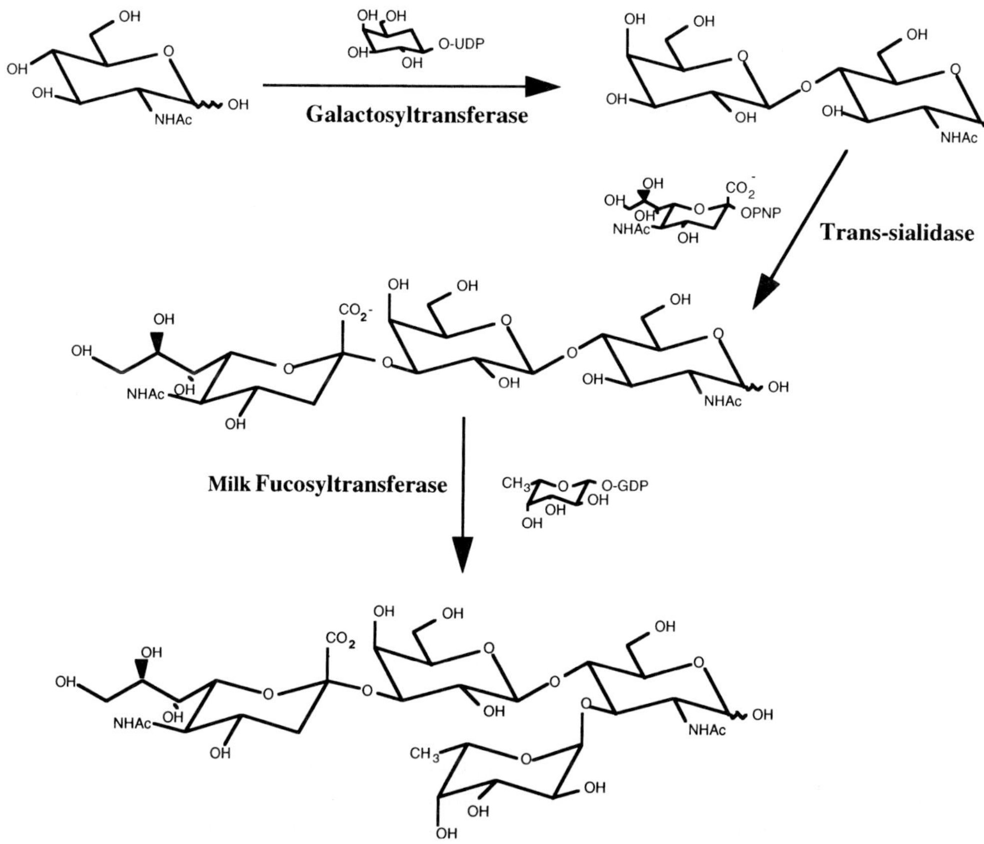

Figure 2. Scheme for the chemoenzymatic synthesis of SLe^x in ^{13}C-enriched form. The donor substrates UDP-Gal, PNP-Neu5Ac and GDP-Fuc are typically prepared synthetically from uniformly ^{13}C-enriched glucose, giving rise to full ^{13}C-enrichment in the glycan moiety of the donor. This scheme works efficiently on the milligram scale providing adequate material for NMR studies.

nm. Hydrogen-bonded hydroxyl protons present significantly smaller exchange rates in comparison with those that are not hydrogen bonded, and also possess significantly smaller temperature coefficients (i.e. the change in chemical shift with temperature). In addition, hydrogen bonded hydroxyl protons often exhibit scalar coupling constants to the adjacent non-exchangeable proton that differ substantially from the ~5.5 Hz coupling that is observed when the hydroxyl group is free to rotate. In the case of Neu5Acα2-6Galβ1-4Glc, Poppe et al. [41] observed anomalous values for these parameters for Neu5Ac OH8, Neu5Ac OH7 and Glc OH3, confirming earlier reports of a hydrogen bond between Neu5Ac OH8 and the ring oxygen or carboxyl group of the Neu5Ac residue [38], and suggesting the presence of a hydrogen bond between Glc OH3 and the ring oxygen of the Gal residue. One further approach for defining hydrogen bond connectivities is to make use of the deuterium isotope effect on ^{13}C chemical shifts. At low temperatures (248–268 K) and neutral pH values, the exchange of hydroxyl protons in water is slow enough to observe these short-range effects, which arise from the influence of the O-D group on the chemical shift of the parent carbon resonance. This effect is most pronounced for hydroxyl groups involved in long-lived hydrogen bonds, and can readily be detected as a 'splitting' of the carbon resonance into two signals in 50%H$_2$0/50%D$_2$O solution [42–44].

Isotopic enrichment is clearly not a prerequisite for the observation of hydroxyl protons in oligosaccharides. However, use of enriched material does simplify the attenuation of the strong H$_2$O resonance and moreover permits spectral editing in a third ^{13}C dimension that can be very useful in overcoming resonance overlap that is typically observed in the hydroxyl region of the ^{1}H spectrum [45].

Heteronuclear Overhauser Effects

It has long been recognized that useful structural information can be obtained from heteronuclear ^{13}C{^{1}H} Overhauser effects [46–48, 30, 49, 50], and it follows that such measurements might be of value in the conformational analysis of oligosaccharides. A principal difficulty with this approach is that NOEs from ^{1}H to ^{13}C are of significant intensity only for quaternary carbons [41]—protonated carbons do not exhibit a substantial NOE by virtue of the efficient dipolar relaxation by the attached proton and consequent leakage of the NOE. A further complication is that an indirect NOE often exists from the source proton through the attached proton to the carbon, and since this NOE is opposite in sign to the direct effect and often larger, ^{13}C{^{1}H} NOEs in carbohydrates are usually negative [51]. This effect can readily be demonstrated from full-relaxation matrix simulations on a simple disaccharide (Figure 3). A suitable means by which the ^{13}C{^{1}H} NOE intensity can be increased to more useful levels is to perdeuterate the aglycon. The indirect effect via protons is thus abolished, and permits the direct trans-glycosidic ^{13}C{^{1}H} NOE to be observed without interference.

Given the substantial potential values of ^{13}C{^{1}H} NOEs for the conformational analysis of oligosaccharides, it is worthwhile considering the appropriate experimental regime for their measurement. Naïvely, one would choose steady-state NOE experiments where the source proton is saturated by an irradiating field, since theoretically it can be shown that these NOEs reach a value more than ten times greater than the transient NOE method. However, steady-state methods involve detection of ^{13}C, whereas by comparison ^{1}H detection offers approximately eight-fold enhancement of signal-to-noise [50]. Given the difficulties with selective saturation of protons in very crowded regions of carbohydrate spectra, the transient method, which is at the heart of two-dimensional methods [46, 47], is the method of choice.

The applications of ^{13}C{^{1}H} NOEs for the conformational analysis of oligosaccharides have thus far been quite limited. However, they have been measured at natural abundance in Neu5Acα2-6Galβ1-4Glc [41] and in ^{13}C, ^{2}H enriched Galβ1-4Glc [51].

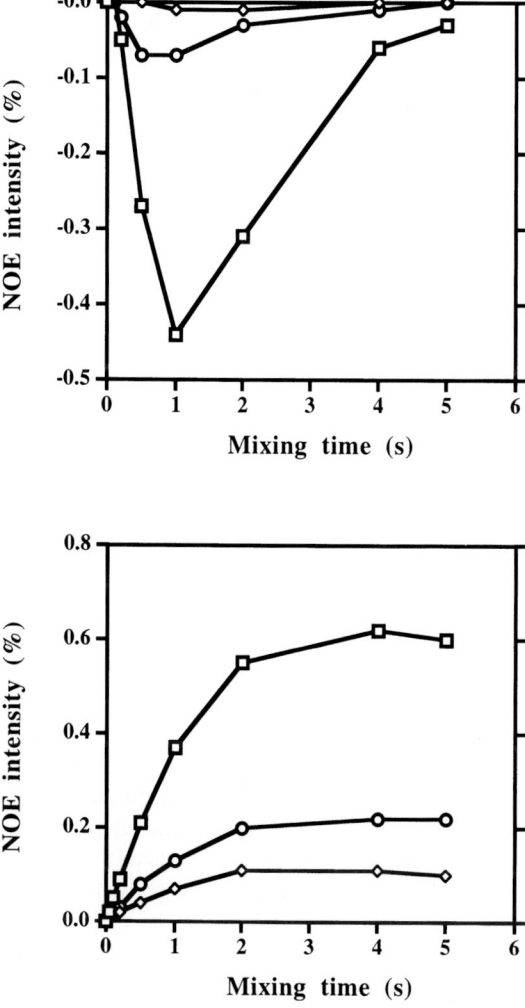

Figure 3. Theoretical full-relaxation matrix simulations of ^{1}H{^{13}C} NOEs across the glycosidic linkage in Galβ1-4[U-^{13}C]Glc (top panel) and Galβ1-4[U-^{13}C, ^{2}H]Glc (bottom panel). Traces show the NOE intensities to Gal H-1 vs mixing time in a HOESY experiment for (□) Glc C-4, (◇) Glc C-5 and (○) Glc C-6.

^{13}C–^{13}C Coupling-Constants

In recent years the value of long-range ^{13}C–^{13}C coupling constants for the conformational analysis of oligosaccharides has been realized [52–55, 56–58]. A principal difficulty with the measurement of these parameters is the requirement for ^{13}C-enriched material—unlike heteronuclear ^{13}C–^{1}H measurements where the ~1% natural abundance of ^{13}C is not a substantial sensitivity barrier, 0.01% of molecules will contain two ^{13}C nuclei. For this reason most applications of ^{13}C–^{13}C coupling constant measurements have concerned bacterial polysaccharides or small oligosaccharides where isotopic enrichment can be achieved either biosynthetically or by chemoenzymatic synthesis. The most convenient approach for the measurement of these couplings appears to be the 'LRCC' method developed by Bax and co-workers for application to proteins [59], and several Karplus parametrizations have been reported with which to convert the relevant coupling constant into angular information for conformational analysis [56, 57, 60].

Dipolar Couplings

Recently, work on the inherent magnetic alignment of proteins in very high magnetic fields (>14 T) has permitted the measurement of residual dipolar couplings between NMR active nuclei [61–63]. These couplings, which average to zero for an isotropically tumbling macromolecule, present a small but finite value due to the small net alignment in the magnetic field. Importantly, residual dipolar couplings provide long-range structural information since their magnitude depends, *inter alia*, on the inverse cube of the distance between the two nuclei and the function $(3\cos^2\theta - 1)$, where θ is the angle between the relevent bond vector and the principal axis of the alignment tensor [61]. The long-range structural information derives from the fact that θ for each bond vector is referenced to the same axis system, and hence the angles in disparate regions of the molecule are directly related. Since the original observation of magnetic alignment in proteins, more recent work has demonstrated that it is possible to obtain alignment two orders of magnitude greater by use of dilute liquid-crystalline solvents [64]. These solvents comprise dihexanoyl-phosphatidylcholine (DHPC) and dimyristoylphosphatidylcholine (DMPC) in aqueous solution, which form disc-like micelles ('bicelles') above the transition temperature (35 °C) that align in a magnetic field [65]. Macromolecules that are not spherically symmetric, when dissolved in these solutions, become partially aligned by virtue of their hydrodynamic properties (Figure 4). Below the transition temperature (27 °C) the solution becomes isotropic, and the dipolar coupling is no longer observed. Thus it is straightforward to measure dipolar couplings for each bond vector by comparison of the splittings at 35 °C, which will be the sum of the scalar and dipolar couplings, with those at 27 °C, which will exhibit only the scalar couplings. An example of the measurement of ^{13}C–^{1}H dipolar couplings from heteronuclear single quantum correlation spectra is shown in Figure 5. The dipolar couplings can be measured to reasonable accuracy in this manner, but for higher accuracy J-modulated or IPAP methods can be used [62, 66].

The measured dipolar couplings are related to θ by the following expression:

$$D_{PQ}(\theta, \phi) = -S\frac{\mu_0}{4\pi}\gamma_P\gamma_Q h \left[A_a(3\cos^2\theta - 1) + \frac{3}{2}A_r \sin^2\theta \cos 2\phi \right] \bigg/ 4\pi^2 r_{PQ}^3,$$

where S is the generalized order parameter for internal motion of the bond vector PQ, μ_0 is the magnetic permeability of vacuum, γ_P and γ_Q are the magnetogyric ratios of P and Q, h is Planck's constant, r_{PQ} is the distance between P and Q, A_a and A_r are the axial and rhombic components of the alignment tensor **A**, and θ and

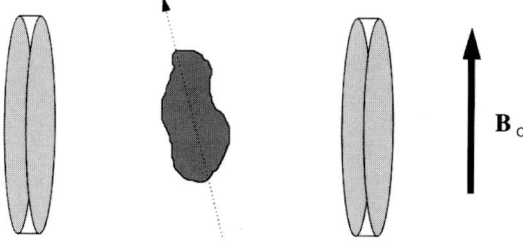

Figure 4. Diagrammatic illustration of the partial alignment of an anisotropic macromolecule in a liquid crystalline solution of phospholipid bicelles. The bicelles align spontaneously with respect to the applied magnetic field (B_0), and the macromolecule gains a small net degree of alignment by virtue of its physicochemical properties in the spaces between the bicelles.

φ are cylindrical coordinates describing the orientation of the vector P-Q in the principal axis system of **A** [64]. It is straightforward to extract the relevant angular information from this dependence, which gives rise to knowledge of the orientations of the relevant bond vectors with respect to a single fixed axis, thus providing long-range structural information that hitherto has been lacking in NMR studies.

33.3 Experimental Restraints in Conformational Analysis

33.3.1 Restraining Protocol

The principal purpose of the collection of the parameters described above is as restraints in conformational analysis. These restraints are incorporated in molecular mechanical simulations as pseudo-energy functions which apply an energy penalty if the theoretical parameter deviates from the experimental value according to some prescribed formula. However, care must be exercized in applying experimental restraints in conformational analysis. This is because in general it is impossible to exclude motional averaging about the glycosidic linkages (see Section 33.4). Under these circumstances, the relevant NMR parameter represents an average over a number of possible different configurations of the molecule, and if the restraints are applied without consideration of this possibility (i.e. assuming the molecule is 'rigid'), there is every possibility that the resulting structure will be a 'virtual conformation' that bears little resemblance to reality [67, 68]. Fortunately this problem has been recognised very early in structural studies of macromolecules, and protocols have been devised to deal with it.

Biharmonic Restraints

In the case of distance restraints derived from NOE measurements or from hydrogen-bond connectivities, it is usual to apply a restraining function in the form of a

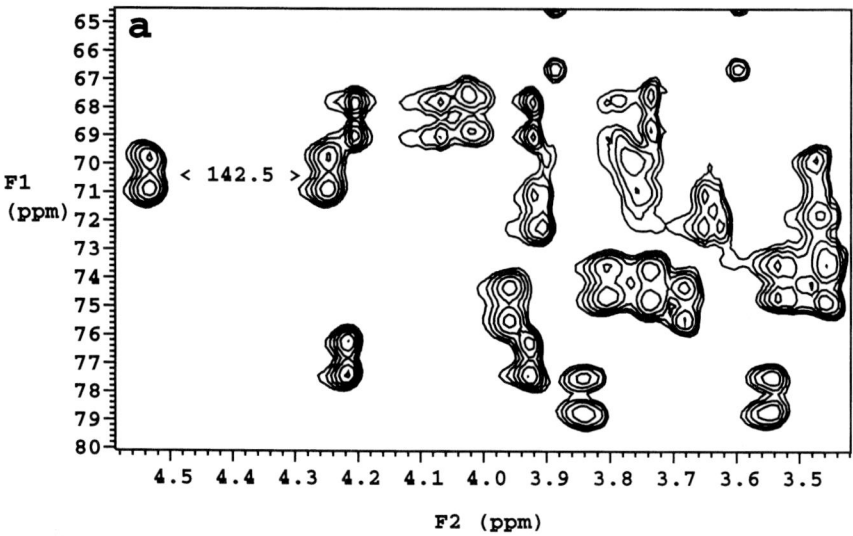

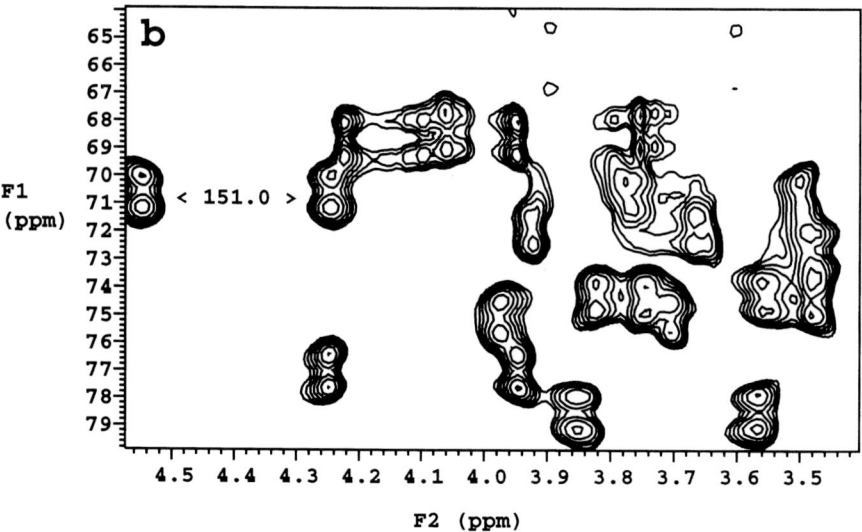

Figure 5. Proton coupled ^{1}H–^{13}C HSQC spectra of uniformly ^{13}C-enriched Galα1-4Galβ1-4Glc in a 7.5% solution of DHPC:DMPC (1:2.9) at (**a**) 27 °C (isotropic phase) and (**b**) 35 °C (liquid crystalline phase). The difference in $^1J_{C,H}$ for Galα H-5 is shown in each case.

biharmonic potential [69]. That is, there is no energy penalty if the theoretical distance lies within prescribed distance bounds, whereas an increasing penalty function is applied if the distance is outwith these bounds. Typically, the NOE restraints are semi-quantitatively categorized into 'weak', 'medium' and 'strong', with corresponding distance bounds of $1.8\,\text{Å} < r < 2.7\,\text{Å}$, $1.8\,\text{Å} < r < 3.3\,\text{Å}$ and $1.8\,\text{Å} < r < 5\,\text{Å}$ respectively. In this manner the molecule is free to adopt any conformation without additional energy penalty provided all theoretical distances are within the relevent bounds. The relatively loose bounds are supposed to take account of the fact that the measured NOE may correspond to a motionally averaged distance.

Incorporation of angular restraints is more difficult. Unlike the NOE, where the measurement of this parameter demonstrates at least that the relevant atoms must be close in space for some of the time, a given spin-coupling constant may correspond with a single, fixed angle in one extreme, or may represent an average over 360° rotation at the other extreme. Although methods have been proposed to overcome this limitation, in general it is undesirable to apply angular terms as primary restraints in conformational analysis.

Time-Dependent Restraints

A principal limitation of conventional biharmonic restraints is that their presence can mask transitions between conformational states that are widely separated on the potential surface. Thus, if during the course of a transition from one low-energy state to another, a distance restraint is violated (the distance between two restrained atoms exceeds 5 Å, for example), then a substantial energy barrier to this transition will be created. In order to overcome this problem, restraints can be imposed in a so-called 'time-dependent manner' [70]. The basic concept is that the theoretical *average* distance over a defined part of the simulation, rather than the theoretical *instantaneous* distance, is required to satisfy certain bounds during the simulation. By use of this protocol, violations of conventional distance bounds are permitted, as long as the running average satisfies these bounds. Thus transitions between different low-energy minima are more likely to occur in simulations using the time-averaging protocol. Remarkably, to our knowledge very few studies on glycoconjugates have utilized time-averaged restraints [71, 72].

33.3.2 Dynamical Simulated Annealing

In principle, conformational restraints can be incorporated into conventional energy minimization algorithms in order to obtain a single, fixed, 'minimum energy' conformation that is consistent with available experimental data. However, this approach is not very efficient, and moreover by virtue of the local minimum problem, it is very likely that the resulting structure will not be the global minimum energy configuration. A much more efficient approach is termed dynamical simulated annealing [69]. This can be thought of as a hybrid between a molecular dynamics simulation and energy minimization. The idea is to simulate the dynamics of the molecule at a relatively high temperature (typically 750 K) over a period of time in order efficiently to sample a substantial region of conformational space. The simulation is continued while the system is then slowly cooled to a very low temperature. Unlike conventional minimization, the system has sufficient thermal energy so there is a tendency for the molecule to escape local minima (provided the cooling is slow enough) and to 'fall' to a lower energy minimum. Once the system has cooled to around 5 K, the simulation is terminated with a conventional energy minimization step. Throughout the simulation the experimental restraints are applied, such that the molecule is constantly driven to a minimum energy conformation that is hopefully consistent with all experimental data. In order that the restraints do not limit the accessible regions of conformational space in the early stages of the simulation at high temperature, their strength is typically scaled for a period of time until the temperature falls to around 300 K. Moreover, a series of simulations is usually performed with a randomized starting structure as input, in order further to explore conformational space and to ensure that the final structure is not dependent on the starting geometry. This typically results in one or more families of structures that are essentially consistent with the applied distance restraints.

33.4 Analysis of Oligosaccharide Dynamics

Oligosaccharides do not exist as 'rigid' entities in solution. While each monosaccharide residue can be thought as being fixed in a certain conformation (at least for pyranoses) on the NMR timescale, there may be substantial torsional oscillations about the glycosidic linkages. In extended structures, these oscillations will be additive, and the manifold of conformations is solution can span considerable regions of conformational space. Characterization of the solution dynamics is very important from the point of view of binding affinity to receptors. To appreciate why this is the case it is convenient to consider a simple ligand–protein association involving a flexible ligand (A) in comparison with a rigid ligand (B). The configurational entropy (i.e. number of available degrees of freedom) of the former is greater than the latter, and hence the Gibbs free energy of the former will be lower by virtue of the relation $G = H - TS$ (where H is the enthalpy and T is the absolute temperature). Thus, the combined free energy of the uncomplexed components will be lower in the case of the flexible ligand and, since the free energy of binding ΔG_b is the difference between the free energy of the complexed (which will be the same irrespective of the dynamics of the ligand, assuming it is rigidly held in the binding site) and the uncomplexed components, the free energy of binding of the flexible ligand will be lower than that of the rigid ligand. This in turn will give rise to a lower affinity by virtue of the relation $\Delta G_b^0 = -RT \ln K_a$, where ΔG_b^0 is the standard free energy of binding, R is the gas constant and K_a is the association constant. Thus in terms of inhibitor design, it is important to consider the dynamics of the ligand, i.e. a rigid ligand that adopts the bound-state conformation in solution is likely to have optimal affinity.

The characterization of oligosaccharide dynamics by NMR is not trivial. All of the conformation-sensitive NMR parameters are averaged by the motion, and it is not possible directly to deconvolute the nature or number of conformational states that contribute to this average. NMR relaxation parameters are directly sensitive to motion if on the appropriate timescale [73, 74], but again cannot be interpreted in the absence of a model for the motion [75]. Thus we are forced to rely on methods that simulate the regions conformational space that an oligosaccharide can access in solution, and two methods are in general use.

33.4.1 Monte-Carlo Simulations

In this method [76–78] the starting point is an arbitrary conformation of the system which is perturbed slightly by, for example, a torsional rotation. The internal energy of the system is calculated and compared with the energy before the perturbation. If the energy has decreased then the new conformation is accepted. If the energy has increased, the increase is compared with the thermal energy kT is order to decide whether the new conformation is accepted. The value of $\exp(-\Delta U/kT)$ is compared with a random number between 0 and 1, where ΔU is the increase in energy. If $\exp(-\Delta U/kT)$ is larger than the random number, the new conformation is accepted. Otherwise a new conformation is generated, and the process is repeated. In this manner the system under investigation explores conformational space. Since perturbations that raise the energy are sometimes allowed, the procedure permits the escape of the molecule from local minima.

33.4.2 Molecular Dynamics Simulations

The molecular dynamics method directly simulates the motions of all the atoms in the system, by using Newton's laws of motion [79]. Since the intenal energy of a given conformation of the system can be derived from a conventional molecular mechanics approach, the force on each atom can be determined from the derivative of the energy along different directions of space. The motion of the atom can in turn be determined from the acceleration which is a known quantity given the force on each atom and its mass. The system is usually started in an arbitrary conformation near 0 K, and the velocities of the atoms are then increased so that the average kinetic energy corresponds with a desired temperature. One of the advantages of the molecular dynamics method is that the temperature can be raised or lowered at will by coupling to a temperature bath, which thus permits energy minimization by simulated annealing (Section 33.3.2). Moreover, since the time-scale of the internal motions is known from the simulation, it is possible to back-calculate time-averaged theoretical NMR parameters for comparison with experiment (Section 33.5.4).

33.5 A Case Study on Neu5Acα2-3Galβ1-4Glc

As an example of the use of some of the techniques described above in the conformational analysis of oligosaccharides, a case study of the solution properties of the trisaccharide Neu5Acα2-3Galβ1-4Glc will be described.

33.5.1 Resonance Assignments in Neu5Acα2-3Galβ1-4Glc

The first step in the conformational analysis of the oligosaccharide is the determination of complete proton and carbon resonance assignments. These have been reported previously for this particular trisaccharide [80]. In systems with unknown assignments, the presence of uniform ^{13}C-enrichment permits the application of conventional HCCH-COSY and HCCH-TOCSY experiments [81, 82], which invariably give complete resonance assignments in an efficient manner.

33.5.2 ROE Connectivities

Although Neu5Acα2-3Galβ1-4Glc contains only 29 non-exchangeable protons, the proton NMR spectrum is remarkably complex, with most resonances concentrated within 0.4 ppm. Despite the fact that complete ^{1}H and ^{13}C resonance assignments are available, this overlap renders impossible analysis of the conventional ^{1}H–^{1}H NOESY spectrum. Isotopic ^{13}C-enrichment permits the acquisition of three-dimensional ^{13}C-edited spectra at high sensitivity. Since the rotational tumbling time of the trisaccharide is close to the point where the homonuclear ^{1}H–^{1}H NOE is zero, it is necessary to acquire three-dimensional ^{13}C edited ROESY-HSQC spectra. It is convenient to examine such spectra as a series of two-dimensional planes. In NMR of proteins, it is usual to examine the F1/F3 (^{1}H/^{1}H) plane, but due to severe overlap of proton resonances, it is more useful to examine F2/F3 (^{13}C/^{1}H) planes in oligosaccharide spectra. A typical F2/F3 plane from the ROESY-HSQC spectrum of Neu5Acα2-3Galβ1-4Glc is shown in Figure 6 [56]. This illustrates all the ROE connectivities derived from Gal H-1. Many of these correspond with intra-residue ROEs that do not contain conformational information across the glycosidic linkages. The remainder (crosspeaks plotted in boldface) represent ROE connectivities across the Galβ1-4Glc glycosidic linkage. The ROE connectivities to Glc H-4 and Glc H-6 are anticipated on the basis of the known conformational preferences about the glycosidic linkage. However, substantial ROEs are also observed to Glc H-3 and H-5. These ROEs are extremely difficult to observe in conventional homonuclear ^{1}H NMR spectra due to extensive overlap, and given their substantial intensity, they cannot arise from a single conformation about the Galβ1-4Glc linkage that simultaneously gives rise to ROEs to Glc H-4 and H-6. Thus these measurements alone indicate that a degree of flexibility exists about the Galβ1-4Glc glycosidic linkage. In particular, the ROEs to Glc H-3 and H-5 demonstrate the existence of the 'anti' conformation about the glycosidic linkage, i.e. where the glycosidic torsion angle ψ adopts a value of ~180° [83–86]. By examination of other F2/F3 planes in the ROESY-HSQC spectrum, a total of seven transglycosidic ROE connectivities is observed for this glycan. These can be used as conformational restraints in the determination of the 'global minimum' energy configuration of the glycan.

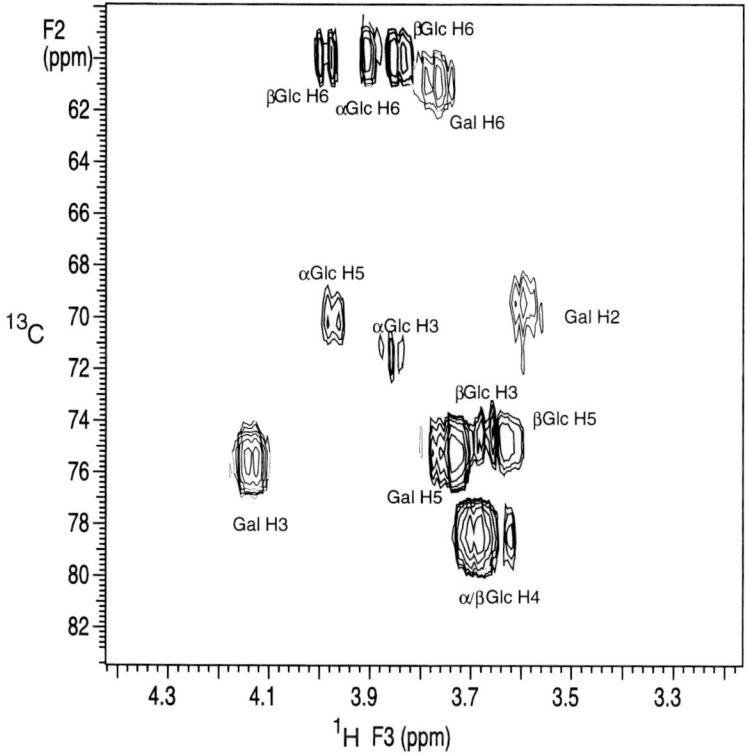

Figure 6. Two-dimensional F2/F3 (^{13}C/^{1}H) plane derived from three-dimensional ROESY-HSQC spectrum of Neu5Acα2-3Galβ1-4Glc. This plane shows all intra- (normal face) and inter-residue (boldface) ROE connectivities to Gal H-1.

33.5.3 'Global Minimum' Conformation of Neu5Acα2-3Galβ1-4Glc

The 'global minimum' energy conformation of the glycan can be determined by dynamical simulated annealing calculations with the ROE connectivities described above as time-averaged conformational restraints. In general it is desirable to compute a series of such calculations with different (pseudo-random) conformations of the glycan as input. In principle, it is possible to begin with random coordinates, but this introduces a substantial complication since it is necessary to ensure that the correct chirality is preserved, and moreover the correct ring geometry must be defined. In practice, the chirality and ring geometry is known, and hence it is more convenient to generate a series of random structures with defined chirality and ring geometry but with pseudo-randomized torsion angles about each glycosidic linkage [87]. A set of such pseudo-random structures can be generated by use of a dynamical quenching procedure.

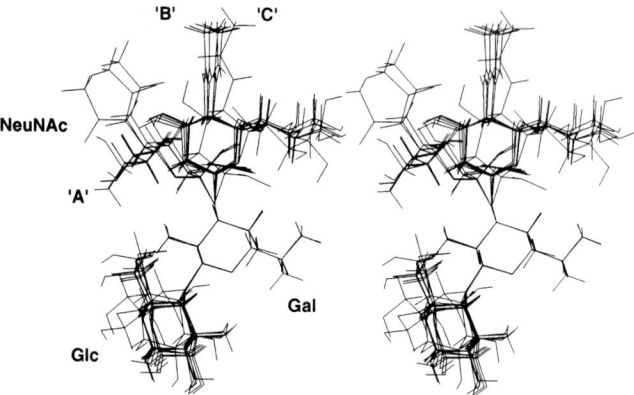

Figure 7. Stereo view of the family of structures derived from a dynamical simulated annealing calculation on Neu5Acα2-3Galβ1-4Glc. The three families are labelled 'A', 'B' and 'C' respectively.

The result of the dynamical simulated annealing calculations is a set of structures most of which satisfy the applied conformational restraints. Those that do not are discarded, and the remaining structures typically form 'families' with similar conformations (Figure 7). These data therefore indicate that different conformations satisfy the experimental ROE restraints simultaneously, and thus provide a crude picture of motional dynamics. A more detailed analysis of dynamics can be obtained from molecular dynamics simulations.

33.5.4 Conformational Dynamics of Neu5Acα2-3Galβ1-4Glc

In order to probe the conformational dynamics of the glycan, one of the low-energy structures derived from the simulated annealing procedure is chosen as input to a restrained molecular dynamics simulation. The conformer with the lowest overall energy can arbitrarily be chosen for this purpose. In principle, it would be desirable to perform free dynamics simulations with explicit inclusion of solvent water in order to obtain an accurate picture of the solution dynamics of the glycan. However, the accuracy of current forcefield parametrizations for oligosaccharides is insufficient over the complete potential surface to generate an accurate model of solution dynamics. Moreover, the inclusion of solvent water molecules severely restricts the length of time over which the molecular dynamics simulation can simulated due to the very large number of computations that are required per step. Thus it is practical only to simulate the effects of solvent water in part by use of an appropriate dielectric constant. In adopting this approach, it must be remembered that a restrained molecular dynamics simulation will be performed which will include data from solution NMR studies, including hydrogen-bonding restraints where appropriate. It is therefore not necessary for the simulation conditions to reproduce accurately these solution properties. It can be argued that the application

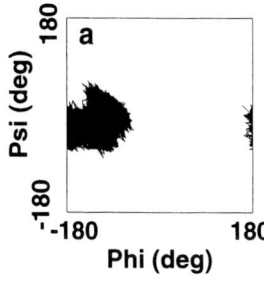 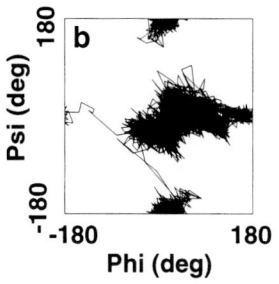

Figure 8. Instantaneous values of φ vs ψ for (**a**) Neu5Acα2-3Gal and (**b**) Galβ1-4Glc glycosidic linkages derived from a 5 ns in vacuo MD simulation of Neu5Acα2-3Galβ1-4Glc.

of experimental restraints will of course modify the available conformational space explored by the glycan during the simulation. However, in the absence of well-parametrized forcefields for oligosaccharides there is little option but to modify the behavior of the forcefield in this manner, and the use of time-dependent restraints minimizes the influence of such restraints in e.g. preventing excursions to other local minima via a substantial energy barrier.

The result of a 5 ns restrained MD simulation for Neu5Acα2-3Galβ1-4Glc is shown in Figure 8. It can be seen that each glycosidic linkage exhibits considerable conformational freedom, and in particular the Galβ1-4Glc glycosidic linkage is seen to adopt the 'anti' conformation for a considerable period of time. The validity of this simulation can be further assessed by back-calculating the theoretical ROEs from the MD simulation for comparison with experimental values, using the appropriate formalism for the computation of the time-averaged ROE. The back-calculation of theoretical parameters that are used as restraints in the simulation is not a circular argument, since such restraints are typically applied as a biharmonic function with lower and upper bounds that differ substantially, such that the theoretical ROE can in principle vary over a very wide range. Moreover, other conformation-sensitive parameters described above such as long range $^{13}C-^{1}H$ and $^{13}C-^{13}C$ coupling constants can be back-calculated from the simulation for comparion with experimental values that are not used as restraints in the simulation. As seen in Table 1, the theoretical results for Neu5Acα2-3Galβ1-4Glc agree well with experimental values, suggesting that the dynamical properties defined by the simulations are a good approximation to the true solution dynamical behavior.

33.5.5 Short-range vs Long-range Restraints

Despite the good agreement between theoretical and experimental parameters in Table 1, a valid criticism of these studies (or indeed all NMR studies on macro-molecular conformation until recently) is that all of the conformational parameters are short-range in nature. Thus by virtue of the inverse sixth power distance dependence of the NOE(ROE), distance restraints are typically limited to 0.5 nm or

Table 1. Experimental ROE intensities and long-range coupling constants for ^{13}C Neu5Acα2-3Galβ1-4Glc vs theoretical values computed from a 5 ns MD simulation.

ROE Connectivity	ROE Intensity (%)[1]		Coupling (Hz)	
	Expt.	Theor.[2]	Expt.[3]	Theor.
ROE				
Neu5Ac H-3ax–Gal H-3	1.0	1.5		
Neu5Ac H-8–Gal H-3	0.4	0.6		
Gal H-1–Glc H-4	4.2	3.9		
Gal H-1–Glc H-6	0.4	0.6		
Gal H-1–Glc H-6′	0.5	0.7		
Gal H-1–Glc H-3	0.4	0.4		
Gal H-1–Glc H-5	0.5	0.6		
$^3J_{CC}$				
Gal C-3–Neu5Ac C-3			1.9	1.6
Gal C-2–Neu5Ac C-2			1.8	1.8
Gal C-4–Neu5Ac C-2			<1	1.1
Gal C-1–Glcβ C-5			1.9	1.6
Gal C-2–Glcα/β C-4			2.6	2.6
Gal C-1–Glcα/β C-3			<1	1.1
$^3J_{HC}$				
Gal H-1–Glc C-4			3.5	3.7
Gal C-1–Glc H-4			5.1	4.9
Neu5Ac C-2–Gal H-3			4.7	4.5

[1] Experimental and theoretical intensities shown are for the Glcβ anomer only, after correction for the mole fraction of this anomer.
[2] Calculated with a rotational correlation time of 0.13 ns.
[3] Error in these measurements estimated as ±0.5 Hz.

less, and the long-range coupling constants provide conformational information on two bond vectors separated by a single covalent bond. It could be argued that such restraints satisfy the local order but do not define long-range order in the molecule. Thus, for extended structures of which polysaccharides are a prime example, the overall geometry of the molecule could be in error. This difficulty has recently been addressed by measurement of residual dipolar couplings as described above (Section 33.2.5). Typical such couplings, measured for Neu5Acα2-3Galβ1-4Glc using the J-modulated HSQC method [88], are given in table 2. These couplings can be applied as restraints in dynamical simulated annealing calculations, in order to obtain a family or families of structures which are consistent with the long-range order imposed by these restraints. The result for Neu5Acα2-3Galβ1-4Glc is two families of structures that differ in the conformation about the Galβ1-4Glc linkage (Figure 9). Importantly, both of these structures have glycosidic torsion angles that map into low-energy regions of conformational space predicted from MD simulations involving 'local' restraints, therefore further suggesting that the latter provide an

Table 2. Residual ^{1}H–^{13}C Dipolar couplings for Neu5Acα2-3Galβ1-4Glc in a 7.5% (w/v) solution of DHPC:DMPC (1:2.9 w/w) in D$_2$O, pD 7.2, containing 100 mM KCl.

Bond Vector	Residual dipolar coupling (Hz)[1]
Glcβ H-1–C-1	+7.4
Glcβ H-2–C-2	+9.5
Glcβ H-3–C-3	+9.6
Glcβ H-4–C-4	+7.7
Galβ H-1–C-1	+11.5
Galβ H-2–C-2	+11.9
Galβ H-3–C-3	+9.2
Galβ H-4–C-4	−1.4
Galβ H-5–C-5	+9.5
Neu5Acα H-4–C-4	+3.2
Neu5Acα H-5–C-5	+4.8
Neu5Acα H-7–C-7	−13.4
Neu5Acα H-8–C-8	−11.1

[1] Values obtained by non-linear least-squares fitting of experimental intensities from J-modulated HSQC experiments. Estimated average error in the measurements is ±0.5 Hz.

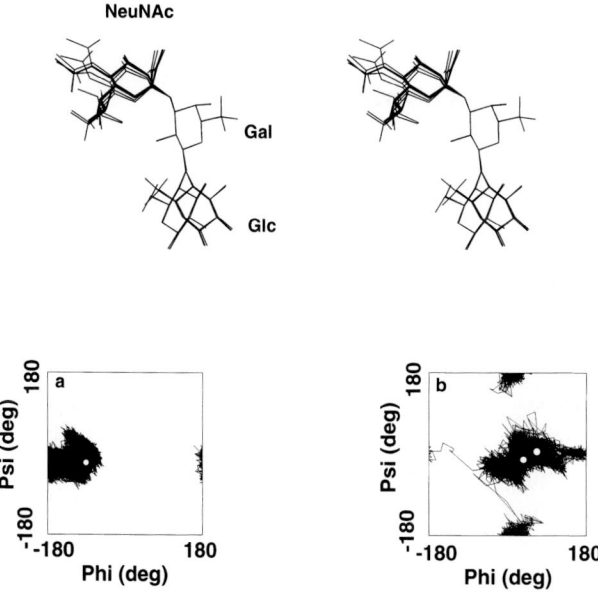

Figure 9. (top) Family of structures derived from dynamical simulated annealing calculation on Neu5Acα2-3Galβ1-4Glc with residual dipolar coupling restraints given in Table 2. (bottom) Resulting values of ϕ and ψ (open circles) for the Neu5Acα2-3Gal and Galβ1-4Glc glycosidic linkages indicated on the plots of Figure 8.

adequate representation of the dynamic behavior of the glycan in solution. Clearly, further work is required to calculate time-averaged residual dipoar couplings from MD simulations for comparison with experimental values.

33.6 Conclusions

Since early studies on oligosaccharides which treated these moieties as essentially 'rigid' bodies [89–93], a wealth of more recent data has indicated that oligosaccharides enjoy considerable motional freedom about the glycosidic linkages in solution. These conclusions are derived from improved computational procedures and better molecular mechanical forcefields, but the availability of additional experimental conformational restraints has also had a particularly significant impact. In particular the possibilities for the synthesis of oligosaccharides in ^{13}C-enriched form using chemoenzymatic methods simplifies considerably not only the task of defining the solution behavior of the free glycan, but also the conformation while bound to a protein receptor. A description of techniques for the study of the latter can be found elsewhere in this volume.

References

1. A. Varki, *Glycobiology*, **1993**, *3*, 97–130.
2. E. Ioffe, P. Stanley, *Proc. Natl Acad. Sci.* U.S.A., **1994**, *91*, 728–732.
3. P. Stanley, E. Ioffe, *Faseb Journal*, **1995**, *9*, 1436–1444.
4. E. Ioffe, Y. Liu, P. Stanley, *Proc. Natl Acad. Sci.* U.S.A., **1996**, *93*, 11041–11046.
5. P. Stanley, T. S. Raju, M. Bhaumik, *Glycobiology*, **1996**, *6*, 695–699.
6. D. Chui, M. OhEda, Y. F. Liao, K. Panneerselvam, A. Lal, K. W. Marek, H. H. Freeze, K. W. Moremen, M. N. Fukuda, J.D. Marth, *Cell*, **1997**, *90*, 157–167.
7. J. J. Priatel, M. Sarkar, H. Schachter, J. D. Marth, *Glycobiology*, **1997**, *7*, 45–56.
8. M. Bhaumik, T. Harris, S. Sundaram, L. Johnson, J. Guttenplan, C. Rogler, P. Stanley, *Cancer Res.*, **1998**, *58*, 2881–2887.
9. T. S. Raju, P. Stanley, *J. Biol. Chem.*, **1998**, *273*, 14090–14098.
10. M. A. J. Ferguson, *Phil. Trans. R. Soc. Lond. Ser. B*, **1997**, *352*, 1295–1302.
11. K. Aktories, *Curr. Topics Microbiol. Immunol.*, **1992**, *175*, 1–143.
12. J. F. G. Vliegenthart, L. Dorland, H. van Halbeek, *Adv. Carbohydr. Chem. Biochem.*, **1983**, *41*, 209–374.
13. S. W. Homans, *Progr. NMR Spectr.*, **1990**, *22*, 55–81.
14. H. van Halbeek, *Meths. Enzymol.*, **1994**, *230*, 132–168.
15. H. van Halbeek, *Curr. Opinion Str. Biol.*, **1994**, *4*, 697–709.
16. J. F. G. Vliegenthart, L. Dorland, H. van Halbeek, *Adv. Carbohydr. Chem. Biochem.*, **1983**, *41*, 209–374.
17. P. E. Jansson, L. Kenne, G. Widmalm, *Carbohydr. Res.*, **1987**, *168*, 67–77.
18. G. M. Lipkind, A. S. Shashkov, Y. A. Knirel, E. V. Vinogradov, N. K. Kochetkov, *Carbohydr. Res.*, **1988**, *175*, 59–75.
19. A. S. Shashkov, G. M. Lipkind, Y. A. Knirel, N. K. Kochetkov, *Mag. Res. Chem.*, **1988**, *26*, 735–747.
20. P. E. Jansson, L. Kenne, G. Widmalm, *Pure Appl. Chem.*, **1989**, *61*, 1181–1192.

21. P. E. Jansson, L. Kenne, G. Widmalm, *Carbohydr. Res.*, **1989**, *193*, 322–325.
22. P. E. Jansson, L. Kenne, G. Widmalm, *Carbohydr. Res.*, **1989**, *188*, 169–191.
23. G. M. Lipkind, A. S. Shashkov, N. K. Kochetkov, *Carbohydr. Res.*, **1990**, *198*, 399–402.
24. H. Baumann, P. E. Jansson, L. Kenne, G. Widmalm, *Carbohydr. Res.*, **1991**, *211*, 183–190.
25. P. E. Jansson, L. Kenne, G. Widmalm, *Anal. Bichem.*, **1991**, *199*, 11–17.
26. P. E. Jansson, L. Kenne, G. Widmalm, *Acta Chem. Scandin.*, **1991**, *45*, 517–522.
27. N. K. Kochetkov, E. V. Vinogradov, Y. A. Knirel, A. S. Shashkov, G. M. Lipkind, *Bioorg. Khim.*, **1992**, *18*, 116–125.
28. N. E. Nifantev, A. S. Shashkov, G. M. Lipkind, N. K. Kochetkov, B. Jann, K. Jann, *Glycoconj. J.*, **1993**, *10*, 335–335.
29. M. Karplus, *J. Am. Chem. Soc.*, **1963**, *85*, 2870–2871.
30. D. Neuhaus, M. P. Williamson, *The Nuclear Overhauser Effect in Structural and Conformational Analysis*, Verlag Chemie, New York 1989.
31. T. J. Rutherford, J. Partridge, C. T. Weller, S. W. Homans, *Biochemistry*, **1993**, *32*, 12715–12724.
32. C. T. Weller, J. Lustbader, K. Seshadri, J. M. Brown, C. A. Chadwick, C. E. Kolthoff, S. Ramnarain, S. Pollak, R. Canfield, S. W. Homans, *Biochemistry*, **1996**, *35*, 8815–8823.
33. Y. Ichikawa, Y.-C. Lin, D. P. Dumas, G.-J. Shen, E. Garcia-Junceda, M. A. Williams, R. Bayer, C. Ketcham, L. E. Walker, J. C. Paulson, C.-H. Wong, *J. Am. Chem. Soc.*, **1992**, *114*, 9283–9298.
34. M. A. Probert, M. J. Milton, R. Harris, S. Schenkman, J. M. Brown, S. W. Homans, R. A. Field, *Tetrahed. Lett.*, **1997**, *38*, 5861–5864.
35. A. K. Misra, J. M. Brown, S. W. Homans, R. A. Field, *Carbohydr. Lett.*, **1998**, *3*, 217–222.
36. H. Shimizu, J. M. Brown, S. W. Homans, R. A. Field, *Tetrahedron*, **1998**, *54*, 9489–9506.
37. T. L. Hwang, A. J. Shaka, *J. Magn. Reson.*, **1995**, *A 112*, 275–279.
38. L. Poppe, H. van Halbeek, *J. Am. Chem. Soc.*, **1991**, *113*, 363–365.
39. L. Poppe, H. van Halbeek, *J. Magn. Reson.*, **1992**, *96*, 185–190.
40. L. Poppe, H. van Halbeek, *Nature Struct. Biol.*, **1994**, *1*, 215–216.
41. L. Poppe, R. Stuikeprill, B. Meyer, H. van Halbeek, *J. Biomol. NMR*, **1992**, *2*, 109–136.
42. J. Reuben, *J. Am. Chem. Soc.*, **1984**, *106*, 6180–6186.
43. J. C. Christofides, D. B. Davies, J. A. Martin, E. B. Rathbone, *J. Am. Chem. Soc.*, **1986**, *108*, 5738–5743.
44. E. Alvarado, T. Nukuda, T. Ogawa, C. E. Ballou, *Biochemistry*, **1991**, *30*, 881–886.
45. R. Harris, T. J. Rutherford, M. J. Milton, S. W. Homans, *J. Biomol. NMR*, **1997**, *9*, 47–54.
46. P. L. Rinaldi, *J. Am. Chem. Soc.*, **1983**, *105*, 5167–5168.
47. C. Yu, G. C. Levy, *J. Am. Chem. Soc.*, **1984**, *106*, 6533–6537.
48. K. E. Kover, G. Batta, *J. Magn. Reson.*, **1988**, *79*, 206–210.
49. D. Neuhaus, C. P. M. Vanmierlo, *J. Magn. Reson.*, **1992**, *100*, 221–228.
50. K. Stott, J. Keeler, *Mag. Res. Chem.*, **1996**, *34*, 554–558.
51. G. R. Kiddle, R. Harris, S. W. Homans, *J. Biomol. NMR*, **1998**, *11*, 289–294.
52. J. M. Duker, A. S. Serianni, *Carbohydr. Res.*, **1993**, *249*, 281–303.
53. T. Church, I. Carmichael, A. S. Serianni, *Carbohydr. Res.*, **1996**, *280*, 177–186.
54. A. S. Serianni, *Abstr. Am. Chem. Soc.*, **1996**, *211*, 10-CARB.
55. T. J. Church, I. Carmichael, A. S. Serianni, *J. Am. Chem. Soc.*, **1997**, *119*, 8946–8964.
56. M. J. Milton, R. Harris, M. Probert, R. A. Field, S. W. Homans, *Glycobiology*, **1998**, *8*, 147–153.
57. Q. W. Xu, C. A. Bush, *Carbohydr. Res.*, **1998**, *306*, 335–339.
58. S. K. Zhao, G. Bondo, J. Zajicek, A. S. Serianni, *Carbohydr. Res.*, **1998**, *309*, 145–152.
59. A. Bax, D. Max, D. Zax, *J. Am. Chem. Soc.*, **1992**, *114*, 6923–6925.
60. B. Bose, S. Zhao, R. Sterutz, F. Clovar, P. B. Bordo, G. Bordo, B. Hertz, I. Carmichael, A. S. Seriarri *J. Am. Chem. Soc.*, **1998**, *120*, 11158–11173.
61. J. R. Tolman, J. M. Flanagan, M. A. Kennedy, J. H. Prestegard, *Proc. Natl. Acad. Sci.*, **1995**, *92*, 9279–9283.
62. N. Tjandra, S. Grzesiek, A. Bax, *J. Am. Chem. Soc.*, **1996**, *118*, 6264–6272.
63. N. Tjandra, A. Bax, *J. Magn. Reson.*, **1997**, *124*, 512–515.
64. N. Tjandra, A. Bax, *Science*, **1997**, *278*, 1111–1114.

65. C. R. Sanders II, J. P. Schwonek, *Biochemistry*, **1992**, *31*, 8898–8905.
66. M. Ottiger, F. Delaglio, A. Bax, *J. Magn. Reson.*, **1998**, *131*, 373–378.
67. O. Jardetzky, *Bioch. Biophys. Acta*, **1980**, *621*, 227–232.
68. D. A. Cumming, J. P. Carver, *Biochemistry*, **1987**, *26*, 6664–6676.
69. M. Nilges, A. M. Gronenborn, A. Brünger, G. M. Clore, *Protein Eng*, **1988**, *2*, 27–38.
70. A. E. Torda, R. M. Scheek, W. F. van Gunsteren, *J. Mol. Biol.*, **1990**, *214*, 223–235.
71. D. G. Low, M. A. Probert, G. Embleton, K. Seshadri, R. A. Field, S. W. Homans, J. Windust, P. J. Davis, *Glycobiology*, **1997**, *7*, 373–381.
72. R. Harris, G. R. Kiddle, R. A. Field, B. Ernst, S. W. Homans, *J. Am. Chem. Soc.*, **1999**, *121*, 2546–2551.
73. M. Vignon, F. Michon, J. P. Joseleau, K. Bock, *Macromolecules*, **1983**, *16*, 835–838.
74. Q. W. Xu, C. A. Bush, *Biochemistry*, **1996**, *35*, 14512–14520.
75. G. Lipari, A. Szabo, *J. Am. Chem. Soc.*, **1982**, *104*, 4546–4559.
76. N. Metropolis, A. W. Rosenbluth, M. N. Rosenbluth, A. H. Teller, E. Teller, *J. Chem. Phys.*, **1953**, *21*, 1087–1092.
77. R. Stuike-Prill, B. Meyer, *Eur. J. Biochem.*, **1990**, *194*, 903–919.
78. T. Peters, T. Weimar, *J. Biomol. NMR*, **1994**, *4*, 97–116.
79. H. J. C. Berendsen, J. P. M. Postma, N. F. van Gunsteren, A. DiNola, J. R. Haak, *J. Chem. Phys.*, **1984**, *81*, 3684–3690.
80. L. Lerner, A. Bax, *Carbohydr. Res.*, **1987**, *166*, 35–46.
81. A. Bax, G. M. Clore, P. C. Driscoll, A. M. Gronenborn, M. Ikura, L. E. Kay, *J. Magn. Reson.*, **1990**, *87*, 620–627.
82. L. Yu, R. Goldman, P. Sullivan, G. F. Walker, S. W. Fesik, *J. Biomol. NMR*, **1993**, *3*, 429–441.
83. G. M. Lipkind, A. S. Shashkov, N. K. Kochetkov, *Carbohydr. Res.*, **1985**, *141*, 191–197.
84. L. Poppe, C.-W. von der Lieth, J. Dabrowski, *J. Am. Chem. Soc.*, **1990**, *112*, 7762–7771.
85. K. Bock, J. O. Duus, S. Refn, *Carbohydr. Res.*, **1994**, *253*, 51–67.
86. J. Dabrowski, T. Kozar, H. Grosskurth, N. E. Nifantev, *J. Am. Chem. Soc.*, **1995**, *117*, 5534–5539.
87. S. W. Homans, M. Forster, *Glycobiology*, **1992**, *2*, 143–151.
88. G. R. Kiddle, S. W. Homans, *FEBS Lett.*, **1998**, *436*, 128–130.
89. K. Bock, J. Arnarp, J. Lonngren, *Eur. J. Biochem.*, **1982**, *129*, 171–178.
90. S. W. Homans, R. A. Dwek, D. L. Fernandes, T. W. Rademacher, *FEBS Lett.*, **1982**, *150*, 503–506.
91. J. R. Brisson, J. P. Carver, *Biochemistry*, **1983**, *22*, 3680–3686.
92. J. R. Brisson, J. P. Carver, *Biochemistry*, **1983**, *22*, 3671–3680.
93. S. W. Homans, R. A. Dwek, D. L. Fernandes, T. W. Rademacher, *FEBS Lett.*, **1983**, *164*, 231–235.

34 Oligosaccharide Conformations by Diffraction Methods

Serge Pérez, Catherine Gautier, and Anne Imberty

34.1 Introduction

Diffraction by single crystals is by far the most powerful experimental method for the characterization of the atomic arrangements in molecules. X-ray and neutron, are used since their wavelengths are of the same order of magnitude as the interatomic distances, typically around an Angstrom unit. The electron and nuclei are the scatterers of the X ray and neutron incident radiation, respectively. The single crystals are usually grown by slow evaporation of saturated solution under controlled environments. Ideally their dimensions should be in the order of 0.2 to 0.5 mm and over 1.0 mm, in the respective case of studies from X-ray and neutron diffraction. Irradiation of the crystal by a monochromatic beam leads to constructive interferences, known as diffraction, of the scattered waves in specific directions. The diffraction patterns, that are recorded on electronic devices, consist of a series of Bragg reflections. Their positions and intensities, respectively, contain the details on the unit cell dimensions, space group symmetry, and atomic positions. Unit cells are the building block of the crystals. In order to determine the crystal structure, both the amplitude and phase of the diffracted wave in every Bragg reflection are required. While the measured intensities are proportional to the square of the amplitude, the phase of the reflection relative to that of the incident beam is unknown as it cannot be recorded experimentally. There are several ways to calculate the phases of some important reflections, among them direct methods which are now used routinely to solve crystal structures. Using Fourier transform, the amplitude and phase information is used to synthesized electron density maps that reveal the atomic content of the unit cell. Least square refinement methods are then used to minimize the discrepancy between the measured and calculated structure amplitudes. From a knowledge of the final atomic coordinates, the molecular shape is revealed and such structural features as bond lengths, bond angles, torsion angles, hydrogen bonds, intermolecular distances are directly computed.

Carbohydrates were among the first organic compounds to be investigated by X-ray crystal structure analysis: pentaerythritol [1] and α-D-glucosamine hydrochloride and hydrobromide [2]. The work developed slowly as only eight crystal structures were described over the next fifteen years; among them to elucidation of sucrose.NaBr.2H$_2$O, by Beevers and Cochran in 1947 [3]. The number of X-ray diffraction studies of carbohydrate increased slowly thereafter and the crystal structures of sucrose and α-D-glucose were analyzed by neutron diffraction in the sixties by Brown and Levy [4, 5]. Very recently, the crystal structure elucidation of the cellodextrins [6–8] were the first example of the use of synchrotron radiation, and exemplifies how single crystals having minute dimensions can be successfully investigated.

34.2 General Analysis

The Cambridge Structural Data Base (CSDB) contains over 160,000 entries in the form of structural data related to geometry, configuration, conformation, and packing of molecular crystals for organic and metal-organic compounds [9]. As a consequence of the automation of experimental measurements along with the development of crystal structure analysis for non-centrosymmetric space groups, there has been a significant increase in the number of reported crystal structures of carbohydrates between the years 1980–1995. The number of entries dealing with carbohydrate crystal structures in the CSDB reaches almost 4000. About 1000 among these entries do not list any atomic coordinates and cannot be considered as structurally informative, in terms of conformations and configurations.

Despite the fact that it is possible to determine the absolute configuration of an optically active molecule from the effects of anomalous scattering, such a level of structural characterization is not performed routinely. Determination of absolute configuration generally involves scattering from relatively heavy atoms. It can nevertheless be determined from molecules containing a significant fraction of oxygen atoms providing that efforts are given to obtain accurate diffracted intensities. Among the 4000 structural investigations of carbohydrates, only 150 dealt with the determination of the absolute configuration. Caution should be taken to the few crystal structures that have been published with the wrong enantiomer. Most of the recently reported crystal structures use the a priori knowledge of the configuration of the constituting monosaccharide units.

Crystalline data can be readily found for many members within the classification of the carbohydrates that encompasses (i) the aldose and kestoses which are the monomers, (ii) the alditol, or sugar alcohol which differ from the aldoses in having members of the series being in *meso*-configurations, (iii) the cyclitols which make up a family of nine compounds, the configurations of which differ by the axial or equatorial disposition of the hydroxyl groups around the chair-shaped cyclohexane ring, (iv) the anhydro sugars which are formed by the elimination of water between hydroxyls of the pyranoses and furanoses thereby forming fused bicylcic or tricyclic ring structures, (v) the carbohydrate acids which are related to the alditols and aldoses, (vi) the oligosaccharides. The disaccharides result from the condensation of a reducing group [C-1 OH in aldose, C-2 OH in ketose] with another hydroxyl.

The family of cyclodextrins is characterized by excellent crystals and therefore many structures analyses have been reported. Typical examples of these compounds are cyclic 1–4 linked α-D-glucopyranosides with 6, 7 and 8 monosaccharides residues. Being shaped like truncated cones, with the inside voids occupied by solvent or guest molecules, these cyclic oligosaccharides are amenable to crystallization, and exhibit only three types of packing modes. A very large number of crystals structures of the cyclodextrins and their complexes have been reported, including some neutron diffraction analysis. Larger cyclodextrins, namely with 10 glucose and 14 glucoses in the ring, have also been crystallized and display distorted ring shape. Very recently, crystal structure of large cycloamylose compound, consisting of 26 glucose residues, have been solved [10]. The large ring is folded into a structure similar to the one of V-amylose. This cyclodextrin field has been extensively reviewed several times [11] and it will now be covered in the present article.

There are 55 crystal structures of unsubsituted disaccharides and a dozen of crystal structures of the fully (or almost fully) acetylated disaccharides. When looking at larger compounds, there are 17 trisaccharide crystals structures and only 4 tetrasaccharides (Table 1). One linear oligomaltose has been crystallized and analyzed as a poly-iodide complex, that of p-nitrophenyl-α-D-maltohexopyranoside [12]. The malto-hexaose molecules form a double helical anti-parallel stranded structure which enclosed the poly-iodide chain in away similar to that found in cyclodextrin inclusion compounds. Both the conformation and the relative orientation of the single stranded chains are completely different from those found in the crystalline arrangement of a small fragment of amylose, the structure of which has been determined from the combined used of fiber X-ray diffraction and electron single crystal diffraction [13].

Crystal structure have been solved for most of the pentoses and hexoses, in one or more of the isomeric forms or as a 1-O-methyl derivatives. A significant amount of crystal structures of carbohydrates have been determined from neutron diffraction experiments. From these highly accurate structural determination, standard molecular dimensions of the constituting units have been established (Table 2). These data are an update of the ones published previously by Jeffrey & Taylor [14] and, in a similar way, they can serve as a basis for parameterizations of molecular mechanics force fields.

There is an obvious reluctance of carbohydrates to crystallize in a form suitable for X-ray or neutron diffraction studies. This is particularly true for aldose and ketose containing carbohydrates as a configurational mixture of four isomers, α- and β-pyranoses and α- and β-furanose, is likely to occur. Configurational heterogeneity in solution tends to inhibit crystallization. For this reason 1-O-methyl derivatives, which cannot epimerize, are likely to crystallize more readily. In some instances, the α and β epimers can co-crystallize, and more than 10 examples of such co-crystallization can be found. The α and β ratio may be dependent on the temperature and solvent of crystallization, and may not be reproducible between independent investigators. One extreme example is provide by the crystal struc-

Table 1. Crystal structures of trisaccharides and tetrasaccharides.

Trisaccharide

Name	Common name	Code	Ref.
αFuc(1–2)βGal(1–3)GlcNAc	Blood group B	nd	[33]
αGlcA-4-O-Me(1–2)βXyl(1–4)βXyl	Aldotriuronic acid	GURXPX10	[34]
βGal(1–4)[αFuc(1–3)]βGlcNAc-O-Me	Lewix X	nd	[29]
Manα(1–3)Manβ(1–4)GlcNAc	Fragment N-glycan	MPYAGL	[35]
αGlc(1–4)αGlc(1–2)βFru	Erlose (mono et trihydrate)	HAHXUJ	[36]
		HEGXOG	[37]
αGlc(1–4)αGlc(1–4)βGlc-O-Me	Maltotrioside	DUDXOP	[38]
αGlc(1–6)αGlc(1–4)α/βGlc	Panose	KOYZAZ	[39]
αGal(1–6)αGlc(1–2)βFru$f$	Raffinose	RAFINO01	[40]
βMan(1–4)βMan(1–4)αMan	Mannotriose	COFMEP10	[41]
βGlc(1–4)βGlc(1–4)βGlc-O-Me	Cellotrioside	TAQYAL	[6]
αGlc(1–2)βFru$f$(1–2)βFru$f$	1-Kestose	KESTOS	[42]
αGlc(1–2)βFru$f$(1–6)βFru$f$	6-Kestose	CELGIJ	[43]
αGlc(1–3)βFru$f$(2–1)αGlc	Melezitose	MELEZT01	[44]
	(2 forms)	MELEZT02	[45]
αGal(1–6)βFru$f$(2–1)αGlc	Planteose	PLANTE10	[46]
βGlc(1–4)βGlc(1–4)βGlc peracetylated	Cellotriose	ACCELL10	[47]

Tetrasaccharides

Name	Common name	Code	Ref
αGal(1–6)αGal(1–6)αGlc(1–2)βFru$f$	Stachyose	STACHY01	[48]
αGlc(1–2)βFru$f$(1–2)βFru$f$(1–2)βFru$f$	Nystose	PEKHES01	[49]
βGlc(1–3)[βGlc(1–6)]βGlc(1–3)βGlc peracetylated	Schyzophyllan fragment	WIMNOV	[50]
βGlc(1–4)βGlc(1–4)βGlc(1–4)βGlc	Cellotetraose		[7]
		ZILTUJ	[8]

ture of the disaccharide lactulose, as the crystals contains an mixture of isomers in which the fructose moiety is β-D-fructofuranose, α-D-fructofuranose and β-D-fructopyranose in the ratio 74.5/10/15.5 [15].

The reluctance of carbohydrate to crystallize in a form suitable for X-ray diffraction studies, is more pronounced for compounds having molecular weights ranging from 1000 to 5000. This is true but for the exception of cyclic compounds such as cyclodextrins and cycloamyloses. One of the reasons may be the lack of sufficient amount of material available for crystal growth. The other reason is that the techniques of growing organic crystals of medium size bio-molecules has not parallel the revolution of growing protein and viruses crystals.

Since the beginning of the 90s an increasing number of crystal structures have been reported for glycoproteins and protein-carbohydrate complexes. The resolu-

Table 2. Standard molecular dimensions for pyranosides. Distances are given in Å and angles in °.

	α-D-$^4C_1$	β-D-$^4C_1$	α-L-$^1C_4$	β-L-$^1C_4$
number	379	184	35	12
C_1-C_2 (Å)	1.52(2)	1.517(19)	1.519(16)	1.515(12)
C_2-C_3 (Å)	1.520(17)	1.522(15)	1.516(17)	1.518(9)
C_3-C_4 (Å)	1.520(18)	1.521(16)	1.514(19)	1.529(9)
C_4-C_5 (Å)	1.526(17)	1.526(16)	1.523(16)	1.521(14)
C_1-O_1 (Å)	1.412(17)	1.394(17)	1.408(15)	1.389(11)
C_1-O_5 (Å)	1.413(17)	1.422(15)	1.413(13)	1.428(9)
C_5-O_5 (Å)	1.440(16)	1.433(14)	1.440(13)	1.437(7)
C_5-O_5-C_1 (°)	113.9(14)	111.9(13)	113.8(16)	111.2(11)
C_5-C_1-O_1 (°)	111.3(11)	107.4(12)	112.0(13)	107.7(6)
C_1-O_1-Cn' (°)	116(3)	115(2)	115(3)	115(2)
C_5-O_5-C_1-O_1 (°)	60(4)	177(3)	−62(4)	177(2)
O_5-C_1-O_1-Cn' (°)	108 (7) *185 frag*	−80(11)	−104(8) *6 frag*	79(8)
	72(10) *194 frag*		−73(12) *29 frag*	

tion of the first reported structures was rarely sufficient to provide reliable conformational information. Significant and rapid progresses arising from the use of synchrotron radiation are providing, now, access to highly resolved structures.

34.3 Crystalline Conformations of Disaccharide Moieties

34.3.1 The Disaccharides

Important in their own right, disaccharides take great importance as the shortest components of the family of oligo, polysaccharides and complex carbohydrates. As such a particular attention is given to their conformational properties because their molecular shapes are considered to be important determinants of their properties and those of their larger parents. The disaccharides result from the condensation of a reducing group [C-1 OH in aldose, C-2 OH in ketose] with another hydroxyl. For hexose in the pyranose shape, this linkage may be 1→n where n is 1 to 6, except 5. If two reducing groups are involved, the disaccharide is non-reducing as in sucroses and trehaloses.

For disaccharide moieties, the main conformational determinants are: the ring shapes, the orientations of the hydroxyl groups, and the relative orientations of the monosaccharide units at the glycosidic linkage. Ring shapes can be defined in terms of reference conformations (chair, C; twist T; boat, B; envelope, E; skew, S) or by the so-called puckering parameters [16]. The exocyclic primary alcohol groups can adopt a number of low-energy conformations. They are usually in staggered ar-

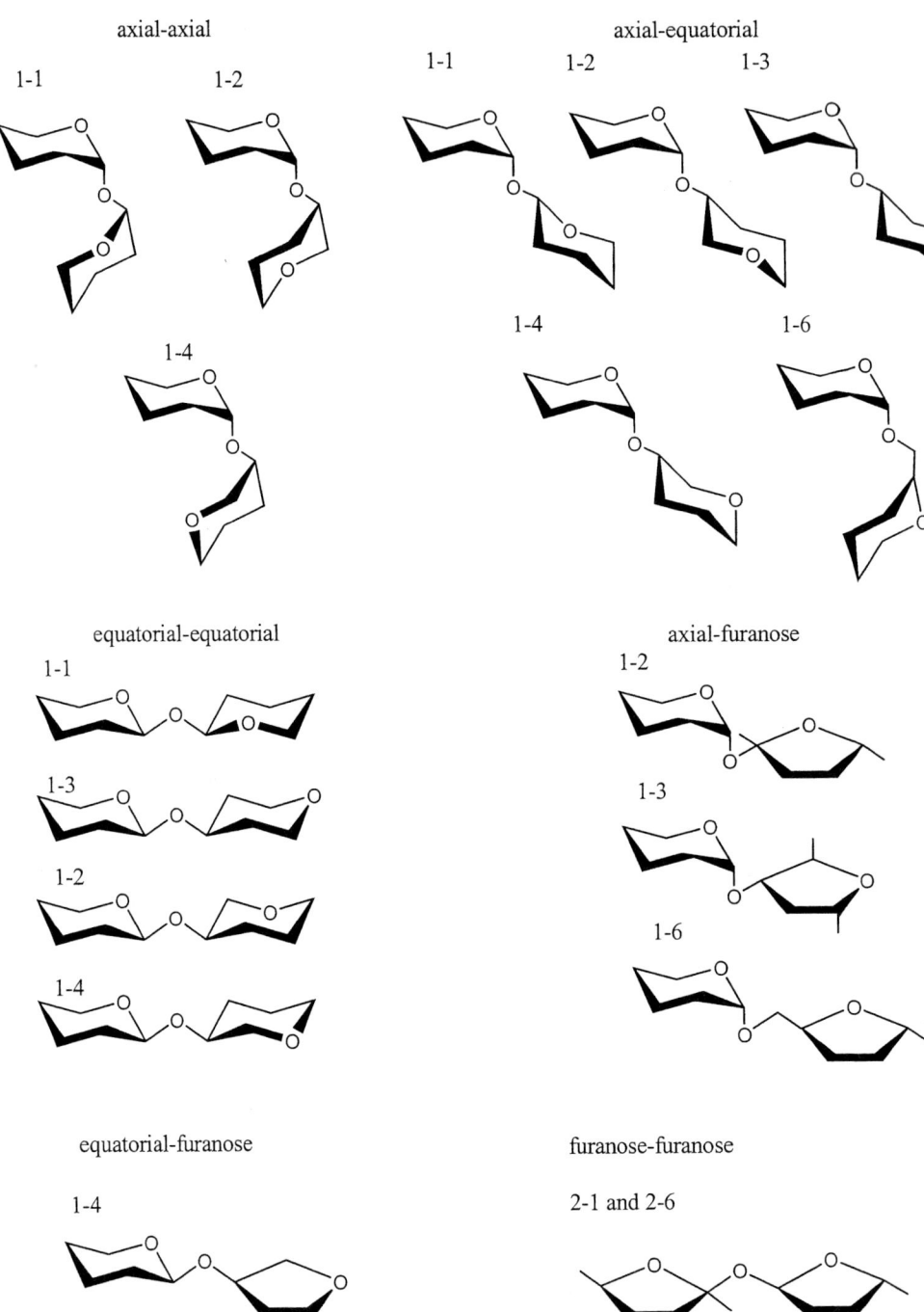

Figure 1. Schematic representation of the different type of linkages observed in crystal structures of disaccharides and oligosaccharides. Sugars with $^1C_4$ ring shape such as L-fucose have not been represented.

rangements that correspond to local minima. In the case of pyranoses, primary hydroxyl groups most frequently occupy two positions, avoiding interactions between O-4 and O-6. However, each of the secondary hydroxyl groups can rotate almost freely. The relative orientation of two consecutive monosaccharide units in a disaccharide is customarily described by the torsion angles Φ and Ψ around the glycosidic bonds. Φ represents the torsion angle about the C(anomeric)–O bond, whereas Ψ represents the torsion angle about the O–Cx bond. The sign of the torsion angles is given in accordance with the IUPAC-IUB Joint Commission of Biochemical Nomenclature [17].

The consideration of the axial/equatorial nature at the glycosidic linkage provides a useful framework for a classification of the disaccharide moieties, independently of the remaining and of the surrounding of the oligosaccharidic molecule. Using such a classification, all the unsubstituted components of the linear oligosaccharide structures have been reported in Table 3. For each class, the nature of the glycosidic linkage: $1 \rightarrow n$ is indicated, along with the REFCODE of the corresponding oligosaccharidic structure, its trivial/usual name, description of the disaccharide, the magnitude of the angles $[(\omega), \Phi, \Psi, \tau]$ at the glycosidic linkage, and the occurrence (if any) of inter-residual hydrogen bonds. As for six membered ring containing disaccharides, axial-axial, axial-equatorial and equatorial-equatorial are found. The families involving axial-furanose, and equatorial-furanose linkage to a hexopyranose are also indicated, along with the furanose-furanose cases. Proper references to the original (or most recent) crystal structure work is also provided.

In order to have a comprehensive vision of the spatial occurrence of all these conformations as a function of their belonging to a given class, a schematic representation of their crystallographic conformations at the glycosidic linkage, has been set with a superimposition onto the low energy contours that have been computed for a prototypical motif using molecular mechanics calculations (Figure 2). Those contours correspond to conformation of the prototypical disaccharide having energy values of 2 kcal/mol and 6 kcal/mol with respect to the lowest energy minimum.

As clearly indicated in Fig. 2a, b, and c, the chosen classification and representations appear to be quite relevant to describe the ensemble of solid state conformations found in crystalline oligosaccharides and their analogs. It is worth noticing that there is no representative of the equatorial-axial class, even though such type of glycosidic linkage can be found in several carbohydrate containing molecules and macromolecules. Whereas all the crystalline conformations lie within the 5 kcal/mol energy contour, it is noteworthy to observe a limited dispersion about the glycosidic Φ angles compared to that observed about Ψ, as an expression of the influence of the exo-anomeric effect on the establishment of preferred conformation in crystalline oligosaccharides.

The exo-anomeric effect arises due to the particular bonding sequence C-5–O-5–C-1–O-g–C-x' in glycopyranosides and disaccharides and influences the conformation about the glycosidic torsion angle Φ [18]. In the case of axial type of linkage (typically as in $^4C_1$ D-α configuration) only one staggered conformation is preferred with Φ being in the vicinity of 60°. As for the equatorial type of linkage (typically as in $^4C_1$ D-β configuration); two staggered conformations are preferred ($\Phi = 60°$ and $-60°$) of which that corresponding to $\Phi = -60°$ is favored due further stabilization occurring from non-bonded interactions. The values of Φ generally varies between

Table 3. Details of the conformations at the glycosidic linkages for disaccharides and oligosaccharides in crystalline state.

a/Axial-axial linkage between two pyranose rings

Atom	Code	Ref	Molecule	Disaccharide	τ	Φ	Ψ	H-bond
1–1	YOXFOG	[51]	α,α-Allo-trehalose, 6H2O	αAllo(1–1)αAllo	110.4	74.7	75.5	
			"		110.2	73.1	74.8	
	ALTRCA	[52]	α,α-Allo-trehalose, CaCl2, 5H2O	αAllo(1–1)αAllo	115.3	45.1	57.1	
	LETTEJ	[53]	3,3′ deoxy-Arabino-trehalose, H2O	αAra(1–1)αAra	114.4	64.5	75.6	
	TREHAL10	[54]	α,α-Trehalose, 2 H2O	αGlc(1–1)αGlc	115.7	75.1	61.6	
	DEKYEX	[55]	α,α-Trehalose	αGlc(1–1)αGlc	113.3	60.8	60.1	
	TRECAB	[56]	α,α-Trehalose, 1/2(CaBr2) H2O	αGlc(1–1)αGlc	113.2	77.0	77.0	
	YOXFUM	[57]	α,α-Galacto-trehalose	αGal(1–1)αGal	116.2	66.3	62.4	
	YODSOZ	[58]	α,α-tetrachloro-Galacto-trehalose	αGal(1–1)αGal	114.8	77.3	77.3	

1-2	FABYOW10	[59]	Man, α(1–2) Manα-O-Me	αMan(1–2)αMan	114.0	64.2	136.1	
1-5	LEUCRO01	[60]	β-Leucrose, 2H2O	αGlc(1–5)βFru$p$	115.4	68.8	-94.0	O-2g...HO-2f
1-4	CITSIH10	[61]	Galabiose	αGal(1–4)αGal	117.5	98.1	157.7	HO-3...O-5'

b/ Axial-equatorial linkage between two pyranose rings

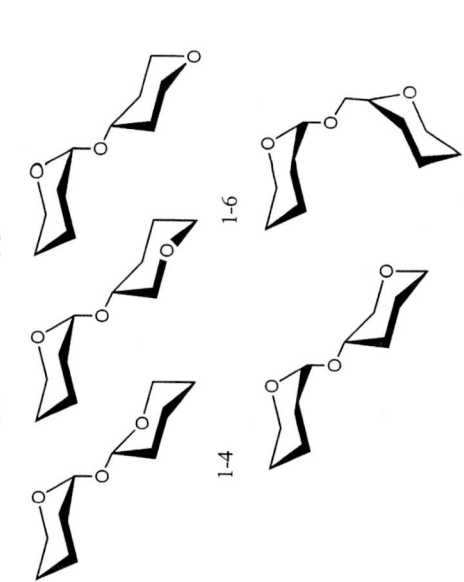

axial-equatorial
1-1 1-2 1-3
1-4 1-6

Atom	Code	Ref	Molecule	Disaccharide	τ	Φ	Ψ	H-bond
1-1	TIQDUS	[25]	α,β-Trehalose, H2O	αGlc(1–1)βGlc	113.7	68.8	-93.4	O-6...O-6
1-2	nd	[33]	Blood group B trisacch.	αFuc(1–2)βGal	115.8	-66.0	-91.3	
	TIYYOP	[62]	H type 1, 1/2 H2O	αFuc(1–2)βGal	116.7	-92.7	-174.8	
				αFuc(1–2)βGal	116.6	-92.6	-174.8	
	RESMOR	[63]	Manα (1–2) βGlc	αMan(1–2)βGlc	118.0	59.0	-148.6	
	GURXPX10	[34]	Aldotriuronic acid trisacch., 3H2O	αGlcA-4-O-Me(1–2)βXyl	116.0	78.8	-81.9	

Table 3 (continued)

	Code	Ref	Name	Bond	τ	φ	ψ	H-bond
1–3	nd	[33]	Blood group B trisacch.	αGal(1–3)βGal	116.4	56.2	60.0	
	nd	[29]	Lewis X trisacch., 9H2O	αFuc(1–3)GlcNAc	115.1	−72.5	139.2	
					115.1	−76.7	139.0	
	MPYAGL	[35]	Manα(1–3)Manβ(1–4)GlcNAc	αMan(1–3)βMan	114.1	60.5	97.0	
	MOGLPR	[64]	Methyl α-Nigeroside	αGlc(1–3)αGlc	116.1	99.9	104.2	HO-4...O-2'
	TURANS	[65]	β-Turanose	αGlc(1–3)βFrup	116.3	98.3	111.6	HO-4...O-2'
1–4	HAHXUJ	[36]	Erlose, H2O	αGlc(1–4)αGlc	117.0	107.2	−129.9	HO-3...O-2'
	HEGXOG	[37]	Erlose 3H2O	αGlc(1–4)αGlc	115.0	68.7	−150.0	
			"		116.7	51.1	−163.5	
	FOXSUG20-b	[12]	p-nitrophenyl-Maltohexaoside, BaI2 27H2O	αGlc(1–4)αGlc	113.3	112.2	−120.8	O-3...O-2'
					119.2	95.7	−117.3	O-3...O-2'
					119.6	93.2	−126.1	O-3...O-2'
					114.1	118.2	−142.6	O-3...O-2'
					113.3	103.1	−128.4	O-3...O-2'
					119.2	45.4	−164.8	O-3...O-2'
					125.5	94.0	−129.5	O-3...O-2'
					112.1	104.7	−133.7	O-3...O-2'
					112.3	79.9	−151.8	O-3...O-2'
					115.7	106.5	−129.6	O-3...O-2'
	MALTOT	[66]	α-Maltose	αGlc(1–4)αGlc	120.1	116.1	−118.0	HO-3...O-2'
	MALTOS11	[67]	β-Maltose, 1H2O	αGlc(1–4)βGlc	117.9	121.7	−107.7	HO-3...O-2'
	PHMALT	[68]	Phenyl α-Maltoside	αGlc(1–4)αGlc	116.9	108.5	−139.4	O-3...HO-2'
			"		116.5	110.0	−138.9	O-3...HO-2'
	IPMALT	[68]	6'-iodo phenyl α-Maltose	αGlc(1–4)αGlc	112.7	72.5	−155.0	
	MMALTS	[69]	Methyl β-Maltoside	αGlc(1–4)βGlc	117.6	109.9	−109.1	
	DUDXOP	[38]	Methyl β-Maltrotrioside, 3H2O	αGlc(1–4)αGlc	114.7	82.3	−148.9	O-3...HO-2'
			"		115.5	82.7	−151.8	
	KOYZAZ	[39]	α/β-Panose	αGlc(1–4)Glc	113.9	96.8	−134.8	

	Code	Ref	Name	Bond	τ	φ	ψ	ω
1–6	MELIBM10	[70]	α/β-Melibiose, H2O	αGal(1–6)Glc	111.7	76.4	−174.2	−63.4
	KOYZAZ	[39]	α/β-Panose	αGlc(1–6)αGlc	112.0	70.7	167.3	74.7

34.3 Crystalline Conformations of Disaccharide Moieties

	Code	Ref	Molecule	Disaccharide					
	RAFINO01	[40]	Raffinose, 5H2O	αGal(1–6)αGlc	−63.1	112.0	71.8	−170.8	
	STACHY01	[48]	Stachyose, 4H2O	αGal(1–6)αGal	87.1	113.1	85.2	−172.4	
				αGal(1–6)αGlc	−62.1	111.1	64.9	−175.1	
2–8	YEPNUC	[71]	KDO disacchac., 2Na+ H2O	αKDO(2–8)αKDO*		115.50	58.1	131.9	HO-7...O1A'

*Two other torsion angles at the linkage: 56.3 63.3

c/ Equatorial-equatorial linkage between two pyranose rings

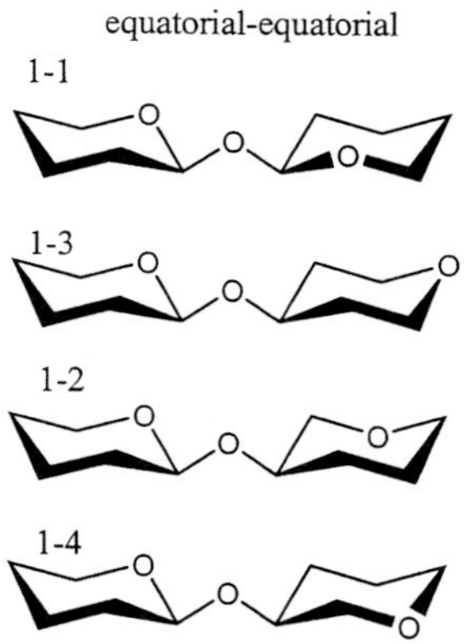

equatorial-equatorial

1-1

1-3

1-2

1-4

	Code	Ref	Molecule	Disaccharide	τ	Φ	Ψ	H-bond
1-1	WACHOX	[72]	β,β-Trehalose, 2H2O	βGlc(1–1)βGlc	115.2	−75.6	−75.5	
1-2	SOPROS	[73]	α-Sophorose, H2O	βGlc(1–2)αGlc	113.7	−78.9	−139.8	
1-3	LAMBIO	[74]	β-Laminarabiose, H2O	βGlc(1–3)βGlc	118.2	−93.6	77.7	HO-4...O5'
	WAGBOV	[75]	Methyl β-Laminarabioside, H2O	βGlc(1–3)βGlc	117.5	−85.7	76.0	HO-4...O5'
	VIZFUF	[76]	Neocarrabiose beta monohydrate	α-L-3,6AnGal(1–3)βGlc	116.5	94.5	141.9	HO-2...O-5'
	CHONDM	[77]	Chondrosine monohydrate	βGlcA(1–3)αGalNAc	116.5	−87.8	55.4	HO-4...O5'

Table 3 (continued)

1–4	GURXPX10	[34]	Aldotriuronic acid trisacch., 3H2O	βXyl(1–4)βXyl	113.8	−145.5	−79.6	
	MAYAGL	[35]	Manα(1–3)Manβ(1–4)GlcNAc	βMan(1–4)αGlcNAc	114.6	−75.7	−130.7	HO-3…O5'
		[29]	Lewis X trisac., 9H2O	βGal(1–4)GlcNAc	117.0	−80.0	−104.6	
					117.7	−70.5	−107.7	
	ACLACT	[78]	α-N-acetyllactosamine, H2O	βGal(1–4)αGlcNAc	116.3	−88.1	−139.5	HO-3…O5'
	LACCCB	[79]	α-Lactose, CaCl2 7H2O	βGal(1–4)αGlc	115.9	−76.9	−136.7	HO-3…O5'
	LACTOS03	[80]	α-Lactose, 1H2O	βGal(1–4)αGlc	117.2	−93.5	−143.5	HO-3…O5'
	BLACTO	[81]	β-Lactose	βGal(1–4)βGlc	116.5	−70.9	−131.5	HO-3…O5'
	REMVUA	[82]	β-Lactosylurea, 2H2O	βGal(1–4)βGlc	117.2	−88.1	−159.4	HO-3…O5'
	DIHTUJ	[83]	α-Mannobiose	βMan(1–4)αMan	115.0	−96.0	−148.1	HO-3…O5'
	COFMEP10	[41]	α-Mannotriose, 3H2O	βMan(1–4)βMan	117.4	−71.9	−131.6	HO-3…O5'
				βMan(1–4)αMan	115.6	−93.9	−150.4	HO-3…O5'
	CELLOB02	[84]	β-Cellobiose	βGlc(1–4)βGlc	116.1	−76.3	−132.3	HO-3…O5'
	WEHTEI	[85]	β-Cellobiosyl-nitromethane	βGlc(1–4)βGlc-	116.5	−93.0	−139.8	HO-3…O5'
	MCELOB	[86]	Methyl β-Cellobioside, methanol	βGlc(1–4)βGlc-O-Me	115.8	−91.1	−160.7	HO-3…O5'
	ZILTUJ	[8]	β-Cellotetraose, 1/2 H2O	βGlc(1–4)βGlc	116.7	−95.3	−143.9	HO-3…O5'
			"	"	115.9	−94.8	−142.9	HO-3…O5'
			"	"	116.2	−93.2	−141.6	HO-3…O5'
			"	"	116.3	−98.3	−150.8	HO-3…O5'
			"	"	117.5	−93.2	−149.0	HO-3…O5'
			"	"	116.3	−91.9	−152.4	HO-3…O5'
	TAQYAL	[6]	Methyl β-Cellotriioside, EtOH,H2O	βGlc(1–4)βGlc	117.3	−93.7	−141.5	HO-3…O5'
			"	"	117.2	−96.7	−141.2	HO-3…O5'
			"	"	116.6	−96.8	−141.9	HO-3…O5'
			"	"	116.8	−93.4	−141.5	HO-3…O5'
			"	"	117.6	−96.3	−150.6	HO-3…O5'
			"	"	117.3	−91.5	−151.8	HO-3…O5'
			"	"	117.2	−91.7	−152.0	HO-3…O5'
			"	"	117.6	−95.5	−150.6	HO-3…O5'
	ACHITM10	[87]	α-Chitobiose	βGlcNAc(1–4)αGlcNAc	116.3	−76.9	−106.9	HO-3…O5'
	BCHITT10	[88]	β-Chitobiose, 3H2O	βGlcNAc(1–4)βGlcNAc	117.1	−90.3	−162.3	HO-3…O5'

	Code	Ref	Molecule	Disaccharide	ω	τ	Φ	Ψ	H-bond
1-6	DICMEH	[89]	Gal β(1-4) Man, 2H2O	βGal(1-4)αMan		117.1	-83.2	-109.9	
1-6	GENTBS	[90]	β-Gentiobiose	βGlc(1-6)βGlc	60.7	113.6	58.6	155.1	

d/linkage between a pyranose ring axial and a furanose ring

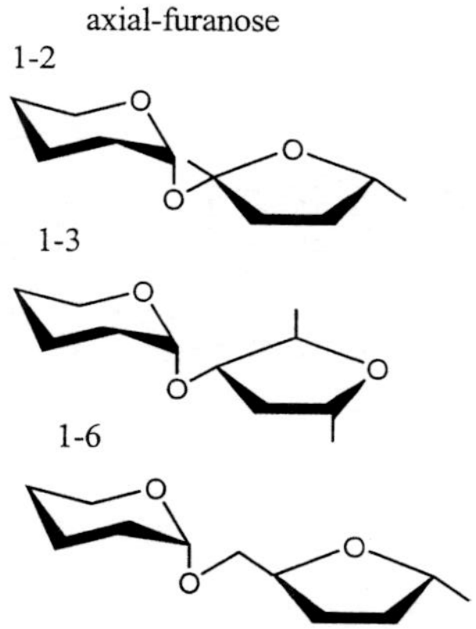

axial-furanose

1-2

1-3

1-6

	Code	Ref	Molecule	Disaccharide	τ	Φ	Ψ	H-bond
1-2	KESTOS	[42]	1-Kestose	αGlc(1-2)βFru	119.4	84.6	-65.9	O-2g...O-1g
	CELGIJ	[43]	6-Kestose, H2O	αGlc(1-2)βFru	121.3	89.6	-168.0	
	MELEZT01	[44]	Melezitose, H2O form I	αGlc(1-2)βFru	119.4	99.8	-30.6	O5-g...HO-6f
	MELEZT02	[45]	Melezitose, H2O form II	αGlc(1-2)βFru	115.3	109.6	-43.4	
	PEKHES01	[49]	Nystose, 4H2O	αGlc(1-2)βFru	118.3	102.3	-18.6	
	PLANTE10	[46]	Planteose, 2H2O	αGlc(1-2)βFru	118.8	108.2	-26.2	
	HAHXUJ	[36]	Erlose, H2O	αGlc(1-2)βFru	116.5	104.5	-32.6	
	HEGXOG	[37]	Erlose, 3H2O	αGlc(1-2)βFru	117.7	98.0	-55.7	
				"	116.2	109.9	-39.8	

Table 3 (continued)

	Code	Ref	Molecule	Disaccharide	ω	τ	Φ	Ψ	H-bond
	RAFINO01	[40]	Raffinose, 5H2O	αGlc(1-2)βFru		120.9	82.1	12.0	
						117.1	109.3	-47.3	O-2g...HO-1f
	STACHY01	[48]	Stachyose, 4H2O	αGlc(1-2)βFru		119.2	91.4	-162.2	
	KANJOY	[91]	Tri-chloro-galactosucrose	4-Cl-Gal(1-2)1,6di-Cl-Fru		114.3	107.8	-44.8	HO-2g...O-3f
	SUCROS04	[92]	Sucrose	αGlc(1-2)βFru		116.7	99.8	-46.1	
	DINYOO10	[93]	Sucrose, NaBr2 2HO	αGlc(1-2)βFru			79.9	-66.6	
	n.d	[94]	Sucrose, ½(2 NaI 3H2O)	αGlc(1-2)βFru			79.6	-61.6	
	NEHCAE	[95]	Sucrose, sarcosine	αGlc(1-2)βFru		118.6	84.0	-20.6	O-2x...HO-1f
	HAHYUK	[96]	Sucroxyl, H2O	αXyl(1-2)βFru		116.6	108.2	-46.9	O-2x...HO-1f
				"		118.1	96.9	-37.2	
1-3	MELEZT01	[44]	Meleziotose monohydrate form I	αGlc(1-3)βFru		115.9	78.4	143.1	
	MELEZT02	[45]	Meleziotose monohydrate form II	αGlc(1-3)βFru		116.9	91.6	153.9	
1-6	IMATUL	[97]	β-Isomaltulose, H2O	αGlc(1-6)βFru	-64.6	115.5	76.9	143.6	
	PLANTE10	[46]	Planteose, 2H2O	αGal(1-6)βFru	63.5	111.2	58.5	172.5	

e/linkage between a pyranose ring equatorial and a furanose ring

equatorial-furanose

1-4

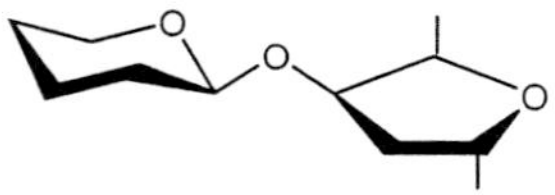

	Code	Ref	Molecule	Disaccharide	τ	Φ	Ψ	H-bond
1-4	BOBKUY10	[15]	β-Lactulose	βGal(1-4)βFru	113.1	-82.1	-164.4	
	PAJNUJ	[98]	β-Lactulose, 3H2O	βGal(1-4)βFru	116.1	-79.9	-170.3	O-5g...HO-3f

f/Linkage between two furanose rings

furanose-furanose

2-1 and 2-6

	Code	Ref	Molecule	Disaccharide	ω	τ	Φ	Ψ	H-bond
1–2	KESTOS	[42]	1-Kestose	βFru(1–2)βFru	179.2	116.4	−41.1	−169.6	
	PEKHES01	[49]	Nystose, 3H2O	βFru(1–2)βFru	−63.3	117.1	55.9	−133.3	
				βFru(1–2)βFru	−175.9	115.3	−47.1	−165.5	
2–6	CELGIJ	[43]	6-Kestose, H2O	βFru(2–6)βFru	67.9	118.0	−55.8	145.7	

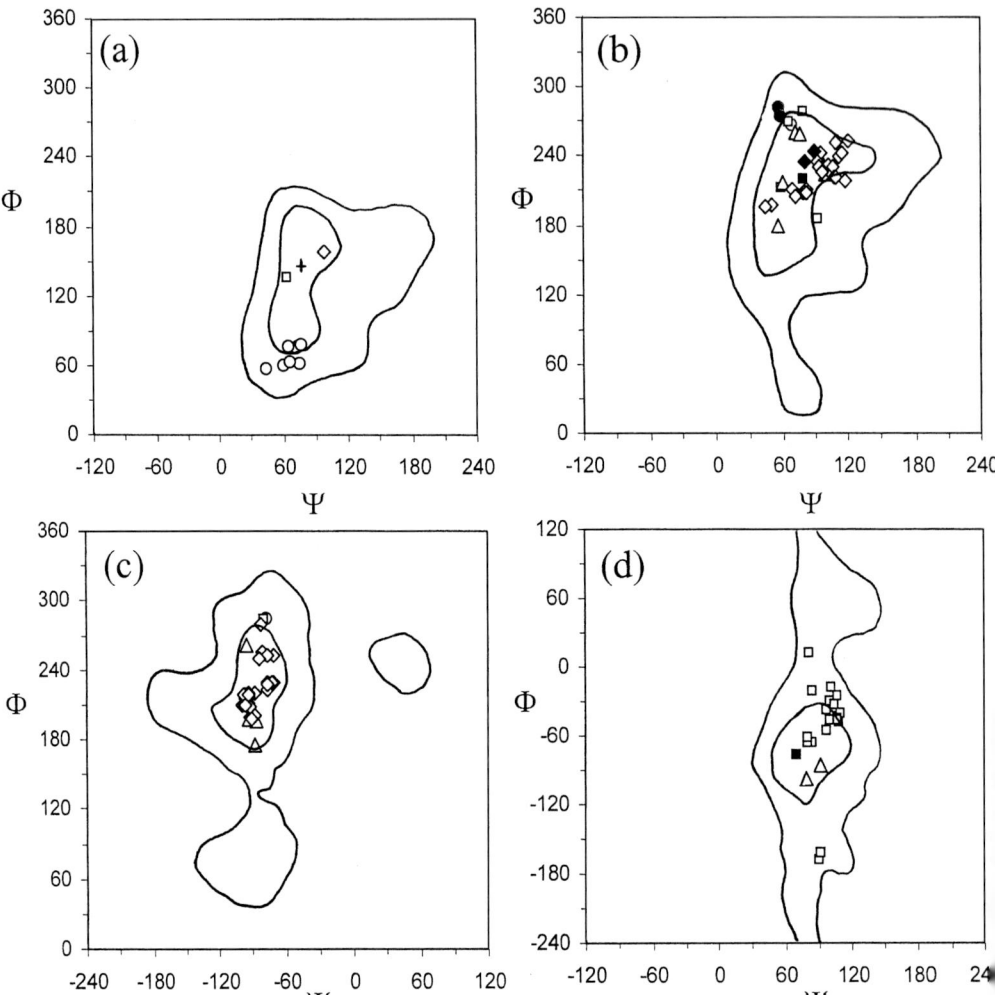

Figure 2. Iso-energy maps for the linkage between two pyranose linked in an axial-axial configuration (a), in a axial-equatorial configuration (b), in a equatorial-equatorial configuration (c) or linkage between a furanose and a pyranose with an axial glycosidic oxygen. The iso-contours represent approximately 2 and 6 kcal/mol as calculated using the MM3 software for αMan(1–2)Man (a), αMan(1–3)Man (b), cellobiose in (c) and sucrose (d). Conformations observed in the crystal structures have been superimposed on the maps using the following code: circle (1-1), square (1-2), triangle (1-3), diamond (1-4) and cross (1-5). The black symbols represent the conformation of an analog with an hetero atom at the position of the glycosidic oxygen. Appropriate transformation has been carried on the value of the Φ and Ψ angle for the aim of comparison: the inverse of the value was taken for L-sugar, for 1–3 linkages, +120° or −120° was added to the value.

40° and 120° for 1→x axial linkages, and between −100° and −65° for 1→x equatorial linkages. Obviously, the magnitude of the exo-anomeric effect is not strong enough to drive the Φ angle to one single conformation. It can be counterbalanced by the occurrence of inter-residue hydrogen bond or other favorable interatomic interactions. Not unexpectedly, the dispersion of the observed conformations about the torsion angle Ψ is much wider.

A somehow similar set of observations can be made in the case of disaccharide segments belonging to the axial-furanose family (Fig. 3d), which encompasses most of the sucrosyl containing molecules. In such cases, a double exo-anomeric effect, occurring from the C-5–O-5–C-1–O-g–C-x'–O-5' sequence could be even more influential in the establishment of glycosidic conformations. Whereas the dispersion in Φ angles goes from 60° to 100°, the Ψ torsion angles span more than 200°. Most of these observations have been rationalized throughout molecular modeling investigations.

34.3.2 The Analogs (S, C, N, ...)

The replacement of the interglycosidic atom in a disaccharide by S, C, NH, ... generates a class of interesting analogues, namely thio-disaccharides, C-disaccharides, diglycosylamines, which all constitute potential non-hydrolyzable epitopes and substrates analogues. Similarly, sulfur-analog can be obtained by substitution of the ring oxygen. The crystal structures of those analogs that have been reported, are listed in Table 4.

The X-ray structure of methyl α-thiomaltoside showed that the effect of replacing the glycosidic oxygen atom by sulfur on the geometry and conformation of the pyranose rings is minimal [19]. However, the magnitude of glycosidic valence angle τ (100.3°) is larger than the value usually found for C–S–C fragments; as such it follows the general trend observed in the corresponding O-glycosides. The same value has been observed when both the glycosidic oxygen and the ring oxygen of the non-reducing unit of maltose are substituted by sulfur [20]. The glycosidic C-1–S and S–C-4' bonds are 1.83 Å length and this results in a larger separation of the two pyranose rings. The torsional angles at the thio-glycosidic linkage are O–C-1–S–C-4' = 89° and C-1–S–C4'–C5' = −116.8°, they correspond to a low energy conformation within a potential energy surface that is similar to that computed for maltose.

In C-glycosides, a methylene group (C-g) replaces the glycosidic oxygen atom and the bonding sequence C-5–O-5–C-1–C-g–Cx' replaces the original sequence present in the parent O-glycosidic molecules. It is therefore expected that in the crystalline conformation, the anomeric and exo-anomeric effects will no longer be driving forces in these molecules. The conformation of C-gentiobioside [21] in the solid state is characterized by torsion angles $\Phi = 55.9°$, $\Psi = 175.1°$ and $\omega = 63.9°$. There is indeed a striking difference between C-gentiobioside and its O-counterpart since there is a 120° difference between the magnitude of the Φ angle in both molecule. The *in vacuo* potential energy calculations for C-gentiobioside indicates that the lack of influence of the *exo*-anomeric effect is not responsible for the occurrence

Table 4. References and selected structural information for the crystal structures of disaccharides analogs.

Code	Ref	Molecule	Disaccharide	ω	τ	φ	φ	H-bond
C-analog								
EABRAA	[22]	C-sucrose octaacetate	αGlc(1–2)βFru		111.3	68.1	–77.2	
	[21]	Methyl-C-gentiobioside	βGlc(1–6)βGlc	–63.9	113.6	55.9	175.1	
RIWZIG	[99]	C-isomaltose heptaacetate	αGlc(1–6)αGlc	61.6	114.6	49.2	64.4	
S-analog								
CIBYAN	[19]	Methyl α-thiomaltoside	αGlc(1–4)αGlc		100.3	89.0	–116.8	HO6...O6'
	[100]	di-thio-maltose, CH3OH	αGlc(1–4)αGlc		100.4	80.7	–125.7	
	[20]	5S-kijobiose	βGlc(1–2)αGlc		114.4	79.3	–140.8	
N-analog								
	[23]	α,β- N-trehalose octaacetate	αGlc(1–1)βGlc		119.9	57.6	–78.8	
					118.9	59.8	–85.9	

of such crystalline conformation; it may simply result from steric interaction of the substituents at the inter-residue linkage. Similarly, the crystalline conformation of C-sucrose, as dictated by torsion angles $\Phi = -54°$ and $\Psi = 44.1°$ is significantly different from that of its O analog [22].

Aminosugars are potent reversible inhibitors of glycosidases and can be used for many therapeutic agents providing that they exhibit sufficient stability. In diglycosylamines, where a NH group replaces the glycosidic oxygen atom, the basicity of the such glycosylamine nitrogen is expected to be decreased by the combined electron withdrawing effect of the two neighboring oxygen atoms. The crystalline and molecular features of the peracetylated α, β-glucopyranosylamine, the N-analog of per-acetylated α, β trehalose, has been solved [23]. The hydrogen atom covalently linked to the nitrogen at the glycosidic linkage is well located, yielding to a R configuration for the nitrogen. The substitution of O by NH at the glycosidic linkage induces a significant perturbation. The glycosidic C–N–C bond angles in 6° larger than that of α, β trehalose, and a lengthening of the O-5–C-1 bond is also observed. This feature was predicted to occur from calculations and is attributed to a stabilizing delocalization of the nitrogen lone pair into an antibonding C–O orbital ($n_N \rightarrow \sigma_{C\text{-}O^*}$) [24]. The torsional angles at the glycosidic linkages (57.6°, −78.8°) and (59.8°, −85.9°) are slightly different from the conformation found in α, β trehalose monohydrate [25] even though the potential energy surface computed for the N-analogue exhibits the same aspects than the one calculated for α, β trehalose, except that the accessible area is larger.

34.4 Hydrogen Bonding in Crystalline Oligosaccharides

The carbohydrates offer rich examples to analyze the hydrogen bonding in crystals, from which rules can be extracted to be further used in molecular modeling situations. Most of the basic rules have been established throughout the analysis of high accuracy X-ray analysis and most evidently from those crystalline structures that have been derived from neutron diffraction investigations.

Neutron diffraction determines nuclear coordinates whereas X-ray diffraction refers to maxima of the electron density distribution. A systematic investigation of the differences in bond lengths from X-ray and neutron diffraction analyses of the same crystal structure gave significant differences for d_X–d_N for C–H and O–H bonds, respectively −0.096(7) and −0.155(10) Å [26]. This is the reason why the X-ray X–H bond lengths have to be normalized to standard, neutron diffraction, values for the interpretation of hydrogen bonding in molecular crystals.

Essentially, the rules governing the establishment of hydrogen bonds obey to two basic concepts [27]:

i) maximize the hydrogen-bond interactions throughout the participation of all hydroxyl groups and as many ring and, to a lesser extend, glycosidic oxygen atoms as possible. These features implies both two and three-center bonds.

ii) maximize cooperativity by forming as many finite and infinite chains of hydrogen bonds as possible.

Several patterns have been identified by Jeffrey [28] and the crystal structure can be roughly divided equally between them. **Class A:** encompasses those structures where the cooperative effect is maximized by the formation of infinite chains or spirals of hydrogen-bonded hydroxyls. In **class B**, a more limited cooperative effect is obtained throughout the occurrence of finite chains with the acetal oxygen atoms acting as chain terminators. **Class C**, in which the maximum cooperative effect is achieved with all the hydroxyl groups, except one, which forms a separate two- or three-centered hydrogen bond to a ring or glycosidic oxygen. In **class D**, the infinite chains are retained and the ring and glycosidic oxygen atoms are included in the hydrogen bonding scheme throughout the occurrence of three-center bonds.

In almost all cases, hydroxyl groups act as both hydrogen bond donors and acceptors. Compared with the other hydroxyl groups, the anomeric hydroxyls tend to be stronger donor and weaker acceptors. As for ring oxygen and glycosidic oxygen, stereochemical and packing reasons may prevent these atoms to be acceptors. Because there are more acceptors than donors in carbohydrates, it is not surprising to find a significant occurrence of three-centered hydrogen bonds.

Hydration is a fairly common features of crystalline oligosaccharides, which may result from their molecular shapes that do not provide efficient packing. Therefore, voids from the water molecules are left in the crystalline structures. It should be clearly stated that the analysis of crystalline oligosaccharides cannot provide a limited insight into the preferred hydration effect. Whereas water molecules may influence the occurrence of a given conformation, these water molecules may be expelled upon formation of the crystalline arrangement. In the hydrate structures, the water molecules always donate two hydrogen bonds; they may accept one or two.

34.5 Packing Features

As for their crystalline structure, more than 95% of the carbohydrates crystallize in non centro-symmetric structures. The rare examples of centro-symmetric space groups deal with some linear or cyclic polyalcohol along with some pyranoses and furanoses. In order to adopt such features, these molecules either have an internal symmetry or crystallize as racemic. Figure 3 displays the histogram of the occurrence of space groups in carbohydrate crystal structures, from which it is obvious that only four space groups, i.e. $P2_12_12_1$, $P2_1$, P_1 and C_2 account for more than 80% of the observed space groups.

Despite the wealth of potential structural information, there is a lack of comprehensive investigation of the molecular principles underlying the formation of the oligosaccharide crystals. Only few reports have tried to analyze the anisotropy of intermolecular interactions in these crystals. A typical way of exploring the packing

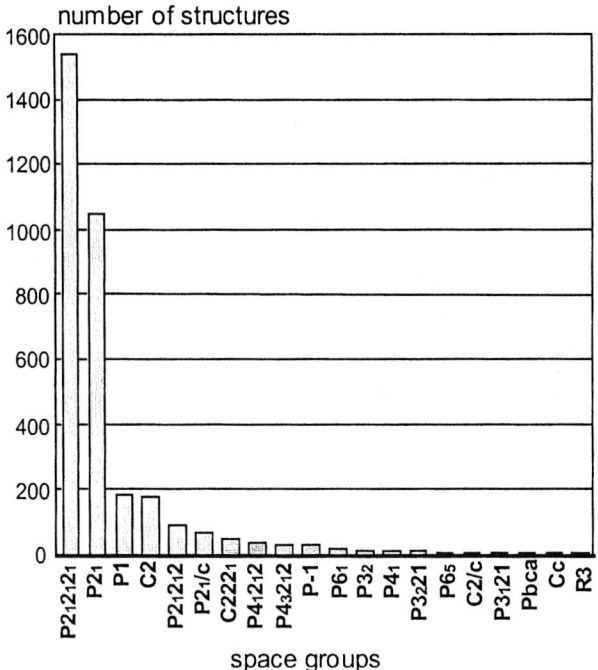

Figure 3. Histogram of the distribution of space groups among the crystal structures of oligosaccharides.

arrangement is to evaluate the intermolecular energy between one molecule, i.e. the reference molecule, and all its neighbors. This energy is evaluated by taking into account the intermolecular hydrogen bonds as well as the non-bonded interactions. The number of 'close' contacts corresponding to interactions distances less than 1.5 times the sum of the van der Waals interacting atoms may be evaluated. The estimation of electrostatic component is more straightforward as it may depend upon the way the atomic charges are evaluated along with the cut-off distance used in the calculations. Since most of the oligosaccharide structures solved up to now concern neutral carbohydrates, such a point has not be given much attention.

The packing arrangements found in crystal structures of oligosaccharides are consistent with a high packing density, as each molecule is usually surrounded by 12 neighbors. Among these neighbors, which occur in pair, a certain anisotropy exist, which in many cases can be correlated to the dimensions of the unit cell, and/or the morphology of the crystals. From the differences in the magnitude of the intermolecular energies can be identity supramolecular elements such as molecular chains and/or molecular layers, that constitute the basis of the molecular layers. From the limited number of packing of oligosaccharide structures which have been investigated so far, the low energy molecular layers correspond to the few which underline the formation of the space groups cited above.

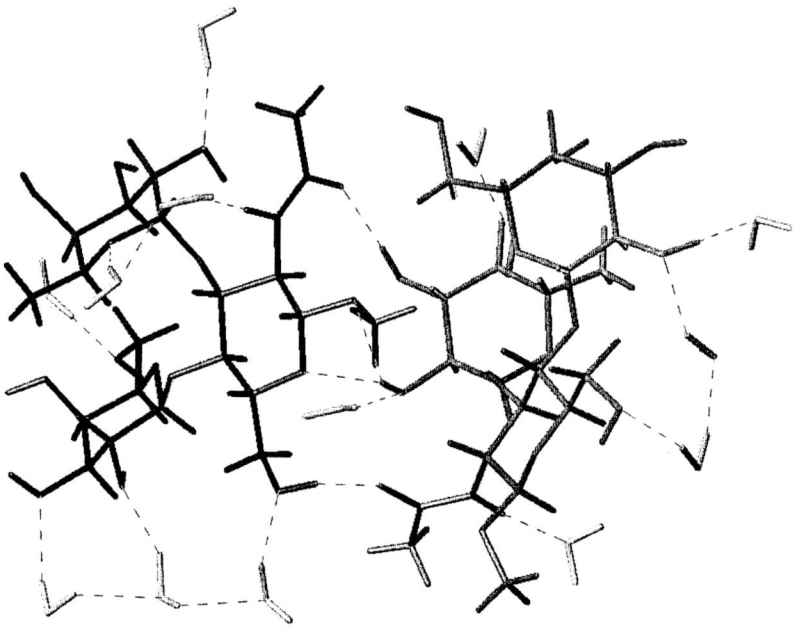

Figure 4. The molecular and crystal structure of Lewis X [29]. Graphical representation of the two trisaccharides constituting the asymmetric unit along with the water molecules hydrating them. Nine of the water molecules are part of the asymmetric unit content and the other ones are related by symmetry element.

34.6 Selected Examples

Lewis^x. Despite their well recognized biological role, the first crystal structure of a histo-blood group carbohydrate-dependent antigen, i.e. Lewis^x [βGall-4[αFucl-3]βGlcNac was reported in 1996 [29] (Figure 4).

Starting from chemically synthesized Le^x trisaccharide methyl glycoside, single crystal could be grown from a slow evaporating solution of Le^x, water/ethanol mixture. Over a period of 2 years, more than 20 crystals having dimensions suitable for X-ray investigations were obtained and investigated. One single crystal having dimensions of $0.5 \times 0.25 \times 0.05$ mm was of a sufficient quality to diffract and yielded enough reflections to solve and refine the structure to a reliability index of 0.05. Le^x crystallizes in the monoclinic space group P2$_1$, with unit cell parameters $a = 12.147(6)$, $b = 27.552(9)$, $c = 8.8662(6)$ A and $\beta = 91.71°(5)$. In such a unit cell, the asymmetric unit contains two independent Le^x molecules, and an unusually high number of water molecules. All hydrogen atoms of the O–H groups and the water molecules could be located in this crystal structure, allowing a straightforward assignment of hydrogen bonding scheme.

The two crystallographically independent Le^x molecules differ in their overall conformations, and these differences are essentially located at the glycosidic torsion angles at the βGalI-4GlcNAc linkage for which angle Φ differs by 10°. Neither of the two trisaccharides exhibits any intramolecular hydrogen bonds. A strong interaction exists between the fucose and galactose residue, but only non polar van der Waals contacts are involved, each ring presenting its most hydrophobic face to the other one. Conformational studies using NMR and/or molecular modeling generally agree on a single conformation for Le^x in solution, corresponding closely to the one found in the crystal structure.

The nine water molecules present in the asymmetric unit are arranged in a cluster-like fashion. They establish hydrogen bonds to other water molecules within the cluster and to the surrounding carbohydrate molecule. They are arranged to fill up an empty space in the crystal packing. Six and seven water molecules, respectively, are involved in the hydration of the two trisaccharides.

The crystal structure displays an extremely dense network of hydrogen bonds. Thirty six hydrogen bonds are observed in the asymmetric unit, 30 of them implying water molecules. Each trisaccharide is involved in 21 hydrogen bonds, the ration per glycosidic residue varying from 5 to 8. Such a high number of hydrogen bonds can be correlated to (i) the high hydration level and (ii) the peculiar folding of the oligosaccharides, burying the hydrophobic faces of the residue and presenting the hydrophilic faces to the external part.

The analysis of the packing indicates a well defined hierarchy of intermolecular contacts, some of them suggesting how Le^x–Le^x interactions may occur in biological conditions providing the molecular basis for cell-cell recognition event.

Cellodextrins. The quest for the structural elucidation of the crystalline allomorphs of cellulose has stimulated interest in the resolution of the crystal structures of the low molecular weight oligomers. In the case of cellulose II, it has been established that the cellodextrins having a degree of polymerization of four and higher, give X-ray powder diffractograms that are similar to those of the polysaccharides. Furthermore, powders of methyl β-cellotrioside, and higher members, give diffraction patterns having the cellulose II features. The crystal structure of methyl β-cellotrioside was solved as a monohydrate 0.25 ethanolate, using synchrotron data collected on crystals of dimensions 0.43 × 0.33 × 0.04 mm [6].

β-D-cellotetraose crystallizes in the triclinic space group P_1 with two independent molecules and one water molecule in the unit cell having dimensions a = 8.045(12), b = 9.003(9), c = 22.51(2) Å, α = 89.66(7), β = 94.83(13), γ = 115.80(4)° [7]. Because of the very small dimensions of the crystals (0.40 × 0.15 × 0.015 mm) the X-ray diffraction data were collected at room temperature, using beam line at the European Synchrotron Radiation Facility in Grenoble. A monochromatic radiation with a wavelength of 0.925 Å was used, and the data collection was achieved on image plate detectors. At about the same time were reported the results of an independent investigation that was performed using a combination of conventional X-ray and synchrotron sources on crystals having dimensions 0.4 × 0. × 0.1 mm, [8]. In that case, the reported unit cell dimensions were a = 8.023(1), b = 8.951(2), c = 22.445(2) Å, α = 89.26(2), β = 85.07(1), γ = 63.93(1)°, this corresponds to a different setting of the crystallographic axis.

The least-squares superposition of the C and O-atoms of these two structural investigations, has a root mean square deviation of 0.05 Å for the non hydrogen atoms. Some several discrepancies are noticed in the positions of O–H hydrogen atoms and consequently in hydrogen bonding scheme. The way H atom positions were located in the two investigations may explained such discrepancies as in one case they were located in 'theoretical' positions whereas in the other one, they were located from differences Fourier analyses.

Among the many structural and conformational insights that such investigations are providing, one would like to emphasize the striking finding that the two crystallographically independent cellotetraose molecules are significantly different. These structural differences are not only observed with the ring puckering parameters, but also with the glycosidic Φ and Ψ torsion angles. Whereas one molecule is in a 'standard' conformation; the other one displays a significant conformational strain. The influence of mode of packing of these two molecules in the crystal has been invoked to explain such a difference, even though no significant molecular mechanic calculations have yet been performed to elucidate the reasons underlying such significant deviations from normality (Figure 5).

When a subcell is constructed of the two central D-glucopyranoses in the two independent molecule, the obtained cell dimensions are identical to those of the cellulose II allomorph. Based on this subcell, a new model for cellulose II has been proposed [8].

34.7 Crystalline Conformations of Oligosaccharides Complexed with Lectins

Since the beginning of the 90's an increasing number of crystal structures have been reported for glycoproteins and protein-carbohydrate complexes. The resolution of the first reported structures was rarely sufficient to provide reliable conformational information. Significant and rapid progresses arising from the use of synchrotron radiation are providing access to highly resolved structures. For example, the protein-carbohydrate literature provides structural data for 15 protein/cellobiose, 10 protein/lactose, 35 protein/maltose 4 protein/sucrose and 1 protein/α-α trehalose complexes crystal structures. Forces in protein-carbohydrate complexes should be more varied than in the fairly homogenous small molecule oligosaccharides crystals. The protein geometry reduces the influence of preferred packing modes, this is particularly true for hydrolytic enzymes that induces significant distortions both of the ring geometry and the conformations at the glycosidic linkages. Other bias may occur, resulting from the low-resolution of the structure, or from the fact that the crystallographic refinements are often guided with force fields. For these reasons, we are restricted the presentation to the class of lectin-oligosaccharide crystal structures, where there is no distortion occurring from hydrolysis and for which the crystallographic resolution is generally very high. Among proteins that interact non covalently with carbohydrates, lectins bind mono- and oligosaccharides reversibly

34.7 Crystalline Conformations of Oligosaccharides Complexed with Lectins

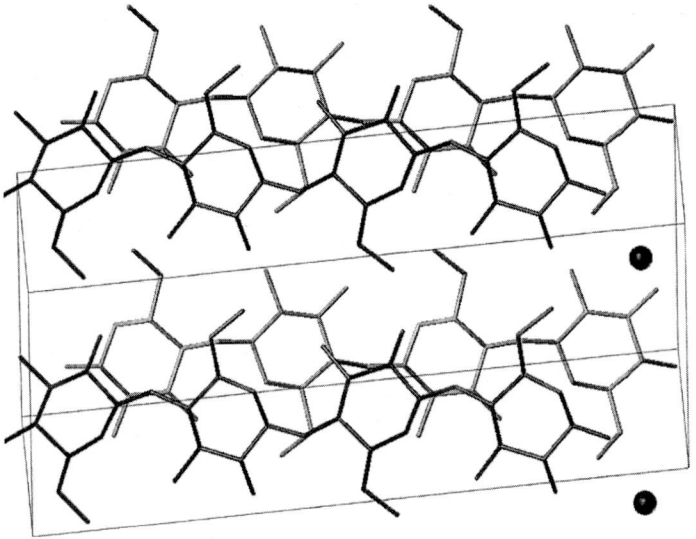

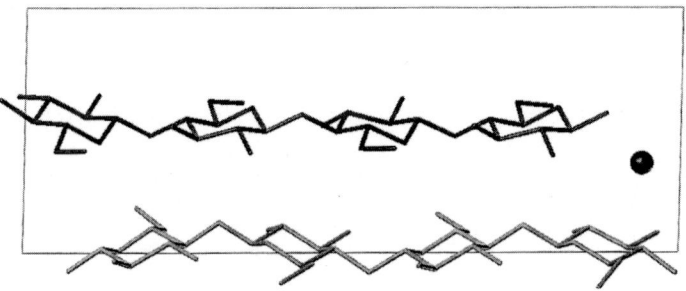

Figure 5. The molecular structure of β-D-cellotetraose [8]. Two orthogonal views of the unit cell.

and specifically, while displaying no catalytic or immunological activity. More than 200 of crystal structures of lectins have been solved, most of them as complexes with carbohydrate ligands. From a data base of three-dimensional structures of lectins (http://www.cermav.cnrs/databank/lectins) Table 5 has been set, to illustrate the extraordinary wealth of information which is available about oligosaccharides.

One can found information about simple disaccharides (maltose, sucrose, ...) and investigate how their 'bio-active' conformation may be similar/different from the one display in molecular crystals. Presumably more informative are those complexes involving biologically important oligosaccharides for which no structural information was available. For example these are: (i) the sialic acid containing oligosaccharides (sialyllactose, sialoglycopeptide, ...), important antigens (T antigen, histo blood group antigen such as Lewis antigens and their sialylated or sulfated

Table 5. References and selected structural information for the crystal structures of lectin-oligosaccharide crystal structures.

Origin	Protein	Glycan	PDB code and ref
Wheat germ	Agglutinin I (WGA)	Sialyllactose	1WGC [101]
		Sialogycopeptide	2CWG [102]
Wheat germ	Agglutinin II(WGA)	Sialyllactose	2WGC [101]
Maclura pomifera	Agglutinin	Galβ1,3GalNAc	1JOT [103]
Ricinus communis	Ricin (RCA)	Lactose	2AAI [104]
Amaranthus caudatus	Agglutinin (ACA)	Gal-β1,3-GalNAc-α-O-benzyl	1JLX [105]
Arachis hypogaea	Lectin (PNA)	Lactose	2PEL [106]
		T-antigen	2TEP [107]
		C-lactose	1BZW [108]
		N-acetyllactosamine	1CIW [109]
Canavalia ensiformis	Concanavalin A (ConA)	αMan(1–3)αMan	1QDO [110]
		αMan(1–6)αMan	1QDC [110]
		αMan(1–6)[αMan(1–3)]αMan	1CVN [111]
			1ONA [112]
		Pentasaccharide from N-glycan	1TEI [113]
Dioclea grandiflora	Lectin	αMan(1–6)[αMan(1–3)]αMan	1DGL [114]
Dolichos biflorus	Lectin (DBL)	Forssman disaccharide	1LU1 [115]
Erythrina corallodendron	Lectin (EcorL)	Lactose	1LTE [116]
			1AX1 [117]
		N-acetyllactosamine	1AX2 [117]
Glycine max	Soybean agglutinin (SBA)	N-acetyllactosamine	2SBA [118]
			1SBD, 1SBE, 1SBF [119]
Griffonia simplicifolia	Isolectin IV (GS4)	Lewis B	1LED [120]
		Lewis Y	1GSL [120]
Lathyrus ochrus	Isolectin I (LOL-I)	αMan(1–3)βMan(1–4)GlcNAc	1LOG [121]
		Biantennary octasaccharide	1LOF [30]
		glycopeptide	1LGC [31]
Lathyrus ochrus	Isolectin II (LOL-II)	N2 fragment of lactotransferine	1LGB [31]
Lens culinaris	Lectine (LcL)	Sucrose	1LES [122]
Galanthus nivalis	Agglutinin (GNA)	αMan(1–3)αMan	1NIV [123]
		Mannopentaose	1JPC [124]

Narcissus pseudonarcissus	Daffodil lectin	αMan(1–3)αMan	1NPL [125]
Rattus rattus	Mannose binding protein A	Biantennary glycopeptide	2MSB [126]
Rattus norvegicus	MBP-A mutant CL-K3	3'sialyl Lewis X	2KMB [127]
		3'sulfo Lewis X	3KMB [127]
		4'sulfo Lewis X	4KMB [127]
Bos taurus	Mannose 6-P receptor	Mannopentaose	1C39 [128]
Bos taurus	Galectin-1	N-acetyllactosamine	1SLT [127]
		Biantennary oligosaccharide	1SLA, 1SLB, 1SLC [129]
Bufo arenarum	Galectin-1	N-acetyllactosamine	1GAN
		Thio-digalactose	1A78
Homo sapiens	Galectin-2	Lactose	1HLC [130]
	Galectin-7	Lactose	4GAL [131]
	Galectin-7	N-acetyllactosamine	5GAL [131]
Conger myriaster	Congerin-I	Lactose	1C1L [132]
Mus Musculus	Sialoadhesin	3' Sialyllactose	1QFO [133]
Bordetella pertussis	Pertussis toxin	3' Sialylgalactose	1PTO [134]
Vibrio cholerae	Cholera toxin	GM1 pentasaccharide	1CHB [135]
			2CHB, 1CT1 [32]
Escherichia coli	Heat-labile Enterotoxin	Lactose	1LTT [136]
	Verotoxin-1	Receptor GB3	1BOS [137]
Staphylococcus aureus	Enterotoxin B	Lactose	1SE4 [138]
		GM3 trisaccharide	1SE3 [138]
Murine Polyomavirus	Coat protein	3' Sialyllactose	1SID [139]
		Disialylated oligosaccharide	1SIE [139]
			1VPS [140]
Influenzae virus	Hemagglutinin	3' Sialyllactose	1HGG [141]
Phage P22	Tailspike protein	Salmonella octasaccharide	1TYU, 1TYX
		Salmonella nonasaccharide	1TYW [142]

forms.), (iii) moieties of glycolipids such as GM1 and GM3, (iv) oligosaccharides belonging to N-linked glycans, biantennary glycopetide, a N2 fragment of lactotransferine, (v) fragments of polysaccharides from the cell wall of pathogenic bacteria (*Salmonella* octa-and nona-saccharide)....

Whereas the study of the conformations about the glycosidic torsion angles indicated a somehow limited flexibility in molecular crystal (*vide infra*) a different pictures emerges for the flexibility of oligosaccharides interacting with lectins. Figure 6 depicts the distributions of glycosidic torsion angles within three disaccharide segments: αMan(1–3)Man, βGlcNAc(1–2) Man and αNeuAc(2–3)Gal. In the case of the αMan(1–3)Man segment, the observed conformations are essentially located around a Φ value of 80°, with an excursion in the vicinity of 140°. May be more interesting is the observation that a remote low energy area (located at $\Phi = 90°$ and $\Psi = 310°$) can be occupied, as observed in the crystalline complex between *Latyrus ochrus* and a biantennary glycan [30]. The study of the dispersion of conformations observed for the disaccharide segment βGlcNAc(1–2) Man, provides another illustration of the occurrence of conformations in remote energy well of the potential energy surfaces. The location of this well is 120° away from what would correspond to the stable conformation driven by the *exo*-anomeric effect in the case of an equatorial type of linkage. Such examples are observed in the crystalline complexes involving the isolectin II of *Latyrus ochrus*, complexed with high molecular oligosaccharides such as a biantennary octasaccharide [30], a glycopeptide or a N2 fragment of lactotransferin [31]. The αNeuAc(2–3)Gal offers an extreme case of conformational flexibility as it can be ten fold more flexible than the other disaccharides. Here again, the conformation corresponding to the establishment of the exo-anomeric effect $\Phi = 60°$ is adopted in several case. Such a stabilizing influence can be easily overridden as exemplified by the occurrence of several low energy conformations having Φ in the vicinity of −60°; this is observed for the GM1 pentasaccharide interacting in the combining site a cholera toxin [32].

34.8 Concluding Remarks

With this article, we have set out to provide enough information to initiate the general reader into the world of oligosaccharide structures as derived from diffraction methods. Oligosaccharide structures, numeric data and detailed reference lists have been presented, which may be useful to those directly involved in structural glycoscience. Some keys conformational features of oligosaccharides, have been dissected using the disaccharide moiety as a basic reference entity. This allowed us to provide a useful classification, both of the disaccharides and their analogs. Some of the general rules which dictate the formation of hydrogen bonding and packing patterns in crystalline oligosaccharides have been presented. They provide an efficient template to decipher those very complex structures. It must certainly be emphasized that despite the wealth of potential structural information available, there is a lack of comprehensive investigation of the molecular principles underlying the formation of oligosaccharide crystals.

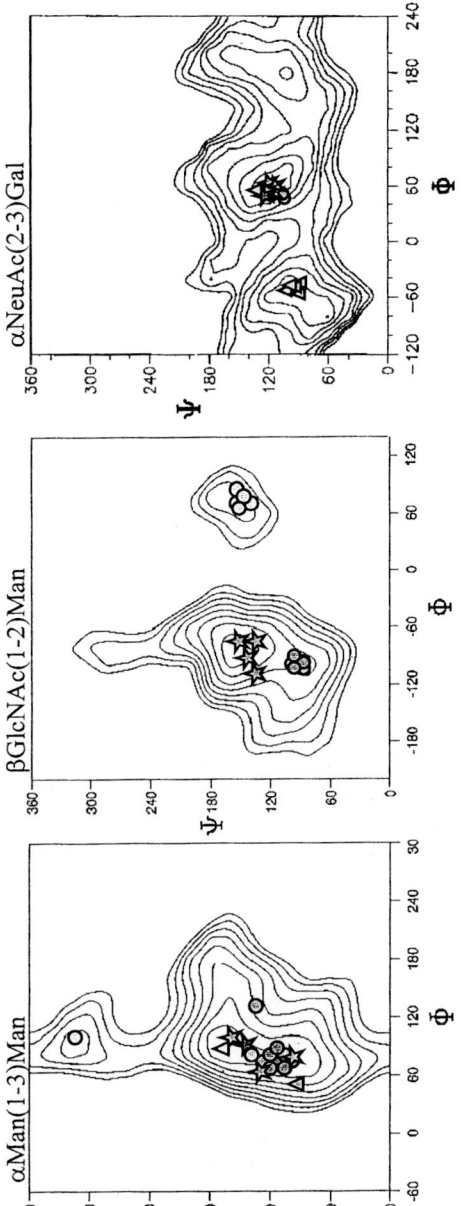

Figure 6. Iso-energy contours of three-disaccharides as calculated with the MM3 program, along with the glycosidic conformations observed in crystalline complexes with lectins.

There has been a significant stagnation of the number of oligosaccharide structure analysis reported over the last decade. Usually, the minute amount of available pure oligosaccharides along with their natural reluctance to crystallize in a form suitable for X-ray diffraction studies may explain such a stagnation. It is nevertheless expected that progress in the synthesis or biosynthesis of carbohydrates as well as development in X-ray diffraction sources will be sufficient to overcome the paucity of structural and crystalline data on oligosaccharides. Interest is focusing on the more complex carbohydrate molecules, as diffraction methods extend to the field of protein-carbohydrate complex. Such a field benefits from the developments of methods for crystallizing these protein-based complexes as well as from the rapid progresses arising from the use of synchrotron radiation which is providing access to highly resolved structures. The conjunction of these new developments is likely to provide exciting new results and a greater insight into the biological mechanisms of this fascinating class of biomolecules.

References

1. F.J. Llewellyn, E.G. Cox, T.H. Goodwin, *J. Chem. Soc.* **1937**, 883–894.
2. E.G. Cox, G.A. Jeffrey, *Nature* **1939**, *143*, 894–895.
3. C.A. Beevers, W. Cochran, *Proc. R. Soc. London Ser. A.* **1947**, *190*, 276–285.
4. G.M. Brown, H.A. Levy, *Science* **1963**, *141*, 921–924.
5. G.M. Brown, H.A. Levy, *Science* **1965**, *147*, 1038–1041.
6. S. Raymond, B. Hennissat, D.T. Qui, A. Kvick, H. Chanzy, *Carbohydr. Res.* **1995**, *277*, 209–229.
7. S. Raymond, A. Heyraud, D. Tran Qui, A. Kvick, H. Chanzy, *Macromolecules* **1995**, *28*, 2096–2100.
8. K. Gessler, N. Krauss, T. Steiner, C. Betzel, A. Sarko, W. Saenger, *J. Am. Chem. Soc.* **1995**, *117*, 11397–11406.
9. F.H. Allen, O. Kennard, *Chemical Design Automation News* **1993**, *8*, 31–37.
10. K. Gessler, I. Uson, T. Takaha, N. Krauss, S.M. SMith, S. Okada, G.M. Sheldrick, W. Saenger, *Proc. Nat. Acad. Sci. USA* **1999**, *96*, 4246–4251.
11. W. Saenger, J. Jacob, K. Gessler, T. Steiner, D. Hoffman, H. Sanbe, K. Koizumi, S.M. Smith, T. Takaha, *Chem. Rev.* **1998**, *98*, 1787–1802.
12. W. Hinrichs, W. Saenger, *J. Am. Chem. Soc.* **1990**, *112*, 2789–2796.
13. A. Imberty, H. Chanzy, S. Pérez, A. Buléon, V. Tran, *J. Mol. Biol.* **1988**, *201*, 365–378.
14. G.A. Jeffrey, R. Taylor, *J. Comput. Chem.* **1980**, *1*, 99–109.
15. G.A. Jeffrey, R.A. Wood, P.E. Pfeffer, K.B. Hicks, *J. Am. Chem. Soc.* **1983**, *105*, 2128–2123.
16. D. Cremer, J.A. Pople, *J. Am. Chem. Soc.* **1975**, *97*, 1354–1358.
17. IUPAC-IUB, *Arch. Biochem. Biophys.* **1971**, *145*, 405–621.
18. R.U. Lemieux, S. Koto, D. Voisin In: *Anomeric effect: origin and consequence*, ACS symposium Series No. 87. Szarek, W. A. and Horton, D., Ed.; Amer. Chem. Soc.: Washington DC, 1979, pp 17–29.
19. S. Perez, C. Vergelati, *Acta Cryst.* **1984**, *B40*, 294–299.
20. T. Weimar, U.C. Keis, J.S. Andrews, B.M. Pinto, *Carbohydr. Res.* **1999**, *315*, 222–233.
21. A. Neuman, F. Longchambon, O. Abbes, H. Gillier-Pandraud, S. Pérez, D. Rouzaud, P. Sinay, *Carbohydr. Res.* **1990**, *195*, 187–197.
22. D.J. O'Leary, Y. Kishi, *J. Org. Chem.* **1993**, *58*, 304–306.
23. A. Imberty, J. Gruza, N. Mouhous-Riou, B. Bachet, S. Pérez, *Carbohydr. Res.* **1998**, *311*, 135–146.
24. I. Tvaroska, J.P. Carver, *J. Chem. Phys.* **1996**, *100*, 11305–11313.
25. T. Taga, Y. Miwa, Z. Min, *Acta Crystallogr* **1997**, *C53*, 234–256.

26. F.H. Allen, *Acta Cryst.* **1986**, *B42*, 515–522.
27. G.A. Jeffrey, *Acta Cryst.* **1990**, *B46*, 89–103.
28. G.A. Jeffrey, W. Saenger *Hydrogen bonding in biological structures*; Springer-Verlag: Berlin, 1991.
29. S. Pérez, N. Mouhous-Riou, N.E. Nifant'ev, Y.E. Tsvetkov, B. Bachet, A. Imberty, *Glycobiology* **1996**, *6*, 537–542.
30. Y. Bourne, P. Rougé, C. Cambillau, *J. Biol. Chem.* **1992**, *267*, 197–203.
31. Y. Bourne, J. Mazurier, D. Legrand, P. Rougé, J. Montreuil, G. Spik, C. Cambillau, *Structure* **1994**, *2*, 209–219.
32. E.A. Merritt, S. Sarfaty, M.G. Jobling, T. Chang, R.K. Holmes, T.R. Hirst, W.G. Hol, *Protein Sci.* **1997**, *6*, 1516–1528.
33. A. Otter, R.U. Lemieux, R.G. Ball, A.P. Venot, O. Hindsgaul, D.R. Bundle, *Eur. J. Biochem.* **1999**, *259*, 295–303.
34. R.A. Moran, C.F. Richards, *Acta Cryst.* **1973**, *B29*, 2770–2783.
35. V. Warin, F. Baert, R. Fouret, C. Strecker, C. Spik, B. Fournet, J. Montreuil, *Carbohydr. Res.* **1979**, *76*, 11–22.
36. T. Taga, E. Inagaki, Y. Fujimori, S. Nakamura, *Carbohydr. Res.* **1993**, *240*, 39–45.
37. T. Taga, E. Inagaki, Y. Fujimori, S. Nakamura, *Carbohydr. Res.* **1994**, *251*, 203–212.
38. W. Pangborn, D. Langs, S. Perez, *Int. J. Biol. Macromol.* **1985**, *7*, 363–369.
39. G.A. Jeffrey, D.-B. Huang, *Carbohydr. Res.* **1991**, *222*, 47–55.
40. G.A. Jeffrey, D.-B. Huang, *Carbohydr. Res.* **1990**, *206*, 173–182.
41. W. Mackie, B. Sheldrick, D. Akrigg, S. Pérez, *Int. J. Biol. Macromol.* **1986**, *8*, 43–51.
42. G.A. Jeffrey, Y.J. Park, *Acta Cryst.* **1972**, *B28*, 257–267.
43. V. Ferretti, V. Bertolasi, G. Gilli, C.A. Accorsi, *Acta Cryst.* **1984**, *40*, 531–535.
44. D. Avenel, A. Neuman, H. Gillier-Pandraud, *Acta Cryst.* **1976**, *B32*, 2598–2605.
45. J. Becquart, A. Newman, H. Gillier-Pandraud, *Carbohydr. Res.* **1982**, *111*, 9–21.
46. D.C. Rohrer, *Acta Cryst.* **1972**, *B28*, 425–433.
47. S. Perez, F. Brisse, *Acta Cryst.* **1977**, *B33*, 2578–2584.
48. G.A. Jeffrey, D.-B. Huang, *Carbohydr. Res.* **1991**, *210*, 89–104.
49. G.A. Jeffrey, D.B. Huang, *Carbohydr. Res.* **1993**, *247*, 37–50.
50. K. Noguchi, E. Kobayashi, K. Okuyama, S. Kitamura, K. Takeo, S. Ohno, *Carbohydr. Res.* **1994**, *258*, 35–47.
51. A. Linden, C.K. Lee, *Acta Cryst.* **1995**, *C51*, 1007–1012.
52. J. Ollis, V.J. James, S.J. Angyal, P.M. Pojer, *Carbohydr. Res.* **1978**, *60*, 219–228.
53. C.-K. Lee, L.L. Koh, *Carbohydr. Res.* **1994**, *254*, 281–287.
54. G.M. Brown, D.C. Rohrer, B. Berking, C.A. Beevers, R.O. Gould, R. Simpson, *Acta Cryst* **1972**, *B28*, 3145–3158.
55. G.A. Jeffrey, R. Nanni, *Carbohydr. Res.* **1985**, *137*, 21–30.
56. W.J. Cook, C.E. Bugg, *Carbohydr. Res.* **1973**, *31*, 265–275.
57. A. Linden, C.K. Lee, *Acta Cryst.* **1995**, *C51*, 1012–1016.
58. C.-K. Lee, A. Linden, *Carbohydr. Res.* **1994**, *264*, 319–325.
59. T. Srikrishnan, M.S. Chowdhary, K.L. Matra, *Carbohydr. Res.* **1989**, *186*, 167–175.
60. J. Thiem, M. Kleeberg, K.-H. Klaska, *Carbohydr. Res.* **1989**, *189*, 65–77.
61. G. Svensson, J. Albertsson, C. Svensson, C. Magnusson, J. Dahmen, *Carbohydr. Res.* **1986**, *146*, 29–38.
62. D.K. Watt, D.J. Brasch, D.S. Larsen, L.D. Melton, J. Simpson, *Carbohydr. Res.* **1996**, *285*, 1–15.
63. L. Eriksson, R. Stenutz, G. Widmalm, *Acta Cryst.* **1997**, *C53*, 1105–1107.
64. A. Neuman, D. Avenel, F. Arene, H. Gillier-Fandraud, J.-R. Pougny, P. Sinay, *Carbohydr. Res.* **1980**, *80*, 15–24.
65. A. Neuman, D. Avenel, H. Gillier-Pandraud, *Acta Cryst.* **1978**, *B34*, 242–248.
66. F. Takusagawa, R.A. Jacobson, *Acta Cryst* **1978**, *B34*, 213–218.
67. M.E. Gress, G.A. Jeffrey, *Acta Cryst.* **1977**, *B33*, 2490–2495.
68. I. Tanaka, N. Tanaka, T. Ashida, M. Kakudo, *Acta Cryst.* **1976**, *B32*, 155–160.
69. S.S.C. Chu, G.A. Jeffrey, *Acta Cryst.* **1967**, *23*, 1038–1049.
70. M.E. Gress, G.A. Jeffrey, D.C. Rohrer, *Acta Crystallogr.* **1978**, *B34*, 508–512.

71. V. Mikol, P. Kosma, H. Brade, *Carbohydr. Res.* **1994**, *263*, 35–42.
72. C.-K. Lee, L.L. Koh, *Acta. Cryst.* **1993**, *C49*, 621–624.
73. J. Ohanessian, F. Longchambon, F. Arene, *Acta Cryst.* **1978**, *BM*, 3666–3671.
74. H. Takeda, N. Yasuoka, N. Kasai, *Carbohydr. Res.* **1977**, *53*, 137–152.
75. K. Noguchi, K. Okuyama, S. Kitamura, K. Takeo, *Carbohydr. Res.* **1992**, *237*, 33–43.
76. D. Lamba, A.L. Segre, S. Clover, W. Mackie, B. Sheldrick, S. Perez, *Carbohydr. Res.* **1990**, *208*, 215–230.
77. M. Senma, T. Taga, K. Osaka, *Chem. Lett.* **1974**, 1415–1418.
78. F. Longchambon, J. Ohanessian, H. Gillier-Pandraud, D. Duchet, J.-C. Jacquinet, P. Sinay, *Acta Cryst.* **1981**, *B37*, 601–607.
79. C.E. Bugg, *J. Am. Chern. Soc.* **1973**, *95*, 908–913.
80. J.H. Noordik, P.T. Beurskens, P. Bennema, R.A. Visser, R.O. Could, *Z. Kristallogr.* **1984**, *168*, 59–65.
81. K. Hirotsu, A. Shimada, *Bull. Chem. Soc. Jpn.* **1974**, *47*, 1872–1879.
82. M.M. Olmstead, M. Hu, M.J. Kurth, J.M. Krochta, Y.L. Hsieh, *Acta Cryst.* **1997**, *C53*, 915–916.
83. B. Sheldrick, W. Mackie, A. D., *Carbohydr. Res.* **1984**, *132*, 1–6.
84. S.S.C. Chu, G.A. Jeffrey, *Acta Cryst.*, **1968**, *B24*, 830–838.
85. P. Koll, M. Petrusova, L. Petrus, B. Zimmer, M. Morf, J. Kopf, *Carbohydr. Res.* **1993**, *248*, 37–43.
86. J.T. Ham, D.G. Williams, *Acta Cryst.* **1970**, *B26*, 1373–1383.
87. F. Mo, L.H. Jensen, *Acta Cryst.* **1978**, *B34*, 1562–1569.
88. F. Mo, *Acta Chem. Scand.* **1979**, *A33*, 207–218.
89. C. Burden, W. Mackie, B. Sheldrick, *Acta Cryst.* **1986**, *C42*, 177–179.
90. F. Arene, A. Neuman, F. Longchambon, *C. R. Acad. Sci. Paris* **1979**, *C288*, 331–334.
91. J.A. Kanters, R.L. Scherrenberg, B.R. Leeflang, J. Kroon, M. Mathlouthi, *Carbohydr. Res.* **1988**, *180*, 175–182.
92. G.M. Brown, H.A. Levy, *Acta Cryst.* **1973**, *B29*, 790–797.
93. C.A. Accorsi, F. Bellucci, V. Bertolasi, V. Ferretti, G. Gilli, *Carbohydr. Res.*, **1989**, *191*, 105–116.
94. C.A. Accorsi, V. Bertolasi, V. Ferretti, G. Gilli, *Carbohydr. Res.* **1989**, *191*, 91–104.
95. R.V. Krishnakumar, S. Natarajan, *Carbohydr. Res.* **1996**, *287*, 117–122.
96. T. Taga, E. Inagaki, Y. Fujimori, K. Fujita, K. Hara, *Carbohydr. Res.* **1992**, *241*, 63–69.
97. W. Dreissig, P. Luger, *Acta Cryst.* **1973**, *B29*, 514–521.
98. G.A. Jeffrey, D.B. Huang, P.E. Pfeffer, R.L. Dudley, K.B. Hicks, E. Nitsch, *Carbohydr. Res.* **1992**, *226*, 29–42.
99. T. Skrydstrup, D. Mazéas, M. Elmouchir, G. Doisneau, C. Riche, A. Chiaroni, J.M. Beau, *Chem. Eur. J.* **1997**, *3*, 1342–1356.
100. S. Mehta, K.L. Jordan, U.C. Kreis, T. Weimar, R.J. Batchelor, F.W.B. Einstein, B.M. Pinto, *Tetrahedron Asym.* **1994**, *5*, 2367–2396.
101. C.S. Wright, *J. Mol. Biol.* **1990**, *215*, 635–651.
102. C.S. Wright, J. Jaeger, *J. Mol. Biol.* **1993**, *232*, 620–638.
103. X. Lee, A. Thompson, Z. Zhang, H. Ton-that, J. Biesterfeldt, C. Ogata, Xu, L., R.A. Johnston, N.M. Young, *J. Biol. Chem.* **1998**, *273*, 6312–6318.
104. E. Rutenber, B.J. Katzin, S. Ernst, E.J. Collins, D. Mlsna, M.P. Ready, R. J.D., *Proteins* **1991**, *10*, 240–250.
105. T.R. Transue, A.K. Smith, H. Mo, I.J. Goldstein, M.A. Saper, *Nat. Struct. Biol.* **1997**, *10*, 779–783.
106. R. Banerjee, K. Das, R. Ravishankar, K. Suguna, A. Surolia, M. Vijayan, *J. Mol. Biol.* **1996**, *259*, 281–296.
107. R. Ravishankar, M. Ravindran, K. Suguna, A. Surolia, M. Vikayan, *Curr. Sci.* **1997**, *72*, 855–861.
108. R. Ravishankar, A. Surolia, M.V.S. Lim, Y. Kishi, *J. Amer. Chem. Soc.* **1998**, *120*, 11297–11303.
109. R. Ravishankar, K. Suguna, A. Surolia, M. Vijayan, *Acta. Crystallogr.* **1999**, *D55*, 1375–1382.

110. J. Bouckaert, T. Hamelryck, L. Wyns, R. Loris, *J. Biol. Chem.* **1999**, *274*, 29188–29195.
111. J.H. Naismith, R.A. Field, *J. Biol. Chem.* **1996**, *271*, 972–976.
112. R. Loris, D. Maes, F. Poortmans, L. Wyns, J. Bouckaert, *J. Biol. Chem.* **1996**, 271, 30614–30618.
113. D.N. Moothoo, J.H. Naismith, *Glycobiology* **1998**, *8*, 173–181.
114. D.A. Rozwarski, B.M. Swami, C.F. Brewer, J.C. Sacchettini, *J. Biol. Chem.* **1998**, *273*, 32818–32825.
115. T.W. Hamelryck, R. Loris, J. Bouckaert, M.-H. Dao-Thi, G. Strecker, A. Imberty, E. Fernandez, L. Wyns, M.E. Etzler, *J. Mol. Biol.* **1999**, *286*, 1161–1177.
116. B. Shaanan, H. Lis, N. Sharon, *Science* **1991**, *254*, 862–866.
117. S. Elgavish, B. Shaanan, *J. Mol. Biol.* **1998**, *277*, 917–932.
118. A. Dessen, D. Gupta, S. Sabesan, C.F. Brewer, J.C. Sacchetini, *Biochemistry* **1995**, *34*, 4933–4942.
119. L.R. Olsen, A. Dessen, D. Gupta, S. Sabesan, J.C. Sacchettini, C.F. Brewer, *Biochemistry* **1997**, *36*, 15073–15080.
120. L.T. Delbaere, M. Vandonselaar, L. Prasad, J.W. Quail, K.S. Wilson, Z. Dauter, *J. Mol. Biol.* **1993**, *230*, 950–965.
121. Y. Bourne, P. Rougé, C. Cambillau, *J. Biol. Chem.* **1990**, *265*, 18161–18165.
122. F. Casset, T. Hamelryck, R. Loris, J.R. Brisson, C. Tellier, M.H. Dao-Thi, L. Wyns, F. Poortmans, S. Perez, A. Imberty, *J. Biol. Chem.* **1995**, *270*, 25619–25628.
123. G. Hester, C.S. Wright, *J. Mol. Biol* **1996**, *262*, 516–531.
124. C.S. Wright, G. Hester, *Structure* **1996**, *4*, 1339–1352.
125. M.K. Sauerborn, L.M. Wright, C.D. Reynolds, J.G. Grossmann, P.J. Rizkallah, *J. Mol. Biol.* **1999**, *290*, 185–199.
126. W.I. Weis, K. Drickamer, W.A. Hendrickson, *Nature* **1992**, *360*, 127–134.
127. K.K. Ng, W.I. Weis, *Biochemistry* **1997**, *36*, 979–988.
128. L.J. Olson, J. Zhang, Y.C. Lee, N.M. Dahms, J.J.P. Kim, *Journal of Biological Chemistry* **1999**, *274*, 29889–29896.
129. Y. Bourne, B. Bolgiano, D.I. Liao, G. Strecker, P. Cantau, O. Herzberg, T. Feizi, C. Cambillau, *Nature Struct. Biol.* **1994**, *1*, 863–870.
130. Y.D. Lobsanov, M.A. Gitt, H. Leffler, S.H. Barondes, J.M. Rini, *J. Biol. Chem.* **1993**, *268*, 27034–27038.
131. D.D. Leonidas, E.H. Vatzaki, H. Vorum, J.E. Celis, P. Madsen, K.R. Acharya, *Biochemistry* **1998**, *37*, 13930–13940.
132. T. Shirai, C. Mitsuyama, Y. Niwa, Y. Matsui, H. Hotta, T. Yamane, H. Kamiya, C. Ishii, T. Ogawa, K. Muramoto, *Structure* **1999**, *7*, 1223–1233.
133. A.P. May, R.C. Robinson, M. Vinson, P.R. Crocker, E.Y. Jones, *Mol. Cell.* **1998**, *1*, 719–728.
134. P.E. Stein, A. Boodhoo, G.D. Armstrong, L. Heerze, S.A. Cockle, M.H. Klein, R.J. Read, *Nat. Struct. Biol.* **1994**, *1*, 591–596.
135. E.A. Merritt, S. Sarfaty, F. Van Den Akker, C. L'hoir, J.A. Martial, W.G.J. Hol, *Protein Sci.* **1994**, *3*, 166–175.
136. T.K. Sixma, S.E. Pronk, K.H. Kalk, B.A.M. Van Zanten, A.M. Berghuis, W.G.J. Hol, *Nature* **1992**, *355*, 561–564.
137. H. Ling, A. Boodhoo, B. Hazes, M.D. Cummings, G.D. Armstrong, J.L. Brunton, R.J. Read, *Biochemistry* **1998**, *37*, 1777–1788.
138. S. Swaminathan, W. Furey, J. Pletcher, M. Sax, *Nature Struct. Biol.* **1995**, *2*, 680–686.
139. T. Stehle, S.C. Harrison, *Structure* **1996**, *4*, 183–194.
140. T. Stehle, S.C. Harrison, *EMBO J.* **1997**, *16*, 5139–5148.
141. N.K. Sauter, J.E. Hanson, G.D. Glick, J.H. Brown, R.L. Crowther, S.J. Park, J.J. Skehel, D.C. Wiley, *Biochemistry* **1992**, *31*, 9609–9621.
142. S. Steinbacher, U. Baxa, S. Miller, A. Weintraub, R. Seckler, R. Huber, *Proc. Nat. Acad. Sci. USA* **1996**, *93*, 10584–10588.

35 Transfer NOE Experiments for the Study of Carbohydrate–Protein Interactions

Thomas Peters

35.1 Introduction

Many biological recognition events are based upon specific protein–carbohydrate interactions. For instance, self non-self discrimination in immune reactions, bacterial and viral adhesion to mammalian target cells as well as key steps in inflammatory processes involve protein–carbohydrate recognition as the heart of the matter at the molecular level [1–5]. The broad interest in a detailed understanding of complexation reactions between carbohydrate ligands and corresponding receptor proteins certainly originates from considerations that go well beyond mere scientific questions. In most cases the biological recognition phenomenon is potentially coupled to pathophysiological conditions that could be cured by inhibition of the particular protein–carbohydrate interaction. Therefore, a strong driving force for research projects that address the molecular nature of protein–carbohydrate recognition reactions stems from potential pharmaceutical applications. A good example that has attracted much attention during recent years is the interaction between selectins and their carbohydrate ligands because these complexation reactions represent the first steps in the inflammatory cascade [6–10]. In pathological situations such as myocardial infarction, transplantation, or rheumatoid arthritis, efficient suppression of the inflammatory cascade is very important. The design of potent inhibitors obviously requires precise knowledge about the underlying binding reaction, and more specifically about the bioactive conformation of the carbohydrate ligand [11–17]. In the case of the selectins, an X-ray structure of the carbohydrate-recognition domain of E-selectin has been published [18]. Attempts to co-crystallize E-selectin with its "natural" ligand, the sialyl Lewis^x tetrasaccharide (Figure 1), failed so far. Knowledge about the bioactive conformation of sialyl Lewis^x came from NMR spectroscopy utilizing experiments that are based upon the so called transfer NOE (trNOE) effect [19–23]. In general, the observation of transfer NOEs allows an analysis of bound ligand conformations without requiring any knowledge about the complex protein NMR spectrum (for a comprehensive explanation of the trNOE effect see [24]).

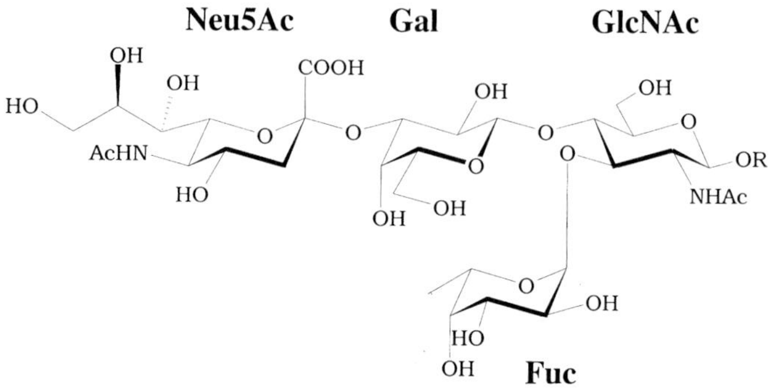

Figure 1. Chemical formula of sialyl Lewis^x (sLe^x), α-D-Neu5Ac-(2→3)-β-D-Gal-(1→4)-[α-L-Fuc-(1→3)]-β-D-GlcNAc.

35.2 The Transfer NOE Experiment

Transferred NOEs were originally observed and their principles described more than twenty years ago [26, 27]. The experiment is based on a chemical exchange between ligand molecules in solution and ligand molecules bound to large proteins. Various experimental implementations have been explored since then, ranging from one dimensional selective saturation transfer experiments [27, 28] to one and two dimensional transient NOE experiments [29, 30]. The chemical exchange between free and bound ligand molecules leads to an averaging of physical properties characteristic for the free and for the bound state, respectively. One important physical parameter that distinguishes free and bound molecules is the so called motional correlation or tumbling time, τ_c. The motional correlation time τ_c is defined as the time that is required for a molecule to advance by one radian during translational and rotational diffusion. It is apparent that small molecules have short correlation times τ_c, whereas large molecules such as high molecular weight proteins have long correlation times τ_c. As an approximation, Equation 1 relates the motional correlation time τ_c to the molecular weight of the molecule in question [24]:

$$\tau_c \approx M_r/4 \; [\text{ns}] \tag{1}$$

with M_r representing the molecular weight in kDa. This equation assumes isotropic overall molecular motion and neglects internal molecular motions but it is usually sufficient for a first estimate of τ_c. It is of fundamental importance to realize that the correlation time τ_c determines sign and size of the NOE [24]. Low or medium molecular weight molecules (MW < 1–2 kDa) have short tumbling times τ_c, and as a consequence such molecules exhibit positive NOEs, no NOEs, or very small negative NOEs depending on their molecular weight, shape and the field strength

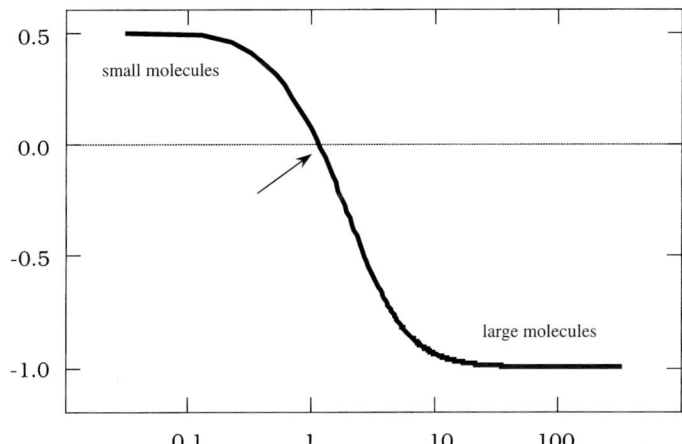

Figure 2. Dependence of the NOE enhancement (η) on the motional correlation τ_c under steady state conditions [24]. The spectrometer frequency ω was set at 500 MHz. The size of tetra- and pentasaccharides usually leads to molecular tumbling times τ_c that cause the NOE to become zero or very close to zero.

applied. Large molecules such as proteins, polysaccharides or nucleic acids have long tumbling times τ_c, and therefore exhibit strong negative NOEs reaching -100% at the extreme. The dependence of the NOE on the tumbling time and the field strength is depicted in Figure 2.

It is obvious that the motion of a small molecule that binds to a large molecule is determined by the long correlation time of the large molecule during the life time of the complex. Upon dissociation of the complex the small ligand tumbles again with its own characteristic correlation time. During a normal NOESY experiment the large negative NOEs of the bound ligand will be transferred to the sharp signals of the free ligand. The NOEs that are observed under such circumstances are called transferred NOEs (trNOEs). It is obvious that trNOEs contain information that is necessary to determine the conformation of the bound ligand, the so called bioactive conformation. One advantage of trNOE experiments is that they usually work at optimum with a ten to twenty fold excess of ligand, making it very easy to observe signals of free ligand molecules against the background of broad protein signals. The reason for the fact that the optimum is reached at excess ligand concentrations is inequality 2 [24]:

$$N_b \times \sigma_b \gg N_f \times \sigma_f \qquad (2)$$

N_b and N_f are the number of ligand molecules in the bound (b) and the free (f) state, respectively. σ_b and σ_f designate the cross relaxation rates of the ligand in the bound and free state. Cross relaxation rates are mainly a function of r^{-6}, with r being the distance between two protons, and of the motional correlation time τ_c (Equation 1). They determine sign and size of the NOE (cf. Figure 2). Since cross relaxation rates of large biomolecules such as proteins are much larger than cross relaxation rates of small molecules, inequality 2 is usually satisfied for a 10–20-fold excess of ligand. In fact, σ_f sometimes is so small that it does not play any significant role. This is the case for the turning point in Figure 2, where no NOEs can be observed. Usually, tetra- and pentasaccharides are of such a molecular weight (Equation 1) that NOEs are very close to zero.

It is important to realize that the observation of trNOEs is only possible under certain conditions. In general, it can be said that trNOEs are only observable for complexes with K_D-values approximately in the range between 10^{-6} to 10^{-3}. It follows that trNOE-experiments are ideally suited for the investigation of protein-carbohydrate complexes which are usually characterized by low binding affinities [31]. The reason for the limitations of the observation of trNOE lies in the nature of the NOE itself. Dipole–dipole relaxation is the source of NOEs, and therefore the observation of trNOEs depends critically on the kinetics of the association-dissociation process of a protein–carbohydrate complex [32–35]. In general, it is required that the dissociation of the complex is fast on the relaxation time scale. Whether the exchange process is fast or slow on the chemical-shift time scale plays no role for the observation of trNOEs, because it is always the signal of the free ligand that carries the information about the bound state. In other words, it is critical that a large amount of ligand molecules samples the binding pocket of the protein frequently enough during the mixing time of the NOESY experiment. This is visualized in Figure 3. In practice, exchange frequencies of 50–100 Hz are sufficient to permit the observation of trNOEs.

35.3 Measurement of trNOEs

The measurement of trNOEs is performed with exactly the same pulse sequences that are used for the measurement of NOEs. Therefore, the NOESY experiment is well suited to record trNOEs. As pointed out above, the right conditions for the measurement of trNOEs are found by varying the carbohydrate-protein ratio. This is usually done in a titration starting at ligand–protein ratios of approximately 4:1 (for an example see [36]) and than increasing the ratio until maximum trNOEs are observed. During titration it is usually sufficient to perform one-dimensional NOE experiments such as 1D-NOESY in order to save time. The mixing times used for the observation of trNOEs are much smaller than the mixing times required to obtain optimum NOEs for di-, tri-, or tetrasaccharides. This is illustrated in Figure 4 that displays typical NOE-curves in comparison to corresponding trNOE curves for the disaccharide α-L-Fuc-(1→6)-β-D-GlcNAc-OMe free and bound to *Aleuria aurantia* agglutinin (AAA) [37].

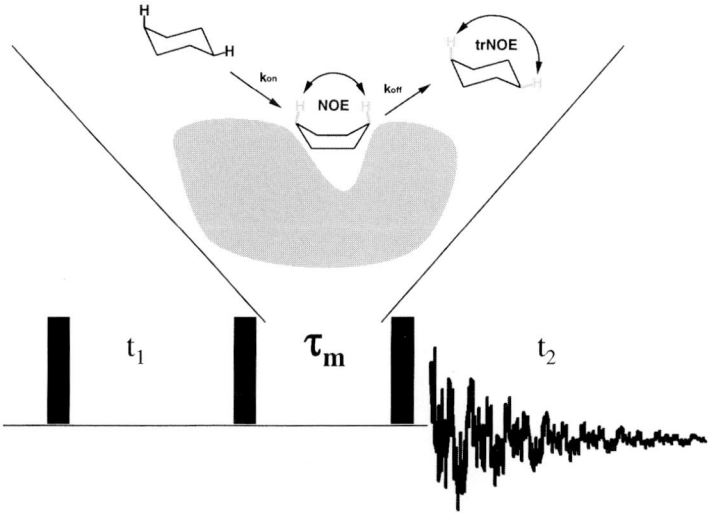

Figure 3. Schematic representation of the transfer NOESY (trNOESY) experiment. Ligands have to sample the binding site of the protein (shown in red) often enough during the mixing time in order to generate a transfer of NOEs. It follows that dissociation of the protein-carbohydrate complex must be fast on the relaxation time scale. With mixing times τ_m usually being 50–300 ms for trNOESY experiments, dissociation rates of ca. 50–100 Hz are sufficient.

Although trNOE-experiments are performed with a large excess of ligand the protein background obstructs the resulting NOESY spectra. Especially if a quantitative evaluation of trNOEs is desired, techniques should be applied to remove unwanted protein signals. The optimum solution is the use of a spin-lock filter that is applied after the first $\pi/2$ pulse in the NOESY sequence [38]. During this time the protein signals undergo fast transverse relaxation whereas signals of the ligand survive because their transverse relaxation rates are negligible. If a quantitative analysis of trNOEs is planned trNOEs should be measured at different mixing times leading to trNOE curves (for an example see [39]). The trNOE step can also be combined with other types of NMR-experiments. For instance, the use of 3D-TOCSY-trNOESY experiments and one-dimensional versions thereof has been demonstrated to relieve signal overlap [20]. Especially powerful is the combination of heteronuclear correlation experiments with trNOESY, such as the 3D-HSQC-trNOESY experiment [23]. Unfortunately, this usually requires the availability of ^{13}C-labelled carbohydrate ligands (see discussion below) that are time consuming and expensive to synthesize [23, 40–42]. In general, any multi-dimensional NMR-experiment that contains a NOESY step can be used to detect trNOEs during this step. Once experimental trNOEs have been measured, a bioactive conformation may be derived.

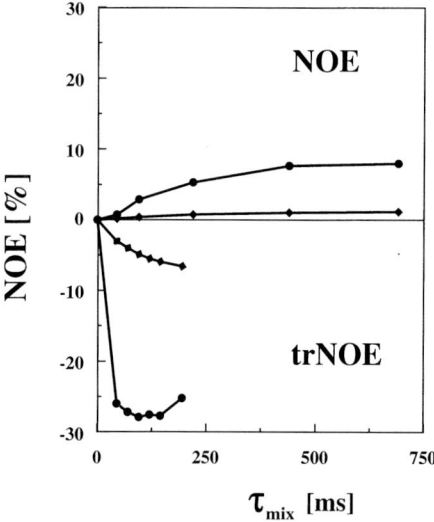

Figure 4. NOEs for free α-L-Fuc-(1→6)-β-D-GlcNAc-OMe (upper half of the graphics), and trNOEs for α-L-Fuc-(1→6)-β-D-GlcNAc-OMe in the presence of *Aleuria aurantia* agglutinin (lower half of the graphics), measured at 600 MHz as a function of the mixing time τ_m. Circles refer to proton pairs H6proR^{GlcNAc}-H6proS^{GlcNAc} and diamonds to H1^{Fuc}-H6proS^{GlcNAc}, respectively. Precise experimental conditions have been described elsewhere [37]. Maximum NOEs for the free disaccharides are found around mixing times τ_m of 1000 ms, whereas maximum trNOEs are reached at mixing times lower than 300 ms.

35.4 Bioactive Conformations of Carbohydrate Ligands From trNOE Experiments

As explained above, trNOEs carry the information about the geometry of the bound state of a ligand molecule binding to a receptor protein. Therefore, the bioactive conformation may be back-calculated from observed trNOEs. This analysis can be performed on different levels ranging from qualitative approaches to highly sophisticated protocols that take into consideration e.g. the binding kinetics and all protons of the receptor–protein binding site and the carbohydrate ligand (for recent reviews on these techniques see: [43, 44]). The most simplistic approach would be to decide on the basis of the absence and/or presence of inter-glycosidic trNOEs which conformation is bound. From the examples presented in the following, it will become clear that this may easily lead to wrong conclusions mainly because of the probability of so called spin-diffusion processes where NOEs or trNOEs occur between protons that are not necessarily close in space. Only if spin-diffusion has been excluded by specific NMR experiments, this simple procedure is allowed. A semi-quantitative method requires the acquisition of trNOE build-up curves that have been corrected for spin-diffusion. Since the initial slope of such buildup-curves is easily translated into inter-proton distances by the so called initial slope approximation (ISPA) [45] bound conformations can be deduced. Two circumstances limit the straightforward translation of initial slopes of trNOE build-up curves into three-dimensional structures. First, inter-glycosidic trNOEs are sparse, and therefore several conformations may fulfill the observed trNOEs. Second, if more than one conformation is bound by the receptor protein "virtual" bioactive conformations may be obtained. In order to interpret data sets that are incomplete in terms of unambiguous structural information experimental restraints may be combined with MD or MMC simulations (see for instance [19, 22, 23, 46]. Following such protocols a range of possible, or "allowed" conformations will be available. If a model for the carbohydrate-protein complex is accessible from X-ray data, or from e.g. homology modelling, extended protocols may be applied. One such approach is implemented in the program CORCEMA ([44] and references cited therein) that is based on full-relaxation matrix calculations taking into account the binding kinetics, and all protein protons that are located in the binding pocket of the protein. Unfortunately, often the structural information required for performing such complex calculations is not available.

35.5 Spin Diffusion may Generate Misleading Distance Constraints

For large molecules such as proteins magnetization transfer in NOESY experiments not only takes place between protons that are close in space. A transfer of magnetization may also occur between protons that are separated by far larger distances than usually required for the observation of NOEs. The reason is a phenomenon called spin diffusion which allows magnetization between two protons to be transferred via relay protons [24]. Spin diffusion becomes of increasing importance with increasing molecular weight and with increasing mixing times. Therefore, spin diffusion also plays an important role for the observation of trNOEs. It is clear that trNOEs that are due to spin diffusion cannot be used as distance constraints in the generation of bioactive conformations. To avoid spin diffusion short mixing times of less than 50 ms are required if no other precautions are taken. But usually at such low mixing times the intensity of trNOEs is so low that no meaningful analysis may be performed. An example will illustrate the misleading effects of spin diffusion artefacts in trNOESY spectra.

The bioactive conformation of a (1→6)-linked disaccharide, methyl 6-*O*-β-D-galactopyranosyl-4-deoxy-2-deutero-4-fluoro-β-D-galactopyranoside bound to the Fab fragment of an anti-(1→6)-β-D-galactopyranan antibody had originally been analyzed on the basis of trNOEs that were not corrected for spin diffusion [47]. It was concluded that a significant conformational change about the (1→6)-glycosidic linkage occurs upon binding. The resulting bound conformation was not found in aqueous solution without the protein present. Later, the authors identified the key trNOEs that lead to the unusual bioactive conformation as spin diffusion artefacts [48]. They reached this conclusion by applying trROESY experiments and comparing the results to the trNOESY spectra. In trNOESY spectra spin diffusion leads to cross peaks that are of the same sign as cross peaks that are due to direct dipolar contacts, and therefore direct dipolar interactions cannot be distinguished from indirect (spin diffusion) interactions. In trROESY spectra cross peaks from indirect magnetization transfer are of opposite sign compared to direct effects, assuming that one "relay" proton mediates the indirect transport of magnetization. If more than one proton mediates spin-diffusion the sign of the corresponding cross peak changes in an alternating manner, but at the same time significantly losing intensity. Therefore, in such cases usually no cross peaks are observed. It follows that the detection of trROEs of opposite sign or of very weak intensity at positions where trNOEs were observed indicates spin diffusion. This is shown schematically in Figure 5a.

In the case described above, trROE experiments showed that the critical cross peaks were due to spin diffusion, and therefore the model for the bound conformation had to be revised. Other groups found similar results [49, 50], and in general it can be stated that testing for spin-diffusion contributions is essential for the eluci-

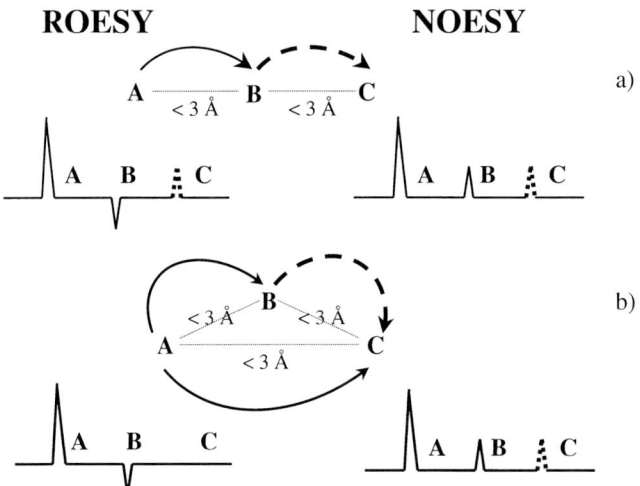

Figure 5. Schematic representation of three spins, A, B, and C interacting via direct dipolar contacts (black arrows) and/or spin diffusion (dashed arrows). **a)** Trace of a 2D NOESY spectrum (right) and a corresponding trace from a 2D ROESY spectrum (left). Spins A and B are close in space, and spin diffusion mediates magnetization transfer between protons A and C, that are not close in space. The 2D ROESY experiment allows unambiguous discrimination between direct and spin diffusion effects. **b)** Protons A and C are close in space, and in addition magnetization is transferred between the two protons via spin diffusion. The effect is a cancellation of the 2D ROESY signal. A cancellation can also occur if indirect magnetization transfer involves more than one relay proton (proton B in this case). Therefore, a discrimination between spin diffusion and direct dipolar interaction is not possible in this case.

dation of bioactive conformations of carbohydrates (and other ligands). TrROESY experiments are a suitable tool to perform this test, but special constellations may exist that require additional experiments. In such cases direct and indirect (spin diffusion) dipolar interactions interfere and lead to a cancellation of trROEs (see Figure 5b) [39]. This observation would mimic the presence of spin diffusion and corresponding distance constraints would be removed, again leading to a false bioactive conformation. The problem here is to separate direct and indirect (spin diffusion) magnetization transfer. It has been shown, that so called QUIET-trNOESY experiments are well suitable for this task [51]. In these type of experiments spins of interest are inverted during the mixing time leaving only their mutual direct dipolar interaction and cancelling all other magnetization transfers [52]. For small ligands such as carbohydrates binding to large receptor proteins such as antibodies the main source of spin diffusion are protons that are attached to the binding site of the protein [35, 39, 48, 51, 53]. Therefore, band-selective inversion of the ring-proton region usually leads to spectra without spin-diffusion contributions from aliphatic or aromatic protein side chain protons [39]. Figure 5 displays a scheme that summarizes the occurrence of cross-peak patterns in the different type of experiments discussed above.

Examples will illustrate the capabilities of the methodology described above. No attempt was made to review the field in total since several very good review articles have appeared during the recent years [54–56].

35.6 The Conformation of Sialyl Lewis^x Bound to E-selectin

Several studies have been performed to elucidate the conformation of the sialyl Lewis^x (sLe^x) epitope bound to E-selectin [19–23]. The selectins are membrane bound glycoproteins that are involved in the initial steps of the inflammatory cascade [6–10]. It has been shown that these proteins mediate rolling of leukocytes by binding to the sLe^x epitope on the surface of leukocytes. Clearly, this provides a key point of therapeutic interaction in the case of inflammatory diseases. In order to design potent mimics of sLe^x the bioactive conformation of the molecule must be known. In solution, the tetrasaccharide sLe^x adopts several conformational states as this has been shown by many conformational analysis studies [19–23, 57–59]. Two main conformational families may be distinguished due to two different major orientations of the α-D-Neu5Ac-(2→3)-β-D-Gal linkage as this is depicted in Figure 6. The two conformational families are characterized by two inter-glycosidic NOEs H3-Gal/H3ax-Neu5Ac and H3-Gal/H8-Neu5Ac, that are mutually exclusive.

The precise distribution of the two orientations and the presence of other local minima in aqueous solution is still a matter of dispute. Upon binding to E-selectin only one of the two major conformational families generated by different orientations of the α-D-Neu5Ac-(2→3)-β-D-Gal linkage is recognized. The bound conformation is characterized by the observation of a trNOE between H3-Gal and H8-Neu5Ac, whereas no interaction is observed between H3-Gal and H3ax-

Neu5Ac (Figure 6). This global result has been found by all conformational analyses of sLe[x] bound to E-selectin performed so far [19–23]. Different protocols have been applied for the analysis of trNOEs, and therefore one would predict that different values are reported for the precise orientation of the α-D-Neu5Ac-(2→3)-β-D-Gal linkage in the bound state of sLe[x]. Indeed, different values were found but nevertheless the gross orientation is similar in all cases.

The latest study reported for the bioactive conformation of sLe[x] utilized fully ^{13}C-labeled sLe[x] [23]. The ^{13}C-enrichment made it possible to perform 3D-NOESY-HSQC spectra. From these spectra additional inter-glycosidic trNOEs were extracted that cannot be obtained from homonuclear spectra because of severe spectral overlap. These additional trNOEs represent extra restraints for the conformations of the β-D-Gal-(1→4)-β-D-GlcNAc linkage and the α-L-Fuc-(1→3)-β-D-GlcNAc linkage but not for the α-D-Neu5Ac-(2→3)-β-D-Gal linkage. Therefore, this study provides more integral data for the bioactive conformation of the Le[x] fragment than previous studies. The glycosidic torsion angles of the different bioactive conformations of sLe[x] bound to E-selectin reported in the studies mentioned above are compiled in Table 1.

Stereo pictures of the different bioactive conformations of sLe[x] from Table 1 are shown in Figure 7. It has been shown that for biological activity the relative orientation of the fucose and the neuraminic acid residue are relevant [11, 12]. Following this guide line, a potent mimic of sLe[x] had been synthesized [11–13]. A conformational analysis of this sLe[x] mimic in the free and bound state [60] clearly showed that the bioactive conformation is very similar to the one derived originally for sLe[x] bound to E-selectin [19].

In none of the studies published so far the protein protons in the binding pocket were taken into account. For a more precise analysis this would be mandatory but it would also require knowledge about the orientation of the tetrasaccharide in the binding pocket. This information is currently not available, although several models for the binding of sLe[x] to E-selectin have been proposed. For a complete understanding of the carbohydrate recognition by E-selectin and consequently as a prerequisite for designing new drugs that block the sLe[x]/E-selectin interaction it would clearly be desirable to have a better knowledge about the orientation of sLe[x] in the binding pocket of E-selectin.

It is interesting to compare the bioactive conformation of sLe[x] when binding to different receptor proteins. For instance, a different orientation of the α-D-Neu5Ac-(2→3)-β-D-Gal linkage was postulated for sLe[x] bound to L-selectin [22]. The biological implications of this finding remain to be elucidated, but certainly it would be interesting to collect more experimental data that demonstrate the presence of distinct bioactive conformations of the sLe[x] ligand when binding to different receptor proteins. For instance, what will happen if totally different (from E-selectin) "carbohydrate recognition" proteins are employed? We therefore studied the binding of sLe[x] to the lectin *Aleuria aurantia* agglutinin (AAA) [61], a lectin from orange peel mushroom, that specifically recognizes fucose residues. Since this lectin is known to have binding specificity for fucose one could well assume that the Neu5Ac part of sLe[x] retains its flexibility even in the bound state, as this had been reported for other saccharide-protein complexes [50, 37]. Surprisingly, this is not the case. Probably

35.6 The Conformation of Sialyl Lewis[x] Bound to E-selectin

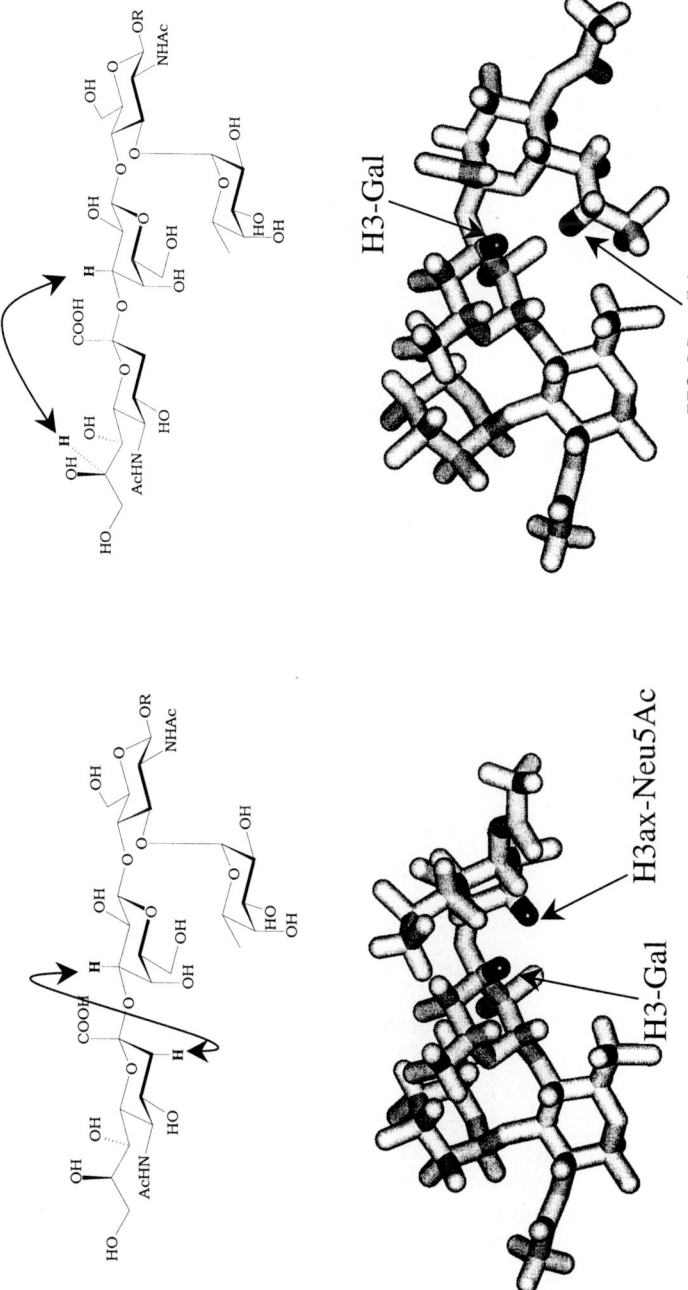

Figure 6. Two conformations of sLe[x] representing the two major conformational families resulting from different orientations around the α-D-Neu5Ac-(2→3)-β-D-Gal glycosidic linkage. Mutually exclusive inter glycosidic NOEs occur.

Table 1. Torsion angles at the glycosidic linkages of bioactive conformations of sLe[x] bound to E-selectin [19, 22, 23] and to *Aleuria aurantia* agglutinin (AAA) [61].

Reference	N(2→3)G	G(1→4)GN	F(1→3)GN
[19]	−76°/+6°	+39°/+12°	+38°/+26°
[22]	−58°/−22°	+24°/+34°	+71°/+14°
[23]	−43°/−12°	+45°/+19°	+29°/+41°
[61]	−61°/−4°	+67°/+17°	−25°/−28°

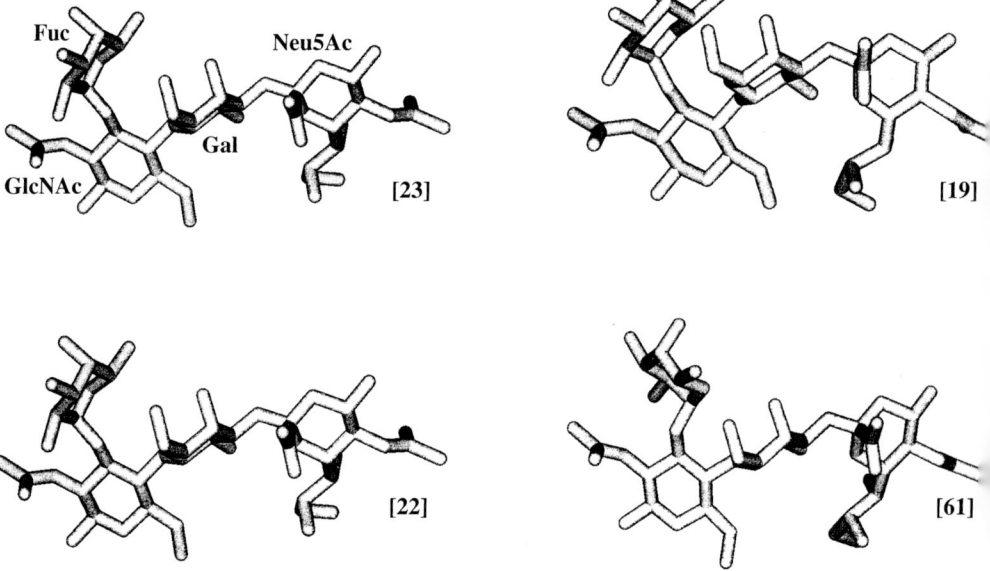

Figure 7. Different bioactive conformations of sLe[x] bound to E-selectin as reported in the literature [19, 22, 23]. The bioactive conformation of sLe[x] bound to *Aleuria aurantia* agglutinin (AAA) is also shown [61]. It is obvious that the latter conformation differs from the other conformations in the orientation of the fucose residue (compare Table 1).

because of unfavourable steric interactions with protein side chains, the Neu5Ac residue adopts a conformation that belongs to the same conformational family that also embraces the conformation of sLe[x] bound to E-selectin. As pointed out above, the inter-glycosidic trNOE H3-Gal/H8-Neu5Ac is characteristic for this conformation [see Figure 6]. Inspecting the data that are available for the bioactive conformation of sLe[x] bound to E-, L- or P-selectin [19–23] it is clear that in all of these conformations the galactose and the fucose pyranose rings have a stacked orientation giving rise to certain characteristic inter-glycosidic NOEs H2-Gal/H5-Fuc, H2-Gal/H6-Fuc, and H1-Fuc/NAc-GlcNAc. This orientation also corresponds to the

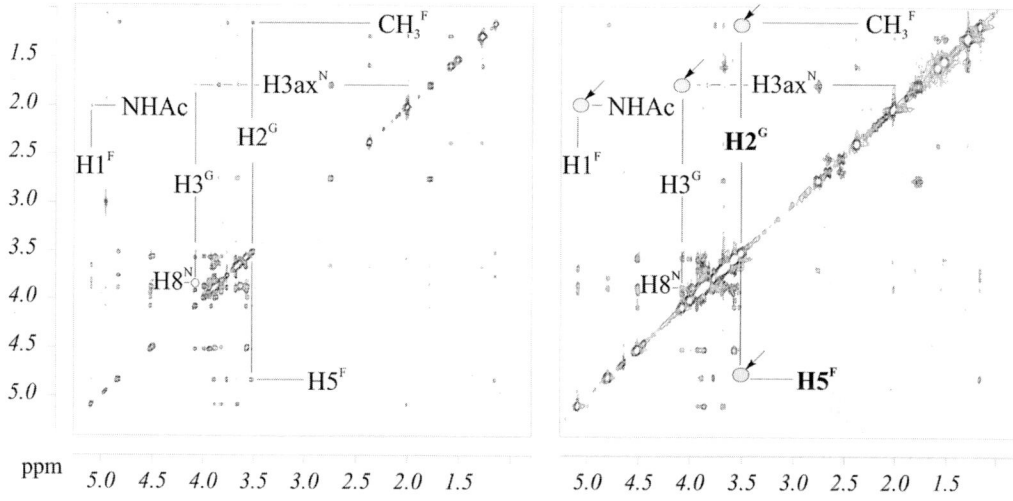

Figure 8. 500 MHz NOESY spectrum of sLe^x (left, 900 ms mixing time) compared to a 500 MHz trNOESY spectrum of sLe^x bound to AAA (right, 150 ms mixing time). It is obvious that a number of inter glycosidic trNOEs are not observed (gray circles) where NOEs were observed (left). Important to notice is the absence of signal intensity for the contact between H2-Gal (H2^G) and H5-Fuc (H5^F), a contact that would indicate stacking of fucose and galactose. An inter glycosidic trNOE is observed between protons H1-Fuc and H2-Gal. This trNOE is difficult to locate at the present expansion of the spectrum but the effect is clearly visible when inspecting the corresponding part of the trNOESY spectrum [61].

global minimum energy region of this glycosidic linkage, and is prevalent in aqueous solution. It follows that the hydrophobic side of the fucose residue is shielded from potential interactions with a protein, and it can be hypothesized that a fucose-binding lectin such as AAA will not recognize a fucose residue that is so well shielded. From trNOE experiments it is evident that AAA recognizes the fucose residue in sLe^x, but in a conformation that is considerably different from the global minimum and that does not obey the exo-anomeric effect. From the arguments given above this is understandable because recognition of the fucose would be hampered by the presence of the galactose-fucose stacking in the global energy minimum. The experimental evidence that led to this conclusion is clear cut. Upon binding to AAA the NOEs characteristic for the global minimum disappear. Instead, a trNOE is observed between H1-Fuc and H2-Gal (a NOESY and a trNOESY spectrum displaying the major features discussed are shown in Figure 8). This indicates that the above mentioned local minimum at the α-L-Fuc-(1→3)-β-D-GlcNAc linkage is selected upon binding to AAA. The torsion angles at the (1→3)-glycosidic linkage are around $\phi = -25°$ and $\psi = -28°$ (compare Table 1 and Figure 7).

Since usually conformations that are recognized by proteins are "preformed" in aqueous solution we reinvestigated the solution conformation of sLe^x and Le^x to find out whether this "non-exo-anomeric" conformation is also present in aqueous solution. The presence of this minimum had been predicted recently on the basis of theoretical MD simulations and has also been discussed in a study that compared the conformational features of the disaccharide α-L-Fuc-(1→3)-β-D-GlcNAc to its thio-analog [46]. But so far, no direct experimental evidence was available for the presence of this local minimum. The minimum is also predicted by MMC simulations that we performed for sLe^x and Le^x. An accurate conformational analysis of Le^x in aqueous solution reveals that the inter-glycosidic NOE between H1-Fuc and H2-Gal is indeed present although with a rather low intensity. For sLe^x we did not detect this NOE, probably because of the larger molecular weight and the charge of sLe^x that renders very weak negative NOEs, as this was reported before [20, 57, 58]. Nevertheless, our experiments allow the conclusion that a small portion of Le^x and most likely of sLe^x in aqueous solution is "preformed" in the local minimum conformation with dihedral angles close to $\phi = -25°$ and $\psi = -28°$. These findings clearly question the concept of rigidity for the Le^x trisaccharide core structure, and suggest that carbohydrate chains in general may carry different biological information that is encoded by different potential bioactive conformations. Depending on the cognate receptor protein different "conformational information" may be read out.

35.7 Interaction of Bacterial Lipopolysaccharide Fragments with Monoclonal Antibodies

Much attention has been paid during the past ten years to the specific recognition of carbohydrate epitopes by monoclonal antibodies (mAbs) [62–66]. Understanding these reactions at a molecular level will certainly help in generating new perspectives in diagnosis and therapy of related diseases. In our laboratory, we studied the interaction of synthetic lipopolysaccharide fragments with corresponding mAbs with NMR [39, 67]. Several mAbs were generated against synthetic and isolated fragments of lipopolysaccharides (LPS) that are present on the surface of chlamydial bacteria [68]. These parasites are responsible for a variety of diseases in humans and animals. During infection, antibodies are expressed against components in the outer membrane, with LPS as one of the major surface antigens. Here, we will focus on the disaccharide element α-Kdo-(2→4)-α-Kdo that constitutes a common structural element of the core region of Gram-negative bacterial LPS in general [69]. The conformational features of the binding of α-Kdo-(2→4)-α-Kdo-(2→O)-allyl (Figure 9) to two mAbs S25-2 and S23-24 will be described [39]. Compared to S25-2, S23-24 binds to disaccharide α-Kdo-(2→4)-α-Kdo-(2→O)-allyl with approximately 50-fold increased affinity, and the question arises if this also is reflected by different bioactive conformations that this disaccharide adopts binding to the two mAbs.

It turned out that the acquisition of QUIET-trNOESY spectra was mandatory because interference of direct and spin-diffusion mediated magnetization transfer was observed. This is demonstrated in Figure 10 that depicts parts of trNOESY, trROESY and QUIET-trNOESY spectra. The critical cross peaks indicate short distances between protons attached to the carbon atom C8 of ring b and protons

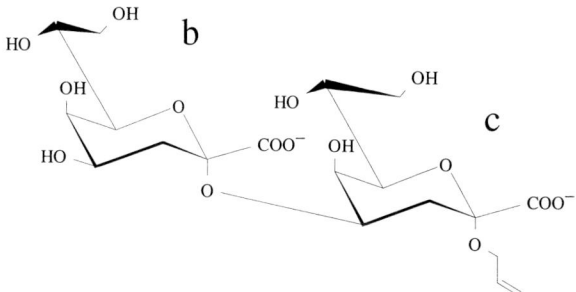

Figure 9. Chemical formula of the disaccharide α-Kdo-(2→4)-α-Kdo-(2→O)-allyl.

attached to C3 of ring c. The trROESY experiment suggests that these interglycosidic trNOEs are due to spin-diffusion because no intensitiy is observed at the critical positions (Figure 10b). In contrast, QUIET-trNOESY experiments clearly show that only part of the interaction is due to spin diffusion (Figure 10c). The experiments show that direct dipolar interactions and spin diffusion occur at the same time, giving misleading results for the trROESY experiments.

From a variety of NMR experiments it was concluded that protein protons are the major source of spin diffusion, and therefore doubly band selective experiments were performed where the regions of ring protons of 4.10–3.60 ppm and 2.17–1.67 ppm were inverted simultaneously during mixing. Buildup curves from such QUIET-trNOESY experiments were measured and translated into bioactive conformations using restrained MMC simulations. For α-Kdo-(2→4)-α-Kdo-(2→O)-allyl binding to S25-2 the analysis results in a rather limited part of conformational space that contains the bioactive conformation as this is shown in Figure 11.

For the other antibody, mAb S23-24, it is impossible to deduce a single bound conformation. Obviously, this antibody recognizes two different bioactive conformations, with one being similar to the global minimum A (Figure 11) of disaccharide α-Kdo-(2→4)-α-Kdo-(2→O)-allyl and the other being rather close to the conformation C (Figure 11) that is bound by the other antibody S25-2. The bioactive conformations found are in good agreement with the binding data. The antibody with higher affinity binds to a conformation that is highly populated in aqueous solution. This study underlines that one carbohydrate epitope may be recognized in different bioactive conformations by different receptor proteins (Figure 12).

These results emphasize that the recognition of different bioactive conformations is an essential component of protein–carbohydrate recognition reactions. Indeed, other examples for this phenomenon have recently been published [66, 70, 71]. As one example, the reader's attention should be drawn to studies that targeted the binding of C-glycosides to glycosidases. Using e.g. C-lactose as a non-hydrolizable substrate for *Escherichia coli*-β-galactosidase, it was shown that the enzyme binds to a high energy local minimum that represents the so called anti-conformation around the pseudo-glycosidic linkage [70]. From crystallographic studies it followed

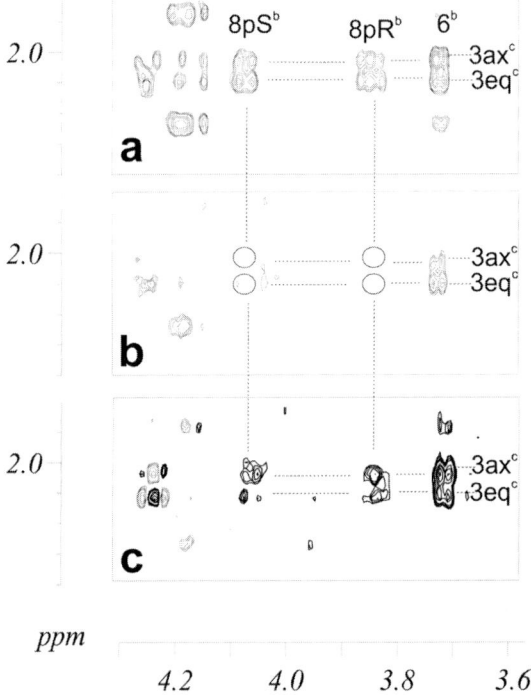

Figure 10. Parts of 2D trNOESY (**a**), 2D trROESY (**b**), and 2D QUIET-trNOESY (**c**) of α-Kdo-(2→4)-α-Kdo-(2→O)-allyl bound to mAb S25-2. The QUIET-trNOESY experiment was recorded with a 15 ms double-band selective Q3 inversion pulse (inversion of regions 4.10–3.60 ppm and 2.17–1.67 ppm). Peaks within the inverted regions show an opposite sign (bold lines, **c**) relative to the other cross peaks outside these regions. The mixing time was 250 ms for all experiments. A comparison of the spectra allows identification of spin-diffusion effects. Cross-peaks that are cancelled in the trROESY spectrum because spin diffusion and direct dipolar interactions take place at the same time (see discussion in the text) are marked with circles in the 2D trROESY spectrum (**b**). Reprinted with permission from Biochemistry [39].

that C-lactose binds to peanut lectin in a conformation that corresponds to the minimum energy conformation of O-lactose [72]. Therefore, the selection of a high energy conformation of C-lactose by *E. coli*-β-galactosidase may promote the cleavage of the glycosidic linkage by the enzyme [70]. At the present, there are too few examples to draw meaningful general conclusions about the biological significance of the recognition of different carbohydrate conformations by different receptor proteins. Certainly, in the future more such data will be collected for other biological relevant cases, and it will be a challenge to link the different bioactive conformations to different biological functions.

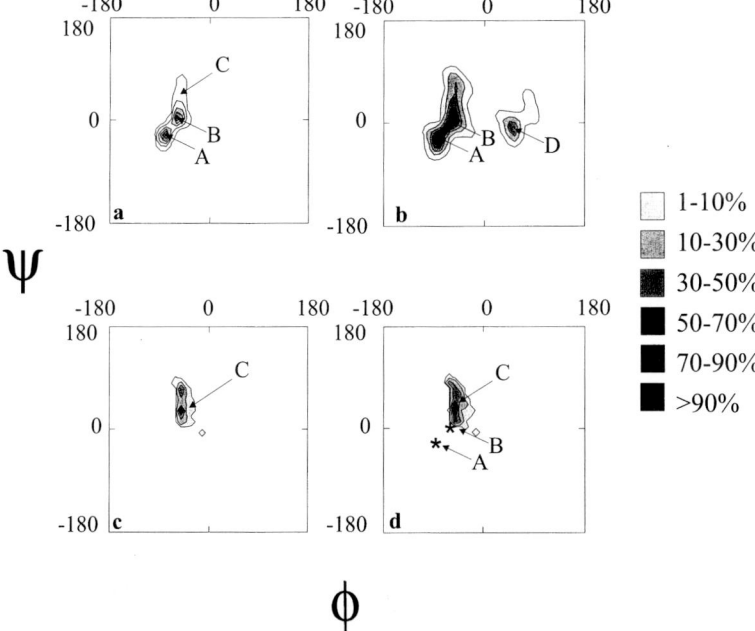

Figure 11. Contour plots showing the relative population of conformational space around the (2→4)-glycosidic linkage in α-Kdo-(2→4)-α-Kdo-(2→O)-allyl. The ϕ/ψ-maps were divided into bins of 10° in ϕ- and ψ-direction, and the number of conformations in each bin was counted. Then, contour levels were calculated relative to the highest populated bin (global minimum). The contour levels are color coded. Magenta encodes 1–10%, dark blue 10–30%, light blue 30–50%, green 50–70%, yellow 70–90%, and red more than 90% of the number of conformations in the highest populated bin. (**a**) MMC simulation at 600 K. (**b**) MMC simulation at 2,000 K. (**c**) and (**d**) represent possible conformations of α-Kdo-(2→4)-α-Kdo-(2→O)-allyl bound to mAb S25-2. (**c**) all conformations from the MMC simulation at 2,000 K that satisfy only positive distance constraints (0.15% of the total number of conformations from the MMC simulation). (**d**) all conformations that satisfy positive and negative distance constraints (0.075% of the total number of conformations). Reprinted with permission from Biochemistry [39].

35.8 Conclusions and Future Directions

A major consequence from the studies reported above, and from other examples from the literature, it can be stated that the inherent flexibility of oligosaccharides as compared to e.g. globular proteins allows to encode different biological information for recognition reactions in a parallel manner. So far it appears that binding affinities for the recognition of a certain conformation parallel with the potential

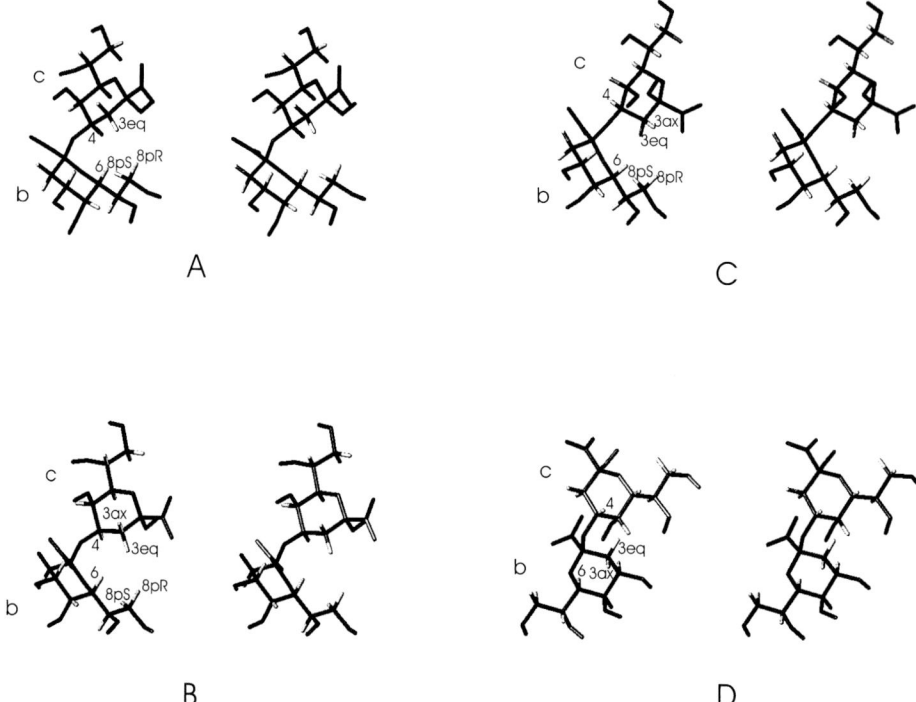

Figure 12. Stereo images of α-Kdo-(2→4)-α-Kdo-(2→O)-allyl (relaxed view); (A) global minimum A; (B) local minimum B; (C) conformation C that is recognized by mAb S25-2; (D) local minimum D. Reprinted with permission from Biochemistry [39].

energy of the conformation in aqueous solution, not bound to a protein. The way how nature takes advantage of this phenomenon is not known yet. In order to establish links between different bioactive conformations and biological functions, it will be important to study the thermodynamics and kinetics of the binding reactions in parallel. Microcalorimetry [31, 65] and surface plasmon resonance [e.g. 73] provide powerful tools to collect such data.

For more detailed structural analyses it will be important to make use of new techniques that have emerged in protein NMR spectroscopy. The main breakthroughs are probably the introduction of TROSY spectroscopy [74], extending the size limit of biological macromolecules that can be subjected to NMR analyses, the creation of so called cross correlation experiments [75] that allow the direct measurement of dihedral angles, and the measurement of residual dipolar couplings [76] that complement data from NOESY spectroscopy. Unfortunately, all of these techniques require, or at least benefit from isotope enrichment which remains a difficult and expensive task for carbohydrates. Consequently, only few examples utilizing e.g. ^{13}C-labelled carbohydrates have been published so far [23, 40–42]. It is obvious that the new tools that have been discovered for NMR spectroscopy will also greatly promote the analysis of protein–carbohydrate interactions.

Interesting developments originate from experiments that use the trNOE effect to screen substance libraries for binding activity against receptor proteins. The first attempts to perform such bio-affinity protocols were rather successful [77, 78], and recently a new experimental strategy was introduced that allows a very robust and powerful screening of libraries, called STD-NMR [79, 80]. The new method is based on the principle of saturation transfer and can be combined with virtually any other NMR technique. Whereas trNOE experiments are well established tools to study bioactive conformations, especially of carbohydrate ligands, the use of bio-affinity NMR methods is still in its infants and promises more surprises in the near future.

References

1. T. A. Springer, *Nature*, **1990**, *346*, 425–434.
2. T. Feizi, *Curr. Opin. Struct. Biol.*, **1993**, *3*, 701–710.
3. A. Varki, *Glycobiology*, **1993**, *3*, 97–130.
4. R. A. Dwek, *Chem. Rev.*, **1996**, *96*, 683–720.
5. P. Sears, C.-H. Wong, *Cell. Mol. Life Sci.* **1998**, *54*, 223–252.
6. M. P. Bevilacqua, S. Stengelin, M. A. Gimbrone, Jr., B. Seed, *Science*, **1989**, *243*, 1160–1165.
7. L. A. Lasky, *Science*, **1992**, *258*, 964–969.
8. A. Levinovitz, J. Muhloff, S. Isenmann, D. Vestweber, *J. Cell. Biol.*, **1993**, *121*, 449–459.
9. M. Lenter, A. Levinovitz, A., S. Isenmann, D. Vestweber, *J. Cell. Biol.*, **1994**, *125*, 471–481.
10. K. L. Moore, N. L. Stults, S. Diaz, D. F. Smith, R. D. Cummings, A. Varki, R. P. McEver, *J. Cell. Biol.*, **1992**, *118*, 445–456.
11. H. C. Kolb, B. Ernst, *Pure Appl. Chem.*, **1997**, *69*, 1879–1884.
12. H. C. Kolb, B. Ernst, *Chem. Eur. J.*, **1997**, *3*, 1571–1578.
13. K. E. Norman, G. P. Anderson, H. C. Kolb, K. Ley, B. Ernst, *Blood*, **1998**, *91*, 475–483.
14. T. F. Lampe, G. Weitz-Schmidt, C.-H. Wong, *Angew. Chem.*, **1998**, *110*, 1761–1764.
15. T. Uchiyama, V. P. Vassilev, T. Kajimoto, W. Wong, H. Huang, C.-H. Wong, *J. Am. Chem. Soc.*, **1995**, *117*, 5395–5396.
16. E. E. Simanek, G. J. McGarvey, J. A. Jablonowski, C.-H. Wong, *Chem. Rev.*, **1998**, *98*, 833–862.
17. P. Sears, C.-H. Wong, *Angew. Chem. Int. Ed.*, **1999**, *38*, 2300–2324; *Angew. Chem.*, **1999**, *111*, 2446–2471.
18. B. J. Graves, R. L. Crowther, C. Chandran, J. M. Rumberger, S. Li, K.-S. Huang, D. H. Presky, P. C. Familletti, B. A. Wolitzky, D. K. Burns, *Nature*, **1994**, *367*, 532–538.
19. K. Scheffler, B. Ernst, A. Katopodis, J. L. Magnani, W. T. Wang, R. Weisemann, T. Peters, *Angew. Chem.*, **1995**, *107*, 2034–2037; *Angew. Chem. Int. Ed. Engl.*, **1995**, *34*, 1841–1844.
20. K. Scheffler, J.-R. Brisson, R. Weisemann, J. L. Magnani, W. T. Wang, B. Ernst, T. Peters, *J. Biomol. NMR*, **1997**, *9*, 423–436.
21. R. M. Cooke, R. S. Hale, S. G. Lister, G. Shah, M. P. Weir, *Biochemistry*, **1994**, *33*, 10591–10596.
22. L. Poppe, G. S. Brown, J. S. Philo, P. V. Nikrad, B. H. Shah, *J. Am. Chem. Soc.*, **1997**, *119*, 1727–1736.
23. R. Harris, G. R. Kiddle, R. A. Field, M. J. Milton, B. Ernst, J. L. Magnani, S. W. Homans, *J. Am. Chem. Soc.*, **1999**, *121*, 2546–2551.
24. D. Neuhaus, M. Williamson, *The Nuclear Overhauser Effect in Strucutral and Conformational Analysis*, VCH Wiley, New York, 1989.
25. P. Balaram, A. A. Bothner-By, J. Dadok, *J. Am. Chem. Soc.*, **1972**, *94*, 4015–4017.
26. P. Balaram, A. A. Bothner-By, E. Breslow, *J. Am. Chem. Soc.*, **1972**, *94*, 4017–4018.

27. G. M. Clore and A. M. Gronenborn, *J. Magn. Reson.*, **1982**, *48*, 402–417.
28. G. M. Clore and A. M. Gronenborn, *J. Magn. Reson.*, **1983**, *53*, 423–442.
29. N. H. Andersen, K. T. Nguyen, H. L. Eaton, *J. Magn. Reson.*, **1985**, *63*, 365–375.
30. N. H. Andersen, H. L. Eaton, K. T. Nguyen, *Magn. Reson. Chem.*, **1987**, *25*, 1025–1034.
31. E. Toone, *Curr. Opin. Struct. Biol.*, **1994**, *4*, 719–728.
32. R. E. London, M. E. Perlman, D. G. Davis, *J. Magn. Reson.*, **1992**, *97*, 79–98.
33. F. Ni, *J. Magn. Reson.*, **1992**, *96*, 651–656.
34. W. Lee, N. R. Krishna, *J. Magn. Reson.*, **1992**, *98*, 36–48.
35. H. N. B. Moseley, E. V. Curto, N. R. Krishna, *J. Magn. Reson. Ser. B*, **1995**, *108*, 243–261.
36. F. Casset, A. Imberty, S. Pérez, M. Etzler, H. Paulsen, T. Peters, *Eur. J. Biochem.*, **1997**, *244*, 242–250.
37. T. Weimar, T. Peters, *Angew. Chem. Int. Ed. Engl.*, **1994**, *33*, 88–91; *Angew. Chem.*, **1994**, *106*, 79–82.
38. T. Scherf, J. Anglister, *Biophys. J.*, **1993**, *64*, 754–761.
39. T. Haselhorst, J.-F. Espinosa, J. Jiménez-Barbero, T. Sokolowski, P. Kosma, H. Brade, L. Brade, T. Peters, *Biochemistry*, **1999**, *38*, 6449–6459.
40. M. A. Probert, M. J. Milton, R. Harris, S. Schenkman, J. M. Brown, S. W. Homans, R. A. Field, *Tetrahedron Lett.*, **1997**, *38*, 5861–5864.
41. D. G. Low, M. A. Probert, G. Embleton, K. Sheshadri, R. A. Field, S. W. Homans, J. Windust, P. J. Davis, *Glycobiology*, **1997**, *7*, 373–381.
42. H. Shimizu, J. M. Brown, S. W. Homans, R. A. Field, *Tetrahedron*, **1998**, *54*, 9489–9506.
43. F. Ni, *Prog. NMR Spectr.*, **1994**, *26*, 517–606.
44. N. R. Krishna, H. N. B. Moseley, in *Biological Magnetic Resonance, 17: Structure and Dynamics in Protein NMR*, Eds. Krishna and Berliner. Kluwer Academic/Plenum Publishers, New York, **1999**, 223–307.
45. A. M. Gronenborn, G. M. Clore, *Progr. Nucl. Magn. Reson. Spectrosc.*, **1985**, *17*, 1–32.
46. B. Aguilera, J. Jiménez-Barbero, A. Feràndez-Mayoralas, *Carbohydr. Res.*, **1998**, *308*, 19–27.
47. C. P. J. Glaudemans, L. Lerner, G. D. Daves Jr., P. Kovác, R. Venable, A. Bax, *Biochemistry*, **1990**, *29*, 10906–10911.
48. S. R. Arepalli, C. P. J. Glaudemans, G. D. Daves Jr., P. Kovác, A. Bax, *J. Magn. Reson.*, **1995**, *106*, 195–198.
49. T. Weimar, S. L. Harris, J. B. Pitner, K. Bock, B. M. Pinto, *Biochemistry*, **1995**, *34*, 13672–13680.
50. J. L. Asensio, F. J. Cañada, J. Jiménez-Barbero, *Eur. J. Biochem.*, **1995**, *233*, 618–630.
51. S. J. F. Vincent, C. Zwahlen, C. B. Post, J. W. Burgner, G. Bodenhausen *Proc. Natl Acad. Sci. USA*, **1997**, *94*, 4383–4388.
52. C. Zwahlen, S. J. F. Vincent, L. Di Bari, M. H. Levitt, G. Bodenhausen, *J. Am. Chem. Soc.*, **1994**, *116*, 362–368.
53. F. Ni, Y. Zhu, *J. Magn. Reson. Ser. B*, **1994**, *102*, 180–184.
54. T. Peters, B. M. Pinto, *Curr. Opin. Struct. Biol.*, **1996**, *6*, 710–720.
55. A. Poveda, J. Jiménez-Barbero, *Chem. Soc. Rev.*, **1998**, *27*, 133–143.
56. J. Jiménez-Barbero, J. L. Asensio, F. J. Cañada, A. Poveda, *Curr. Opin. Struct. Biol.*, **1999**, *9*, 549–555.
57. J. Breg, L. M. J. Kroon-Batenburg, G. Strecker, J. Montreuil, J. F. G. Vliegenthart, *Eur. J. Biochem.* **1989**, *178*, 727–739.
58. Y. Ichikawa, Y.-C. Lin, D. P. Dumas, G-J. Shen, E. Garcia-Junceda, M. A. Williams, R. Bayer, C. Ketcham, L. E. Walker, J. C. Paulson, C.-H. Wong, *J. Am. Chem. Soc.* **1992**, *114*, 9283–9298.
59. C. Mukhopadhyay, K. E. Miller, C. A. Bush, *Biopolymers* **1994**, *34*, 21–29.
60. W. Jahnke, H. C. Kolb, M. J. J. Blommers, J. L. Magnani, B. Ernst, *Angew. Chem.*, **1997**, *109*, 2715–2719.
61. T. Haselhorst, T. Peters, unpublished results.
62. D. R. Bundle, N. M. Young, *Curr. Opin. Struct. Biol.*, **1992**, *2*, 666–673.
63. D. R. Bundle, E. Eichler, M. A. J. Gidney, M. Meldal, A. Ragauskas, B. W. Sigurskjold, B. Sinnott, C. D. Watson, M. Yaguchi, N. M. Young, *Biochemistry*, **1994**, *33*, 5172–5182.

64. D. R. Bundle, H. Baumann, J.-R. Brisson, S. M. Gagne, A. Zdanov, M. Cygler, *Biochemistry*, **1994**, *33*, 5183–5192.
65. D. R. Bundle, B. W. Sigurskjold, *Methods Enzymol.*, **1994**, *247*, 288–305.
66. M. J. Milton, D. R. Bundle, *J. Am. Chem. Soc.*, **1998**, *120*, 10547–10548.
67. T. Sokolowski, T. Haselhorst, K. Scheffler, R. Weisemann, P. Kosma, H. Brade, L. Brade, T. Peters, *J. Biomol. NMR*, **1998**, *12*, 123–133.
68. Y. Fu, M. Baumann, P. Kosma, L. Brade, H. Brade, *Infect. Immun.*, **1992**, *60*, 1314–1321.
69. L. Brade, P. Kosma, B. J. Appelmelk, H. Paulsen, H. Brade, *Infect. Immun.*, **1987**, *55*, 462–466.
70. J. F. Espinosa, E. Montero, A. Vian, J. L. García, H. Dietrich, R. R. Schmidt, M. Martín-Lomas, A. Imberty, J. Cañada, J. Jiménez-Barbero, *J. Am. Chem. Soc.*, **1998**, *120*, 1309–1318.
71. M. Gilleron, H. C. Siebert, H. Kaltner, C. W. von der Lieth, T. Kozar, K. M. Halkes, E. Y. Korchagina, N. V. Bovin, H. J. Gabius, J. F. G. Vliegenthart, *Eur. J. Biochem.*, **1998**, *252*, 416–427.
72. R. Ravishankar, A. Surolia, M. Vijayan, S. Lim, Y. Kishi, *J. Am. Chem. Soc.*, **1998**, *120*, 11297–11303.
73. C. R. MacKenzie, T. Hirama, J. T. Buckley, *J. Biol. Chem.*, **1999**, *274*, 22604–22609.
74. K. Pervushin, R. Riek, G. Wider, K. Wüthrich, *Proc. Natl Acad. Sci. USA*, **1997**, *94*, 12366–12371.
75. S. J. Glaser, T. Schulte-Herbrüggen, M. Sieveking, O. Schedletzky, N. C. Nielsen, O. W. Sørensen, C. Griesinger, *Science*, **1998**, *280*, 421–424.
76. N. Tjandra, A. Bax, *Science*, **1997**, *278*, 1111–1114.
77. B. Meyer, T. Weimar, T. Peters, *Eur. J. Biochem.*, **1997**, *246*, 705–709.
78. D. Henrichsen, B. Ernst, J. L. Magnani, W.-T. Wang, B. Meyer, T. Peters, *Angew. Chem. Int. Ed.*, **1999**, *38*, 98–102.
79. M. Mayer, B. Meyer, *Angew. Chem. Int. Ed.*, **1999**, *38*, 1784–1788.
80. J. Klein, R. Meinecke, M. Mayer, B. Meyer, *J. Am. Chem. Soc.*, **1999**, *121*, 5336–5337.

36 Carbohydrate–Protein Interactions: Use of the Laser Photo Chemically Induced Dynamic Nuclear Polarization (CIDNP)-NMR Technique

Hans-Christian Siebert and Johannes F. G. Vliegenthart

36.1 Introduction

The laser photo chemically induced dynamic nuclear polarization (CIDNP) method is a sophisticated NMR technique [1], which can be used for the detection of surface exposed Tyr-, His- and Trp-residues of a protein. It is possible to correlate the intensity of a CIDNP signal with the degree of accessibility of the corresponding CIDNP-sensitive amino acid residue in a protein-structure [2–7]. To obtain CIDNP signals it is necessary that a laser-light excited dye undergoes radical reactions with one of the three CIDNP-sensitive amino acids. Two spectra are recorded: one with and one without laser-light irradiation at each CIDNP experiment. The resulting light spectrum is subtracted from the dark spectrum, thereby establishing the CIDNP difference spectrum, containing the signals of polarized residues only. Trp- and His-signals occur in positive, Tyr-signals in negative direction in the aromatic part of a CIDNP-difference spectrum [3, 8]. Trp- and His-signals can be discriminated by the strong pH-dependence of the His-signals resulting in alterations of chemical shift and signal intensity. The pH-dependence of the intensities of Tyr-, His- and Trp-CIDNP-signals have been described [3, 9, 10]. CIDNP-measurements are usually carried out in D_2O solution, therefore one has to consider the relation: $pD = pH_{\text{meter reading}} + 0.4$. In general, it is possible to distinguish highly, partly and non-accessible Tyr-, His- and Trp-residues. In case a CIDNP-reactive amino acid is located in or in the vicinity of the binding pocket of a protein, the corresponding CIDNP signals can in many cases be suppressed by the addition of a specific ligand. In the bound form this ligand hinders the excited flavin dye to undergo radical reactions with the CIDNP-reactive residues. Provided that the alternative interpretation can be excluded that ligand binding alters the general conformation of a receptor thereby changing the surface accessibilities of various amino acid residues. For this reason, CIDNP results have always to be evaluated in close correlation to a computer supported molecular modelling procedure. The CIDNP-technique is a rather quick method, but restricted to special problems, namely the detection and analysis of Tyr-, His and Trp-residues on the surface of a protein. CIDNP data provide their optimal benefit when they are combined with molecular modelling data based on X-ray and NMR structures of similar molecules [5–7]. The CIDNP method sometimes also allows to detect an influence on the conformation caused by ligand interaction or by site-directed mutagenesis.

The validity of CIDNP data can be checked when protein-structures are completely examined by multi-dimensional NMR experiments. In contrast to multi-dimensional NMR experiments the CIDNP technique is not limited by the size of the molecule [11].

36.2 The CIDNP Method

The CIDNP radical reaction is initiated by the flavin I mononucleotide (N3 of the isoalloxazine ring substituted with $-CH_2COOH$, and N10 with $-CH_3$) as laser-responsive dye. The laser-light used is generated by a continuous-wave argon ion laser which operates in the multi-line mode with principal emission wavelengths of 488.0 and 514.5 nm, close to the edge of the 450 nm absorption band of the dye. By an optical fibre the laser light is directed to the sample and chopped by means of a mechanical shutter controlled by the spectrometer to prevent heating of the protein-containing solution. The irradiation leads to the generation of protein–dye radical pairs by surface-exposed Tyr-, Trp-, and His-side chains.

Nuclear spin-polarization is obtained from back-reactions of the radical pairs. The irradiated dye converts from an excited singlet state to an excited triplet state where it can undergo radical reactions with the corresponding groups of the CIDNP-sensitive amino acids. Recombination and escape reactions are possible. The recombination, which only occurs in the singlet state, depends on the spin of the nuclei leading to a nuclear spin polarization which can be detected by NMR-spectroscopy [3, 12–14].

The Tyr-dependent CIDNP effect corresponds to a spin-density distribution of the intermediate phenoxy radical with strong negative signals of the protons H3, H5 and less intense positive signals of the protons H2, H6. The CIDNP signals of Trp are generated by an intermediate radical with strong spin densities at the positions H2, H4, H6 of the indole unit which all yield positive CIDNP signals. The CIDNP responses of H2, H4 of His occur as positive singlets [3].

A typical experimental setting consists of a presaturation pulse of 1 s, a light pulse of 0.5 s (5 W), a resonance frequency pulse (90° flip angle) of 5 µs, an acquisition time of 1 s and a delay of 5 s. In general, an adequate signal-to-noise ratio is obtained after 16, 32 or 64 light scans [3]. The individual mixtures consisted of 0.1 mM protein, 0.1 to 2 mM ligand and 0.4 mM flavin derivative. Chemical shifts were assessed relative to acetone (2.225 ppm) and HDO (4.76 ppm at a defined temperature and pH/pD-value).

36.3 CIDNP-related Molecular Modelling

As mentioned before CIDNP-data should whenever possible be correlated with model structures. In case no further NMR-data are available, X-ray coordinates from the studied molecule or even from homologous structures can be used. The values of the surface accessibility of CIDNP-reactive side chains have shown to be suited parameters for the combination of experimentally and theoretically derived results. Those values can be calculated with the help of the Connolly program in InsightII following an established method [5–7, 15–19]. The calculated values correspond to the accessible exterior part of the molecule obtained by smoothening the van der Waals surface with a test sphere that has the average radius of a water molecule (1.5 Å). Additionally, the average radius of flavin (4 Å) can be used without yielding different conclusions concerning the degree of accessibility of the studied residues. The average radii of water ($r = 1.5$ Å) and flavin ($r = 4$ Å) are calculated according to a published formula [20]. The dot density of the van der Waals-like spheres representing the Connolly surfaces is generally set to a value of 1 corresponding to a distribution of calculated surface coordinate values.

36.4 Applications

Beside successful studies of protein–nucleic acid interactions [19, 21, 22] and protein-folding processes [23] the method has shown to be a proper tool in glycosciences since Tyr-, His- and Trp-residues are often constituents of the carbohydrate binding domain [5–7, 15, 24, 25]. Many plant and animal lectins harbour CIDNP-reactive amino acid residues as an essential part of their binding pocket architecture [26–28].

The presence of aromatic amino acids in the binding site of various lectins raises the question why such hydrophophic amino acids are important for the binding of the hydrophilic carbohydrates. An answer may be found in Figure 1a and b which documents the hydrophilic and the hydrophobic parts of a monosaccharide and its position between polar and non-polar amino acids. The CIDNP-reactive amino acid residues can form hydrophobic contacts with the unpolar parts of the carbohydrate in the binding pocket. Lectins are interesting objects to study this aspect of binding in detail. These molecules are carbohydrate binding proteins from non-immunological origin and without enzymatic activity. They play important roles in various regulation processes in the plant as well as in the animal organism and are therefore of great interest in the field of glycosciences [25, 26, 28–30].

Since Tyr-, His- and Trp-residues are CIDNP-responsive and often involved in carbohydrate binding, alterations of the respective signal intensities of lectins can be detected after addition of a specific saccharide. It is possible to correlate such findings with the theoretical results derived by molecular modelling techniques. CIDNP

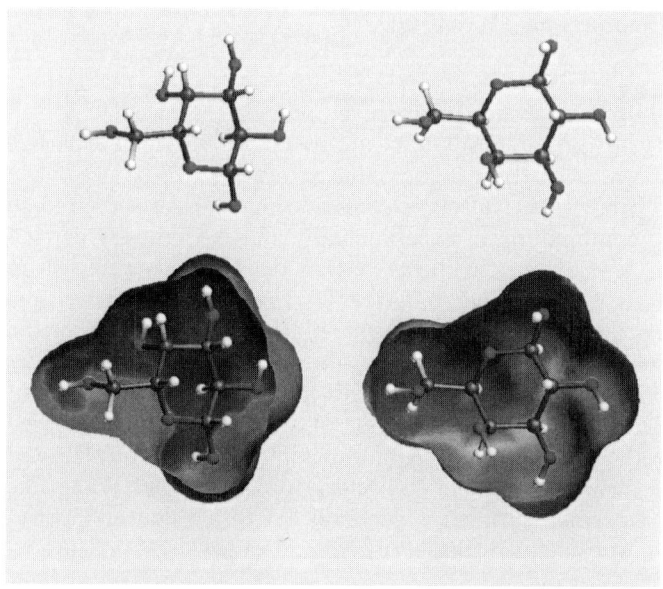

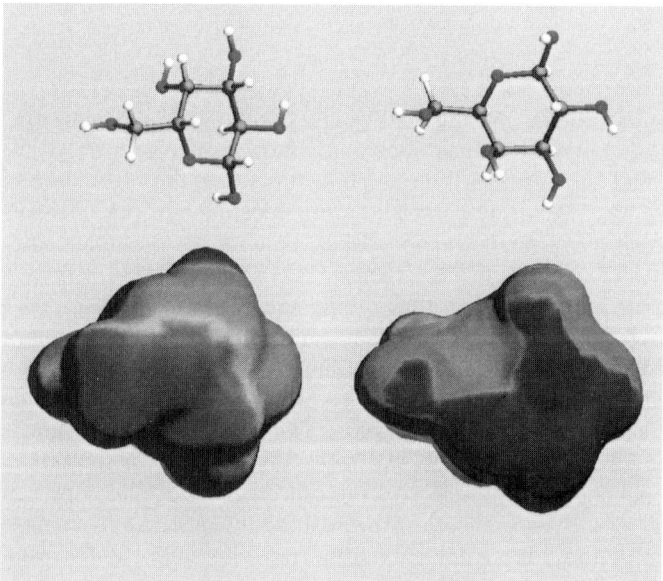

Figure 1a. Hydrophobic an hydrophilic site of a monosaccharide.

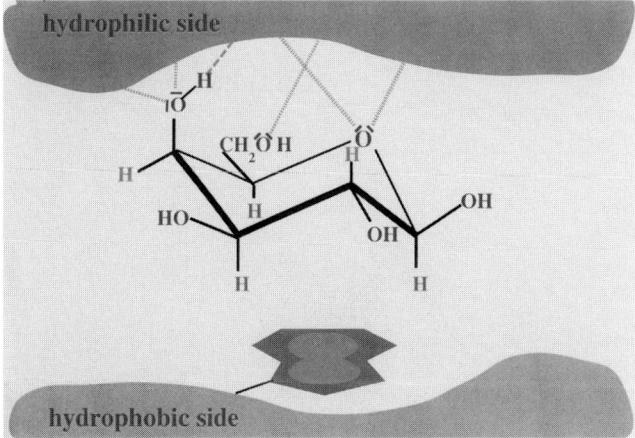

Figure 1b. Schematic representation of a monosaccharide in a binding pocket with a hydrophobic and a hydrophilic site.

data provide furthermore complementary information to structural data derived by other biochemical and biophysical methods [31]. Molecular dynamics simulations in combination with CIDNP experiments make it possible to refine the models obtained by interpretation of the rigid X-ray coordinates alone. The success of such a study can be investigated by isoforms and/or mutants of carbohydrate-binding proteins, when available. In case X-ray data from one isoform or the wild type are known, information about important structural features can already be obtained by CIDNP-experiments.

36.5 Hevein-like Lectins

The role of Tyr-, His- and Trp-residues can be exemplarily studied in the small lectins hevein, pseudo-hevein (two isolectins from the rubber tree) and the B-domain of wheat germ agglutinin (WGA) which consist of 43 amino acid residues [6, 32, 33]. All hevein-like lectins have a binding specificity for D-N-acetylglucosamine oligomers (GlcNAc$_n$). The CIDNP signals of the corresponding amino acids in the binding pocket show a significantly weakened intensity after complexation with a ligand, as demonstrated for hevein at a low pD-value of 4.4 (Figure 2).

The lectin *Urtica dioica* agglutinin (UDA) from the stinging nettle and wheat germ agglutinin (WGA) have also an affinity to GlcNAc$_n$. UDA consists of two and WGA of four hevein-like domains. Beside its affinity to GlcNAc$_n$ WGA also binds oligosaccharides containing sialic acids. In all mentioned examples the corresponding signals of Tyr- and Trp-residues of the binding pocket are affected by the addition of GlcNAc$_n$ indicated by CIDNP signals with lower intensity after ligand complexation. These results were confirmed by two-dimensional NMR methods

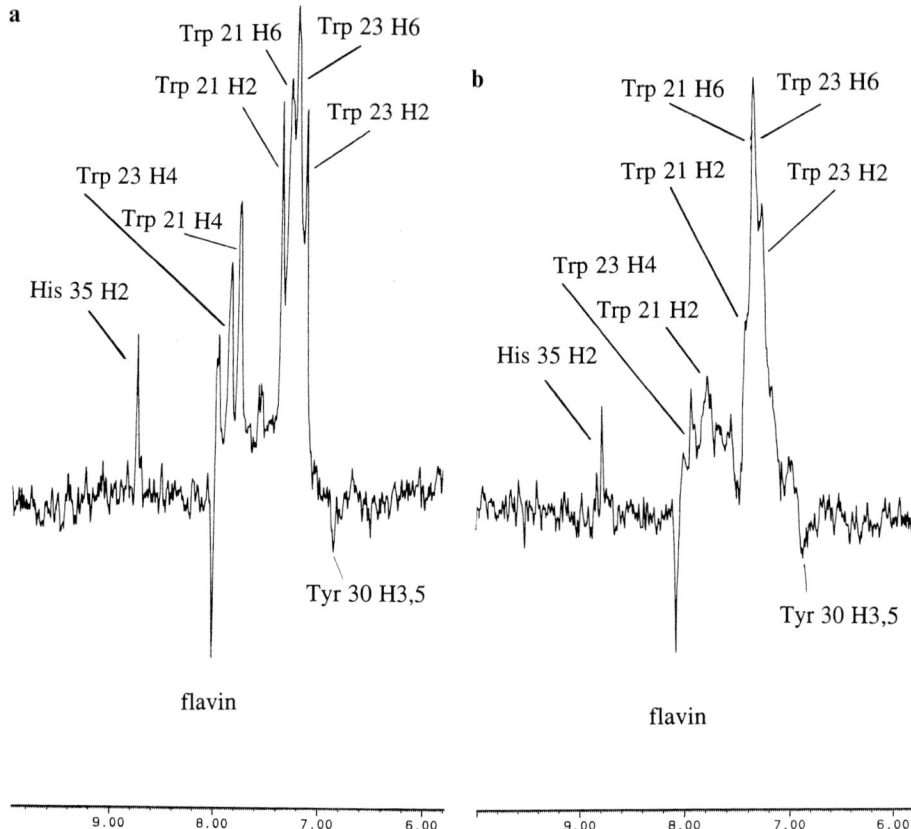

Figure 2. Laser photo CIDNP difference spectra (aromatic part) of hevein and hevein-GlcNAc$_4$ complexes at pD = 3.5. **a.** ligand-free hevein; **b.** 0.5 mmol hevein + 1 mmol GlcNAc$_4$.

[13, 34, 35]. The degree of suppression for WGA and the hevein-like monomers was significantly stronger than for UDA. Interestingly, complexation of UDA with a small amount of GlcNAc$_3$ leads to an altered CIDNP spectrum in which a new small Trp-signal has occurred. This finding argues in favour for a small conformational change of the binding pocket during ligand binding (Figure 3) [6].

In order to obtain further information about the architecture of the hevein-domains the CIDNP data were correlated with X-ray crystallographic coordinates from WGA [36] and completely NMR-obtained conformations from hevein and hevein-ligand complexes [13, 32, 33, 37]. Comparison of CIDNP-derived surface accessibilities with computationally obtained values on the basis of refined X-ray or NMR-structures leads to the following results: Buried Tyr-residues have a surface accessibility below 30 Å^2, partly buried ones have accessibilities of 30–80 Å^2, whereas a value above 80 Å^2 is calculated for Tyr-residues which are considered to be the surface exposed ones causing intense CIDNP signals. Corresponding values

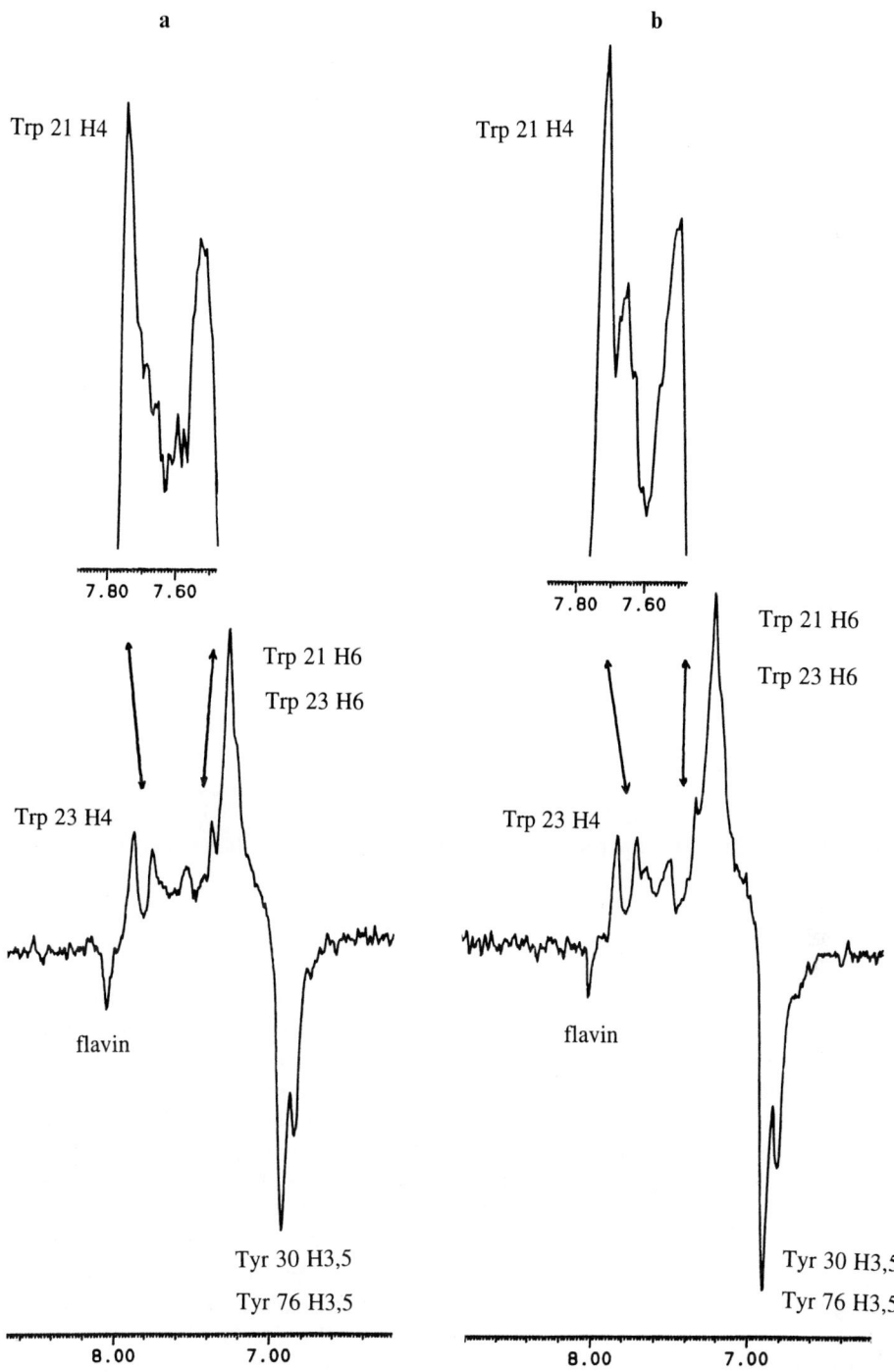

Figure 3. Laser photo CIDNP difference spectra (aromatic part) of UDA and UDA-GlcNAc$_4$ complexes. **a.** ligand-free UDA; **b.** 0.5 mmol UDA + 1 mmol GlcNAc$_4$.

are estimated for Trp- and His-residues. The calculations are carried out with a dot density of 1 and a test sphere radius of 1.5 Å [5–7, 15–19].

These data make it possible to use CIDNP-derived results for quality control of modelled structural data [5–7]. The surface accessibility values of CIDNP-reactive amino acid residues of hevein-like lectins can be compared with the intensities of the corresponding CIDNP signals. CIDNP and modelling data indicate that the architecture of the binding pockets of hevein, pseudo-hevein and the B-domain of WGA1 are similar [6]. In detail: different ways in designing the models have been used. In one case the B-domain of WGA1 was directly taken from the X-ray structure of WGA1. In the other case the B-domain of WGA1 has been constructed by amino acid replacement in the hevein-NMR-structure. The initial structures have been energetically minimized before starting the MD (molecular dynamics) simulations. Values of the surface accessibility of CIDNP-reactive amino acid residues are obtained during the MD-simulation and listed in a table [6]. Modelled structures of pseudo-hevein and the B-domain of WGA1 which are in optimal agreement with the CIDNP-results are shown in Figure 4a, b.

36.6 Galactoside-binding Lectins from Plant and Animal Origin

NMR-spectra of *Erythrina corallodendron* lectin (EcorL) are not resolved (Figure 5) without the use of special techniques, like CIDNP. Resolved CIDNP signals from EcorL, however, can be assigned by comparison of wild-type- and mutant-EcorL-CIDNP-spectra, as published [7]. For completion, galectins, which are galactoside-binding lectins from animal origin which do not need any Ca^{2+}-ions for saccharide binding and which share structural homologies with EcorL, have been included in this study.

From X-ray data as well as from the results of the CIDNP experiments one can conclude that the highly accessible amino acid residues Tyr106 in EcorL and Trp68 in galectins are involved in the binding process. However, X-ray data of EcorL and bovine galectin also indicate that the extent of surface accessibility of other CIDNP-reactive moieties besides Tyr106 and Trp68 can be examined. Since the crystallographic data sets provide the starting point for knowledge-based homology modelling, MD (molecular dynamics)-derived conformational parameters for the related proteins can likewise be correlated to the results of the CIDNP spectroscopic measurements. This comparison is also facilitated by the availability of mutants for the legume lectin, in which Tyr-residues at position 106, 108 or 229, respectively, were replaced by Thr-, Gly- or Ala-residues [38, 39]. Furthermore, a mutant was studied in which Trp is replaced by Ala at position 135. The introduction of the single-site mutations in the legume protein can cause non-uniform impacts on the calculated conformational parameters beyond the immediate vicinity of the site of mutation [7, 40]. These results deserve attention for the interpretation of comparative data sets of wild-type and single-site mutant proteins. The structural predictions of molecular modelling are in agreement with the CIDNP-spectra. Regarding the

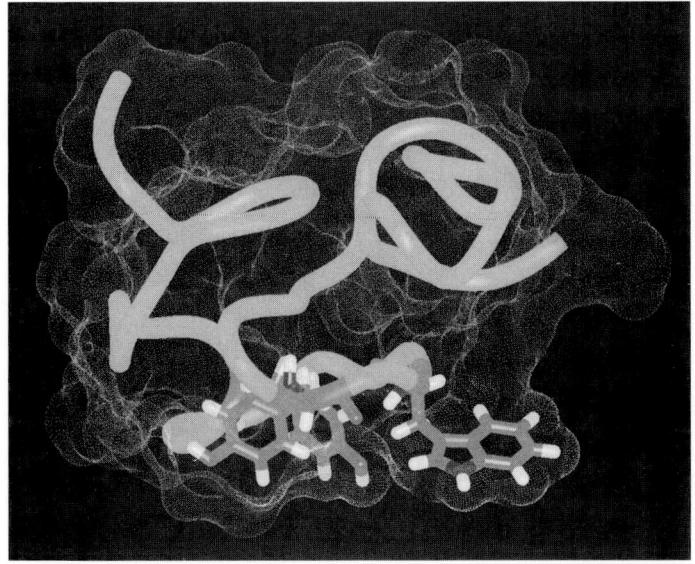

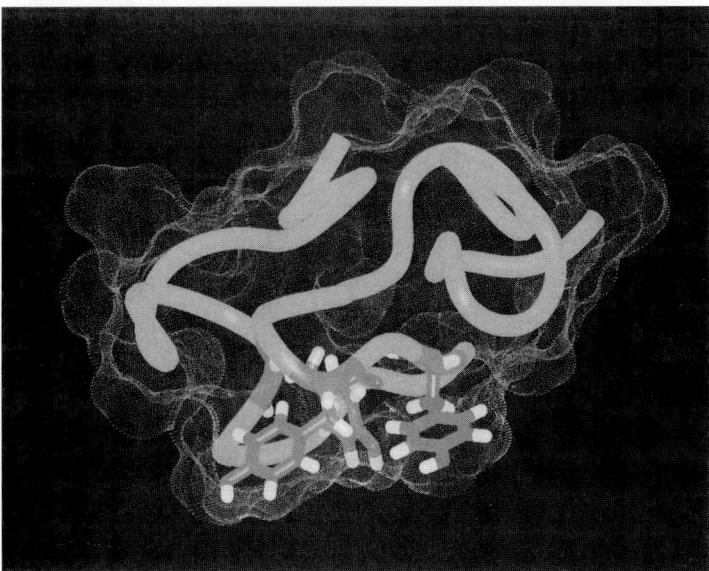

Figure 4a. Energy minimum conformation of pseudo-hevein, emphasising the surface exposition of Tyr21 and Trp23 in the upper part of the picture; **b.** energy minimum conformation of domB of WGA1, emphasising the surface exposition of Tyr64 and Tyr66 in the lower part of the picture.

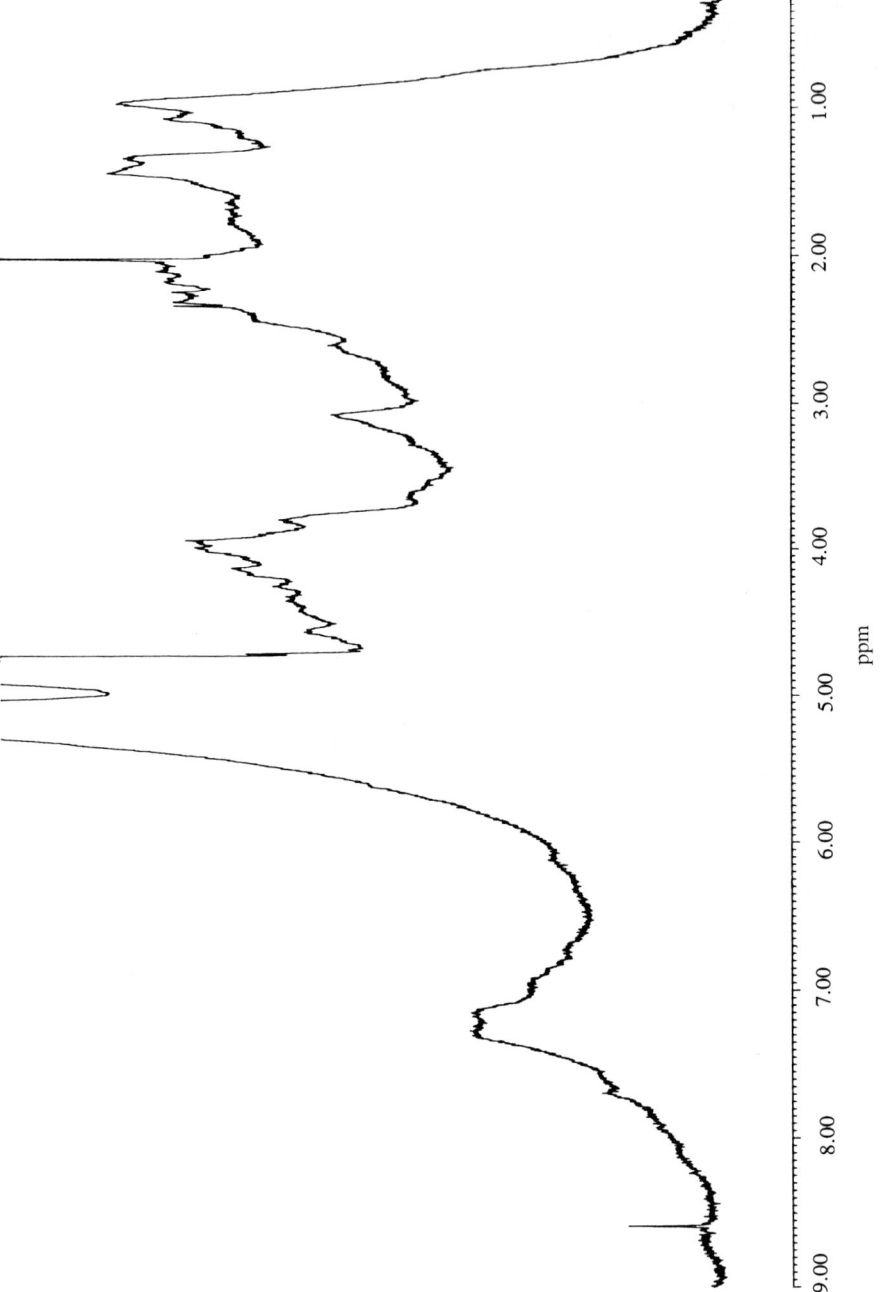

Figure 5. One dimensional ^{1}H-NMR spectrum of EcorL (60 kDa).

intensity of the Tyr-signal, the CIDNP-spectrum of the mutant Trp135Ala (Figure 6a) does not alter remarkably from that of the wild type [7]. A single site mutation of the highly accessible Tyr 106 leads to the complete disappearance of the Tyr-signal in the CIDNP difference spectrum, Figure 6b. However, also the CIDNP-spectrum of an EcorL mutant wherein Tyr108 is exchanged against a CIDNP-inert residue does not show any Tyr-CIDNP-signal, Figure 6c. This can only be the case when Tyr106 loses its surface accessibility due to the replacement of Tyr108. Our experimental finding is exactly reproduced by molecular modelling calculations which are based on the X-ray structure data of EcorL [41]. The derived MD data [7, 42] confirm the experimental results, as demonstrated in Table 1. Furthermore, as can also be seen from Table 1, the single site mutations of Tyr106 or Tyr108 have an impact on the surface accessibilities of Tyr192 and Tyr229 which are lowered significantly. On the other hand a replacement of Tyr by Ala at position 229 influences the surface presentation of Tyr106 and Tyr108 [7] but no impact could be detected on the binding specificity [38, 39]. The conformational rearrangement in this mutant is shown in Figure 7a, b.

The CIDNP method was also applied to five members of the galectin family, namely bovine heart galectin, human lung galectin, CG-14—the galectin from chicken intestine, CG-16—the galectin from chicken liver and the recombinant murine galectin-3. The experimental data show a significant covering of Trp68 after addition of lactose whereas the Tyr-residues of the studied galectins are unaffected by the ligand. The X-ray structure from bovine heart galectin and the CIDNP data of all studied galectins have been used as valuable information for the design of four model structures [7, 43].

CIDNP-responsive amino acids are therefore important constituents in the binding site of the β-galactoside specific galectins. One galectin-domain with a bound ligand and the involvement of Trp is shown in Figure 8.

The availability of well-defined X-ray structures [44–46] and sequence homology between various galectins from different species allow the prediction of structures by use of the knowledge-based protein modelling approach [47, 48]. The results obtained from CIDNP experiments provide valuable information concerning the reliability of structures obtained by different modelling techniques. The occurrence of an additional Trp-CIDNP signal in a spectrum obtained from a saccharide–human galectin1 complex argues in favour of a change in orientation of the Trp-ring. Furthermore, effects on the binding strength of galectins caused by site-directed mutagenesis [49] can be analysed by a combination of CIDNP and modelling techniques. Such a protein–carbohydrate interaction study can be completed when the results from transferred nuclear Overhauser experiments (trNOE) are integrated (50). TrNOE-experiments are a proper tool in glycosciences [50–58]. These kinds of experiments are used when the conformation of a small ligand in the presence of a much bigger receptor is studied. Since the mentioned biophysical methods are complementary to each other, a combination of X-ray-data, NMR methods (multi-dimensional experiments, trNOE, CIDNP) and molecular modelling techniques is in general the best way to analyse protein-carbohydrate complexes on an atomic scale [48, 58, 59].

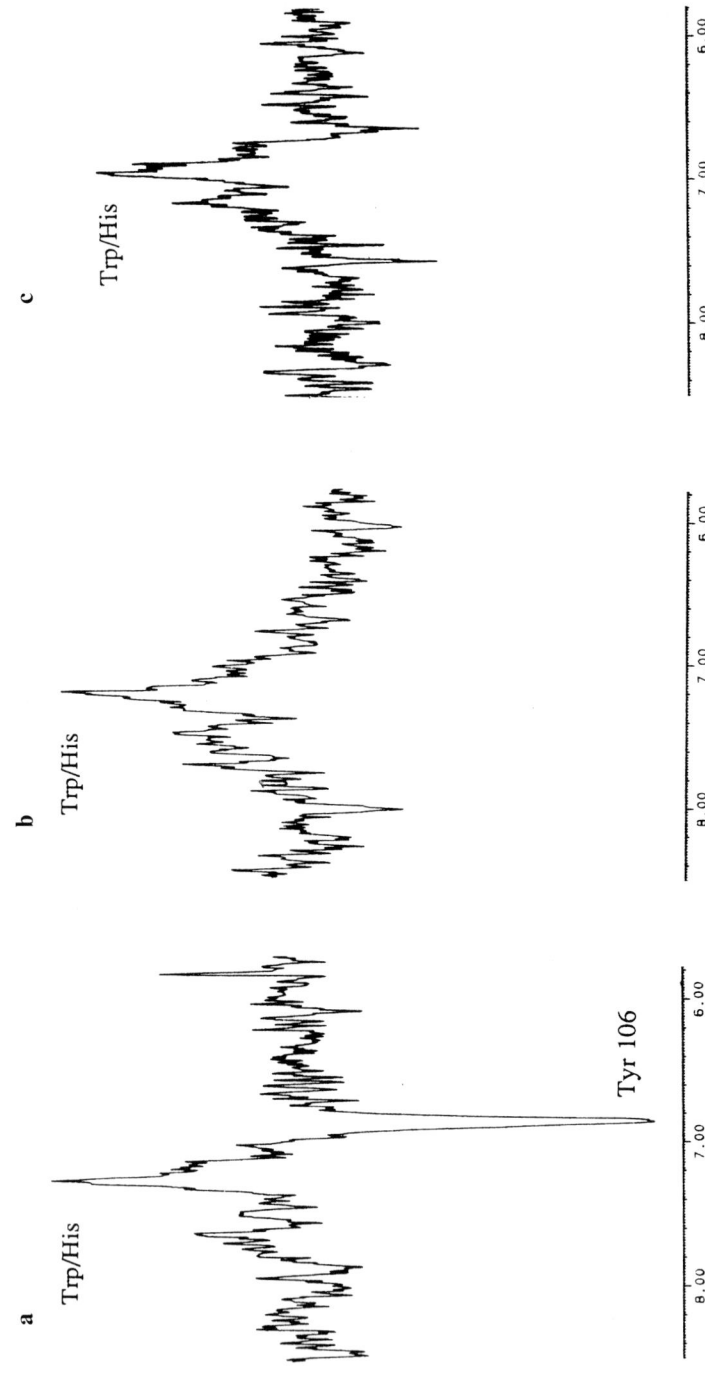

Figure 6. CIDNP-difference spectra (aromatic part) of three single-site mutants of recombinant EcorL. **a.** [W135A]EcorL (ligand-free); **b.** [Y106G]EcorL (ligand-free); **c.** [Y108T]EcorL (ligand-free).

Table 1. Average surface accessibilities ($\bar{x}$) and the differences of the surface accessibility areas of CIDNP-sensitive residues derived from molecular dynamics calculations between the wild-type of EcorL ($\bar{x}$) and the three single-site mutants [7]. All area values are given in Å^2. Dot density: 1; test sphere radius: 1.5 Å. **** denotes the site of mutation substituting the respective tyrosine residue with CIDNP-inert alanine.

	$\bar{x}$	ΔY106	ΔY108	ΔY229
Tyr 53	10.0	+1.2	−10.0	+15.5
Tyr 82	69.3	−10.7	+7.0	+4.4
Tyr 106	67.7	****	−14.4	+14.1
Tyr 108	75.3	−30.7	****	−32.6
Tyr 121	49.4	+12.2	+15.7	+11.6
Tyr 172	7.9	−7.4	+7.4	−7.9
Tyr 185	6.1	+3.6	−1.6	+16.3
Tyr 192	65.8	−18.6	9.0	−30.4
Tyr 229	81.0	−11.9	−25.4	****
Trp 45	100.9	+5.2	−29.0	−15.5
Trp 60	30.9	+2.8	−17.6	−26.3
Trp 135	69.1	−20.6	+23.6	−2.8
Trp 207	113.3	−5.5	−42.2	−20.6
Trp 231	3.0	−1.0	−3.0	−3.0
His 58	28.7	−10.0	−3.6	−14.9
His 142	0.0	0.0	0.0	0.0
His 180	45.0	+13.0	+9.4	+17.2
His 226	0.0	+2.1	0.0	0.0

36.7 Sialidase from *Clostridium Perfringens* (Wild Type and Mutants)

The presence of a number of CIDNP-sensitive amino acids on the protein surface can enable the detection of conformational alterations resulting from single site mutations. This has been demonstrated for wild-type forms and various mutants of EcorL and of the sialidase of *Clostridium perfringens* [5, 7]. Sialidase of *C. perfringens* is larded with CIDNP-responsive amino acid residues. Therefore, these residues have been used as valuable sensors in NMR and modelling studies in which CIDNP-spectra of the sialidase wild type and various mutants are compared and correlated with modelled structures.

Starting with the crystal structure of the bacterial sialidase of *Salmonella typhimurium* [60] which is used as a framework, the knowledge based homology modelling produces a set of energy-minimised conformations for the small sialidase of *C.*

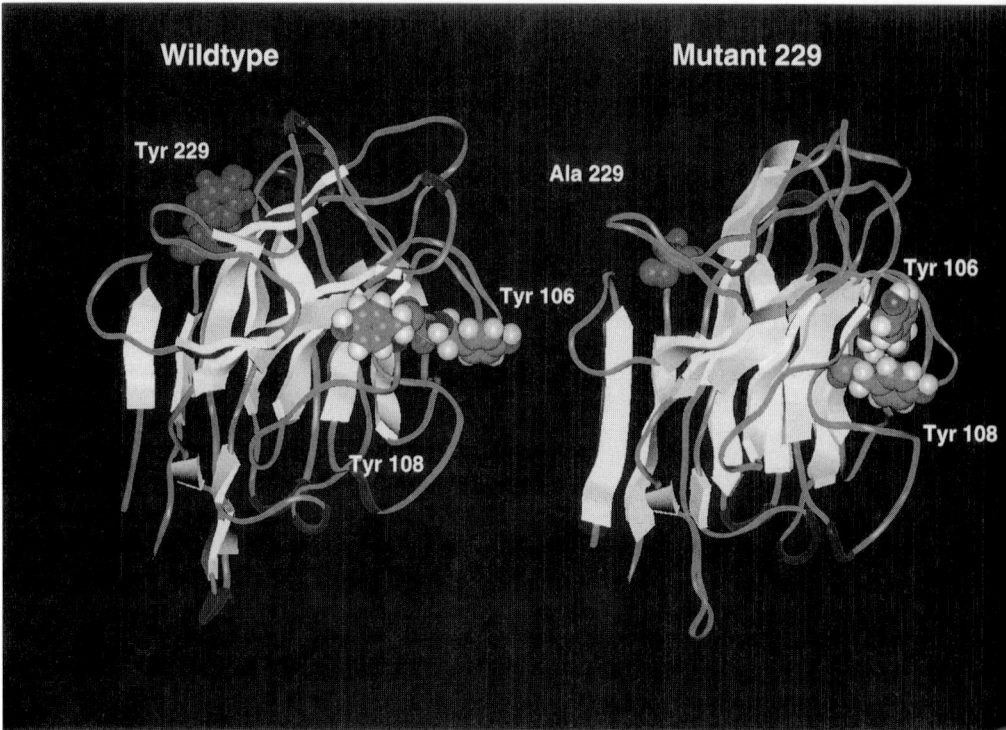

Figure 7. MD-derived energy minimum structures of wild-type EcorL and single-site mutants. **a.** wild-type EcorL; **b.** mutant [Y229A]EcorL.

perfringens. Although the number of CIDNP-reactive amino acids is to large to unequivocally assign signals to defined residues, as has been feasible for hevein [6], the signal intensity can be set into correlation with the model-derived expectations.

In addition to the wild-type enzyme the impact of introducing amino acid substitutions by site-directed mutagenesis has been theoretically and experimentally delineated as a further test of the model structure. Changes in surface accessibilities of widely separated residues affected by a Tyr/Phe- or a Cys/Ser-substitution could be measured and calculated. This leads to the conclusion that conformational changes of mutant enzymes relative to the wild-type form should not be underestimated.

However, not only CIDNP-spectra and modelling data argue in favour of an influence of single site mutations on the overall conformation. The CIDNP- and modelling-results obtained for the sialidase of *C. perfringens* and several of its mutants are in perfect agreement with observations concerning the enzymatic activities of these molecules determined by biochemical methods. The differences in enzymatic activities between the wild type and various mutants confirm the conclusions drawn from CIDNP experiments and molecular modelling [5].

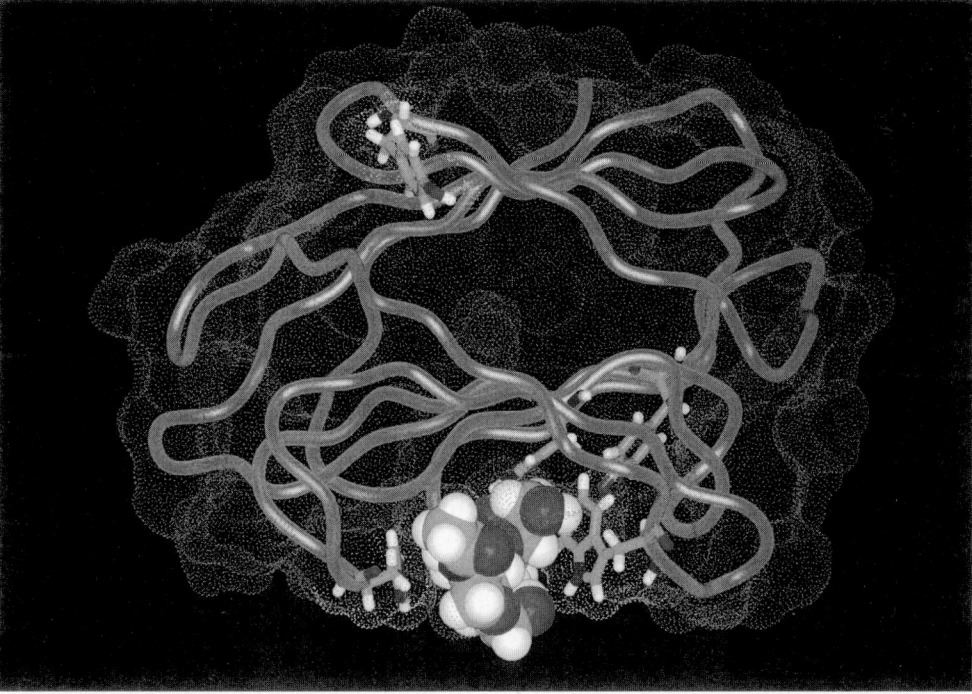

Figure 8. Galectin-1 monomer with a specific disaccharide in its binding pocket.

The success of knowledge-based homology modelling is critically dependent on the predictive potency of the program of structure-based calculations, which attempts to translate homologous sequences into three-dimensional structures. In order to evaluate the actual relevance of molecular-modelling-supported crystal structures for the protein topology in solution, CIDNP data providing selected parameters of the protein's conformation can be used for quality ranking.

36.8 CIDNP Analysis of Glycoproteins

When CIDNP-reactive amino acids are accessible on the surface of a glycoprotein, a possible impact of the covalently bound oligosaccharide chain on the protein part can be scrutinized [14, 15, 61] by this method. Therefore, CIDNP-experiments are also suitable to improve structural models of glycoproteins. On the basis of a X-ray structure of a glycoprotein, very often only the coordinates for the protein part are available since the flexibility of the oligosaccharide chain hampers a complete crystallographic structure. NMR measurements of the oligosaccharide chain [62–66] in combination with molecular mechanic and molecular dynamic calculations make it possible to gain the needed structural data. CIDNP experiments were applied on the glycoprotein SAP (serum amyloid P-component) from human serum [67, 68]. In one series of experiments the oligosaccharide chains were intact and in another series they were desialylated [14, 15].

The respective spectra of SAP and enzymatically desialylated SAP were determined. Six Trp/His signals and one Tyr signal are present in the aromatic part of the CIDNP difference spectrum of SAP. The corresponding spectrum of desialylated SAP shows remarkable alterations. The chemical shift of one Trp/His-characteristic signal is decreased by 0.1 ppm. One Trp/His-signal disappeared and a new one was formed in the CIDNP difference spectrum of desialylated SAP, while the other signals were unaffected. The Tyr signal has a clearly enhanced intensity in desialylated SAP. Therefore, the removal of sialic acid residues from the single *N*-glycan of each monomer apparently affects the surface presentation of distinct CIDNP-reactive amino acids of SAP [14, 15]. A conformational change of the protein part of SAP in relation with a different orientation of the desialylated oligosaccharide chain in comparison to the sialylated chains or a covering of CIDNP-responsive amino acids by the oligosaccharide chain itself are possible explanations of the CIDNP results. In order to prove these experimental data molecular dynamics simulations of SAP with an intact and an asialo-saccharide chain were carried out. The structural data of the protein part of SAP were obtained by using X-ray crystallographic coordinates of SAP [69, 70] as a start structure for the molecular dynamics simulations. The computationally constructed oligosaccharide chains were energetically minimised with the SWEET program [71] and linked to the protein part (Figure 9) before the MD-run. The MD-simulations address the question in which way an interaction between the oligosaccharide and protein part is possible. Small long range interactions within oligosaccharide chains of *O*-acetylated and non-*O*-acetylated gangliosides (9OAc-GD1a and GD1a) have already been identified with help of NMR-supported MD-simulations [72].

36.9 Conclusions

In summary, the CIDNP method is a potent addition to the arsenal of biophysical methods which are successfully used in the field of structural glycosciences. The combination of X-ray data, molecular modelling, trNOE-methods and laser-photo CIDNP provides valuable structural information about protein-carbohydrate complexes. This is of particular importance, when multidimensional NMR-data can not be recorded due to the size of the molecule. Furthermore, the analysis of similar carbohydrate-binding proteins sharing a sufficient sequence homology is possible when complete structural data are available only for a few of them and CIDNP- and molecular modelling data are present for all of them. The successful role, which CIDNP-experiments can play for the solution of important structural glyco-

Figure 9. X-ray structure of human serum amyloid P component with modelled oligosaccharide chains.

biochemical problems, is based on the important function of Tyr-, His- and Trp-residues in constituting the hydrophobic interactions between the corresponding combining sites of a carbohydrate and a protein.

Acknowledgments

This work was supported by the Human Capital and Mobility Program of the European Community, the Netherlands Foundation for Chemical Research (SON) and the Netherlands Organisation for Scientific Research (NWO).

References

1. D. Neuhaus, M. P. Williamsen, *The Nuclear Overhauser Effect*, **1989**, VCH, Weinheim–New York.
2. R. Kaptein, K. Dijkstra, K. Nicolay, *Nature*, **1978**, *274*, 293–294.
3. R. Kaptein in *Biological Magnetic Resonance*, **1982**, *4*, (L. J. Berliner, ed.) pp. 145–191, Plenum Press, New York.
4. P. J. Hore, R. W. Broadhurst, *Progr. NMR Spectrosc.*, **1993**, *25*, 345–402.
5. H.-C. Siebert, E. Tajkhorshid, C.-W. von der Lieth, R. G. Kleineidam, S. Kruse, R. Schauer, R. Kaptein, J. F. G. Vliegenthart, H.-J. Gabius, *J. Mol. Model.*, **1996**, *2*, 446–455.
6. H.-C. Siebert, C.-W. von der Lieth, R. Kaptein, J. J. Beintema, K. Dijkstra, N. van Nuland, U.M.S. Soedjanaatmadja, A. Rice, J.F.G. Vliegenthart, C.S. Wright, H.-J. Gabius, *Proteins*, **1997**, *28*, 268–284.
7. H.-C. Siebert, R. Adar, R. Arango, M. Burchert, H. Kaltner, G. Kayser, E. Tajkhorshid, C.-W. von der Lieth, R. Kaptein, N. Sharon, J. F. G. Vliegenthart, H.-J. Gabius, *Eur. J. Biochem.*, **1997**, *249*, 27–38.
8. S. Stob, R. Kaptein, *Photochem. Photobiol.*, **1989**, *49*, 565–577.
9. P. F. Heelis, B. J. Parsons, G. O. Phillips, *Biochim. Biophys. Acta*, **1979**, *587*, 455–462.
10. K. A. Muszkat, T. Wismontski-Knittel, *Biochemistry*, **1985**, *24*, 5416–5421.
11. R. M. Scheek, R. Kaptein, J. W. Verhoven, *FEBS Lett.*, **1979**, *107*, 288–290.
12. Y. N. Molin (ed.), *Spin Polarization and Magnetic Effects in Radical Reactions*, **1984**, Elsevier, Amsterdam.
13. H.-C. Siebert, R. Kaptein, J. J. Beintema, U. M. S. Soedjanaatmadja, C. S. Wright, A. Rice, R. G. Kleineidam, S. Kruse, R. Schauer, P. J. W. Pouwels, J. P. Kamerling, H.-J. Gabius, J. F. G. Vliegenthart, *Glycoconj. J.*, **1997**, *14*, 531–534.
14. H.-C. Siebert, S. André, G. Reuter, R. Kaptein, J. F. G. Vliegenthart, H.-J. Gabius, *Glycoconj. J.*, **1997**, *14*, 945–949.
15. H.-C. Siebert, S. André, G. Reuter, H.-J. Gabius, *FEBS Lett.*, **1995**, *371*, 13–16.
16. B. Lee, F. M. Richards, *J. Mol. Biol.*, **1971**, *55*, 379–400.
17. M. L. Connolly, *J. Appl. Cryst.*, **1983**, *16*, 548–558.
18. M. L. Connolly, *Science*, **1983**, *221*, 709–713.
19. E. Kellenbach, T. Härd, R. Boelens, K. Dahlman, J. Carlstedt-Duke, J.-Å. Gustafsson, G. A. van der Marel, J. H. van Boom, B. Maler, K. R. Yamamoto, R. Kaptein, *J. Biomol. NMR*, **1991**, *1*, 105–110.
20. J. T. Edward, *J. Chem. Educ.*, **1970**, *47*, 261–270.
21. F. Buck, H. Rüterjans, R. Kaptein, K. Beyreuter, *Proc. Natl Acad. Sci. USA*, **1980**, *77*, 5145–5148.
22. S. Stob, R. M. Scheek, R. Boelens, R. Kaptein, *FEBS Lett.*, **1988**, *239*, 99–104.
23. R. W. Broadhurst, C. M. Dobson, P. J. Hore, S. E. Radford, M. L. Rees, *Biochemistry*, **1991**, *30*, 405–412.
24. F. Quiocho, *Pure Appl. Chem.*, **1989**, *61*, 1293–1306.
25. N. Sharon, H. Lis, H., *Science*, **1989**, *246*, 227–246.
26. N. Sharon, *Trends Biochem. Sci.*, **1993**, *18*, 221–226.
27. J. M. Rini, *Annu. Rev. Biophys. Biomol. Struct.*, **1995**, *24*, 551–577.
28. H.-J. Gabius, S. Gabius, *Glycosciences: Status and Perspectives*, **1997**, Chapman & Hall, Weinheim–London.
29. R. A. Dwek, *Chem. Rev.*, **1996**, *96*, 683–720.
30. H.-J. Gabius, *Eur. J. Biochem.*, **1997**, *243*, 543–576.
31. T. Díaz-Mauriño, D. Solís, J. Jiménez-Barbero, M. Martín-Lomas, H.-C. Siebert, J. F. G. Vliegenthart, *Carbohydr. Eur.*, **1998**, in press.
32. J. L. Asensio, F. J. Cañada, M. Bruix, A. Rodríguez-Romero, J. Jiménez-Barbero, *Eur. J. Biochem.*, **1994**, *230*, 621–633.
33. J. L. Asensio, F. J. Cañada, M. Bruix, C. Gonzáles, N. Khiar, A. Rodríguez-Romero, J. Jiménez-Barbero, *Glycobiology*, **1998**, *8*, 569–577.
34. K. Hom, M. Gochin, W. J. Peumans, N. Shine, *FEBS Lett.*, **1995**, *361*, 157–161.

35. H.-C. Siebert, R. Kaptein, J. F. G. Vliegenthart, in *Lectins and Glycobiology* (H.-J. Gabius, S. Gabius, eds), **1993**, pp. 105–116, Springer Verlag, Heidelberg, New York.
36. C. S. Wright, *J. Mol. Biol.*, **1989**, *209*, 475–487.
37. N. H. Andersen, B. Cao, A. Rodriguez-Romero, B. Arreguin, *Biochemistry*, **1993**, *32*, 1407–1422.
38. R. Arango, E. Rodriguez-Arango, R. Adar, D. Belenky, F. G. Loontiens, S. Rozenblatt, N. Sharon, *FEBS Lett.*, **1993**, *330*, 133–136.
39. R. Adar, N. Sharon, N., *Eur. J. Biochem.*, **1996**, *239*, 668–674.
40. T. P. Kogan, B. M. Revelle, S. Tapp, D. Scott, P. J. Beck, *J. Biol. Chem.*, **1995**, *270*, 14047–14055.
41. B. Shaanan, H. Lis, N. Sharon, *Science*, **1991**, *254*, 862–866.
42. E. Moreno, S. Teneberg, R. Adar, N. Sharon, K.-A. Karlsson, J. Ångström, *Biochemistry*, **1997**, *36*, 4429–4437.
43. E. Tajkhorshid, H.-C. Siebert, M. Burchert, H. Kaltner, G. Kayser, C.-W. von der Lieth, R. Kaptein, J. F. G. Vliegenthart, H.-J. Gabius, *J. Mol. Model.*, **1997**, *3*, 325–331.
44. Y. D. Lobsanov, M. A. Gitt, H. Leffler, S. H. Barondes, J. M. Rini, *J. Biol. Chem.*, **1993**, *26*, 27034–27038.
45. Y. Bourne, B. Bolgiano, D. Liao, G. Strecker, P. Cantau, O. Herzberg, T. Feizi, C. Cambillau, *Nature Struct. Biol.*, **1994**, *1*, 863–870.
46. D. Liao, G. Kapadia, H. Ahmed, G. R. Vasta, O. Herzberg, *Proc. Natl Acad. Sci. USA*, **1994**, *91*, 1428–1432.
47. M. S. Johnson, N. Srinivasan, R. Sowdhamini, T. L. Blundell, *Crit. Rev. Biochem. Mol. Biol.*, **1994**, *29*, 1–68.
48. M. W. MacArthur, R. A. Laskowski, J. M. Thornton, J. M., *Curr. Opin. Struct. Biol.*, **1994**, *4*, 731–737.
49. J. Hirabayashi, in *Glycosciences: Status and Perspectives* (H.-J. Gabius, S. Gabius, eds.), **1997**, pp. 355–368, Chapman & Hall, London-Weinheim.
50. H.-C. Siebert, M. Gilleron, H. Kaltner, C.-W. von der Lieth, T. Kozár, N. V. Bovin, E. Y. Korchagina, J. F. G. Vliegenthart, H.-J. Gabius, *Biochem. Biophys. Res. Commun.*, **1996**, *219*, 205–212.
51. V. L. Bevilacqua, D. S. Thomson, J. H. Prestegard, *Biochemistry*, **1990**, *29*, 5529–5537.
52. V. L. Bevilacqua, Y. Kim, J. H. Prestegard, *Biochemistry*, **1992**, *31*, 9339–9349.
53. J. L. Asensio, F. J. Cañada, J. Jiménez-Barbero, *Eur. J. Biochem.*, **1995**, *233*, 618–630.
54. T. Weimar, T. Peters, *Angew. Chem. Int. Ed. Engl.*, **1994**, *33*, 88–91.
55. L. Poppe, G. S. Brown, J. S. Philo, P. V. Nikrad, B. H. Shah, *J. Am. Chem. Soc.*, **1997**, *119*, 1727–1736.
56. M. Gilleron, H.-C. Siebert, H. Kaltner, C.-W. von der Lieth, T. Kozár, N. V. Bovin, E. Y. Korchagina, H.-J. Gabius, J. F. G. Vliegenthart, *Eur. J. Biochem.*, **1998**, *252*, 416–427.
57. H.-C. Siebert, C.-W. von der Lieth, M. Gilleron, G. Reuter, J. Wittmann, J. F. G. Vliegenthart, H.-J. Gabius, in *Glycosciences: Status and Perspectives* (H.-J. Gabius, S. Gabius, eds.), **1997**, pp. 291–310, Chapman & Hall, London-Weinheim.
58. A. Poveda, J. Jimenez-Barbero, *Chem. Soc. Rev.*, **1998**, *27*, 133–143.
59. G. Wagner, S. G. Hyberts, T. F. Havel, *Annu. Rev. Biophys. Biomol. Struct.*, **1992**, *21*, 167–198.
60. S. J. Crennel, E. F. Garman, W. Graeme Laver, E. R. Vimr, G. L. Taylor, *Proc. Natl Acad. Sci. USA*, **1993**, *90*, 9852–9856.
61. K. Hård, J. P. Kamerling, J. F. G. Vliegenthart, *Carbohydr. Res.*, **1992**, *236*, 315–320.
62. K. Hård, J. F. G. Vliegenthart, in *Glycobiology. A Practical Approach* (M. Fukuda, A. Kobata, eds.), **1993**, pp. 223–242, Oxford University Press, Oxford, New York, Tokyo.
63. T. de Beer, C. W. E. M. van Zuylen, K. Hård, R. Boelens, R. Kaptein, J. P. Kamerling, J. F. G. Vliegenthart, *FEBS Lett.*, **1994**, *348*, 1–6.
64. T. de Beer, C. W. E. M. van Zuylen, B. R. Leeflang, K. Hård, R. Boelens, R. Kaptein, J. P. Kamerling, J. F. G. Vliegenthart, *Eur. J. Biochem.*, **1996**, *241*, 229–242.
65. J. P. M. Lommerse, L. M. J. Kroon-Batenburg, J. Kroon, J. P. Kamerling, J. F. G. Vliegenthart, *J. Biomol. NMR*, **1995**, *5*, 79–94.
66. J. P. M. Lommerse, L. M. J. Kroon-Batenburg, J. P. Kamerling, J. F. G. Vliegenthart, *Biochemistry*, **1995**, *34*, 8196–8206.

67. G. A. Tennent, M. B. Pepys, *Biochem. Soc. Trans.*, **1994**, *22*, 74–79
68. M. B. Pepys, T. W. Rademacher, S. Amatayakul-Chantler, P. Williams, G. E. Noble, W. L. Hutchinson, P. N. Hawkins, S. R. Nelson, J. R. Gallimore, J. Herbert, T. Hutton, R. A. Dwek, *Proc. Natl Acad. Sci. USA*, **1994**, *91*, 5602–5606.
69. J. Emsley, H. E. White, B.P. O'Hara, G. Oliva, N. Srinivasan, I.J. Tickle, T. L. Blundell, M. B. Pepys, S. P. Wood, *Nature*, **1994**, *367*, 338–345.
70. A. K. Shrive, G. M. T. Cheetham, D. Holden, D. A. A. Myles, W. G. Turnell, J. E. Volanakis, M. B. Pepys, A. C. Bloomer, T. J. Greenhough, *Nature Struct. Biol.*, **1996**, *3*, 346–354.
71. A. Bohne, E. Lang, C.-W. von der Lieth, *J. Mol. Model.*, **1997**, *3*, 1–5.
72. H.-C. Siebert, C.-W. von der Lieth, X. Dong, G. Reuter, R. Schauer, H.-J. Gabius, J. F. G. Vliegenthart, *Glycobiology*, **1996**, *6*, 561–572.

37 Biacore

Wolfgang Jäger

37.1 Introduction

Since the first commercially available Biacore system (Biacore AB, Uppsala, Sweden) was introduced back in 1990, the technology has advanced to become a standard method in studying biomolecular interactions. This Chapter describes the function and practical utilization of this technology, discusses typical user questions and offers suggestions for problem solving. Studies of oligosaccharide interactions are presented, while the whole application range includes all kinds of biomolecules.

Since applications are continually optimized and new reagents and consumables are being developed, detailed protocols will not be given. The reader is referred to updated protocols provided by the supplier, to the literature cited in the text, and to the current reference list on the Internet under http://www.biacore.com.

37.1.1 Real-time Analysis by Surface Plasmon Resonance

The detection principal of Biacore relies on the physical phenomenon of surface plasmon resonance (SPR: Figure 1). Specifically, it is based on an optical unit which measures the refractive index located on the matrix side of the flat sensor chips (Figure 2). One of the interacting partners is immobilized on the chip matrix. Substance accumulation by molecule binding to the immobilized partner directly influences the refractive index, which is detected in real-time and indicated immediately [1].

Inside the instrument, the chip matrix forms a roof of small flow cells. In current instrument configurations, each sensor chip serves two or four of these flow channels, giving two or four separate measuring surfaces. The integrated flow system provides a continuous flow of buffer or samples over the chip surface. Sample injections and measurements run automatically: only specimen loading to a sample loop can be

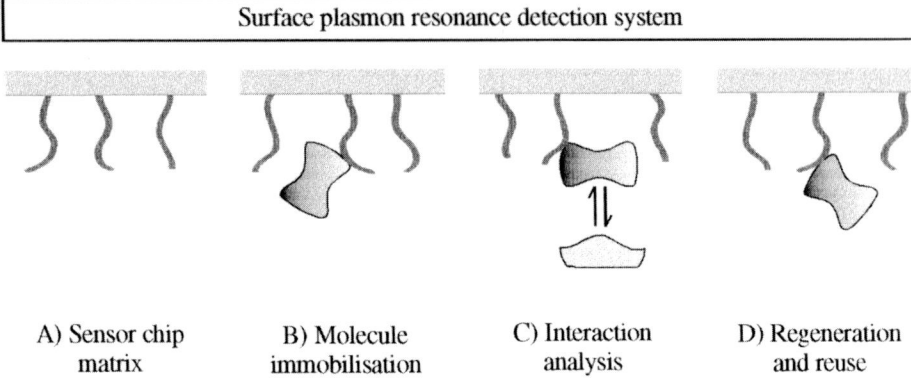

Figure 1. Different steps for applying the biosensor are displayed. **A)** A suitable sensor chip is chosen. **B)** Immobilization of biomolecules in the chip matrix generates a specific measuring surface. **C)** Interaction is analyzed by injecting binding partners along the surface. **D)** After the regeneration the surface is used for the next analysis.

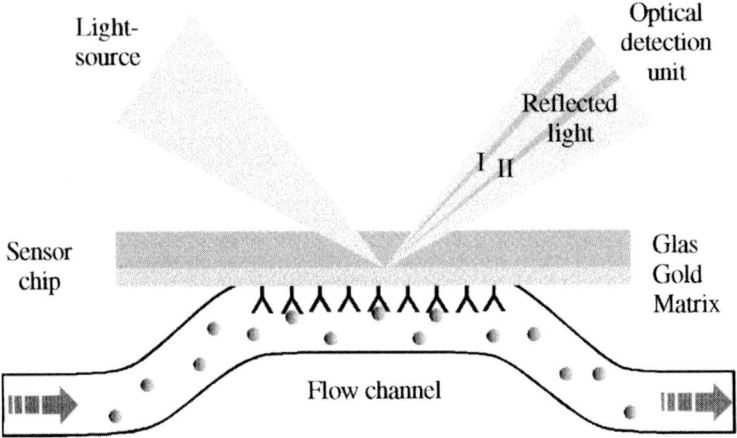

Figure 2. The sensor chip is composed of a glass carrier coated with a thin gold layer on which a biocompatible matrix is attached. An evanescent field is generated when light under the condition of total internal reflection is directed to the glass side. This field extends about 300 nm into the solution on the other side (the matrix side) and interacts with the refractive index of the solution close to the gold surface. The electromagnetic phenomenon of surface plasmon resonance (SPR) arises in the gold film, resulting in extinction of the reflected light at a specific angle. The angle of minimum reflected intensity (the resonance angle) varies with the refractive index on the matrix side. Changes in the resonance angle are directly proportional to changes in mass concentration due to binding or dissociation of biomolecules. The signal is presented in resonance units (RU), where 1 RU corresponds approximately to 0.8 pg carbohydrate or 1 pg protein bound per mm^2.

either automatic or manual, depending on the system used. Signal detection occurs simultaneously in all channels and a serial flow along different surfaces allows monitoring of separate interactions with just one sample injection, or measuring the in-line control signal at a reference surface simultaneously.

Because detection directly reacts to substance accumulation, the signal depends both on the molecular weight and concentration of the injected partner, and on the number of binding sites present on the matrix. However, detectable limits are picomolar concentrations or a minimal mass of about 180 Da [2].

37.1.2 Information in a Sensorgram

In the sensorgram the actual signal corresponding to mass concentration at the sensor surface is plotted against time (Figure 3). Interaction with the immobilized partner during sample injection causes an increase in signal. At the end of the injection, the system automatically switches to buffer flow and a decrease in signal now reflects the dissociation of interactants. The matrix is regenerated and reused.

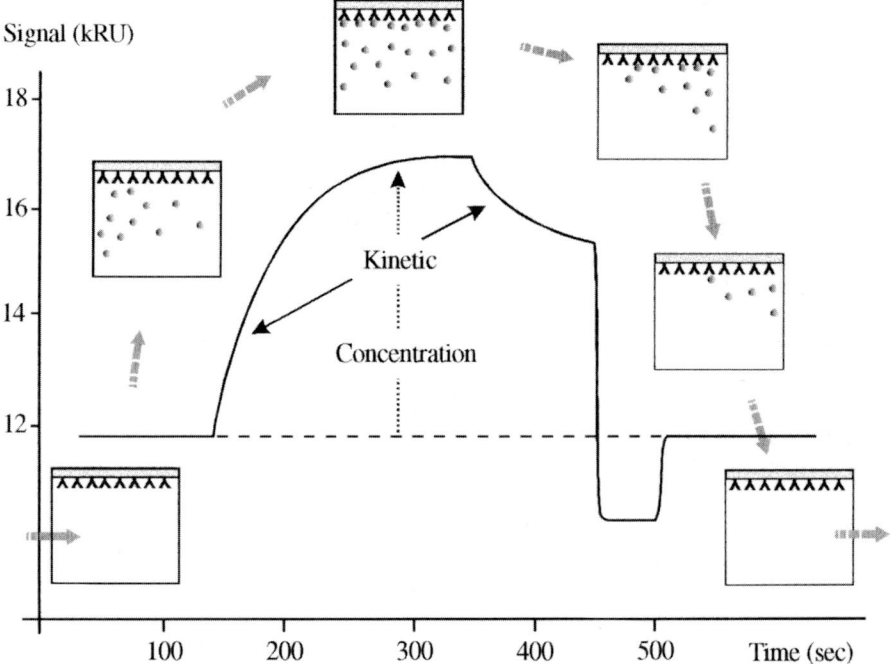

Figure 3. Interaction progress is monitored in a sensorgram. Binding of molecules during sample injection and dissociation during buffer flow are indicated. The height or the initial slope of the signal are directly correlated to sample concentration, while the shape of the curves reflect association and dissociation kinetics. After each measurement, remaining material is washed away by injecting an appropriate regeneration solution leaving only the immobilized partner at the surface, and a new analysis cycle can be done.

Multiple usage of one surface greatly enhances reproducibility of results and, therefore, is especially valuable for comparative studies.

Analysis can be purely qualitative (yes or no), in order to determine the presence or absence of binding activity in a particular sample, or under selected conditions. Whole complexes can be built up on the matrix, step by step, and simultaneously viewed on the screen. The concentration of biologically active molecules can be determined by use of a standard curve. Association and dissociation kinetics as well as the affinity of an interaction are calculated by the fitting routines of the evaluation software [3].

37.2 Experimental Procedures

This biosensor allows experiments to be organized in different ways. The choice can be made as to which interacting partner to immobilize on the chip matrix and which to inject in solution. Different immobilization strategies may be applied, and both purified samples or extracts can be used for the measurements. Interaction analyses are either performed by direct binding to the immobilized molecule, or as an inhibition assay with both partners free in solution.

With this flexibility, the easy handling of the instruments, and a wide diversity of standardized protocols, each user has the opportunity to immediately start their analysis. On the other hand, many assays can be carried out in different ways. Therefore, it is recommended to obtain a general view of the technology and to clearly define the goal for each measurement, in order to structure an analysis in the best way.

37.2.1 Immobilization of Biomolecules at the Sensor Surface

Before the actual measurement, a sensor chip is specifically prepared by immobilizing the biomolecule of interest in the matrix. There is a variety of sensor chips available with different surface properties, and there are standardized methods for coupling just about every biomolecule. A summary of the currently available sensor surfaces and the corresponding immobilization strategies is presented in Table 1. The immobilization level can be controlled on the screen and varied to fit any particular application. It is recommended to use purified samples for the coupling in order to create a homogeneous measuring surface, while the total amount of material consumed is usually only 1–5 µg.

Covalent coupling via amine groups is the standard immobilization procedure [21], which includes three steps. First, the carboxymethyl (CM)-groups on the sensor chip matrix are activated by a mixture of EDC (*N*-ethyl-*N'*-(3-dimethylaminopropyl)-carbodiimide) and NHS (*N*-hydroxysuccinimide), then covalently bound to free amine groups of the biomolecules, and finally all remaining activated

Table 1. Sensor surfaces.

Sensor chip matrix	Coupling via*	Immobilized molecules/comment
Standard CM-dextran	NH_2-, SH-, CHO- or COOH-groups	Proteins or small compounds
Streptavidin-dextran	Biotin	Biotinylated molecules like Oligosaccharides or nucleic acids
NTA-dextran**	Histidin-6mer	His-tag fusion proteins
Flat hydrophobic surface**	Hydrophobic adsorption	Lipid-monolayers
New sensor chips (continually developed, surfaces might change)		
Low carboxylated dextran	As standard CM-dextran	Reduced non-specific binding of culture medium or cell homogenate
Shortened dextran	As standard CM-dextran	Reduced non-specific binding of serum, injection of phages or cells
Flat CM-surface	As standard CM-dextran	When dextran is not required, injection of phages or cells
Lipid-anchor-dextran**	Liposome capture	Liposomes (lipid bilayer)
Pure gold surface	User-defined	Customized surfaces, spin-coating, self-assembled monolayers

* for detailed protocols please refer to information supplied by the manufacturer
** surface can be completely regenerated

groups are deactivated by an excess of ethanolamine. In the same manner, CM-surfaces can be chemically modified, in order to bind proteins or other molecules via thiol-, aldehyde-, or carboxyl-functions [4]. In some cases, covalent coupling can result in loss of biological activity. Then, the immobilization strategy has to be changed, for instance from amine to thiol coupling or to indirect binding by high affinity capture. Covalent immobilization ensures a stable baseline, which is important whenever low signals are expected in the measurements.

Molecule capture is an indirect immobilization, carried out for instance by covalent coupling of an antibody for capturing the corresponding antigen. This method allows removal of the captured interaction partner after measurement, and the surface can be re-loaded with the same or a new molecule. Moreover, specifically captured molecules need not be purified, but can be present in extracts, so that the danger of protein inactivation by a purification protocol is eliminated [5].

Carbohydrates are usually biotinylated and bound to a streptavidin matrix. A simple procedure has been published for the preparation of 4-(biotinamido)-phenylacetylhydrazine and the employment of this substance to biotinylate oligosaccharides at the reducing end [6]. An additional incubation step at 4 °C at pH 3.5 greatly reduces the mixture of stereoisomers resulting in an excess of the cyclic β-glycoside form, which also exists in natural *N*-linked oligosaccharides. This aspect is important in that the type of modification can influence the activity of bio-

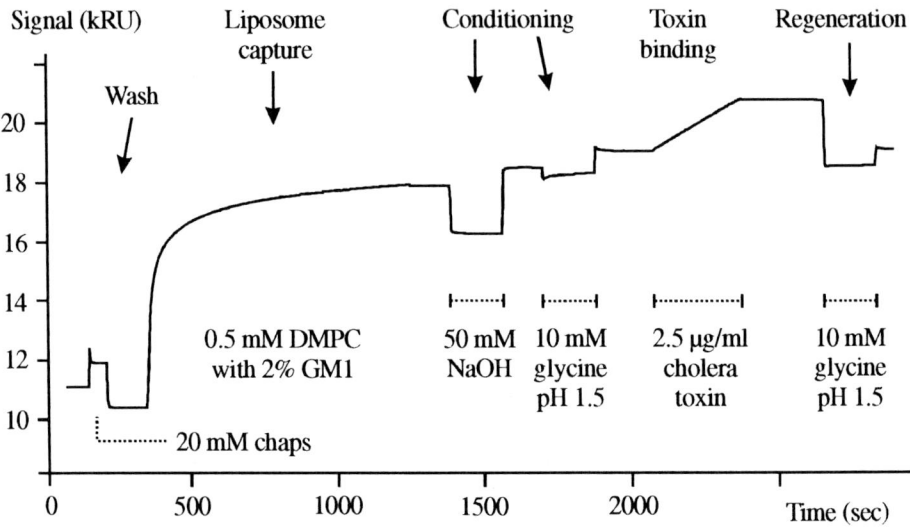

Figure 4. The sensorgram shows the capture of liposomes on a sensor chip matrix derivatized with hydrophobic groups and the interaction response of cholera toxin. Different injections are indicated.

molecules. Heparin, for instance, showed slightly changed binding characteristics depending on whether amino or oxidized cis-diol groups were biotinylated [7].

Glycolipids can be incorporated in liposomes, directly captured on the matrix of the sensor chip (Figure 4), and completely removed after a measurement [22]. In a similar approach using immobilized antibodies for liposome capture, bacterial toxin affinity and binding specificity for glycolipid receptors were determined [8].

37.2.2 Surface Regeneration

After immobilization, surface activity and regeneration conditions are tested. Repeated injections with the same sample and regeneration of the surface must lead to identical binding signals and to a stable baseline. Regeneration is needed, whenever the injected partner does not dissociate in buffer flow within an acceptable time. Care should be taken while testing the conditions, in order not to inactivate the surface-bound molecules. Injections of 1 min pulses are recommended, using successively harsher solutions, until all complexes have dissociated and the signal is back on the original baseline level. Acidic solutions down to pH 1 (100 mM HCl), basic solutions up to pH 13 (100 mM NaOH), hydrophobic solutions (up to 50% ethylene glycol), chaotropic ions, detergents, nonpolar solvents, and also chelating agents may be applied. If the inactivation of the matrix-bound molecule by the regeneration procedure cannot be avoided, then a capturing strategy should be applied, by completely regenerating the surface after each analysis and re-loading it with an active interaction partner.

37.2.3 Interaction Analysis and Controls

Sample injections are typically performed at flow rates of 5–30 µl min^{-1} using sample concentrations corresponding to the affinity of the interaction (K_D, affinity dissociation constant). A measurement will immediately give an answer as to whether binding occurs or not. Nevertheless, each response should be verified through a surface and a sample control. A second trace on the same sensor chip, chemically modified in the same manner as the active surface (e.g. activated/deactivated), usually serves as a control channel. However, an optimized reference surface carries an inactive molecule, which is similar to the active molecule and immobilized at the same amount. Negative samples like non-specific proteins or carbohydrates, or inactive extracts, serve as a control in solution.

The interaction signal during sample injection is divided into two main parts, the so-called bulk response at the beginning and the end of the injection and the actual binding signal (Figure 5). Since the detector reacts to changes in the refractive index, every solution which is optically denser or thinner than the running buffer causes this signal jump. Using the sample buffer as running buffer greatly minimizes this effect and improves the data quality.

Possible background binding to the chip surface can be minimized by choosing a different sensor chip (see Table 1). If background is based on ionic interaction with the CM-groups, then the negative charge at the matrix should be reduced by expanding activation and deactivation times to about 15 min each. Increasing the ionic strength in the injected sample up to 500 mM sodium chloride can also help.

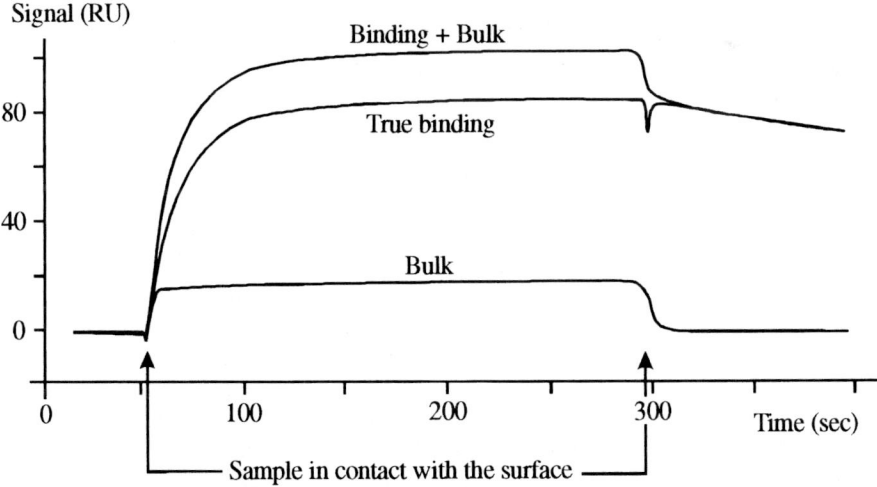

Figure 5. The measured response is usually the sum of binding and bulk effect. The latter occurs when the injected solution does not exactly match the refractive index of the running buffer. Bulk signal is measured on a control surface without interacting molecules and subtracted from the response on the active surface giving the true binding signal.

For the analysis of crude samples, like extracts or serum, a 1:10 dilution with running buffer is often sufficient to decrease any background to an acceptable level. Non-specifically bound material or aggregated proteins can usually be removed from the surface by using detergent or 5–50 mM sodium hydroxide.

37.2.4 Determination of Kinetic Rate Constants

The dynamics of an interaction are characterized by the speed of complex building and decay and described by the rate constants of association and dissociation. Quantitative kinetic analysis is performed by using a set of curves obtained by injecting samples in concentrations of about 0.1–100 times K_D. Based on the mathematical rate equation for the assumed interaction model, a global fitting routine directly evaluates the measured curves. 1:1 binding and reactions, like heterogeneity, conformational changes, or bivalence, are predefined in the software, and customized models can be created by the model editor. Published dissociation rate constants range from 6×10^{-7} sec^{-1} [9] to 10 sec^{-1} [10]. Association rate constants, which can be determined by global fitting, are shown in Figure 6.

The quality of a fit is assessed by means of the corresponding residual plot. Differences between measured and calculated curves should be near the noise level and regularly dispersed. Any trend in the residuals might indicate use of a wrong interaction model or inaccuracy during the analysis. The choice of the correct model is important as well as proof that the binding partners react in the way the model assumes. For example, a monovalent molecule will bind multivalently if it ag-

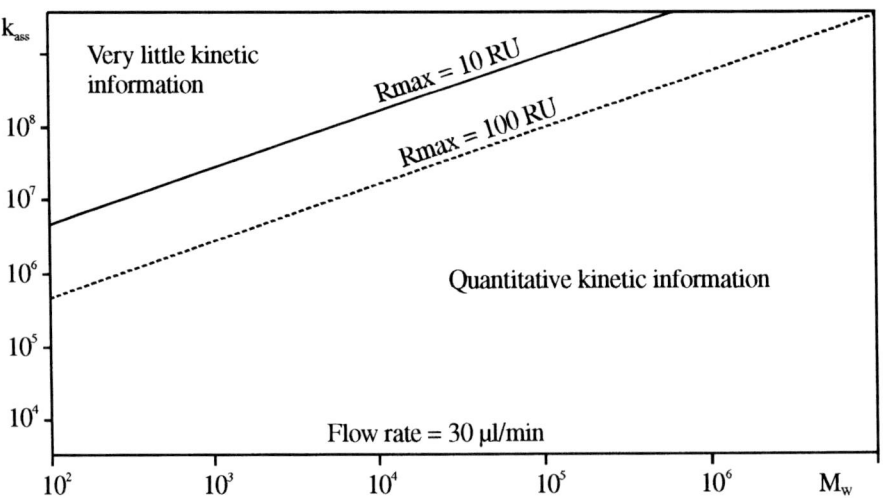

Figure 6. Theoretical limits for kinetic analysis as a function of association rate constant (k_{ass}), molecular weight (M_W) of the injected molecule, and surface binding capacity ($Rmax$) are shown.

gregates in solution. This was in fact found during the kinetic analysis of a single chain Fv antibody raised against *Salmonella O*-polysaccharide [11]. Deviations from a 1:1 interaction are indicated by altered dissociation curves following injections with different contact times [3]. Mass transport to the surface can affect binding signals of interactions with fast association rates, since the last distance to the immobilized partner must be overcome by diffusion. Nevertheless, due to the defined laminar flow mass transport can be mathematically described and, if necessary, taken into account during the data evaluation [12]. Transport has no influence on the interaction, when the initial slopes of binding curves obtained at flow rates of 15 and 75 µl min^{-1} do not deviate more than 10% from each other.

In general, kinetic analyses should be carried out at flow rates of ≥ 30 µl min^{-1} on a surface with a low immobilization level corresponding to a maximum surface binding capacity (R_{max}) of 10–100 RU. The experimental arrangement should be as simple as possible, in order to ensure robust and secure data evaluation. It can be achieved, for instance, by coupling a bivalent molecule to the surface and injecting the monovalent partner. This results in a 1:1 binding scheme, whereas in an opposite construction a bivalent interaction would occur.

37.2.5 Affinity Determination

Affinity can be directly determined from the equilibrium response, which is reached during an injection, when as many complexes are formed as decay and the signal no longer shows any change. The sample concentration leading to 50% surface saturation directly corresponds to the affinity dissociation constant (K_D). The data can be evaluated by the existing software program or a standard Scatchard plot [13]. Affinity determination with both partners free in solution can be performed by an inhibition assay [14, 15]. Interaction partners are mixed and injected after equilibrium binding has been reached in solution. Biacore specifically detects free molecules of one of the interactants. This concentration is inversely proportional to the amount of formed complexes.

Low molecular weight carbohydrates often show weak and rapid interactions, reaching equilibrium within seconds after injection and dissociating rapidly during buffer flow. The corresponding response looks like a square pulse, similar to bulk signals. Nevertheless, even transient binding can be quantitatively analyzed by real-time detection (Figure 7). When working with small sugars, active and reference surfaces should be immobilized at the same density, in order to avoid differences in the bulk signal [16], or a correction factor should be included [23].

A comparison of affinity and kinetic data ensures consistency of the analysis. For a 1:1 interaction, affinity is equal to the quotient of dissociation and association rate constants ($K_D = k_{diss}/k_{ass}$). However, some reactions do not follow a 1:1 interaction, and it must be noted that the apparent affinity of bi- or multivalent reactions can be directly affected by the density of immobilized interaction partners. Nevertheless, the chip surface resembles the situation on a cell surface and possibly reflects the in vivo environment better than an assay free in solution [17].

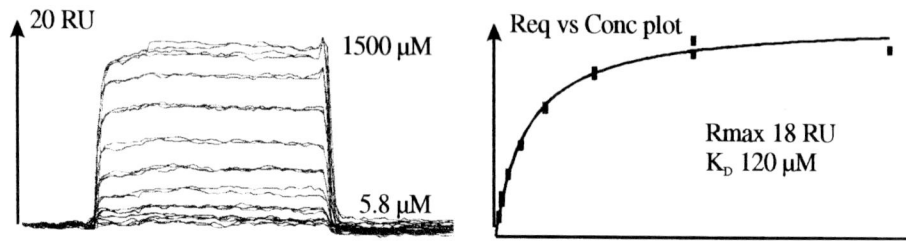

Figure 7. Maltose binding to immobilized anti-maltose antibody is shown. The concentration of maltose ranges over 5.8–1500 μM with duplicate injections of each concentration. Depicted are the binding signals after subtraction of the background response measured on a surface, which was coated with an unspecific antibody. Affinity determination was done by fitting data from the Plot of equilibrium response (*Req*) against concentration to the Langmuir 1:1 binding isotherm using BIA evaluation software 3.0.

37.3 Application Areas

The application range of Biacore for studying biomolecular binding is quite diverse. Some of the references describing analyses of oligosaccharide interactions have already been cited in the text, while three publications are presented below in more detail.

37.3.1 Selectin Binding to a Glycoprotein Ligand

Selectins are transmembrane proteins with membrane-distal Ca^{2+}-dependent lectin domains. This family of cell adhesion molecules is involved in the tethering and rolling of leukocytes on the blood vessel endothelium. Nicholson and colleagues [10] published the analysis of leukocyte selectin, CD62L, binding to the mucin-like glycoprotein, Gly-CAM-1.

The extracellular portion of CD62L was expressed as a fusion protein with two domains of CD4. The resulting soluble monomeric form of CD62L was used to measure the monovalent affinity and kinetics of its interaction with the native Gly-CAM-1. Injection of CD62L-CD4 along a matrix coated with GlyCAM-1 resulted in an immediate increase in signal, attaining equilibrium within seconds. Different concentrations of CD62L-CD4 were applied and subtracting the signals on the control surface carrying no GlyCAM-1 from the signals on the active surface resulted in the actual binding response. Fitting the graph of equilibrium binding level against concentration to a 1:1 binding model showed a low affinity of $K_D \approx 108$ μM. No binding occurred when CD62L-CD4 was applied in the presence of EDTA, or when the control chimera sCD48-CD4 was injected.

The binding response dropped with a half time of 0.07 sec after switching to buffer, corresponding to a dissociation rate constant of $k_{diss} \approx 10$ sec^{-1}. But, since it also takes this time to wash the sample out of the flow cell, dissociation of the complex may even be faster and was therefore given as $k_{diss} \geq 10$ sec^{-1}. Direct measurement of the association rate constant was not possible because the reaction reached equilibrium within one sec. However, with the values for affinity, 108 µM, and dissociation rate, ≥ 10 sec^{-1}, the association rate could be calculated as $k_{ass} \geq 10^6$ M^{-1} sec^{-1}.

A higher affinity was found when the analysis arrangement was changed by immobilizing CD62L-CD4 and injecting GlyCAM-1 along the surface. Binding was apparently detectable at concentrations as low as 30 nM GlyCAM-1, and the dissociation rate constant was reduced approximately by a factor of 10,000 when compared to the data obtained in the above 1:1 binding assay. This strongly suggested that GlyCAM-1 interacts in a multivalent manner to immobilized CD62L-CD4.

The data supported the hypothesis that the highly dynamic cell adhesion process is based on the low affinity and fast kinetics of selectin interaction. GlyCAM-1 was shown to bind multivalently to immobilized CD62L at concentrations just above its mean serum level, indicating a physiological role by interacting with CD62L embedded in leukocyte membranes.

37.3.2 Oligosaccharide Characterization

An analytical method was described by Blikstad et al. [18] to detect unlabelled oligosaccharides or glycopeptides in column effluents. *Sambucus nigra* agglutinin (SNA) specific for sialic acid, and *Ricinus communis* agglutinin (RCA) specific for terminal galactose were covalently immobilized and used to identify different bi- and triantennary N-linked oligosaccharides. After each measurement, the lectin surfaces were regenerated with a 3 min pulse of 100 mM HCl. Surface stability was found for at least 300 measurements over a period of 2 weeks. For 25 consecutive analysis cycles the coefficient of variation was calculated to be as low as 1.21%.

Detection was highly specific, since Sialylated oligosaccharide bound to SNA but not to RCA, whereas desialylated oligosaccharide with a terminal galactose interacted only with RCA. The sensitivity of the analysis was determined by injecting a concentration series of trisialyated triantennary N-linked oligosaccharide (A3) and disialylated biantennary N-linked oligosaccharide (A2) over the SNA surface. Both oligosaccharides could be detected at concentrations of less than 1 nM, while above 200 nM no further increase in signal occurred, indicating surface saturation. Similarly, asialo-galacto biantennary N-linked oligosaccharide (NA2) was detectable at a concentration of 10 nM, using immobilized RCA as detector. During the measurements only 35 µl of sample were consumed for each injection.

The feasibility of this assay was demonstrated by using a pronase digest of transferrin. After a chromatographic separation, individual fractions from the column were diluted and injected over the SNA surface. While measuring binding activity to SNA, one major peak showed the presence of responding glycopeptides. These fractions showed practically no UV absorbance. By using other lectins, the method can be extended to detect oligosaccharides containing other structures.

Combining gel filtration, different glycosidases and different lectins for an analysis, the overall structures of oligosaccharides could be deduced.

37.3.3 *In situ* Modification of Immobilized Carbohydrates

Beside qualitative detection of oligosaccharides and quantitative affinity or kinetic analyses, the *in situ* modification of carbohydrates immobilized at a sensor chip surface has been published [6, 19]. Van der Merwe et al. [20] used this approach to analyze the binding of B-lymphocyte antigen, CD22, to the highly glycosylated leukocyte surface protein CD45.

Native CD45 (CD45-thy) purified from rat thymus was covalently coupled to sensor chip CM5. CD22Fc (domains 1–3 of mouse CD22 fused to the Fc portion of human IgG) interacted with immobilized CD45-thy. The *in situ* modification of the sialoglycoconjugates present on CD45-thy by desialylation and resialylation provided the basis for a detailed characterization of this interaction.

Maackia amurensis agglutinin (MAA) and *Sambucus nigra* agglutinin (SNA) lectins, which are specific for α2-3- and α2-6-linked sialic acids, respectively, bound to CD45-thy, indicating the presence of the corresponding sialic acids. Treatment of immobilized CD45-thy by a 30 min injection of *Vibrio cholerae* sialidase abolished CD22Fc interaction and substantially decreased both MAA and SNA binding. Resialylation of CD45 with Galβ1-4GlcNAc α2-6-sialyltransferase using NeuAc as substrate increased SNA binding, while CD22Fc did not bind. CD22Fc interaction was fully restored after α2-6-resialylating CD45-thy using NeuGc as a substrate. Binding of SNA and the lack of MAA binding confirmed the specificity of the α2-6-resialylation.

The results showed that CD22Fc binds to NeuGcα2-6Galβ1-4GlcNAc, which are carried on CD45-thy *N*-glycans. Moreover, the experiments demonstrated the potential for analyzing binding specificities to selectively modified carbohydrates immobilized at a sensor chip.

References

1. Jönsson U. and Malmqvist M. (1992) Real time biospecific interaction analysis. The integration of surface plasmon resonance. Detection, general biospecific interface chemistry and microfluidics into one analytical system, Adv. Biosensors 2:291–336.
2. Karlsson R. and Ståhlberg R. (1995) Surface plasmon resonance detection and multi-sensing for direct monitoring of interactions involving low molecular weight analytes and for determination of low affinities, Anal. Biochem. 228:274–280.
3. Karlsson R. and Fält A. (1997) Experimental design for kinetic analysis of protein protein interactions with surface plasmon resonance biosensors, J. Immunol. Methods 200:121–133.
4. Johnsson B., Löfås S., Lindquist G., Edström A., Müller-Hilgren R.-M. and Hansson A. (1995) Comparison of methods for immobilization to carboxymethyl dextran sensor surfaces by analysis of the specific activity of monoclonal antibodies, J. Mol. Recognit. 8:125–131.
5. Nath D., van der Merwe P. A., Kelm S., Bradfield P. and Crocker P. R. (1995) The amino-terminal immunoglobulin-like domain of sialoadhesin contains the sialic acid binding site. Comparison with CD22, J. Biol. Chem. 270:26184–26191.

6. Shinohara Y., Sota H., Gotoh M., Hasebe M., Tosu M., Nakao J. and Hasegawa Y. (1996) Bifunctional labeling reagent for oligosaccharides to incorporate both chromophore and biotin groups, Anal. Chem. 68:2573–2579.
7. Mach H., Volkin D. B., Burke C. J., Middaugh C. R., Lindhardt R. J., Fromm J. R., Loganathan D. and Mattsson L. (1993) Nature of the interaction of heparin with acidic fibroblast growth factor, Biochemistry 32:5480–5489.
8. MacKenzie C. R., Hirama T., Deng S., Bundle D. R., Narang S. A. and Young N. M. (1997) Quantitative analysis of bacterial toxin affinity and specificity for glycolipid receptors by surface plasmon resonance, J. Biol. Chem. 272:5533–5538.
9. Deka J., Kuhlmann J. and Müller O. (1998) A domain within the tumor suppressor protein APC shows very similar biochemical properties as the microtubule-associated protein tau, Eur. J. Biochem. 253:591–597.
10. Nicholson M. W., Barclay A. N., Singer M. S., Rosen S. D. and van der Merwe P. A. (1998) Affinity and kinetic analysis of L-selectin (CD62L) binding to glycosylation-dependent cell-adhesion molecule-1, J. Biol. Chem. 273:763–770.
11. MacKenzie C. R., Hirama T., Lee K. K., Altmann E. and Young N. M. (1996) Analysis by surface plasmon resonance of the influence of valence on the ligand binding affinity and kinetics of an anti-carbohydrate antibody, J. Biol. Chem. 271:1527–1533.
12. Myszka D. G., He X., Dembo M., Morton T. A. and Goldstein B. (1998) Extending tha range of rate constants available from BIACORE: Interpreting mass transport-influenced binding data, Biophys. J. 75:583–594.
13. Karlsson R., Fägerstam L., Nilshans H. and Persson B. (1993) Analysis of active antibody concentration. Separation of affinity and concentration parameters, J. Immunol. Methods 166:75–84.
14. Karlsson R. (1994) Real-time competitive kinetic analysis of interactions between low-molecular-weight ligands in solution and surface-immobilized receptors, Anal. Biochem. 221:142–151.
15. Nieba L., Krebber A. and Plückthun A. (1996) Competition BIAcore for measuring true affinities: Large differences from values determined from binding kinetics, Anal. Biochem. 234:155–165.
16. Ohlson S., Strandh M. and Nilshans H. (1997) Detection and characterization of weak affinity antibody antigen recognition with biomolecular interaction analysis, J. Mol. Recognit. 10:135–138.
17. Shinohara Y., Hasegawa Y., Kaku H. and Shibuya N. (1997) Elucidation of the mechanism enhancing the avidity of lectin with oligosaccharides on the solid phase surface, Glycobiology 7:1201–1208.
18. Blikstad I., Fagerstam L. G., Bhikhabhai R. and Lindblom H. (1996) Detection and characterization of oligosaccharides in column effluents using surface plasmon resonance, Anal. Biochem. 233:42–49.
19. Hutchinson A. M. (1994) Characterization of glycoprotein oligosaccharides using surface plasmon resonance, Anal. Biochem. 220:303–307.
20. van der Merwe P. A., Crocker P. R., Vinson M., Barclay A. N., Schauer R. and Kelm S. (1996) Localization of the putative sialic acid-binding site on the immunoglobulin superfamily cell-surface molecule CD22, J. Biol. Chem. 271:9273–9280.
21. Johnsson B. and Löfås S. (1991) Immobilization of proteins to a carboxymethyldextran modified gold surface for biospecific interaction analysis in surface plasmon resonance, Anal. Biochem. 198:268–277.
22. Cooper, M. A., Hansson, A., Lofas, S. and Williams, D. H. (2000) A vesicle capture sensor chip for kinetic analysis of interactions with membrane-bound receptors, Anal. Biochem. 277:196–205.
23. Karlsson, R., Kullmann-Magnusson, M., Hämäläinen, M., Remaeus, A., Andersson, K., Borg, P., Gyzander, E. and Deinum, J. (2000) Biosensor analysis of drug target interactions, Anal. Biochem. 278:1–13.

Part I

Volume 2

V

Carbohydrate–Carbohydrate Interactions

38 Carbohydrate–Carbohydrate Interactions

Dorothe Spillmann and Max M. Burger

38.1 Introduction

Cell recognition and adhesion are important events in the formation and maintenance of functional tissues. These processes must be selective and flexible in order to guarantee proper development as in embryogenesis, during the development of the nervous system, for continuous activity of the lymphoid system, but can also be misused by parasitic intruders or aberrant cells like tumor cells. How are glycans involved in these recognition and adhesion phenomena so important for all living organisms?

Carbohydrates are the most ubiquitous and prominently exposed molecules on the surface of living cells. The combination of a few building blocks in a seemingly countless array of structures is a principle characteristic of carbohydrate chains that distinguishes carbohydrate sequences from all other biomolecules whether nucleic acids or peptides. Such, these chains offer simultaneously flexible, ordered and easily modulatable motives for cellular interactions. The specific arrangement of individual carbohydrate sequences on surfaces and within matrices creates yet another dimension of diversity: Well defined structures or patches of carbohydrate motives can be created within linear stretches of glycans and glycosaminoglycans (GAG), attached to core protein backbones or located on freely movable lipid anchors for interaction with other molecules [1, 2]. Glycans are therefore predestined to serve as crucial molecules for recognition and attachment at the moment of cell encounter.

Carbohydrates have been recognized as interaction sites in many different instances. Many of these processes are mediated by lectin- or lectin-like molecules that recognise specific carbohydrate motifs. Lymphocyte homing is a multistep process with increasing interaction strengths. A first rolling stage of circulating lymphocytes is dependent on recognition of glycan ligands by protein receptors, the selectins, on the endothelium as well as on the lymphocytes [2, 3]. Neuronal development depends on reversible cell extensions and connections followed by disruption which requires the proper interaction with cells, extracellular matrix and soluble factors like growth factors. Many of these interactions have glycans implicated [4–6]. Plant and mammalian lectins have defined roles in tissue maintenance and defense. Glycan ligands are also used as primary receptors by microbial pathogens [7, 8]. Many of these interactions are characterized by fairly weak forces which, however, are easily potentiated by orders of magnitude through multimerization of ligand and receptor [2, 9, 10]. Further along this conceptual line is the idea of carbohydrate–carbohydrate interactions mediated by specific carbohydrate sequences in an ordered polyvalent array [11]. Instead of the protein receptor for a glycan ligand, a second glycan is receptor for the first. Such interactions could be both specific and easily controlled by external influences like ionic conditions, shear forces, availability of substrates and biosynthetic activity of genes involved [12] and would therefore be ideal for early steps in cell interactions where links should have modest affinities in order to guarantee a quick and flexible reversion.

The property of potentially weak interactions, on the other hand, raises one major difficulty, namely the question how to measure such interactions. In cases of structural components as found for instance in plants where direct carbohydrate–carbohydrate interactions have been postulated first, the number of binding sites is extremely large and therefore the avidity reasonably high to be measurable by classic methods of biophysics and biochemistry. In more dynamic systems, like in mammalian tissues that undergo continuous restructuring or underlay shear forces from body fluids the possibilities of carbohydrate–carbohydrate interactions are obvious, but hard to measure. The number of publications dealing with this special aspect of glycobiology is therefore limited. However, the advent and continuous refinement of modern methods like atomic force microscopy, nuclear magnetic resonance, surface plasmon resonance or the increasing availability of tools like anti-

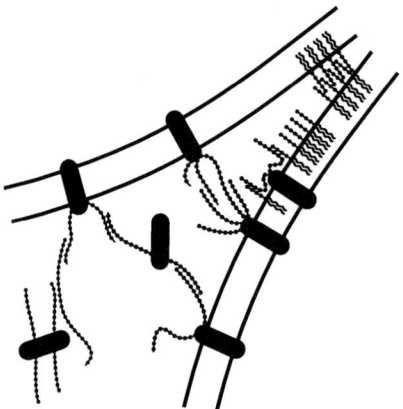

Figure 1. Schematic overview of carbohydrate–carbohydrate interactions. Binding of carbohydrate sequences to one another can be envisaged both in *cis* and *trans* mode at the cell surface between different or identical glycoconjugates and between identical or different types of sequences. Also cell–matrix interactions might be stabilized in this way. ▌ protein core, ┃ lipid core, ⸎ carbohydrate chains

bodies and homogeneous, synthetic ligands should allow it to study such phenomena equally well as other types of interactions where carbohydrates are involved.

This Chapter is divided into a first general part with an overview of the biological background for the different models and is followed by a discussion of the molecular details assumed to play a role in such interactions. Finally, methods in use for assessing carbohydrate–carbohydrate interactions are described. An extensive description of "cook-book" recipes is omitted as the use of most methods is discussed in detail in other parts of this series and does not differ for different applications whether carbohydrate–carbohydrate or carbohydrate–protein interactions are analyzed.

38.2 From Structural Components to Cell Recognition

38.2.1 Carbohydrate–Carbohydrate Interactions as Part of Structural Components

Extracellular Matrix of Seaweeds—Agarose, Carrageenan and Alginate

The classical view on carbohydrates is the one of the energy store and the structural component, whether extracellular space filler and cushion or as part of a cell wall. In this latter form, carbohydrate–carbohydrate interactions have been proposed for the first time. Agarose, carrageenan and alginate, are examples of carbohydrate networks which are produced by different species of algae as intercellular matrix. They are examples for structural networks created from regular and random stretches of carbohydrate chains which form hydrated, elastic gels, that are stabilized by interchain hydrogen-bonding and ionic complexing thereby providing a very basic form of carbohydrate–carbohydrate interaction. Agarose and carrageenan are each built of two alternating sugar residues that form a repeating unit ([3,6-anhydro-L-Galα1-3D-Galβ1-4] and [3,6 anhydro-D-Galα1-3D-Gal(4-OSO$_3$)β1-4] for agarose and κ-carrageenan, respectively). The chains can form double helices which are interrupted at sites where the chains kink due to further modification of the basic units by sulfation or dehydration. The extent of modification, which is under biosynthetic control, determines the structural behaviour of the chains and thereby affects the functional properties of the gelling polysaccharides [13]. Alginates are another group of glycan networks. They are composed of homo- and heteropolymers of β-D-mannuronic acid and its C5 epimer α-L-guluronic acid, and are produced by brown algae and certain strains of *Pseudomonas* and *Azotobacter* bacteria [14]. The special characteristics of alginates are determined by the sequence of building blocks in the single chain. Blocks of homogeneous mannuronate or guluronate polymers alternate with heteropolymers containing both units in variable order of sequence. Due to the conformation of the two uronic acids, the different types of sequences adopt different three-dimensional conformations that in turn determine the extent of chain-interaction. While the mannuronic acid is in a preferred $^4C_1$ chair conformation and the mannuronan-homopolymer therefore adopts an extended structure of either a ribbon in its acid form or a threefold helix in its salt form, the guluronic acid conformation of $^1C_4$ drives the guluronan-homopolymer to adopt a twofold buckled ribbon, that is more condensed [14]. In presence of Ca^{2+} ions these so called G-blocks complex to egg-box like structures, that lead to intra- and interchain cross-linking, for the first time suggested by Rees [13]. A high content of guluronic acid is consequently resulting in much stiffer gels due to carbohydrate–carbohydrate binding while high mannuronic acid contents provide more loosely, flexible gels that have a low degree of chain contacts. The extent of interaction can be modulated not only by the ratio and distribution of guluronic and mannuronic acid along the chain, but also by post-polymerization modification. Some of the bacterial alginates for instance, though identical in basic structure to the algae structures, are O-acetylated on mannuronate which changes their Ca^{2+}-uptake and complexing characteristics [15].

Cell Walls

Many cell walls from bacteria to plants are predominantly composed of carbohydrates. Especially capsular polysaccharides of bacteria have attracted the interest because of their properties being similar to other polysaccharide structures found in mammalians and as virulence factors. In the case of *Bacteroides fragilis* the identity of a virulent capsular polysaccharide has been identified and reveal an unusual dimeric carbohydrate complex of a polysaccharide chains A and B. The two polysaccharides are linked by non-covalent electrostatic interactions between negatively and positively charged groups within both polysaccharide chains providing a carbohydrate–carbohydrate interaction [16]. Considering findings with isolated polysaccharides as bacterial colominic acid [17] or the behavior of GAGs [18, 19], one might anticipate similar phenomena to occur within capsular structures in more than the *Bacteroides* example.

Another form of very simple carbohydrate–carbohydrate interactions can be found in plant cell walls. These walls contain a high content of carbohydrates of which approximately 20% are cellulose microfibrils, some 70–80% are non-cellulose polysaccharides, hemicelluloses and pectins, and up to 10% of the mass is glycoproteins. While some of these structures are very conserved over a wide range of organisms (like cellulose), others are not, as detected by specific antibodies [20]. Cellulose, the homogeneous polymer of β-4-linked D-glucose and the polymer with the largest biomass on earth, forms the basic frame of plant cell walls with rigid microfibrils. Hydrogen-bonds and hydrophobic interaction between the individual, parallel chains stabilize the assembly to a near crystalline form [21, 22]. These homogeneous fibrillar structures are cross-linked by an uncounted number of different heteropolysaccharides. Hemicelluloses (xylans, gluco- and galactoglucomannans, xyloglucans and branched glucans), on the one hand, and pectins (galacturonans, rhamnogalacturonans, arabinans, galactans and arabinogalactans), on the other hand, create a fine network between different cellulose fibres. The non-cellulose chains are all more heterogeneous and linear and branched elements create even more versatile structural properties. In this way a three-dimensional network is created that guarantees the physical properties of the plant cell wall. Covalent ester

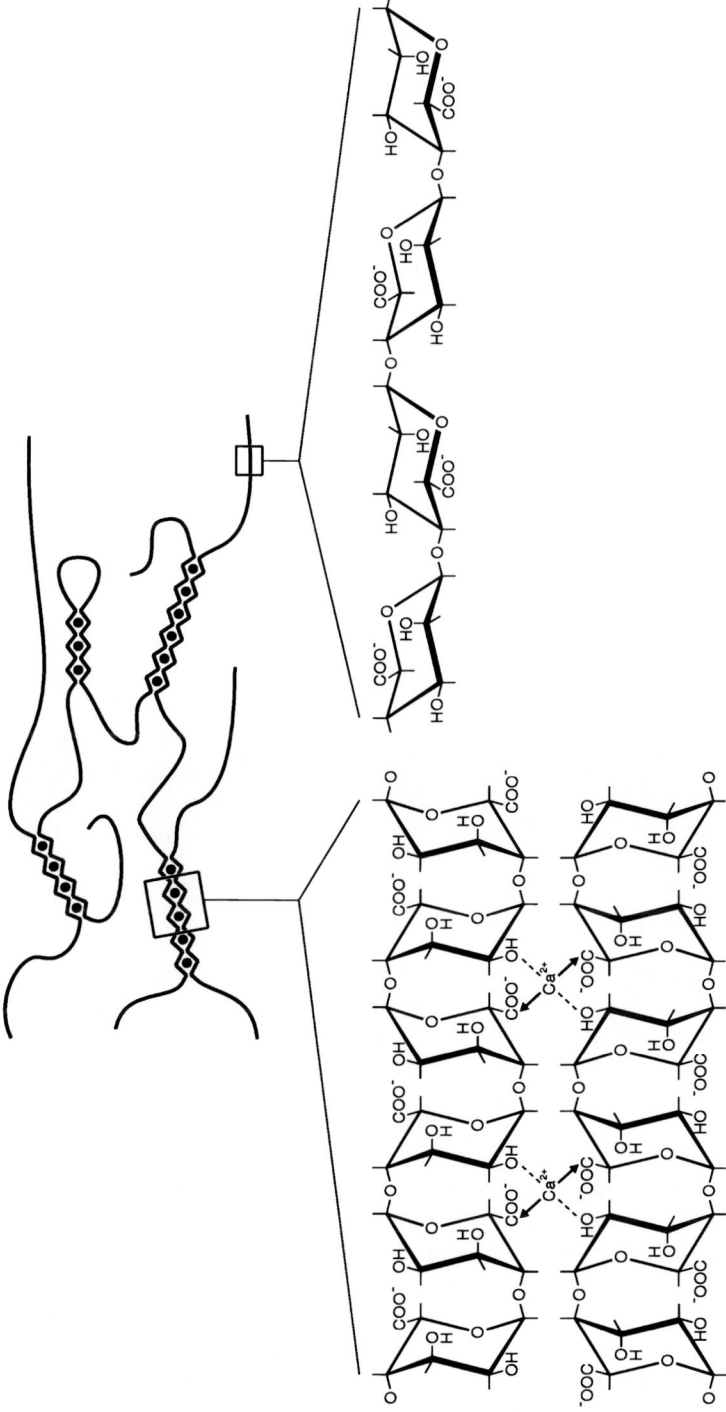

Figure 2. Model of an alginate network. Stretches of poly-mannuronan (detail view on the left) and poly-guluronan (detail view on the right) are interrupted by alternating mannuronan–guluronan sequences. The lengths of each of these sequences varies depending on content and distribution of the individual components and thereby determines the viscosity of the three-dimensional network. The condensed ribbon conformation of poly-guluronan allows "egg-box" piling in the presence of Ca^{2+} (adapted from [14]).

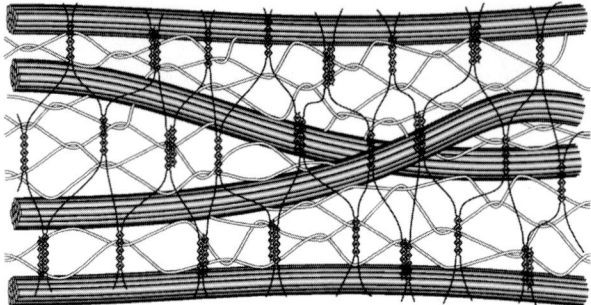

Figure 3. Segment of a plant cell wall (schematic drawing). Bundles of cellulose microfibers in near crystalline packing (thick bundles) are crosslinked by hemicelluloses (drawn as double-lined intercalated network) and pectins (fine lines with egg-box interaction sites). Carbohydrate–carbohydrate interaction occurs between individual cellulose chains, between cellulose and xyloglucan (a hemicellulose) and between different hemicelluloses and pectins. Besides different forms of non-covalent interactions (Ca^{2+} mediated "egg-box" structures as well as ion-independent forms of molecular interactions) also covalent ester linkages between carbohydrate chains are found (adapted from [23]).

linkages between individual polysaccharides are found as well as ion-complexed "egg-box" structures and non-covalent carbohydrate–carbohydrate hydrogen-bonding [21, 23, 24]. Xyloglucan for instance associates in carbohydrate–carbohydrate manner to cellulose, both within the cell wall as well as extracted in in vitro assays [21, 25]. Still there are plenty of questions open concerning the three-dimensional structure of different cell walls especially considering the needs of the plant during development and growth, when the cells should overcome the rather rigid network and be able to extend themselves [20].

Mammalian Extracellular Matrix Components

Although the number of different molecules present in matrices of higher organisms is clearly larger than in more simple organisms similar concepts of structural organization of carbohydrate networks might still serve as a basic fundament. The extracellular matrix and cell surfaces of higher eukaryotic organisms consist of a rich assembly of highly glycosylated molecules [26–28]. Among those, proteoglycans with their long GAG chains provide motifs that could serve not only for binding proteins [1, 29] but might also show a higher order of structural organisation. Matrix constituents are glycoproteins, proteoglycans and the only type of free carbohydrate chains, hyaluronan (HA). This last one is not only an important component of extracellular matrices, but also essential component of the vitreous body of the eye and of synovial fluid. HA chains are the most simple type of GAG chains with the basic structure of an alternating glucuronic acid and glucosamine unit to form long, uniform chains of the structure $[GlcA\beta1\text{-}3GlcNAc\beta1\text{-}4]_n$. HA chains in solution are highly hydrated randomly kinked coils, which are entangled at concentrations of less than 1 mg/ml as seen by sedimentation analysis and viscosity measurements, function as lubricator and as controller of liquid and macromolecular flow (for review see [30]). Although they interact with proteins in the matrix and on the cell surface their gel forming properties are also occurring in protein-free preparations [31], by direct interaction of chains with one another. Fibrillarization of HA chains with one another has been observed by electron microscopy [19, 32] as well as other techniques like spectroscopic methods [33] and viscosity measurements [34]. This purely carbohydrate mediated network might well serve as basic cushion into which other molecules are integrated. In the crystal an antiparallel helix has been described [35, 36]. Driving force for the complexing could be the exposure of hydrophobic stretches stabilized by intramolecular hydrogen bonding [37] which would lead to hydrophobic association of the chains.

The protein-bound forms of GAG chains, namely the heparan sulfates (HS) and chondroitin sulfates (CS) that are parts of the ubiquitous proteoglycans have been tested for direct carbohydrate–carbohydrate interaction in vitro. Isolated GAG chains free from their core proteins were found to form double-helical structures with other GAGs [38]. GAG chains immobilised on beads to mimic their natural, polyvalent appearance on protein cores, showed GAG specific interactions [39, 40]. HA modified beads aggregate with chondroitin-6-sulfate (CSC) beads and less so with chondroitin-4-sulfate (CSA) but only very modestly with heparin or dermatan sulfate (DS) modified beads. Neither of the beads showed homotypic aggregation. These interactions were independent of Ca^{2+} ions and required intact, full-length chains, as neither oxidized nor fragmented chains were mediating bead aggregation, thus indicating a strong need for polyvalent interaction [39]. Similarly, rotary shadowing and electron microscopy showed aggregation to meshes of CSC with itself or HS, but not with CSA, and CSA did not form networks with itself [18]. A potential explanation could be the interaction of hydrophobic sites within CSC to aline with hydrophobic stretches within HA, or in another molecule of CSC, whereas the 4-sulfation of N-acetyl-galactosamine in CSA would disturb these hydrophobic patches explaining the reduced self-complexing of CSA compared to CSC [18]. HS and DS chains, on the other hand, were proposed to bind in a homotypical way to the respective type of GAG either as free chains or multimerized on the core protein [41, 42] when tested by affinity chromatography on HS-modified matrices. This affinity interaction was dependent on intact sulfation and carboxylation of the chains, desulfation or reduction of the carboxyl groups did abolish the binding [43], and on the cooperative interactions of several sites within an intact chain [44]. Such GAG–GAG interactions as measured in vitro could be of value at the interface of cells and matrix in tissues with a high content of GAGs (i.e. cartilage) while they are probably of rather negligible importance in others considering the large number of potentially GAG-binding proteins present in the tissue. Of relevance they might be in particular pathogenic situations where GAG-chains become accumulated [45] as for instance in the deposition of plaques in amyloidoses.

Glycolipids have been postulated to assume structural purposes via carbohydrate–carbohydrate mediated interactions for instance by stabilizing myelin layers around axons.

Galactosylceramide (GalCer) and cerebroside sulfate (CBS) are the predominant glycolipids found in myelin in higher vertebrates and it has been suggested that the

Table 1. Different classes of glycans suggested in carbohydrate–carbohydrate interactions.

Structure[a]	Type of interaction[b]	Organism/System[c]	Reference
κ-Carrageenan	homo	Algae	[13]
Agarose	homo	Algae	[13]
Alginate	homo	Algae, bacteria	[14, 15]
Cellulose	homo	Plants, bacteria	[21, 23]
Cellulose/Hemicellulose	hetero	Plants	[21, 23]
Neutral glycans	hetero	In vitro	[121]
Glycosaminoglycans	homo/hetero	In vitro	[39, 42]
Acidic glycans	homo/hetero?	Marine sponge, In vitro	[11, 17]
Zwitterionic glycans	hetero	*Bacteroides fragilis*	[16]
Glycolipids	homo/hetero	Different cell types	[47]

[a] Classes of glycan structures associated with carbohydrate–carbohydrate binding
[b] Indicates whether identical or different sequences of the same class of glycans are interacting with one another
[c] Organism/model in which these interactions have been described

amount of these ceramides correlates with the stability and compactness of the myelin sheet [46]. In vitro liposome assays have demonstrated a preferred heterogeneous interaction of GalCer-liposomes with CBS-liposomes in presence of Ca^{2+} over a homogeneous interaction of either glycolipid type with itself [46, 47].

38.2.2 Carbohydrate–Carbohydrate Interactions as Part of Recognition Keys?

In all of the examples described so far, often homogeneous types of carbohydrate sequences interact with neighboring chains. Stable structures are created by cooperative action of a multitude of low affinity binding sites that accumulate to an avidity keeping for instance cell walls together. In none of these examples, however, is there flexibility tolerated unless on the expense of stability. A hyaluronan-pad can be compacted or extended by exchange of H_2O and ions, but it is stable unless it becomes degraded by enzymatic or chemical means. Cell walls, once created, are rigid and flexibility must be "bought" by the action of reconstructive enzymes, sofisticated hydrolysis processes and biosynthetic adaptation to new requirements [23]. A totally different situation arises, if carbohydrate–carbohydrate are considered in recognition and adhesion process of tissues that are dependent on flexibility. Here, advantage could be taken from the fact that carbohydrate–carbohydrate interactions are rather low affinity interactions and be used as primary, reversible contact glues that allow the organism to quickly release or reinforce adhesion between two cells. Mainly two lines of research have been followed testing these possibilities: The properties of an extracellular aggregation molecule that mediates aggregation of marine sponges, on the one hand, and cell surface glycoconjugates that participate in the adhesion of embryonic and tumor cells, on the other hand.

Carbohydrate Interactions in Invertebrates—The Marine Sponge *Microciona prolifera* as a Model System

Sponges are the most simple multicellular organisms. They do not have any defined tissues or organs instead their pluripotent cells remain motile within the animal and maintain different functions. For this simplicity and for the fact that mechanically isolated cells have the property to reaggregate and reform animals [48, 49] the marine sponges have been model systems for studying recognition and adhesion in multicellular organisms since the beginning of the century. Early it was recognized that cell suspensions from two different sponges do sort out when mixed and reaggregate in a species-specific fashion. As the sponge is feeding on filtration basis, the animal is continuously in contact with foreign particles a fact which probably has helped to let the animal evolve a modus for recognising self *vs* foreign. Equally well has this sponge model been one of the first examples where carbohydrate–carbohydrate interactions have been postulated in context of a recognition phenomenon [50]. The extracellular matrix of the sponge is similarly composed as higher eukaryotic tissues, containing both proteoglycan-like complexes as well as collagens and other glycoproteins. A major adhesion mediating factor of the sponge tissue has been isolated in form of a proteoglycan-like complex from *Microciona prolifera* [51]. This large complex, the *Microciona* aggregation factor (MAF), with a carbohydrate content of more than half the molecular mass of the macromolecule has been shown to mediate species-specific aggregation of dissociated sponge cells [52–54] in a Ca^{2+}-rich environment ($\sim$10 mM, as in sea water). In the electron microscope, the macromolecule appears as a sunburst [55] with a ring of $\sim$200 nm diameter to which up to 20 arms are attached, a feature confirmed by more physiological assessments of the structure by atomic force microscopy (AFM) in its hydrated form [56]. EDTA-treatment of the complex disrupts the arms from the central ring core, indicating that the cation is important for the structural integrity of the complex [54]. Several hundred Ca^{2+}-binding sites are also important for the Ca^{2+}-dependent self-interaction [57]. Two functionally distinct binding sites have been identified for the complex, one of which is a Ca^{2+}-independent binding site to cellular receptors and the second, the Ca^{2+}-dependent self-association site which is providing the intercellular adhesion force [57]. Cellular receptors may include membrane-associated glycoproteins of 210 kDa and 68 kDa [58] and have a high affinity to both, the cells as well as the aggregation factor [59, 60].

The carbohydrates are essential for the adhesion and recognition process as either glycosidase treatment [61, 62] or periodate oxidation abolish the aggregation of isolated sponge cells [57]. A carbohydrate–carbohydrate interaction of the glycans by themselves was suggested by experiments with protein-free glycan preparations from MAF, where a Ca^{2+}-dependent aggregation of beads could be demonstrated with multimerized glycans [11]. Species-specific binding could be achieved by either the intact aggregation factor or the polymerized glycans to create *de novo* polyvalence [50, 63]. Ca^{2+} is essential for aggregation of the glycans and cannot be replaced by other divalent cations [64]. At much lower charge concentrations, polycations as polybrene or polylysine are able to replace Ca^{2+} indicating a cooperative effect of the polycations with the polyanionic carbohydrate chains. The quantitative

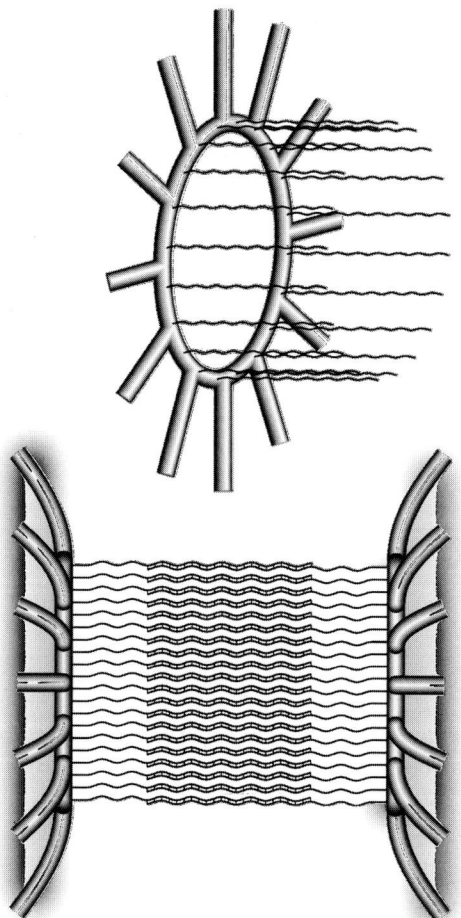

Figure 4. Model of polyvalent interaction between two *Microciona prolifera* cells. The high molecular weight proteoglycan-like complex of MAF is composed of ~20 arms attached to a central ring each composed both of protein and carbohydrate components. The small, N-linked g6 glycans are predominantly localized in the arms of the molecule [55] and supposed to interact with the cell surface [67]. The large glycans associated with the self-association property of the complex [11] have not been unambiguously localized but might correspond to the larger glycans identified in the central ring ([55] and J. Jarchow personal communication) and are recognised by anti-MAF-antibodies blocking the self-aggregation [11, 68, 69], adapted from [12].

differences in interactive strength of various polycations could reflect the variations around an optimal fit of the polycations to the polyanionic glycan chains of MAF as demonstrated by phase-partition assays [65]. Whereas many different polycations are able to precipitate the intact factor glycans, the self-interaction of glycans under physiological conditions, i.e. in the presence of Ca^{2+} ions, is restricted to glycan chains of the same species [50, 63]. These findings indicate that the self-interaction of the glycan chains is not due to unspecific charge interaction of polyionic chains. Rather, the glycans contain a species-specific arrangement of residues in a proper spacing of interactive sites, whether these are single-charged residues in a defined distance [65] or sequences of several residues [66]. Monoclonal antibodies which inhibit the glycan mediated self-aggregation of factor molecules [11, 67] are directed against distinct carbohydrate motives [68, 69] favoring the sequence model. Observation of cross-reactivity of some of the monoclonal antibodies with glycans from different species [11] and the observation that different sponge type cells may first aggregate randomly before sorting out [88] could easily be explained by the different arrangement of otherwise identical or practically identical motives.

Carbohydrate Interactions in Vertebrates—Embryonal and Tumor Cells

Eukaryotic cells express, besides glycoproteins, high amounts of different glycosphingolipids (GSL) on their plasma membranes and many of them are subjected to developmental regulation. Not only embryonic cells but also tumor cells undergo changes in the expression pattern of GSL and therefore these glycoconjugates became classified as onco-developmental antigens [47, 70]. Early mouse embryos at the 8- to 16-cell stage when expression of the stage specific embryonic antigen 1 (SSEA-1) is highest, are susceptible to decompaction by the analog free hapten, namely the blood group antigen Lewis^x (Le^x: Galβ1→4[Fucα1→3]GlcNAcβ1→3Galβ1→4Glcβ) [71]. The embryonal carcinoma cell line F9, which shows a self-aggregation capacity, expresses also a large number of Le^x epitopes, whereas the more differentiated embryonal cell line PYS-2 does not self-aggregate and does not express Le^x epitopes [72]. The correlation of self-aggregation capacity with Le^x expression level was corroborated by binding properties of these cells to Le^x coated culture dishes indicating that F9 cells depend on a Le^x based binding mechanism. F9 cells were therefore used to analyze the mechanism behind the Le^x mediated decompaction of either morula stage embryos or aggregated tumor cells. Indeed, also F9 cell aggregation turned out to be sensitive for competition by Le^x containing oligosaccharides. In isolates from F9 cell membranes both Le^x-containing glycoproteins [72, 73] as well as GSL having this hapten were found to bind in a Ca^{2+}-dependent fashion to Le^x-modified matrices [73]. The glycan nature of the aggregation mediator was further supported by the observation that protein-free embryoglycan chains from the F9 cells precipitated in presence of Ca^{2+}, but not EDTA and were sensitive towards removal of fucose residues, suggesting that the Le^x epitopes can mediate a Ca^{2+}-dependent carbohydrate–carbohydrate interaction [72]. Not only Le^x epitopes can mediate cell interaction but also diverse other glycans. Undifferentiated human embryonal carcinoma 2102 cells express high amounts of the Le^x precursor nLc4 and SSEA-3 (with the major epitope GalGb4) and these cells bind well to Gg3 (GalNAcβ1→4Galβ1→4Glcβ1→1Cer) and Gb4 (GalNAcβ1→3Galα1→4Galβ1→4Glcβ1→1Cer) coated dishes while they do not adhere to several other tested surfaces. The binding is dependent on the differentiation state of the cells as retinoic acid or bromodeoxyuridine induced differentiation alters both, the expression level of these GSL on the cell surface and the adhesion phenotype of the cells [74]. Furthermore, binding to Gb4 was reported to induce activation of the transcription factors AP1 and CREB, while binding to Gg3 did not result in any cell activation [74] indicating that there is also some qualitative difference in binding to different GSL layers. A similar observation was made for B16 melanoma cells either plated on non-coated, G_{M3} (NeuAcα2→3Galβ1→4Glcβ1→1Cer) or Gg3 coated surfaces. Only on the preferential coating, the Gg3 surface, adhesion went along with enhanced tyrosine phosphorylation of FAK [75]. That direct binding between G_{M3} on the cell surface to Gg3 on the culture dish is the activating mechanism, is suggested by the ability to activate FAK also with antibodies against G_{M3} [75].

When cells interact with other cells or with the extracellular matrix, not only glycolipids may be involved in the interaction with the surroundings but equally well glycoproteins. In what context do these different molecules act? What is the role of other adhesion molecules that are well known mediators of cell interaction and what forces could be tolerated by the different combinations? Work with the mouse melanoma cell line B16 addressed these questions. As seen before with the embryonal carcinoma cell lines, also the melanoma cells could adhere in a GSL-dependent manner to either other cells or glycolipid coated culture dishes [76]. The adhesion and spreading on coated culture dishes was most obvious at early stages of cell plating indicating that GSL mediated interactions are very early phenomena in cell interactions and will be overtaken by protein-mediated binding [77]. This fact was also confirmed by plating cells on dishes coated both with GSL and extracellular matrix proteins. There, melanoma cells could adhere and spread faster on glycolipid coated plates while the adhesion and spreading on laminin or fibronectin coated plates were slower [78]. Different clones of the melanoma line B16 with different expression levels of the predominant GSL were tested for their binding behavior to non-activated endothelial cells. Binding of melanoma cells was dependent on their G_{M3} expression level in the static system and faster, but weaker than binding to laminin or fibronectin. Under shear forces, binding strength of the mutant B16 cells was still correlated to the expression level of G_{M3}, but also stronger than the binding via the proteins. From these findings the hypothesis was raised that metastatic tumor cells make use of the high expression rates of certain glycolipids to attach to the unstimulated endothelium, before the next steps in cell activation and transmigration are mediated by protein–protein interactions [79].

Repulsive Carbohydrate–Carbohydrate Interactions

Where there are adhesive carbohydrate–carbohydrate interactions, there must also exist anti-adhesive or repulsive ones. Aggregation factors and glycans from different sponge species show only little or no cross-reactivity [11] or similarity in glycan composition [80], and beads coated with aggregation factors from different species sort out [80, 81] as cells have been known to do since long ago. These properties might suggest that there are sequence compositions that are not promoting adhesion as explained by the zipper model [12] and therefore participate in the self vs non-self recognition phenomenon of sponge cells.

Examples of non-adhesive carbohydrate–carbohydrate interactions are also seen between different GSLs. GSL–GSL interactions have been found to promote

Table 2. Carbohydrate motifs suggested to interact in carbohydrate–carbohydrate interactions.

Motif[a]	Organism/System[b]	Reference
$[GulA]_n$–$[GulA]_n$	Algae, bacteria	[14]
PSA-PSA	In vitro	[17]
Le^x–Le^x	Early mouse embryo, teratocarincoma cells	[71, 122]
G_{M3}-Gg_3	Melanoma/lymphoma cells	[76]
G_{M3}-lacCer	Melanoma/endothelial cells	[79]
Gb4-nLc4	Teratocarcinoma cells	[74]
Gb4-GalGb4	Teratocarcinoma cells	[74]
GalCer-Gal(SO_3)Cer	In vitro liposome assay	[46]
H-H	In vitro liposome assay	[90]
H-Le^Y	In vitro liposome assay	[90]

[a] Identified structure or epitope suggested to interact in a carbohydrate–carbohydrate interaction
[b] Describes in what context these observations were made

binding, to be neutral or to be anti-adhesive [82]. An example of an anti-adhesive GSL interaction has been observed for B16 melanoma cells coated on G_{M3}-plates. While cells coated on Gg3 were nicely spreading, less than 1% of all cells did so on G_{M3}. Sialidase treatment of the dish could revert the effect, so that cells did spread to control levels, indicating that a G_{M3}–G_{M3} interaction is of repulsive nature [77]. Control of GSL expression could therefore easily allow the switch from a non-adhesive to an adhesive phenotype or *vice versa* by the activation/inactivation of a sialyltransferase.

A fascinating oligosaccharide modulating neuronal plasticity by repulsive carbohydrate–carbohydrate interactions is polysialic acid (PSA). PSA is a homogeneous polymer of α2-8 linked sialic acid residues [83], which in neuronal tissue is expressed almost exclusively on the neuronal cell adhesion molecule (N-CAM) attached to a *N*-linked core glycan [84]. The role of PSA on embryonal forms of N-CAM, but also on adult forms as on axons that undergo plasticity or synaptic remodelling, is to prevent axon fasciculation and thereby guaranteeing easier axon migration towards the target, and higher accessibility of the axon for signalling molecules from the target. The injection of endo-neuraminidase during embryonal development leads for instance to serious pathfinding problems for the motorneurons as they should brake away from their neighbors in the plexus once they have emerged from the neural tube. In adult tissue, where the amount of PSA on N-CAM is drastically reduced this repulsive effect is lost. The mechanism behind this anti-adhesive carbohydrate–carbohydrate interaction could be pure ionic repulsion or physical hindrance due to the large hydration volume of the polymer. In either way, cell contacts can be affected in *trans* or *cis* fashion, hindering either receptor–receptor interaction between two cells or affecting receptor mobility and interaction with other molecules within the cell membrane which ultimately also leads to altered cell–cell contacts [84]. An unexpected finding with isolated PSA from either bacterial or mammalian source, however, demonstrates an adhesive capacity of the

Table 3. Carbohydrate motifs suggested to function in anti-adhesive mode.

Motif	Organism/System	Reference
PSA><PSA	Embryonal brain	[6]
G_{M3}><G_{M3}	Melanoma cells	[77]
Le^Y><Le^Y	In vitro liposome assay	[90]
MAF><HAF[a]	Marine sponges	[80, 81]

[a] A carbohydrate–carbohydrate interaction is only indirectly suggested as the sponge glycans are postulated to bind in a homophilic way [11, 67]. By bead-aggregation and AFM, however, sorting out of MAF and HAF modified beads could be demonstrated, whereas a direct proof of MAF-glycan–HAF-glycan interaction has not yet been established.

chains as observed by atomic force microscopy. Similar to HA, PSA forms bundles in the presence of Ca^{2+} beyond a chain length of 12 units and longer fragments have a tendency to form branched bundles [17]. These findings refresh the discussion about the mechanism and functioning of PSA in the developing brain and in bacterial cell walls [17].

38.3 Molecular Aspects of Carbohydrate Interactions

38.3.1 Polyvalence to Inforce Weak Interactions

Molecular interactions where carbohydrates are involved are usually weak interactions. Proteins usually recognize and bind to just a few carbohydrate residues within an oligosaccharide, whether it is an antibody or a lectin. Nature adopts to the needs of higher affinity requirements by the multimerization of either protein or carbohydrate or both units in order to reach avidities that can hold molecules and cells under physiological conditions [2, 66].

Even more true is the need for polyvalence in the case of carbohydrate–carbohydrate interactions. Association of glycoconjugates as small as a mono- or disaccharide with another sugar molecule can be observed by methods as mass spectrometry [85, 86] or crystallography [87]. The affinities are, however, so low that the stability of the complexes would not even survive the measuring time in NMR [72] not to think of a situation in vivo. Polyvalence, i.e. the repetition of a binding motif in either of different modes, allows these molecules to bind together and gives the scientist the chance to measure them by different methods. Sponge cell aggregation mediated by isolated glycans needs the *de novo* polymerization of the isolated chains into longer, multivalent complexes to achieve approximately the size of the native aggregation factor [50, 63]. Polyvalence could also be achieved by attachment of glycans to beads which could aggregate in a species-specific manner [11, 50] confirming earlier proposals and data [65].

The same needs hold true in the GSL-mediated interactions of tumour cell lines. The adhesion capacity of cells as well as liposome aggregation and binding are dependent on the concentration of glycolipids on the receptor side. Correlation of binding strength was both seen for the amount of a specific GSL expressed on a certain cell line as well as to the concentration of the glycolipid on the substrate side [74, 77]. Also the length and degree of hydroxylation of the fatty acid portion has an influence on the interaction of glycolipids with one another, suggesting that developmental control or pathological changes can affect carbohydrate–carbohydrate mediated cell interactions [46].

In the carbohydrate–carbohydrate model of cell interaction advantage is taken from single low affinity sites. Control of the polyvalence by various means (surface density of presented structures, ionic strength to modulate attractive vs repulsive forces, subtle changes in biosynthesis of the carbohydrate sequences, etc.) to change the affinity of the interactive molecules provides an adaptable system beyond a structural scaffold. Such recognition systems are created which allow different cells to test surrounding surfaces and release or reinforce interactions [88]. Species-specific sorting is one example of the value for moderate interaction strengths which achieve biological relevance through polyvalence. Moderate strength is also required within one species, e.g. within an individual sponge, otherwise the cells could not migrate past each other. In an experiment where factor-mediated aggregation was induced by polybrene instead of Ca^{2+}, secondary migration of cells was inhibited. This observation demonstrates the drastic effect of "locking" the molecules in a permanent tight interaction mode [65].

38.3.2 Arrangement of Motifs and the Possibility to Control Specificity

A velcro pad or a zipper can be used as simple models to highlight the value and simplicity by which nature may create specific and lasting binding sites by such carbohydrate–carbohydrate interactions from compositionally rather similar structures [66]. The creation of repetitive, interacting glycan sequences is feasible in different modes. Oligosaccharide motives can be repeated along the primary glycan sequence as seen in the examples of plant cell wall carbohydrates. Glycans can also be arranged in a repetitive pattern along a backbone structure which is not directly participating in binding, as illustrated by mucin structures on protein backbones [2] or by branching carbohydrates on a glycan scaffold as for instance blood group antigens on poly-N-acetyllactosamine type glycan backbones [70, 89]. Finally, proteoglycans, glycoproteins or glycolipids can be presented in clusters or surface superstructures partly due to mobile anchors. Only in this last version advantage is taken of the fluidity of the biomembrane for control of the surface density of these glycans and therefore their avidity [2, 66, 90]. There are indications that GSL form clusters within membranes and are not randomly distributed [75]. Indeed, GSL accumulate on apical surfaces of epithelial cells [91, 92], are observed as patches in erythrocyte membranes [93] and peripheral lymphocytes [94] or as morphological clusters in vitro in liposomes as detected by electron microscopy of freeze fractured samples [95].

Primary glycan sequences determine whether there is the possibility of interaction, but how these sequences will be arranged in a three-dimensional context will define the specificity of interaction. Even though sequences could be similar, different spacing in the three-dimensional architecture could create a "non-binding" phenotype as illustrated with two half-chains of a zipper being of different dimensions. These architectural differences could even be gradual, allowing for instance two different sponges to first slightly complex before sorting out again [88]. Or a tumor cell with a high density of a certain GSL be an effective metastatic colonizer, whereas the one with a lower concentration of the same GSL be less efficient due to reduced primary binding capacity [96]. The next step in binding via carbohydrates is the question how for instance a certain combination of GSL would trigger the cells to activate signalling molecules while another combination, although still mediating binding would not activate the cell [75].

38.3.3 Molecular Basis of Carbohydrate–Carbohydrate Interactions

The forces playing between carbohydrates are no different from those acting between other biomolecules. Most carbohydrates are neutral or negatively charged due to carboxy- and sulfate groups, although positively charged glycans also occur [16, 97] and ionic interactions can be anticipated. Except for rare ionic interactions between oppositely charged glycans [16], direct carbohydrate–carbohydrate interaction between two anionic carbohydrate rather leads to repulsive forces as nicely illustrated for PSA [84]. With the aid of divalent cations, this effect can be overcome as demonstrated for sponge cell aggregation, where both carboxyl and sulfate groups as well as Ca^{2+}-ions are important for proper species-specific aggregation of cells [57, 98]. However, not just any kind of acidic sponge glycan is interacting in the presence of Ca^{2+} indicating that the effect of Ca^{2+} is not merely a charge effect. Another possibility of the action of Ca^{2+} is the property to complex carbohydrates via a suitably arranged combination of sugar hydroxyl groups. Single hydroxyl groups in sugars are too weak to coordinate cations in the presence of water molecules. However, in the combination of two to three well positioned hydroxyls on one sugar residue or over two adjacent residues, cations can become coordinated to the carbohydrate which can force the chain into a specific conformation or lock it there as seen for instance with guluronic acid sequences in alginate [14] or pectins [23]. The preference of Ca^{2+} over other cations lies in the molecular dimensions and complexing properties: Complexing strength raises from mono- to trivalent cations, while the ionic radii determine how well the ions fit into molecular dimensions of the molecules which are optimal around 100–110 pm as seen for Na^+, Ca^{2+}, La^{3+} [99]. Some, but not all, of the GSL interactions have been shown to depend on Ca^{2+} and it is not clear what is the difference between those. For the interactions of G_{M3} with Gg3 and Le^x–Le^x an association of the molecules via their hydrophobic sides that complement each other followed by a locking through Ca^{2+} has been proposed based on molecular modelling [72, 76]. However, a similar molecular modelling approach has led to the suggestion that Ca^{2+} is the interlinking force between the two Le^x molecules [85]. No Ca^{2+} was present in the Le^x crystal [87], why further studies are required to resolve the exact role of Ca^{2+} in this interaction.

Carbohydrates offer a rich source for hydrogen-bonds due to hydroxyl-, amine- and carboxy-groups. Hydrogen bonding can be seen intramolecular as well as intermolecular or in combination with the solvent. An extremely high number of hydrogen bonds could be seen in the crystal of Le^x [87] with bonds between the trisaccharide and water and between the carbohydrates themselves. An indirect consequence of intramolecular hydrogen bonding has been suggested for HA. As with the locking properties of Ca^{2+} a similar fixing the conformation has been proposed for neighbouring residues in HA, which would lead to exposure of a larger hydrophobic patch in the chain, which in turn would favor hydrophobic interactions between different chains explaining chain interactions [37]. An opposite interpretation of similar data from molecular dynamics models for short saccharide sequences would suggest the rapid exchange of different intramolecular H-bonds in favor of a prolonged solubility of even high concentrations of HA [100]. Improved measuring methods must be awaited for deciding which ones are the driving forces in vivo.

Also hydrophobic interactions and Van der Waals forces are likely to occur as not only HA-chains present hydrophobic patches [18, 101]. Hydrophobicity of the

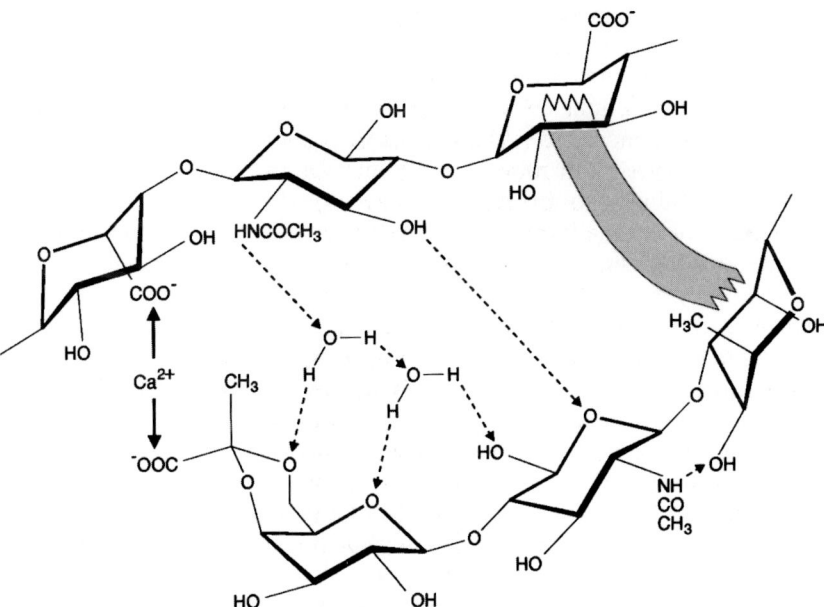

Figure 5. Schematic drawing of forces that might act between two carbohydrate chains in sponge glycans. Possible forces are sketched as ionic interaction via Ca^{2+} ions between two carboxyl groups (Ca^{2+} between arrows), hydrogen-bonds (arrows with dashed lines) might occur within the chain, between chains and between a chain and the solvent, and hydrophobic interactions (broad interaction band) between apolar surfaces of the sugar rings (adapted from [12]).

carbohydrates depends on the conformation of the rings, the epimer conformation and the glycosidic linkages and therefore not all carbohydrate chains are equally amphiphilic [102].

38.4 Experimental Approaches

38.4.1 General Considerations

Low affinity interactions are difficult to measure and the intrinsic problem is to distinguish specific from non-specific interaction. In this context, it is important to realise that not all biological interactions are equally "specific", in other words, to understand specificity not merely as a term of high affinity, unique binding site, but rather as an adaptation to the physiological possibilities and needs. The coexistence in space and time of potential partners in an interaction scheme is therefore a more important requirement than the existence of unique structural motifs recognized by individual molecules.

A measure for the strength of two molecules interacting with one another is the affinity between them. This can be expressed in terms of an association constant which can differ for many orders of magnitude depending on the kind of interaction. Association strengths between two sugar molecules can be below 10^2 M^{-1}, whereas low affinity interactions have a range of approximately 10^2–10^4 M^{-1} and high affinity interactions more than 10^5 M^{-1}, already indicating that the possibilities of detection and choices to measure such interactions can be limited by the inherent technical possibilities of different approaches.

As measurements of interactions often happen in dissected systems where molecules are tested in other than the original context, it should be crucial to test a model in conceptually different assay set-ups to validate findings and try to eliminate experimental artefacts. These requirements can be difficult to achieve for the carbohydrate–carbohydrate interaction; even more difficult than to find the natural components of protein–carbohydrate interaction which is already a difficult task as demonstrated by the hunt for the "real" selectin ligands [103]. Much of older literature on carbohydrate–carbohydrate interaction does not necessarily hold the requirements for unambiguous testing in different systems, but due to lack of possible alternatives at the time they were performed they have been included in this summary. Improvement of methods and technology should give the chance to test these concepts in future.

The collection of methods presented below is chosen on the basis of experiments described in the literature used for studies of carbohydrate–carbohydrate interactions. Rather than being a collection of ready to use recipes the different methods are discussed considering important points as most of the techniques are discussed in full extent in other parts of these volumes and do not differ conceptually whether measuring carbohydrate–carbohydrate interactions or any other type of interaction.

38.4.2 Affinity Interactions

In principle all studies to measure interaction between two molecules are affinity interactions. What is described under this heading could also be described as traditional cell biological or biochemical approaches to measure binding between two systems. It was historically started with the observation of whole cells interacting, before more sophisticated methods of cell fractionation and artificial reconstruction have progressed to the point that almost any molecule can be isolated and packed into a *de novo* assembly to test for its role in the intact system.

Cell Binding Studies

The advantage to use intact cells for binding studies is that the molecular assembly is in its original form and effects are less likely due to altered exposure of molecules, artificial combinations, lack or changed arrangement of components. Obviously that creates also a disadvantage in itself, as designing which of the components does what is less ambiguous in an intact system. Cell binding studies can be performed in several different approaches. Single-cell suspensions are allowed to aggregate with themselves or another cell type, to bind to a supporting cell layer or to an artificial substrate.

Cell aggregation

Cell aggregation is a classical type of experimental set-up and has been used extensively, hemagglutination just being the most prominent one. Mechanically dissociated and well washed cells are allowed to aggregate in the presence of a potential adhesion mediator or inhibitor. Cells can be left untouched or exposed to shear forces for instance on a gyratory shaker.

To exemplify this approach, sponge cells can be mechanically dissociated from the sponge matrix, washed extensively and allowed to reaggregate in the presence of Ca^{2+} and an aggregation factor [51]. This approach can be further extended to test the role of different components of the adhesion molecule. Chemical treatment of the factor or enzymatic removal of certain sugars [57, 61, 62], fragmentation into smaller subunits [63] or chemical cross-linking of factor glycans to a multimeric complex [11] aid to substantiate the mode of interaction between two cells. The reversed scenario can also be applied: A cell complex is broken up by adding competitors from which one assumes that they have the capacity to inhibit the interaction. This can be antibody molecules against different surface molecules or glycan motifs to be tested for their effect as done with early embryos [71].

Cell adhesion to substrate

The binding capacity of cells to different substrates is a first step in dissecting the system and testing individual components for binding. While the cell is still an intact entity and serves as a tool to screen with, the complementary adhesion site is modified and theoretically can be built from almost any kind of combination of

Table 4. Cell lines applied to GSL-mediated binding studies.

Cell line[a]	Expression[b]	Binding[c]	Substrate[d]	Reference
B16 melanoma	G_{M3} high	Strong	LacCer-coated plate, non-stimulated endothelial cells (LacCer expression)	[75, 77, 79]
B16 melanoma	G_{M3} high	Strong	Gg3 coated plate, T-cell lymphoma L5178 AA12 (high Gg3 expression)	[75, 76]
B16 melanoma	G_{M3} high	Weak	T-cell lymphoma L5178 AV27 (low Gg3 expression)	[76]
B16-F10 clone	G_{M3} high	Strong	Non-stimulated endothelial cells (LacCer expression)	[79]
B16-F1 clone	G_{M3} medium	Medium	Non-stimulated endothelial cells (LacCer expression)	[79]
B16-WA4 clone	G_{M3} low	Weak	Non-stimulated endothelial cells (LacCer expression)	[79]
F9 teratocarcinoma	Le^x high	Yes	Le^x coated plate	[72]
PYS-2 teratocarcinoma	Le^x low	No	Le^x coated plate	[72]
2102 embryonal carcinoma	High nLc4 and GalGb4, moderate Gb4, no Gg3	Yes	Gg3 and Gb4-coated plate	[74]
2102 embryonal carcinoma	High nLc4 and GalGb4, moderate Gb4, no Gg3	No	Gb3, Lc3, Le^a, Le^b, G_{M1}, G_{M3}-coated plate	[74]

[a] Cell lines and clones used for studying GSL-mediated binding
[b] Expression levels are described with low, moderate or high according to the original description
[c] The binding strength is graded as described in the original publication

molecules. The advantage lies in the simplicity of the test and the better control of components, whether concerning amounts or combinations. The *caveat* arises from the same point as mentioned earlier, namely the artificial combination of components and the uncertainty as to the steric properties of the molecules when immobilized on a surface. Nevertheless this type of approach has been used extensively in the study of GSL mediated cell binding. Individual GSL alone or in combination with different glycoproteins have been immobilized to culture dishes and testing occurred with different cell lines under static or flow conditions. Conclusions drawn from this kind of assays are that different GSL have different adhesion promoting/preventing potentials for a given cell line, an effect which is clearly dependent on both the concentration of GSL expressed on the cell surface and that coated on the culture dish [74, 75, 77]. Combination of GSL with proteins and testing cell adhesion under both static and flow conditions indicate that GSL–GSL interactions are primarily mediating cell adhesion in the early phases of cell coating and are replaced by protein–carbohydrate and protein–protein interactions after that, but are more prominent under shear forces than under static conditions [78, 79].

Aggregation of *de novo* Complexes

Once a cell is dissected into single components and these should be tested individually, different options can be tested. Aggregation or complexing are easily observed and do not need too much of technical equipment. Macromolecules that allow such observations have been favored due to simplicity and easiness to create such polyvalent assemblies. Modification of inert beads and integration of molecules into liposomes have been two different approaches used extensively in the study of carbohydrate–carbohydrate interaction. Aggregation of beads can be followed macroscopically or with a normal light microscope, or, when using fluorescent beads, by fluorescent microscopy. Liposome aggregation can be followed simply by measuring absorption with a spectrometer.

Modification of inert beads

The choice for beads is merely empirical. Two important points must be considered before an experiment: First, how shall the molecules be coupled to the beads in order to exert a minimal effect of the immobilization method on the assay. Is non-covalent adsorption possible and stable enough or does it need a covalent-linking scheme. The chemistry of linking molecules to beads depends on molecular characteristics (which chemical groups are available for linking, which ones should not be used to conserve the biological activity) and can affect the properties of the linked molecule in terms of its binding properties, i.e. its functional integrity and its way of presentation (steric factor, architecture on the bead). Integrity and access of the molecule on the bead should therefore be tested with a suitable probe as could be an antibody, a lectin or another molecule for which the binding behavior is known. Second, the chemical groups on the beads used for covalent coupling of the sample must not affect the binding properties of the beads. It must be possible to neutralize them without changing the binding characteristics of the bead. This problem is usually rather easily circumvented as control beads can be modified and blocked in parallel without any ligand or an alternative ligand and be used as negative (or positive) control.

Beads have been used to demonstrate that either intact aggregation factor or immobilized factor glycans can mediate the sponge cell aggregation [57, 67], that GAGs can bind to one another [39, 44] and to demonstrate GSL–GSL interaction [72].

Integration of glycans into liposomes

Glycolipids are highly suited to be integrated into liposomes for testing glycan-dependent liposome aggregation and liposome binding to surfaces. No chemical modification is needed, as the quality of liposomes be controlled by the choice of lipids used to create the liposomes. The integration of radioactively marked glycolipids can help to control the level of incorporation. Different glycan motifs can be integrated and tested in parallel. The possibility to create neoglycolipids from short glycans also opens the possibility to test glycans that originally are not isolated as glycolipids but may originate from a glycoprotein and could be tested independent of the protein core [104].

Hypotheses of GSL-mediated cell interactions have been tested by demonstration of the aggregability of GSL-liposomes and adhesion of such liposomes to glycolipid coated surfaces [46, 82, 90].

Affinity Chromatography

Alternative to using immobilized glycans in aggregation assays, cell or liposome binding and similar experiments, they can also be used in chromatography. Whereas high affinity interactions are easily seen by affinity chromatography, low affinity interactions might be overlooked due to insufficient retention. For larger glycans that are inherently polyvalent, retention on a glycan modified column could be manifested on the basis of carbohydrate–carbohydrate [42]. As holds true for the bead coating, the inclusion of an identically modified column matrix without the ligate of choice attached should be considered as a control or precolumn to avoid looking at non-specific matrix–ligand interactions. At least the theoretical possibilities of weak affinity chromatography are impressive if one succeeds to immobilize the ligate at high enough densities to cope with the low affinity of carbohydrate–carbohydrate interactions [105].

Distribution between Compartments

Boyden chamber

The distribution of molecules between two phases or two liquid chambers separated by a membrane has been occasionally used for the demonstration of carbohydrate–carbohydrate interaction. In the Boyden chamber, two compartments with identical buffer content are separated by a membrane with suitable molecular weight cut-off. To one chamber the molecules to be tested, for instance a soluble oligosaccharide and a potential macromolecular ligand are added, and the diffusion of the oligosaccharide into the other chamber is followed. If it does not interact with the macromolecule, the distribution of the oligosaccharide should be under free diffusion and an equilibrium must be reached. Similarly, two freely diffusible species can be observed, if their interaction creates complexes that become too big for free diffusion (by correct choice of the membrane pore size), and will remain in the chamber they have originally been placed in. Binding of radiolabelled oligosaccharides to stationary liposomes containing different glycolipid species was observed in the Boyden chamber. Oligosaccharides that bound to the GSL in the liposomes did not pass the membrane, whereas oligosaccharides that could not bind to the GSL were recovered also from the other compartment [73].

Phase partitioning

In a similar type of approach, phase partitioning, the distribution of molecules between two different liquid phases, is analyzed without any separating membrane. In this assay, the solubility and interaction with the solvent in the two different phases is driving force for the separation behavior of the analytes. The distribution of ^{125}I-MAF between a lower, dextran-rich and an upper, polyethylene glycol phase had been analyzed as a function of cations added. While the non-complexed MAF distributed to the dextran-rich phase due to its polyanionic character, aggregation of MAF in the presence of cations reduced the net charge of the aggregate and resulted in a shift to the polyethylene glycol phase [65]. The limitation of this assay system is, however, that the effects of pure charge neutralization by the addition of cations to the polyanion cannot be distinguished from the effect of cation-mediated complexing of the polyanion to a less charged macrocomplex. The change of distribution properties of liposomes modified with neutral glycans like dextran, pullulan or mannan between a carbohydrate-rich and a carbohydrate-poor phase, might rather been taken as a sign of direct carbohydrate–carbohydrate interaction between the carbohydrate on the liposome and the bulk carbohydrate [106].

38.4.3 Microscopy

Electron Microscopy

Tertiary structures beyond the cellular level can be observed by electron microscopy (EM). Both, macromolecular complexes of proteoglycan-like aggregation factors [55] and isolated glycosaminoglycan chains forming fibrillar structures have been observed by EM [19, 32].

Many of the observations have been made on replicas of rotary shadowed structures that had been dried on the carrier. These dehydrated samples have most obviously lost some of their physiological properties beside the fact that the preparation as such could be deleterious to the sample and result in artefacts. The development of atomic force microscopy (AFM) in the 1980s is therefore a valuable improvement of technology towards more physiological sample application possibilities.

Atomic Force Microscopy

AFM is an attractive development which has resulted from scanning tunneling microscope technology. In AFM a sensor is raster scanned in nanometer distance over a surface with the immobilized sample giving a picture of the specimen. In this imaging mode, measurements can be achieved for single molecules as well as whole cells allowing the observation of biological phenomena in a more physiological context. In contrast to EM, samples do not need to be fixed or shadowed and they can be observed in hydrated form. What was anticipated a few years ago to be a general method for imaging from molecules to cells [107] has already been successfully used for imaging cells [108], proteins [109], ribonucleic acid [108] and carbohydrates [56, 110] but even more to measure forces acting between molecules [56, 110]. In the force-distance approach, the tip with an immobilized molecule is moved in an approach-and-retract cycle towards the sample on the substrate surface. The forces acting on the sensor are recorded under all the approach-and-retract cycle and give a picture of singular binding events, the distance at which the binding ruptures, the adhesion probability and, at least theoretically, also the energy dissipated during the process [111]. In order to really be able to measure forces acting between the two samples and not those between the samples and their respective immobilization surface, it is important that the molecules are immobilized so well as to resist forces larger than the ones acting between the molecules. As discussed with other immobilization techniques, it is also important with AFM, that the immobilization procedure has to conserve the right orientation and binding activity of the molecules to be analyzed. And equally well is it important to discriminate against any unspecific interaction with the surface and sensor [111].

Applying AFM technology, imaging demonstrated the same contours of the aggregation complex from sponges as had earlier been observed by EM [56]. In the force mode, an average molecular contour length of ~220 nm and arm lengths of ~150 nm for MAF could be determined [111]. Furthermore, the average forces of individual interaction sites between two MAF molecules were estimated to ~40–50 pN [56, 111]. As multiple jump-off events of ~40 pN ± 15 pN were seen in a singular approach-and-retract cycle, but the average forces measured between the molecules amounted to ~125 pN with maxima up to ~400 pN, this result seems to favor multivalent binding of 3–10 binding pairs [56]. As it is, however, still unresolved how the single glycan species are integrated in the whole aggregation factor complex a conclusive picture of this interaction awaits further experimentation. An unexpected finding by AFM is the observation of PSA filaments in the presence of Ca^{2+}-ions [17].

38.4.4 Crystallography

Crystallography is one method of choice to receive detailed conformational information about molecules. The flexibility and heterogeneity of oligosaccharides, however, sets serious limitations to the analysis of either glycoconjugates or carbohydrate ligands bound to other macromolecules [112]. Crystallization of carbohydrates might be more difficult and even impossible for certain structures, but still has a considerable value in cases where the crystallization succeeds. Not surprisingly HA, which is very homogeneous in sequence, has been described as a crystal and assumes a double helical form [35, 36]. More recently, crystallization of a blood group antigen succeeded for the first time ever [87]. This Le^x crystal is characterized by a high degree of hydration with water molecules filling all spaces in the crystal and participating in the intermolecular hydrogen bonding pattern. Hydrophobic interaction within a single trisaccharide can be seen between fucose and galactose rings, but no intramolecular hydrogen bonding. The conformation of the trisaccharides in the crystal fit to the conformations predicted from NMR and molecular modelling data [113], and the order of the trisaccharide in a head-to-head arrangement in the crystal would suggest that a tight packing of glycolipids could occur in a similar fashion [87].

38.4.5 Mass Spectrometry

Mass spectrometry (MS) has been used since its beginnings for assisting the characterization of carbohydrate structures. More recently it has gained much of value due to improved ionization tools that provide milder treatment of the sample without fragmentation risk. Whereas the advent of matrix assisted laser desorption ionization (MALDI) MS has improved the characterization of even larger compounds, electrospray ionization (ESI) MS has opened the field to studies of interactions between molecules. In ESI-MS highly charged liquid droplets in a strong electric field are dispersed and evaporated into the mass analyzer [114]. Due to this soft ionization technique even non-covalent interactions can therefore be observed.

By ESI-MS homodimers could be observed for both Le^x and Le^x-LacCer in presence of divalent, but not monovalent cations. A heterospecific interaction in presence of divalent cations could also be observed for Lex-LacCer in interaction with LacCer, GalCer and even Cer with a decreasing affinity in this order [85] confirming findings realized by other methods. Another example for the value of ESI-MS is provided by the study of GalCer with CBS, that shows a Ca^{2+} dependent oligomerization and confirms results from liposome assays. A qualitative picture of the affinity between different oligomers can be gained by measuring the stability at different declustering potentials. Under increasing declustering potentials GalCer-Ca^{2+} was the most stabile complex which contrasts to findings with the liposome assay. The divergence might, however, be explained by the choice of the solvent used for the ESI-MS measurements [86]. Collision induced decomposition of the complexes, on the other hand, showed that the non-covalent interactions between the carbohydrate portions and Ca^{2+} are more stable than covalent linkages, as the ceramide portion could be split off before the complex broke. Combining the results from changing declustering potentials and performing collision induced decomposition suggests therefore, that the GalCer-CBS-Ca^{2+} complex is the most stabile form [86] which corresponds to the liposome assays. Similar observations were also made with the Lewis antigens where ceramide and fucose were split off before the Ca^{2+}-stabilized carbohydrate–carbohydrate interaction was disintegrated [85].

38.4.6 Nuclear Magnetic Resonance

The advantage of nuclear magnetic resonance (NMR) lies in the fact that measurements occur in aqueous solutions, i.e. the most natural form if one considers biological materials. NMR is widely used for structural characterizations of carbohydrates even though the requirements for homogeneous probes and relative insensitivity can pose limitations. The flexibility of carbohydrates sets another factor of limitation and therefore structural and dynamic properties must be determined in a combination of different methods [112]. Conformations of bound carbohydrates *vs* non-bound carbohydrates can be measured by transfer nuclear Overhauser effect (trNOE) experiments. These trNOEs can only be determined for a carbohydrate complexed to a large molecular weight molecule, for instance a protein, when the relaxation is governed by the tumbling time of the protein. The only information gained in this way is the difference between non-bound and bound carbohydrate. To elucidate the structure of the entire complex needs also the parallel analysis of the protein structure by either crystallography or NMR and the combination of the data acquired in the different experiments with molecular modelling techniques [112, 113]. As carbohydrate–carbohydrate interactions are usually weak and require polymerization of both binding partners resulting in larger complexes, trNOEs are not feasible for measuring carbohydrate–carbohydrate interactions. The only possibility for analyzing characteristics of two carbohydrate molecules against one another is by modelling interactions with data obtained for the individual components from either NMR or crystallographic approaches.

38.4.7 Molecular Modelling

Due to the nature of carbohydrates, appropriate force-fields must be used to describe the conformation and dynamic properties of these structures. A number of different programs have been adapted or created especially for modelling carbohydrates (for review see [113, 115]). Due to the limited number of data for larger carbohydrates, models only exist for a few carbohydrate–carbohydrate interactions mediated by small oligosaccharides. The interactions of the GSL headgroups Gg3 and G_{M3} have been modelled by creating minimum energy conformation models of the two head groups based on hard sphere exoanomeric calculations, suggesting the interaction via an exposed hydrophobic patch in the two molecules [76]. Similarly, Le^x–Le^x interaction was modelled and proposed to assume a head-to-head alignment of the molecules [72] as also suggested by crystal data [87]. The increasing potential of calculation power and development of enhanced measuring capacity with NMR equipment will probably add information to the field of carbohydrate–carbohydrate interactions.

38.4.8 Tools

Important tools for the study of carbohydrate interactions have been antibodies directed against carbohydrate epitopes. Even more important is the synthesis of carbohydrate sequences to create homogeneous, well defined sequences which allow the study of interactions without problems created by heterogeneous populations of glycans in functional and structural tests.

Synthetic Oligosaccharides

The synthesis of well defined and homogeneous structures does allow the study of interaction phenomena in different ways. For obvious reasons oligosaccharide synthesis involves more effort than peptide- or oligonucleotide-synthesis, but nevertheless the number of structures synthesized for different purposes is increasing. Sponge epitopes [116], GSL-epitopes [117] and, the most prominent examples, Lewis blood group antigens have been synthesized and applied for direct measurement of interaction by NMR, crystallography [87], ESI-MS [85] and other methods [118].

Antibodies against Carbohydrate Motifs

Antibodies against carbohydrate determinants have been widely used to study the functional aspects of carbohydrate interactions. They are applied to control the expression level of a given carbohydrate epitope as GSLs [77], to block binding sites [11] or to affinity purify and detect oligosaccharide structures containing the epitope [68]. One important aspect with antibodies is the fact that their reactivity can depend both on the conformation of the epitope due to different surroundings and on the degree of polyvalence due to clustering [119]. The recognition of the same antenna structure on glycopeptides and glycolipids has been shown to differ for various antibodies as a consequence of different conformation of these epitopes depending on the respective anchoring sequences [119].

Two monoclonal antibodies inhibiting *Microciona* aggregation are directed against the carbohydrate portion of the aggregation factor (Block 1 and Block 2), while a third one does not block the aggregation although recognizing the carbohydrate portion [11]. All three have been characterized by a combination of chemical degradation and characterization of the recovered structures by enzyme susceptibility, NMR and MS [68, 69]. Several antibodies directed against glycolipid structures have been used in the characterization of GSL–GSL interaction to

Table 5. Monoclonal antibodies applied for studying carbohydrate–carbohydrate mediated interaction.

Antibody	Epitope recognized	Assay system[a]	Reference
Block 1	Pyr-4,6Galβ1-4GlcNAcβ1-3Fuc	*Microciona* aggregation	[68]
Block 2	GlcNAc(3OSO$_3$)β1-3Fuc	*Microciona* aggregation	[69]
DH2	Anti-G$_{M3}$	Melanoma adhesion to T-cell lymphoma	[76]
2D4	Anti-Gg3	Melanoma adhesion to T-cell lymphoma	[76]
T5A7	Anti-LacCer	B16 melanoma adhesion	[79]
9G7	Anti-Gb4	2102 lymphoma cell adhesion	[74]
1B2	Anti-nLc4	2102 lymphoma cell adhesion	[74]
MC631	Anti-SSEA-3 (GalGb4 main epitope)	2102 lymphoma cell adhesion	[74]
SNH3	Anti-SLex	B16 melanoma adhesion	[79]
SH1	Anti-Lex	GSL-GSL in vitro	[73]
BE2	Anti-H	B16 melanoma adhesion	[79]

[a] Assay system indicates in which context the antibodies were used for binding studies

quantitate the GSL-expression on cell lines and to interfere with the binding of cells and liposomes to one another [76, 77, 79, 82, 117, 120].

Cells

What started in the beginning of the century with the observation of cells taken from different sponges to allow the study of recognition processes [48, 49] has developed into a major research tool in biology with the use of both, primary and immortalized cells from many different sources. The selection of different clones from the same parent line has further stimulated the application and direct comparison of cells with similar but not identical surface composition. Examples of such cell lines that have been used for the study of carbohydrate–carbohydrate mediated phenomena are variants of the mouse melanoma cell line B16 expressing different levels of G_{M3}-ceramides [79] or T-cell lymphoma clones with different levels of Gg3-ceramides on the cell surface [76] (see also Table 4).

References

1. D. Spillmann and U. Lindahl, *Curr. Opin. Struct. Biol.*, **1994**, *4*, 677–682
2. A. Varki, *Proc. Natl Acad. Sci. USA*, **1994**, *91*, 7390–7397
3. T. A. Springer, *Cell*, **1994**, *76*, 301–314
4. H. Rauvala and H. B. Peng, *Prog. Neurobiol.*, **1997**, *52*, 127–144
5. M. Schachner, *Curr. Opin. Cell Biol.*, **1997**, *9*, 627–634
6. U. Rutishauser, *J. Cell. Biochem.*, **1998**, *70*, 304–312
7. K.-A. Karlsson, *Curr. Opin. Struct. Biol.*, **1995**, *5*, 622–635
8. K. S. Rostand and J. D. Esko, *Infect. Immun.*, **1997**, *65*, 1–8
9. N. Sharon and H. Lis, *Science*, **1989**, *246*, 227–234
10. T. Feizi, *Trends Biochem. Sci.*, **1994**, *19*, 233–234
11. G. N. Misevic, J. Finne and M. M. Burger, *J. Biol. Chem.*, **1987**, *262*, 5870–5877
12. D. Spillmann and M. M. Burger, *J. Cell. Biochem.*, **1996**, *61*, 562–568
13. D. A. Rees, *Biochem. J.*, **1972**, *126*, 257–273
14. P. Gacesa, *Carbohydr. Pol.*, **1988**, *8*, 161–182
15. F. Clementi, *Crit. Rev. Biotech.*, **1997**, *17*, 327–361
16. A. O. Tzianabos, A. Pantosti, H. Baumann, J.-R. Brisson, H. J. Jennings and D. L. Kasper, *J. Biol. Chem.*, **1992**, *267*, 18230–18235
17. J. Toikka, J. Aalto, J. Häyrinen, L. J. Pelliniemi and J. Finne, *J. Biol. Chem.*, **1998**, *273*, 28557–28559
18. J. E. Scott, Y. Chen and A. Brass, *Eur. J. Biochem.*, **1992**, *209*, 675–680
19. J. E. Scott, C. Cummings, A. Brass and Y. Chen, *Biochem. J.*, **1991**, *274*, 699–705
20. P. Albersheim, J. An, G. Frshour, M. S. Fuller, R. Guillen, K.-S. Ham, M. G. Hahn, J. Huang, M. O'Neill, A. Withcombe, M. V. Williams, W. S. York and A. Darvill, *Biochem. Soc. Trans.*, **1994**, *22*, 374–378
21. M. McNeil, A. G. Darvill, S. C. Fry and P. Albersheim, *Annu. Rev. Biochem.*, **1984**, *53*, 625–663
22. L. M. J. Kroon-Batenburg and J. Kroon, *Glycoconj. J.*, **1997**, *14*, 677–690
23. N. C. Carpita and D. M. Gibeaut, *Plant Cell*, **1993**, *3*, 1–30
24. N. Carpita, M. McCann and L. R. Griffing, *Plant Cell*, **1996**, *8*, 1451–1463
25. G. O. Aspinall In *The Biochemistry of Plants*; Academic Press: New York, 1980; pp 473–500.
26. T. N. Wight, M. G. Kinsella and E. E. Qwarnström, *Curr. Opin. Cell Biol.*, **1992**, *4*, 793–801
27. H. Lis and N. Sharon, *Eur. J. Biochem.*, **1993**, *218*, 1–27
28. A. Varki, *Glycobiology*, **1993**, *3*, 97–130

29. L. Kjellén and U. Lindahl, *Annu. Rev. Biochem.*, **1991**, *60*, 443–475
30. T. C. Laurent, U. B. G. Laurent and J. R. E. Fraser, *Imm. Cell. Biol.*, **1996**, *74*, A3–A7
31. L. A. Sellers and A. Allen In *Mucus and Related Topics*; University Press Corporation: 1989; pp 65–71.
32. M. Mörgelin, M. Paulsson, T. E. Hardingham, D. Heinegård and J. Engel, *Biochem. J.*, **1988**, *253*, 175–185
33. R. E. Turner, P. Lin and M. K. Cowman, *Arch. Biochem. Biophys.*, **1988**, *265*, 484–495
34. E. R. Morris, R. D. A. and E. J. Welsh, *J. Mol. Biol.*, **1980**, *138*, 383–400
35. J. K. Sheehan, K. H. Gardner and E. D. T. Atkins, *J. Mol. Biol.*, **1977**, *117*, 113–135
36. S. Arnott, A. K. Mitra and S. Raghunathan, *J. Mol. Biol.*, **1983**, *169*, 861–872
37. J. E. Scott, *Ciba Found. Symp.*, **1989**, *143*, 6–20
38. L.-Å. Fransson and L. Cöster, *Biochim. Biophys. Acta*, **1979**, *582*, 132–144
39. E. A. Turley and S. Roth, *Nature*, **1980**, *283*, 268–271
40. L.-Å. Fransson, I. Sjöberg and V. P. Chiarugi, *J. Biol. Chem.*, **1981**, *256*, 13044–13047
41. L.-Å. Fransson, L. Cöster, A. Malmström and J. K. Sheehan, *J. Biol. Chem.*, **1982**, *257*, 6333–6338
42. L.-Å. Fransson, I. Carlstedt, L. Cöster and A. Malmström, *J. Biol. Chem.*, **1983**, *258*, 14342–14345
43. L.-Å. Fransson and B. Havsmark, *Carbohyd. Res.*, **1982**, *110*, 135–144
44. L.-Å. Fransson, B. Havsmark and J. K. Sheehan, *J. Biol. Chem.*, **1981**, *256*, 13039–43
45. A. D. Snow, H. Mar, D. Nochlin, R. T. Sekiguchi, K. Kimata, Y. Koike and T. N. Wight, *Am. J. Pathol.*, **1990**, *137*, 1253–1270
46. R. J. Stewart and J. M. Boggs, *Biochemistry*, **1993**, *32*, 10666–10674
47. S.-i. Hakomori, *Pure Appl. Chem.*, **1991**, *63*, 473–482
48. H. V. Wilson, *J. Exp. Zool.*, **1907**, *5*, 245–258
49. P. S. Galtsoff, *J. Exp. Zool.*, **1925**, *42*, 223–251
50. G. N. Misevic and M. M. Burger, *J. Biol. Chem.*, **1986**, *261*, 2853–2859
51. T. Humphreys, *Dev. Biol.*, **1963**, *8*, 27–47
52. T. Humphreys In *The Specificity of Cell Surfaces*; Englewood Cliffs: Prentice Hall, NJ, 1967; pp 195–210.
53. P. Henkart, S. Humphreys and T. Humphreys, *Biochemistry*, **1973**, *12*, 3045–3050
54. C. B. Cauldwell, P. Henkart and T. Humphreys, *Biochemistry*, **1973**, *12*, 3051–3055
55. S. Humphreys, T. Humphreys and J. Sano, *J. Supramol. Struct.*, **1977**, *7*, 339–351
56. U. Dammer, O. Popescu, P. Wagner, D. Anselmetti, H.-J. Güntherodt and G. N. Misevic, *Science*, **1995**, *267*, 1173–1175
57. J. E. Jumblatt, V. Schlup and M. M. Burger, *Biochemistry*, **1980**, *19*, 1038–1042
58. J. A. Varner, M. M. Burger and J. F. Kaufman, *J. Biol. Chem.*, **1988**, *263*, 8498–8508
59. J. A. Varner, *J. Cell Sci.*, **1995**, *108*, 3119–3126
60. J. A. Varner, *J. Biol. Chem.*, **1996**, *271*, 16119–16125
61. S. R. Turner and M. M. Burger, *Nature*, **1973**, *244*, 509–510
62. G. N. Misevic and M. M. Burger, *J. Cell. Biochem.*, **1990**, *43*, 1–8
63. G. N. Misevic, J. E. Jumblatt and M. M. Burger, *J. Biol. Chem.*, **1982**, *257*, 6931–6936
64. D. J. Rice and T. Humphreys, *J. Biol. Chem.*, **1983**, *258*, 6394–6399
65. W. Burkart and M. M. Burger, *J. Supramol. Struct. Cell. Biochem.*, **1981**, *16*, 179–192
66. D. Spillmann, *Glycoconj. J.*, **1994**, *11*, 169–171
67. G. N. Misevic and M. M. Burger, *J. Biol. Chem.*, **1993**, *268*, 4922–4929
68. D. Spillmann, K. Hård, J. Thomas-Oates, J. F. G. Vliegenthart, G. Misevic, M. M. Burger and J. Finne, *J. Biol. Chem.*, **1993**, *268*, 13378–13387
69. D. Spillmann, J. E. Thomas-Oates, J. A. van Kuik, J. F. G. Vliegenthart, G. Misevic, M. M. Burger and J. Finne, *J. Biol. Chem.*, **1995**, *270*, 5089–5097
70. T. Feizi, *Nature*, **1985**, *314*, 53–57
71. B. A. Fenderson, U. Zehavi and S.-i. Hakomori, *J. Exp. Med.*, **1984**, *160*, 1591–1596
72. N. Kojima, B. A. Fenderson, M. R. Stroud, R. I. Goldberg, R. Habermann, T. Toyokuni and S.-i. Hakomori, *Glycoconj. J.*, **1994**, *11*, 238–248
73. I. Eggens, B. Fenderson, T. Toyokuni, B. Dean, M. Stroud and S.-i. Hakomori, *J. Biol. Chem.*, **1989**, *264*, 9476–9484

74. Y. Song, D. A. Withers and S.-i. Hakomori, *J. Biol. Chem.*, **1998**, *273*, 2517–2525
75. K. Iwabuchi, S. Yamamura, A. Prinetti, K. Handa and S.-i. Hakomori, *J. Biol. Chem.*, **1998**, *273*, 9130–9138
76. N. Kojima and S.-i. Hakomori, *J. Biol. Chem.*, **1989**, *264*, 20159–20162
77. N. Kojima and S.-i. Hakomori, *J. Biol. Chem.*, **1991**, *266*, 17552–17558
78. N. Kojima and S.-i. Hakomori, *Glycobiology*, **1991**, *1*, 623–630
79. N. Kojima, M. Shiota, Y. Sadahira and S.-i. Hakomori, *J. Biol. Chem.*, **1992**, *267*, 17262–17270
80. J. Jarchow and M. M. Burger, *Cell Adhes. Comm.*, **1998**.
81. O. Popescu and G. N. Misevic, *Nature*, **1997**, *386*, 231–232
82. S.-i. Hakomori, *Biochem. Soc. Trans.*, **1993**, *21*, 583–595
83. J. Finne, *J. Biol. Chem.*, **1982**, *257*, 11966–11970
84. U. Rutishauser and L. Landmesser, *Trends Neurosci.*, **1996**, *19*, 422–427
85. G. Siuzdak, Y. Ichikawa, T. J. Caulfiedl, B. Munoz, C.-H. Wong and K. C. Nicolaou, *J. Am. Chem. Soc.*, **1993**, *115*, 2877–2881
86. K. M. Koshy and J. M. Boggs, *J. Biol. Chem.*, **1996**, *271*, 3496–3499
87. S. Pérez, N. Mouhous-Riou, N. E. Nifant'ev, Y. E. Tsvetkov, B. Bachet and A. Imberty, *Glycobiology*, **1996**, *6*, 537–542
88. M. M. Burger In *The Role of Intercellular Signals: Navigation, encounter, outcome*; Verlag Chemie: Berlin, 1979; pp 119–134.
89. M. Fukuda, *Biochim. Biophys. Acta*, **1985**, *780*, 119–150
90. N. Kojima, *Trends Glycosci. Glycotechnol.*, **1992**, *4*, 491–503
91. K. Simons and G. van Meer, *Biochemistry*, **1988**, *27*, 6197–6702
92. D. A. Brown and J. K. Rose, *Cell*, **1992**, *68*, 533–544
93. T. W. Tillack, M. Allietta, R. E. Moran and W. W. J. Young, *Biochim. Biophys. Acta*, **1983**, *733*, 15–24
94. M. Sorice, I. Parolini, T. Sansolini, T. Garofalo, V. Dolo, M. Sargiacomo, T. Tai, C. Peschle, M. R. Torrisi and A. Pavan, *J. Lipid Res.*, **1997**, *38*, 969–980
95. P. Rock, M. Allietta, W. W. j. Young, T. E. Thompson and T. W. Tillack, *Biochemistry*, **1990**, *29*, 8484–8490
96. S.-i. Hakomori, *Cancer Res.*, **1996**, *56*, 5309–5318
97. M. Salmivirta, K. Lidholt and U. Lindahl, *FASEB J.*, **1996**, *10*, 1270–1279
98. W. J. Kuhns, G. Misevic and M. M. Burger, *Biol. Bull.*, **1990**, *179*, 358–365
99. S. J. Angyal, *Adv. Carbohydr. Chem. Biochem.*, **1989**, *47*, 1–43
100. A. Almond, A. Brass and J. K. Sheehan, *Glycobiology*, **1998**, *8*, 973–980
101. R. U. Lemieux, *Chem. Soc. Rev.*, **1989**, *18*, 347–374
102. D. Balasubramanian, B. Raman and C. S. Sundari, *J. Am. Chem. Soc.*, **1993**, *115*, 74–77
103. A. Varki, *J. Clin. Invest.*, **1997**, *99*, 158–162
104. T. Feizi, M. S. Stoll, C. T. Yuen, W. Chai and A. M. Lawson, *Methods Enzymol.*, **1994**, *230*, 484–519
105. D. Zopf and S. Ohlson, *Nature*, **1990**, *346*, 87–88
106. E.-C. Kang, K. Akiyoshi and J. Sunamoto, *Int. J. Biol. Macromol.*, **1994**, *16*, 348–353
107. M. Radmacher, R. W. Tillmann, M. Fritz and H. E. Gaub, *Science*, **1992**, *257*, 1900–1905
108. H. G. Hansma, K. J. Kim, D. E. Laney, R. A. Garcia, M. Argaman, M. J. Allen and S. M. Parsons, *J. Struct. Biol.*, **1997**, *119*, 99–108
109. D. J. Müller, C.-A. Schoenenberger, F. Schabert and A. Engel, *J. Struct. Biol.*, **1997**, *119*, 149–157
110. M. Rief, F. Oesterhelt, B. Heymann and H. E. Gaub, *Science*, **1997**, *275*, 1295–1297
111. J. Fritz, D. Anselmetti, J. Jarchow and X. Fernandez-Busquets, *J. Struct. Biol.*, **1997**, *119*, 165–171
112. T. Peters and B. M. Pinto, *Curr. Opin. Struct. Biol.*, **1996**, *6*, 710–720
113. A. Imberty, *Curr. Opin. Struct. Biol.*, **1997**, *7*, 617–623
114. G. Siuzdak In *Mass Spectrometry for Biotechnology*; Academic Press, Inc.: San Diego, New York, Boston, London, Sydney, Tokyo, Toronto, 1996; pp 1–161.
115. R. J. Woods, *Glycoconj. J.*, **1998**, *15*, 209–216
116. T. Ziegler, *Carbohyd. Res.*, **1994**, *262*, 195–212

117. M. R. Stroud, S. B. Levery, S. Mårtensson, M. E. K. Salyan, H. Clausen and S.-i. Hakomori, *Biochemistry*, **1994**, *33*, 10672–10680
118. T. Le Bouar, F. Pincet, J.-M. Mallet, J. Esnault, Y. M. Zhang, E. Perez and P. Sinay, *Glycoconj. J.*, **1997**, *14*, S72–73
119. S. Saito, S. B. Levery, M. E. K. Salyan, R. I. Goldberg and S.-i. Hakomori, *J. Biol. Chem.*, **1994**, *269*, 5644–5652
120. M. R. Stroud, S. B. Levery, E. D. Nudelman, M. E. K. Salyan, J. A. Towell, C. E. Roberts, M. Watanabe and S.-i. Hakomori, *J. Biol. Chem.*, **1991**, *266*, 8439–8446
121. D. Gupta, R. Arango, N. Sharon and C. F. Brewer, *Biochemistry*, **1994**, *33*, 2503–2508
122. I. Eggens, B. A. Fenderson, T. Toyokuni and S.-i. Hakomori, *Biochem. Biophys. Res. Commun.*, **1989**, *158*, 913–920

Part I
Volume 2

VI

Carbohydrate–Nucleic Acid Interactions

39 Carbohydrate–Nucleic Acid Interactions

Heinz E. Moser

39.1 Introduction

Among the various classes of biomolecules, nucleic acids play a central role not only as carrier of genetic information but also in the form of RNA as important link for the production of proteins. Specific interactions both on DNA and RNA are essential to retrieve the information required at various stages of life cycles. In general this is largely achieved by a variety of DNA or RNA binding proteins to regulate accessibility and transcription of DNA as well as modification, transport and translational processes of messenger RNA. Molecular interference with these processes can result in more or less specific disturbance of normal cell functions and evolution produced a variety of relatively small ligands that bind with high affinity to DNA or RNA, thereby disrupting crucial functions for life. This often translates in cytostatic or cytotoxic activity and depending on specificity, quite a few of these molecules found a therapeutic application as antiinfective or antitumor agents. Only a limited number of these compounds can be considered as pure carbohydrates. The binding to highly charged nucleic acids required structural properties that are usually not common for carbohydrates. Features like reduction of hydrophilicity and incorporation of positive charges are often found in carbohydrate containing molecules that tightly bind nucleic acids and they indicate how nature managed to optimally benefit from the richness of structural information in carbohydrates by adopting required physicochemical properties.

Only a few structures of pure carbohydrate nucleic acid complexes are known thereby limiting our understanding of these interactions. However, many of the DNA ligands known to date carry one or more carbohydrate residues (recognition elements [1]). The increasing availability of x-ray and NMR structures helps to understand their influence on interaction with either DNA or RNA at the molecular level.

A small number of review articles have previously addressed carbohydrate nucleic acid interactions [2–5]. The focus of this overview is the discussion of selected examples from recently published structures with the aim to better understand common rules that govern interactions between nucleic acids and modified carbohydrates.

39.2 Carbohydrates Binding to DNA

39.2.1 Ene-Diyne Antibiotics and Antitumor Agents

A family of extremely potent antitumor antibiotics were discovered over the past decades with many common features [3, 6]. These molecules all bind double stranded DNA in the minor groove and consist of a bicyclic core containing a strained 9- or 10-membered ring with an ene-diyne functionality and at least one sp$_2$-like bridgehead carbon atom. Either an intra- or intermolecular nucleophilic attack at this central ring triggers a conformational change allowing a spontaneous cyclization to take place. The resulting, highly reactive aryl diradical was demonstrated to initiate single or double strand DNA cleavage based on H-abstraction from the deoxyribose backbone. In addition, most compounds carry one or more carbohydrate side chains and/or an intercalating group, adding to structural complexity and tuning their biological function. Selected members of this family include neocarzinostatin (**1**), esperamicin A$_1$ (**2**), calicheamicin γ_1 (**3**), and dynemicin A (**4**) (Scheme 1). Recent publications of high resolution NMR structures of such ligands to short DNA duplexes [7, 8] allow a much better understanding of DNA interaction and modification by this complex class of molecules, including the important function contributed by carbohydrate moieties.

The incredibly potent family of calicheamicins were discovered independently almost 20 years ago at Lederle Laboatories and Brystol Myers [9–12] based on a focused effort to identify natural products with the ability to inhibit tumour growth. Esperamicin A$_1$ (**2**) and calicheamicin γ_1 (**3**) were among the most potent members isolated (approximately 1000-fold more potent than adriamycin (**8b**) against murine tumors, Scheme 3), and they both share the same aglycon but vary in their side chains influencing their specificity to interact with and damage DNA. Their biological activity triggered extensive research efforts to understand structure and function of aglycon and side chains to gain better insight at the molecular level of this fascinating and complex class of compounds. A key component to unravel this puzzle was the synthetic accessibility of the core ene-diyne structure that was successfully tackled by various groups [13–20]. Mastering the synthetic accessibility to these complex molecules paved the way to explore structure activity relationship of analogs.

Esperamicins

Esperamicin A$_1$ (**2**) and related members of this family, originating mainly from chemical hydrolysis or nuclephilic attack inducing the cycloaromatization, were

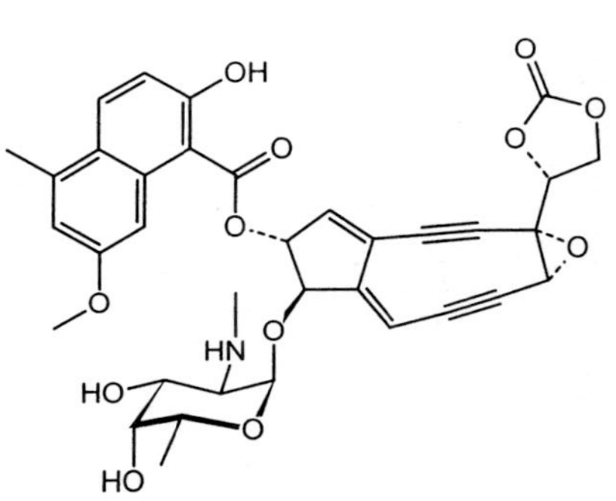

Scheme 1. Structures of selected ene-diyne antitumor antibiotics.

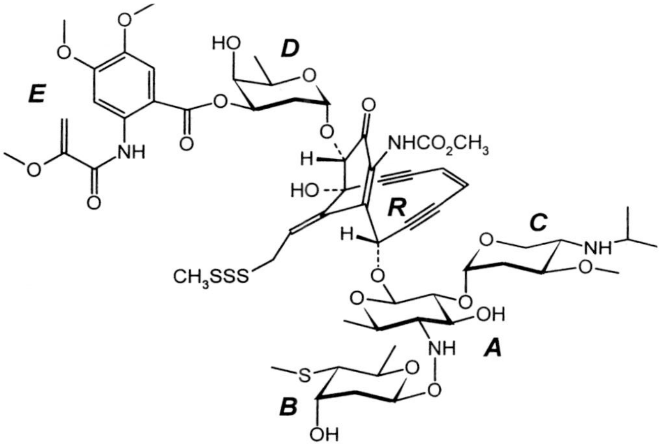

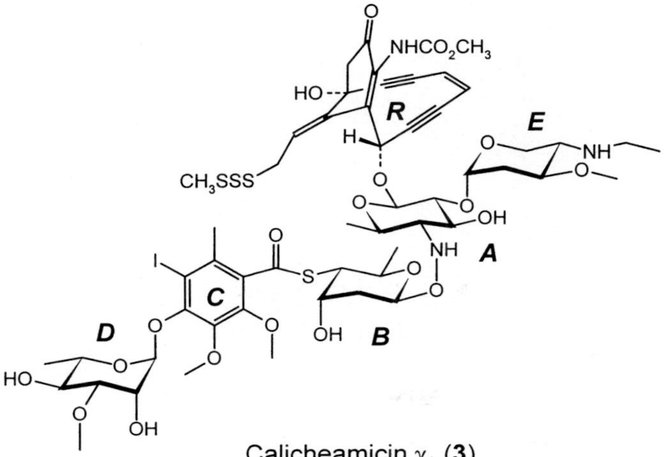

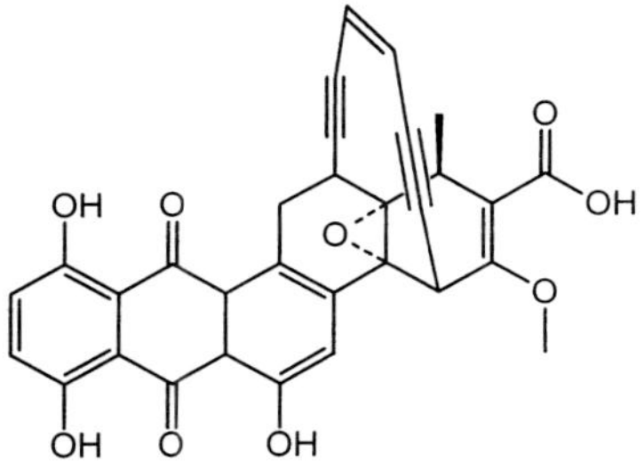

studied extensively to understand their mechanism of action in relation to biological activity [21–24]. Esperamicin A_1 (**2**) was shown to be highly cytotoxic in various eukaryotic cell lines with IC_{50}s of 0.3–4.5 ng/ml. Removal of parts **D** and **E** (esperamicin C) reduced the cytotoxic effect by roughly 200-fold. Additional removal of carbohydrate moiety **B** (esperamicin D) further reduced the potency by approximately five-fold. Interestingly, this trend was also found for their ability to induce DNA breaks in HCT116 cells; however, the increase of concentration needed to trigger a similar amount of strand breaks varied 5000-fold for removal of **D** and **E** and additional 30-fold for esperamicin D, indicating that additional functions might be involved in the observed cytotoxicity [21]. Surprisingly examination of DNA scission in vitro showed equipotent behaviour of both esperamicin A_1 and C, indicating a role of the deoxyfucose-anthranilate moiety **D** and **E** for drug accumulation in cells. However, esperamicin A_1 predominantly produced single strand breaks at low concentration, pointing out the importance of the deoxyfucose-anthranilate moiety to position the ene-diyne aglycon and ultimately the aryl diradical in the minor groove of DNA, facilitating H-abstraction mainly from one of the two sugar phosphate backbones. These findings as well as sequence preferences and the proposed intercalation of the methoxyacrylyl-anthranilate **E** [24, 25] were experimentally supported by the NMR solution structure of esperamicin A_1 (**2**) bound to the self complementary DNA duplex d(CGGATCCG) [8] (Figure 1a). Upon closer examination of these published structures some of the previously observed properties of the various esperamicins can be much better understood and indicate the role of the carbohydrate side chains. The ene-diyne aglycon **R** and the methoxyacrylyl-anthranilate **E** are linked by the deoxyfucose **D** which serves as ideal spacer accommodating the necessary 90° turn and distance between these residues. In addition, a hydrogen bond between the axial hydroxyl group and the carbonyl group of cytosine (C7′, see Figure 1b) contributes to the overall stability of the complex. Remarkably, this hydrogen bond could hypothetically be formed also with the three other bases replacing cytosine as they all contain a hydrogen bond acceptor at this position (thymine: O-2; adenine: N-3; guanosine: N-3), thereby supporting the lack of a clear sequence preference for this particular interaction.

The trisaccharide **A-B-C** on the opposite side of aglycon **R** has a key role to help fixing the ene-diyne moiety deep in the minor groove (Figure 1b). All sugar residues are deoxygenated and carry rather unusual groups optimizing favorable interactions (hydrophobic and/or opposite charges) with the DNA duplex. Carbohydrate moiety **A** serves as connection point between the aglycon **R** and residues **B** and **C**, facing the relatively wide minor groove with the 6-membered ring and interacting with the phosphodiester backbone via its equatorial hydroxyl group. Residue **B** is placed through the unusual NH–O linkage deep into the minor groove with the ring parallel to the rims of the groove, establishing extensive hydrophobic contacts in addition to a hydrogen bond with N-3 of adenine via its only axial hydroxyl group. As mentioned above, this specific interaction could in principle take place with the other three nucleic acid bases as well. An additional interaction was pointed out between the large polarizable sulfur atom on ring **B** and the exposed exocyclic amino proton of the proximal guanine base, detectable by the large upfield shift for this proton. This observation would indicate a sequence pref-

39.2 Carbohydrates Binding to DNA 1099

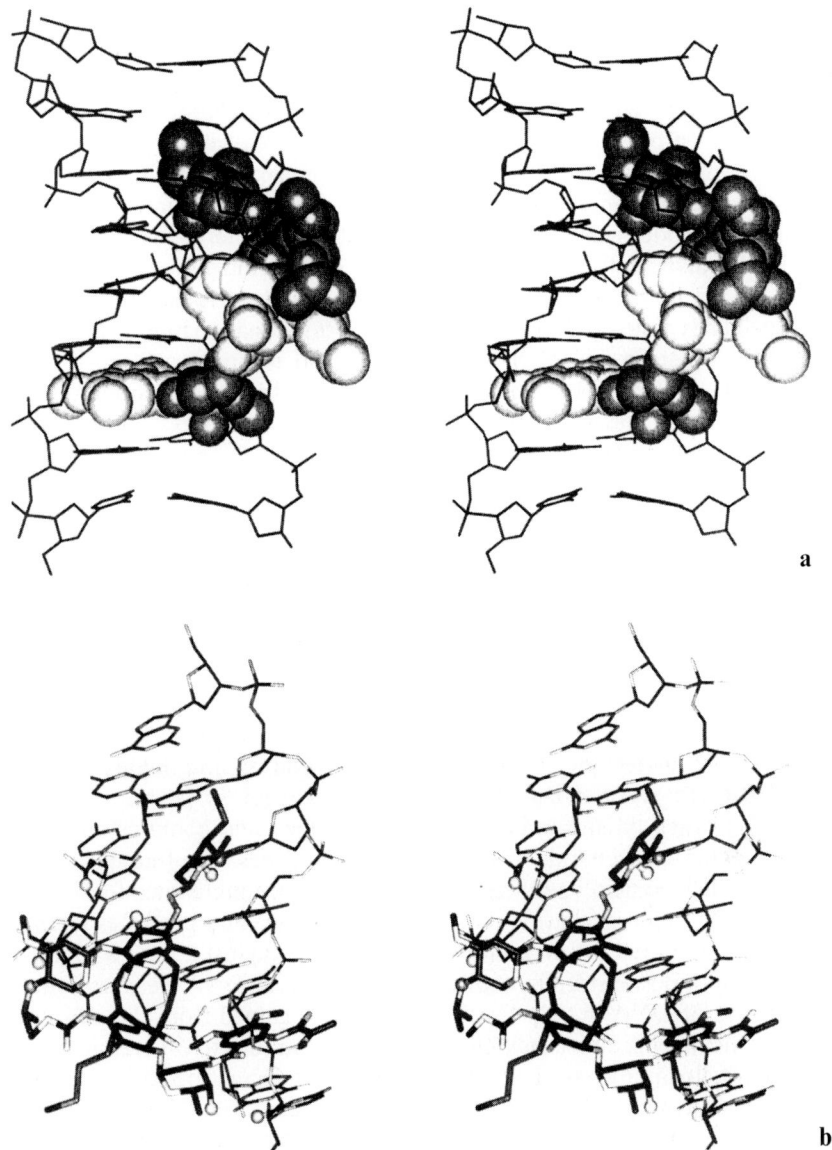

Figure 1. Stereo representation of esperamicin A_1 bound to the self complementary oligodeoxyribonucleotide duplex d(CGGATCCG)$_2$ ([8], structures were retrieved from the Brookhaven database: 1pik; to simplify the discussion only structure #2 is represented, with the hydrogen atoms omitted). **a)** Different parts of esperamicin A_1 are colored as follows: aglycon ***R*** and methoxyacrylyl-anthranilate ***E*** (light grey), carbohydrate residues ***A***, ***B***, ***C***, and ***D*** (dark grey). **b)** Same structure with one base pair omitted at each end. Specific heteroatoms intermolecularly interacting between esperamicin A_1 and the DNA via hydrogen bonds are highlighted as small balls. Oxygen atoms are represented in light grey and both, nitrogen and sulfur, are shown in dark grey; all the other elements are black.

erence for a guanine at this position; even though such sites are found upon close examination of esperamicin A_1 induced cleavage patterns, overall this interaction does not seem to play a dominant role. Residue **C** is placed over the sugar phosphate backbone of the DNA duplex forming a hydrogen bond between its charged isopropyl ammonium group and one of the phosphodiester oxygens.

All the interactions discussed above are in excellent agreement with the observed, relatively weak sequence preferences for DNA cleavage by esperamicin A_1. The reported structure provides a plausible explanation for the preferred break of only one strand. The ene-diyne aglycon **R** is tightly placed deep in the minor groove fixed on both sides by the anthranilate-deoxyfucose **D-E** and the trisaccharide **A-B-C** sidechains, respectively, allowing only minimal conformational changes relative to the duplex DNA. Whereas one of the potential radical positions will be ideally placed to abstract H_1' from the bottom strand (Figure 1b), the other position on the opposite site is not only further away from H_5' of the other strand but also not ideally directed at this position. It is important to note that these differences have to be read with caution as in particular the position of H_5' is not as well defined and the interpretation of this structure in terms of cleavage prediction is limited. Nevertheless, the hypothesis that esperamicin A_1 predominantly cleaves one strand based on conformational fixation is supported by the cleavage behavior of esperamicin C: removal of the intercalating part **E** and linking sugar **D** yields an esperamicin analog that exclusively produces double strand scissions [21].

Calicheamicins

Calicheamicin γ_1 (**3**) shares the same ene-diyne aglycon **R** and highly similar trisaccharide **A-B-C** (for **2**) or **A-B-E** (for **3**) with esperamicin A_1 (**2**) but lacks the intercalating methoxyacrylyl-anthranilate deoxyfucose part and contains an additional extension on the trisaccharide [6]. Two major differences were found between these antibiotics: calicheamicin γ_1 (**3**) shows a clear sequence preference for TCCT (found as well for the oligosaccharide moiety alone [26–28]) and produces mainly double strand cleavages even at low concentrations with a typical 3 base 3'-shift of the cleavage sites on opposite strands, commonly observed for DNA cleavers bound to the minor groove of DNA [6, 29]. Further insight was gained from the cleavage behaviour of calicheamicins either lacking rings **D** or **E**: Similar to esperamicin A_1, omission of the 4-ethylamino sugar **E** did not cause a change in the observed specificity patterns but dropped the efficiency by 2–3 orders of magnitude indicating a function for binding affinity but not specificity. The calicheamicin lacking the terminal rhamnose **D** exhibited a similar cleavage specificity with a roughly 100-fold lower efficacy [30]. Many groups initiated various studies to better understand the molecular interaction between DNA and calicheamicin γ_1 (**3**); in particular NMR proved to be extremely valuable [31–34]. Finally, a high resolution structure of calicheamicin γ_1 (**3**) bound to the 23-mer hairpin duplex d(CACTCCTGGTTTTTCCAG$_{18}$G$_{19}$AGTG) was obtained and published that is used here as basis for the discussion of specific DNA carbohydrate interactions [7].

Calicheamicin γ_1 (**3**) binds in an extended conformation to the minor groove of DNA, placing the terminal ene-diyne aglycon deep into the groove (Figure 2a).

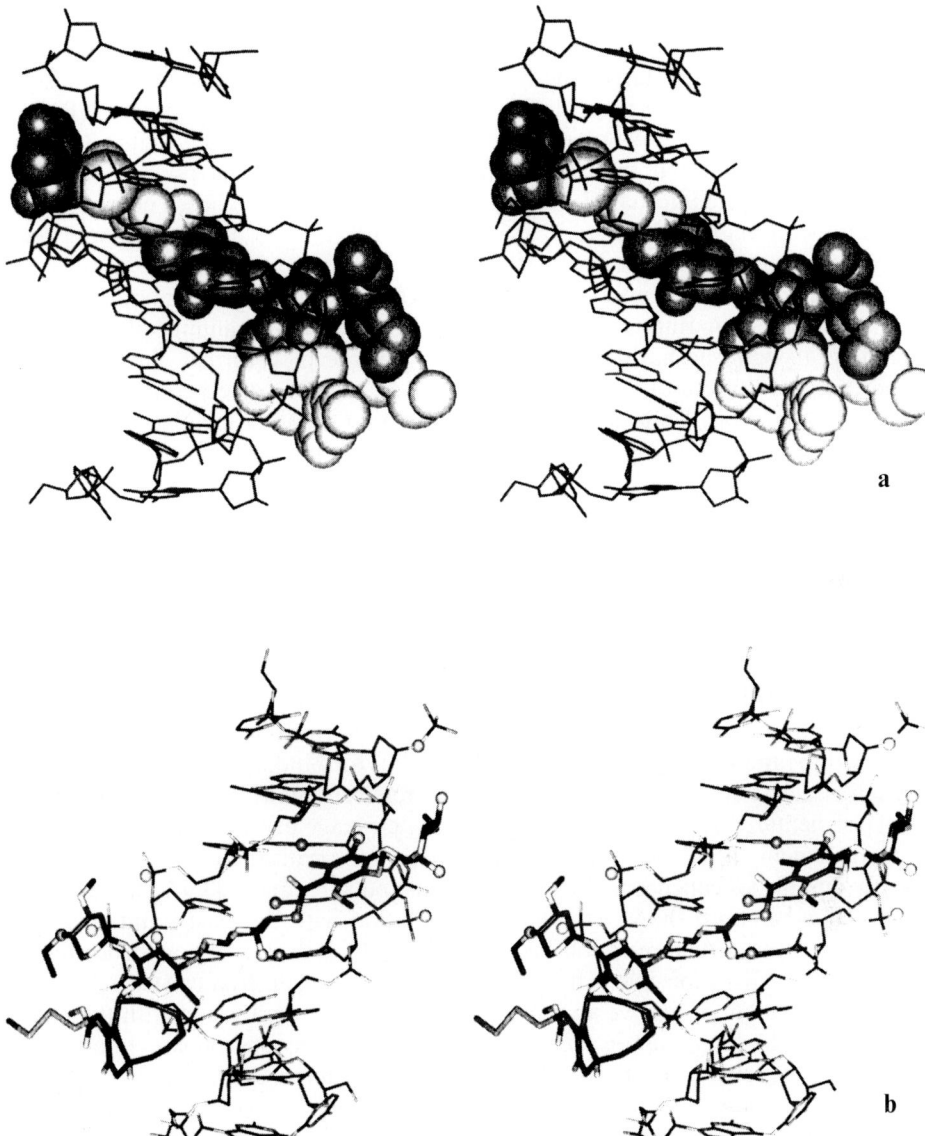

Figure 2. Stereo representation of calicheamicin γ_1 (**3**) bound to the 23-mer hairpin duplex d(CACTCCTGGTTTTTCCAGGAGTG) ([7], the structure was retrieved from the Brookhaven database as 2pik and due to simplicity only the first structure is represented omitting hydrogen atoms and the T_5 hairpin loop. **a**) Different parts of calicheamicin γ_1 (**3**) are coloured as follows: aglycon **R** and thiobenzoate ring **C** (light grey), carbohydrate residues **A**, **B**, **D**, and **E** (dark grey). **b**) Same structure with one base-pair omitted at the top end. Specific heteroatoms intermolecularly interacting between calicheamicin γ_1 (**3**) and the DNA via hydrogen bonds are highlighted as small balls. Oxygen and iodine atoms are represented in light grey and both nitrogen and sulfur, are shown in dark grey; all the other elements are black.

Determination of the structure without target DNA under various conditions indicated a high degree of conformational preorganization [35], allowing this rather big molecule to bind its target site in an extended conformation with high affinity. The interactions for sugar rings *A*, *B*, and *E* closely resemble those discussed for the similar counterparts in esperamicin A_1 (**2**) discussed previously (see Figure 2b). It is worthwhile to point out that the required orientation between rings *A* and *B* is maintained with the unusual N–O bond, ideally adopting an eclipsed conformation with a torsion angle of close to 120° [36, 37] ($-128.4 \pm 2.6°$ [7]) providing a relatively rigid, gentle curvature in the middle of the molecule to optimally achieve the overall shape to bind the minor groove. A change of the natural configuration at the anomeric center on ring *A* completely disrupts the binding affinity of the corresponding aryltetrasaccharide to the target DNA site [28], further supporting the importance of structural preorganization. As mentioned for esperamicin A_1, the sulfur atom is in proximity with the exocyclic amino group of guanine 19 causing a large difference in the chemical shift of the corresponding proton.

However, replacing this guanine by inosine lacking the exocyclic amino group caused only a modest change in the binding affinity of the aryltetrasaccharide and therefore indicates this interaction to be of minor importance [38]. The thiobenzoate ring *C* is buried similar to sugar *B* deep in the minor groove with its methyl and iodine substituents facing the bottom of the groove and establishing stacking interactions with the deoxyribose rings of T_7 and A_{20} on opposite strands. The large polarizable iodine atom is in proximity to the exocyclic amino group of guanine 18 favorably contributing to the sequence preference of calicheamicin γ_1 (**3**) [1]. Substitution of the iodine atom by bromine, chlorine, fluorine, methyl, or hydrogen [38] indicated the importance of iodine in this position for binding of the corresponding aryltetrasaccharide as these substitutions progressively reduced the binding affinity as experimentally determined by a competition cleavage experiment.

In agreement with these findings, substitution of the critical guanine 18 by inosine greatly diminishes calicheamicin γ_1 (**3**) binding and subsequent DNA cleavage [38]. Rings *B* and *D* both showed the highest resolution with the least standard deviations of the four structures published [7] giving a high degree of confidence in this central part of the molecule. The terminal rhamnose sugar *D* is twisted relative to the rims of the minor groove and interacts via its 2-hydroxyl group with the proximal phosphodiester group on the purine strand of the recognition site. In addition, the second hydroxyl group is within distance to interact with a phosphodiester oxygen on the opposite strand. Finally, the ene-diyne aglycon *R* lacking further substitution on the opposite side of its aryltetrasaccharide substituent, interacts via its unsubstituted hydroxy group with the proximal oxygen atom on the terminal phosphodiester of the complex.

The calicheamicin oligosaccharide **5** (Scheme 2) was shown to bind with low micromolar affinity to its target site with similar selectivity as compared to calicheamicin γ_1 (**3**) [38] (see also references in [39]). This unique property could be used to compete off specific transcripton factors from their target sequence at the expected concentrations [39]. The limited selectivity and potency, however, were not sufficient for a meaningful biological application and the recent progress in the synthesis of complex oligosaccharides allowed this approach to be explored further.

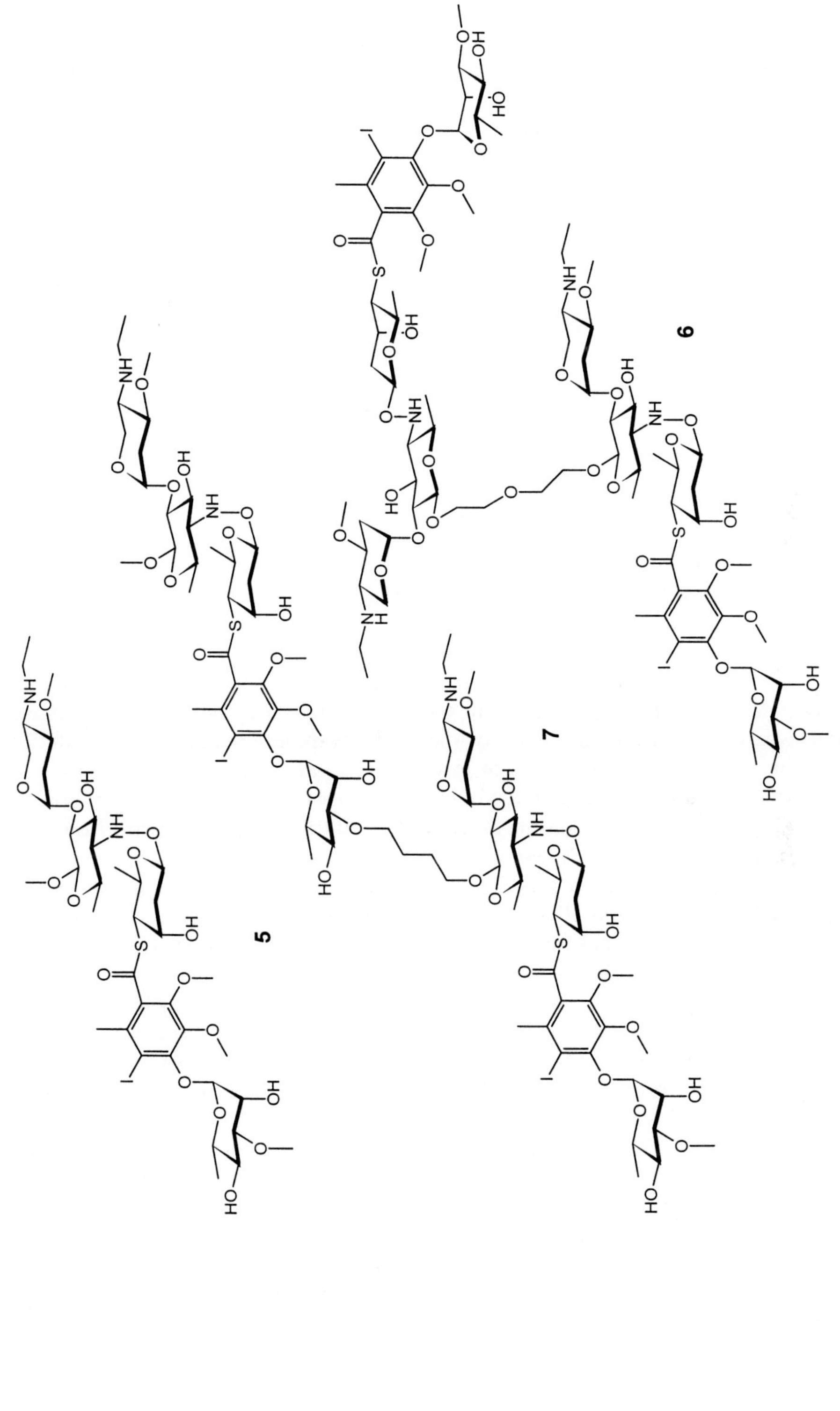

39.2 *Carbohydrates Binding to DNA* 1103

Scheme 2. Structures of the calicheamicin γ_1 tetrasaccharide **5** and the corresponding head to head and head to tail dimers **6** and **7**.

Based on structural knowledge of the calicheamicin oligosaccharide bound to its target DNA tetramer site, head to head and head to tail dimers **6** and **7** were designed and synthesized [40–42]. Examination of the inhibitory activity on calicheamicin γ_1 (**3**) induced DNA cleavage revealed an estimated 100-fold higher affinity of the head to head dimer **6** and more than 1000-fold improved binding of dimer **7** for the target sites as compared to monomer **5** [43]. Both dimers bound with >100-fold selectivity to the corresponding target site but dimer **6** interacted differently with two distinct target sites indicating the influence of neighboring sequences on the overall activity. This dimer was subjected to structure elucidation by NMR as complex with the self complementary oligonucleotide duplex d(CGTAGGATATCCTACG)$_2$ [34]. Close examination of the head to head dimer **6** interaction with the DNA duplex essentially revealed a highly similar binding mode of each carbohydrate unit as compared to the monomer **5**, reproducing all the individual interactions discussed previously (see above). The recently published structure of the head to tail dimer **7** confirmed the previous findings and added an interesting insight on sequence selectivity [44]: the structure was resolved with the target duplex d(GCACCTTCCTGC) × d(GCAGGAAGGTGC) containing two slightly different affinity sites. As discussed with calicheamicin γ_1 (**3**) itself, the 3-OH group of ring **B** hydrogen bonds with N-3 of A$_{20}$ [7] (see Figure 2b). Flipping this base pair in the duplex positions the O-2 carbonyl function in almost exactly the same place allowing the hydrogen bond to be formed as evident from the NMR structure. The calicheamicin γ_1 (**3**) and calicheamicin oligosaccharide structures **6** and **7** discussed above [7, 34, 44] revealed an almost perfect fit between duplex DNA and bound carbohydrate residues. Structural analysis of the DNA binding sites revealed no major conformational changes at the sites of carbohydrate interactions whereas the accommodation of the en-diyne moiety required a widening of the minor groove. These findings are in contrast to an earlier hypothesis [31, 33] that assumed the conformational flexibility of the DNA binding site to be of crucial importance for the selective binding event. The head to head dimer **6** was also tested for its ability to interfere with transcriptional events [45]. At low micromolar concentrations oligosaccharide **6** was able to disrupt binding of transcription factors (AP-1, STAT-3, NFI, and PU.1) to the corresponding target sites containing a overlapping binding site as shown by gel mobility assays and in vitro transcription by polymerase II.

These experiments demonstrate the validity of modified carbohydrate units as minor groove recognition elements, not limiting their function as originally believed to tune the pharmacokinetic behavior. The required degree of conformational preorganization with an evolutionary adaption of the individual units by decreasing hydrophobicity and adding positive charges make the functionally rich carbohydrates a truly unique scaffold for minor groove recognition of DNA. Whereas the work described above was mainly based on further elaboration of oligosaccharides from natural sources, the increasing structural understanding on molecular level paired with powerful technologies like combinatorial (carbohydrate) chemistry, improved synthetic methodologies, and experimental binding optimization by NMR [46–48] might lead to new horizons for DNA—carbohydrate interactions. In particular, probing preformed DNA-carbohydrate complexes by NMR for adjacent

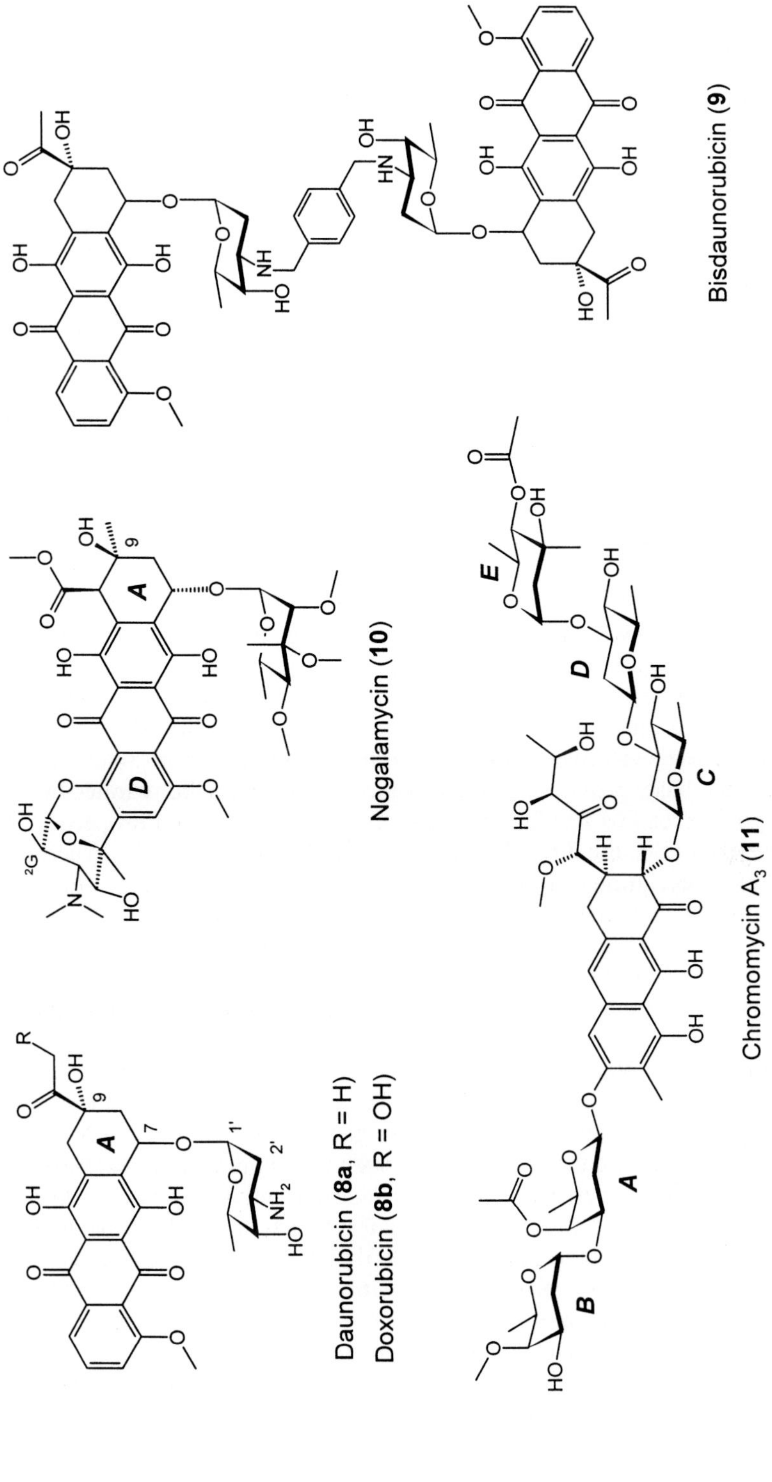

Daunorubicin (**8a**, R = H)
Doxorubicin (**8b**, R = OH)

Bisdaunorubicin (**9**)

Nogalamycin (**10**)

Chromomycin A$_3$ (**11**)

Scheme 3. Structures of representative antibiotics from the anthracycline (**8–10**) and aureolic acid (**11**) families.

binding events of structurally diverse carbohydrate residues (or other low molecular weight compounds) could lead after appropriate linkage to new classes of high affinity DNA ligands that will go beyond the scope of currently available molecules from natural sources.

39.2.2 Anthracyclins

Anthracycline antibiotics or antitumor agents are a class of biologically active compounds isolated from various *Streptomyces* strains. They have indicator-like properties and consist of a core tetracyclic anthraquinone chromophore carrying one or multiple carbohydrate substituents. The first members of this family were isolated roughly half a century ago and shown to display potent antibacterial activities (for recent reviews see [49–52]). The high toxicity in mice, however, precluded their further application in man and this changed in the sixties with the discovery of the structurally closely related daunorubicin (or daunomycin, **8a**) [53, 54] and doxorubicin (or adriamycin, **8b**) [55], both currently used in the clinical treatment of leukaemias and solid tumors, respectively. They interfere with DNA replication and transcription by tightly binding to double stranded DNA via intercalation, preferably at alternating pyrimidine–purine tracts, placing the carbohydrate moiety in the minor groove. In addition, they interfere with topoisomerase II activity [56]. Due to an intense interest in their biological behaviour and shortcomings in the clinical application (cardiotoxic side effects and development of resistance) efforts were initiated to understand the interaction with DNA on the molecular level. To date well over 40 high resolution structures are available from the Brookhaven database, providing insight in the role of the carbohydrate–DNA interaction that might lead to the design of derivatives with improved properties.

Daunorubicin (**8a**) was the first anthracycline antibiotic for which structural information at high resolution became available of its DNA-bound state [57]. This structure was refined to remarkable 1.2 Å [58] and, later on, was followed by additional structures using various DNA templates [59–61]. In the highly refined 2:1 complex with d(CGTACG)$_2$ [58], the tetracyclic aglycon is intercalated between the terminal CG base pairs, oriented almost at a right angles to the long axis between the base pairs (Figure 3a). In this complexed form the conformation of ring *A* varies from the x-ray structure obtained from daunorubicin (**8a**) alone [62]: only C-9 lies outside the plane defined by the fused aromatic rings, positioning O-9 almost perfectly perpendicular to this plane and within hydrogen bonding distance to N-3 and N-2 of G$_2$ (O–N distances: 2.6 and 2.9 Å, respectively; see Figure 3b). Interestingly, this hydroxy group at C-9 is essential for the biological activity of daunorubicin, supporting the importance of this intermolecular interaction. Daunorubicin (**8a**) was reported to preferentially bind to 5'-A/T-CG-3' or 5'-A/T-GC-3' triplets, albeit with only modest preference over other sequences [63, 64]. This is in agreement with the hydrogen bonding interaction discussed above, as the more important interaction with N-3 of G$_2$ can take place with the other three bases as well (N-3 of adenosine or O-2 of either thymine or cytosine). The second main difference between the two x-ray structures is the torsion angle of the glycosyl linkage (C7-O7-C1'-C2')

39.2 Carbohydrates Binding to DNA   1107

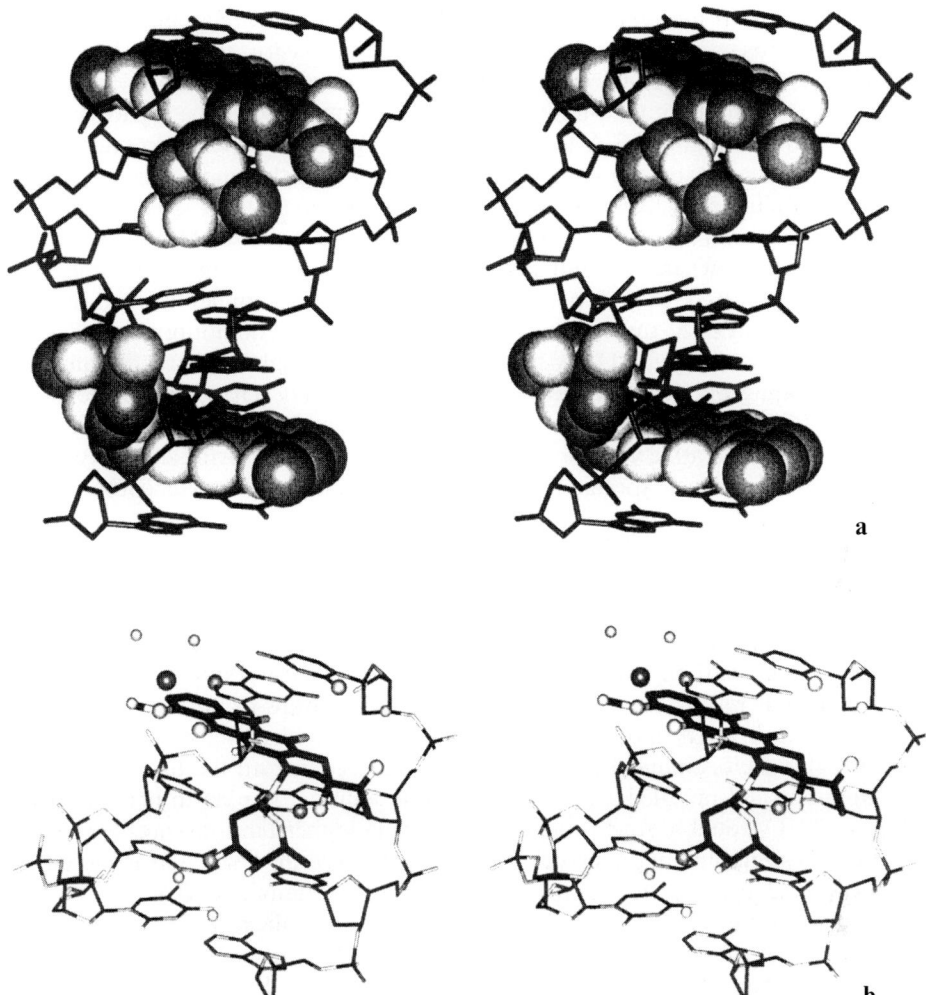

Figure 3. Stereo pictures of daunorubicin (**8a**) bound to d(CGTACG)$_2$ [58]. The co-ordinates have been retrieved from the Brookhaven database (11d1) and for clarity, hydrogens are omitted. **a**) Representation of the DNA duplex with two daunorubicin molecules intercalated; the tetracyclic anthraquinone aglycon is represented as CPK model with carbon atoms in dark grey and oxygen/nitrogen atoms in light grey. **b**) Part of the same structure including selected interacting water molecules (small balls in light grey) and one sodium atom (ball in dark grey). Heteroatoms involved in hydrogen bonding are drawn as small balls. Carbon and phosphorous atoms are shown in black, oxygen in light grey and nitrogen in dark grey.

which is smaller in the DNA-bound molecule. This small change allows the aminosugar (daunosamine) to adopt the correct orientation for an optimal fit in the minor groove of the right-handed DNA duplex. In order to optimally accommodate the carbohydrate in the minor groove the proximal G_2-C_{11} base pair needs to be dislocated by roughly 1.3 Å towards the major groove. Opposite to interactions seen in other complexes, the amino group does not directly bind to the negatively charged phosphodiester backbone but hydrogen bonds to O-2 of C-5 and two water molecules. The axial hydroxy group at C-4' points away from the DNA duplex into the solvent region where it might play an important role for the interaction with topoisomerase II. As shown previously by footprinting studies of various anthracyclines with DNA, the attached carbohydrate residues do not largely influence the sequence selectivity for DNA binding [65], even though they greatly influence both the binding affinity [66] and biological activity [49]. These observations are supported by the high resolution structure of the complex: the binding affinity is increased by co-operative van der Waals contacts of the carbohydrate moiety in the minor groove. However, specific hydrogen bond interactions between carbohydrate and bases are lacking.

The wealth of detailed information of the complex discussed above served as starting point for structure-based design of new bis-intercalating anthracycline antibiotics [67]. In all structures reported to date, a 2:1 ratio of daunorubicin (**8a**) to target DNA was observed. Both drug molecules intercalate at the end of the duplex positioning their carbohydrate groups in the minor groove pointing at each other (see Figure 3a). In this bound form, the two amino groups of both daunorubicins are roughly 7 Å apart and therefore could easily be bridged with an appropriate linker to give the bis-intercalator **9**. Binding studies confirmed the bis-intercalating binding mode and revealed a picomolar binding affinity (27 pM) to herring sperm DNA at 20 °C, roughly four orders of magnitude higher than daunorubicin (**8a**). The hypothesized binding mode of dimer **9** was later confirmed by structure determination using x-ray crystallography [68] and high resolution NMR [69].

The structural information on daunorubicin (**8a**) and calicheamicin γ_1 (**3**) bound to duplex DNA guided the design of a hybrid molecule between the daunorubicin aglycon and the calicheamicin γ_1 tetrasaccharide [70]. The main goal of this approach was the creation of an intercalating daunorubicin analog with improved sequence selectivity by replacing the carbohydrate moiety. Examination of models indicated the requirement of a β-glycosidic linkage and a five-atom spacer ($CH_2CH_2OCH_2CH_2$) between aglycon and tetrasaccharide. The corresponding "calichearubicin" was indeed shown to intercalate double stranded DNA, whereas an analog lacking the five atom spacer failed to intercalate. Footprinting studies on a 155 base pair fragment of pBR322 with methidium propyl [71] indicated a gain in binding specificity compared to daunorubicin. Due to the difficult interpretation of these results and the lack of discrete binding affinities for this hybrid molecule, a clear picture on sequence preference could not be obtained.

During the late sixties the structurally more complex nogalamycin (**10**) [72] was isolated and characterized. This anthracycline is unique as two carbohydrate groups (nogalose and aminoglucose) are linked to the opposite ends of the tetracyclic aglycon, yielding a dumbbell shaped molecule. As a consequence, one of the bulky sugar residues needs to move through the helix to reach the opposite side, most likely requiring a local melt of base pairs. This special feature is reflected in the slow on- and off-rates for binding. The publication of various high resolution structures of nogalamycin (**9**) to short DNA duplexes [73–81] confirmed the proposed binding mode (Figure 4a) which is discussed in more detail below.

In all the structures reported to date nogalamycin **10** is intercalating DNA, preferentially between purine–pyrimidine tracts, with the doubly connected aminoglucose on ring **D** positioned in the major groove. The conformation of the DNA bound ligand [77] closely resembles the ligand itself [82], indicating a well preorganized structure. A subtle difference is the slightly closer proximity between nogalose and aminoglucose in the bound state that allows nogalamycin (**10**) to optimize its interactions with the DNA in the complex (see below). The aglycon intercalates almost perpendicular to both flanking GC base pairs, similar to the orientation reported for daunorubicin (**8a**). The main difference, however, is the shift towards the minor groove by $\sim$2 Å, caused by interactions of the aminoglucose in the major groove and the large nogalose in the minor groove. The aminoglucose faces the GC base pair with its flat surface of the six-membered ring, positioning both hydroxy groups (O_2G and O_4G) in hydrogen bonding distance to N-7 of G_2 and N-4 of m^5C_{11}, respectively (Figure 4b). The dimethylamino group is not directly interacting with the DNA or its negatively charged phosphodiester backbone but is bridged through a water molecule to N-6 of A_{10}. Ring **A** of the aglycon adopts a similar conformation as reported for daunorubicin with C_9 dislocated from the mean plane by 0.53 Å, placing the hydroxy group in an equatorial position pointing away from the minor groove. The axial methyl ester at C-10 points in direction of the minor groove and interacts with N-2 of G12 through a hydrogen bond. The large hydrophobic nogalose inserts sideways in the minor groove and primarily interacts by Van der Waals contacts, thereby widening the groove by roughly 3 Å. Only one direct interaction was proposed between the glycosidic O-7 between nogalose and ring **A** and the hydrogen donating N-2 of G2. The distance between these two atoms (3.5 Å) and the lower electron density of the involved oxygen lone pair due to the stereoelectronic conjugation with the neighboring C–O bond, however, suggest that this interaction is only weak.

As seen previously for the calicheamicin antibiotics, the anthracyclins represent another family of DNA binding ligands carrying modified carbohydrates. To gain energetic benefits from binding to DNA, carbohydrates have to be less hydrophilic and/or contribute to the overall charge neutralization. This allows optimal interaction at the bottom and sidewalls of the lipophilic minor groove allowing excellent shape recognition through van der Waals contacts. In addition, the conformationally rigid six-membered ring is an excellent scaffold to place hydroxy or amino groups such that they can specifically interact with the functional groups of nucleotide bases on the floor of both grooves. Nature chose a variety of approaches to make sugar residues successful DNA or RNA ligands: often the carbohydrate is deoxygenated and some or all of the remaining hydroxy groups are methylated as seen in nogalose. The aminoglucose found in nogalamycin (**10**) is deoxygenated as well and, in addition, contains a dimethylamino functionality that energetically contributes to the interaction with DNA by charge neutralization. Both the remain-

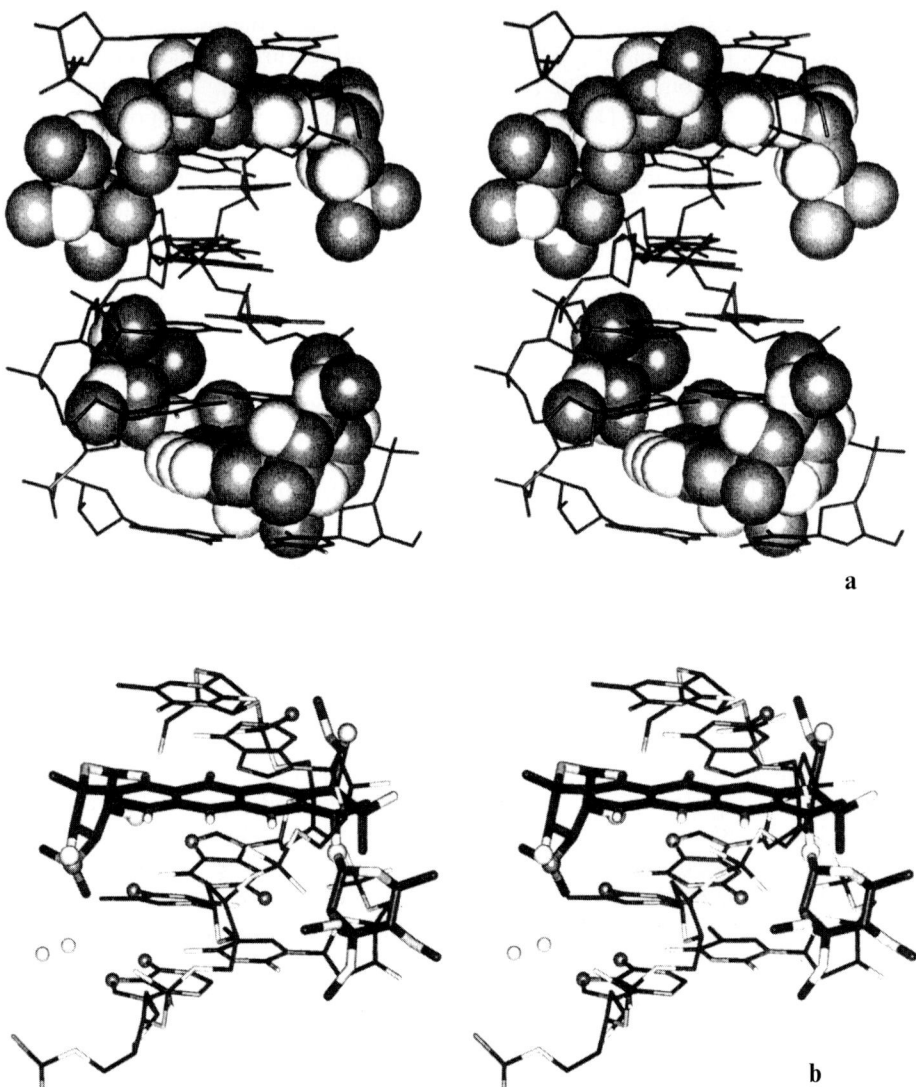

Figure 4. Stereo representation of the x-ray structure of nogalamycin (**9**) (Brookhaven: 1d21; [77]) bound to d[m^{5}CGT(pS)Am^{5}CG]$_2$. The hydrogen atoms are omitted to simplify the representation of the complex. **a**) 2 nogalamycin molecules are shown in space filling models intercalated in the modified duplex with water molecules removed, carbon atoms are shown in dark grey and oxygen/nitrogen atoms in light grey. **b**) Part of the same structure including selected interacting water molecules (small balls in light grey). Heteroatoms involved in hydrogen bonding are drawn as small balls. Carbon and phosphorous atoms are shown in black, oxygen in light grey and nitrogen in dark grey.

ing hydroxyl groups are perfectly positioned for direct interaction with hydrogen bond donor or acceptor sites at the binding site.

39.2.3 Pluramycins and Aureolic Acids

Pluramycins are closely related to nogalamycin (**10**): they consist of a central, tetracyclic chromophore with up to three substituents (altromycins) at different positions. Besides carbohydrates they contain epoxy functionalities that provide them with alkylating properties in the bound state. This class of antibiotics was recently reviewed [83] and will not be discussed further in this review.

Chromomycin A_3 (**11**) is an antitumor antibiotic belonging to the group of aureolic acids along with related analogs like mithramycin and olivomycin. Chromomycin is believed to bind DNA, thereby inhibiting RNA transcription and regulation [84, 85]. It was demonstrated, that it binds in the presence of Mg^{2+} to double-stranded DNA [86] with a sequence specificity to GC rich sites [87–90]. Even though the central, tricyclic aglycon indicated an intercalative binding mode, NMR structure elucidation revealed a unique binding mode of the magnesium bound dimer in the minor groove of DNA [91, 92]. The complex is bound symmetrically around the central $(GC)_2$ sequence of the target duplex d(AAGGCCTT)$_2$ (Figure 5, [91]) with its unligated hydroxy group of the aglycon directly interacting with the exocyclic amino group of the central guanosines. The binding of the dimer requires a widening of the minor groove to accommodate both chromophores and

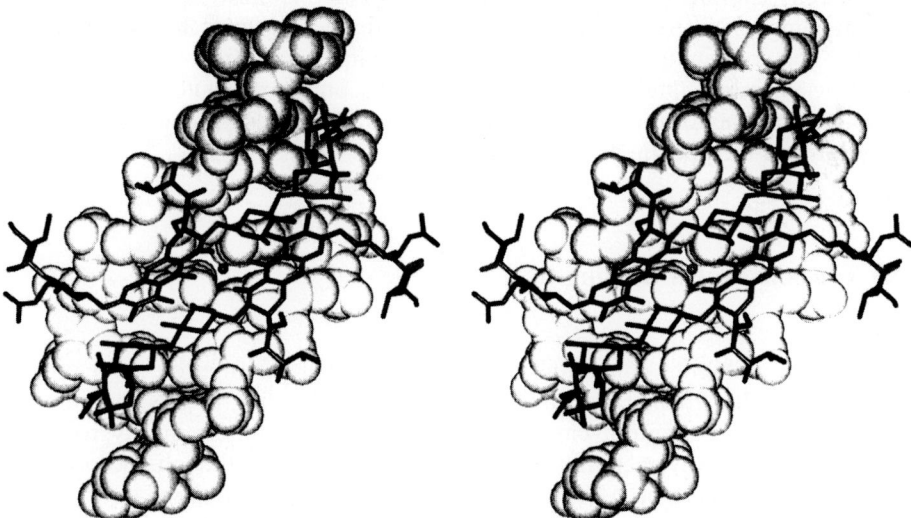

Figure 5. Stereo picture of a Mg^{2+} chelated chromomycin A_3 (**11**) dimer bound to d(AAGGCCTT)$_2$ [91] as retrieved from the Brookhaven database (1d83). The hydrogen atoms are omitted to simplify the representation of the complex. The DNA duplex is represented as a CPK model (grey) and both the chromomycin dimer and magnesium ion are shown in black.

sugar moieties *C* and *D* side by side in a parallel orientation to the walls of the groove. Carbohydrates *E* extend on both sides in the minor groove, facing the bottom of the groove. The trisaccharide *C-D-E* along with mainly with the sugar moiety *A*, primarily stabilize the complex by van der Waals intermolecular interactions within the minor groove of the target duplex. As seen with the previous antibiotics, the hydrophilic nature of all carbohydrate moieties has been altered by deoxygenation (*ABCDE*), methylation (*BE*) and acetylation (*AE*) to improve the overall binding characteristics. This complex reveals a different molecular recognition approach for DNA binding molecules that is related to the recent finding of alternate binding modes for distamycin A or netropsin related compounds [93, 94]. However, the metal ion plays a crucial role indicating new possibilities in the design of specific ligands that achieve higher molecular weight by chelating to physiologically present metal ions. A different but nevertheless related approach was described for protease inhibitors that bind the target enzyme as dimers chelated to Zn^{2+} [95].

In summary, the properly functionalized, six-membered carbohydrates can fit into the minor groove of DNA, depending on the groove width either facing the floor or squeezed between the phosphodiester backbone facing the walls. Besides changing the overall properties like solubility and pharmacokinetic behavior, these carbohydrates usually improve the overall binding affinity either in a selective or non-selective manner. As discussed previously, these carbohydrates are modified to increase their lipophilic properties and in some cases carry cationic functionalities to contribute to the overall charge neutralization. Such scaffolds seem to be ideally suited as DNA recognition elements due to size, conformational preorganization, and the high degree of potential functionalization. The synthetic difficulties, however, to fully explore this potential explain why there is currently only a limited understanding on energetic contributions of individual groups to overall DNA binding. The combination of combinatorial techniques with improvements in carbohydrate chemistry and biosynthesis might not only contribute to our understanding of interactions at the molecular level but could also provide compounds with improved therapeutic potential.

39.3 Carbohydrates Binding to RNA

Quite a few antibiotics containing modified carbohydrates interfere with protein biosynthesis rather than transcriptional processes by binding to DNA. Examples are erythromycin, streptomycin (**12**), tetracycline, spectinomycin, hygromycin, edeine and the neomycin family of aminoglycosides. Even though the observed biological effects have been correlated with perturbation of particular ribosomal events, the understanding of specific interactions on molecular level between such antibiotics and ribosomal components remained unknown for a long time. In analogy to enzyme inhibitors, it was assumed for these molecules to directly and specifically interact with functional sites of ribosomal components but until quite recently, no convincing evidence was found. The hypothesis that rRNA is the essential determinant of ribosome function lead to close examination of the interaction between a variety of antibiotics and rRNA by chemical footprinting [96, 97]. Over the past decade increasing information has been provided to support specific and direct interaction with rRNA but size and complexity of this target [98] rendered structural elucidation extremely difficult. Two high resolution structures were reported only recently [99, 100]. They provide for the first time structural insight in molecular recognition between aminoglycosides and RNA. Further information was obtained by attempts to optimize either the low molecular weight aminoglycosides or the target RNA using chemical synthesis or affinity selection and amplification approaches, respectively. In addition to ribosomal RNA, aminoglycosides have been shown to interact with various, biologically relevant RNA structures. Among those are catalytic group I introns [101, 102], hammerhead ribozymes [103–106], the rev-binding element of HIV [107–109], and Hepatitis delta virus ribozyme [110].

39.3.1. Aminoglycosides

Aminoglycosides are a class of aminosugar and aminocyclitol containing antibiotics that inhibit prokaryotic cell growth by interfering with ribosomal protein biosynthesis [111–114] (Scheme 4). They have been in clinical use for several decades and lack of oral bioavailability, toxicity and frequent occurrence of resistance limit their current use in humans. They are known to perturb translational events, causing a marked decrease in the fidelity of translation and/or inhibiting translocation. A variety of experimental findings led to an increased understanding of aminoglycoside function. Antibiotic resistance was not only found by enzymatic modifications of the aminoglycosides or mutations in ribosomal proteins but also by direct changes of the rRNA sequence [115], indicating points of direct contact. The small 30S ribosomal subunit of procaryotes consists of 16S ribosomal RNA (1540 nucleotides in length) and 21 proteins. Sequence analysis of this RNA from aminoglycoside resistant strains indicated functionally important areas for binding. Streptomycin resistance in *Escherichia coli* was originally shown to be governed by ribosomal protein S12. More recently, however, mutants have been found with a C to U transition at position 912 of 16S rRNA that was directly correlated with the observed aminoglycoside resistance [116, 117]. This finding indicated that streptomycin (**12**) directly binds to RNA rather than proteins. Experimental support for this assumption was reported from RNA footprinting studies of intact 70S ribosomes and sequence analysis of the 16S rRNA: chemical modification by dimethylsulfate was strongly suppressed in the presence of streptomycin (**12**) at adenosines 913–915 and weak but significant protection was noted for both, uracil 911 and cytosine 912. This protection or shielding from chemical modification surrounds and includes the same nucleotide that was mutated in a resistant strain thereby indicating the importance of streptomycin binding at this RNA site for antibiotic function. The neomycin family of aminoglycosides displays a different pattern of RNA protection and mutations associated with resistance. For example paromomycin resistance has been shown to result from single mutations in both

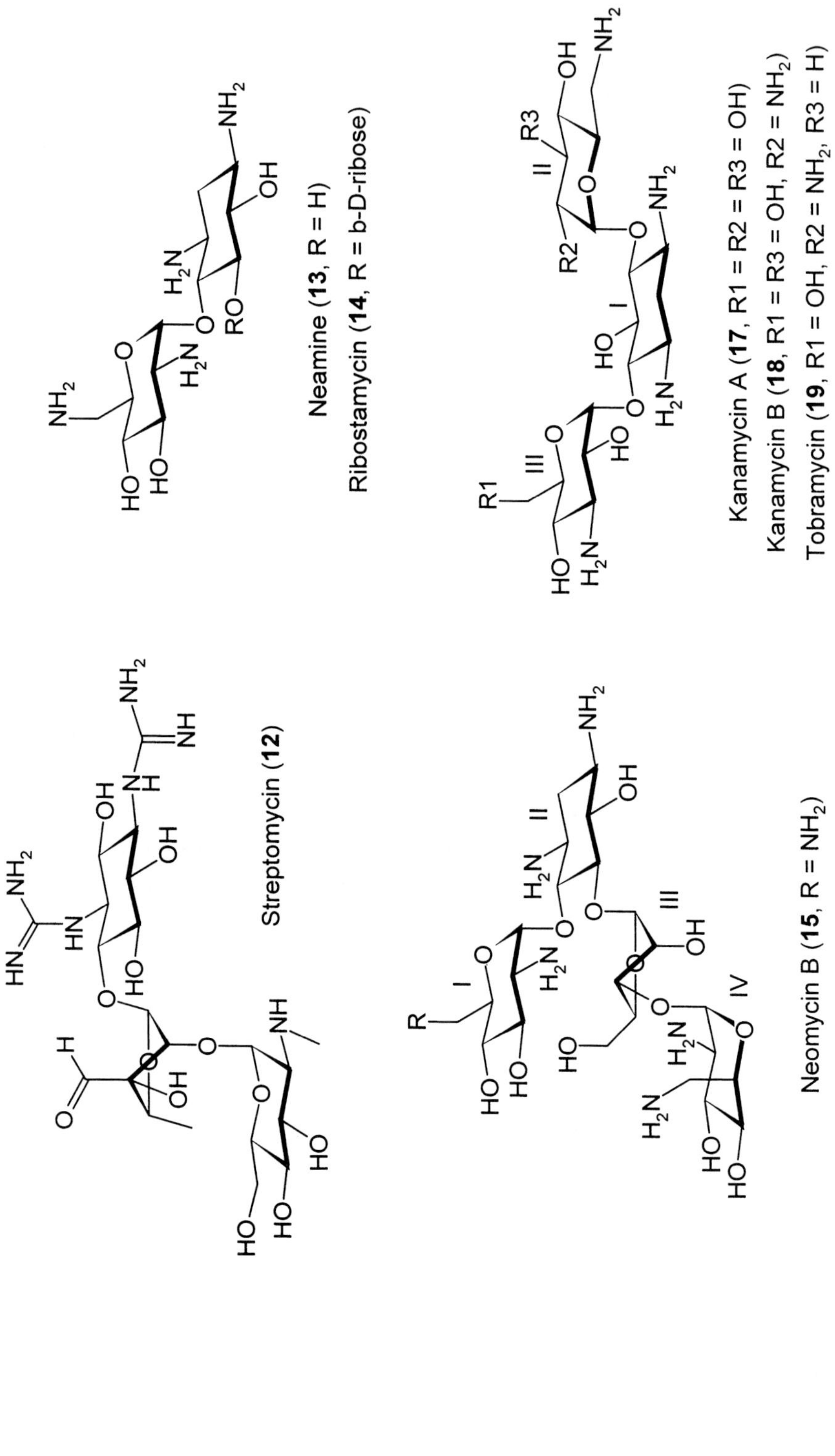

Neamine (**13**, R = H)
Ribostamycin (**14**, R = b-D-ribose)

Kanamycin A (**17**, R1 = R2 = R3 = OH)
Kanamycin B (**18**, R1 = R3 = OH, R2 = NH$_2$)
Tobramycin (**19**, R1 = OH, R2 = NH$_2$, R3 = H)

Streptomycin (**12**)

Neomycin B (**15**, R = NH$_2$)
Paromomycin (**16**, R = OH)

Scheme 4. Structures of selected aminoglycosides. For discussion refer to text.

yeast mitochondrial [118] and Tetrahymena rRNA [119, 120]. In accordance to the numbering of *E. coli* 16S ribosomal RNA, both identified mutations aim at the disruption of the base pair between positions 1409 and 1491 (Scheme 5) that is part of the *A*-site involved in the decoding by binding aminoacyl-tRNAs. As described for streptomycin (**12**), an excellent correlation was found with chemical footprinting pointing at the A-site as binding site for paromomycin (**16**): adenosine 1408 and guanosine 1494 are strongly protected from alkylation at N-1 and N-7, respectively. Similar behavior was detected for neomycin and gentamycin, whereas the protection for the smaller kanamycin is less complete at both nucleotides.

An additional clue is provided by antibiotic producing microorganisms in which the resistance is achieved by methylation of specific nucleotides on rRNA. The cloned methylases can be expressed in heterologous organisms like *E. coli* which they render resistant towards selected aminoglycosides by transforming either guanosine 1405 or adenosine 1408 in the *A*-site of 16S rRNA [121]. This site was mimicked by a self-structured oligoribonucleotide that bound both antibiotic and RNA ligands of the 30S subunit in a similar manner as compared to the normal subunit [122]. This oligoribonucleotide was further reduced in size to the 27-mer **21** (Scheme 5) that was shown to specifically bind paromomycin (**16**) with similar binding affinity and protection pattern as found for *E. coli* 16S rRNA in the 30S subunit [123, 124]. This complex turned out to be suitable for structure elucidation by NMR spectroscopy, providing insight in molecular interactions between aminoglycoside and target RNA [99, 125, 126] (Figure 6a).

Paromomycin (**16**) binds in the major groove of the *A*-site within a pocket that is formed by the asymmetric internal loop, thereby changing the local structure of the bound RNA [126]. Not bound to RNA, paromomycin (**16**) has in solution rather rigid ring conformations with certain flexibility at the glycosidic linkages connecting the four rings. In the bound form, however, paromomycin adopts a L-shaped conformation with rings **II**, **III** and **IV** linearly arranged. Both, rings **I** and **II** are conformationally highly conserved; they adopt chair conformations placing all the substituents at the 6-membered ring in equatorial positions with a number of direct contacts to the RNA (see Figure 6b). Ring **I** stacks on the purine ring of guanosine 1491 and its substituents interact with the phosphodiesters of guanosine 1491 and adenosines 1492 and 1493. Both hydroxy groups, 3'-OH and 4'-OH, are not essential for aminoglycoside function [127] but the replacement of the pro-R oxygens of the phosphates which point into the minor groove at positions 1492 and 1493 with sulfur interfere with paromomycin binding [128]. The 6'-OH group might possibly not only interact with the proximal phosphodiester of adenosine 1493 but also with its N-7 of the purine base. The only amino group at C-2' strongly forms an intramolecular hydrogen bond with ring **III** and might additionally interact with the phosphodiester of guanosine 1491. Both amino groups of ring **II** (2-deoxystreptamine) directly interact with N-7 of guanosine 1494 and O-4 of uracil 1495 and possibly with the phosphodiester between adenosines 1492 and 1493. The CG base pair between 1407 and 1494 is essential for aminoglycoside binding [123], and carboxyethylation at N-7 of guanosine 1494 interferes with paromomycin binding. These crucial interactions between paromomycin (**16**) and the A-site model oligoribonucleotide **21** have been largely confirmed through an affinity chromatography

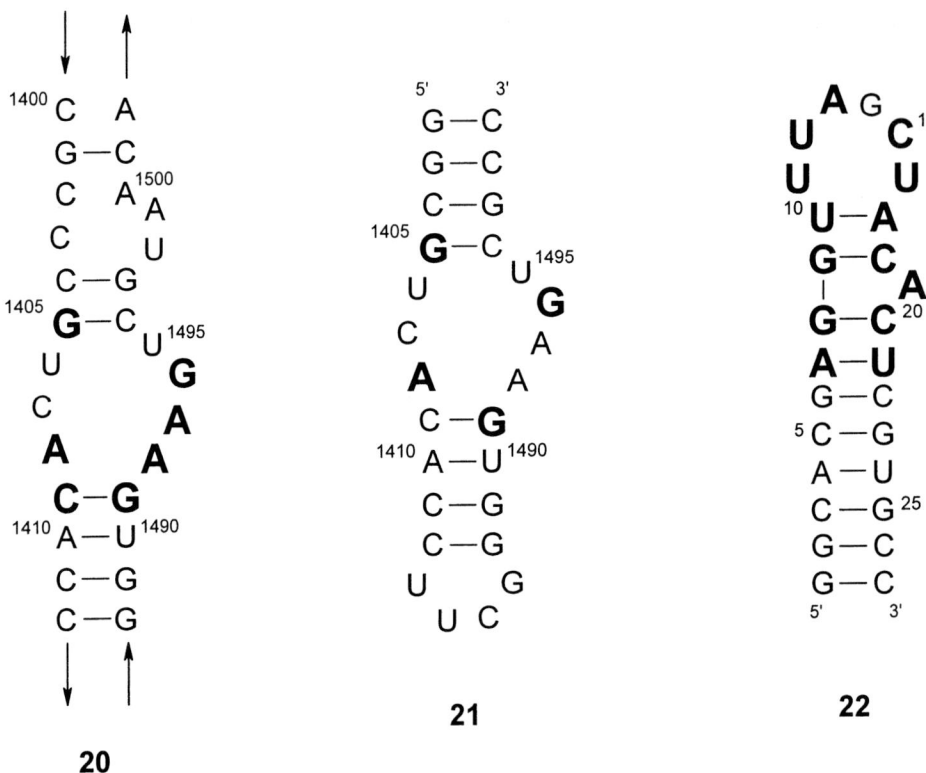

Scheme 5. Secondary structure on *Escherichia coli* 16S RNA in the region of decoding A-site **20** and its shortened hairpin version (27-mer model oligoribonucleotide **21**) used for the structure determination with bound paromomycin (**16**) [125]. Nucleotides which are protected from chemical modification by bound aminoglycosides, methylated in resistant strains or essential for aminoglycoside binding are shown in bold. For more detailed discussion refer to text. The sequence of the 27-mer RNA aptamer **22** with its secondary structure is shown. The consensus sequence identified for this family of aptamers is highlighted in bold [141].

based assay looking at binding interference between immobilized paromomycin (**16**) and various, base or backbone modified RNAs [128].

Rings **III** and **IV** of paromomycin (**16**) extend into the minor groove and in particular ring **IV** contributes through positive charges to the overall binding affinity without adding much to the specificity. This is supported by the fact that both, ribostamycin (**14**) and neamine (**13**), lacking ring **IV** and rings **III** and **IV**, respectively, bind to 16S rRNA and cause miscoding, albeit at higher concentration. Measured dissociation constants by surface plasmon resonance [108] are 19 nM, 25 µM, and 7.8 µM for neomycin B (**15**), ribostamycin (**14**), and neamine (**13**), respectively (paromomycin (**16**): 0.20 µM) [129]. In addition, ring **IV** is partially disordered in the ensemble of 20 NMR structures providing the least confidence for conformational determination. Exchanging ring **IV** (2,6-dideoxy-2,6-diamino-β-L-

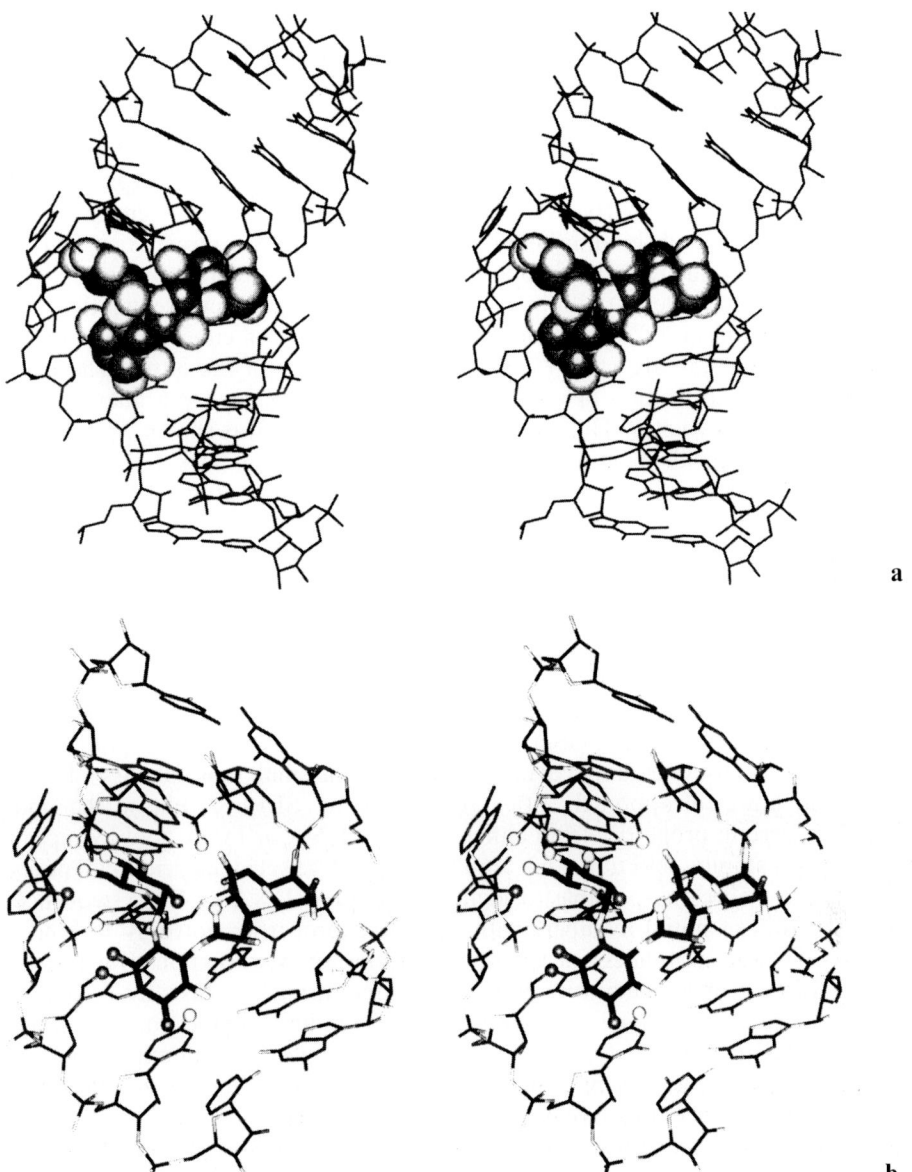

Figure 6. Minimized average stereo representation of paromomycin (**16**) binding to the model 27-mer oligoribonucleotide **21** (Scheme 5) [99, 125] as retrieved from the Brookhaven database (1pbr). The hydrogen atoms are omitted to simplify the representation of the complex. a) Complex with DNA in black and paromomycin (**16**) as a CPK model with carbon atoms shown in dark grey and oxygen/nitrogen atoms in light grey. b) Part of the same structure with the non-interacting base pairs and the loop removed. Carbon and phosphorous atoms are shown in black, oxygen in light grey and nitrogen in dark grey. Heteroatoms potentially involved in hydrogen bonding are drawn as small balls.

idopyranose) by either simple amines or deaminated idoses revealed the importance of charges for high affinity binding even though flexible amines with identical charge were not fully able to substitute for the hydroxy group bearing and preorganized ring **IV** [130]. A library approach was chosen by Wong and co-workers to identify new A-site binding molecules based on the modification of ring **I** of neomycin B (**15**), containing a core 1,3-hydroxyamine motif [131, 132]. Despite various binding affinities (between 10 µM and >500 µM), no molecules were identified containing either a high binding affinity or selectivity for the target A-site model RNA, presumably due to lack of sufficient conformational preorganization.

These aminoglycoside antibiotics of the neomycin family act preferentially on prokaryotic organisms. Comparison between ribosomes of either eukaryotic or prokaryotic origin reveals a 10–15-fold higher sensitivity of the latter. The main difference is the substitution of adenosine 1408 in prokaryotes for a guanosine at this position in eukaryotes. The adenosine pair 1408–1491 was shown to be essential for aminoglycoside binding, causing a specific binding pocket that accommodates ring **I**. As mentioned before, the base pair between positions 1409 and 1491 is essential for aminoglycoside binding as it forms the floor of the antibiotic binding pocket. Disruption of this base pair in prokaryotes leads to aminoglycoside resistance. Higher eukaryotic organisms, including humans, have both disruptions in their cytoplasmic rRNA sequences and as a consequence these aminoglycosides cannot bind the ribosomal target with high affinity.

The perfect optimization of a RNA binding ligand is tedious and as experience demonstrates is not that straightforward. Independently, the groups of Gold [133], Joyce [134], and Szostak [135] described a new technique that allowed optimization of the RNA (or DNA) binding partner rather than the small molecule ligand. This elegant and powerful approach, often referred to as 'SELEX', allows the optimization of certain properties from either a diverse RNA or DNA pool by repeating cycles of affinity selection and amplification of selected nucleic acid pools. Within weeks it became possible to identify optimized DNA or RNA sequences from 10^{14}–10^{16} of initial members with optimized properties not only to bind a variety of different molecules but also for catalyzing specific reactions (for reviews see [136–138]). Under conditions of high selection stringency, usually conserved RNA sequences are identified with low nanomolar binding affinities for their target molecule. This in vitro selection process has been applied to optimize RNA sequences for binding neomycin [139], kanamycin [140], and tobramycin (**19**) [141]. For the latter target aminoglycoside, on particular clone (named J6) was isolated coding for a 60-mer oligoribonucleotide insert with a dissociation constant for tobramycin (**19**) of 0.77 nM [142]. Key structural component of this RNA was shown to be a small stem loop that was incorporated in a 27-mer oligoribonucleotide **22** subjected to structure elucidation by NMR in the tobramycin (**19**) bound form (Figure 7a, [100]). This structure is of particular interest as the affinity of tobramycin (**19**) to this RNA aptamer is roughly three orders of magnitude greater as compared to the ribosomal RNA target site and structural insight should provide the answer of how tighter binding to RNA can be achieved by aminoglycosides. Unfortunately, the resolution of the reported structure is not sufficient to locate specific intermolecular interactions at atomic level with possibly the exception of a

39.3 Carbohydrates Binding to RNA   1119

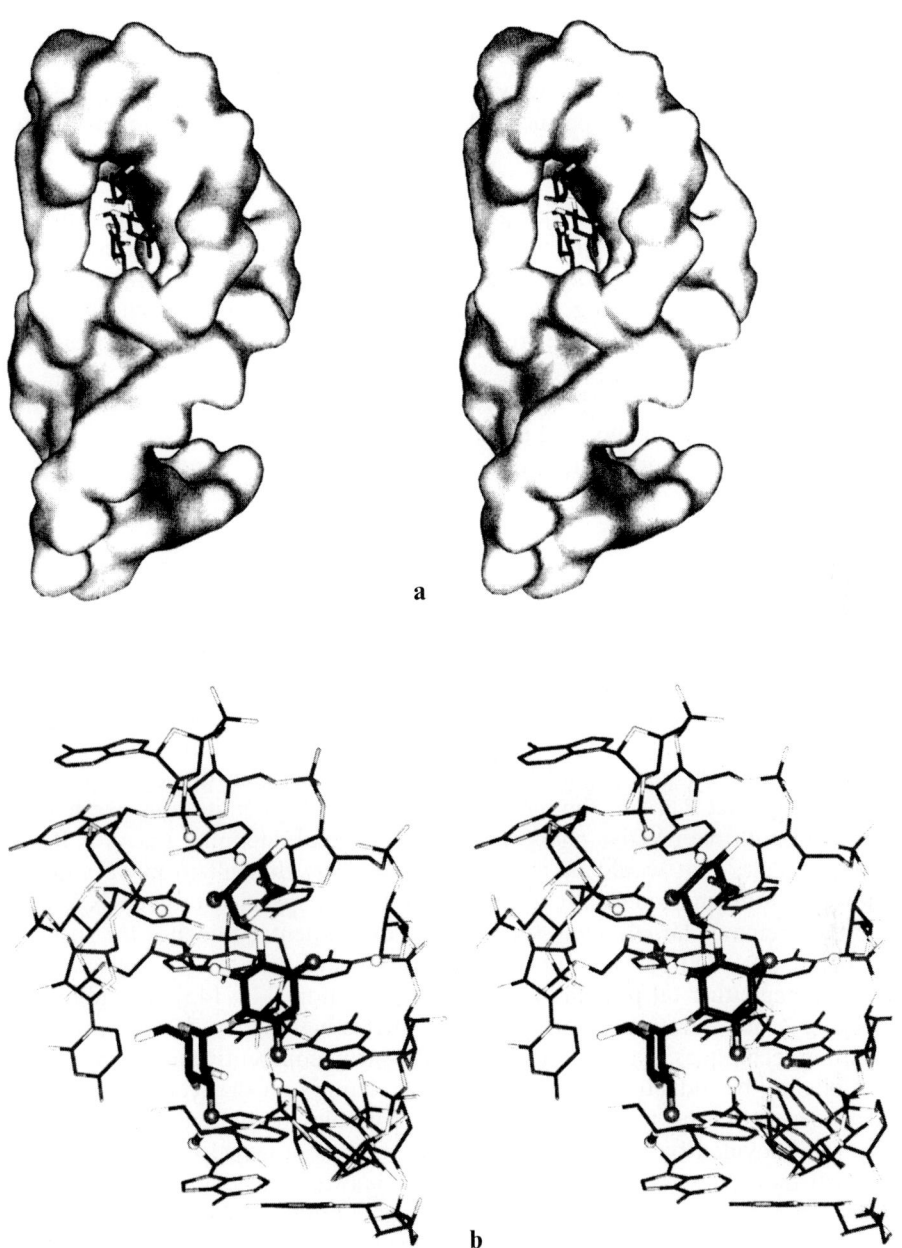

Figure 7. Structure of tobramycin (**19**) bound to RNA apatmer **22** [100]. From the co-ordinates of the 7 structures deposited in the Brookhaven database (1tob), only the last was chosen to visualize interactions. **a)** Representation of the whole complex with the RNA displayed in surface view. **b)** Stick model of the actual binding sites with direct interactions by potential hydrogen bonds highlighted. Carbon and phosphorous atoms are shown in black, oxygen in light grey and nitrogen in dark grey.

hydrogen bond between one of the amino groups at ring **I** and N-7 of guanosine 9 (Figure 7b). However, tobramycin (**19**) is well resolved with all three rings adopting a chair conformation connected by well-defined torsion angles. Some of the key features responsible for this specific and tight interaction can be summarized as follows:

i) the increased width of the major groove at the binding site is governed by the mismatch between uracils 11 and 16 as part of a well defined and conserved loop that is critical for tobramycin binding;
ii) cytosine 15 that is flapping over ring **III** of the bound tobramycin (**19**) and contributes to the large surface area (52%) that is covered by the RNA ligand in the complex;
iii) the defined floor of the major groove that is formed by the edges of base pairs interacting with the ligand; and finally
iv) the overall shape complementarity between tobramycin (**19**) and RNA aptamer **22**.

The work discussed above clearly indicates the importance of RNA as molecular target for small molecules to specifically (or non-specifically) interfere with crucial biological functions, an aspect that was recently reviewed by Michael and Tor [143]. A standard RNA duplex or stem with its wide, shallow minor groove and deep, narrow major groove is not ideally shaped to accommodate a low molecular weight compound. Distortions like bulges or internal loops often create the necessary binding pockets required for the binding event. The modified carbohydrates discussed here greatly helped to understand interactions between this class of molecules and structured RNA at the molecular level. Based on limitations like hydrophilicity and multiple positive charges, however, general designing rules for RNA binding molecules that would be of therapeutic use are difficult to make and the question if neutral molecules might specifically bind a specified target RNA with high affinity remains largely unanswered. First experimental evidence for such an approach are provided by Mei and Czarnik with the identification of neutral molecules disrupting either tat protein–TAR RNA interaction [144, 145] or inhibiting a self-splicing group I intron ribozyme [144]. Interestingly, they demonstrated by footprinting techniques that binding is not necessarily competitive and can disrupt protein binding by induction of conformational changes on the RNA level. Additional evidence for the feasibility of neutral molecules to bind RNA with high affinity was provided by the identification of aptamers designed to bind a variety of different molecules. This approach, however, is extremely powerful as 'perfect' sequences can be selected from large pools containing roughly 10^{15} individual members and it might prove to be more difficult designing small molecules with drug-like properties for a given RNA target of biological relevance.

References

1. R.C. Hawley, L.L. Kiessling, S.L. Schreiber. *Proc. Natl Acad. Sci. U.S.A.* **1989**, 86, 1105–1109.

2. D. Kahne. *Chem. Biol.* **1995**, *2*, 7–12.
3. J. Hunziker. *Chimia* **1996**, *50*, 248–256.
4. P. Sears, C.H. Wong. *Proc. Natl Acad. Sci. U.S.A.* **1996**, *93*, 12086–12093.
5. P. Sears, C.H. Wong. *Chem. Commun.* **1998**, 1161–1170.
6. M.D. Lee, G.A. Ellestad, D.B. Borders. *Acc. Chem. Res.* **1991**, *24*, 235–243.
7. R.A. Kumar, N. Ikemoto, D.J. Patel. *J. Mol. Biol.* **1997**, *265*, 187–201.
8. R.A. Kumar, N. Ikemoto, D.J. Patel. *J. Mol. Biol.* **1998**, *265*, 173–186.
9. J. Golik, J. Clardy, G. Dubay, G. Groenewold, H. Kawaguchi, M. Konishi, B. Krishnan, H. Ohkuma, K. Saitoh, T.W. Doyle. *J. Am. Chem. Soc.* **1987**, *109*, 3461–3462.
10. J. Golik, G. Dubay, G. Groenewold, H. Kawaguchi, M. Konishi, B. Krishnan, H. Ohkuma, K. Saitoh, T.W. Doyle. *J. Am. Chem. Soc.* **1987**, *109*, 3462–3464.
11. M.D. Lee, T.S. Dunne, M.M. Siegel, C.C. Chang, G.O. Morton, D.B. Borders. *J. Am. Chem. Soc.* **1987**, *109*, 3464–3466.
12. M.D. Lee, T.S. Dunne, C.C. Chang, G.A. Ellestad, M.M. Siegel, G.O. Morton, W.J. McGahren, D.B. Borders. *J. Am. Chem. Soc.* **1987**, *109*, 3466–3468.
13. S.L. Schreiber, L.L. Kiessling. *J. Am. Chem. Soc.* **1988**, *110*, 631–633.
14. S.J. Danishefsky, N.B. Mantlo, D.S. Yamashita, G. Schulte. *J. Am. Chem. Soc.* **1988**, *110*, 6890–6891.
15. W.L. Jorgensen. *Chemtracts: Org. Chem.* **1988**, *1*, 485–487.
16. A.S. Kende, C.A. Smith. *Tetrahedron Lett.* **1988**, *29*, 4217–4220.
17. K.C. Nicolaou, Y. Ogawa, G. Zuccarello, E.J. Schweiger, T. Kumazawa. *J. Am. Chem. Soc.* **1988**, *110*, 4866–4868.
18. P. Magnus, P.A. Carter. *J. Am. Chem. Soc.* **1988**, *110*, 1626–1628.
19. K. Tomioka, H. Fujita, K. Koga. *Tetrahedron Lett.* **1989**, *30*, 851–854.
20. S.M. Kerwin, C.H. Heathcock. *Chemtracts: Org. Chem.* **1989**, *2*, 21–23.
21. B.H. Long, J. Golik, S. Forenza, B. Ward, R. Rehfuss, J.C. Dabrowiak, J.J. Catino, S.T. Musial, K.W. Brookshire, T.W. Doyle. *Proc. Natl Acad. Sci. U.S.A.* **1989**, *86*, 2–6.
22. Y. Sugiura, Y. Uesawa, Y. Takahashi, J. Kuwahara, J. Golik, T.W. Doyle. *Proc. Natl Acad. Sci. U.S.A.* **1989**, *86*, 7672–7676.
23. M. Uesugi, Y. Sugiura. *Biochemistry* **1993**, *32*, 4622–4627.
24. L. Yu, J. Golik, R. Harrison, P. Dedon. *J. Am. Chem. Soc.* **1994**, *116*, 9733–9738.
25. N. Ikemoto, R.A. Kumar, P.C. Dedon, S.J. Danishefsky, D.J. Patel. *J. Am. Chem. Soc.* **1994**, *116*, 9387–9388.
26. J. Drak, N. Iwasawa, S. Danishefsky, D.M. Crothers. *Proc. Natl Acad. Sci. U.S.A.* **1991**, *88*, 7464–7468.
27. J. Aiyar, S.J. Danishefsky, D.M. Crothers. *J. Am. Chem. Soc.* **1992**, *114*, 7552–7554.
28. K.C. Nicolaou, S.C. Tsay, T. Suzuki, G.F. Joyce. *J. Am. Chem. Soc.* **1992**, *114*, 7555–7557.
29. N. Zein, A.M. Sinha, W.J. McGahren, G.A. Ellestad. *Science (Washington, D.C.)* **1988**, *240*, 1198–1201.
30. N. Zein, M. Poncin, R. Nilakantan, G.A. Ellestad. *Science (Washington, D.C.)* **1989**, *244*, 697–699.
31. S. Walker, J. Murnick, D. Kahne. *J. Am. Chem. Soc.* **1993**, *115*, 7954–7961.
32. S.L. Walker, A.H. Andreotti, D.E. Kahne. *Tetrahedron* **1994**, *50*, 1351–1360.
33. L.G. Paloma, J.A. Smith, W.J. Chazin, K.C. Nicolaou. *J. Am. Chem. Soc.* **1994**, *116*, 3697–3708.
34. G. Bifulco, A. Galeone, L. Gomez-Paloma, K.C. Nicolaou, W.J. Chazin. *J. Am. Chem. Soc.* **1996**, *118*, 8817–8824.
35. S. Walker, K.G. Valentine, D. Kahne. *J. Am. Chem. Soc.* **1990**, *112*, 6428–6429.
36. S. Walker, D. Yang, D. Kahne, D. Gange. *J. Am. Chem. Soc.* **1991**, *113*, 4716–4717.
37. S. Walker, V. Gupta, D. Kahne, D. Gange. *J. Am. Chem. Soc.* **1994**, *116*, 3197–3206.
38. T. Li, Z. Zeng, V.A. Estevez, K.U. Baldenius, K.C. Nicolaou, G.F. Joyce. *J. Am. Chem. Soc.* **1994**, *116*, 3709–3715.
39. S.N. Ho, S.H. Boyer, S.L. Schreiber, S.J. Danishefsky, G.R. Crabtree. *Proc. Natl Acad. Sci. U.S.A.* **1994**, *91*, 9203–9207.
40. K.C. Nicolaou, K. Ajito, H. Komatsu, B.M. Smith, T. Li, M.G. Egan, L. Gomez-Paloma. *Angew. Chem., Int. Ed. Engl.* **1995**, *34*, 576–578.

41. K.C. Nicolaou, K. Ajito, H. Komatsu, B.M. Smith, P. Bertinato, L. Gomez-Paloma. *Chem. Commun.* **1996**, 1495–1496.
42. K.C. Nicolaou, B.M. Smith, J. Pastor, Y. Watanabe, D.S. Weinstein. *Synletters* **1997**, 401–410.
43. K.C. Nicolaou, B.M. Smith, K. Ajito, H. Komatsu, L. Gomez-Paloma, Y. Tor. *J. Am. Chem. Soc.* **1996**, *118*, 2303–2304.
44. G. Bifulco, A. Galeone, K.C. Nicolaou, W.J. Chazin, L. Gomezpaloma. *J. Am. Chem. Soc.* **1998**, *120*, 7183–7191.
45. C. Liu, B.M. Smith, K. Ajito, H. Komatsu, L. Gomez-Paloma, T. Li, E.A. Theodorakis, K.C. Nicolaou, P.K. Vogt. *Proc. Natl Acad. Sci. U.S.A.* **1996**, *93*, 940–944.
46. S.B. Shuker, P.J. Hajduk, R.P. Meadows, S.W. Fesik. *Science (Washington, D.C.)* **1996**, *274*, 1531–1534.
47. P.J. Hajduk, et al. *J. Am. Chem. Soc.* **1997**, *119*, 5818–5827.
48. P.J. Hajduk, R.P. Meadows, S.W. Fesik. *Science (Washington, D.C.)* **1997**, *278*, 497–499.
49. J.W. Lown. *Chem. Soc. Rev.* **1993**, *22*, 165–176.
50. W.R. Strohl, M.L. Dickens, V.B. Rajgarhia, A.J. Woo, N.D. Priestley. *Drugs Pharm. Sci.* **1997**, *82*, 577–657.
51. F.M. Arcamone. *Biochimie* **1998**, *80*, 201–206.
52. A.H. Wang. *Adv. DNA Sequence Specific Agents* **1996**, *2*, 59–100.
53. F. Arcamone, G. Franceschi, P. Orezzi, S. Penco, R. Mondelli. *Tetrahedron Lett.* **1968**, 3349–3352.
54. R.H. Iwamoto, P. Lim, N.S. Bhacca. *Tetrahedron Lett.* **1968**, 3891–3894.
55. F. Arcamone, G. Franceschi, S. Penco, A. Selva. *Tetrahedron Lett.* **1969**, 1007–1010.
56. P.B. Jensen, B.S. Soerensen, M. Sephested, E.J. Demant, E. Kjeldsen, E. Friche, H.H. Hansen. *Biochem. Pharmacol.* **1993**, *45*, 2025–2035.
57. G.J. Quigley, A.H. Wang, G. Ughetto, G. van der Marel, J.H. van Boom, A. Rich. *Proc. Natl Acad. Sci. U.S.A.* **1980**, *77*, 7204–7208.
58. A.H. Wang, G. Ughetto, G.J. Quigley, A. Rich. *Biochemistry* **1987**, *26*, 1152–1163.
59. M.H. Moore, W.N. Hunter, D. Langlois, O. Kennard. *J. Mol. Biol.* **1989**, *206*, 693–705.
60. G.A. Leonard, T.W. Hambley, K. McAuley-Hecht, T. Brown, W.N. Hunter. *Acta Crystallogr., Sect. D: Biol. Crystallogr.* **1993**, *D49*, 458–467.
61. Y. Gao, H. Robinson, E.R. Wijsman, G. van der Marel, J.H. van Boom, A.H. Wang. *J. Am. Chem. Soc.* **1997**, *119*, 1496–1497.
62. S. Neidle, G. Taylor. *Biochim. Biophys. Acta* **1977**, *479*, 450–459.
63. J.B. Chaires, J.E. Herrera, M.J. Waring. *Biochemistry* **1990**, *29*, 6145–6153.
64. C.J. Shelton, M.M. Harding, A.S. Prakash. *Biochemistry* **1996**, *35*, 7974–7982.
65. K.R. Fox. *Anti-Cancer Drug Des.* **1988**, *3*, 157–168.
66. S. Kunimoto, Y. Takahashi, T. Uchidsa, T. Takeuchi, H. Umezawa. *J. Antibiot.* **1988**, *41*, 655–659.
67. J.B. Chaires, F. Leng, T. Przewloka, I. Fokt, Y.H. Ling, R. Perez-Soler, W. Priebe. *J. Med. Chem.* **1997**, *40*, 261–266.
68. G.G. Hu, X. Shui, F. Leng, W. Priebe, J.B. Chaires, L.D. Williams. *Biochemistry* **1997**, *36*, 5940–5946.
69. H. Robinson, W. Priebe, J.B. Chaires, A.H. Wang. *Biochemistry* **1997**, *36*, 8663–8670.
70. K.M. Depew, S.M. Zeman, S.H. Boyer, D.J. Denhart, N. Ikemoto, S.J. Danishefsky, D.M. Crothers. *Angew. Chem., Int. Ed. Engl.* **1997**, *35*, 2797–2800.
71. R.P. Hertzberg, P.B. Dervan. *Biochemistry* **1984**, *23*, 3934–3945.
72. P.F. Wiley, F.A. MacKellar, E.L. Caron, R.B. Kelly. *Tetrahedron Lett.* **1968**, 663–668.
73. M.S. Searle, J.G. Hall, W.A. Denny, L.P. Wakelin. *Biochemistry* **1988**, *27*, 4340–4349.
74. Y.C. Liaw, Y.G. Gao, H. Robinson, G. van der Marel, J.H. van Boom, A.H. Wang. *Biochemistry* **1989**, *28*, 9913–9918.
75. X. Zhang, D.J. Patel. *Biochemistry* **1990**, *29*, 9451–9466.
76. L.D. Williams, M. Egli, Q. Gao, P. Bash, G. van der Marel, J.H. van Boom, A. Rich, C.A. Frederick. *Proc. Natl Acad. Sci. U.S.A.* **1990**, *87*, 2225–2229.
77. Y.G. Gao, Y.C. Liaw, H. Robinson, A.H. Wang. *Biochemistry* **1990**, *29*, 10307–10316.
78. M. Egli, L.D. Williams, C.A. Frederick, A. Rich. *Biochemistry* **1991**, *30*, 1364–1372.

79. H. Robinson, D. Yang, A.H. Wang. *Gene* **1994**, *149*, 179–188.
80. C.K. Smith, G.J. Davies, E.J. Dodson, M.H. Moore. *Biochemistry* **1995**, *34*, 415–425.
81. C.K. Smith, J.A. Brannigan, M.H. Moore. *J. Mol. Biol.* **1996**, *263*, 237–258.
82. S.K. Arora. *J. Am. Chem. Soc.* **1983**, *105*, 1328–1332.
83. M.R. Hansen, L.H. Hurley. *Acc. Chem. Res.* **1996**, *29*, 249–258.
84. W. Kersten, H. Kersten, F.E. Steiner, B. Emmerich. *Z. Physiol. Chem.* **1967**, *348*, 1415–1423.
85. W.E. Mueller. *Pharmacol. Ther., Part A* **1977**, *1*, 457–474.
86. Ward, D. C., Reich, E., and Goldberg, Irving H. *Science (Washington, D.C.)* **1965**, 1259–1263.
87. M.W. Van Dyke, P.B. Dervan. *Biochemistry* **1983**, *22*, 2373–2377.
88. K.R. Fox, N.R. Howarth. *Nucleic Acids Res.* **1985**, *13*, 8695–8714.
89. B.M. Cons, K.R. Fox. *Nucleic Acids Res.* **1989**, *17*, 5447–5459.
90. A. Stankus, J. Goodisman, J.C. Dabrowiak. *Biochemistry* **1992**, *31*, 9310–9318.
91. X. Gao, P. Mirau, D.J. Patel. *J. Mol. Biol.* **1992**, *223*, 259–279.
92. D.L. Banville, M.A. Keniry, M. Kam, R.H. Shafer. *Biochemistry* **1990**, *29*, 6521–6534.
93. J.G. Pelton, D.E. Wemmer. *Proc. Natl Acad. Sci. U.S.A.* **1989**, *86*, 5723–5727.
94. C.L. Kielkopf, S. White, J.W. Sewczyk, J.M. Turner, E.E. Baird, P.B. Dervan, D.C. Rees. *Science (Washington, D.C.)* **1998**, *282*, 111–115.
95. K.D. Rice, R.D. Tanaka, B.A. Katz, R.P. Numerof, W.R. Moore. *Curr. Pharm. Des.* **1998**, *4*, 381–396.
96. D. Moazed, H.F. Noller. *Nature (London)* **1987**, *327*, 389–394.
97. J. Woodcock, D. Moazed, M. Cannon, J. Davies, H.F. Noller. *EMBO J.* **1991**, *10*, 3099–3103.
98. E.V. Puglisi, J.D. Puglisi. *Mod. Cell Biol.* **1997**, *17*, 1–21.
99. D. Fourmy, M.I. Recht, S.C. Blanchard, J.D. Puglisi. *Science (Washington, D.C.)* **1996**, *274*, 1367–1371.
100. L. Jiang, A.K. Suri, R. Fiala, D.J. Patel. *Chem. Biol.* **1997**, *4*, 35–50.
101. U. von Ahsen, J. Davies, R. Schroeder. *J. Mol. Biol.* **1992**, *226*, 935–941.
102. U. von Ahsen, H.F. Noller. *Science (Washington, D.C.)* **1993**, *260*, 1500–1503.
103. T.K. Stage, K.J. Hertel, O.C. Uhlenbeck. *RNA* **1995**, *1*, 95–101.
104. H. Wang, Y. Tor. *Angew. Chem., Int. Ed.* **1998**, *37*, 109–111.
105. H. Wang, Y. Tor. *J. Am. Chem. Soc.* **1997**, *119*, 8734–8735.
106. T. Hermann, E. Westhof. *J. Mol. Biol.* **1998**, *276*, 903–912.
107. M.L. Zapp, S. Stern, M.R. Green. *Cell (Cambridge, Mass.)* **1993**, *74*, 969–978.
108. M. Hendrix, E.S. Priestley, G.F. Joyce, C.H. Wong. *J. Am. Chem. Soc.* **1997**, *119*, 3641–3648.
109. S.H. Wang, P.W. Huber, M. Cui, A.W. Czarnik, H.Y. Mei. *Biochemistry* **1998**, 37, 5549–5557.
110. J.S. Chia, H.L. Wu, H.W. Wang, D.S. Chen, P.J. Chen. *J. Biomed. Sci. (Basel)* **1997**, *4*, 208–216.
111. Puglisi, Joseph D. *'Structural basis for aminoglycoside antibiotic action'* in Many Faces RNA, (8th SmithKline Beecham Pharm. Res. Symp.), ed. by D.S. Eggleston, Ed. Academic Press, San Diego, CA **1998**, p. 97.
112. Schroeder, U. von Ahsen. *Nucleic Acids Mol. Biol.* **1996**, *10*, 53–74.
113. Wank, Herbert and Schroeder, Renee. *'Antibiotics that interfere with RNA/RNA interactions'* in Ribosomal RNA Group I Introns, ed. by R. Green and R. Schroeder, Eds. Landes, Austin, TX **1996**, p. 129.
114. C. Bochaton, S. Rochegude, R. Roubille. *Lyon Pharm.* **1997**, *48*, 226–239.
115. P. Chakrabarti. *Proc. Natl Acad. Sci., India, Sect. B* **1997**, *67*, 169–179.
116. P.E. Montandon, R. Wagner, E. Stutz. *EMBO J.* **1986**, *5*, 3705–3708.
117. P.E. Montandon, P. Nicolas, P. Schuermann, E. Stutz. *Nucleic Acids Res.* **1985**, *13*, 4299–4310.
118. M. Li, A. Tzagoloff, K. Underbrink-Lyon, N.C. Martin. *J. Biol. Chem.* **1982**, *257*, 5921–5928.
119. E.A. Spangler, E.H. Blackburn. *J. Biol. Chem.* **1985**, *260*, 6334–6340.
120. M. Li, A. Tzagoloff, K. Underbrink-Lyon, N.C. Martin. *J. Biol. Chem.* **1982**, *257*, 5921–5928.
121. A.A. Beauclerk, E. Cundliffe. *J. Mol. Biol.* **1987**, *193*, 661–671.
122. P. Purohit, S. Stern. *Nature (London)* **1994**, *370*, 659–662.

123. M.I. Recht, D. Fourmy, S.C. Blanchard, K.D. Dahlquist, J.D. Puglisi. *J. Mol. Biol.* **1996**, *262*, 421–436.
124. H. Miyaguchi, H. Narita, K. Sakamoto, S. Yokoyama. *Nucleic Acids Res.* **1996**, *24*, 3700–3706.
125. D. Fourmy, M.I. Recht, J.D. Puglisi. *J. Mol. Biol.* **1998**, *277*, 347–362.
126. D. Fourmy, S. Yoshizawa, J.D. Puglisi. *J. Mol. Biol.* **1998**, *277*, 333–345.
127. R. Benveniste, J. Davies. *Antimicrob. Agents Chemother.* **1973**, *4*, 402–409.
128. S.C. Blanchard, D. Fourmy, R.G. Eason, J.D. Puglisi. *Biochemistry* **1998**, *37*, 7716–7724.
129. C.H. Wong, M. Hendrix, E.S. Priestley, W.A. Greenberg. *Chem. Biol.* **1998**, 5, 397–406.
130. P.B. Alper, M. Hendrix, P. Sears, C.H. Wong. *J. Am. Chem. Soc.* **1998**, *120*, 1965–1978.
131. M. Hendrix, P.B. Alper, E.S. Priestley, C.H. Wong. *Angew. Chem., Int. Ed. Engl.* **1997**, *36*, 95–98.
132. C.H. Wong, M. Hendrix, D.D. Manning, C. Rosenbohm, W.A. Greenberg. *J. Am. Chem. Soc.* **1998**, *120*, 8319–8327.
133. C. Tuerk, L. Gold. *Science (Washington, D.C.)* **1990**, *249*, 505–510.
134. D.L. Robertson, G.F. Joyce. *Nature (London)* **1990**, *344*, 467–468.
135. A.D. Ellington, J.W. Szostak. *Nature (London)* **1990**, *346*, 818–822.
136. G.F. Joyce. *Curr. Opin. Struct. Biol.* **1994**, *4*, 331–336.
137. L. Gold, B. Polisky, O. Uhlenbeck, M. Yarus. *Annu. Rev. Biochem.* **1995**, *64*, 763–797.
138. J.R. Lorsch, J.W. Szostak. *Acc. Chem. Res.* **1996**, *29*, 103–110.
139. M.G. Wallis, U. von Ahsen, R. Schroeder, M. Famulok. *Chem. Biol.* **1995**, *2*, 543–552.
140. S.M. Lato, A.R. Boles, A.D. Ellington. *Chem. Biol.* **1995**, *2*, 291–303.
141. Y. Wang, R.R. Rando. *Chem. Biol.* **1995**, *2*, 281–290.
142. Y. Wang, J. Killian, K. Hamasaki, R.R. Rando. *Biochemistry* **1996**, *35*, 12338–12346.
143. K. Michael, Y. Tor. *Chem. Eur. J.* **1998**, *4*, 2091–2098.
144. H.Y. Mei, M. Cui, S.M. Lemrow, A.W. Czarnik. *Bioorg. Med. Chem.* **1997**, *5*, 1185–1195.
145. H.Y. Mei, M. Cui, A. Heldsinger, S.M. Lemrow, J.A. Loo, K.A. Sannes-Lowery, L. Sharmeen, A.W. Czarnik. *Biochemistry* **1998**, *37*, 14204–14212.

Index

Roman figures attached to the page numbers refer to Part I (vols. 1 and 2) and Part II (vols. 3 and 4) respectively: eg. II/478 refers to page 478 in Part II: Biology of Saccharides (volume 3 or 4).

ABAKAN II/412
ABO blood groups II/314
Acanthamoeba II/878
acceptor analog II/300, II/304 f
acceptor mapping II/300, II/308
accumulation in Tay-Sachs and Sandhoff disease II/951
2-acetamido-2-deoxy-β-D-mannopyranosides I/335
acetimidyl-2-deoxy-hexopyranoses I/394
acetobacter II/794
- xylinum II/793
- - cellulose synthase genes II/793
N-acetyl-4-O-acetylneuraminic acid II/235
N-acetyl-9-O-acetylneuraminic acid II/228
O-acetylation II/323
Di-N-acetylchitobiase II/475, II/479 ff
- evolutionary deficiency II/954
acetylCoA:α-glucosaminidase acetyltransferase
- in mucopolysaccharidoses II/949
acetyl-coenzyme A:sialate-4-O-acetyltransferase II/234
acetyl-coenzyme A:sialate-7(9)-O-acetyltransferase II/234
3,4-di-O-acetyl-L-fucal I/374
- NIS-promoted glycosylation I/374
- ciclamicin 0 I/376
- synthesis I/376
N-acetylgalactosamine (Tn-antigen) I/274, I/287
N-acetylgalactosamine-6-sulfatase
- in mucopolysaccharidoses II/949
N-acetyl-α-D-galactosaminidase
- in lysosomal disorder II/954
α-N-acetylgalactosaminidase II/501
β4-acetylgalactosaminyltransferase II/265
β4-N-acetylgalactosaminyltransferase II/263, II/266 ff
- family I/615

N-acetylgalactosaminyltransferase II/334, II/855
9-O-acetyl GD3 II/238
3,4,6-tri-O-acetyl-D-glucal I/369 f
- addition of alcohol I/368
- chlorine addition I/370
N-acetylglucosamine II/717
- bisecting II/853
- 4-sulfatase
- - in mucopolysaccharidoses II/949
- 6-sulfatase
- - in mucopolysaccharides II/949
N-acetylglucosamine, O-linked I/279, II/652 ff, II/661
- dynamic nature II/653 f
- galactosyltransferase II/652
- hexosamine biosynthetic pathway II/657
- insulin resistance II/657
- O-GlcNAcase II/653
- O-GlcNAc transferase II/653
- protein stability II/655 f
- protein translation II/656
- RNA polymerase II II/655, II/657
- transcriptional regulation II/657
- transcription factors II/652, II/655 f
- type II diabetes II/657
N-acetylglucosaminidase II/904
N-acetyl-β-D-glucosaminidase
- lysosomal II/475 f
- HexA and HexB deficient mice II/948, II/951 ff
- in lysosomal disorder II/954
- mucopolysaccharidoses in null mice II/952
N-acetylglucosaminyltransferase II/146, II/262, II/336, II/412
N-acetylneuraminic acid-9phosphate II/228
N-acetyl-β-D-glucosaminidase B
- degradation of G_{M2} by II/952
β-1,2-N-acetylglucosaminyltransferase I II/148
β-1,2-N-acetylglucosaminyltransferase II (GnT-II)

II/152, II/962
β-1,4-*N*-acetylglucosaminyltransferase III   II/155
β-1,4-*N*-acetylglucosaminyltransferase IV   II/157
β-1,4-*N*-acetylglucosaminyltransferase VI   II/161
β-1,6-*N*-acetylglucosaminyltransferase   II/337
β-1,6-*N*-acetylglucosaminyltransferase V   II/158
β-3-acetylglucosaminyltransferase (i-enzyme)   II/262
β-3-*N*-acetylglucosaminyltransferase (i-enzyme)   II/268
β-4-*N*-acetylylglucosaminyltransferase   II/265 ff
β-6-*N*-acetylglucosaminyltransferase family   I/613
N-acetyl-α-D-hexosaminidase
- in mucopolysaccharidoses   II/949
N-acetyl- or *N*-glycolyl-8-*O*-sulfoneuraminic acid   II/239
1-*O*-acyl-2-deoxy-hexopyranoses   I/394
β-*N*-acetylhexosaminidase   II/501
N-acetyllactosamine   II/265, II/717
- galectins   II/639
- lacNAc   II/261
N-acetyllactosaminoglycans   I/656
- α3-sialylation   I/656
N-acetylmannosamine   II/229
- formation of   II/1044 f
- metabolic labeling sialic acid   II/1046 f
N-acetylneuraminic acid   II/228, II/485
α$_1$-acid glycoprotein
- lysosomal degradation   II/476, II/479
acid phosphatase
- lysosomal disorder   II/948, II/954
acremonium α-*N*-acetylgalactosaminidase   II/502
acrosome reaction   II/896
- gamete interaction   II/896
acrosome reaction   II/904
acrosome reaction   II/906 f
activation   II/318
active specific immunotherapy   II/679 f
- animal models   II/679
- carbohydrate antigens   II/680
- clinical studies   II/680
- MUC1 tandem repeat   II/680
- vacinnia virus   II/680
acute phase proteins   II/1030
acyl carrier protein (ACP)   II/439
acyloxy   II/441 ff
acyloxyacyl units   II/437
acyl-protected nucleophiles   I/181
- anomeric O-alkylation with primary triflates   I/181
lipoteichoic acid fragment   I/182
- synthesis *via* anomeric O-alkylation   I/182
acyltransferases   II/441
addition of carbohydrate side chains   II/424
- Gal   II/424
- GalNAcβ1-4   II/424
- mammalian cells   II/424
- Man   II/424
- Taxoplasma gondii   II/424
- T. brucei   II/424
- yeast   II/424
S-adenosyl-L-methionine:sialate-9-*O*-methyl-transferase   II/238
Ad$_E$2-mechanism   I/369
adhesin   II/969 f, II/973, II/1030
adhesion molecule   II/590 f
- acessory molecule   II/591
- COS cells   II/590
- length of Sn extracellular region   II/590
- macrophages   II/590
adhesion, cell   II/721
affinity   II/553
- chromatography   I/1082
- enhancement   II/554
- labeling   II/457
- tagging   II/971
agalactosyl-IgG and rheumatoid factor binding   II/982
aggrecan   II/375, II/377 ff, II/382, II/384, II/386, II/390, II/409, II/411, II/722
aggregation of *de novo* complexes   I/1081
- integration of glycans into liposomes   I/1081
- modification of inert beads   I/1081
aglycon   II/455
AGP   II/1030
agrin   II/712
-α-dystroglycan   II/712
- neuromuscular junction   II/712
- renal tubular Bms   II/712
- synaptic differentiation   II/712
AIDS, acquired immuno deficiency syndrome   II/851, II/862
alanes   I/171
- formation using diazirines   I/171
albumen gland   II/266
aldolase   I/638
alg (asparagine-linked glycosylation   II/47
ALG genes   II/134 ff
alginate network   I/1065
alkaloids, naturally occuring   II/514
alkyl glycosidases   I/811 ff
- enzymatic synthesis (table)   I/811
alkylations of reducing sugars   I/177
- anomeric O-alkylation   I/177
allantoic   II/722
allergenicity   II/1032
allergens   II/1036
allosamidin   I/66
- synthesis via iterative assembly   I/66
alternative splicing   II/155
Alzheimer´s disease   II/721
amide-linked carbohydrate oligomers   I/574
- solution synthesis   I/574
amino acids   I/40

- glycosylation I/40
aminoglycosides I/1113
aminosugar trichloroacetimidates I/8 ff
amipurimycin I/567
amylases II/498
α-amylases II/498
β-amylases II/498
anarchimeric stabilization I/436
anemia II/963
2,6-anhydro-2-thio sugars I/389
- glycosyl donors I/389
2,7-anhydro-N-acelylneuraminic acid II/228
animal models II/467
ankylosing spondylitis II/990
anomeric α,β-S-phosphorodithioates I/396
- from glycalepoxide I/396
anomeric allyl carbamate I/223
anomeric bromides I/484
- preparation I/484
anomeric radicals I/402
- glycosylation reactions I/402
anomeric sulfates I/232
antagonists II/446
anthracyclins I/1106
antiadhesion therapy II/967
antibacterial agents II/430
antibiotics II/967
antibodies II/560, II/737 ff
- anti-T II/858
- anti-Tn II/858, II/860
- blood groups II/858, II/859
- cross-reactions II/854
- O-glycan recognizing II/857, II/858
- hybridoma II/854, II/858
- mannose-related II/857
- natural II/853, II/858, II/859
- neutralizing II/855
- serum II/851, II/859
anti-cancer agents II/935
- carbohydrate processing inhibitors II/935
antigen
- masking by carbohydrate II/856
- recognition by T lymphocytes II/856
- carbohydrate II/851
O-antigen II/968
- repeat II/435 ff
O2-antigen I/226
antigenicity and glycosylation of gp 120 II/856
antigens
- blood group II/858, II/859, II/860
- Ii II/856
- T and Tn II/860
anti-infection therapy II/973
anti-influenza drugs II/235
antitumor
- agents I/1096
- antibiotics I/1097

apical surface II/776
apoprotein B II/746
apoptosis II/40, II/1034
- galectins II/638, II/640, II/641
Matrix assisted laser desorption ionization-time
   of flight-mass spectrometry (MALDI-TOF-MS)
   I/917, I/926
- application in glycobiology I/926
APS kinase II/245 f
AR see acrosome reaction II/896 ff
arabidopsis II/786, II/793
- cellulose synthesis II/793
arabinans II/790
- RG-I II/790
arabinogalactans II/790
- RG-I II/790
arabinoxylans II/787
- cross-linking II/787
- ferulic acid II/787
- glucuronarabinoxylans II/787
arachidonic acid II/446
armed-disarmed
- concept I/230
- principle I/203, I/230
arterial smooth muscle II/746
Arthrobacter sialidase II/504
arylβ-D-mannopyranosides I/336
arylsulfatase A II/460, II/465
- in metachromatic leukodystrophy II/947, II/953
ascidian II/896, II/903, II/899, II/902
- sperm-egg coat binding II/899
ascidran II/896
asialofetuin
- lysosomal digestion II/479
asialoglycoprotein receptor II/476, II/549 ff
asialoglycoproteins II/262
- catabolism II/475 f, II/479
asialoorosomucoid
- lysosomal digestion II/475 f, II/479
ASN-GlcNAc
- hydrolysis by glycosylasparaginase II/477
ASN-linked glycoproteins
- lysosomal degradation II/473, II/475 ff
Asn-linked oligosaccharides
- biological roles II/82
- processing II/82
- quality II/85
- control II/85
- mannose trimming II/85
- ER mannosidase I II/85
Asn-X-Thr/Ser II/48
asparagine N-glycosides I/271
- formation I/271
asparagine-linked glycoproteins II/37
asparagine-linked glycosylation see
   N-glycosylation
aspartylglycosaminuria II/477

- animal model   II/948
- in Finland   II/953
Asp-box motif   II/490
assignment of mass values   I/921
Asx-turn   II/49
atherosclerosis   II/743
- extracellular matrix   II/743
ATP sulfurylase   II/245 f
attachment of the GPI precursor to a protein
- ω site   II/425
- transamidase   II/425
aureolic acids   I/1111
Austin disease   II/465
autoantibodies   II/968
autoantigen   II/856
autoimmune arthritis   II/979
avermectin α disaccharide   I/203
- synthesis with exo-enol ethers   I/203
avidity   II/861
azatrisaccharide   I/203
2-azido-_-imidates   I/475
- reactivity   I/475
1-azi-tetra-*O*-benzylglucopyranose 7   I/162
- glycosidation of phenols   I/162

Bacillus circulans β-galactosidase   II/499
bacteria   II/1029
bacterial
- lipopolysaccharide fragments   I/1016
- interaction with monoclonal antibodies   I/1016
- sialidases   II/490
baculovirus   II/1053
- enzymes needed to improve recombinant protein production   II/1048, II/1054
- deleterious carbohydrate structures   II/1056
baculovirus/Sf9 insect cell system   II/151
bamacan   II/377, II/381
band 3   II/963
Bandeiraea simplicifolia lectin II   II/980
Barton deoxygenation   I/73
basalioma   II/238
basolateral surface   II/776
batroboxin   II/267
betaglycan   II/377, II/381 f II/707
- TGFβ type III receptor   II/707
biacore   I/1046 ff
- affinity determination   I/1053
- application areas   I/1054 ff
- determination of kinetic rate constants   I/1052
- experimental procedures   I/1048
- immobilization of biomolecules at the sensor surface   I/1048
- *in situ* modification of immobilized carbohydrates   I/1056
- interaction analysis and controls   I/1051
- oligosaccharide characterization   I/1055

- real-time analysis by surface plasmon resonance   I/1045
- selectin binding to a glycoprotein ligand   I/1054
- sensor chip   I/1046
- sensorgram   I/1047
- steps for applying the biosensors   I/1046
- surface regeneration   I/1050
- system   I/1045
bidentate ligand   I/223
- as a leaving group   I/223
biglycan   II/377, II/380, II/384, II/748
BI-GP   II/554 ff
binding specificity   II/553
biological recognition   II/230
biosynthesis   II/791 ff, II/968
- cellulose   II/791, II/793
- cellulose synthase genes   II/793
- glucuronoarabinoxylan   II/791
- glycosyl transferase   II/792
- Golgi apparatus   II/791 f
- methyl esterified homogalacturonan   II/791
- nucleotide sugars   II/792
- of hyaluronan   II/363 ff
- of nod factors   I/847
- pectic polysaccharides   II/792
- plasma membrane   II/791
- RG-I   II/791
- RG-II   II/791
- wall glycoproteins   II/791
- xyloglucans   II/791 f
biosynthetic
- mechanism of HA   II/367
- precursors   II/364
- - nucleotide sugars   II/364
bis(monoacylglycero)phosphate   II/461, II/464
bisected oligosaccharides   II/148
bisecting GlcNAc   II/148
- residue   II/155
bisubstrate analog   II/300
blastocyst   II/723
B-lipoproteins   II/745
blood-group   II/1029
- antigens   II/853, II/856, II/857, II/858
- i-type   I/652
- - enzymatic synthesis   I/652
- A
- - active glycoproteins   II/502
- - galectins   II/639
- A and B   II/321
blood group Lewis: FucT III, V and VI   II/204
- cancers   II/205
- chronic liver diseases   II/205
- digestive tract   II/205
- Lewis enzyme   II/204
- lung   II/205
- plasma-type enzymes   II/204
- leukocyte enzyme: FucT VII   II/206

- controll by cytokines II/206
- trafficking of leukocytes II/206
blood-type-related oligosaccharides I/419
- synthesis by the ortogonal strategy I/419
boranes I/171
- formation using diazirines I/171
boron II/795
- cross-link II/795
- deficiency II/795
- wall pectin II/795
- wall pore size II/795
bovine mammary gland II/264, II/266 f, II/269
bovine milk glycoprotein II/269
brain II/412
- KS proteoglycans II/410
- tumor II/238
branch specificity II/157
branches II/317
breast cancer II/675 ff
- cells II/318, II/323
- cell lines II/676
breast carcinomas II/238
brefeldin A II/386 f
brevican II/377, II/380
bromine addition to D-glucal I/371
- influence of protecting groups I/371
bromoalkoxylation I/372
N Butyldeoxynojirimycin
- inhibition of G_{M2} biosynthesis II/953

C. elegans II/154
cadherin *see* E-cadherin
caenorhabditis elegans II/276
- galectins II/625, II/627, II/632, II/635
calicheamicins I/1100
calmegin II/997
calnexin II/835, II/997
- role in glycoprotein folding II/123 ff
calorimetric data I/882 ff
- interpretation I/882
- solvation/desolvation I/882
- solvation entropy I/883
- translational/rotational entropy I/884
calreticulin II/997
- role in glycoprotein folding II/123 ff
calustrin II/412
calystegines II/514
- A_3 II/514
- B_1 II/514
- B_2 II/514
- C_1 II/514
Camillo Golgi II/977
cancer II/157, II/696 f
- associated antigens II/315
- cells II/318
- galectins II/637, II/641 , II/642

- immunotherapy II/1029
- initiation and progression II/926
- metastasis II/637, II/641 , II/642
carbamate family of participatinggroups I/475
carbohydrate I/269 ff
carbohydrate amino acids I/566 ff, I/572
- as important construction elements I/566
- conensation I/572
- natural I/566
- synthetic I/567
carbohydrate-antigen II/851
- xeno-antigen II/853
-binding proteins II/853, II/857, II/861, II/862
- in parasitic protozoans II/869
carbohydrate-binding specificities II/861 f,
        II/968 f,
carbohydrate-carbohydrate interactions I/1061 ff
- affinity interactions I/1079 ff
- agarose I/1063
- alginate I/1063
- arrangement of motifs I/1075
- atomic force microscopy I/1083
- carbohydrate motifs I/1073
- carrageenan I/1063
- cells I/1088
- cell walls I/1064
- crystallography I/1084
- electron microscopy I/1083
- experimental approaches I/1078
- extracellular matrix of seaweeds I/1063
- mammalian extracellular matrix components
        I/1066
- mass spectrometry I/1085
- measurements of interactions I/1078
- microscopy I/1083
- molecular aspects I/1074
- molecular basis I/1076
- molecular modelling I/1086
- nuclear magnetic resonance I/1085
- part of recognition keys I/1068
- polyvalence I/1074
- recognition I/1063
- repulsion I/1072
- schematic overview I/1062
- structural components to cell recognition
        I/1063
- tools I/1086
- weak interactions I/1074
carbohydrate-deficiency glycoprotein syndrome
        II/20, II/29
carbohydrate-deficient glycoprotein syndrome
        (CDGS) II/154, II/959
carbohydrate-interactions I/1069, I/1071
- embryonal cells I/1071
- invertebrates I/1069
- sponges I/1069
- tumor cells I/1071

- vertebrates I/1071
carbohydrate-nucleic acid interactions I/1096 ff
- carbohydrates binding to DNA I/1096
- carbohydrates binding to RNA I/1112
carbohydrate-polymers I/572
carbohydrate-protecting groups I/269
carbohydrate-receptor analogs II/967
carbohydrate-recognition II/584 ff
- A crystal structure II/584
- binding site II/584
- NMR II/586
- 3´-sialyllactose II/584
- V-set domain II/584
- X-ray crystallography II/584
-recognition domain (CRD) II/551, II/599 ff
- galectins II/626, II/627, II/629, II/630, II/635
- ligand binding II/600
carbohydrate-specificity II/537 f
- concanavalin A II/537
- D. biflorus lectin II/537
- EcorL lectin II/537
- G. simplicifolia lectin II/537
- L. tetragonolobus lectin II/538
- pea lectin II/537
- peanut lectin II/537
- phaseolus vulgaris lectin II/538
- potato lectin II/538
- ricinus communis agglutinin II/537
- soybean agglutinin II/537
- S. nigra lectin II/538
- snow drop lectin II/537
- U. europaeus lectin II/538
- wheat germ agglutinin II/537
carbohydrate-sulfotransferases II/249, II/252 ff
- cDNAs II/250
- - 3-OSTs II/250
- - chondroitin 6-sulfotransferase II/250
- - GlcA-3-O- sulfotransferase II/250
- - IduA-2-O- sulfotransferase II/250
- - keratan sulfate sulfotransferase II/250
- galactosyl ceramide 3-O- sulfotransferase II/252
- GSTs II/252 ff
- HEC-GlcNAc6ST II/255
- HNK-1 sulfotransferase II/252
- NSTs II/252
- 3-OSTs II/252
- 3-OST-1 II/255
- 3-OST-2 II/255
- 3-OST-3A II/255
- purification II/249
- SPLAG domain II/252
- stem domain II/252
- 2-O-sulfotransferases II/252
- 2-O-sulfotransferase domain II/252
carboxylic acids I/44
- glycosylation I/44
carcinomas II/285, II/560

carnitine
- formation by proteolytic release of trimethyllysines II/474
cartilage II/409, II/722, II/731, II/733, II/735 f
castanospermine II/480, II/935, II/1000
cathepsins
- A
- - in galactosialidosis II/957
- - protective protein, for β-galactosidase and α-neuraminidase II/946
- D
- - deficiency in mice II/948, II/950
- K
- - in pycnodysostosis II/945, II/948 ff
- role in degradation of Asn-linked glycoproteins II/475 ff
- role in degradation of thyroglobulin II/478 f
caveolae II/764
- GPI II/764, II/766
caveolae II/772 ff
CCAAT II/151
CD4 II/857, II/860, II/862
CD14 II/446, II/1030
CD22 II/233
CD33 II/231
CD43
- galectins II/639
CD44 II/156, II/377, II/381, II/411, II/687 f, II/694, II/696, II/708, II/721, II/747
- CD44E II/708
- hematopoiesis II/708
- hyaluronan receptor II/708
- lymphocyte homing II/708
- inflammation II/708
- metastasis II/708
- RANTES II/708
- tissue inhibitor of metalloproteases-1 (TIMP-1) II/708
- tumor progression II/708
CD44H II/721
CD45
- galectins II/639
CDA II/963
- type II II/963
CDGS II/88
- type I II/959 ff
- type II II/959, II/961 ff
CD-MPR II/565, II/567, II/572
- crystal structure II/572
- - disulfide bond pairing II/572
- - interaction with Man-6-P II/572
- - oligomeric state II/572
- - polypeptide fold II/572
- intracellular trafficking pathways II/565
- - cellular machinery II/565
- - targeting sequences II/565
- co and post-translational modifications II/568

- mutant forms II/571
- oligomeric state II/568
- structure II/567
CDw60 II/237
cell
- adhesion
- - molecules, see also selectins II/862, II/912 ff
- - galectins II/638, II/640, II/641
- - integrin ligand ($\alpha_6\beta_1$) II/916
- - L1 II/914
- - laminin II/916
- - NCAM II/912 ff
- - PSA-NCAM II/913
- binding studies I/1097
- - cell adhesion to substrate I/1097
- - cell aggregation I/1097
- cycle II/139
- fusion activity of HIV-1, see also syncytium II/855, II/857
- migration II/930
- - galectins II/640
- proliferation
- - galectins II/638, II/640
- recognition and adhesion I/1061, II/1013
- surface heparan sulfate proteoglycans II/705
- -cell recognition II/1014, II/1017, II/1019
cellobioside I/734
- glucosynthase-catalyzed synthesis I/734
α-cellobiosyl phosphate I/629
- synthesis I/629
cellodextrins I/991
- conformations I/991
β-D-cellotetraose I/993
- molecular structure I/993
cellular
- adhesion II/237 ff
- aspects II/552
- communication II/229
cellulose II/787
- xyloglucan interactions II/793 f
- - arabinosylated xyloglucans II/794
- - computer modeling II/793
- - cross-linkages II/794
- - fucosylated xyloglucans II/794
- - hydrogen-bonding II/793
- - in muro II/794
- - in vitro II/794
- - microfibrils II/793
- - modification
- xyloglucan interactions II/801
- - cell elongation II/801
- xyloglucan network II/798
ceramidase II/465
ceramide II/455 ff, II/465, II/972
- β-galactosyltransferase II/187
cerebrosides II/340
cervix II/722

chain elongation II/368
- non-reducing end II/368
- reducing end II/368
chaperone proteins, calnexin and calreticulin II/852
chemotaxis
- galectins II/640
chinese hamster ovary (CHO) cells II/853, II/854
chitin I/731
- enzymatic synthesis I/731
- oligosaccharides I/852
- - production in E. coli I/852
chitobiase see-Di-N-acetylchitobiase
chlaydia trachomatis II/441
chlorate II/245
chol-1 II/1018
cholesterol II/761
- detergent insolubility II/761 f, II/764
- GPI II/761
- membrane domains II/761 ff
chondroitin sulfate II/1021
chondroitin 4-sulfate
- digestive pathway in lysosomes II/947 f, II/950
chondroitin 6-sulfate
- digestive pathway in lysosomes II/947 f, II/950
- proteoglycan-LDL complexes II/746
chromomycin I/1111
ciclamycin 0 I/64
- synthesis via iterative assembly I/64
CIDNP
- analysis of glycoproteins I/1040
- method I/1025 ff
- - applications I/1027
- - detection of surface exposed Tyr-, His- and Trp-residues of a protein I/1025
- related molecular modelling I/1027
CI-MPR II/565 ff
- co- and post-translational modifications II/567
- intracellular trafficking pathways II/565
- - cellular machinery II/565
- - targeting sequences II/565
- mutant forms II/571
- oligomeric state II/567
- structure II/566
circulating anodic antigen II/881
- Schistosoma mansoni II/881
class I histocompatibility II/1000
class 1 mannosidases
- ER mannosidase I II/93
- golgi mannosidase I II/95
- fungal secreted mannosidases II/97
- new genes with unknown functions II/98
class 2 mannosidases II/98 ff
- golgi mannosidase II II/99
- lysosomal mannosidase II/101
- epididymal/sperm mannosidase II/103
- heterogeneous cluster of mannosidase homologs

II/104
classification of mannosidases II/89 ff
- class 1 mannosidases (family 47 glycosyl-hydrolases) II/92
- class 2 mannosidases (family 38 glycosyl-hydrolases) II/92
claustrin II/720
cloning strategy I/849
cluster glycoside effect I/910
- molecular basis I/910
CMP-N-acetylneuraminic acid II/1044 f
- Golgi transporter II/1045
CMP-Neu5,9Ac$_2$ II/235
CMP-Neu5Ac hydroxylase II/230 f
CMP-Neu5Ac monooxygenase II/232
- gene cloning II/232
CMP-Neu5Ge II/231
CMP-sialic acid I/631
- synthesis I/631
- derivatives I/631
- synthesis I/631
CoA transporter II/235
collagen II/983
- fibrillogenesis II/748
- type I, digestion by cathepsin K II/951
collectin II/601 ff
- domain organisation II/601
collectin-43 II/608
colon cancer tissue II/319
colon tumor II/238
competition II/267 ff
competitive inhibitors II/151, II/305 f
complement, activity, pathway II/859, II/862.
complement, glycoproteins C3 and iC3b II/862
complementation group II/23
- Lec2 II/23
- Lec8 II/23
- Had-1 II/23
complex N-glycans II/145
complex-type N-glycans II/884
- Acanthocheilonema vitae II/884
- Trichinella spiralis II/884
complex-type oligosaccharide I/607
- lacdiNAc pathway I/607
complex-type oligosaccharide II/261
- synthesis I/602, II/261, II/264 f, II/267
- - chitobio pathway II/265, II/267
- - lacdiNAc pathway II/264 f
- - lacNAc pathway II/261, II/265
- - terminal reactions I/602
concanvalin A II/536, II/542
- Con A II/857
congenital dyserythropoietic anemia (CDA) II/963
conglutinin II/608, II/857, II/861
consensus sequon II/46 ff
control of sugar nucleotide levels II/13

conversion of 6-OH into 6-NH$_2$ II/558
co-operative effects of GalNAc-transferase isoforms II/281
corbohydrate-protein interactions I/1025
- laser photo chemically induced dynamic nuclear polarization (CIDNP)-NMR technique I/1025
core α-1,6-fucosyltransferase II/148
core 1 II/323 f
- β3-Gal-transferase II/316
core 2 β6-GlcNAc-transferase II/317 f
core 3 II/319
- β3-GlcNAc-transferase II/319
core 4 II/319
- β6-GlcNAc-transferase II/319
core 5 II/320
- α3-GlcNAc-transferase activity II/320
core 6 II/320
core 7...II/320
core 8 II/320
core glycosylation II/45
core protein II/731, II/734 f
- hyalectans II/731
- proteoglycans, small leucine-rich II/731, II/733
core region of O-glycans I/648
- in vitro synthesis I/648
core sugars II/435 ff
cornea II/407, II/718
coronaviruses II/238
coulombic stabilization I/870
coupling of sugar phosphates I/630
Creutzfeldt-Jakob II/1036
cryotosporidium parvum II/878
crystalline oligosaccharides I/987
- hydrogen bonding I/987
Csk II/772
CSPGs II/720
cyclic sulfites I/233
cyclization I/401
- of acyclic sugars I/401
- Wittig reactions I/401
cystinosis II/474, II/945
cytokeratins
- galectins II/639
cytokines II/370, II/446, II/852
- IFN-γ II/370
- IL-1 II/370
- TNF-α II/370
cytopathogenic effects, HIV-1 II/852
cytoplasmic II/658, II/661
- glycosylation II/658
cytosolic sulfotransferase II/245
- neurotransmitters II/245
- steroids II/245

5D4 II/722
DAD1 (defender against apoptotic death) II/57

daunorubicin I/1107
deacetylase II/439
deacylase I/638
decarboxylative glycosylation I/221
- method I/221
decorin II/375, II/377 ff, II/390
decorin/biglycan (DSPGs) II/748
delivery II/559
dendritic cells II/856
3-deoxy-C-cellobiose I/507
- preparation I/507
2′-deoxy-β-disaccharide I/378
- synthesis with a glycal epoxide as glycosyl donor I/378
2′-deoxy-β-glycosides I/385
- synthesis with cycloadduct glycosyl donors I/385
2-deoxyglycopyranosides I/198
- enol ether synthesis I/198
2-deoxyglycosides I/13, I/202, I/367
- synthesis I/13, I/367
- - with endo-glycals I/202
2-deoxyglycosyl bromides I/395
- glycosylation reactions I/395
2-deoxyglycosyl donors I/393
2-deoxyglycosyl fluorides I/395
- glycosylation reactions I/395
2-deoxyglycosyl phosphate I/397
- for glycosylations I/397
2-deoxyglycosyl phosphites I/127, I/397
- for glycosylations I/397
- glycosylation I/127
2-deoxyglycosyl phosphoramidites I/397
- for glycosylations I/397
S-(2-deoxyglycosyl)phosphorodithioates I/396
- glycosylation reactions I/395
2-deoxyglycosyl sulfoxides I/399
- armed/disarmed concept I/399
- for glycosylations I/399
2-deoxyhexopranoses I/394
2-deoxyhexopyranoside I/369
- synthesis I/369
3-deoxy-D-*manno*-octul II/441 ff
3-deoxy-D-*manno*-octulosonic acid (Kdo) II/437
deoxynojirimycin II/408
1-deoxynojirimycin II/1000
2′-deoxy 2′-phenylseleno-α- and β-disaccharides I/383
- stereoselective syntheses I/383
2-deoxy-2-phtalimido glycosides I/472
2-deoxy sugars I/367 ff
2-deoxy thioglycosides I/398
- glycosylation reactions I/398
dermatan sulfate
- digestive pathway in lysosomes II/947 f, II/950
development II/729, II/735 ff, II/911 ff
- auditory organ II/913

- axonal growth and synaptogenesis II/916
- blastocyst, trophnectoderm and inner cell mass II/917
- brain microvasculature II/912
- capillary morphogenesis II/917
- cartilage and mesenchyme II/916
- chick embryo II/915
- ciliary ganglion motoneurons in chicken II/915
- cultured rat calvarial osteoblasts II/916
- early-postnatal mouse cerebellum II/914
- embryo development II/912
- fertilization II/915
- fetoembryonic defense mechanisms II/914
- formation of myotubes II/917
- hamster embryos II/916
- heart, kidney and hair-follicule II/914
- human neuroblastoma cell line SK-N-MC II/917
- human prosencephalon II/911
- human sublingual salivary glands II/911
- immune systems II/914
- implantation II/917
- kidney II/916
- malformed tissues II/912
- motoneuron and myotube II/913
- mouse brain II/915
- mouse embryo II/915
- mouse Leydig cell II/912
- mouse olfactory system II/916
- mouse neuroblastoma cells II/913
- nervous systems II/914, II/916
- neurogenesis II/913
- of multi-cellular animals II/154
- preimplantation embryos II/917
- rabbit fallopian tubes, uterus and blastocysts II/912
- rat brain II/913
- rat cerebellum II/917
- sensory systems and neuronal plasticity II/912
- thymus of fetal and newborn animals II/917
N,N'-diacetylchitobios II/265
N,N'-diacetyllactosediamine II/265
N,N'-diacetyllactosediamine, lacdiNAc II/261, II/263
diaziridine I/157
- formation I/157
- oxidation I/157
- oxidation with I_2
diazirines I/171
- exploratory use I/171
Dictyostelium discoideum II/419
2,2′-dideoxy-β-disaccharide B-C I/387
- synthesis I/387
differentation II/318
diffraction methods I/969
1,1-difluorides I/171
- formation using diazirines I/171
dimethyldioxirane epoxidation I/205

2,2-dimethyldioxirane (DMDO) I/377
- for epoxidation of glycals I/377
disaccharide synthase II/440
disaccharides I/973
- crystalline conformations I/973
dispersive interactions I/872
distribution between compartments I/1082
boyden chamber I/1082
dog pancreas II/51
dolichol II/47, II/960 f
- chain length II/132
- phosphate glucose II/39
- phosphate mannose II/39, II/131f, II/135f
- - synthase II/132, II/135
- pyrophosphate II/131, II/134
- - oligosaccharides
- - - degradation II/480
Dol-P-Man see dolichol phosphate mannose
Dol-PP see dolichol pyrophosphate
domain structure II/148
domains, N-linked II/1035
DPM synthase II/41
dual glycon specificity II/501
dylinositol (GPI)-anchored variable surface glycoprotein (VSG) II/873

E-64 II/951
Ebola virus II/832
E-cadherin II/156
echinoderms II/238 ff
EGCases II/507
egg cells II/239
electrospray-mass spectrometry (ES-MS) I/918, I/926
- application in glycobiology I/926
embryonic fibroblasts II/932
- focal adhesions II/932
embryonic lethality II/154
endo D II/505 f
endo H II/505 f
endo-β-galactosidase II/507
endo-β-galactosidase II/963
endo-α–N-acetylgalactosaminidase II/506
endo-β-N-acetylglucosaminidase II/505
endocytosis II/456 ff, II/552, II/761
endoglucosaminidase
- degradaton of polymannose oligosaccharides in ER II/480 f
endoglycoceramidase II/507
endo-glycosidase II/505
endomannosidase
- absence in Chinese hamster ovary cells II/71
- alternate deglucosylation pathway II/73 f
- assay II/72
- cloning II/71 f
- concerted action with other deglucosylating

enzymes II/72 ff
- inhibitors II/67, II/72
- molecular characteristics II/67, II/71
- phylogenetic distribution II/67, II/71
- purification II/67, II/71
- processing of N-linked oligosaccharides II/66
- properties II/67, II/70 ff
- specificity II/66f, II/69 ff
- subcellular location II/67, II/71
endomembrane assembly line II/146
endometrial eoithelial II/723
endometrium II/722
endoplasmic reticulum (ER) II/134, II/376, II/997
- role in degradation of polymannose oligosaccharides II/479 ff
endothelial cells II/862
endotoxin II/435
enediyne antibiotics I/1096
Entamoeba histolytica II/878
Enterobacter II/439
envelope glycoprotein, see also gp4l, gp120, gp 160 II/851, II/852, II/861
- viral II/832
- - intracellular transport II/832
- - pathogenicity II/832
- retroviral II/830
- - carbohydrate structure II/830
enzymatic glycosylations I/647 ff, I/663, I/685 ff
- N-acetylglucosaminylation I/696
- N-acetylglucosaminyltransferase I, II and III I/696
- N-acetylglucosaminyltransferase V I/698
- fucosylations I/690
- FucT V I/692
- FucT III and IV I/692
- galactosylations I/686
- α1,3-galactosyltransferase I/688
- β1,4-galactosyltransferase I/686
- human milk _1,3/4-fucosyltransferase I/690
- non-natural acceptors I/658
- non-natural donors I/658
- recycling of sugar necleotides I/663
- sialylations I/692
- α2,3-sialyltransferase I/692
- β2,6-sialyltransferase I/692
enzymes
- catalyzed additions I/370
- glycals I/370
- linked lectin assay (ELLA) I/878
- modifying Neu5Ac II/230
- replacement II/469
E-PHA II/156
epidermal growth factor (EGF) II/156
epiphycan II/377, II/380
epithelia II/860
epithelial membrane mucins II/674 ff
- MUC1 II/675

epithelial mucins II/670, II/672
- expression of epithelial mucins II/672
epithelium II/972
epitope II/968, II/970 f
epoxidation I/200
Epstein Barr virus II/856
ER see also endoplasmic reticulum II/378 f, II/383 ff
- mannosidase I II/85
- retrieval
- - sequence II/28
ERGIC-53 II/86, II/545
Erp57 II/1001
erythroleukemia II/830
erythromycin A I/391
- synthesis I/391
erythropoietin receptor II/831
Escherichia coli II/437
- K26 antigen I/226
E-selectins II/322, II/613 f, II/619, II/1030
ES-MS see electrospray-mass spectrometry (ES-MS)
esperamicins I/1096
17-β-estradiol II/722
Ets transcription factors II/153
eubacteria
- lipopolysaccharide II/130, II/139
- mannosyltransferase II/139
- mycobacterium II/130, II/135, II/139
everninomicin I/388
- synthesis I/388
evolution II/154, II/239 ff, II/266, II/276
- of GPI biosynthesis II/426
- - archaea II/426
- - GlcNα1-6myo-inositol-1-phosphodialkyl-glycerol II/426
- - Methanosarcina barkeri II/426
evolutionary considerations II/588 f
- disulphide bond II/589
- hypervariable interstrand loop regions II/589
- immunoglobulins II/589
- primitive multicellular organisms II/589
- primordial Ig-like domain II/589
- prokaryotes II/589
- protein-protein interactions II/589
exo-glycosidases II/497
exon II/266 f
expansins II/799
- cell expansion II/799
- wall stress relaxation II/799
expression patterns II/286
expression systems I/849
extracellular HSPGs, other. II/713
- collagen XVIII II/713
- endostatin II/713
- heparin II/713
extracellular matrix II/701

extracellular matrix heparan sulfate proteoglycans II/710
- aggrecan family II/710
- agrin II/710
- basement membrane II/710
- decorin II/710
- leucine-rich repeats II/710
- perlecan II/710
extrahepatic tissues II/550
ezomycin A$_1$ I/567

FAB-MS see fast atom bombardment-mass spectrometry
fabry disease
- animal model II/953
FAK II/772, II/777
Farber disease II/465
Fasciola hepatica II/868
fast atom bombardment-mass spectrometry (FAB-MS) I/915, I/926
- application in glycobiology I/926
fertilization II/895, II/901 ff
- sperm-egg binding II/901 f
fibril, collagen II/719
fibromodulin II/409, II/717
fibronectin
- galectins II/638, II/640
filarial parasites II/267
fluorophore-coupled II/992
F-MCFV see Friend mink cell focus-inducing virus
F-MuLV see Friend murine leukemia virus
formation of complex-type I/594
- branching I/594
- committed steps I/594
forskolin II/156
Forsmann
- galectins II/639
Forssman glycolipid synthetase II/191
free GPIs II/417, II/758, II/760
- trafficking II/760
Friend mink cell focus-inducing virus II/830
Friend murine leukemia virus II/825, II/831
- glycosylation mutants II/831
Friend spleen focus-forming virus II/830
frog II/899 f
- integumentary mucin II/673
fructosides I/201
- enol ether synthesis I/201
fruit ripening II/794, II/801
- enzymic activities II/801
- gene expression II/801
- pectins II/801
- wall loosening II/801
- xyloglucan II/801
F-SFFV see Friend spleen focus-forming virus

fucose II/410
- -dependent binding II/971
- inhibition of glycosylasparaginase II/476 f
- N-linked I/295
a-L-fucosidase II/502, II/902, II/903
- lysosomal II/475, II/477
- lysosomal disorder II/948, II/954
β-D-fucosidase II/503
fucosidosis II/477
α-fucosidosis
- animal model II/948
α-fucosylation I/271
α3-fucosylation of α1-acid glycoprotein II/981
fucosyl phosphites I/125
- glycosylation I/125
fucosyltransferases II/197 ff, II/621
- enzymatic reaction mechanism II/201 f
- - deoxyfuconojirimiycin II/201
- - ethylmaleimide (NEM) II/201
- GDP-Fucose: Fucα1(Fucα1,2Fuc)α2-fucosylransferase...II/203
- - Schistosoma mansoni II/203
- - Tricobilharzia occelata II/203
- - general characteristics II/198
fucosyltransferase
- nomenclature II/198
- protein structure and topology II/200
- - structural organization II/200
- sequence peptide motifs II/199
- - catalyc site II/199
- - GDP binding domain II/199
- specificity II/199
- - hypervariable region II/199
- unconventional types of fucosylation: Fucβ1-P-Ser and cytoplasmic Fucα1,2-Galβ-1,3-GlcNAc-Pro (Dictyostelium discoideum) II/207
- - Dictyosteliumdiscoideum II/207
- - Fucβ1-P-Ser II/207
- - Fucα1,2-Galβ1,3-GlcNAc-Pro II/208
- - - cytosolic II/208
α1,3-fucosyltransferase II/148, II/331
α2-fucosyltransferase II/264, II/268
- family I/611
α3-fucosyltransferase I/602, II/264, II/268
α3/4-fucosyltransferase family I/612
β4-fucosyltransferase II/266

G_{A2}
- accumulation in Tay-Sachs and Sandhoff disease II/951
- intermediate in G_{M2} degradation II/952
GAG II/717
- chains II/376 ff, II/381 ff, II/388 f
- - biosynthesis II/379, II/388 ff
- - disaccharides II/388

- - epimerization II/376, II/383, II/389 ff
- - formation, repeating II/388
- - formation, repeating disaccharides II/376, II/378
- - functions II/382
- - heterogeneity II/382 f
- - initation II/379, II/381
- - lyases II/379
- - microstructures II/379, II/383
- - precursors II/376, II/384
- - - nucleotide sugars II/384
- - regulation II/390
- - specification II/382, II/385, II/388
- - structure II/379, II/382
- - sulfation II/376, II/378, II/382, II/390 f
- - types II/376 ff, II/385, II/388 f
- - - CS II/376 ff, II/381 ff, II/389 ff
- - - DS II/376 ff, II/381 ff, II/389
- - - heparin II/381 ff
- - - HS II/377 ff, II/385, II/388 ff
- - - HS/heparin II/389 ff
GaINAc
- galectins II/639
galactans II/790
- RG-I II/790
galactocerebrosidase II/464, II/466
galactopyranoses I/180
- anomeric O-alkylation with primary triflates I/180
- result of the reactive anomeric β-anion I/180
galactose
- addition to cultured cells II/1050
- biosynthesis in animal cells II/1049
- effect on complement fixation II/1049
- levels on recombinant proteins II/1050 f
galactose-mediated uptake of antigens II/857
galactose-recognizing receptors II/230
galactosialidosis II/467
- animal model II/946, II/948
α-galactosidase II/498
β-galactosidase II/464, II/499, I/727
β-D-galactosidase
- in galactosialidosis II/946
- in G_{M1} gangliosidosis II/948, II/953
- in mucopolysaccharidoses II/949
- in degradation of globoside II/952
- in degradation of G_{M1} II/952
- lysosomal II/475 f
β-galactosides
- galectins II/625
α3-galactosyl/N-acetylgalactosaminyltransferase family I/613
galactosyl phosphites I/125
- glycosylation I/125
galactosylation of IgG II/980
galactosylceramide II/466, II/1013 ff
galactosylsphingosine II/466

galactosyltransferases II/175, II/977 f
- database inquiries II/177
- nomenclature II/177
- reaction catalyzed II/175
α(1-3)galactosyltransferase I/642
α3-galactosyltransferase I/602, II/175
- gene family II/188
- - evolution of II/191
- α-gal epitope II/190
- - xenotransplantation II/190
- ABO blood group II/190
α4-galactosyltransferase II/192
β-galactosyltransferase II/412
β(1-4)galactosyltransferase I/642
β3-galactosyltransferase II/264
- family I/617
β4-galactosyltransferase II/175, II/261, II/263, II/265, II/267, II/269
- family I/615, II/267
- β4-galactosyltransferase-I II/178
- - biosynthetic reactions II/178
- - lactose biosynthesis II/178, II/182
- - protein domain structure II/181
- - gene organization II/181
- - gene regulation II/181
- gene family II/182
- - evolution of II/184
- gene family II/185
- protein domain structure II/186
- acceptor sugar substrates II/186
galectins II/1023
- candidates II/627, II/629, II/630ff, II/635
- expression II/637, II/638
- functions II/638, II/640, II/641
- gene knockout II/638, II/641
- genes II/629, II/630, II/632, II/633, II/635, II/637
- ligands II/639 ff
- regulation II/637
- specificity II/639
- structure
- - acetylation II/628, II/635
- - carbohydrate recognition domain II/626, II/627
- - crystallography II/626
- - linker II/626
- - phosphorylation II/628
- - repetitive domain II/628, II/634
- subcellular targeting
- - nucleus II/628, II/636, II/642
- - secretion II/635, II/636
GalNAc
- benzyl II/316
- glycosylation II/276
- peptide II/314
- transferases II/274, II/315
- - catalytic domains II/274
- - - DxD-like II/274

- - - sequon II/274
GalNAcα-Ser/Thr-linked oligosaccharides
    (O-glycans) II/313
α3-Gal-transferase II/321
β3-Gal-transferase II/321
gamete interaction *see* sperm-egg interaction
ganglio-series gangliosides I/305 ff
- retrosynthetic analysis I/305
- synthesis I/305
gangliosides I/305, I/345, II/456 ff, II/1014, II/1016 f
- classification I/305
- comparison with GD1a I/311
- GD3 II/235, II/237 ff
- GM1 II/458,
- - galectins II/639
- GM2 II/458, II/462
- GQ1b I/311
- synthesis I/305 ff
gangliotetraosylceramide II/969, II/971
gastric disease II/967 ff
- atrophic gastritis II/967
- duodenal ulcer II/967
- gastric cancer II/967
- gastritis II/967
- Helicobacter pylori II/967
- iniflammation II/967
- peptic ulcer II/967
- therapy II/967
gastric mucosa II/1030
gastritis II/967
Gaucher disease II/464, II/460
- animal model II/948, II/953
- enzyme replacement II/946
- in Australian population II/954
GD1a II/1016 ff, II/1022
GD1b II/1016 f
GDP-deoxyfucoses I/634
- preparations I/634
GDP-Fucose: GlcNAc-N(Fucα1,6GlcNAc)
    α6-fucosylransferasesII/207
- hepatocellular carcinoma II/207
GDP-Fucose: O-Ser(Fucα1-O-Ser)GlcNAc
    polypeptide fucosyltransferases...II/207
- EGF domains II/207
- O-glycosidic bond II/207
GDP-Fucose:
    Galβ1(Fucα1,2Gal)α2-fucosyltransferase...II/204
- xenograft rejection II/204
GDP-Fucose: Galβ1,3GlcNAc(Fucα1,3GlcNAc)
    bacterial (Helicobacter pylori) α3-fucosyl-
    transferase...II/207
- Helicobacter pylori II/207
- leucine zipper II/207
GDP-Fucose: Galβ1,4/3GlcNAc
    (Fucα1,3/4GlcNAc)α3/4-fucosylransferases
    II/204

gel-forming mucins II/672 ff
- processing and function II/672 ff
gene
- disease II/959
- knockout II/964
- therapy II/469
- transfer II/229
gene structure
- evolution II/199
- Caenorhabditis elegans II/199
genetic manipulation II/973
genome sequence II/973
genomic organization II/266
Giardia lamblia II/878
GlcNAc I/183
- anomeric O-alkylation I/183
- -benzyl II/322
- -PI deacetylation II/421
- - GlcNAc-PI II/421
- - GlcN-PIs II/421
- - GTP II/421
- - PIG-L II/421
- - sub-compartment of the ER II/421
- -PI synthesis II/420
- - GlcNAc transferase II/420 f
- - GPII II/420
- - PI II/420
- - PIG-A II/420
- - PIG-C II/420
- - PIG-H II/420
- - PIG-a II/420
- - UDP-GlcNAc II/420
- transferase II II/962
- -2 epimerase
- - in animal cells II/1045
- - in insect cells II/1054
- - in plants II/1057
- -TV gene expression II/929
- - up-regulation II/929
O-GlcNAc I/279
- synthesis I/279
- side chains I/294
β1,3-GlcNActransferase II/337
β6-GlcNAc-transferase II/320 f
- family II/319
GlcT II/961
GlcT-1 II/342
globo H I/82
- synthesis via iterative assembly I/82
globoside
- lysosomal digestive pathway II/952
D-glucal I/368
- addition of alcohol I/368
α-glucan-active enzymes I/560
glucoamylases II/498
glucocerebrosidase II/457, II/464
glucopyranoses I/180 f

- anomeric O-alkylation I/181
- anomeric O-alkylation with primary triflates I/180
- result of the reactive anomeric β-anion I/180
β-glucopyranosides I/329
- epimerization at C-2 I/329
α-D-glucopyranosyl 1-thio-α-D-mannopyranose I/557
glucosamine II/437
glucose I/188
- anomeric O-alkylation with cyclic sulfates I/188
glucosidase I II/1000
- assay II/68
- cloning II/67 f
- concerted action with other deglucosylating enzymes II/72 ff
- inhibitors II/67 f
- molecular characteristics II/67 f
- mutants II/67, II/74 f
- phylogenetic distribution II/67 f
- processing of N-linked oligosaccharides II/66
- properties II/66 ff
- purification II/67 f
- specificity II/66 ff
- subcellular location II/67 f
glucosidase II II/1000
- assay II/70
- cloning II/67, II/70
- concerted action with other deglucosylating enzymes II/72 ff
- inhibitors II/67, II/69 f
- molecular characteristics II/67; II/69 ff
- mutants II/67, II/74 f
- phylogenetic distribution II/67, II/69
- processing of N-linked oligosaccharides II/66
- properties II/67 ff
- purification II/67, II/69 f
- role in glycoprotein processing II/119
- role in glycoprotein folding II/123 ff
- sequence homology II/67, II/70
- specificity II/66 ff
- subcellular location II/67, II/69
glucosidase inhibitors II/514, II/522 ff
- australine II/514, II/524
- castanospermine II/522
- deoxynojirimycin II/524
- 2,6-diamino-2,6-imino-7-O-(β-D-glucopyranosyl)-D-glycero-gulohepitol (MDL 25,637) II/524
- 2,5-dihydroxymethyl-3,4-dihydroxypyrrolidine II/514
- DMDP (2,5dihydroxymethyl-3,4-dihydroxypyrrolidine) II/524
- glucosidase I II/522
- glucosidase II II/522
- lentiginosine II/514

- MDL 25,637 II/514
- processing glucosidases II/522
- trehazolin II/524
α-glucosidase II/497
α-D-glucosidase
- in biosynthetic quality control II/480 f
- in pompe diasease II/954
β-glucosidase II/498
glucosylation II/47
glucosylceramide II/464
glucosyltransferase (Glc-T) II/41, II/341, II/961
glucuronic acid-2-sulfatase
- in mucopolysaccharidoses II/949
β-D-glucuronidase
- in mucopolysaccharidoses II/949
glucuronyltransferases II/340
β3-glucuronyltransferase family I/615
glycals I/61, I/369 f, I/376 ff
- addition of a phenylsulfenate ester I/382
- addition of PhSCl to D-glucal derivatives I/380
- addition of selenium based electrophiles I/382
- addition of sulfonium salts and sulfenates I/381
- addition of sulfur based electrophiles I/379
- assembly I/62 ff
- conversion into spiro-ortholactones I/383
- derivatives I/61
- - iterative assembly I/61
- epoxidation I/377
- epoxides I/377
- - nucleophilic attack I/377
- fluoroglycosylation I/385
- for the assembly of oligosaccharides and glycoconjugates I/376
- halogenation I/370
- iterative assembly I/61
- phenylsulfenyl chloride addition I/379
- protonation I/369
- stereochemistry I/380
GlyCAM-1 II/616
glycans I/1068, II/881
- biosynthesis II/238
- - N-linked II/925
- carbohydrate-carbohydrate interactions I/1068
- processing II/924
- Schistosoma sp. II/881
N-glycans II/20, II/314, II/856, II/859, II/861, II/880, II/1033
- Schistosoma sp. II/880
N-glycans, di-, tri-, tetra-antennary II/853
N- and O-glycan changes in cancer cells II/924
- summary
O-glycan II/20, II/273, II/314, II/322, II/315, II/319 f, II/323, II/676, II/854, II/856
- biosynthesis I/648, II/323
- - extension of core 3 and core 4 glycans I/651
- - initialization I/648
- - in vitro extension of core 1 glycans I/650

- - in vitro extension of core 2 glycans I/651
- - synthesis of core 1 I/648
- - synthesis of core 2 I/649
- - synthesis of core 3 and core 4 I/650
- core 1 II/315
- core 2 II/316 ff
- core 3 II/316
- core 4 II/316
- core 5 II/316
- core 6 II/316
- core 7 II/316
- core 8 II/316
- in cerclerial glycocalyx II/881
- - Schistosoma mansoni II/881
- Schistosoma sp. II/880
glycocalix II/455
glycoconjugates I/715, II/239, II/911 ff
- N-acetylgalactosamine II/912
- N-acetylglucosamine II/912
- acetylglucosaminoside II/912
- β-N-acetylglucosaminoside II/912
- amine II/912
- diversification II/239
- enzymatic synthesis on soluble supports I/715 ff
- galactose II/912
- β-D-galactose-BSA
- β-galactoside II/911 f
- _-galactoside-_2,6-sialyltransferase II/913
- ganglioside sialidase mediated GM(1) II/916
- lactose-BSA II/912
- N-linked oligosaccharides II/914
- 2,8-linked sialic acid units II/913
- long-chain polysialic acid (PSA) II/913
- mannosides II/912
- NOC-3 II/913
- NOC-4 II/913
- sialyl Lewis-A II/917
- sialyl Lewis-X II/917
glycodelin A II/262 ff
glycoforms I/591, II/979
glycogen II/349
- biosynthesis II/349
- branching enzyme II/349 f, II/356
- bulk synthesis II/354
- glycogenin II/349 ff
- initiation II/349
- intermediates II/357 f
- proglycogen II/349, II/357 f
- synthase II/349 f, II/354 f
- synthesis II/350 f
glycoglycerolipids I/305
- synthesis I/305
glycolipids I/305, II/229, II/455 ff, II/809, II/963, II/970 ff, II/1013
- bacterial pathogenesis II/809
- fucosyltransferases II/330
- GA2 II/462

- galactosyltransferases II/333
- receptors II/812
- - stress response II/812
- receptor function II/810
- - modulation II/810
- recognition II/810
- - via the stress response II/812
- sialyltransferases II/338
- synthesis I/305
N-glycolylmannosamine II/233
N-glycolylneuraminic II/228
N-glycolylneuraminic acid II/231 ff, II/323
- biological roles II/231
- great apes II/232
- human tumours II/233
glycomimetics I/495 ff
- definition I/495
glycopeptides I/267 ff, I/712
- biological role I/267 ff
- definition I/267 ff
- enzymatic solid-phase synthesis I/712
- formation I/271
- O-glycopeptides I/274
- N-glycopeptides I/280
- glycosylation methods I/271
- structural diversity I/267 ff
- synthesis I/274, I/280
- - in solution I/274
N-glycopeptide libraries I/298
N-glycopeptides I/297, I/280, I/285
 - Lewis-type saccharide side-chains I/285
- natural saccharide side-chains I/280
- synthesis on the solid phase I/297
O-glycopeptides I/274, I/287
- synthesis I/274
- synthesis on the solid phase I/287
glycopolymer mass spectrometry I/923
- fragmentation pathways I/923
glycoproteins I/268, II/229, II/669, II/791
- arabinose II/791
- catabolism
- - oligosaccharide release from the ER II/87
- - failure of quality control II/87
- - cytosalic oligosaccharide degradation II/87
- cytoplasmic II/658, II/661
- - glycogenin II/659
- - parafusin II/659
- - phosphoglucomutase II/659
- - SKP1 II/658- galactose II/791
- hydroxypyroline II/791
- hydroxypyroline-rich glycoproteins II/791
- linkage regions I/268
- N-linked II/37
- lysosomal degradation II/473 ff
- mitochondrial II/661
- serine II/791
- viral II/821 ff

- - biosynthesis II/826
- - carbohydrate substituents II/826, II/828
- - functions II/825 f
- - - enzymatic activities II/826
- - - immune response II/826
- - - membrane fusion II/825
- - - receptor binding II/825
- - - virion assembly II/825
- - glycosylation II/822, II/826 f
- - - epitopes, antigenic II/827
- - - folding of proteins II/826
- - - oligosaccharide diversity II/827
- - - oligosaccharides, N-linked II/827
- - - oligosaccharides, O-linked II/827
- - - structure analysis II/828
N-glycoproteins
- endoplasrnic reticulum processing II/119
- quality control of folding II/123 ff
- synthesis II/119
glycosaminoglycans (GAG) II/375, II/701, II/717 ff, II/729 ff, II/744, II/1020
- catabolism in lysosomes II/947 ff
- chains II/385 ff
- - precursors II/385
- - - nucleotide sugars II/385 ff
- - - phosphoadenosine 5´-phosphosulfate (PAPS) II/385
- - specification II/387
- - types II/386 f
- - - CS II/387
- - - DS II/386
- - - HS II/387
- chondroitin 4 II/744
- chondroitin sulfate II/729
- dermatan sulfate II/729 ff, II/734, II/737 ff, II/744
- disaccharides II/730 f
- - chondroitin sulfate D II/730, II/737
- - chondroitin sulfate E II/730
- - chondroitin 4-sulfate II/736
- - chondroitin 6-sulfate II/730, II/736 f
- - 2-sulfated iduronic acid II/739
- - unsulfated chondroitin II/730
- heparin II/729, II/739, II/745
- heparan sulfate II/729, II/739, II/744
- heterogeneity II/375
- hyaluronan II/729, II/731, II/733, II/736
- hyaluronic acid II/744
- keratan sulfate II/729, II/732, II/734, II/736, II/745
- nuckar II/660
- 6 sulfate (C-4-S, C-6-S) II/744
- types II/375
- - chondroitin sulfate (CS) II/375
- - dermatan sulfate (DS) II/375
- - heparan sulfate (HS)/heparin II/375
glycosidases I/723, II/456

- amyloglucosidase II/517
- aryl-glycosidases II/517
- cytosolic II/480 f
- ER II/480 f
- ER mannosidase II/517
- families I/723
- β-glucosidase II/517
- glucosidase I II/517
- glucosidase II II/517
- glycoprotein processing enzymes II/517
- indolizidines II/515
- inhibitors II/515
- inhibitory activity II/517
- intestinal maltase II/517
- known number I/723
- mannosidase I II/517
- nortropanes II/515
- piperidines II/515
- pyrrolidines II/515
- pyrrolizidines II/515
- sucrase II/517
- trehalase II/517
glycoside
- bond formation I/5
- - through anomeric oxygen-exchange reactions I/5
- - through retention of the anomeric oxygen I/5
- cluster II/549
- preparation I/197
- - via enol ethers I/197
C-glycosides I/51
N-glycosides I/51
glycosidic bond formation I/408
- control of stereochemistry I/408
glycosidic mechanism I/428 ff
- Lemieux´s glycosidic mechanism I/428
glycosignaling domain (GSD) II/772 ff
glycosolic stereocontrol I/408
- axial glycoside formation I/408
- axial glycoside formation by S_N^2 displacement I/408
- equatorial glycoside formation I/408
- β-mannosylation by intramolecular aglycone delivery I/408
- β-mannosylation by means of insoluble silver salts I/408
- stereochemistry I/408
- 1,2-trans glycoside formation by means of a C-2 participating group I/408
glycosphingolipids (GSLs) I/305, II/329, II/455 ff, II/761, II/810, II/1014, II/1016 f
- functional domains II/810
- synthesis I/305
- detergent insolubility II/761
- GPI II/761
- membrane domains II/761
glycosphingolipidoses

- substrate depravation therapy II/953
α-glycosylation I/355
glycosyl
- donors with a C-2 heteroatom I/386 ff
- - 2,6-anhydro-2-thio-glycosyl donors I/388
- - 2-bromo-2-deoxyglycosyl bromides I/386
- - 2-deoxy-2-(thiophenyl)-glycosyl fluorides I/387
- esters I/216
- - precursors for other glycosyl donors I/216
- phosphatidylinositols (GPIs) II/417, II/757
- phosphines I/171
- - formation using diazirines I/171
- phosphorodithioates I/233
- - preparation I/233
- phosphoroselenoates I/233
- - preparation I/233
- phosphorothioates I/233
- - preparation I/233
- transfer I/384
- - via cycloaddition I/384
- transferases II/382, II/385 ff
- - GacA transferases II/387
- - GalNAc transferases II/388
- - Gal transferases II/386
- - GlcA transferases II/389
- - GlcAT-P II/387
glycosylation I/724
- basic mechanisms I/724
- N-linked II/840 f, II/844
- synthesis vs. hydrolysis I/724
N-glycosylation II/45, II/134 ff, II/315, I/593
- gp 120, HIV-1 II/851
- HIV-1 II/851, II/854
- host-cell dependent II/853, II/854
- inhibitors, antiviral effects II/852
- reactions in the rough endoplasmic reticulum I/593
- site II/851, II/855
O-glycosylation II/273, II/315, II/671, II/676 f
- chain extension II/671
- chain termination II/671
- changes in cancer II/677
- core 1 II/671
- core 2 II/671, II/676
- terminal sugars II/671
- inhibitors II/316
1,2 cis glycosyl donors I/108
α-glycosyltransferase II/133
β-glycosyltransferase II/133
glycosylasparaginase II/475 ff, II/479
- knock-out mouse II/949
- lysosomal disorder II/954
glycosylated natural products I/61
- synthesis I/61
glycosylation I/443, II/912 ff, II/1029
- N-acetylglucosamine II/912

- acetylglucosaminoside II/912
- β-N-acetylglucosaminoside II/912
- amine II/912
- galactose II/912
- β-D-galactose-BSA
- β-galactoside II/911 f
- β-galactoside-α2,6-sialyltransferase II/913
- ganglioside sialidase mediated GM(1) II/916
- heterogenity II/859
- homeostasis II/985
- lactose-BSA II/912
- 2,8-linked sialic acid units II/913
- N-linked oligosaccharides II/914
- long-chain polysialic acid (PSA) II/913
- mannosides II/912
- methods I/178, I/196 ff
- - conventional I/178
- - endo-enol ethers I/198, I/201
- - endo-glycals I/202
- - esters I/216
- - exo-glycals I/204
- - heterogeneous acid catalysis I/196
- - 1-hydroxy sugars I/209
- - iodoetherification reaction I/201
- - isopropenyl glycosides I/207
- - mild activation of O-glycosyl N-allyl carbamates with soft electrophiles I/223
- - new I/178
- - NIS-mediated I/204
- - orthoesters I/223
- - oxazolines I/223
- - phosphorus I/229
- - sugar carbonates I/221
- - sulfur derivatives I/229
- - use of ortho esters as intermediates I/196
- - vinyl glycosides I/206
- NOC-3 II/913
- NOC-4 II/913
- phenotype II/861
- product determining step I/443
- rate-limiting step I/443
- sites II/316, II/851, II/855, II/860
- strategy I/411
- - stepwise synthesis I/411
- with a 2-thiophenyl-α-D-glucopyranosyl donor I/381
- - stereochemistry of I/381
- with thioglycoside donors I/97 ff
glycosylidene diazirines I/155 ff
- addition to aldehydes I/170
- course of the glycosidation I/158
- effect of intermolecular hydrogen-bonding from the diazirine to the acceptor I/166
- glycosidation I/158
- glycosidation of anomeric N-phthaloylated alosamine I/166
- glycosidation of diols I/164

- glycosidation of fluorinated alcohols I/163
- glycosidation of monovalent alcohols I/163
- glycosidation of phenols I/162
- glycosidation of strongly acidic hydroxy compounds I/162
- glycosidation of triols I/164
- glycosidation of weakly acidic hydroxy compounds I/163
- ketones I/170
- precursors of glycosylidene carbenes I/155
- stability I/158
- synthesis I/155
- synthesis of spirocyclopropanes I/168
glycosylphosphatidylinositol (GPI) II/130, II/135 f, II/138, II/872
- anchored II/872
glycosyltransferase I/589 ff, I/647 ff, I/705, II/20, II/29 f, II/264, II/266, II/314, II/382, II/671, II/676 f, II/914 f, II/977, II/1036
- β-N-acetylglucosaminyltransferase IV II/915
- biological role I/589 ff
- families I/608 ff, II/267
- β-galactosides II/915
- Gal transferases II/382
- galactosyltransferase II/915
- GleA transferase II/382
- Golgi apparatus II/914
- inhibitors II/293 ff
- - N-acetylglucosaminyltransferases II/305
- - blood group A and B glycosyltransferase II/306
- - design II/294
- - donor analogs II/296
- - α1,2-fucosyltransferase II/300
- - α1,3-fucosyltransferase II/300
- - α1,3-galactosyltransferase II/297 f
- - - acceptor analogs II/298
- - β1,4-galactosyltransferase II/296 f
- - - acceptor mapping II/297
- - natural II/293
- in solid-phase synthesis I/705
- Kyl transferase II/382
- polypeptide glycosyltransferases II/671
- polysialyltransferases (PST) II/914 f
- sialyltransferase (STX) II/915
- α2,3 sialyltransferase II/676
- C2GnT II/676
- core 2β1,6GlcNAc T II/676
- ST3Gal I II/677
glypians II/706 f
- dally II/707
- glycosylphosphatidyinositol II/706
- glypican-3 II/707
- HS binding proteins b II/706
- ordered microdomains II/706
- rafts II/706
- Simpson-Golabi-Behmel Syndrome II/707

- wingless II/707
- X-linked overgrowth II/707
GM1 II/1016 f
G$_{M1}$ ganglioside
- lysosomal digestive pathway II/952
G$_{M1}$ gangliosidosis
- animal model II/948, II/953
G$_{M2}$ ganglioside
GM2 gangliosidoses II/462
G$_{M2}$
- storage in HexA deficient mice II/952
GM2-activator II/458 ff, II/462
GM3-dependent cell adhesion II/778
GnT I-null mouse II/154
GnT-II II/962 f
GnT-II II/963 f
Golgi II/323, II/376, II/378 f, II/383, II/385, II/386 ff, II/389 ff, II/413, II/962, II/964, II/977
- apparatus II/146, II/274, II/855
- biosynthetic pathways II/925
- compartment II/231, II/234 ff, II/925
- glycosyltransferases I/597
- localization II/150
- membrane II/315
- sulfotransferases II/245
- - carbohydrates II/245
- - tyrosine residues II/245
- - tyrosine sulfation II/246
Gp 120
- binding II/857
- HIV-1 II/851, II/852, II/855, II/860, II/861, II/862
- - V3 loop II/855, II/860
- - oligosaccharides II/851 ff
- - shedding II/853
Gp 160, HIV-1 II/852, II/856
Gp 41, HIV-1 II/852, II/855
GP1bα II/246
GPI see also glycosylphosphatidylinositol II/417 ff, II/757 ff
- acylated inositol II/420
- 1-alkyl-2-acyl glycerol II/419
- anchored glycoproteins II/1033
- Anchors II/417
- biosynthesis II/427, II/759
- - drug development II/427
- - - Plasmodium falciparum II/427
- - regulation II/427
- caveolae II/764, II/766
- ceramide II/419
- detergent insolubility II/761 f, II/764 f
- diacylglycerol II/419
- disease II/759
- dimyristoyl glycerol II/419
- endocytic pathways II/760 f
- fyn II/762 f
- glycosphingolipis (GSLs) II/761

- - detergent insolubility II/761
- - membrane domains II/761
- inositol phosphoglycan (IPG) II/766
- insulin II/766 f
- leishmania II/758
- lyso-glycerolipid II/419
- mannosylation II/422
- - Dol-P-Man II/422
- - L- major II/422
- - PIG-B II/422
- - substrate channeling II/422
- - T. brucei II/422
- membrane anchors II/1029
- membrane domains II/761 ff
- membrane release II/765 f
- parasite coasts II/758
- paroxysmal nocturnal hemoglobinuria (PNH) II/757, II/759
- phospholipases II/765 ff
- precursors II/420
- - glycolipid A II/420
- - synthesis II/419 f
- - - cell free systems II/420
- - - ER II/419
- - - human II/419
- - - mutants II/420
- - - T. brucei II/419
- - - yeast II/419
- prion protein II/757, II/762
- second messenger II/766
- secretory pathway II/760 f
- signaling II/764 ff
- T. brucei II/758
- - African sleeping sickness II/758
- - trypanosomiasis II/758
- - variant surface glycoprotein (VSG) II/758
- T. cruzi II/758
- trafficking II/760
- transport II/760
- trypanosoma brucei II/757
- trypanosoma cruzi II/765
- yeast II/758 ff
- - cell wall II/758 f
- - Gas1p II/759
- - Gce1p II/759
GQ1b I/313 f
- construction of oligosaccharide I/314
- preparation of building block I/313
- retrosynthetic analysis I/313
growth factor II/370, II/732 f, II/736, II/739
- bFGF II/370
- EGF II/370, II/732 f,
- FGF-2 II/739
- PDGF-BB II/370
- receptors II/774, II/776
- TGF β II/734
- TGF-β1 II/370

GSLs II/762
- clusters II/772
- detergent insolubility II/762, II/764
- -GSL interaction II/773
- membrane domains II/762 ff
GT1b II/1016 ff, II/1022
guanosine diphosphate mannose II/131, II/133 ff, II/137 ff
- and cell cycle II/139 f
- transporter II/139

H-1 II/968, II/971
HA II/363
- concentrations II/363
- invertebrates II/363
- network II/363
- physiological function II/363
- synthesizing enzymes (HAS) II/365
Haemophilus influenzae II/437
2-halo-3-β-substituted sialic acid derivatives I/347
- sialy donors I/347
Hansch parameters I/874
HbsAg *see* hepatitis surface antigen
HBV *see* hepatitis B virus
heamagglutinin (Hag) II/839 ff
HEF II/839
- esterase domain II/844
Helicobacter pylori II/967 ff
- antiadhesion therapy II/967
- antibiotics II/967
- atrophic gastritis II/967
- carbohydrate receptor analogs II/967
- duodenal ulcer II/967
- gastric cancer II/967
- gastritis II/967
- inflammnation II/967
- peptic ulcer II/967
- therapy II/967
- vaccines II/967
helminthic parasites II/879
helminths II/868
hemagglutinin II/235, II/486
- neuraminidase II/487
hemorrhagic fever II/832
HEMPAS (Hereditary Erythroblastic Multinuclearity with Positive Acidified Serum lysis test II/959, II/963
- HEMPAS disease II/89
heparan *N* sulfatase
- in mucopolysaccharidoses II/949
heparan sulfate II/395, II/701, II/969 ff
- biosynthesis II/395 ff, II/701 f
- - highly sulfated domains, HSDs II/702
- - macroscopic structure II/702
- - unmodified domains, UMDs II/702

- - galactosyltransferases II/396
- - GlcA C5-epimerase II/398 ff
- - GlcA/GlcNAc copolymerase II/397
- - GlcNAc (HexNAc) transferase II/397
- - GlcNAc *N*-deacetylase/*N*-sulfotransferase (NDST) II/398 ff
- - glucuronyl transferase I II/397
- - linkage region II/396 ff
- - regulation II/402 ff
- - 2-*O*-sulfotransferase II/398 ff
- - 3-*O*-sulfotransferase II/398 ff
- - 6-*O*-sulfotransferase II/398 ff
- - xylosyltransferase II/396
- functions, physiological II/396
- interactions II/395 ff
- - antithrombin-binding pentasaccharide II/402
- - facilitated diffusion II/401
- - fibroblast growth factor II/403
- - platelet-derived growth factor II/403
- - specificity II/401
- lysosomal digestive pathway II/947 ff
- proteoglycans II/248, II/396, II/750
- - blood coagulation components II/248
- - cell-cell adhesion molecules II/248
- - chemokines II/248
- - cytokines II/248
- - extracellular II/248
- - growth factor II/248
- - lipid carrier molecules II/248
- - matrix proteins II/248
- - viral attachment receptors II/248
- structure II/400 ff
- - anti-HS monoclonal antibodies II/400
- - domains II/400 ff
- - fine structure II/401
heparin II/395 ff, II/701, II/750
- biosynthesis II/395 ff
- - galactosyltransferases II/396
- - GlcA C5-epimerase II/398 ff
- - GlcA/GlcNAc copolymerase II/397
- - GlcNAc (HexNAc) transferase II/397
- - GlcNAc *N*-deacetylase/*N*-sulfotransferase (NDST) II/398 ff
- - glucuronyl transferase I II/397
- - linkage region II/396 ff
- - regulation II/402 ff
- - 2-*O*-sulfotransferase II/398 ff
- - 3-*O*-sulfotransferase II/398 ff
- - 6-*O*-sulfotransferase II/398 ff
- - xylosyltransferase II/396
- functions, physiological II/396
- /HS chains II/249
- - antithrombin II/249
- - FGFR1 II/249
- - GlcNS 6-*O*-sulfate II/249
- - IduA-2-*O* sulfate II/249
- - 6-*O*-sulfation of GlcN(Ac/S) II/249

- interactions II/395 ff
- - antithrombin-binding pentasaccharide II/402
- - facilitated diffusion II/401
- - fibroblast growth factor II/403
- - platelet-derived growth factor II/403
- - specificity II/401
- pentasaccharide sequence I/482 ff
- - structure I/482
- - synthesis I/482
- proteoglycans II/396
- structure II/400 ff
- - anti-HS monoclonal antibodies II/400
- - domains II/400 ff
- - fine structure II/401
- synthesis I/482
- - antithrrombin-binding pentasaccharide sequence I/482
hepatitis B II/156
- virus II/833 ff
- envelope (glyco)proteins II/833 f
- - N-glycosylation II/834
- - O-glycosylation II/834
- - L-protein II/834
- - M-protein II/834
- - S-protein II/834
- subviral particles II/834
- virion formation II/835
- - glycosylation inhibitors II/835
hepatitis surface antigen II/833
chaperone II/835
hepatocellular carcinoma II/157
hepatoma II/157
heterologous ('recombinant') oligosaccharide production I/847
- concept I/847 ff
- *E. coli* I/847
- methodology I/847
heterologous oligosaccharides I/845
hetero-oligomer synthesis I/418
HexA genes II/951 ff
HexB genes II/951 ff
hexosaminidases II/462, II/469
β-hexosaminidases II/457
β-D-hexosaminidase *see* -
N-Acetyl-_-D-glucosaminidase
HIV gp120 II/280
HIV-1 II/851
- disease, pathogenesis II/862
- dissemination, spread II/862
- glycosylation variants II/862
- molecular clones II/860
- infected lymphocytes II/853, II/854
- transmissibility II/853
HIV-2 II/853
HNK-1 II/387, II/1018, II/1020, II/1022 f
homogalacturonan II/788, II/795 f
- cross-linked by calcium II/795

- esterified II/795
- localization II/795
- methyl esters II/795
- methyl esterified II/788
- rheological properties II/795
hormone non-specific
β4-N-acetylgalactosaminyltransferase II/264
hormones II/370
- follicle stimulating hormone II/370
- biosynthesis of thyroid II/477 ff
host cell II/860, II/862
host surfaces II/968 f
housekeeping gene II/151
HPAEC methods II/980
HPLC II/980
hsp110 II/905
htrB II/441
human II/419 f
- CD52 II/419
- paroxysmal nocturnal hemoglobinuria II/420
- PIG-A II/420
Hunter disease *see* mucopolysaccharidoses, MSP II
- animal model II/948
Hurler disease see mucopolysaccharidoses, MSP I
- animal model II/948
hyaladherins II/687
hyaluronan II/685, II/744, II/745, II/747
- binding proteins II/687 ff
- in vascular disease II/743
- receptors II/687 f, II/694, II/696
- synthase II/691, II/693
hyaluronate II/156, II/721
hyaluronic acid
- digestive pathway in lysosomes II/947, II/950
hybrid N-glycans II/146
hybridoma-derived antibodies II/854
hydrogen bonding I/871
hydrophobicity
- analysis II/25
- plot II/25
hydroximolactones I/157
- O-Sulfonylation I/157
6-hydroxyhexyl β-D-galactoside I/727
- enzymatic synthesis I/727-
2-D-hydroxyl fatty acids II/972
1-hydroxy sugars I/209 f
- acidic activation I/210
- acidic activation with additional reagents I/211
- dehydrative glycosylation I/212

I β3-GlcNAc-transferase II/320
I β6-GlcNAc-transferase II/318, II/321
I antigen II/321
I,i antigens II/856
I22 II/722

I-cell disease
- animal model II/948, II/954
iduronate-2-sulfatase
- in mucopolysaccharidoses II/949
- in degradation of glycosaminoglycans II/948 ff
α-L-iduronidase
- in mucopolysaccharidoses II/949
- in degradation of glycosaminoglycans II/948 ff
- α-lactalbumin II/266, II/269
IgG glycosylation II/978
IgG sugars II/979
IL-6 II/446
1-imidazolylcarbonyl-glycosides I/222
1-imidazolylthiocarbonyl-glycosides I/222
immobilized glycosyl acceptor I/240
- polymer-supported oligosaccharide synthesis I/240
immobilized glycosyl donors I/242
- anomeric specificity I/242
- for the synthesis of oligosaccharides I/242
- on polystyrene support I/242
immune response II/968
- galectins II/640
immune system II/678
- adaptive II/678 f
- - MHC class I II/679
- - MHC class II II/679
- innate II/678
immunogenic carbohydrates II/1043
- in insect cells II/1054, II/1056
- in plants II/1056, II/1059
immunogenicity and glycosylation of gp 120 II/856
immunoglobulins II/1036
immunoglobulin G II/977 f
- superfamily II/579
- - cell recognition II/579
- - cell surface II/579
- - signal transduction II/579
immuno-suppression II/859
in vivo II/279 f
inflammation II/862, II/967, II/972, II/1029
influenza II/839, II/973
- A and B viruses II/233
- C virus II/234, II/238
- - HEF envelope glycoprotein II/238
- hemagglutinin II/1000
- virus II/968
- - neuraminidase II/486
inhibitors of NAm II/847
inhibitors of N-linked oligosaccharide processing II/522
inhibitory potency II/553 ff
innate immune system II/1030
inositol
- acylation II/421
- - mammals II/421
- - T. brucei II/421
- - yeast II/421
- glycosides I/36
- - synthesis I/36
- phosphoceramide II/130, II/137
- phosphoglycan (IPG) II/766
- - GPI II/766
- - insulin II/766
- - second messenger II/766
- - signaling II/766
Insect cells
- as a source for recombinant proteins II/1053
insects II/262
integrins
- galectins II/639, II/640
cis interactions II/589 f
- on cell surfaces II/590
- purified proteins II/590
- sialidase treatment II/590
interest cancer II/323
interleukin 1 (IL-1) II/446
intracellular proteoglycans II/703
intracellular sources of sugars
- salvage II/5
- activation and interconversion of monosaccharides
- - glycogen II/6
- - glucose II/7
- - glucuronic acid II/8
- - iduronic acid II/8
- - xylose II/8
- - mannose II/8
- - fucose II/9
- - galactose II/10
- - N-acetylglucosamine II/10
- - N-acetylgalactosamine II/10
- - sialic acids II/11
intramolecular glycosidation reactions I/449 ff
- carbon thethers I/450
- definition I/449
- silicon tethers I/454
- tether does not participate in the reaction I/459
- tethering to the leaving group I/459
- tether participation I/450
invertebrate II/269
β-inverting glucosidases I/725
iodoalkoxylation I/372
IPG II/767
- insulin II/767
- signaling II/767
isoantigen II/856
isoforms II/280
- overlapping substrate specificities II/283
isolation...II/516
- glycosidase inhibitor II/516
- radial chromatography II/516
- thin layer chromatography II/516

- GC-MS II/516
- ^{1}H and ^{13}C NMR II/516
- high resolution mass spectrometry II/516
isothermal titration microcalorimetry I/878
iterative glycal assembly strategy I/62 ff
jelly coat II/896 ff, II/906
juvenile chronic arthritis II/986

KDNase II/504
KDO I/185
- anomeric O-alkylation with primary triflates I/184
- [3] glycosides I/199, I/201
- - enol ether synthesis I/198, I/201
Kdo disaccharides I/201
- enol ether synthesis I/201
- glycosides I/199, I/202, I/205
- transferase II/441
keratan sulfate II/717
- digestive pathway in lysosomes II/947, II/950
keratin II/412, II/722
keratinocytes II/722
keratocan II/717
keratocytes, stromal II/720
keratosulfate II/407
2-keto-3-deoxy-nononic acid II/227
ketofuranosides I/201
- enol ether synthesis I/201
KH-1 I/86
- synthesis via iterative assembly I/86
kin recognition II/150
kinases II/370
- cAMP-dependent protein kinase II/370
- protein kinase C II/370
kinetic analysis II/151
Klebsiella II/439
knock-out II/718, II/970
Koenigs-Knorr method I/345
kojitriose I/459
Krabbe disease II/466
- animal model II/948, II/953
KS II/717 ff
- 502 I/69
- - synthesis via iterative assembly I/69
KSPGs II/717 ff

β-lactosamine linkage I/272
- formation I/272
α-D-lannosidase
- lysosomal disorder II/948
β-D-lannosidase
- lysosomal disorder II/948
α-lannosidosis
- animal model II/948
β-lannosidosis

- animal model II/948
lectin, C-type II/549 ff, II/597 ff
- function II/598
- domain organization II/598
lectin, D. biflorus II/542
lectin, I-type II/579
L enzyme II/318
L. major II/418, II/421
- gp63 II/418
L1 II/720
lacdiNAc
- analog Lewis^x II/263, II/264
- synthase II/264
lactosamine II/408
lactosamine
- saccharides I/657
- - α3-fucosylation I/657
- galectins II/639
lactosylceramide I/183, II/464, II/969, II/972
- synthesis from acyl-protected lactose via anomeric O-alkylation I/183
laminin II/720
- galectins II/638 ff
LAMPs
- galectins II/639 ff
laser photo chemically induced dynamic nuclear polarization (CIDNP)-NMR technique I/1025
latent-active glycosylation I/208
lateral pressure II/460
LBP II/446. II/1030
LDL II/745
LEC4 II/159
LEC4A II/159
LEC14 II/161
LEC18 II/161
LEC1 CHO cell mutant II/152
lecloir pathway I/663
- glycosyltransferases I/664
- - sugar nucleotide substrates I/664
lecticans II/1021 f
lectins II/491, II/535 f, , II/539 ff, II/853, II/857, II/860 f, II/911 ff, II/999, II/1021
- anomeric preference II/539
- applications II/535
- apyrase II/540
- artificial neolectins II/912
- association constants II/539
- availability II/535
- carbohydrate crystal structures I/994 ff
- carbohydrate specificity II/536
- Con A II/912
- crystal structure II/539 f, II/543
- endogenous lectin, called R1 II/917
- galectin-1 II/916 f
- galectin-3 II/916 f
- galectin-4 II/916
- galectin-5 II/917

- galectin-6   II/915
- galectin-9   II/916
- homezygous mutant mice null for all three selectins   II/917
- hydrophobic binding site   II/543 f
- hydrophobic sites   II/540
- ligand interactions   I/562
- LNP   II/540
- LPA   II/912
- mice carrying a null mutation in the gene encoding galectin 1 or other galectins   II/917
- monosaccharide specificities   II/536
- multivalence   II/539
- nuclear and cytoplasmic   II/661
- - galectins   II/661
- - CBP70   II/661
- pentraxias   II/915
- PNA   II/911 f
- P-type   II/563, II/569
- - cation-independent mannose 6-phosphate receptor (CI-MPR)   II/563 ff
- - cation-dependent mannose 6-phosphate receptor (CD-MPR)   II/563 ff
- - lysosomal enzyme recognition   II/569 f
- - - cation dependence   II/569
- - - pH dependence   II/569
- - - recognition of phosphodiesters   II/570
- - - recognition of diphosphorylated oligosaccharides   II/570
- quaternary structures   II/541, II/543
- RCA I   II/912
- SBA   II/912
- selectins   II/917
- E-selectin   II/917
- P-selectins   II/917
- SNA   II/912
- specific carbohydrate ligands   II/911
- C-type lectins   II/915
- P-type lectins   II/915
- usual abbreviations   II/911
- VVA   II/911
- WGA   II/912
leech sialidase L   II/504
Legionella pneumophila   II/437
Leishmania   II/868, II/876 f
- donovani   II/875
- major   II/417
- - gp63   II/417
- sp   II/876
Leloir glycosyltransferases   I/846
letachromatic leukodystrophy
- animal model   II/947 f
leucine zipper   II/25
leukemia   II/157
- cells   II/318, II/323
leukocyte, extravasation   II/862
leukotrienes   II/446

leupeptin   II/476, II/479, II/951
Lewis a   II/968, II/971
Lewis b   II/968, II/971
Lewis c   II/968
Lewis d   II/968
Lewis gene   II/971
Lewis x   II/968
Lewis y   II/968
Lewis^y I/76, I/990, II/261
- conformations   I/990
- synthesis via iterative assembly   I/76
Le^x II/332, II/882, II/1030
- schistosomes   II/881
ligand
- coreceptors   II/709
- - dimeric receptor state   II/709
- - immobile complex   II/709
- - cell-cell adhesion   II/709
- - cell-ECM adhesion   II/709
- - ligand activity   II/709
- - microbial pathogenesis   II/709
- - seven pass transmembrane receptors   II/709
- - single pass transmembrane receptor tyrosine kinases   II/709
- linkage   I/905
- receptors   II/708
- - clearance/internalization receptors   II/708
- - ligand localization   II/709
linear B determinant   II/321
linear polystyrene   I/249
- polymer-supported synthesis   I/249
α-linked trisaccharide   I/378
- synthesis with glycal epoxides as glycosyl donors   I/378
linkers   I/250 ff
- *p*-acylaminobenzyl   I/257
- *p*-alkoxybenzyl   I/258
- benzylphenol   I/251
- cleavable by photolysis   I/260
- crosslinked polystyrene resin   I/255
- diaryl(alkyl)silyl   I/258
- diethylsilyl   I/258
- dioxyxylyl diether (DOX)   I/252
- DOX   I/251
- ethoxydimethylsilylbutylcarbonamide   I/257
- fluorene   I/251
- functionalized polystyrene cross-linked with divinylbenzene   I/255
- in two-phase (solid state) oligosaccharide synthesis   I/257
- 4-mercaptophenol   I/257
- 4-mercaptophenol-acetamide   I/257
- mercaptopropanol   I/258
- mercaptopropyl   I/257
- mercaptothiopropanol   I/258
- nitrobenylphenolbenzoate   I/258
- nitrobenzyl   I/257

- one-phase (solution) oligosaccharide synthesis I/251
- *p*-oxybenzyl I/258
- pentenyl I/257
- phenylaccetamide I/251
- photochemically removable I/260
- pore glass (CPG) I/255- dialkyl-or diary-silyl I/259
- Rich´s linker I/257
- succinoyl I/251, I/257
- succinoyl diester I/250
- succinoyl glycine I/257
- sulfonyl I/258
- thiglycoside linkers I/259
- thio...I/251
- thioethanol I/251
- two-phase systems I/255
lipase
- lysosomal disorder II/954
lipid II/963
- IV$_A$ II/440
- A II/435 ff
- linked oligosaccharide biosynthesis II/83
- remodeling II/423 f
- - ceramide II/424
- - distearoyl glycerol II/423
- - human CD52-1 II/424
- - L. mexicana II/423
- - mucin-like proteins II/424
- - myristoyl CoA II/423
- - porcine membrane dipeptidase II/424
- - *sn*-1-alkyl group II/424
- - T. brucei II/423
- - T. cruzi II/424
- - yeast II/423
- saccharide donor II/47
- sorting based on chain length II/815
- X II/440
lipoarabinomannan II/130, II/139
lipogranulomatosis II/465
lipopolysaccharide (LPS) II/131, II/435 ff, II/1030
- *Escherichia coli* II/131
- *Rhizobium* II/131
lipoprotein II/969
- oxidation II/746
Liver MBP II/607
LNP II/546
- apyrase II/546
- carbohydrate binding activity II/546
- root hairs II/546
locked anomeric configuration method I/189 f
- epimerization at C-2 I/189
- isomerization of acetal I/190
- mannosyl stannylene acetal I/190
- rhamnosyl stannylene acetal I/190
- synthesis of methyl, allyl, benzyl glycosidases I/189
L-PHA II/156
LPS *see* lipopolysaccharide
lpxA II/439
lpxB II/440
lpxC II/439
lpxD II/440
lpxH II/440
lpxK II/441
lpxP II/442 ff
lpxY II/442 ff
LSC II/315
L-selectin II/613, II/620
lucopolysaccharidoses
- MSP I II/949
- MSP II II/949
- MSP IIIA II/949
- MSP IIIB II/949
- MSP IIIC II/949
- MSP IIID II/948 f
- MSP IVA II/949
- MSP IVB II/949
- MSP VI II/949
- MSP VII II/948 f
lumican II/717 f
lutropin II/262
Lymnaea stagnalis II/263 f
lymphocyte *see* also T cell lymphotropic virus II/862
- homing II/249
- Gal6-*O*-sulfotransferate II/249
- GlcNAc-6-*O*-sulfotransferate II/249
- L-selectin ligands II/249
lyosomal
- enzymes II/564
- - synthesis II/564
- - acquisition of Man-6-P II/564
- hydrolases II/473 ff
- membrane transporters
- - for cystine
- - - lysosomal disorder II/946, II/954
- - for sialic acid
- - - lysosomal disorder II/954
- sialidases II/486
- storage diseases *see* aspartylglycosaminuria, cystinosis, fucosidosis, α-mannosidosis, β-lysosomal storage diseases II/953 f
- in Australia II/953 f
lysosomes II/456 ff, II/564
- role in degradation of cytosolic oligosaccharides II/473 ff
- role in degradation of glycoproteins II/473 ff

Mac-2-BP
- galectins II/639, II/640
macrocyclic glycosides I/42

macrophages II/721
- infection II/860, II/861
- receptor, endocytosis II/852, II/861
macular corneal dystrophy II/719
malaria parasites II/238
MALDI-TOF (matrix-assisted-laser desorption iorllzation -time of flight mass spectrometry) for monitoring glycans on recombinant proteins II/1050
MALDI-TOF-MS see matrix assisted laser desorption ionization-time of flight-mass spectrometry
mammalian
- development II/153
- hepatic lectin II/549 ff
mammals II/262
mammary carcinoma cells II/930
α-D-Man(1-2)-D-Man I/727
- enzymatic synthesis I/727
mannan II/134, II/136 ff, II/858
- binding protein II/861
mannobioses II/500
mannofuranose I/179
- anomeric O-alkylation with primary triflates I/179
mannopyranose I/184 f
- anomeric O-alkylation with primary triflates I/184
- complex formation I/185
- coupling reaction I/185
β-D-mannopyranosylamines I/337
mannose binding protein (MBP) II/603 ff
mannose
- ligand binding II/603
- complement activation II/604
- C-linked II/130
- O-linked II/136 ff
- MBP-associated serine proteases (MASPs) II/604
- MBP deficiency II/605
- on recombinant proteins II/1052
- phosphate II/137f, II/140
- 6-phosphate receptor II/457
- recognizing proteins, and antibodies II/857, II/861
mannosidase inhibitors II/514, II/525 ff
- deoxymannojirimycin (DMJ) II/514, II/526
- DIM (1,4-dideoxy-1,4-imrno-D-mannitol II/527
- 1,4-dideoxy-1,4-imino-D-mannitol II/515
- kifunensine II/514, II/527
- mannonolactam amidrazone II/527
- mannostatin II/528
- mannostatin A II/514
- swainsonine II/514 II/525
mannosidase inhibitors
α-mannosidase I/727, II/500, II/962
α-mannosidase II (MII) II/963

α-mannosidase IIx (MIIx) II/963
α-3/6-mannosidase II II/148
α-D-mannosidase
- α (1 → 6) specific II/477, II/479
- cystolic II/480 f
- lysosomal II/475 ff, II/479, II/481
β-mannosidase II/500
β-D-mannosidase
- lysosomal II/475 f, II/481
mannosidosis, salla disease II/88
- animal models
- - N-acetyl-β-D-glucosaminidase (α-subunit) II/948, II/951 ff
- - N-acetyl-β-D-glucosaminidase (β-subunit) II/948, II/951 ff
- - N-acetygalactosamine-4-sulfatase II/948
- - N-acetylglucosamine-1-phosphotransferase II/948, II/954
- - N-acetylglucosamine-6-sulfatase II/948
- - acid phosphatase II/948
- - arylsulfatase A II/947 f
- - cahepsin K II/948 ff
- - cathepsin A (protective protein) II/946, II/948, II/950
- - cathepsin D II/948
- - α-L-fucosidase II/948
- - α-D-galactosidase II/948
- - β-D-galactosidase II/948
- - β-D-galactocerebrosidase II/948
- - β-D-glucuronidase II/948
- - β-D-glucocerebrosidase II/948
- - α-D-glucosidase II/948
- - glycosylasparaginase II/948
- - G_{M2} activator II/948
- - α-L-iduronidase II/948
- - α-D-mannosidase II/948
- - β-D-mannosidase II/948
- - α-neuraminidase II/948
- - sphingolipid activator protein II/948
- - sphingomyelinase II/948
- pathology II/945 ff
- storage disorders II/564
- - I-cell disease (mucolipidosis II) II/564
- - pseudo-Hurler polydystrophy (mucolipidosis III) II/564 f
β-mannosides I/13, I/319 f, I/329 ff
- alkylation of 1-O-metal complexes I/332
- anomeric O-alkylation I/332
- de novo syntheses I/334
- direct inversion I/329
- enzymatic synthesis I/337
- exomannosidases I/337
- from non-carbohydrate precursors I/334
- mannosyl transferases I/337
- oxidation-reduction approach I/329
- radical inversion of the anomeric chirality of α-D-mannopyranosides I/333

- reductive cleavage of cyclic orthoesters I/334
- stannylene acetal method I/332
- 2-ulosyl donor method I/331
- glycosylation with mannosyl donors I/320
- stereoselective synthesis I/319
- synthesis I/13
α-mannosidosis II/477, II/479
β-mannosidosis II/478
mannosylation I/320 ff
- donor based methods I/327
- insoluble promoters I/320
- intramolecular I/324
- silver oxide activation I/320 ff
- with sulfonates I/322
C-mannosylation see mannose, C-linked
O-mannosylation, see mannose, O-linked
mannosysltransferase II/41m II/129
- acceptor specificity II/132
- archaeal II/139
- cell cycle II/139
- donor specificity II/131
- dolichol phosphate mannose-utilizing II/135 f
- endoplasmic reticulum II/134 ff
- eubacterial II/139
- families II/133 ff
- golgi II/136 ff
- structural features II/131 ff
- yeast II/133 ff
Marburg virus II/832 f
- glycoprotein II/832 f
- - function II/833
- - N-glycosylation II/832
- - O-glycosylation II/832
- - sialylation II/833
Maroteaux-Lamy disease
- animal model II/948 f
mass spectra I/922
- fragmentation pathways I/922
mass spectrometry II/972
MBGV see Marburg virus
MBL II/1030
MCD I II/719
MCD II II/719
mechanism of sialic acid cleavage II/845
melanoma II/238
membrane protein
- type III II/23
menstrual cycle II/723
M enzyme II/318
metabolic engineering I/858
- definition II/1043
- in animal cells II/1044
- in insect cells II/1053
- in plants II/1056
- production of more complex oligosaccharides I/858
metachromatic leukodystrophy II/460, II/465

metastatic potential II/156
methyl α-thiogentiobioside I/534
methyl 4,4^{II}-dithiomaltotrioside I/547
methyl glucopyranosides I/430
- rates of hydrolysis I/430
N-(O-methyl)glycolylneuraminic acid II/234, II/239
MGAT1 II/151
MGAT2 II/153
MHC antigens
- processing II/473 f
Michael addition to unsaturated acceptors I/549
microbial enzymes II/365
- glycosyltransferase II/365
- Pasteurella multocida II/366
- recombinant enzyme II/365
- Streptococcus equisimilis II/365
- Streptococcus pyogenes II/365
microbial glycosidases II/497
microcal omega titration microcalorimeter I/880
microglia II/720, II/860
microheterogeneity I/591
Micromonospora viridifaciens II/490
MII II/964
MIIx II/964
mimecan II/717 f
mimicry II/968
mitsunobu
- glycosylation I/214
- reaction I/210, I/214
model of arthritis II/983
moenomycin-type analogs I/12
- synthesis I/12
molecular
- chaperones II/997
- dynamics simulation II/970
- dynamics simulations I/959
- modeling II/972
- molecular diversity II/923
- - generated by protein glycosylation II/923
monoclonal antibodies II/287
monocyte II/856, II/860
monoglcosylated oligosaccharides II/999
mononuclear cells II/858
monosaccharide polymerization I/227
monosaccharide transporters II/4
Monte-Carlo simulations I/959
Morquio A disease
- MSP IVA II/949
Morquio B disease (G$_{MI}$ gangliosidosis)
- MSP IVB II/949
mouse
- hepatitis virus II/235
- hepatocarcinomas II/930
MPS IIID
- animal model II/948
MPS VII

- animal model II/948
MS analysis I/924 f
- protocols for I/924
- sample loading for FAB-MS analysis I/924
- sample loading for MS-MS analysis on the Q-TOF I/925
- sample loading for LC-ES-MS on the Q-TOF I/925
- sample loading for LC-ES-MS-MS on the Q-TOF I/925
- sample loading for MALDI-MS analysis I/925
- sample loading for NanoES-MS analysis on the Q-TOF I/925
MsbA II/441
msbB II/441
MUC1 I/275, II/284, II/411; II/669, II/672, II/675 ff, II/723
- as an antigen II/679
- cell-cell interactions II/678
- different glycoforms II/678
- effectson behavioral properties II/677
- expression II/677
- O-glycosylation II/675
- immune responses II/677
- key building blocks I/275
- synthesis I/275
- tandem repeat II/675
MUC2 II/283, II/673
MUC2, 3 etc. II/669 f
MUC4 II/672, II/675
MUC5AC II/673
MUC5B II/673
mucins II/237, II/313 f, II/316, II/322 f, II/410, II/970, II/1030
- colonic II/237
- human nasal
- genes II/313, II/670
- like glycoproteins II/313
- type O-glycosylation II/273 ff, II/669 ff
mucus II/672
- barrier II/237
Mukaiyama promotor system I/73
multiple
- binding specificities II/972
- sulfatase deficiency II/465
multivalent binding II/588
- clustering II/588
- high avidity II/588
- ligands I/903
- very low affinity II/588
muramic acid I/566
mycoplasma pneumoniae II/856
myelin II/1013 ff, II/1021
- associated glycoprotein II/233, II/1017, II/1021 f
myeloid enzyme: FucT IV II/205
- atherosclerosis II/206

- lung tumors II/206

NAD(P)H cytochrome b₅ reductase II/231
NANA II/485
nanomelia II/384
NAP II/969 f
nascent polypeptides II/46 ff
naturally occuring alkaloids II/515
- indolizidines II/515
- nortropanes II/515
- piperidines II/515
- pyrrolidine alkaloids II/515
- pyrrolizidines II/515
NB-DNJ see N-butyldeoxynojirimycin
NCAM II/1020
Neisseria meningitidis II/439
neoglycoproteins II/553
nerve growth cone guidance II/719 ff
nervous system II/1013, II/1015, II/1019
Neu5Ac II/485
- de-N-acetylase II/230
neural cell adhesion molecule (NCAM) II/1020
neuraminic acid II/227 ff
neuraminidases (Nam) II/485, II/839
- lysosomal II/476 f
α-neuraminidase
- in galactosialidosis II/946, II/948, II/950
- degradaton of G_{M2} by II/952
neurite II/720 ff
- outgrowth II/720
neurocan II/377, II/380
neuronal II/719 ff
neuronal enzyme: FucT IX II/206
- neurons II/206
neutrophil-
- activating protein, NAP II/969 f
- binding protein II/972
neutrophils II/972
Newcastle disease virus II/487
- sialidase II/503
NG2 II/377, II/381
Nicholas reaction I/219
Niemann-Pick disease II/465
- animal model II/948, II/952
nod factors I/467 ff, II/546
- Beau synthesis I/471
- Hui synthesis I/477
- metabolic pathway I/854
- precursors I/854
- Nicolaou synthesis I/468
- Ogawa snthesis I/475
- structure I/468
- syntheses I/467
nogalamycin I/1110
nonhydroxyl fatty acids II/972
non-natural CMP-'Sia' derivatives I/638

northern analysis II/287
NPD-sugars I/666
- recycling system I/666
NPG *see* pentenyl glycosides
nuclear factor-κB (NF—κB) II/446
nucleoside diphosphatase II/21
nucleotide
- α-diphosphate sugars I/628
- activated donor sugars I/627
- sugar
- - CMP-sialic acid II/21
- - GDP-fucose II/22, II/29
- - synthesis II/21
- - UDP-galactose II/21
- - UDP-GlcNAc II/22
- - transporter II/21 f, II/24
- - - cell mutants II/19
- - - cloning II/22 f, II/25, II/28
- - - CMP-sialic acid II/19, II/26, II/29
- - - defects II/29
- - - deudrogram II/24
- - - function II/20
- - - mechanism II/20
- - - membrane topology II/19, II/25 f, II/30
- - - molecular defects II/28
- - - structure II/25
- - - subcellular distribution II/27
- - - UDP-galactose II/19
- - - UDP-galactose transporter II/28

oligomannose *N*-glycans II/145
oligosaccharides I/202, I/239, I/310, I/706, I/715,
    I/723, I/736 ff, I/915, I/947 ff,
- accessibility II/861
- amino-functionalized - backbone, type 1 and
    type 2 II/858
- biosynthesis I/589 ff
- - folding and quality control I/593
- chains I/594
- - *N*-linked I/606
- - - invariable core I/606
- - - protein-specific processing I/606
- - - site-specific processing I/606
- clustering II/861
- complex type II/851, II/853, II/854, II/857
- conformation I/988
- - packing features I/988
- conformational analysis in solution by NMR
    I/947
- conformational parameters I/948
- ^{13}C-^{13}C coupling-constants I/953
- display II/860, II/861
- dynamics I/958
- - analysis in solution by NMR I/958
- electrophoresis II/992
- elucidention of conformation by NMR I/947

- enzymatic synthesis
- - on insoluble supports I/707 ff
- - on soluble supports I/715 ff
- - with *endo*-glycosidases (table) I/791 ff
- - with *exo*-glycosidases (table) I/736 ff
- glycosidase-catalysed snthesis I/723
- high mannose II/851, II/852, II/853, II/854,
    II/857, II/861, II/862
- in plant defense and cell signalling II/801 f
- - chitin II/802
- - defense response II/802
- - developmental regulation II/802
- - elicitors II/802
- - endoglycanases II/802
- - formation of II/802
- - glucan II/802
- - growth regulation II/802
- - inhibitor proteins II/802
- - oligogalacturonides II/802
- - receptors II/802
- into glycolipids I/310
- ligand II/861, II/862
- lipid-linked II/134 ff
- of gp120 II/851, II/852, II/853
- of gp160 II/856
- phenotype II/861
- polymer-supported synthesis I/239
- presentation II/861
- processing II/147
- - *N*-linked II/519 ff
- - - *endo* α1,2-mannosidase II/520
- - - glucosidase I II/519
- - - glucosidase II II/519
- - - α1,2-mannosidase II/519
- - - mannosidase IA II/520
- - - mannosidase IB II/520
- - - mannosidase II II/520
- signaling II/545
- solid-phase synthesis with glycosyltransferases
    I/706
- solution conformations I/947
- structural analysis I/915
- synthesis I/407 ff, I/412
- - chemoselective glycosylations I/407
- - efficiency I/408
- - efficient chemical synthesis I/407
- - iterative reactions I/409
- - orthogonal coupling concept I/412
- - orthogonal strategy I/407
- - polymer-supported synthesis I/407
- - strategic aspects I/408
- - with endo-glycals I/202
- transformation I/310
oligosaccharides, *N*-linked II/854, II/856
- on gp 120 II/855
- synthesis I/609
- - major pathways I/609

C-oligosaccharides I/495
- anionic pathways I/496 ff
- by applying a cycloaddition reaction I/496 ff
- by direct coupling I/496 ff
- by the *de novo* synthesis I/496 ff
- radical pathways I/496 ff
- synthesis I/495 ff
- synthetic approaches I/496 ff
C-oligosaccharides, synthesis of I/496 ff
- cycloaddition and rearrangement I/527
- *de Novo* approach I/518
- from smaller building blocks I/518
- intermolecular anomeric radical addition I/511, I/513
- via C1-glycal carbanions I/500
- via C5-alkynyl anions I/496
- with anomeric samarium species I/502
- with C-branched carbanions I/506
- with C6-phosphoranes I/508
- Wittig olefination I/508
S-oligosaccharides I/531, I/560 f
- α-glucan-active enzymes I/560
- β-glucan-active enzymes I/560
- enzyme-substrate interactions I/560
- synthesis I/531 ff
S-oligosaccharides, synthesis of I/548
- chemoenzymatic methods I/548
- solid-support synthesis I/550
oligosaccharyl
- dolichol II/37, II/43
- - catabolism II/43
- transferase II/37, II/147
N-oligosaccharyltransferase II/45
oligosialyl chains II/239
oligosialylglycosides I/353
- synthesis I/353
olivomycin A I/391
- synthesis I/391
OMIM (online Mendelian Inheritance in Man) II/945 f
oncogenesis II/1034
opaque, corneas II/719
ordered sequential Bi-Bi mechanism II/151
organ abscission II/794, II/801
- wall loosening II/801
organ transplantation II/469
orosomucoid see α$_1$-acid glycoprotein
orthoester rearrangement I/225
orthogonal
- chain elongation of homo-oligosaccharides I/414
- coupling I/418
- glycosylation I/410 ff
- - current concepts I/410
- - solid-phase oligosaccharide synthesis I/414
- - strategy I/414
- oligosaccharide synthesis I/420

- - polymer-supported I/420
- protecting groups I/413
orthogonality I/413
- definition I/413
orthomyxoviridae II/839
orthomyxoviruses II/486
OST complex II/51
OST1 II/55 ff
OST2 II/57 ff
OST3 II/55 ff
OST4 II/56 ff
OST5 II/56 ff
OST6 II/55 ff
osteoadherin II/409, II/717
osteoclasts
- role in pycnodysostosis II/951
osteoglycin II/717
osteoporosis
- anti-cathepsin K drugs II/951
outgrowth, neurite II/720
overexpression II/156
oxonium ion I/431
- torsional strain I/431

PAMPs II/678
Panstrongylus II/873
Pantibody II/1057 f
PAPS II/245 f, II/254 f, II/719
- synthetase polypeptide II/246
- translocase II/246
parainfluenza II/487
paramyxoviridae II/839 ff
paramyxoviruses II/486
parasites II/867
paromomycin I/1117
paroxysmal nocturnal hemoglobinuria (PNH) II/757, II/759
- GPI II/757, II/759
- PIG-A gene II/759
partially protected nucleophiles I/187
- anomeric O-alkylation I/187
pathogen-associated molecular patterns II/678
pea lectin II/536
pectin interactions II/794 f
- covalently cross-linked II/795
pentasaccharide core II/145
n-pentenyl glycosides (NPG) I/135 ff
- as protecting groups I/146
- chemistry of I/137
- glycosylation I/135
- orthogonal protecting groups I/148
- relative reactivities I/138
- solid-phase iterative couple-deprotect-couple strategy I/146
n-pentenyl orthoesters I/141, I/144
- latent C2 esters I/144

peptic ulcer II/967
peptide
- antigens II/314
- *N*-glycosidase
- - degradation of polymannose oligosaccharides in ER II/480 f
- *N*-glycanase F II/506
pericellular matrices II/693
pericellular matrix II/690 f, II/694
perlecan II/710 ff
- Arg-Gly-Asp motif II/711
- chondrogenesis II/711
- Englebreth-Holm-Swarm murine tumor II/710
- Leu-Arg-Glu (LRE) cell adhesive motif II/711
- rotary shadowing electron microscopy II/710
- SEA region II/711
- sperm protein, enterokinase, agrin II/711
- /syndecans (HSPGs) II/749
P-glycosides I/51
phagocytosis II/230, II/970, II/972
- erythrocytes II/230
- lymphocytes II/230
- thrombocytes II/230
phase partitioning I/1083
3´-phosphoadenosine-5´-phosphosulfate II/719
phosphocan II/410, II/412
phosphokinase I/638
phosphomannoisomerase (PMI) II/961
phosphomannomutase (PMM) II/960 f
phosphomannose *see* mannose phosphate
phosphoric acids I/44
- glycosylation I/44
photoaffinity labeling II/457, II/557 ff
phthalic acid-tethered glycosidation I/463
phylogenetic tree II/277
physiological role II/552
phytosphingosine II/972
pig, organ transplantation II/859
pigxenotransplants II/156
pituitary hormones II/264
plants
- as a source for recombinant proteins II/1056
- cell expansion II/800 f
- - cell elongation II/801
- - cellulose microfibrils II/801
- - environmental stimuli II/800
- - hormones II/800
- - morphogenesis and differentiation II/800
- - organogenesis II/801
- - rate of II/801
- - regulation II/800
- cell walls I/1066, II/786, II/791
- - angiosperms II/786
- - arabinoxylan II/786
- - biosynthesis II/800
- - - cell expansion II/800
- - - differentiation II/800
- - cellulose II/786
- - dicotyledons II/786
- - diversity II/786
- - glycoproteins II/796 f
- - - arabinogalactan proteins II/796
- - - cross-linking II/796
- - - extensin II/796
- - - GPI membrane anchors II/796
- - - hydroxyproline-rich glycoproteins II/796
- - - glycine-rich proteins II/796
- - - load-bearing structure II/799
- - - proline-rich proteins II/796
- - - viscoelastic deformation II/799
- - - wall rigidification II/796
- - - wall stress relaxation II/799
- - - wall synthesis II/799
- - - water potential II/799
- - homogalacturonan II/786
- - monocotyledons II/786
- - rhamnogalacturonan I II/786
- - rhamnogalacturonan II II/786
- - structural components II/786
- - type I II/786
- - type II II/786
- - xyloglucan II/786
- enzymes needed to improve recombinant protein production II/1048
- kingdom II/515
- - glycosidase inhibitors II/515
- primary cell walls II/798
- - function II/798
- - metabolism II/798
- - structural support II/798
- - turgor pressure II/798
- - cell expansion II/798
plant lectins, role II/543 f
- cytokinins II/545
- hormone binding site II/546
- oligsaccharide signaling II/546
- rhizobium-legume symbiosis II/546
- plant defense II/544
plasma lipids II/744
plasmodium II/868
- genus II/868
P-loop motif II/254
PLP (proteolipid protein) II/774
pluramycins I/1111
PMI II/961
PMM II/960 f
PMM1 II/960
PMM2 II/960
Pneumocystis carinii II/879
PNGase
- degradation of polymannose oligosaccharides in ER II/480 f
PNH II/760
- PIG-A gene

polyacrylamide gel I/707
- amino-functionalized I/707
- water-compatible I/707
polylactosamine backbones, I-type I/653 ff
- central branching I/654
- distal branching I/653
- β4-galactosylation I/656
polyadenylation II/718
polycyclic glycosides I/42
polyethylene glycol polyacrylamide (PEGA) I/715
polyethyleneglycol_-monomethylether (MPEG) I/248
- polymer-supported synthesis I/248
polyglycosylceramides II/970
polyhydroxylated alkaloids II/513
- nitrogen, in the ring II/513
- glycosidase, inhibitors of II/513
- imino-sugars, alkaloids II/513
polyisoprenoid *see* polyprenol
polylactosamines II/407, II/963
- backbones biosynthesis I/651
- - enzymatic in vitro synthesis I/651
- galectins II/635, II/641
polylactosaminoglycans I/603, II/265, II/507
- chains I/605
- - specific modification I/605
polylactosediaminoglycans II/265, II/268
polymannose oligosaccharides
- degradative pathway II/473, II/479 ff
polymer-supported
- glycosylations I/48
- synthesis I/239
- - of oligosaccharides I/241, I/246 ff, I/260
- - - automation I/246
- - - combinatorial libraries I/261
- - - examples of syntheses I/260
- - - linkers I/250
- - - one-phase I/241, I/247
- - - polymer supports I/246
- - - two-phase I/241
poly-*N*-acetyllactosamine II/320, II/853
polyoma virus II/159
polyoxin J I/567
polypeptide
- *N*-acetylgalactosaminyltransferase family I/614
- bound oligosaccharide I/592
- α-GalNAc-transferases II/315
- GalNAc-transferase II/273
- GalNAc-transferase II/273 ff
- GalNAc-transferase gene family II/275
- - GalNAc-T1 II/276 f
- - GalNAc-T2 II/277
- - GalNAc-T3 II/277
- - GalNAc-T4 II/277
- - GalNAc-T6 II/277
- - GalNAc-T8 II/277

- - kinetic properties II/278
polyprenols II/47
- reductase II/38
- chain length II/132
- pyrophosphate II/139
polysaccharide II/785
- fractionation II/785
polysialic acid II/23, II/721, II/1020
polysialo ganglio-series gangliosides I/311
- synthesis I/311
polyvalent interaction I/1070
pompe disease
- animal model II/948, II/954
posttranslational modification II/466
precipitin assays I/877
predictive value of in vitro *O*-glycosylation II/288
pregnancy II/980
cis-prenyltransferase II/38
primary cell walls II/793, II/797, II/784 f, II/803
- biotechnology II/803
- Fourier transform infra-red (FTIR) spectroscopy II/797
- fractionation II/785
- glycoprotein II/785
- hemicelluloses II/785
- heterogeneity II/797
- immunocytochemical methods II/797
- interacting networks II/793
- intercellular transport and storage II/803
- isolation II/784 f
- fruit ripening II/797
- organization II/793
- pectic polysaccharides II/785
- plasmodesmata II/803
- pore size II/803
- purification II/784
- structural domains II/797
- structural networks II/797
primary walls II/791
- apiogalacturonans II/791
- callose II/791
- coumaric II/791
- galactoglucomannans II/791
- β-D-glucans II/791
- ferulic II/791
- silicon II/791
- spezialized tissues II/791
- xylogalacturonans II/791
primates, New world, Old World II/857, II/859
prion protein II/757, II/762
- GPI II/757
processing of *N*-glycans II/146
progeroid syndrome II/386
promotor region II/371
- transcription factors II/371
prosaposin II/458
proteases

- cathepsin A
- - lysosomal disorder II/946, II/948, II/950, II/954
- cathepsin K
- - lysosomal disorder II/948 ff, II/954
proteasomes
- degradation of deglycosylated proteins II/480
protecting groups I/147, I/427, I/430 ff
- at the anomeric position I/195
- - transformation into a good leaving group I/195
- coupling efficiency I/427
- effects on reactivity I/427
- electronic and torsional effects I/430
- glycosylation stereoselectivity I/427
- influence on donor reactivity I/431
- influence on the acceptor I/441
- neighboring-group participation I/436
- n-pentenyl-based
- reactivity control in stereoselectivity I/437
- steric effects on glycosylation I/443
- stereoselectivity I/436
- strategy I/427
protective protein II/467
protein
- C II/262
- carbohydrate
- - binding I/876
- - - evaluation I/876
protein-carbohydrate interaction I/863, I/887 ff
- - association in aqueous solution I/864
- - cluster glycoside effect 901
- - degradation II/1002
- - dipole-dipole interactions I/864
- - dipole-induced dipole interactions I/866
- - fundamental considerations I/863
- - gas phase non covalent interactions I/864
- - glycosylation and cancer II/923
- - N-glycosylation I/591 f
- - - initial steps I/592
- - O-glycosylation I/608
- - hydrogen bonding and n-_ bonding I/868
- - 'hydrophobic' interactions I/872
- - machinery for GPI addition II/426
- - - GAA1 II/426
- - - GP18 II/426
- - O-mannosyltransferase II/136
- - misfolding II/852, II/855
- - role of multivalency 901
- - thermodynamic data (table) I/888 ff
- - thermodynamics I/887- conformation II/861
proteoglycans II/375 ff, II/379 ff, II/703, II/717 ff, II/729, II/731 ff, II/743 f
- aggrecan II/731 ff, II/735, II/738
- attachment sequence II/703
- biglycan II/734 ff
- biosynthesis II/375 f, II/378 ff
- core proteins II/375 f, II/378 ff, II/383 f, II/703

- - folding II/378, II/384
- - post-translational processing II/384
- - structure II/377, II/380 ff
- - sugar nucleotides II/376
- - trafficking II/384
- - translation II/376, II/380, II/383
- - translocation II/376, II/383
- distribution II/375
- decorin II/734 ff
- epiphycan II/734 f
- families II/380
- function II/375 f
- in vascular disease II/743
- keratan sulfate II/717 ff
- linkage region II/376, II/378, II/381 f, II/385 ff
- - biosynthesis II/385 ff
- - DS II/387
- - galactosylation II/386
- - phosphorylation II/381. II/386
- - structure II/381
- - sulfation II/387
- - xybsylation II/378
- - xylosylation II/381, II/385
- part-time II/376, II/381
- small leucine-rich repeat proteoglycan (SLRP) II/380
- structure II/375 f, II/379
- versican II/731 ff, II/735, II/739
proteolipid protein (PLP) II/775
proteoliposome II/21 f
Proteus II/439
protioglycosylation I/369
protozoans II/868
P-selectin II/246, II/613 f, II/619 f, II/1030
- glycoprotein ligand-1 (PSGL-1) I/267
- - synthesis I/267
Pseudomonas aureginosa II/439
PSGL-1 II/246, II/616, II/670, II/677
psicosides I/201
- enol ether synthesis I/201
psychosine II/467
Pulmonary surfactant protein A (SP-A) II/608
Pulmonary surfactant protein D (SP-D) II/608
pycnodysostosis
- animal model II/945, II/948 ff, II/954
2-pyridincarboxyl-glycosides I/222
pyrophosphatase I/638
_-pyrophosphate sugars I/626
pyruvate II/229

quality control II/1002
- of glycoproteins
- - effect of castanospermine II/75
- - role of calreticulin and calnexin as lectin-like chaperones II/69, II/75 ff
- - role of endomannosidase II/75 ff

- - role of glucosidases I and II II/75 ff
- - role of monoglucosylated N-linked oligosaccharides II/75 ff
- role of UDP-Glyglycoprotein glucosyltransferase II/76
- schematic representation II/76

R-3-hydroxymyristate II/437
radical substitution I/201
Raf kinase II/160
Rar II/980
Ras II/772, II/777
RAS pathway II/929
rat hepatocarcinomas II/930
rebeccamycin I/69
- synthesis via iterative assembly I/69
receptor
- binding proteins II/840
- destroying activity II/839
- destroying enzymes II/235, II/237 ff
recognition II/968 f
recombinant
- GnT I II/151
- GnT II II/153
- oligosaccharides I/845, I/852
- - purification I/852
- proteins see also Baculovirus, Insect cells, Plants, Chinese hamster ovary cells) II/1043, II/1044
reducing sugars I/188
- anomeric O-alkylation I/188
regeneration systems I/665 ff
- application in carbohydrate synthesis I/666
- carbohydrate-based I/680
- CMP-NeuAc I/671
- GDP-sugars I/676
- UDP-galactose I/666
- UDP-sugars I/669
regulation of HA synthesis II/370
regulatory elements II/285
remote activation I/223
- concept I/223
repertoire of GalNAc-transferases II/287
reptiles II/262
restonic lesions II/747
β-retaining glucosidases I/725
retrovirusses II/830, II/851, II/855, II/857, II/858, II/859
RG-II II/795
- 1:2 borate-diol ester II/795
- cross-linked II/795
- dimer II/795
- self assembly II/795
RHAMM (Receptor for hyaluronan mediated motility) II/687 f, II/694, II/747
rhamnogalacturonan I II/786, II/788 f

- composition II/788
- ferulic acid II/789
α-L-rhamnosidase II/504
repeating disaccharide II/788
rhamnogalacturonan II II/786, II/789 f
- backbone II/790
- borate ester II/790
- composition II/789
- methyl esterified II/790
rheumatoid arthritis II/977, II/979
rheumatoid factors II/983
Rhizobium
- legume symbiosis II/546
- leguminosarum II/437
- Rhizobium meliloti II/250
- - nod factors II/250, II/252
- - - lipooligosaccharide II/252
Rho II/772, II/777
Rhodnius II/873
Rhodobacter capsulatus II/446
β(1-5)ribf I/179
- stereoselective syntheses I/179
ribfa I/179
- stereoselective syntheses I/179
ribofuranose I/178
- anomeric O-alkylation with primary triflates I/178
ribophorins II/51
- I II/55 ff
- II II/54 ff
Ricinus communis agglutinin II/980
Ric^{R}14 BHK cell mutant II/152
Rieske iron-sulfur protein II/232
RIP-motif II/490
Rnase B II/1001
rotaviruses II/238
Rous sarcoma virus II/159

S. cerevisiae II/418
- Gas1p II/418
S. typhimurium sialidase II/504
saccharide-peptide hybrids (SPHs) I/565, I/574 ff
- biological activity I/579
- conformation I/582
- definition I/565
- directed synthesis I/574
- solid-phase synthesis I/578
Saccharomyces cerevisiae II/51
SA-Le^x II/330
salidases II/485
salla disease II/474. II/954
Salmonella II/442 ff
- typhimurium II/490
Sandhoff disease II/462, II/467, II/948, II/951 ff
Sanfilippo A disease see MSP IIIA
Sanfilippo B disease see MSP IIIB

Sanfilippo C disease *see* MSP IIIC
Sanfilippo D disease *see* MSP IIID
SAP-A II/466
SAP-B II/460, II/465
SAP-C II/465 f
SAP-D II/465
saposins II/459
SAP-precursor protein II/457
Sarcocystis spp II/879
SAT-3 II/339
saturation of II/132
Schindler disease II/954
Schistosoma II/262, II/868
- haematobium II/879
- japonicum II/879
- LDN glycans II/883
- LDNF glycans II/883
- mansoni II/263 f, II/879
sclera II/718
sea urchins II/239, II/896 f, II/899, II/904, II/906 f
secondary walls II/798
- cellulose II/798
- lignin II/798
secretory pathway II/378 ff, II/384
selectfluor I/385 f
- fluoroglycosylation of
3,4,6-tri-*O*-acetyl-D-glucal I/386
selectins II/229, II/613
- E and P II/862
sensory nerves II/721
sepharose matrix I/708
septic shock II/446
sequential *N*-glycopeptide synthesis I/299
sequon II/147
serglycin II/375, II/378 ff, II/381 f
- and heparin II/704
- - anticoagulant II/705
- - basophils II/704
- - low molecular weight heparin II/705
- - mast cells II/704
- - NK II/704
- - pentasaccharide sequence II/705
- - protein-free heparin II/705
- - secretory granules II/704
- - serine glycine repeats II/704
Serratia II/439
serum proteins II/960
SGGL II/1020, II/1022 f
S-glycosides I/49
shed effectors II/709 f
- agonists II/709
- antagonists II/709
- HS oligosaccharides II/710
- tethered PG modulators II/709
shedding II/369
sialate
- 4-*O*-acetylesterase II/230, II/235

- - mouse hepatitis virus II/237 ff
- 9-*O*-acetylesterase II/230, II/235, II/238 ff, II/839
- - influenza C virus II/237 ff
- *O*-acetylisomerase II/236 ff
- *O*-acetylmigrase II/235
- 7,8,9-*O*-acetylmigrase II/230
- 4-*O*-acetyltransferase II/230
- 7(9)-*O*-acetyltransferase II/236 ff
- 7(9?)-*O*-acetyltransferase II/230
- 9-*O*-acetyltransferase II/230
- 9-*O*-lactyltransferase II/230
- lyasa II/229
- 8-*O*-methyltransferase II/230
sialic acid I/199, I/566, II/227 ff, II/234 ff,
    II/238, II/485 f, II/583, II/721, II/839 ff, II/853,
    II/969 f, II/972, II/1016
- *O*-acetylated II/234
- *O*-acetyl migration II/235
- *O*-acetylation, Golgi-membranes, bovine
submandibular glands II/236
- 4- and 9-*O*-acetylesterases II/234
- biological functions II/229 ff
- biosynthesis II/227 f
- diversity II/227 ff
- echinoderms II/228
- enol ether synthesis I/198
- evolution II/229
- formation in liver II/1044
- from *N*-acetylmannosamine II/1044 f, II/1047, II/1050
- in cultured aortic cells II/1044
- in human fibroblasts II/1044
- in insect cells l054 f
- in plants II/1059
- insects II/229 ff
- lactones II/239
- man II/228
- masking effect II/230
- *O*-methylated II/238
- microorganisms II/229
- modification II/229 ff
- occurence II/227 ff
- purification II/234
- recognition II/843
- - sites II/231 ff
- substituents II/233
- - biological significance II/233
- *O*-sulfated II/238
- synthetase I/638
- 2,3-unsaturated II/228
sialidase II/228 f, II/235, II/238 ff, II/503
- superfamily II/493
sialidation I/345 ff
- auxiliary group at C-3 I/347
- Koenigs-Knorr method I/345
- mechanism I/356

- products I/351
- - structures of I/351
- reaction mechanism I/357
- special methods I/359 ff
- with phosphites I/356
- with phosphites of sialic acids I/349
- with 2-thioglycosides I/349
- with xanthates I/349, I/356
sialidosis
- animal model II/948
- in Australian population II/954
sialoadhesins II/231, II/233 ff
sialoglycoconjugates I/345, II/229, II/232
- life-time II/232
sialoglycoproteins I/345
sialomucin complex II/672, II/674
sialyl
- donnors I/350
- glycosides I/199
- 6´-sialyl-LacNAc I/674
- - one-pot synthesis I/674
- Lewis antigen I/296
- Lewis^x (sLe^x) II/261, II/318, II/321, II/323, II/614, II/977
- - bound to E-selectin I/1011
- - - conformation I/1011
- oligosaccharide I/731
- - one-pot synthesis I/731
- l-6-sulfo Lewis^x II/229
- T antigen I/278, II/854
- - synthesis I/278
- Tn II/323
- Tn antigen I/291, II/314, II/321
- transferase II/213 ff, II/229, II/234, II/236 ff, II/322
- - classification and nomenclature II/216
- - family I/610
- - α2,3-ST family II/216
- - α2,6-ST family II/217
- - α2,8-ST family II/218
- - regulation and functionality II/219
2,3-sialyl T antigen I/293
α2,3-sialyltransferase II/304
α2,6-sialyltransferase II/302
- transition state analogs II/295 f
- donor analogs II/302
α3-sialyltransferase I/602, II/322 f
α6-sialyltransferase I/602, II/264, II/268, II/322 f
sialylation II/322
- lation and metastasis II/933
α-sialylation I/272
α3-sialylation I/656
siglecs II/231, II/233 ff, II/237 ff. II/580 f
- CD33 II/581
- CD22 II/581
- MAG II/581
- sialoadhesin (Sn) II/581

signal
- sequence for GPI addition II/426
- - hydrophobicity II/426
- - mutagenesis II/426
- - ω spacer II/426
- - ω + 1 spacer II/426
- - ω + 2 spacer II/426
- transducers II/772
- transduction II/235, II/815
- - glycosphingolipids II/815
signalling II/591 ff
- B cell activation II/592
- CD22-deficient mice II/592
- ITIM-like motif II/592
- negative regulatory functions II/592
- SH2-containing II/591
- SHP-1 tyrosine phosphatase II/592
- tyrosine residues II/591
silylated acceptors I/216
1-O-silyl glycosides I/215
- as glycosyl donor I/215
simian immunodeficiency virus, SIV II/856
site of biosynthesis II/364
- eucaryotic II/364
- pericellular space II/364
- streptococci II/364
SIV II/856
Sjogren´s syndrome II/986, II/990
skin II/734 ff, II/739
Sly disease see MSP VII
Small Leucine-Rich Proteoglycan (SLRP) II/410
SMC II/674 f
SMC II/677
smooth muscle cell II/746
snail II/263 f, II/264, II/266 f
solid
- oligosaccharides I/45
- phase synthesis I/45, I/239, I/705
- supports I/255
- - crosslinked polystyrene resin I/255
- - functionalized polystyrene cross-linked with divinylbenzene I/255
- - pore glass (CPG) I/255
soybean agglutinin II/542
spatial arrangement II/556
species-specific II/897, II/899, II/904, II/906 f, II/905
α1→2-specific fucosidases II/502
specificities II/318, II/323
spermatozoa II/239
sperm-egg
- binding see also sperm-egg interaction II/904, II/907
- interaction II/895
sphingolipid activator protein II/458
sphingolipidoses II/466 ff
sphingolipids

- lysosomal digestive pathway II/951 ff
sphingomyelin II/465
sphingomyelinase II/465
sphingornyelin
- lysosomal digestive pathway II/952
sphingosine II/972
- derivatives and mimics I/36
- - glycosylation I/36
spirocyclopropanes, synthesis of I/168
splicing, alternative II/718
Src family kinases II/772
Src kinase II/160
c-Src II/772, II/777
ST6Gal-1 II/339
stannanes I/171
- formation using diazirines I/171
starfish II/896, II/899
- Asterias rubens II/238
Stenotrophomonas maltophilia I/226
stomach pit cells II/971
Streptococcus pneumoniae II/493
Streptomyces I/93
stress protein II/970
structural determination II/516
- glycosidase inhibitor II/516
- radial chromatography II/516
- thin layer chromatography II/516
- GC-MS II/516
- ^{1}H and ^{13}C NMR II/516
- high resolution mass spectrometry II/516
structure-activity relationship II/518
- molecular modeling II/518
- molecular orbital calculations II/518
structure-based drug design II/487, II/969
STT3 II/58
STZ II/339
subcellular Gb$_3$ trafficking II/813
submaxillary mucin II/674
substrate specificities II/151, II/205, II/279 f, II/280
subunit organization II/558
succinic
- acid tether I/461
- succinic acid-tethered I/462
- - stereodifferentiation I/462
sucrose synthase I/642
sugar
- acceptors I/350
- combining site II/559
- metabolism
- - basic principles II/3
- - nucleotide transporters II/11
- nucleotides I/625
- - synthesis I/625
- nucleotides I/641
- - in *situ* generation I/641
- nucleotides I/665

- - biosynthetic pathways I/665
- nucleotides, synthesis I/629 ff
- - CMP-activated sugars I/629
- - GDP-activated donors I/632
- printing II/990
- specificities II/551
sugar nucleotides, synthesis of I/626, I/635 ff
- chemical synthesis I/626
- chemo-enzymatic synthesis I/635
- CMP-activated sugars I/637
- GDP-activated sugars I/639
- UDP-activated donors I/626
- uridine diphosphate-activated donor sugars
  I/635
sulfate polysaccharide II/897
sulfated
- carbohydrates II/247
- - monosaccharide modifications II/247
- *O*-glycans II/323
sulfatide II/465, II/969, II/971 f, II/1013, II/1015
- lysosomal digestive pathway II/952
sulfatide (3´ sulfogalactosyl ceramide) II/813
- recognition II/813
sulfation II/382
- linkage region II/382
- - structure II/382 ff
N-sulfation II/408
sulfatransferase II/268
sulfoglucuronyl glycolipids (SGGL) II/1018
sulfonamidoglycosylation I/68
N-sulfonylamines I/171
- esters I/171
- formation using diazirines I/171
sulfotransferase II/264, II/314, II/324, II/390 f, II/413
- chondroitin II/390 f
- KS II/719
- urnic acid II/391
6-sulfotransferase II/390
summary and future prospects II/528
- high degree of potency II/528
- new glycosidases II/528
- substrate specificity II/528
surface IgM II/446
SV1 II/720
SV2 II/720
swainsonine II/935
SWPl (suppressor of *WBP*) II/54 ff
syncytium formation by HIV-1 II/857
syndecan II/378, II/749, II/376 f, II/381, II/705 f
- cell scaffolding proteins II/706
- cytoplasmic domains II/706
- dominant negative inhibitors II/706
- ERM II/706
- GAG attachment II/706
- MAGUK II/706
- oligomerization domains II/706

- paracrine effectors II/706
- PDZ II/706
- phosphorylation II/706
- proteolytic cleavage II/706
- putative cell interaction II/706
- shedding II/706
- syntenin II/706
syndecan-1 II/382
synergistic inhibitors II/300
synthetic di- and tri-valent glycopeptides II/555 ff
systemic lupus erythematosus II/990

T antigen (Gal-GalNAc) I/290, II/854
T. brucei II/418 ff
- glycolipid A II/420
- glycolipid C II/422
- inositol acyltransferase II/422
- inositol deacylase II/422
- PARP II/418 ff
- VSG II/419, II/421
- VSG anchor II/418
T cell infections with HIV-1 II/862
T cell lymphotropic virus II/861
T. cruzi II/419
T24 H-ras II/159
T_3-(3,3',5-triiodo-L-thyronine)
- formation via lysosomal degradation of thyroglobulin II/477 ff
T_4-(3,3',5,5',-tetraiodo-L-thyronine)
- formation via lysosomal degradation of thyroglobulin II/477 ff
tandem repeats II/670, II/672
- sequences II/313
T-antigen (Gal-GalNAc) I/276
- formation I/276
targeting II/559
- to Golgi Apparatus II/150
TATA II/151
Tay Sachs
- /Sandhoff disease II/951, II/952
- disease II/462, II/467
- - animal model II/948, II/951 ff
- - mouse models II/951 ff
tentaGel-N I/261
- solid phase I/267
- solution I/267
tert-butyldimethylsilyl 2-deoxyglycosides I/394
tether I/449
- stable I/449
- temporary I/449
tetrasaccharides I/972
- crystal structures I/972
therapeutics II/860
therapy II/469, II/967
thermodynamics of association I/885
- binding site reorganization I/885

- proton transfer I/885
- salt effects I/885
4-thiocellobiose I/544
thiocyclodextrins I/538
1,2-thiodisaccharides I/553
1,3-thiodisaccharides I/551
1,4-thiodisaccharides I/541
thioethyl donors I/74
- synthesis via iterative assembly I/74
thiogalactosides I/435
- relative reactivities I/435
thioglycoses I/532
- selective S-deprotection I/533
- synthesis I/532
1-thioglycoses I/532
2-, 3-, 4-, 5-, or 6-thioglycoses I/532
thioglycosides I/93 ff
- applications I/110
- as acceptors I/110
- - thioglycosides as both donors and acceptors I/111
- - glycosyl phosphites I/117 f
- - for glycosylation I/117 ff
- - formation of glycosyl phosphonate I/131
- - glycosylation mechanism I/119
- - glycosylation of sialyl phosphites I/122
- - glycosylation of C-2-acylated glycosyl phosphites I/123
- - glycosylation with C-2-O-benzylated glycosyl phosphites I/124
- - glycosylation with α-selectivity I/124
- - glycosylation with β-selectivity I/124
- - preparation I/118
- - stereoselectivity in glycosylation I/121
- - synthesis of CMP-NeuAc I/129
- - synthesis of GDP-fucose I/129
- - transformation of glycosyl donor I/131
- block syntheses I/110
- donors I/99
- - carbonium type I/100
- - direct activation I/99
- - halonium type I/100
- - heavy metal salt promotors I/99
- - single-electron activation I/106
- - sulfonium type I/100
- donors with an anomeric sulfur I/108
- from anomeric acetates I/94
- from glycosyl halides I/95
- glycosyl donor qualities I/93
- intramolecular glycosidations I/112
- orthogonal glycosylations I/110
- protecting group manipulations I/96
- solid phase synthesis I/113
- synthesis I/94
2-thioglycosides I/349
- for sialyl glycoside syntheses I/349
thio group migration I/388

1,1-thio linkages I/557
1,2-thio linkages I/553 ff
1,3-thio linkages I/551 ff
1,4-thio linkages I/541
1,6-thio linkages I/534
4-thiomaltose I/544
1-thio-β-D-mannopyranosides I/336
thiooligosaccharides I/558
- conformation I/558
1,3-thiooligosaccharides I/552
1,4-thiooligosaccharides I/546
6-thiooligosaccharides I/538
6-thiosaccharides I/534
1-thiosucrose I/553
α,α-thiotrehalose I/557
threshold theory II/466
thrombomodulin II/376 ff, II/381, II/388
thyroglobulin
- degradation in lysosomes II/474, II/477 ff
- degradation of carbohydrate chains II/478 f
- iodination II/477 f
- proteolysis II/478 f
thyroid hormones
- biosynthesis II/474
- formation via lysosomal degradation of thyroglobulin II/474, II/477 ff
tirosuine sulfation II/411
tissue-specific II/718
tlc plates II/970
TLR2 II/446
TLR4 II/446
Tn antigen II/315
to cis to II/319
tobramycin I/1119
Toll-like receptor proteins II/446
topology II/457
- of GPI biosynthetic pathways II/424 f
- - flippase II/425
- - L. major II/425
- - PIG-A II/424
- - PIG-B II/425
- - PIG-L II/424
- - Trypanosoma brucei II/425
Toulouse-Lautrec II/950 f
Toxocara II/868
Toxoplasma gondii II/879
transcription
- factor Sp1 II/151
- start sites II/161
transfer NOE experiments I/1003
- study of carbohydrate-protein interactions I/1003
transfer of EtN-P II/423
- aminoethylphosphonate II/423
- PIG-F II/423
- T. brucei II/423
- T. cruzi II/423

transferrin II/960, II/962 f
transgenic mouse II/971
transglycosidation II/228
- intramolecular II/228
trans-Golgi I/597
- elongation I/597
- reaction I/597
- termination reactions I/597
transition-state analogs II/303
translocation II/42, II/369
transmembrane domain II/150
transmissibility
- of HIV-1 II/851
- of retroviruses II/851
transparency, corneal II/718
transplantation, organ II/859
trans-sialidases II/229, II/491
C-β,β-trehalose I/507
- synthesis I/507
triantennary structure II/553 ff
Triatoma II/873
trichloroacetimidates I/5 f
- formation I/6
- reaction I/6
- method I/5, I/7
- - basic principle I/5
- - glycosylation of carbohydrate acceptors I/7
- - synthesis of oligosaccharides I/7
Trichobilharzia ocellata II/263 f
triflates I/541
- S_N2-displacement I/541
TRI-GP II/554 ff
trimannan I/143
- synthesis from n-pentenyl orthoester
trisaccharides I/972
- crystal structures I/972
tritylated acceptors I/224
Trypanosoma
- brucei II/868, II/873
- - gambiense II/868, II/873
- - rhodesiense II/868, II/873
trypanosoma cruzi II/486, II/765, II/868, II/873 ff
- GPI II/765
TSG-6 II/747
tumor
- associated changes II/1036
- cells II/932
- - adhesion II/934
- - -endogenous lectins II/934
- - focal adhesions II/932
- - proliferation II/927
- necrosis factor-α II/446
tunicamycin II/40, II/408, II/1000
β-turn structure II/48 ff
type I
- arabinogalactans II/791
- chains II/968

- collagen II/734, II/748
(Gal(β1-3)GlcNAcβ-R) chains I/607
- - synthesis I/607
type II
- arabinogalactans II/791
- collagen II/733
- hydroxyproline-containing proteins II/791
- (Gal(β1-4)GlcNAcβ-R) chains I/607
- - synthesis I/607
- membrane protein II/315, II/320
type III collagen II/748
type IX collagen II/377
type IX collagen II/381
tyrosine
- phosphorylation II/156
- sulfation II/617 f
tyrosylsulfotransferases II/246, II/252 f

UDP-Glyglycoprotein glucosyltransferase
- amino acid sequence II/122
- calcium requirement II/120
- distribution in nature II/120
- donor specificity II/120
- localization, subcellular II/120
- role in glycoprotein folding II/123 ff
- reaction catalyzed II/119
- sensor of glycoprotein conformation II/121 f
- sequence analysis II/122 f
- size II/120
UDP-GlcNAc:N-acetylglucosamine-I-
    phosphotransferase
- in I-Cell disease II/954
UDP-GlcNAc-2 epimerase
- in animal cells II/1044, II/1046
UDP-glicose-4´-epimerase I/642
UDP-glucose: glycoprotein glucosyltransferase
    II/1003
UDP-N-acetylglucosamine II/439
UDP-N-acetylglucosaminyl:dolichyl phospate
    N-acetylglucosaminyl phosphoryl transferase
    II/40
unprotected aldoses I/187
- anomeric O-alkylation with bromides I/187
- anomeric O-alkylation with cyclic sulfates I/187
- anomeric O-alkylation with primary triflates
    I/187

virus shedding II/862
vaccines II/967
valency II/553 ff
vascular
- disease II/743
- endothelial cells II/749
- lessions II/744

verotoxin (VT) mediated pathology II/811
versican (CSPGs) II/377, II/380, II/745
vertebrate synthases II/366
- HAS genes II/366 ff
- HAS isoenzyme from mouse II/366
- human II/366
- structural features II/366
- DG42 protein from Xenopus laevis II/366
vibrio cholerae II/491
VIM-2 ganglioside I/485
- total synthesis I/485 ff
VIP36 II/545
viral
- evolution II/235
- proliferation
- - effect of glucosidase I and II inhibitors II/77 f
- - role of endomannosidase pathway II/78
- receptor destroying enzymes II/847
virion, virus particle II/859, II/862
virus II/821 f
- classification II/822
- dissemination II/852
- entry II/860
- envelope II/851, II/862
- Epstein Barr II/856
- glycoproteins
- - glycosylation II/853, II/859, II/862
- - macrophage-derived II/853
- inactivation II/851, II/859
- infection II/852, II/859, II/860, II/862
- infectivity II/851, II/855, II/860, II/862
- neutralization II/857, II/858
- pig, porcine II/859
- resistance II/859
- surface glycoproteins II/822
- transmissibility II/858
- transmission II/851, II/859
- tropism II/852, II/860, II/861
visceral Leishmaniasis II/875
vitelline coat *see also* vitelline envelope II/902 f
vitelline envelope II/899, II/901
VMP-Neu5Ac II/229
von Willebrand factor II/246, II/279, II/673

WBPl (wheat germ agglutinin protein) II/54 ff
wheat germ agglutinin II/536, II/543
Wolman disease II/954
worm development II/155

xanthan II/139
Xanthomonas manihotis II/499
xeno
- antibodies II/857, II/859
- antigen II/853
- graft II/859

Xenopus *see also* frog
- laevis II/900
xenotransplantation II/859, II/1030
X-ray
- crystal structures II/845
- crystalography II/1034
m-xylene-tethered glycosidation I/464
xyloglucans II/787, II/799
- cell expansion II/799
- cellular differentiation II/799
- exoglycanases II/799
- fruit ripening II/799
- hydrolytic enzymes II/799
- load-bearing structure II/799
- poaceae II/787
- solanaceae II/787
- structural variation II/787
- transglycosylases II/799

β-xylosidase II/505
β-D-xylosides II/386
β1-2-xylosyltransferase II/148

yeast II/419, II/758 ff
- GPI II/758, II/760
- cell wall II/758 f
- Gas1p II/759
- Gce1p II/759

zona pellucida (ZP) II/409, II/902, II/896, II/900, II/906